Handbook of Polymer Science and Technology

Volume 1

Basic Concepts of Polymers, Polymerisation Techniques, Characterisation and Processing of Polymers and Industrial Polymers

Handbook of Polymer Science and Technology

Volume 1

Basic Concepts of Polymers, Polymerisation Techniques, Characterisation and Processing of Polymers and Industrial Polymers

M.H. FERRY
C.Eng.

A.V. BECKER
M.Sc., Ph.D.

Edited by

S.C. BHATIA
B.E. (Chem), M.B.A.

CBS

CBS PUBLISHERS & DISTRIBUTORS PVT. LTD.

NEW DELHI • BENGALURU • PUNE • KOCHI • CHENNAI

ISBN : 81-239-1132-7

First Edition : 2004
Reprint : 2012

Published by Satish Kumar Jain and produced by V.K. Jain for
CBS Publishers & Distributors Pvt. Ltd.,
CBS Plaza, 4819/XI Prahlad Street, 24 Ansari Road, Daryaganj,
New Delhi - 110002, India. • Website: www.cbspd.com
e-mail: delhi@cbspd.com, cbspubs@airtelmail.in
Ph.: 23289259, 23266861, 23266867 • Fax: 011-23243014

Branches:

- ***Bengaluru:*** Seema House, 2975, 17th Cross, K.R. Road,
  Bansankari 2nd Stage, Bengaluru - 560070 Ph.: +91-80-26771678/79
  • E-mail: cbsbng@gmail.com, bangalore@cbspd.com
- ***Pune:*** Bhuruk Prestige, Sr. No. 52/12/2+1+3/2,
  Narhe, Haveli (Near Katraj-Dehu Road By-pass), Pune - 411051
  Ph.: +91-20-64704058/59, 32342277 • E-mail: pune@cbspd.com
- ***Kochi:*** 36/14, Kalluvilakam, Lissie Hospital Road,
  Kochi - 682018, Kerala • Ph.: +91-484-4059061-65
  Fax: +91-484-4059065 • E-mail: cochin@cbspd.com
- ***Chennai:*** 20, West Park Road, Shenoy Nagar, Chennai - 600030
  Ph.: +91-44-26260666, 26208620 E-mail: chennai@cbspd.com

Printed at :
India Binding House, Noida (UP)

Preface

This volume, dealing with basic concepts of polymer, polymerisation techniques, characterisation and processing of polymers and industrial polymers, is the first volume of 'Handbook of Polymer Science and Technology'.

Every industry uses rubber for conveyor belts, hoses, tubings or as insulation in heavy cables. Every vehicle, such as motor car, bus and aeroplane moves on rubber tyres and consists of other parts made out of rubber such as anti-vibration mountings, window and door seals, radiation hoses or fan belts, and every home uses polymer-based household equipments.

Recognising the great importance of polymers, this reference text book is intended for the practising polymer technologists, rubber and plastic technologists, chemical and mechanical engineers, industrial chemists and research students. This book also provides an upto date information for the users of polymer products in other industries.

This volume is divided into six sections. Section I deals with basic concepts of polymers. Chapter 1 focuses on introduction to polymer science, while chapter 2 is devoted to natural and synthetic polymers. Block and graft co-polymer is undergoing a very rapid growth, keeping this is mind chapter 3 deals with chemistry of block and graft co-polymers.

Section II deals with various polymerisation techniques which refer to the physical aspects of polymerisation reactions and which decide whether the monomer is polymerised in its condensed or gaseous state, or whether it is polymerised as such or along with other inert components such as solvents and non-solvents. Chapter 4 provides valuable information on step reaction (condensation) polymerisation, while chapter 5 focuses on radical chain (addition) polymerisation. Chapter 6 deals with ionic and co-ordination chain (addition) polymerisation and chapter 7 focuses on co-polymerisation, various types of co-polymerisations and their kinetics are also discussed in detail. Chapter 8 concentrates on co-polymers of ethylene and propylene oxides, their classifications, manufacturing aspects and applications. In chapter 9, the latest polymer conditions and reactions including micro-emulsions–a novel medium for polymerisation reactions are discussed. Chapter 10 focuses on chemical modifications of polymers with special emphasis on esterification, hydrolysis and hydrogenation of polymers.

Section III deals with characterisation of polymers. Knowledge of the solubility of various polymers in different solvents is important in assessing their chemical resistance and their application potentialities in the field of paints, fibres and casting films. Keeping this in mind chapter 11 describes the criteria of polymer solubility, thermodynamics and phase equilibrium in polymer solutions. Polymers are non-homogenous with respect to molecular sizes and individual characteristics of varying lengths etc. Hence a concept of an average molecular weight is adopted, keeping this in mind chapter 12 focuses on measurement of molecular weight and size of polymers. Chapter 13 deals with structure of polymeric materials, structure, polymer crystallinity, thermal transitions and glass transition temperature of various polymers etc.

Section IV explains in detail various industrial polymers, their preparation, properties and applications. Chapter 14 is devoted to basic concept of industrial polymers. Chapter 15 provides valuable information on addition polymers, which are produced in large quantities by cracking or refining of crude oil. Chapter 16 is devoted to condensation polymers which are formed from bio-or polyfunctional monomers by reactions which involve elimination of some smaller molecules.

Section V deals with polymer processing - that is fabrication of polymers. In a very broad sense, polymer processing may be defined as an engineering speciality which is used to convert polymeric materials into useful products. In fact, the need to understand the finer aspects of polymer processing surfaced as conventional materials were required to be substituted by those with better finish and elegance. Chapter 17 is devoted to plastic processing and fabrication processes, while chapter 18 deals with rubber compounding and processing technology. Chapter 19 is devoted to miscellaneous processing techniques such as coating processes, powder moulding, adhesive bonding, welding etc. Chapter 20 concentrates on decoration of plastics. Various commercial techniques for decorating plastics such as painting, printing, hot stamping, embossing and electroplating are discussed in detail.

Section VI is devoted to some special topics which are of great importance to polymer, rubber, and plastic technologists. Chapter 21 focuses on polymer blends. The commercial utility of polymers can be greatly extended through the use of blends and alloys. Various blending aspects have been discussed such as, rubber-plastic blends, PVC-nitrile rubber blend, rubber-rubber blends etc. Chapter 22 concentrates on some current problems related to polymerisation reaction engineering while chapter 23 is devoted to development of a software sensor for control of bulk polymerisation of methyl methacrylate. In chapter 24, discussed topics are water soluble polymers, their kinetic behaviour and synthesis from divinyl monomers.

The text in this book is throughout supplemented with diagrams and tables. The treatment of all topics is in a cogent, lucid style aimed at enabling the reader to grasp the information quickly and easily. The references and index are also given at the end of the book.

While painstaking care has gone into producing a useful and exhaustive reference textbook, the authors would welcome any constructive criticism and creative feedbacks from the students, teachers and professionals in the industry.

We wish to thank all those colleagues who provided figures and photographs and offered important advice during its preparation.

M.H. Ferry
A.V. Becker

Contents at a Glance

SECTION V

SECTION VI

Contents

SECTION IV

SECTION V

SECTION VI

SECTION I

Basic Concepts of Polymers

CHAPTER 1

Introduction to Polymer Science

INTRODUCTION

Polymer science is concerned with the composition and properties of a large number of substances classed as 'polymers', which include rubbers, plastics, and fibres. The meaning of the term 'polymer' will become clear during the course of this chapter, but it does involve some understanding of basic chemistry. The science of chemistry concerns itself with the composition of matter and with the changes that it undergoes. To the layman, the methods and processes of chemistry appear quite baffling and difficult to comprehend. Yet these are essentially based on a simple, logical development of knowledge about the substances which comprise our physical world.

ATOMS AND MOLECULES

In this 'atomic' age, it is certainly common knowledge that the atom is the basic unit of all matter. There are as many different kinds of atoms as there are elementary substances, i.e., elements, and these number about 100. It is the myriads of combinations of these atoms which make possible the hundreds of thousands of different substances comprising our world. Just as in the case of the 100-odd elements it is the atom which is the smallest unit, so too in the case of all other substances (chemical compounds), it is the particular group of atoms, or molecule, which is the smallest possible unit. If the molecule of a compound substance is broken up, the substance ceases to exist and may be reduced to the elementary substances which comprise it. It is the various possible combinations of element atoms to form different molecules which is the basis of chemistry.

Now, in this 'nuclear' age, we hear much about 'atomic fission' processes, in which atoms are split apart, with the accompaniment of considerable energy release. How then can an atom be split if it is supposedly a basic unit of matter? Well, we know today that atoms are not the indivisible particles they were originally believed to be. Instead, they, too, are made up of even smaller particles such as protons, electrons, neutrons, and others. The differences between atoms can be explained by the different numbers and arrangements of these sub-atomic particles. The thing to remember, however, is that the different substances of our world make their appearance once atoms have been formed. This becomes obvious when we consider that all protons are the same, all neutrons are the same, and so on. There is no such thing as a 'proton of iron' being different from a 'proton of oxygen'. However, an atom of iron is considerably different from an atom of oxygen. That is why the chemist is concerned mainly with the arrangement of atoms into molecules and considers the elemental atoms as the basic building blocks.

Valence

So far nothing has been said about the manner in which the sub-atomic particles are held together or what holds atoms together in molecules. Although this is not our immediate concern here, it should be understood that there are electromagnetic forces which hold all these particles together. Thus most of the sub-atomic particles carry either positive or negative electrical charges, which act as binding forces within the atoms as well as between different atoms. Because each atom contains a different number and arrangement of these charged particles, the forces acting between different atoms also are different. These forces, therefore, result in certain well-defined rules which govern the number and type of atoms which can combine with each other to form molecules. These rules are called the rules of valence, and refer to the 'combining power' of the various atoms.

From the beginning of the nineteenth century, chemist have been carefully weighing, measuring, and speculating about the proportions in which the element substances combine to form compounds. Since they found that these proportions were always the same for the same substances, this was most easily explained by assuming combinations of fixed numbers of atoms of each element. This was the basis for *Dalton's* Atomic Theory, first proposed in 1803. In accordance with this hypothesis, an atom from the element hydrogen is never found to combine with more than one atom of any other kind. Hence, hydrogen has the lowest combining power, and is assigned a valence of 1. On the other hand, an atom of the element oxygen is found capable of combining with two atoms of hydrogen, hence can be said to have a valence of 2.

In this way, the kinds and numbers of atoms that make up a molecule of a given compound can be deduced and expressed as a formula for that compound. Thus the compound water, whose molecules are each found to contain 2 atoms of hydrogen and 1 atom of oxygen, is denoted by the formula H_2O, where the letters represent the kinds of atoms, and the numbers tell how many of each atom are present within one molecule. This formula can also be written as H—O—H to show how 1 oxygen atom is bonded to 2 hydrogen atoms. In the case of the well-known gas, carbon dioxide, the molecule would be represented as CO_2, where the carbon atoms would show a valence of 4, since it is capable of combining with 2 oxygen atoms (each of which has a valence of 2). This formula could also be written as O=C=O, which is known as a structural formula, since it shows exactly how the carbon and oxygen atoms are bonded. Such structural formulae must be written in a way which will satisfy the valence of all atoms involved.

MOLECULES AND MACROMOLECULES

We have seen how the chemist is concerned with the structure of the molecules which comprise the substances in our world. At this point it might be of interest to consider the actual size of these molecules. It is now known that a molecule of water, for instance, measures about ten billionths of an inch, that is, one million such molecules could be laid side-by-side to make up the thickness of the type used on this page. Putting it another way, one drop of water contains 1,500,000,000,000,000,000,000 molecules. Most ordinary substances have molecules of this approximate size. Thus the molecule of ordinary sugar, which has the rather impressive chemical formula $C_{12}H_{22}O_{11}$, is still only a few times larger than the water molecule.

Such substances as water, sugar, ammonia, gasoline, baking soda, and so forth, have relatively simple molecules, whose size is about the same as indicated above. However, when chemists turned their attention to the chemical structure of some of the 'materials' like wood, leather, rubber, and so forth, they encountered some rather baffling questions. For instance, cellulose is the basic fibrous material which comprises wood, paper, cotton, linen, and other fibres. Chemists knew long ago that cellulose has the formula $C_6H_{10}O_5$, as shown by chemical analysis, yet their measurements did not indicate that the molecules of cellulose existed in the form shown by the formula. Instead, they found evidence that the cellulose molecule was unbelievably large. Hence they did not really believe that it was a molecule, but considered the true molecule to be $C_6H_{10}O_5$, and that hundreds of such molecules were bunched together in aggregates.

Rubber, like cellulose, was also classed as a colloid, i.e., a substance which contains large aggregates of molecules. Over 150 years ago, the formula for rubber was established as C_5H_8, but here again the evidence indicated that the molecule was much larger than shown by the formula. It was only as recently as the 1920's that *Staudinger* advanced the revolutionary idea that the rubber molecule was indeed a giant molecule, or macromolecule. Today, it is generally accepted that the class of materials which form the useful fibres, plastics, and rubbers are all composed of such macromolecules. Hence the formula for cellulose, if it is to represent one molecule, should more correctly be written as $(C_6H_{10}O_5)_{2000}$, while rubber should be represented as $(C_5H_8)_{20,000}$.

Such molecules are, of course, many thousands times larger than the molecules of ordinary chemical substances, hence the name 'macromolecules'. However, it should be remembered that even these giant molecules are far from visible in the best microscopes available. Thus the cellulose or rubber molecule may have a diameter several hundred times larger than that of a water or sugar molecule, but its size is still only several millionths of an inch. Even so, however, their relatively enormous dimensions, in comparison with ordinary molecules, give them the unusual properties observed in plastics, rubbers, and fibres.

POLYMERS AND MONOMERS

The molecular formula for rubber, $(C_5H_8)_{20,000}$, at first glance appears formidable indeed. There can obviously be a vast number of possible arrangements of 100,000 carbon atoms and 160,000 hydrogen atoms. To deduce the exact way in which they are arranged seems an impossible task. Yet it is not as difficult as it would seem. Fortunately—for the chemists—these huge molecules are generally composed of a large number of simple repeating units attached to each other in long chains. Thus the strucutral formula for the molecule of natural rubber may be represented by the simple unit C_5H_8 multiplied many thousand times. It is actually shown thus:

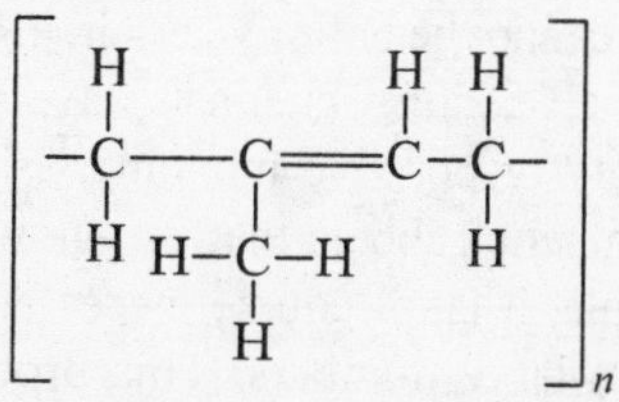

In the above structure, *n* represents a value of about 20,000, as stated previously. The arrangement of carbon and hydrogen atoms obeys the valence rules of 4 for carbon and 1 for hydrogen. The unattached bonds at each end of the units are, of course, meant to show attachments to adjacent units.

Because these giant molecules consist of a large number of repeating units, they have been named polymers, from the Greek 'poly' (many) and 'meros' (parts). The repeating unit shown above would, therefore, be called the monomer. It is actually very similar to a simple compound, known as isoprene, a low-boiling liquid. *Greville Williams* first isolated this compound as a decomposition product obtained when rubber was heated at elevated temperatures, out of contact with air. The relationship between the monomer and polymer structures, in this case, is illustrated below.

Monomer Polymer

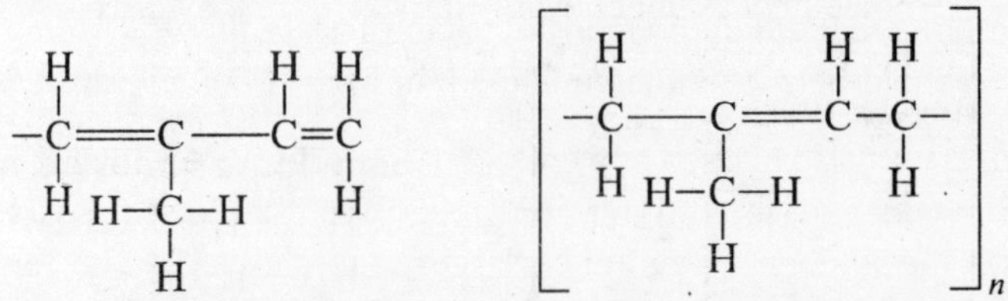

Isoprene Polyisoprene (natural rubber)

It can be seen here that the unit in the chain molecule corresponds exactly with the uncombined monomer molecule except for the necessary rearrangement of the bonds between the carbon atoms. Other examples are also shown below:

Monomer		*Polymer*	
Propylene	$CH_3CH{=}CH_2$	Polypropylene	$\left[CH_2-\underset{\displaystyle CH_3}{\underset{\vert}{CH}}\right]_n$
Vinyl chloride	$CH_2{=}CHCl$	Polyvinyl chloride	$\left[CH_2-\underset{\displaystyle Cl}{\underset{\vert}{CH}}-\right]_n$
Tetrafluoroethylene	$CF_2{=}CF_2$	Teflon	
Styrene	$H_5CH{=}CH_2$	Polystyrene	$\left[-CH_2-\underset{\displaystyle C_6H_5}{\underset{\vert}{CH}}-\right]_n$

The subscript n in the formula indicates the number of monomeric units contained in the polymer.

SYNTHESIS OF MACROMOLECULES

So far we have been discussing the macromolecules, or polymers, which occur ready-made in nature. Although the chemist may surmise, he cannot be entirely certain just how these large molecules are built up in nature. However, over the past half-century, chemists have become very proficient in producing synthetic polymers from a wide variety of possible monomers. It is from these processes that there have arisen all the new and varied synthetic rubbers that we know today.

In order to form a polymer molecule, it is, of course, necessary to find a way in which individual small molecules may join together in large numbers. This process is, for obvious reasons, called polymerisation.

Polymerisation is the combination of several simple molecules without the formation of by-products. In the general form, polymerisation can be represented by the equation:

$$nA \longrightarrow (A)_n$$

The simplest examples of polymerisation are olefin and their derivatives which polymerise as a result of opening of their double bond

$$nCH_2{=}CHX \longrightarrow -CH_2-CHX-CH_2-CHX-CH_2-CHX-$$

Polymerisation is always accompanied by a decrease in unsaturation of the reactants. When a compound with more than one double bond, e.g., a drene, polymerises the resulting macromolecular compound is less unsaturated than the monomer. Not only compounds containing multiple carbon-carbon bonds are capable of polymerisation but also compounds with multiple bonds between a carbon and some other atom (N,O,S).

Degree of Polymerisation

Number of units or mers, composing a chain is known as degree of polymerisation. Polymer mixtures are usually composed of chains of varying lengths, the term average degree of polymerisation $\overline{DP}$ is often used with the bar over the DP designating the value as an average.

There are a number of ways in which this process can be effected, but these can all be broadly classed under two main types of chemical reactions. One of these is known as addition polymerisation, since it involves a simple addition of monomer molecules to each other, without the loss of any atoms from the original molecules.

The other type is called condensation polymerisation, since it involves a reaction between monomer molecules during the course of which a bond is established between the monomers, but some of the atoms present are lost in the form of a by-product compound.

In this case, then, the polymer molecule is a condensed version of the original monomer molecules which reacted.

Backbone of Polymers

The principal chain in the polymer is known as backbone.

For polystyrene

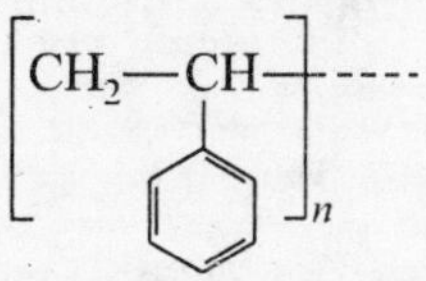

the backbone is simply ~C—C—C—C~ ...

For Nylon-6, $—(CH_2)_6—\underset{\underset{O}{\|}}{C}—NH$, the backbone is ~C—C—C—C—C—C—N~.

Linear Polymers

Macromolecules of linear polymers are long chains with a high degree of asymmetry (Fig. 1.1).

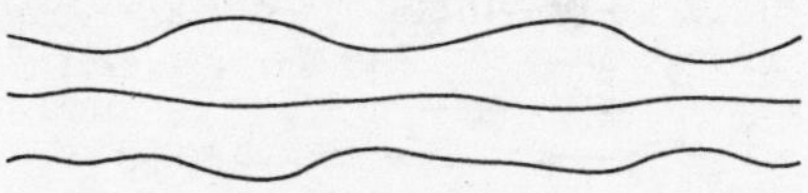

Fig. 1.1. Structure of linear polymer.

Some linear polymers are cellulose, natural rubber, certain proteins e.g., casein, Zein, amylose etc.

Branched Polymers

The macromolecules of a branched polymer are chains with branches. The number of branches and the ratio of the length of the main chain to that of the side chains may vary (Fig. 1.2).

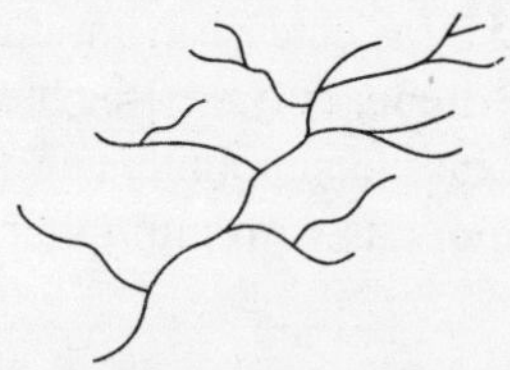

Fig. 1.2. Structure of branched polymers.

Amylopectin, glycogen and some mixed natural high molecular mass compounds show such structures. The synthesis of branched polymers has developed extensively in recent years.

Cross-linked Polymers

Such polymers are composed of macromolecular chain formed together with covalent bonds or cross-links as shown in Fig. 1.3.

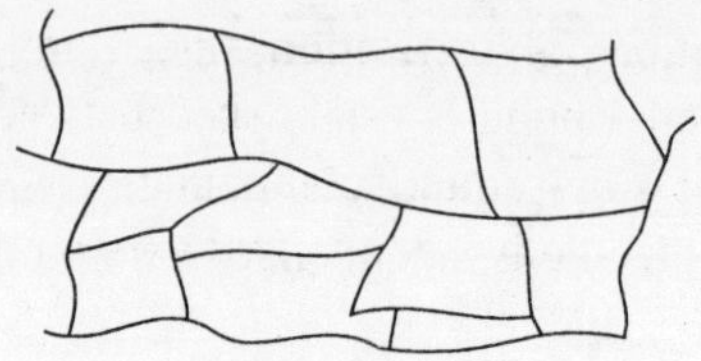

Fig. 1.3. Structure of cross-linked polymer.

Cross-linked polymers developed in three directions at right angles are called spatial or three dimensional polymers. Classical examples are diamond and quartz. Cross-linked polymers of two dimensional structure such as graphite are called lamellar polymers.

Co-polymers

Polymers having monomer units of different chemical composition are known as co-polymers e.g.,

A—B—B—C—A—C—C—B—A—B—C—

where A, B and C are monomer units of different chemical composition. Co-polymers include many proteins, lignin, nucleic acids, mixed polysaccharides and many synthetic high molecular mass compounds. In some co-polymers, monomeric units are arranged either at random or as alternate runs of each of them.

A—A—A—A—B—B—B—B—A—A—A—A

Such co-polymers are termed block co-polymers.

Graft Co-polymers

In some cases different monomer units can be added or grafted onto the linear chain of a macromolecule, such polymers are referred to as graft co-polymers, e.g.,

```
                          B
                          |
                          B
                          |
—A—A—A—A—A—A—A—A—...
   |        |
   B        B
   |        |
   B        B
```

Crystalline Polymers

Polymers may be amorphous (completely disordered) or partially ordered.

The amorphous phase (Fig. 1.4) consists of linear, branched or cross-linked forming tridimensional networks. Ordered structures have a tendency to crystallisation. Crystalline forms include, fringed, micelles, fibrils, folded lamellae, spherulites.

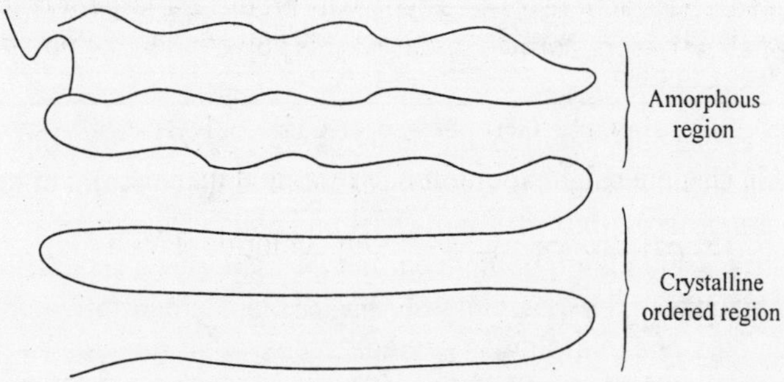

Fig. 1.4. Crystalline polymers.

In order to improve mechanical properties, crystallinity is induced in amorphous polymers. Crystalline polymers include polyethylene, polycaproamide, polyethylene terephthalate, gutta-percha etc. These are widely used in fabricating various articles.

ADDITION AND CONDENSATION POLYMERS

Polymer molecules are often classified as being of the addition or of the condensation type. This is somewhat misleading, as there is little difference between the resultant materials. The difference lies in the chemical reactions involved in the synthesis of the material, and a more accurate classification would be polymers formed by addition polymerisation and polymers formed by condensation polymerisation.

Carbon can exist in other valency states than the tetrahedral one already mentioned. In one of these the carbon atom forms a double bond with another carbon atom and two single bonds with other atoms or groups. These bonds are in one plane and make an angel of 120° with one another. The simplest substance, with this type of bonding, is ethylene.

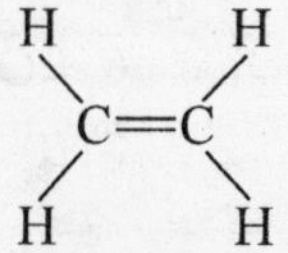

There is a large family of ethylenic compounds in which one or more of the hydrogen atoms have been replaced by other atoms or organic groups. The carbon-carbon double bond in this type of molecule will often break during a reaction. The two carbon atoms revert to the tetrahedral valency state and stay joined by a single bond. The fourth bond to each carbon atom is made with the other reactant. This is illustrated by a reaction of ethylene with chlorine.

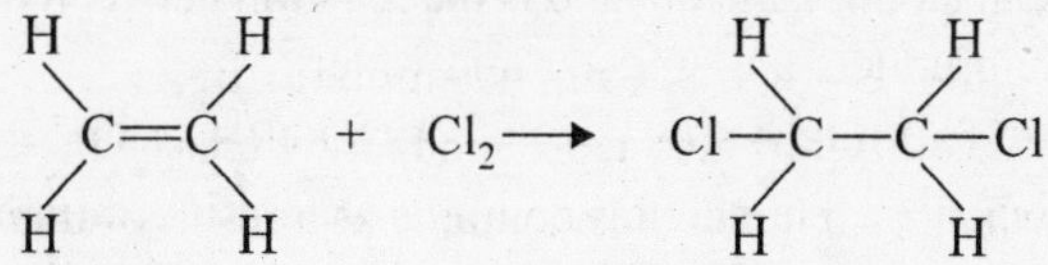

Reactions of this type are known as addition reactions.

Simple addition reactions with other reactants give rise to a new small molecule. It is sometimes possible to make molecules of this type react with themselves. When this occurs, the new bonds formed by the carbon atoms in one molecule, are with carbon atoms of other molecules, e.g.,

$$CH_2{=}CH_2 + CH_2{=}CH_2 \longrightarrow -CH_2-CH_2-CH_2-CH_2-$$

A large number of addition reactions of this type will produce a large molecule, which is a linear polymer. The simplest of all polymers, formed in this way, is polyethylene or polythene as it is commonly called.

$$nCH_2{=}CH_2 \longrightarrow -(CH_2-CH_2)_n-$$

Polymers made in this way are called addition polymers and include

$-(CH_2-CHCl)_n-$ Polyvinyl chloride (PVC)

$-(CH_2-CH(C_6H_5))_n-$ Polystyrene

and $-(CF_2-CF_2)_n-$ Polytetrafluoroethylene (PTFE)

Values of n may run into thousands.

Organic molecules exist in which there are two carbon-carbon double bonds, e.g., $CH_2{=}CH-CH{=}CH_2$, butadiene. The presence of two double bonds, which can take part in addition polymerisation, can lead to more complicated reactions and network polymers.

Many organic molecules contain groups, which are much more reactive than the carbon and hydrogen which make up the remainder of the molecule. When such materials react, it is these active groups which determine the nature of the reaction. They are called functional groups. Organic compounds can be classified in families, all the members of which contain the same functional groups, and are therefore chemically similar, e.g.,

Alcohols	R–O–H	such as	CH_3OH methyl alcohol
			C_2H_5OH ethyl alcohol
			$C_6H_5CH_2OH$ benzyl alcohol
Acids	$R-C(OH){=}O$	such as	CH_3COOH acetic acid
			$CH_2{=}CHCOOH$ acrylic acid
Amines	$R-N(H)-H$	such as	CH_3NH_2 methylamine
			$C_6H_5NH_2$ aniline

(R is used to denote an organic grouping containing carbon and hydrogen only).

Reactions between organic substances are often reactions involving only the functional groups. For instance, the reaction between an alcohol and an acid is the reaction between the hydroxyl group and the carboxylic acid group, and the products are an ester and water.

$$RCOOH + ROH \longrightarrow RCOOR + H_2O$$

A reaction of this type, in which two molecules combine as a result of interaction of their functional groups with the elimination of a simple molecule, is known as a condensation reaction.

The above reaction between monofunctional acid and alcohol gives an ester, which contains neither acid nor alcohol functional groups. Many organic compounds contain more than one functional group. If a dibasic acid is reacted with a dihydric alcohol, a series of condensation reactions occur. The first step is

$$\mathrm{HOOCRCOOH + HOR'OH \longrightarrow HOOCRCOOR'OH + H_2O}$$

then

$$\mathrm{HOOCRCOOR'OH + HOOCRCOOH \longrightarrow HOOCRCOOR'OOCRCOOH + H_2O}$$

or

$$\mathrm{HOOCRCOOR'OH + HOR'OH \longrightarrow HOR'OOCRCOOR'OH + H_2O}$$

The result is a new dibasic acid, or a new dihydric alcohol of increased size. These then continue to react with one another until a polymer molecule results. This has the form

$$\mathrm{-\!\!\left(O-\overset{O}{\overset{\|}{C}}-R-\overset{O}{\overset{\|}{C}}-O-R'\right)\!\!_n-}$$

Polymers made in this way are called condensation polymers.

Other pairs of functional groups undergo condensation reactions which can be used in polymer synthesis. These include acids with amines and acid chlorides with amines or alcohols. A molecule, which contains two different functional groups, can form a condensation polymer by a similar series of reactions, e.g., an amino acid

$$n\,\mathrm{H_2NRCOOH} \longrightarrow \mathrm{-\!\!\left(R-\overset{O}{\overset{\|}{C}}-\overset{H}{\overset{|}{N}}\right)\!\!_n-} + n\,\mathrm{H_2O}$$

Condensation reactions between molecules with more than two functional groups per molecule can lead to the formation of network polymers.

Details of the methods used to bring about these polymerisation reactions, and of the synthesis of polymeric materials are also discussed in various chapters of this book.

MECHANISM OF POLYMERISATION

Mechanism of polymerisation of single monomer is given below.

If M represents the monomer and $Init_2$ represents a free radical initiator then we can write,

$$Init_2 \xrightarrow{k_i} 2\ Init \qquad \text{Chain initiation}$$

$$\left.\begin{array}{l} Init + M \xrightarrow{k_p} Init\ M \\ \cdots \quad \cdots \quad \cdots \\ \cdots \quad \cdots \quad \cdots \\ Init\ M_n + M \xrightarrow{k_p} Init\ M_{n+1} \end{array}\right] \qquad \text{Chain propagation}$$

$$Init\ M_n + Init\ M_m \xrightarrow{k_t} Init\ M_{m+n}\ Init$$

or

$$\xrightarrow{k_t} Init\ M_n + Init\ M_m$$

Chain termination

Using a steady-state approximation on the concentration of all types of radicals $[R^{\bullet}]$ we obtain,

$$[R^{\bullet}] = \left(\frac{k_i}{k_t}\right)^{1/2} [Init]^{1/2}$$

and rate of propagation

$$\begin{aligned} &= k_p\,[R^{\bullet}]\,[M] \\ &= k_p\,(k_i/k_t)^{1/2}\,[Init_2]^{1/2}\,[M] \\ &= k\,[Init_2]^{1/2}\,[M] \end{aligned}$$

The above mechanism is also consistent with the kinetics of co-polymerisation.

POLYMER DEGRADATION

Degradation process involves breakage of bonds and results in a decrease in molecular mass of polymers.

$$\underset{\text{Starch}}{(C_6H_{10}O_5)_x} \longrightarrow \underset{\text{Maltose}}{\frac{x}{2}\,C_{12}H_{22}O_{11}} \longrightarrow \underset{\text{Glucose}}{C_6H_{12}O_6}$$

Degradation of polymers is a very important phenomenon. It is used for determining the structure of polymeric compounds and for obtaining valuable low molecular mass substances from natural polymers, e.g., glucose from cellulose and starch.

Types of Polymer Degradation

Polymer degradation may take place in several ways.

Chemical degradation

Polymer degradation may take place under the action of chemical agents such as water, acid, alcohol etc. The most common type of chemical degradation is by hydrolysis, i.e., cleavage of chemical bond with the addition of water molecules in presence of H^+ or OH^- ions.

Examples

1. Hydrolysis of the amide bond results in an amino group and a carboxyl group

$$-NH \xrightarrow{HOH} CO- \xrightarrow{\overset{+}{H}\text{ or }\overset{-}{OH}} -NH_2 + -COOH$$

2. Hydrolysis of cellulose: Cellulose degrades rapidly under the action of gaseous HCl.

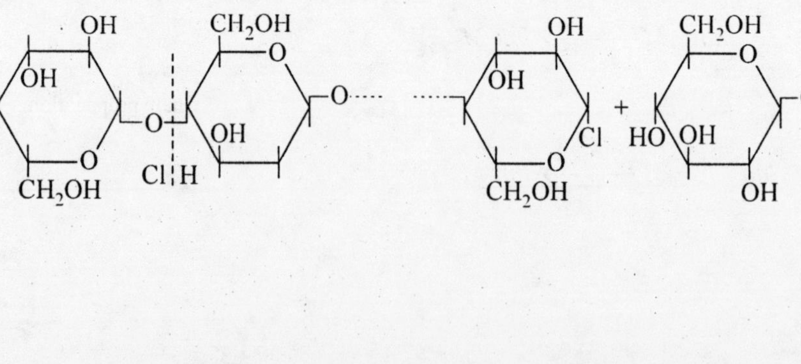

Photochemical Degradation

Degradation of polymers also occurs upon irradiation of light. Such degradation reactions play an important role in determining the life of natural and synthetic fibres, plastics and rubber goods.

$$A + h\nu \longrightarrow B$$

where $M_B < M_A$ (M = molecular mass).

Polyethylene decomposes in presence of light within two years whereas in the dark at ordinary temperature it does not decomposes at all.

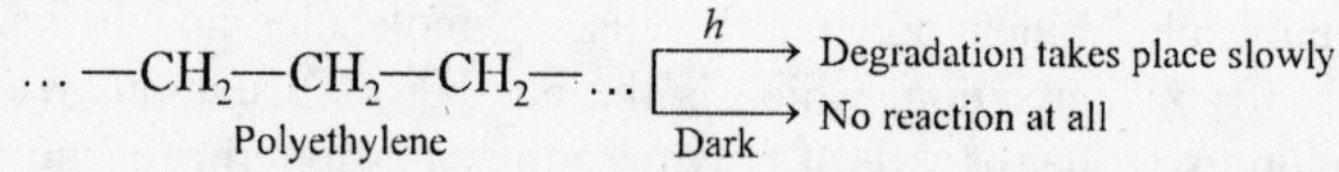

Besides the chemical (including photochemical) degradation, polymers also undergo degradation under the action of luminous energy, high energy radiation and thermal energy. The degradative reactions which result in the cleavages of carbon-carbon bonds in the polymer backbone are accelerated in the presence of oxygen. All these processes proceed according to a chain mechanism with the intermediate formation of free radicals.

The oxidation of a branched poly(vinyl chloride) proceeds according to the following equations.

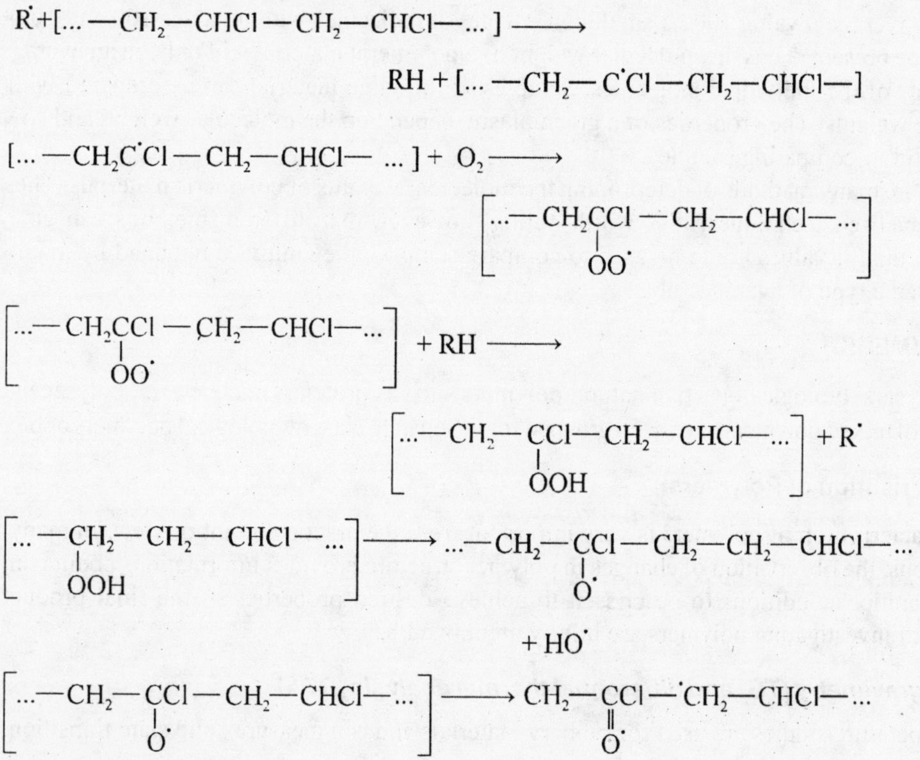

DILUTE POLYMER SOLUTIONS

Polymer solutions are considered dilute if practically no collisions occur between the macromolecules. For dilute solutions, the activity of the solute in solution is directly related to its mole fraction as the solute concentration goes to zero. When the activity of an added material is directly related to the mole fraction of material present, the property is referred to as a colligative property. Colligative property methods employed for the determination of molecular weights include boiling point elevation, freezing point depression, osmotic pressure etc.

MOLECULAR WEIGHT

For a cross-linked material, the term molecular weight has no significance, as the material does not exist in the form of discrete molecules. Linear polymers are discrete molecules, and their size can be expressed in terms of the degree of polymerisation, which is the number of structurally recurring units in the chain. It is more common to define the size of polymer molecules in terms of molecular weight.

In a substance which consists of small molecules, the molecular weight of each molecule is the same, because each one has the same composition. For instance, every molecule in a sample of TNT will have the formula $C_7H_5N_3O_6$ and the molecular weight 227. This is not the case with polymeric materials, which may have molecular weights thousands of times as large as this. The formula for a linear polymer is $-(M)_{\overline{n}}$ where M is the repeating unit and n the degree of polymerisation. In a given sample of a polymeric material, molecules with many different n values will be present. The spread of these values will depend on the nature of the material, and on the method of polymerisation, but in sample or average n value 1000 a small percentage of molecules with n values 100 and 10,000 would normally be present. Thus the molecular weight of a polymeric material will only be an average value of the weights of the individual molecules. Samples of the same material can be prepared with different molecular weights. The properties of a given plastic depend on the molecular weight, and so values are often useful in comparing samples.

There are many methods of determining the molecular weights of polymeric materials. These involve fairly specialised techniques. It is worth noting, however, that different methods can give different values, so that, if values are to be used to compare samples, they must be obtained by methods which give the same type of average value.

BIOPOLYMERS

In recent years, biologically active natural polymers such as proteins, nucleic acid, polysaccharides and a number of mixed polymers have been grouped into a separate class of biological polymers or biopolymers.

Characterisation of Polymers

The characterisation of polymers is very important from the practical point of view. In many technical applications, the observation of changes in polymer structure provides informations about compositions and preparation conditions to be chosen to achieve desired properties of the final product. Several methods of investigating polymers are briefly mentioned below:

Thermogravimetry (TG) and differential thermal analysis (DTA)

Low temperature studies are used for rubbery materials and can measure solid state transitions in them.

Differential scanning calorimetry (DSC)

It is used to study the influence of the composition and thermal history on crystallinity, melting point, glass transition temperature etc. Glass transition T_g is a characteristic temperature where the glassy, amorphous regions of polymers become flexible or rubber like when the temperature is increased.

Polarised light optical microscopy

Polarised light optical microscopy is used for the investigation of crystalline polymers. Spherulites show the typical maltese cross in polarised light which is caused by the birefringence effects.

Scanning electron microscopy (SEM)

Scanning electron microscopy (SEM) is a very powerful tool for studying the morphology of surfaces. The surface of non-conductive samples must be coated with a thin layer (20–50 nm) of gold or platinum. Following applications of SEM have been made: (i) study of surface microstructure of polymer films; (ii) investigation of liquid crystals; (iii) structure of co-polymer; (iv) polymer networks; (v) structural defects; (vi) sample roughness; (vii) fracture surface morphology; and (viii) grafting processes etc.

Transmission electron microscopy (TEM)

It is used to observe the internal microstructure of polymers.

Scanning tunnelling microscopy (STM) and atomic force microscopy (AFM)

These techniques allow examination of polymer samples from the scale of micrometers to angstroms.

ESCA spectrometry

Using this technique, electrons can not penetrate deeply into the material studied, thus surface roughness can be studied.

Nuclear magnetic resonance (NMR) spectrometry

This technique is an excellent one for revealing the environment of protons in molecules. Best results are often found by warming the polymer solution to increase the rate of movement of chain segments.

X-ray fluoroscence (XRF) spectrometry

The technique is used to determine Cl, S, P and Si quantitatively in polymers.

Mass spectrometry (MS)

This is well known technique that determines the molecular weight of polymer fragments. Study of these fragments can be used to suggest the structure of the original polymer.

Gel permeation chromatography

Gel permenation chromatography provides a method to obtain the molecular weight distribution from which molecular weight average can be estimated.

Infrared (IR) analysis

The co-polymer composition can be determined conveniently by IR spectrometry of thin films deposited on NaCl hull plates.

CHAPTER 2

Natural and Synthetic Polymers: An Overview

INTRODUCTION

Polymer is a generic name given to a vast number of materials of high molecular weight. These materials exist in countless forms and numbers because of a very large number and types of atoms present in their molecules. Polymers can have different chemical structures, physical properties, mechanical behaviour, thermal characteristics, etc., and can be classified in different ways, as now discussed.

Natural and synthetic polymers: Depending on their origin, can be grouped as natural or synthetic. Those isolated from natural materials are called natural polymers. Typical examples are: cotton, silk, wool and rubber. Cellophane, cellulose rayon, leather and so on are, in fact, chemical modifications of natural polymers.

Polymers synthesised from low molecular weight compounds are called synthetic polymers. Typical examples are: polyethylene, PVC, nylon and terylene.

Organic and inorganic polymers: A polymer whose backbone chain is essentially made of carbon atoms is termed an organic polymer. The atoms attached to the side valencies of the backbone carbon atoms are, however, usually those of hydrogen, oxygen, nitrogen, etc. The majority of synthetic polymers are organic and they are very extensively studied. In fact, the number and variety of organic polymers are so large that when we refer to 'polymers', we normally mean organic polymers. The molecules of inorganic polymers, on the other hand, generally contain no carbon atom in their chain backbone. Glass and silicone rubber are examples of inorganic polymers.

Thermoplastic and thermosetting polymers are those polymers which soften on heating and can be converted into any shape that they can retain on cooling. The process of heating, re-shaping and retaining the same on cooling can be repeated several times. Such polymers, which soften on heating and stiffen on cooling, are termed 'thermoplastics'. Polyethylene, PVC, nylon and sealing wax are examples of thermoplastic polymers. Some polymers, on the other hand, undergo some chemical change on heating and convert themselves into an infusible mass. They are like the yolk of the egg, which on heating sets into a mass, and, once set, cannot be re-shaped. Such polymers, which become an infusible and insoluble mass on heating, are called 'thermosetting' polymers.

Plastics, elastomers, fibres and liquid resins: Depending on its ultimate form and use, a polymer can be classified as plastic, elastomer, fibre or liquid resin. When, for instance, a polymer is shaped into hard and tough utility articles by the application of heat and pressure, it is used as a 'plastic'. Typical examples are polystyrene, PVC and polymethyl methacrylate. When vulcanised into rubbery products exhibiting good strength and elongation, polymers are used as 'elastomers'. Typical examples are natural rubber, synthetic rubber, silicone rubber. If drawn into long filament-like materials, whose length is at least 100 times its diameter, polymers are said to have been converted into 'fibres'. Typical examples are nylon and terylene. Polymers used as adhesives, potting compounds, sealants, etc., in a liquid form are described as liquid resins. Commercially available epoxy adhesives and polysulphide sealants are typical examples.

NATURALLY OCCURING POLYMERS

Carbohydrates are widely occuring compound, containing C, H and O with the empirical formula $(CH_2O)_n$. These products may be named as monosaccharides, oligoscharides and polysaccharides. Monosaccharides are the building blocks in polysaccharide. The principal polysaccharides are cellulose and starch.

Cellulose is the principal polysaccharide in living plants. It is a polydisperse linear poly (1 → 4)-β (anhydroglucose) or poly (1 → 4)-β-D-(glucopyranose). Starch, the second most abundant polysaccharide is obtained in various parts of plants such as seeds, roots, lubers and stems. This homopolysaccharide is formed by photosynthesis in corn, white potatoes, wheat, rice, millet, barley, cassava etc. Highly branched polysaccharides called dextrans are produced by the fermentation of sucrose by bacteria. Heparin, and anticoagulant which helps to prevent thrombosis, is a polysaccharide consisting of D-glucoronic acid and D-glucosamine in which the amino groups and an occasional hydroxyl group are sulphated. Antibiotics such as streptomycin are tri and tetrasaccharides.

Protein, like polysaccharides are heteropolymers . It is essential to all forms of life and are of prime importance for the life processes in plants and animals. Proteins may be water soluble as well as water insoluble. The water insoluble class includes silk, wool and hair. Polyamide is a simple protein and all simple proteins contain peptide linkage 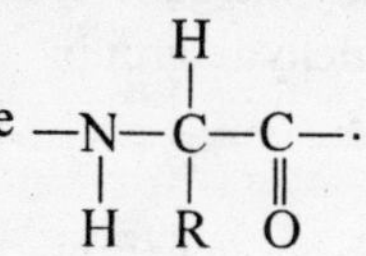.

Glycyl glycine 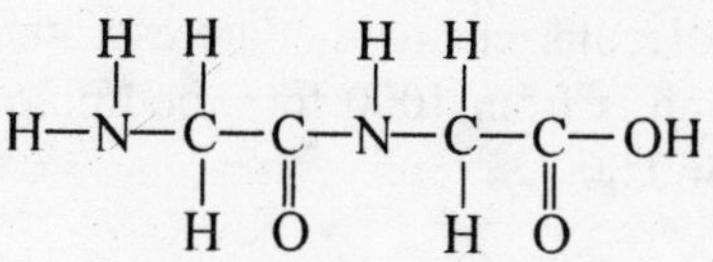 is a simplest peptide molecule containing only one peptide link.

An α-amino acid 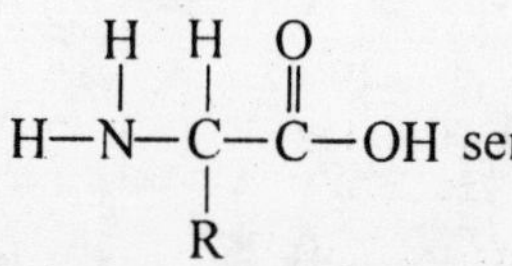 serves as building blocks for proteins.

Natural rubber differs from *gutta-percha* in the spatial arrangement but having the same monomeric units. In rubber the first and fourth carbon atoms are in *cis*-configuration (Fig. 2.1) relative to the double bond whereas in *gutta-percha* they are in the *trans*-configuration (Fig. 2.2).

Besides the naturally occurring polymers, a wide variety of polymers of technological importance have also been synthesised. The output of synthetic polymers obtained from low molecular mass starting material, keeps growing from year to year. Industries producing plastics, synthetic fibres, rubber, paints, electrical insulators are developing at a rapid pace.

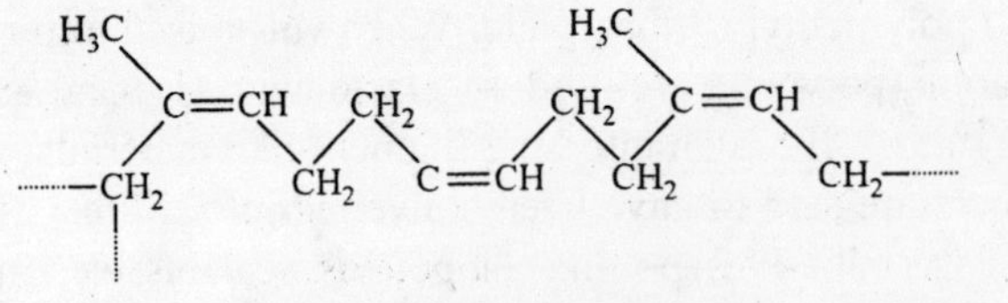

Fig. 2.1. *Cis*-isomer (rubber).

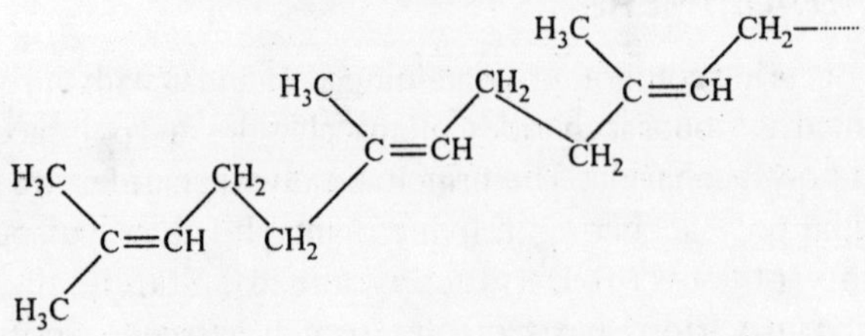

Fig. 2.2. *Trans*-isomer (*gutta-percha*).

Polysaccharides

Polysaccharides are natural polymers. These are high molecular weight molecules made up of monosaccharides (D-glucose) linked together through oxygen atoms to form chain of varying lengths. The monosaccharide units are linked by glycoside linkages. As a class they occur widely in plants and animals; and some serve as reserve for food supplies. Two of the important polysaccharides are starch and cellulose both of which can be represented by the general formula $(C_5H_{10}O_5)_n$. Polysaccharides do not exhibit the characteristic reactions of the aldehyde group.

Cellulose

Cellulose $(C_6H_{10}O_5)_n$, a natural carbohydrate high polymer (polysaccharide) consisting of anhydroglucose units joined by an oxygen linkage to form long molecular chains that are essentially linear. It can be hydrolysed to glucose. The degree of polymerisation is from 1000 for wood pulp to 3500 for cotton fibre, giving a molecular weight from 160,000 to 560,000.

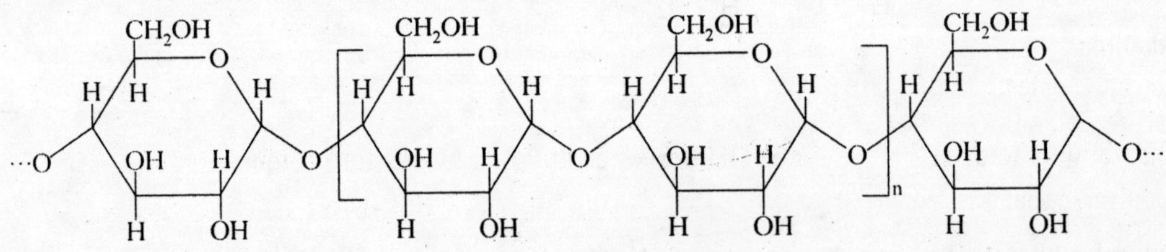

Cellulose (β-linkage)
n = ~ 750–3000

Cellulose is a colourless solid, specific gravity about 1.50, insoluble in water and organic solvents; it will swell in sodium hydroxide solution and is soluble in *Schweitzer's* reagent. It is the fundamental constituent of all vegetable tissues (wood, grass, cotton, etc.) and is the most abundant organic material in the world. Cotton fibres are almost pure cellulose; wood contains about 50 per cent.

The physical structure of cellulose is unusual in that it is not a single crystal, but consists of crystalline areas embedded in amorphous areas. Chemical reagents penetrate the latter more easily than the former.

Cellulose is virtually odourless and tasteless, but it is combustible, with an ignition point of about 450°F; in some forms it is flammable. For example, railroad shipping regulations require a flammable label on such items as 'burnt fibre', 'burnt cotton', 'wet waste paper' and 'wet textiles'. These were undoubtedly due to heat build-up in the paper caused by microbial activity and self-sustaining oxidation.

The most important uses of cellulose are—bulk woods of many kinds; paper, most of which is made from wood pulp; cotton products (clothing, sheeting, industrial fabrics); packaging, ranging from wooden barrels to candy pats; and as a source of ethyl alcohol (enzymatic hydrolysis) and methyl alcohol (destructive distillation of wood). Specialised uses include non-woven fabrics, medical equipment (artificial kidney), insulation and soundproofing, sausage casings, etc. Cellulose has about 60 per cent of the energy content of bituminous coal; its use as a fuel has increased, especially in rural locations.

There are many chemical modifications of cellulose, including its esters (cellulose acetate), ethers (methylcellulose), the nitrated product (nitrocellulose), as well as rayon and cellophane (from cellulose xanthate). Thus it is the basis of many plastics, fibres, coatings, lacquers, explosives, and emulsion stabilisers. Alkali cellulose is an intermediate made by the action of sodium hydroxide solution on cellulose and is used for making cellulose ethers and viscose.

Cellulose exists in three forms—*alpha*, *beta*, and *gamma*. *Alpha*-cellulose has the highest degree of polymerisation and is the chief constituent of paper pulp. It is insoluble in strong caustic soda solution. The *beta* and *gamma* forms have much lower DP and are known as hemicellulose.

Cellulose can be decomposed to glucose by the enzyme cellulase or by hydrolysis. Cellulosic materials (paper, cotton and textile wastes) when wet with water are a dangerous fire hazard and subject to shipping regulations.

Determination of molecular weight

Determination of the molecular weight of highly polymerised substances by physical methods is not easy. The usual cryoscopic and ebullioscopic methods are not applicable, since excessively high concentrations are required in order to obtain measurable effects. Three more sensitive methods are based upon osmotic pressure, viscosity, and the equilibrium or rate of sedimentation in a centrifugal field. The determinations required in the first method are by no means easy, and several measurements are necessary in order to extrapolate to infinite dilution, where molecular weight can be assumed to be inversely proportional to osmotic pressure. The method is limited because low molecular weight fractions exert a greater influence than high molecular weight fractions and accurate measurements can be obtained only with substances of molecular weight less than 500,000. The viscosity method, as applied to linear macromolecules, is based upon a relationship, developed experimentally by Staudinger and later theoretically interpreted by *W. Kuhn*, according to which viscosity is proportional to molecular weight. On theroretical grounds determinations by the ultracentrifuge method are probably the soundest. All three methods are uncertain in that some degradation may attend solution of cellulose. Although cellulose is insoluble in the usual solvents, it can be dissolved in ammoniacal cupric hydroxide (*Schweitzer's* solution). Various nitro and acetyl derivatives are more soluble and are often used for check determinations. Values obtained by the three methods indicate an average molecular weight of 300,000–500,000.

The formula for cellulose derived from chemical evidence has been confirmed by X-ray examination, which reveals the fine structures of the cellulose molecule. The length of the unit cell along the chain axis (frequency period) is 10.3Å, whereas the value calculated for one cellobiose unit is 10.25Å; consequently the chains must be nearly straight with respect to the fibre axis. The fibres show crystalline regions that are explained on the basis of the micellar theory; this postulates that the crystalline supermolecular unit is made up of bundles of parallel orientated chains (micelles). The width of a micellar unit is about 60Å, corresponding to 100–200 cellulose chains; the length is at least 600Å, corresponding to a minimal chain length of 200 glucose units. The remarkable mechanical strength and chemical stability are considered to be a result of micellar structure. Cellulose swells to some extent in water or dilute alkali, but since the crystal lattice is not destroyed, the reaction is said to be intermicellar. The lattice is destroyed whenever solution is effected.

Man and other carnivorous animals are unable to utilise cellulose as food, since they lack enzymes required for hydrolysis. Many micro-organisms, some protozoa, and the snail can decompose cellulose. Digestion of cellulose by ruminants (cud-chewing animals) is due to the presence of micro-organisms within the specially constructed alimentary system.

Hemicelluloses

Of several polysaccharides that occur in association with cellulose, two are pentosans; that is, they yield pentoses on hydrolysis. The more common one, xylan, is built from D-xylose units linked in the 1- and 4-positions. It is closely related to polyglucuronic acid with which it is associated in nature, and from which it may be produced by decarboxylation. The pentosans occur in large amount (20–40 per cent) in cereal straws and brans, and are used for the large-scale industrial preparation of furfural. The pentosans are hydrolysed by sulphuric acid, which then dehydrates the liberated pentoses.

Chitin

Chitin is a polysaccharide that forms the hard shell of crustaceans and insects. Chitin is analogous to cellulose; for example, the identity period along the fibre axis is 10.4Å, and hence the units probably are linked as in cellulose.

Starch

Starch is a polymer of glucose and occurs naturally as granules in the roots, seeds and stems of various plants. It comprises a large percentage of potatoes, cereals, rice, and barley. Starch constitutes the main carbohydrate reserve of plants. The animal starch is called glycogen. Partial hydrolysis of starch yields maltose while complete hydrolysis either enzymatically or chemically forms D-glucose. The hydrolysis takes place through dextrin (the polysaccharides of intermediate chain length that are formed from starch components by the action of amylase are called dextrins). Starch is not a single substance but consists of a mixture of two main types of polysaccharides, amylose and amylopectin. These two portions can be separated by fractionating starch with hot water.

$$\text{Starch} \xrightarrow{H_3O^+} \text{dextrin} \xrightarrow{H_3O^+} \text{maltose} \xrightarrow{H_3O^+} \text{glucose}$$

Amylose constitutes 20 per cent of starch and is sparingly soluble in hot water. It is a linear polysaccharide, and the glucose molecules are linked by α-glucosidic linkage at C-1 and C-4 atoms.

Amylose is responsible for the blue colour of the starch iodine test.

Formally, the only difference between amylose and cellulose is the stereochemistry of the glycosidic bond.

The molecular weight of amylose ranges from 10,000–50,000 (osmotic pressure). The value derived by end-group assay is in the same range; therefore amylose contains only one end-group per molecule and is a long-chain molecule structurally related to cellulose.

Molecular weights of amylopectin samples are in the range 50,000–1,000,000 (osmotic pressure). Since end-group assay (periodate procedure) indicates one end-group for each 25–27 units, the structure must consist of branched chains. Hydrolysis of exhaustively methylated amylopectin furnishes 2,3,6-trimethylglucose, and in addition small amounts of 2,3,4,6-tetramethylglucose derived from end-groups and of 2,3-dimethylglucose corresponding to branching points. Therefore the main chains are composed of glucose units joined by 1,4-links and the branches are linked to it by 1,6-links.

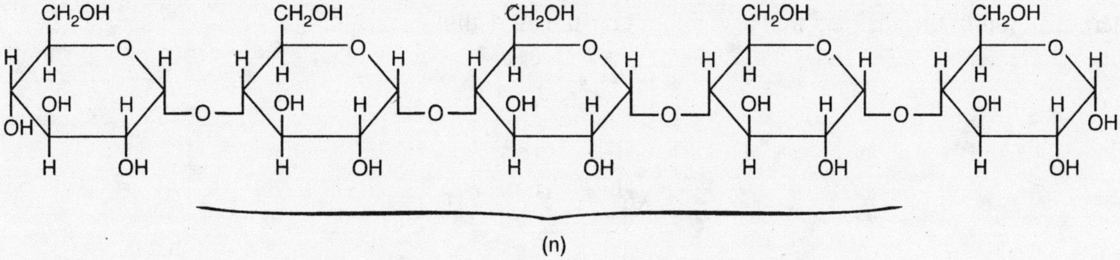

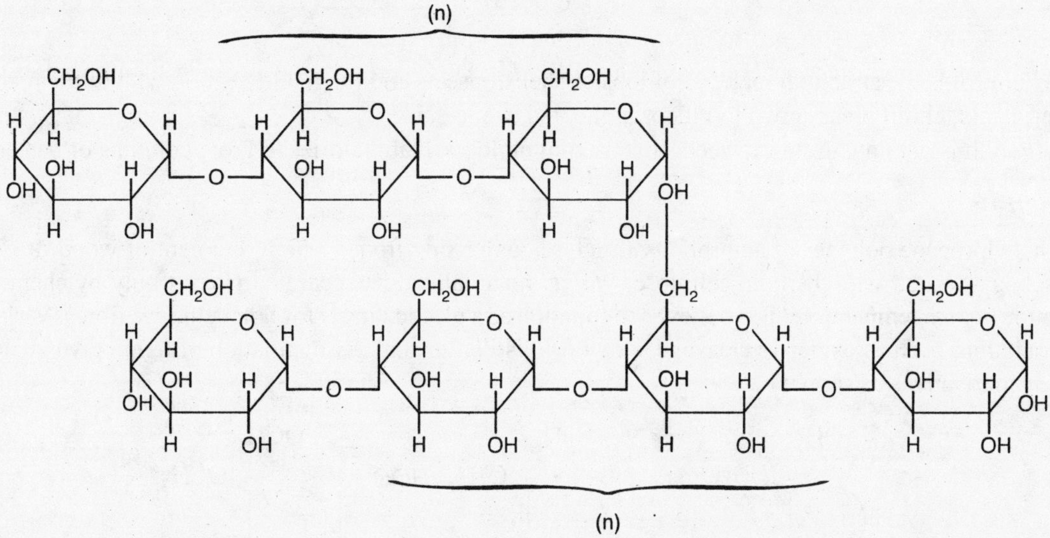

Amylopectin: Starch granules comprise about 13 per cent of amylopectin. It is a highly branched molecule and has an average molecular weight of over one million. A partial structure is given above.

Because of its branched structure, starch is not suitable for plastics and fibres like cellulose. It is, however, useful as food because animals have the enzymes necessary to break the α-glucoside linkages.

Other important polysaccharides

Glycogen: Starch is a major nutrient for animals. It is converted to (+)-glucose by animal metabolism. Glucose is re-polymerised to form glycogen in the liver. It serves as a reserve energy source. Glycogen consists of branched chain of glucose molecules. It gives a brown to violet colour with iodine, resembling that of grenzdextrin. It is composed entirely of glucose units joined as in maltose, a product of enzymatic hydrolysis; and the end-group method indicates one end-group for every 12–18 glucose units. Glycogen is very similar to amylopectin, since methylated glycogen yields the same three methylated glucose derivatives as methylated amylopectin. One difference, indicated by an end-group content of 9 per cent as compared with 5 per cent for amylopectin, in that the degree of branching is about twice as great. Another is that the proportion of the molecule hydrolysed by amylase is slightly smaller (55–60 per cent).

It occurs in almost all animal cells and some plants and is the main carbohydrate food reservoir of animals.

Inulin: It is another reserve carbohydrate of plants. It occurs in dahlia bulbs and in Jerusalem artichokes. An example is inulin, the polysaccharide of the *Compositae* (dahlia, dandelion). It was originally believed to consist entirely of fructose units but more recent results have shown that glucose is a terminal unit of the chain as well as a minor constituent within the chain.

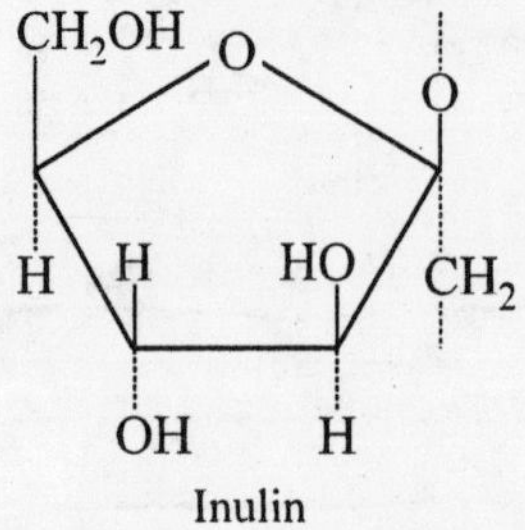

Inulin

Chitin: It is the structural material of lobster shells, insects and fungi. On heating with dilute hydrochloric acid, chitin break down to glucosamine and acetic acid.

Agar: It is obtained from seaweed. It is used in biological laboratories to grow columns of bacteria.

Lignin

A phenylpropane polymer of amorphous structure comprising from 17 to 30 per cent of wood. It is so closely associated with the holo cellulose. Woody material can be separated from it only by chemical reaction at high temperature. It is believed to function as a plastic binder for holocellulose fibres. Various degradations such as oxidative cleavage, zinc dust distillation and dry distillation of lignin give guaicol and its derivatives as shown below:

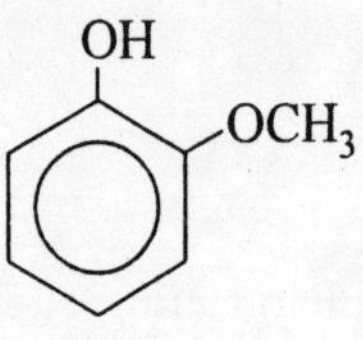

(Guaicol)

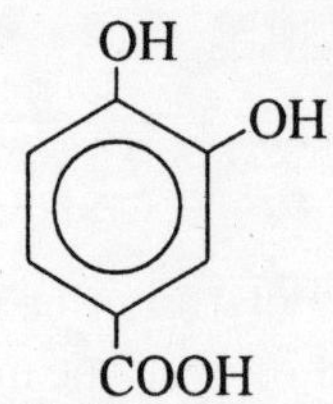

(Protocatechuic acid)

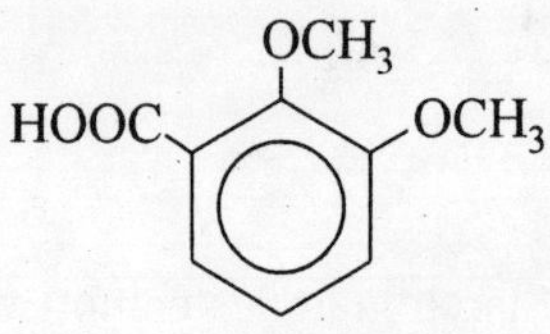

(Isohempinic acid)

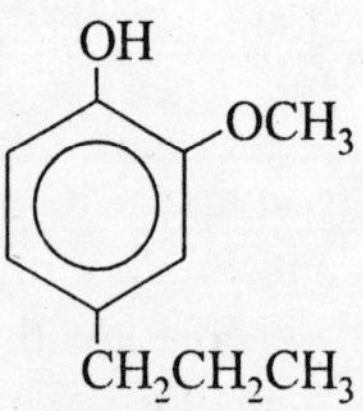

(4-*n*-propyl guaicol)

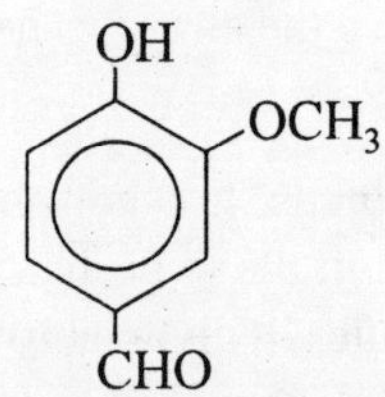

(Vanillin)

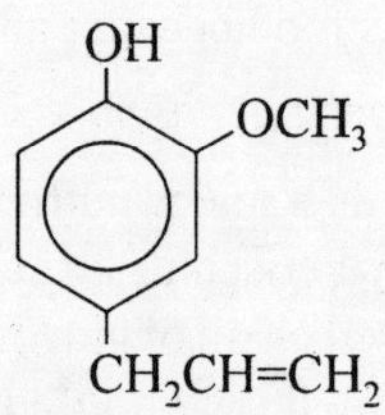

(Eugenol)

It is recovered from wood processing wastes in limited amounts. Lignin finds numerous uses; as a dispersant and wetting agent used in oil well digging muds, in adhesives, road binders, industrial cleaners and leather tanning. It is also used in the industrial preparation of vanillin.

Glycogen

Glycogen is a highly branched poly(anhydroglucose) with a structure somewhat similar to that of amylopectin. It has a chain length of about 10–14 glucose units.

PROTEIN

A complex high polymer containing carbon, hydrogen, oxygen, nitrogen, and usually sulphur, and comprised of chains of amino acids connected by peptide linkages (–CO.NH—).

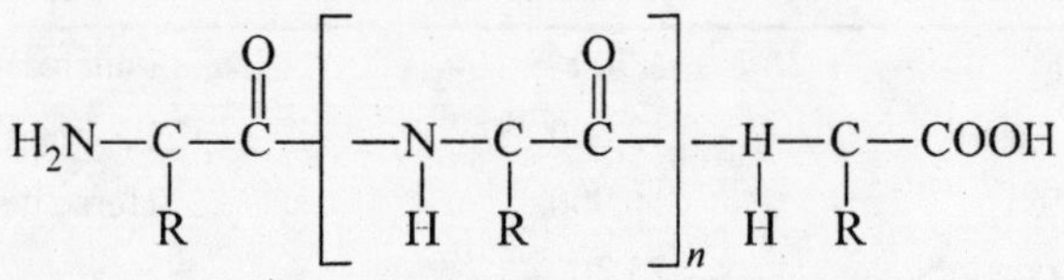

Proteins occur in the cells of all living organisms and in biological fluids (blood plasma, protoplasm). They are synthesised by plants largely because of the nitrogen-fixing ability of certain soil bacteria. Their molecular weight may be as high as 40 million (*Tobacco mosaic* virus). They have many important functional forms; enzymes, haemoglobin, hormones, viruses, genes, antibodies, and nucleic acids. They also comprise the basic component of connective tissue (collagen), hair (keratin), nails, feathers, skin, etc. Some have been synthesised in the laboratory.

The sequence of amino acids in the polypeptide chain is of critical importance in genetics. Proteins can be hydrolysed to their constituent amino acids and can be broken down into simpler forms by proteolytic enzymes. They form colloidal solutions, and behave chemically as both acids and bases simultaneously. They are denatured by changes in pH, by heat, ultraviolet radiation, and many organic solvents.

Simple proteins contain only amino acids; conjugated proteins contain amino acids plus nucleic acids, carbohydrates, lipids, etc. On the basis of solubility they can be classified as albumins (water-soluble); globulins (insoluble in water but soluble in aqueous salt solutions), and prolamines (soluble in alcohol-water mixtures but not in alcohol or water alone). A number of proteins have been synthesised, notably the hormone insulin. Proteins are an essential component of the diet, occuring chiefly in meat, eggs, milk and fish. Edible proteins suitable for human food as well as cattle feed can be produced from micro-organisms grown in carbonaceous or nitrogenous media to form yeast-like materials. Paraffinic hydrocarbons (methane) and petroleum-derived ethyl alcohol can be used as growth media for protein biosynthesis.

Physical Properties of Proteins

Most of the proteins are colourless, tasteless, and insoluble in water, alcohol and ether. They posses high molecular weights and are optically active (laevorotatory). The proteins structure can be disrupted on heating, on treatment with mineral acids or on exposure to light. This process is called denaturation. It is the sensitivity of proteins to denaturation that makes its isolation difficult.

Chemical properties of proteins

1. Proteins form salts with bases. Casein, for instance, is present in milk as its calcium salt.
2. On acid or enzymatic hydrolysis of proteins, amino acids are formed. Digestive enzymes hydrolyse proteins to peptides and amino acids which are absorbed and transferred by the blood to various body tissues.

3. Proteins respond to several colour tests:
   (a) With ninhydrin, they give a purple colour.
   (b) Millon's reaction yields a pink colour.
   (c) A yellow colour is obtained with iodine.

Examples of some selected proteins along with their sources of extraction are given in Table 2.1.

Table 2.1. Selected proteins and their sources.

Protein	*Mol. weight*	*Source*
Insulin	5,700	Pancreas (beef)
Ribonuclease	12,600	Pancreas (beef)
Myoglobin	16,900	Horse heart
Chymotrypsinogen	23,200	Bovine
β-Lactoglobulin	35,400	Milk
Zein	40,000	Corn
Fibroin	56,000	Silk
Haemoglobin	64,500	Human
Hexokinase	102,000	Yeast
Glutenin	120,000	Wheat

Structure of proteins

There are four types of structures that can be associated with proteins.

Primary structure

This refers to the covalently bonded structure of an amino acid and the order in which the amino acids are joined along the chain. The determination of primary structure employs the same experimental procedure as are used for the peptides. The elucidation of the amino acid sequence of a protein was first successfully established for insulin by *F. Sanger*. The establishment of the primary structure of other proteins such as myoglobin trypsin, haemoglobin, and others quickly followed (Fig. 2.3).

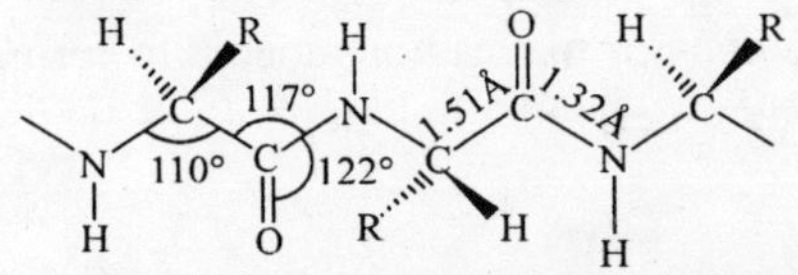

Fig. 2.3. Primary structure of protein.

Secondary structure

This structure results from interaction by H-bonding between the carbonyl and the imide group of the peptide bonds which appear in a regular fashion along the chain. By a careful X-ray crystallographic study showed that the peptide bond was a rigid planar structure.

The hydrogen atom of —N(H)— group is nearly always *trans* to the Fig. 2.4.

Moreover, there is no freedom of rotation about the C—N bond of the peptide, as this bond has a partial double bond character. It has been investigated that a protein can fold itself in two ways; (i) α-helix; and (ii) a pleated sheet.

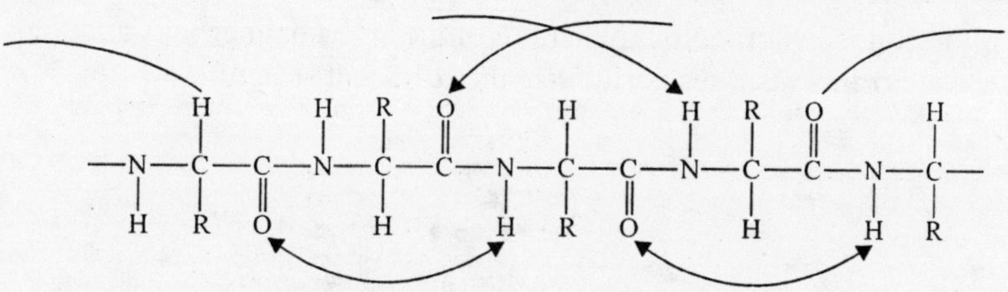

Fig. 2.4. Secondary structure of protein.

The α-helix is a spiral that can be either right-handed or left-handed (Fig. 2.5). The tightly coiled polypeptide main chain forms the inner part of the rod and the side-chains extend outward in a spiral form.

The α-helix is stabilised (Fig. 2.6) by H-bonding between $-\underset{\text{H}}{\underset{|}{\text{N}}}-$ and $-\overset{\text{O}}{\overset{\|}{\text{C}}}-$ groups of the main chain. The $-\overset{\text{O}}{\overset{\|}{\text{C}}}-$ group of each amino acid is hydrogen-bonded to the $-\underset{\text{H}}{\underset{|}{\text{N}}}-$ group of the amino acid that is situated for residues ahead in the liner sequence.

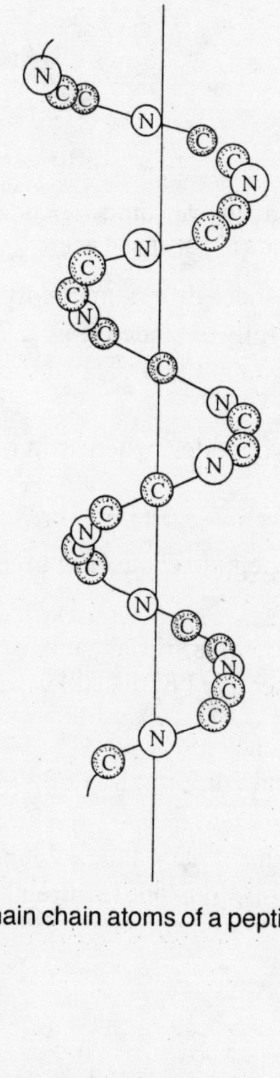

Fig. 2.5. Orientation of the main chain atoms of a peptide about the axis of an α-helix.

The stability of the coil is governed by the nature of the side-chain groups and their sequence along the chain, for example, proline residues destabilise the α-helical structure.

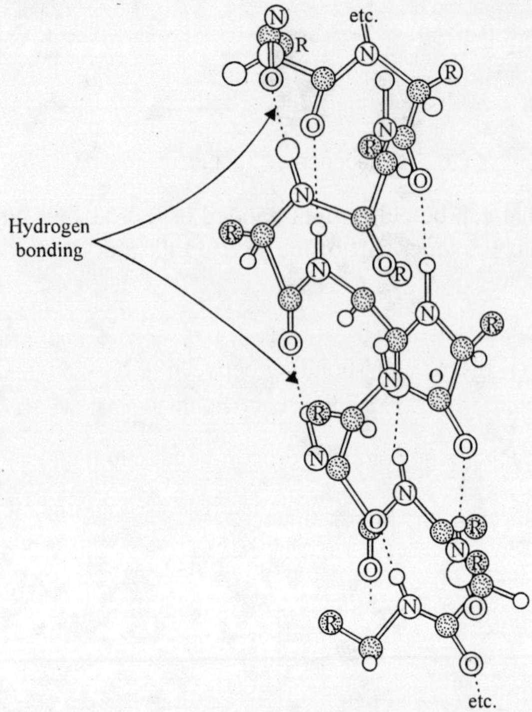

Fig. 2.6. Stabilisation of an α-helical configuration by hydrogen bonding to yield a secondary structure. All the shaded balls represent carbon atoms or residues (R) of amino acids.

An alternative structure the β-pleated sheet differs markedly from the α-helix in that it is a sheet rather than a rod; the polypeptide chain is fully extended (Fig. 2.7) rather than being tightly coiled as in α-helix. The second important difference is that the β-pleated sheet is stabilised by H-bonding between $-\overset{\overset{\displaystyle O}{\|}}{C}-$ and $-\underset{\underset{\displaystyle H}{|}}{N}-$ groups in different polypeptides, whereas in the α-helix the H-bonds occur between $-\overset{\overset{\displaystyle O}{\|}}{C}-$and $-\underset{\underset{\displaystyle H}{|}}{N}-$ groups in the same polypeptide chain. The secondary α-helix structure is particularly common in keratins which are fibrous proteins of hair and wool. Extension of the β-pleated sheet structure has been found in silk fibrogin secreted by the silkworm *Bombyse mori.*

Tertiary structure

This refers to a complete three-dimensional conformation of a protein to a compact and tightly folded structure.

This structure is far more complex than primary and secondary structures because the polypeptide chains of proteins are not arranged along one axis but are irregularly and compactly folded.

Quaternary structure

This type of structure refers to how individual polypeptide chains of a protein having two more chains are arranged in relation to each other. For example, haemoglobin is composed of four polypeptide chains linked together non-covalently.

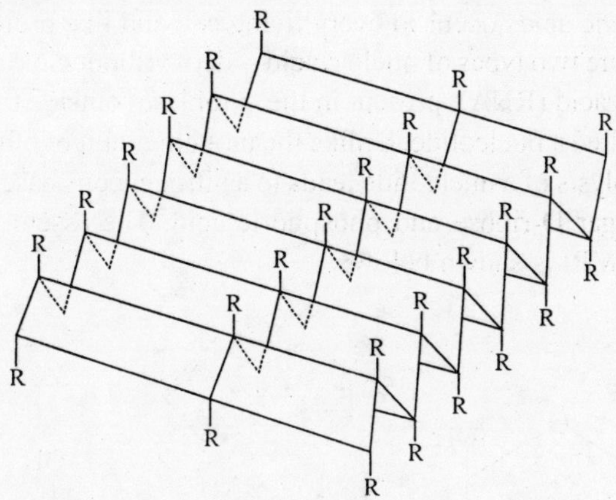

Fig. 2.7. An extended polypeptide chain of β-pleated sheet.

Role of proteins and enzymes in biological processes

The biological processes are extremely diverse and various proteins have a different role to play. Proteins are vital for the living organisms. They constitute an essential component of animal and human diet. A few functions of some specific proteins may be described here. Haemoglobin acts as an oxygen carrier from lungs to the tissues, nucleoproteins form the important constituents of the cells that supply and transmit genetic message in cells, sperum albumin binds free fatty acids tightly, liporotein of blood plasma transport liquids between the intestine, liver and adipose (fatty) tissues. The blood proteins thrombin and fibrinogen participate in blood clotting and thus prevent the loss of blood. Many of the hormones such as insulin and thyroxins which are proteins regulate various metabolic processes in the body. An enzyme is a protein that is synthesised in a living cell and catalyses or speeds up a thermodynamically possible reaction. A catalyst is a substance which provides energy and thus makes the reactions go faster. *Louis Pasteur* recognised that fermentation is catalysed by enzymes. Today approximately 2,000 different enzymes are known, many of them have been isolated in pure state. Many enzymes are named by putting a suffix *ase* to the name of the substrate which they catalyse. For instance, urease, catalyses urea to ammonia and carbon dioxide, phosphates catalyse the hydrolysis of phosphate esters. Some enzymes are given chemical names such as trypsin or pepsin. Many enzymes, in addition, to a protein part contain a non-protein component called the prosthetic groups or co-enzyme.

Enzymes work under mild conditions and are extremely specific about the reaction that they catalyse. In other words, a given enzyme attacks a particular substrate only. The enzyme catalysed reactions are much more faster than the corresponding uncatalysed ones.

Industrial applications of proteins

Industrial applications of proteins include plastics, adhesives and fibres derived from casein and soybean protein, but these have been declining in recent years. Special forms in which proteins are commercially available include textured proteins for food products, protein hydrolysate and liquid pre-digested protein, both for medical use.

Nucleic Acids

Nucleic acids are biomolecules and play an important role in the storage of genetic information and in the biosynthesis of proteins. Nucleic acids occur in every living cell and like proteins they are polymers of high molecular weight. There are two types of nucleic acid—deoxyribonucleic acid (DNA), a constituent of the nucleus and ribonucleic acid (RNA), present in the cytoplasm outside the nucleus. The monomer unit of these biopolymers is called a nucleotide. Unlike the monomer unit of other polymers, a nucleotide can be hydrolysed. Base hydrolysis of a nucleotide leads to a nitrogeneous base (a derivative of purine or pyrimidine), a five-carbon sugar D-ribose and phosphoric acid. The assembly of typical nucleotide, adenosine monophosphate (AMP) is shown below:

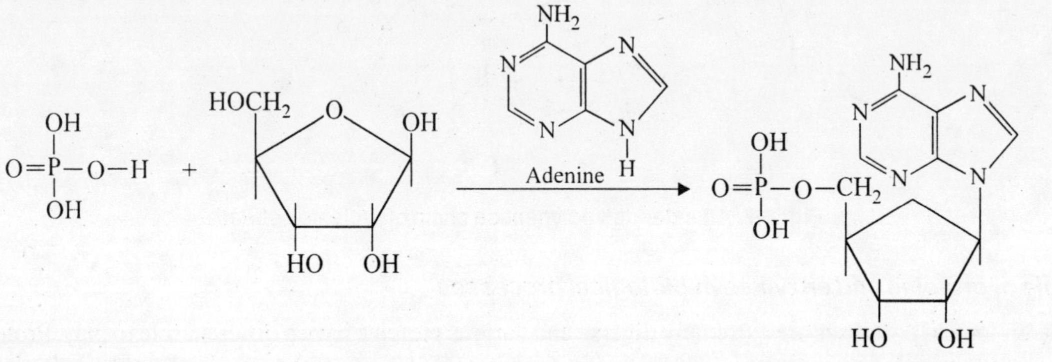

Phosphoric acid Ribose AMP

The nucleotides of DNA are built from deoxyribose (*deoxy* means lacking in an oxygen atom) and those of RNA, from ribose.

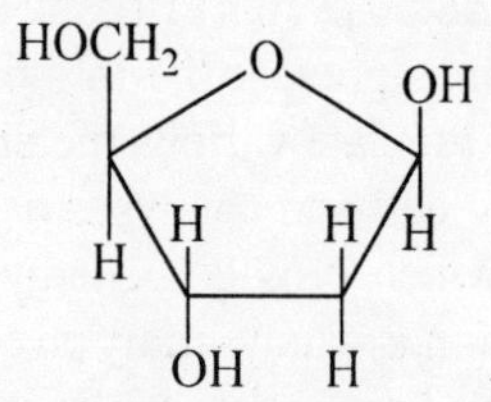

Deoxyribose

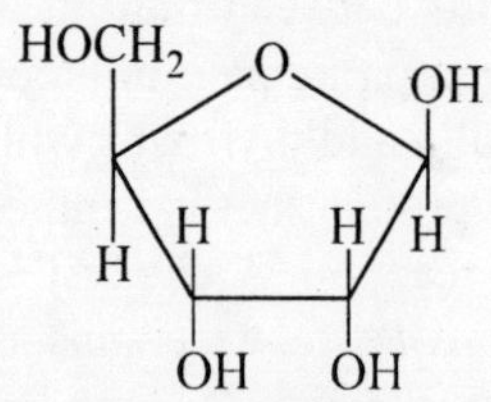

Ribose

Both DNA and RNA consist of a long chain alternating sugar and phosphate residues.

Hydrolysis of nucleic acids yields five heterocyclic bases. These are common to both DNA and RNA—adenine, guanine and cytosine.

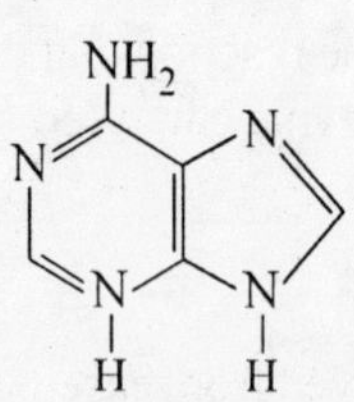

Adenine

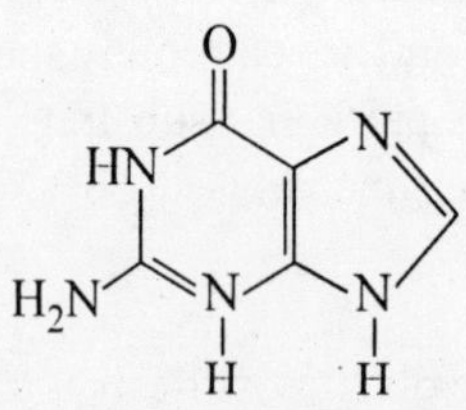

Guanine

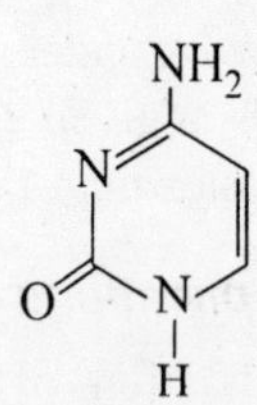

Cytosine

Of the remaining two-thymine is found only in DNA, while uracil only in RNA.

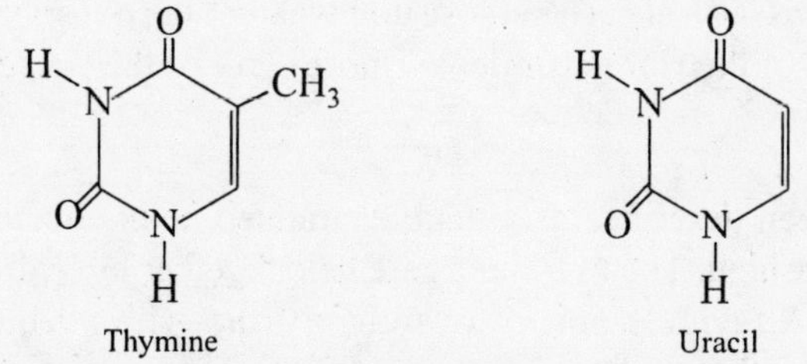

Various nucleotides combine by the loss of a molecule of water from the phosphoric acid unit of one nucleotide with an alcohol unit of the sugar residue of the next molecule. This gives rise to a phosphate ester linkage. This process is repeated until hundreds of nucleotide units are incorporated into the polymer. A segment of deoxyribonucleic acid (DNA) is shown in Fig. 2.8.

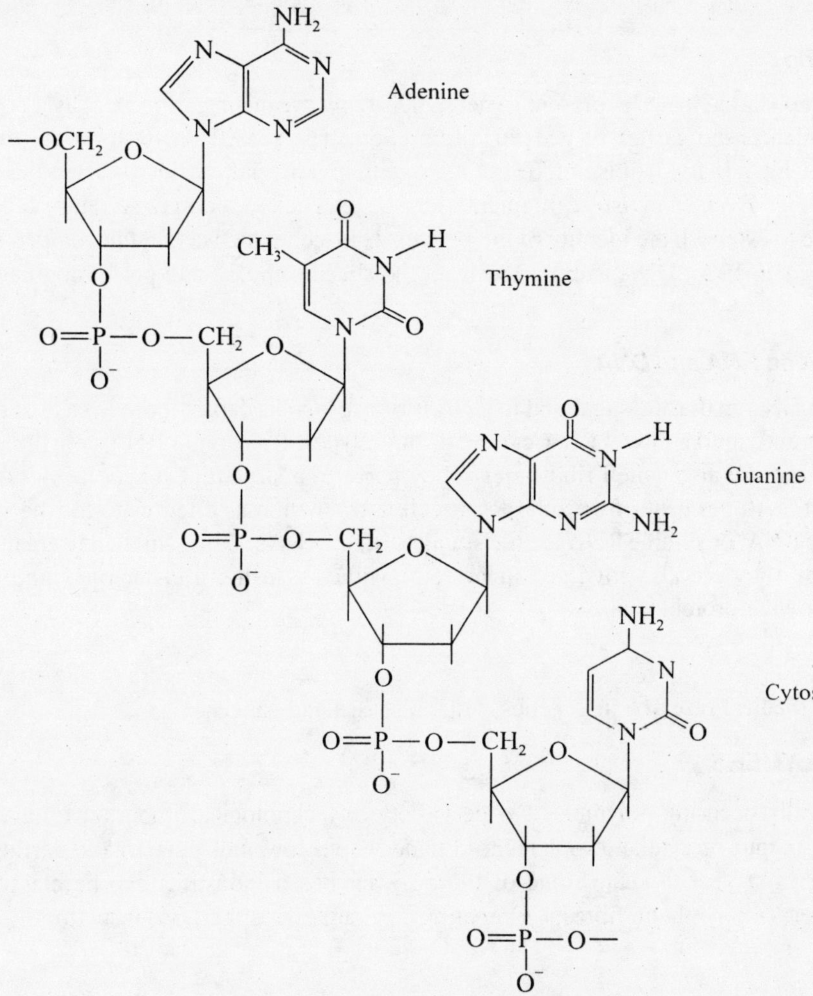

Fig. 2.8. A segment of deoxyribonucleic acid (DNA).

The backbone of nucleic acid consists of regularly alternating phosphate and pentose units. The heterocyclic bases are linked to C—1 of each pentose unit whereas the phosphate group of one nucleotide is joined with the C—3 atom of the ribose (or deoxyribose) unit of the succeeding nucleotide.

Structure of DNA

The structure of DNA has been described as a double-stranded helix comprising of two DNA chains twisted together in a helical fashion. The two chains are held together by hydrogen-bonds between pairs of bases in the two chains. A thymine unit of one chain pairs with adenine of the other chain via hydrogen bonds and guanine with cytosine. Genetic information is passed from cell to cell in a process called DNA replication. A gene is a section of a DNA molecule that is 'coded' for a specific genetic message. Since the DNA molecules have the same backbones, the differences in life functions related to it are based on the sequence and kind of bases on the nucleic acids. It is this sequence of bases, which is generally accepted as a genetic code in which biological information is recorded in nucleic acid molecule.

DNA finger printing

DNA is the self-replicating material present in nearly all living organisms, especially as a constituent of chromosomes which is the carrier of genetic information. It is passed on from parent to a child in identifiable codes. Each individual is unique but a progeny inherits patterns partly from the mother and partly from the father. Even an invisible fragment of tissue or speck of blood or serum may be sufficient to clinch evidence to establish the identity of the person. This technique was first developed by a British geneticist *A. Jeffrey* in 1985. DNA is extracted from the sliced sampled and put through a radioactive probe.

Difference between RNA and DNA

RNA differs from DNA in that the sugar unit in RNA is ribose but it is deoxyribose in case of DNA. RNA exists as single strand, unlike DNA. RNA can be of three types. Ribosomal RNA constitute the major amount of RNA and are also called ribosomes. They perform a structural function. Messenger RNA account for about 5–10 per cent of the total RNA cell and function as a template for the synthesis of proteins. Transfer RNA, or soluble RNA are the smallest of the RNAs. Being small they are more soluble and more mobile. They are also involved in protein synthesis, in the transfer of amino acids from solution to the growing protein chain.

Uses

Biochemical and medical research in genetics, virus diseases and cancer.

SYNTHETIC POLYMERS

Besides the naturally occuring polymers, a wide variety of technological importance have also been synthesised. The output of synthetic polymers obtained from low molecular mass starting material, keep on growing from year to year. Some of the polymer based industries can be classified as: (i) plastics; (ii) rubber; (iii) synthetic fibres; (iv) coatings; (v) adhesives; and (vi) insulators.

Plastics

It is well-known fact that plastics have played a very vital role in the growth phase of the Indian economy during the recent past. Every vital sector of the economy starting from agriculture to packaging, automobile, building construction, communication or infotech have been virtually revolutionised by the applications of plastics. The plastic processing industry which made a modest beginning during the 1970's, had really taken off only during the post-liberalisation era. The current business environment and abundant domestic availability of plastic raw materials have resulted in a double-digit growth rate consistently during the recent years.

The success story of plastics would not have been possible but for the proven advantages that it possesses over its nearest substitutes like metal, glass or paper. Today, the packaging industry is, by far the major user of plastics. They are used to pack cosmetics, toiletries, milk, edible oil and food products. Plastics in the packaging sectors provides the convenience, ease of handling and packaging efficiency. It definitely increases the shelf life of food products.

In the transport sector, plastics are increasingly replacing or have replaced traditional material in automobiles, aircrafts and boats; for example, fibre glass, reinforced plastics, fuel tanks, interiors, dash boards, bumpers and so on. In these applications, the functional superiority of plastics has been established.

The contribution of plastics as an aid to modern agriculture is immense. Drip and sprinkler irrigation systems, mulch films, green house films, pond and canal lining films are proven products to conserve water, protect crops from vagaries of weather, thus resulting in better productivity.

It is a high polymer, usually synthetic, combined with other ingredients, such as curatives, fillers, reinforcing agents, colorants, plasticisers, etc. The mixture can be formed or moulded under heat and pressure in its raw state, and machined to high dimensional accuracy, trimmed and finished in its hardened state. The thermoplastic type can be re-softened to its original condition by heat; the thermosetting type cannot.

Plastics in general (including all forms) are sensitive to high temperatures, among the more resistant being fluorocarbon resins, nylon, phenolics, polyimides, and silicones, though even these soften or melt above 260°C. Other types are combustible when exposed to flame for a short time (polyethylene, acrylic polymers, polystyrene), and still others burn with evolution of toxic fumes (polyurethane).

Engineering plastics are those to which standard metal engineering equations can be applied; they are capable of sustaining high loads and stresses, and are machinable and dimensionally stable. They are used in construction, as machine parts, automobile components, etc. Among the more important are nylon, acetals, polycarbonates, ABS resins, PPO/styrene, and polybutylene terephthalate.

Fibres, films and bristles are examples of extruded forms. Plastics may be shaped by either compression moulding (direct pressure on solid material in a hydraulic press) or injection moulding (ejection of a measured amount of material into a mould in liquid form). The latter is most generally used, and articles of considerably size can be produced. Because of their dielectric and non-toxic properties, plastics are essential components of electrical and electronic equipment (especially for use within the human body).

Plastics can be made into flexible and rigid foams by use of a blowing agent; they are light and strong, and the rigid type is machinable. These forms are collectively called cellular plastics. Plastics can also be reinforced, usually with glass or metallic fibre for added strength. They are laminated to paper, cloth, wood, etc. for many uses in the packaging, electrical and furniture industries; they also can be metal-plated. Plastic pipe is widely used for underground transportation of gases and liquids over long distances as well as intraplant. Several natural materials (waxes, clays, and asphalts) have rheological properties similar to synthetic products, but as they are not polymeric, they are not considered true plastics. Certain proteins (casein, zein) are natural high polymers from which plastics are made (buttons, and other small items), but they are of decreasing importance.

Plastics have permeated industrial technology. Not only have they replaced and improved upon many materials formerly used, but also have made possible industrial and medical applications that would have been impracticable with older technologies.

Their major application areas are:

1. Automobile bodies and components; boat hulls.
2. Building and construction (siding, piping, insulation, flooring).
3. Packaging (vapour-proof barriers; display cartons; bottles; drum linings).
4. Textiles (carpets, cordage, suiting, hosiery, drip-dry fabrics, etc.).
5. Organic coatings (paint and varnish vehicles).
6. Adhesives (plywood, reinforced plastics, laminated structures).
7. Pipelines.
8. Electrical and electronic components.
9. Surgical implants.
10. Miscellaneous (luggage, toys, tableware, brushes, furniture, etc.).

Rubber

Rubber is the name given to elastic, high-molecular compounds capable of changing their shape (deforming) when acted upon by an external force and of rapidly resuming their original shape when the force is removed. The elastic properties of rubber are due to the fact that the linear macromolecules of the material which normally are of a bent, curled or curved shape, easily stretch out of contract under the effect of an applied force; when the force is removed, the molecules return to their initial shape and length.

If sulphur is added to natural rubber which is then heated, the sulphur will form links between the linear macromolecules of the rubber, i.e., form bridges between them; by this larger macromolecules of a web structure will be produced. This process is called vulcanisation of the rubber, and the product of reaction between the rubber and sulphur is vulcanised rubber. Vulcanisation improves the mechanical properties of the rubber and increases the range of temperatures within which the elasticity of the material is retained; it also increases the resistance of the rubber to organic solvents, etc.

Rubbers are classified as natural rubber and synthetic rubber. Until the nineteen thirties only natural rubber, obtained from the latex of rubber plants which grow in countries with a damp tropical climate (mainly in Malayasia, Indonesia, etc.), was used.

Natural and synthetic rubber is the basic material in the fabrication of rubber-textile and rubber-metal articles which are so extensively used in various branches of the economy. These are motor car and aircraft tyres, drive belts and conveyer belts, flexible hoses and sleeves, machine parts and various gaskets, electric insulating materials, rubberised fabrics and rubber footwear, articles of personal hygiene and many other rubber products. The consumption of rubber has increased year by year. This explains the attention devoted to developing the rubber industry in India and in many other countries.

Natural rubber

Natural rubber can be produced from many plants, however, actually the latex is obtained from a small number of so-called rubber plants. The most important of these is a tree-like plant *Hevea brasilliensis*, whose latex contains 20 to 35 per cent of a rubber constituent, formed by biochemical synthesis. The latex is drawn from notches in the tree bark in the same way as turpentine. This is called tapping.

Latex contains the rubber in the form of tiny pear-like particles or globules, which are suspended in water and bear a negative electric charge; this prevents them from coagulating. Latex also contains a small amount of protein substances (2 per cent), resins (up to 2 per cent), sugary substances (0.4 per cent), and ash (0.5 per cent). To remove the charge from the rubber particles formic or acetic acid diluted with water is added to the latex; this causes the latex to coagulate and the rubber is separated in the form of doughy, bulky mass. The rubber is filtered off, thoroughly washed with water and passed through corrugated rollers. The rubber sheets produced are dried and smoked in a smoke chamber so that they will keep without spoiling. The sheets are packed in bales. Natural rubber is a polymer of isoprene, $CH_2{=}C{-}CH{=}CH_2$. The macromolecules of natural rubber are made up of

$$\begin{array}{l} \quad\;\, | \\ CH_3{-}CH_2{-}C{=}CH{-}CH_2{-}\text{structural units.} \\ \quad\;\, | \\ \quad CH_3 \end{array}$$

The degree of polymerisation exceeds 2,000 and the molecular weight is from 136,000 to 340,000. Natural rubber possesses high strength and elasticity within a wide range of temperatures.

Natural latex is also used directly, without separating the rubber from it. It is used as a glue, for impregnating textiles and making microporous rubbers.

Synthetic rubber

Mainly synthetic rubber is used in the rubber industry. At the present time the chemical industry turns out a wide range of synthetic rubbers, both made from one individual monomer by addition polymerisation or from a blend of two or three monomers by co-polymerisation. The synthetic rubbers can be divided into the two groups of general-purpose and special-purpose rubbers. This division is purely nominal. General-purpose rubbers are used for making tyres and other rubber articles of mass consumption. Special rubbers have certain improved properties. Such rubbers can operate under severe conditions, for instance at low or at high temperatures resist highly corrosive media (acids, alkalies, oxidants, solvents, etc.)

Synthetic Fibres

Synthetic fibre is a fundamental form of solid (usually crystalline), characterised by relatively high tenacity and an extremely high ratio of length to diameter (several hundred to one). Natural fibres are animal, such as wool and silk (proteins); vegetable, such as cotton (cellulose); and mineral (asbestos). Cotton fibre is called staple, and rarely exceed 2″ in length.

Semi-synthetic fibres include rayon and inorganic substances extruded in fibrous form, such as glass, boron, boron carbide, boron nitride, carbon, graphite, aluminium silicate, fused silica and some metals (steel). Synthetic fibres are made from high polymers (polyamides, polyesters, acrylics, and polyolefins) by extruding from spinnerets (nylon, 'Orlon', etc.) Some are being used in speciality papers, though the primary use is in textile fabrics.

Metal fibres are used in several ways; (i) as 'whiskers', which are single-crystal fibres up to 2 inch. long having extremely high tensile strength; they are made from tungsten, cobalt, tantalum and other metals, and are used largely in composite structures for specialised functions; (ii) as filaments, which are alloys drawn through diamond dies to diameters as small as 0.002 cm; steel for tyre cord and antistatic devices has been developed for such applications; and (iii) in biconstituent structures composed of a metal and a polymeric material, for example, aluminium filament covered with cellulose acetate butyrate.

Hollow fibres of cellulose acetate and nylon are used as membrances in the reverse osmosis method of water purification.

Coatings

A film or thin layer applied to a base material called a substrate. The coatings most commonly used in industry are metals, alloys, resin solutions, and solid/liquid suspension on various substrates (metals, plastics, wood, paper, leather, etc.). They may be applied by electrolysis, vapour deposition, vacuum, or mechanical means such as brushing, spraying, calendering, and roller coating. Products such as cables and power cord are coated by extrusion. Such thermosetting resins as acrylics, epoxies, and polyesters appropriately compounded are available in powder form and are applied to auto bodies, machinery, and other industrial products by electrostatic spraying techniques. This process is claimed to minimise the pollution problems and waste encountered with solvent-based sprayed coatings.

Adhesives

Adhesives, often referred to as cements, glues, or pastes, are defined as substances 'capable of holding materials together by surface attachment'. Adhesive technologists do not all agree as to what is and what is not an adhesive. Substances may attach to surfaces and develop the internal or cohesive strength necessary to hold materials together while cooling from liquid to solid state, while losing solvent, or during chemical reaction. Pressure-sensitive adhesives do not undergo a phase change in order to hold materials together.

Many of the substances designated as adhesives may be called paints, finishes, or coatings when applied in thin films to only one surface, or may be called caulking, potting, casting, or encapsulating compounds when employed in thick masses.

To be termed an adhesive, a substance must be a liquid or tacky semi-solid, at least for an instant to contact and wet a surface, and be used in relatively thin layer to form a useful joint capable of transmitting stresses from one substrate to another. Adhesives and sealants include cement, glue, paste and waxes.

Adhesion is an interfacial phenomenon: wetting of substrates is essential. The electrical theory presumes that the adhesive and the substrates are like the plates of a capacitor that become charged due to the contact of two substances. The theory fails to predict the strong joints that result when a layer of water is frozen to join two blocks of ice, or when an epoxy adhesive is used to join two previously cured blocks of cast epoxy. Diffusion theory presumes the penetration of the substrates by the adhesive prior to its solidification; this is easily applied to many porous plastics; however, not to metal, glass, or glazed ceramic. Adsorption theory specifies the concept of forces, such as van der waals' forces, acting across the space between molecules in a material. Rheological theory suggests that the removal of weak boundary layers in plastics leaves the joint's mechanical properties determined by the mechanical properties of the materials making up the joining and local stresses. In the absence of weak boundary layers, joint failure must be cohesive within the bondline of the substrates.

Types and application

Adhesives have been categorised by suitability for bonding various substrates by physical form, by method of application, and by temperature-resistance recommendations for many substrates when using a particular type of bonding agent.

Much of the equipment used to apply adhesives is the same as used to apply surface coatings such as roller, flow, curtain, and knife coaters.

A brief classification is give below:

Inorganic

1. Soluble silicates (water glass).
2. Phosphate cements.
3. Portland cement (calcium oxide-silica).
4. Other hydraulic cements (mortar, gypsum).
5. Ceramic (silica-boric acid).
6. Thermosetting powdered glasses ('Pyroceram').

Organic

1. Natural
   (a) Animal hide and bone glue; fish glue blood and casein glues
   (b) Vegetable soyabean, starch, cellulosics, rubber latex and rubber-solvent (pressure-sensitive). Gums, terpene resins (rosin), mucilages.
   (c) Mineral asphalt, pitches, hydrocarbon resins.
2. Synthetic
   (a) Elastomer-solvent cements.
   (b) Polysulphide sealants.
   (c) Thermoplastic resins (for hot-melts). Polyethylene, isobutylene, polyamides, polyvinyl acetate.
   (d) Thermosetting resins. Epoxy, phenolformaldehyde, polyvinyl butyral, cyanoacrylates.
   (e) Silicone polymers and cements.

High-temperature adhesive

1. Organic polymers, e.g., polybenzimidazoles, that retain bonding strength up to 260°C for a relatively long time (500–1000 hours); above 260°C strength drops rapidly, 80 per cent being lost after 10 minutes at 535°C.
2. Inorganic (ceramic), e.g., silica-boric acid mixtures produce bonds having high strength above 2000°F; adhesive lap-bond strengths can be over 2000 psi at 1000°F. These adhesives are used largely for aerospace service, and metal/metal and glass/metal seals.

Hot-melt adhesive

A solid, thermoplastic material which quickly melts upon heating, and then sets to a firm bond on cooling. Most other types of adhesives set by evaporation of solvent. Hot-melt types offer the possibility of almost instantaneous bonding, making them well-suited to automated operation. In general, they are low-cost, low-strength products, but are entirely adequate for bonding cellulosic materials. Ingredients of hot-melts are polyethylene, polyvinyl acetate, polyamides, hydrocarbon resins, as well as natural asphalts, bitumens, resinous materials, and waxes.

Rubber-based (cement, rubber) adhesive

1. A solution of natural or synthetic rubber in a suitable organic solvent, without sulphur or other curing agent.
2. A mixture of rubber (often reclaimed), filler, and tackifier (pine tar, liquid asphalt) applied to fabric backing (pressure-sensitive friction tape).
3. A room-temperature curing rubber-solvent-curative mixture, often made up in two parts, which are blended just before use.

Hazards those containing organic solvents, (1) and (3) above, are flammable.

Setting and curing

Only the reactive adhesives truly cure. For example, acrylics, unsaturated polyesters, and other monomer adhesives containing ethylenic unsaturation cure by formation of free radicals when catalysed by peroxides and accelerated by metal-ion donors. Epoxies cure by addition mechanisms. Polysulphides react with active oxidising agents to form rubbery polymers particularly useful in joining low expansion glasses to metals. Many urethane and silicone adhesive cure in the presence of moisture. Most of the reactive adhesive cures are accelerated by heat. Cyanoacrylate adhesives cure only in thin bondlines, i.e. 1.0 μm to 0.1 mm, without the addition of a catalyst or hardener. The basicity present on many substrates initiates cure when inhibitors present in the adhesive are overcome.

Uses

Adhesives are used widely in a variety of industries including automotive; construction; electronic bonding, with special applications to computer memories and both active and inactive micro- and macro-electronic elements; packaging; plastics; and in the textile products and apparel industry.

Insulators

Insulators are substances or mixtures that have an extremely low dielectric constant, low thermal conductivity, or both. Electrical insulators are either solid or liquid, the latter being used in transformers (mineral oils, silicone oils). A wide variety of solid types includes porcelains, glass, mica, alumina, various high polymers (epoxies, polyethylene, polystyrene, phenolics), cellulosic materials, nylon, and silicone resins. All these may be used alone or combined with other insulators as composites.

Thermal insulators comprise an equally broad range of materials. Such inorganics as mineral fibres, magnesia, aluminium silicate, cellulose and glass fibres are widely used for steam and hot-water pipes, furnaces, and blown-in home insulation. Organic products that are effective include plastics foams (polyurethane, polyvinyl chloride, polystyrene) and cellular rubber. There are a number of materials that may be called double insulators, since they have both electrical and thermal insulating properties, e.g., polystyrene, PVC, cellulose, glass, magnesia and aluminium silicate.

Air is a unique case, as it is the only gaseous material in actual use as an insulator. Its dielectric constant is 1.00058, far less than that of any other dielectric material, and it has low thermal conductivity as well. It is particularly effective when trapped within a solid network, as in wool, cellular plastics, or glass fibres, or as an interlayer between wall panels.

BICYCLOHEPTENE AND HEPTADIENE POLYMERS

In the recent years there has been considerable interest in polymers derived from the unusual hydrocarbon monomer bicyclohept-2-ene (1) and related substances such as bicyclohepta-2,5-diene (2), now more commonly known as 2-*norbornene* and 2,5-*nonbornadiene*, respectively. Facile polymerisation of ring systems of this type are accomplished by co-ordinate polymerisation, but polymerisations have also been reported using free-radical initiators, reduced metal oxides, and some hydrated transition metal halides in polar media.

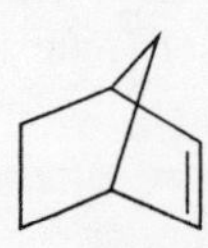

(1)

Bicyclohept -2-ene

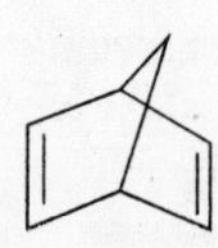

(2)

Bicylohepta-2,5-diene

Not only homopolymerisations, but co-polymerisations of the mono- and diunsaturated bicyclic hydrocarbons as well have been studied. For example, 2-norbornene has been co-polymerised with ethylene, propylene, or sulphur dioxides; 2,5-norbornadiene has been co-polymerised with an even wide variety of olefinic materials.

Bicyclohept-2-ene

Physical properties

2-Norbornene, formula weight 94.150, is a white crystalline solid boiling at 94–97°C at 740 mm pressure and exhibiting a melting point of 44–44.5°C in a sealed capillary.

Manufacture

2-Norbornene most conveniently is prepared by the addition of ethylene to cyclopentadiene under pressure (Eq. 2.1). The reaction is preferably carried out at 190–200°C; therefore dicyclopentadiene may also be employed in place of monomeric cyclopentadiene

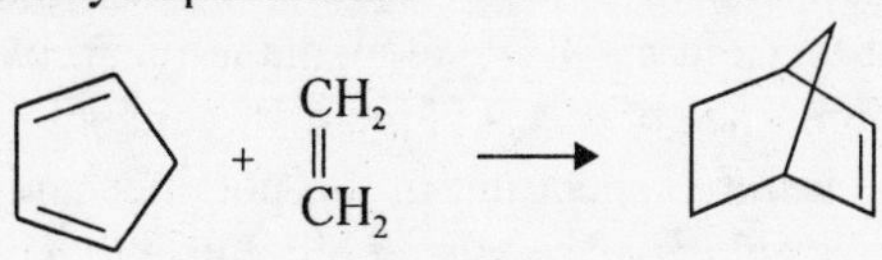

... (2.1)

since dimer is pyrolysed to monomer at elevated temperatures.

Chemical properties

2-Norbornene has been noted to exhibit some unusual olefinic chemical properties because of the fact that both carbon atoms alpha to the double bond are bridgehead carbon atoms in the bicyclic ring system. For example, bromination of 2-norbornene with *N*-bromosuccinimide results in a rearrangement and formation of the tricyclic derivative 3-bromotricyclo heptane, more commonly known as 3-*bromonortricyclene* (Eq. 2.2). However, this novel rearrangement is not limited to the reaction with *N*-bromosuccinimide; the rearrangement has been demonstrated by the same investigators, who

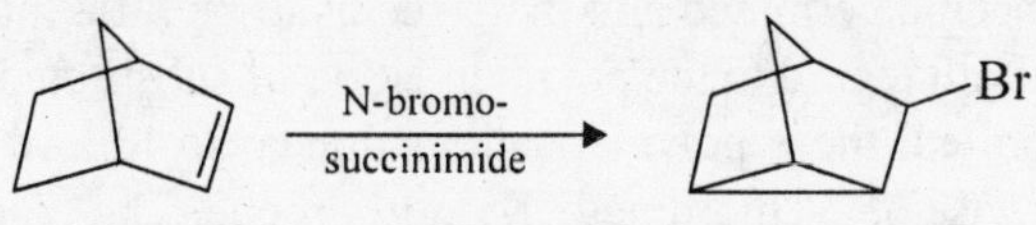

... (2.2)

also obtained the tricyclene derivative upon substituting bromine or bromine and pyridine for the *N*-bromosuccinimide.

2-Norbornene, like 2,5-norbornadiene is a highly strained structure, and many reactions, including polymerisation, are probably facilitated by the resulting release of this strain.

Polymerisation

2-Norbornene has been polymerised using a variety of agents including co-ordinate polymerisation catalysts, reduced metal oxides, ruthenium, osmium, and iridium halides, and free-radical catalysts. One of the most facile methods of producing polymers from 2-norbornene is by means of co-ordinate polymerisation. A catalyst derived from lithium aluminium tetraheptyl and titanium tetrachloride is mixed with 2-norbornene in an inert hydrocarbon medium and allowed to stand for 24 hr. at room temperature and atmospheric pressure. The solid polymeric product is washed free of hydrocarbon diluent, catalyst, and low-molecular-weight polymer by means of acetone. The dried polymers are crystalline in nature, and may vary in crystallinity from 5 to 55 per cent. A sample with 20 per cent crystallinity has been shown to have a crystalline melting point of 202°C by X-ray diffraction. The structure of the polymer has been determined by use of infrared spectroscopy and chemical degradation. The most prominent feature of the infrared spectrum of poly-2-norbornene is the absorption band at 10.37 μ (964 cm^{-1}), which suggests the presence of a major amount of *trans* unsaturation. The fact that a semiquantitative measure of the amount of the unsaturation in the polymer demonstrated the presence of a major amount of unsaturation has led to the hypothesis that polymerisation proceeds via a ring-opening reaction. A plausible mechanism of polymerisation involves a stereospecific ring opening and inversion of a *cis* to a *trans* double bond (Eq. 2.3).

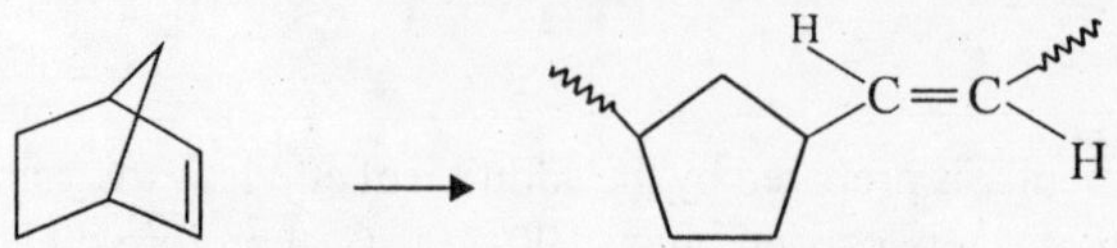

... (2.3)

Ozonisation of the polymer comprised of repeating units having the postulated structure indicated in equation 2.2 would be expected to yield *cis*-1,3-cyclopentane-dicarboxylic acid as a product. Actual experiments have confirmed the presence of this acid in an 18 per cent yield. The absence of bands due to *cis*-ethylenic absorption in the infrared spectrum of the polymer and the absence of any detectable *trans*-1,3-cyclopentanedicarboxylic acid after ozonisation indicate that co-ordination polymerisation of 2-norbornene proceeds via a highly stereospecific reaction.

Two methods using reduced valence states of metals for the polymerisation of 2-norbornene have also been developed. Polymerisation of 2-norbornene to the *trans*-unsaturated structure shown in equation 2.2 utilising reduced oxides of heavy metals such as chromium, molybdenum, and tungsten has been reported. A recent report provides details for the polymerisation of 2-norbornene using hydrated trichlorides of ruthenium (III) or related metals in a polar medium. The polymers are substantially similar to those previously described, except that an appreciable number of the olefinic bonds are *cis* rather than *trans*, as determined by infrared spectra showing absorption in the 13.5 μ (742 cm^{-1}) region. Several investigators have described the formation of polymers of 2-norbornene using free-radical catalysis; however, no structures have been determined and there seems to be little interest in this area.

There is little information with regard to the polymerisation of substituted norbornenes. Based upon the work which has been reported, these polymerisations appear to follow the expected course. This has been demonstrated in the case of 2-chloromethyl-5-norbornene. The co-ordinate polymerisation of 5-(3-butenyl)-2-norbornene (3) has been noted to yield a mixture of polymers formed either by a ring-opening polymerisation of the bicyclic nucleus or by a polymerisation across the terminal unsaturation of the alkenyl group.

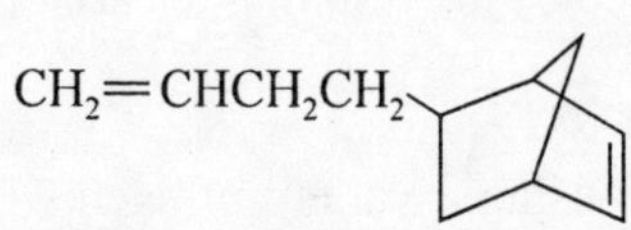

(3)

Co-ordinate polymerisation of 5-(3-butenyl)-2-norbornene

Non-allylic alcohols containing the norbornene nucleus have been graft co-polymerised with acrylic ester polymers and with acrylic ester polymers and diepoxides using free-radical catalysts.

The co-polymerisation of 2-norbornene has been reported to proceed with either co-ordinate polymerisation or free-radical catalysts. Co-ordinate polymerisation co-polymers have been formed from 2-norbornene (3) and ethylene and propylene, although no detailed structure analyses have been given. Co-polymerisation of 2-norbornene and sulphur dioxide has been effected with free-radical catalysts. Recently, the polymerisation was found to proceed rapidly even in the absence of added initiator.

Properties of polymers

Polymers of 2-norbornene formed via co-ordinate polymerisation vary from brittle, resinous, inhomogeneous materials melting above 300°C to tough homogenous products melting over the range 180–205°C. The type of polymer produced depends upon the ratio of organometallic reducing agent to transition metal compound employed in the catalyst. The crystallinities of both types of polymer are in the range 5–55 per cent. A tensile impact in excess of 40 ft-Ib/in. has been reported for the tough variety of polymers. A broad spectrum of physical properties is available in the co-polymers of 2-norbornene and either ethylene or propylene. Co-polymerisation of the bicyclic derivative with either of the latter monomers provides polymers which have exceptionally good high-temperature properties. Moreover, it has been noted that the low-temperature properties of polypropylene may be improved by co-polymerising small amounts of 2-norbornene with the propylene.

Processing of polymers

Poly-2-norbornene may be compression moulded into shaped structures at 200–250°C. The co-polymers for example with ethylene or propylene, may be processed at lower temperatures by means of a wider variety of techniques depending upon the amount of the bicyclic monomer incorporated into the co-polymer. Co-polymer comprised of major amounts of ethylene or propylene may be fabricated into moulded or extruded articles at 170–200°C employing conventional processing techniques.

Bicyclohepta-2,5-diene

Physical properties

2,5-Norbornadiene, formula weight 92.134, has a melting point of about 25°C, a boiling point at atmospheric pressure of 89–90°C, a density d_4^{20} of 0.9056, and a refractive index n_D^{20} of 1.4699.

Chemical properties

2,5-Norbornadiene is also a highly strained structure and, many reactions, including polymerisation, are probably facilitated by the resulting release of this strain.

Manufacture

The synthesis of 2,5-norbornadiene may be achieved by the Diels-Alder reaction of cyclopentadiene and acetylene, or by the dechlorination of 5,6-dichloro-2-norbornene employing magnesium iodide in ether.

Polymerisation

The homopolymers obtained from this unusual ring system have been given little attention. Although no definitive work has been published on the polymerisation of this monomer by means of co-ordinate polymerisation catalysts, the formation of homopolymers from 2,5-norbornadiene using these catalysts has been disclosed. The free-radical polymerisation of this monomer have been studied and the properties of soluble poly-2,5-norbornadiene obtained at low conversions noted, as well as the tractability of the polymer to casting. Russian workers have disclosed the formation of linear polymers containing unsaturation from this monomer via Diels-Alder reaction. Some very precise work has been done to elucidate the structure of the dimer formed from the diene upon irradiation with iron carbonyl. The infrared absorption and the nuclear magnetic resonance spectra of 2,5-norbornadine polymer obtained by ^{60}Cr γ-radiation-induced polymerisation have also been studied.

2,5-Norbornadiene has been co-polymerised with a variety of monomers employing virtually all available polymerisation techniques. The co-polymerisation of 2,5-norbornadiene with ethylene via co-ordinate polymerisation to yield a polymer containing unsaturation available for subsequent cross-linking has been described. Sulphur-curable elastomers from a mixture of alpha-olefins and methyl-substituted 2,5-norbornadiene have also been prepared with co-ordination polymerisation catalysts. Friedel-Crafts or Lewis acid catalysts have been employed in the co-polymerisation of 2,5-norbornadiene with $C_{10}H_{16}$-terpenes, and the co-polymerisation of the diene with unsaturated petroleum feeds by means of Freidel-Crafts catalysts to increase the softening point of petroleum resins has also been reported. Several investigators have been active in the area of free-radical co-polymerisations employing 2,5-norbornadine. *Schmerling* has disclosed the formation of low-molecular-weight resins from the diene and aliphatic or cyclic olefins; Russian workers have reproted the co-polymerisation of the diene and styrene or vinyl acetate, and of the diene and 1,1,2-trichloro-1,3-butadiene; other investigators have reported its co-polymerisation characteristics with vinyl acetate and with *p*-chlorostyrene. An extensive study has been made of the free-radical co-polymerisation of 2,5-norbornadiene with vinyl chloride, vinylidene chloride, acrylonitrile, ethyl acrylate, and methyl methacrylate. Infrared studies have shown that the diene polymerises, with rearrangement, across the 2,6 positions to yield co-polymers containing 3,5-disubstituted nortricyclene units.

POLYMER CONCRETE MATERIALS

For the past 30 years, progress in a wide range of research and development concerning concrete-polymer composites has been made throughout developed countries like Japan, USA, Canada, UK, Germany, etc. Concrete is considered to be a two-phase composite material whose constituents are binder and aggregate and the use of polymers for the binder produces concrete-polymer composites with the upgraded performance. This section gives brief outline of some of the materials involved.

Commercially available binders for polymer mortar and concrete are various thermosetting resins, fat-modified resins, resin-modified asphalts and vinyl monomers, epoxy resin, unsaturated polyester resin, methylmethacrylate (MMA) monomer etc. On account of the good workability and low temperature curing characteristics, the methyl methacrylate monomer is widely used as a binder in polymer concrete materials. Modified PP (vinylbond) also could be used as binder.

Fillers such as ground calcium carbonate, silica are used to extend binders and to fill the very fine particles for aggregates, coarse and fine aggregates used for polymer mortar and concrete are river gravels and sands, crushed stones, silica sands etc. In the USA, electric utility industries are developing polymer concrete materials in place of conventional porcelain material because of its economy and ready availability.

The electric porcelain industry requires an extremely high capitalisation for plant construction. Porcelain is highly energy-intensive and in future, there is a question whether fuel will be available for firing the kilns necessary to produce electrical porcelain.

A number of plants are opening in the USA to produce a number of precast polymer concrete products such as electrical insulators, various pads and supports for electrical applications, underground utility boxes, channels and trenches, machine bases, various panels, tiles. The development in the USA of the electrical applications for polymer concrete (PC) is particularly note-worthy.

Polymer Concrete Materials

Monomer systems used for binder consists of methylmethacrylate (MMA) monomer, polymethyl methacrylate (PMMA), trimethylol propane trimethacrylate, benzoyl peroxide as an initiator, dimethyl aniline as promoters. Some of the application of polymer concrete materials are given in Table 2.2.

Table 2.2. Applications of polymer concrete materials.

Application	*Location of work*
Floorings (including decorative finishing)	Floors for warehouses, offices, hospitals, factories and shops, railway platforms, train floors, etc.
Anti-corrosive linings	Effluent drains, chemical and machinery floors, grauts for acid proof tiles, floors for chemical laboratories and pharmaceutical warehouses, septic tanks, electrolyte baths, hot spring baths, offshore structures.
Adhesive	Adhesives for flooring, heat insulating materials, tile adhesives, embedment of anchor bolts, etc.
Repairing materials	Grout for repairing cracks and delaminations of concrete structure, patching materials for damaged concrete structures, protective coatings for corroded reinforcing bars, etc.
Structural precast products	Manholes and handholes for tele-communication cable lines, electric power cable lines and gas application (pipe lines), tunnel liner segments for telephone cables and sewage systems, piles for port, manufacture of high voltage electric insulators, bushes, etc.
Non-structural precast products	Gutter covers, U-shaped gutters, footpath panels, terrazzo tiles and panels and large-sized or curved decorative panels, for building partition wall panels, sink, prefabricated cellars, counters, wash stands, speakers, etc.

Filler and Aggregates

Commercially available calcium carbonate (heavy grade), coarse and fine silica sands are used as an aggregates.

Manufacturing and Moulding

The manufacturing process for polymer concrete material insulators is simple and it requires only 10–15 percent of the consumption of energy as does porcelain, The necessary equipment for production consists of mixer, vibrating tables, vacuum systems, laboratory equipment, moulds, etc.

The process used is a batch process that consists of weighing aggregates and organics separately, mixing the aggregates, adding the organics and mixing again, powering the mix into the mould and vibrating under vacuum, curing, demoulding and inspecting for quality control. Test results of methyl methacrylate polymer concrete materials are given in Table 2.3.

Table 2.3. Test results of (methyl methacrylate polymer concrete materials).

Test	*Temperature, °C*	*Results*
Compressive strength	–26	171 N/mm^2
	21	135 N/mm^2
	49	109 N/mm^2
	88	97 N/mm^2

(Contd ...)

Test	*Temperature, °C*	*Results*
Modulus elasticity	–26	46×10^3 N/mm^2
	21	36×10^3 N/mm^2
	88	31×10^3 N/mm^2
Poisson's Ratio	–26	0.24
	21	0.23
	88	0.22
Tensile splitting strength	–26	10.4 N/mm^2
	21	9.9 N/mm^2
	88	9.4 N/mm^2
Modulus of rupture	21	21 N/mm^2
Shear strength	21	26 N/mm^2
Sp. Gravity		2.4
Water absorption percent		0.6

Comparison of properties of polymer materials versus porcelain materials are given in Table 2.4.

Table 2.4. Comparison of polymer materials versus porcelain materials.

Properties	*Polysil*	*Porcelain*
Dielectric strength, v/mil	450–650	55–450
Dielectric constant	4.5	5.4–7.0
Energy required, BTU/tb	10,000	20,000
Compressive strength	22,226	16,450
Modulus of elasticity, $\times 10^6$ 16/in^2	4.4	10
Impact strength 24°C	0.52	0.4–0.6
40°C	0.56	—
Sp. Gravity	135	143
Thermal resistivity	50	58.8
Centilever, lb	1,825	1500
Tensiles, lb	26,780	12,000

Equipment for Manufacturing Polymer Concrete materials

Some of important equipment are: (i) concrete mixer; (ii) vibrating table; (iii) vacuum pump, pipes etc.; (iv) weighing scale, table, etc.; (v) different moulds for electrical insulators switch, etc.

Raw Materials/Chemicals for Manufacturing Polymer Concrete Materials

Some of the important chemicals raw materials are: (i) silica sand; (ii) MMA-monomer; (iii) PMMA powder; (iv) Catalyst/promoters; and (v) other chemicals.

BOUNCING PUTTY

Bouncing putty is a soft, plastic-like material which is perhaps best known as a novelty item sold commercially under the names of Nutty Putty, Crazy Clay, or Silly Putty. Chemically, it is a boron-containing silicone, and it has the surprising property of re-bounding when dropped on a hard surface.

Manufacture

The equipment generally used in the manufacture of bouncing putty consists of a reaction vessel equipped with a high-powered stirrer which scrapes the inside surface of the vessel. Dough mixers and Banbury mixers are good examples.

There are a number of alternative methods which can be used to manufacture bouncing putty. Practically all of these processes embody two major steps: polymerisation, and compounding the polymer with the appropriate plasticisers, fillers, and pigments to give the product its desired properties and colour.

Polymerisation

The polymer can be made by mixing about 5 parts of boron oxide with 100 parts of dimethylsilicone oil and heating the mixture at a temperature of 150–250°C until a rubber-like mass is formed. Another procedure for preparing the polymer consists of heating a mixture of silicone oil and a boron compound such as pyroboric acid, boron oxide, boric acid, ethyl borate, and other esters of boric acid in the presence of ferric chloride for a period of 2–4 hr. at a temperature of 100–200°C. This method yields a comparatively hard, brittle polymer having a characteristic yellowish-brown to brown colour. This material is very resilient, having a rebound of 80–90 per cent or greater, and exhibits cold flow properties.

Compounding

At this stage of manufacture the product is not in its optimum condition for many commercial applications due to its hardness and brittleness. Therefore, it is often necessary to soften it by kneading, usually in the presence of a plasticiser. The plasticiser is usually a fatty acid such as oleic acid, stearic acid, or lauric acid, or a metal salt thereof, for example, ferric stearate, zinc stearate, zinc laurate, etc. The amount of plasticiser to be added is determined by the degree of softness desired. A decrease in the resiliency of the polymer results from plasticisation, but it can be partially regained by mixing hydroxyl-containing materials, such as glycerol, zinc hydroxide, and moist bentonite into the putty. Finely divided red iron oxide, potassium alum, titanium dioxide, lithopone, a wide range of silicas, and other fillers can be used to alter the flow properties of the bouncing putty and to extend it. Colouring pigments which are ordinarily used in silicones can be used in the product. Some of them are red iron oxide, white titanium dioxide, white zinc oxide, and green chromium oxide. Caution must be taken to select non-toxic materials in the manufacture of bouncing putty for use in novelty items which are likely to be handled by small children.

Properties

Bouncing putty is noted for its most unusual properties. The type that is sold commercially as a novelty is soft, readily mouldable, and easily drawn into threads. Surprisingly, this material will re-bound as much as 80 per cent when it is dropped on a hard surface. When struck suddenly with a hammer it will shatter into many small pieces and when pulled rapidly it will break. Bouncing putty will also pick up printing ink and has achieved wide popularity with children because of its ability to lift various figures from the newspaper cartoons which can later be distorted into many odd shapes. The material has cold-flow properties, which in some recent commercial products has caused it to penetrate clothing and fabric upholstery.

This feature has made it difficult to remove from fabrics. The best solvents to use to remove it are 2-propanol and carbon tetrachloride. However, new bouncing putties have recently been introduced on the market which have varying degrees of cold flow. Some of these materials will not penetrate or strain textiles under the pressure of their own weight.

The bouncing putty is composed of linear dimethyl polysiloxanes with Si—O—B bonds at intervals along the chain. Its properties of high elasticity and very viscous flow are due to temporary cross-linking by means of a transfer of electrons from the oxygen atom of a siloxane to the boron atom of a borosiloxane illustrated as follows:

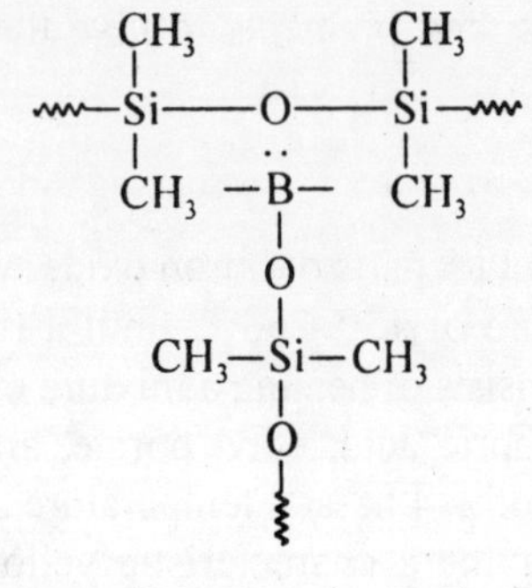

Uses

Bouncing putty has been used in the centres of golf balls to increase their resiliency, in automatic leveling devices, in novelty items, in athletic mouthpieces for the purpose of absorbing sudden shocks to the teeth, for reducing corona around electrical connections, in damping devices, and therapeutically for the exercise of injured arm muscles, a ball of it being worked in the hand.

FEEDSTOCKS FOR POLYMERS

Even when considered only as the monomer for the homopolymer, ethylene is a very important commodity. More than that, it is an essential ingredient in vinyl chloride and styrene, the other two largest volume monomers for plastics (Fig. 2.9). Polymer-grade propylene and butadiene are made in large measure as by-products of ethylene manufacture. As seen in Fig. 2.9, these two raw materials enter into the production of virtually all the other major plastic, rubber, and fibre polymers.

Natural gas is a diminishing resource, so almost all new plants are made to use heavier feeds such as heavy naphtha (boiling range of 100 to 220°C) or atmospheric gas oil, (AGO) (boiling range of 200 to 350°C). Almost any other hydrocarbon source is a possible feed, but matters of aromaticity and sulphur content change the desirability from one refinery stream to another.

The heavier feeds yield a greater variety of by-products that must be sold or used to make the process economical. Some chemical companies have been exploring the direct production of ethylene from crude oil, by-passing the usual oil refinery. By recycling various streams, the yield of ethylene can be optimised without having to market gasoline or fuel oil in competition with the traditional oil refiners. No full-scale plants have yet been built using a crude oil feed.

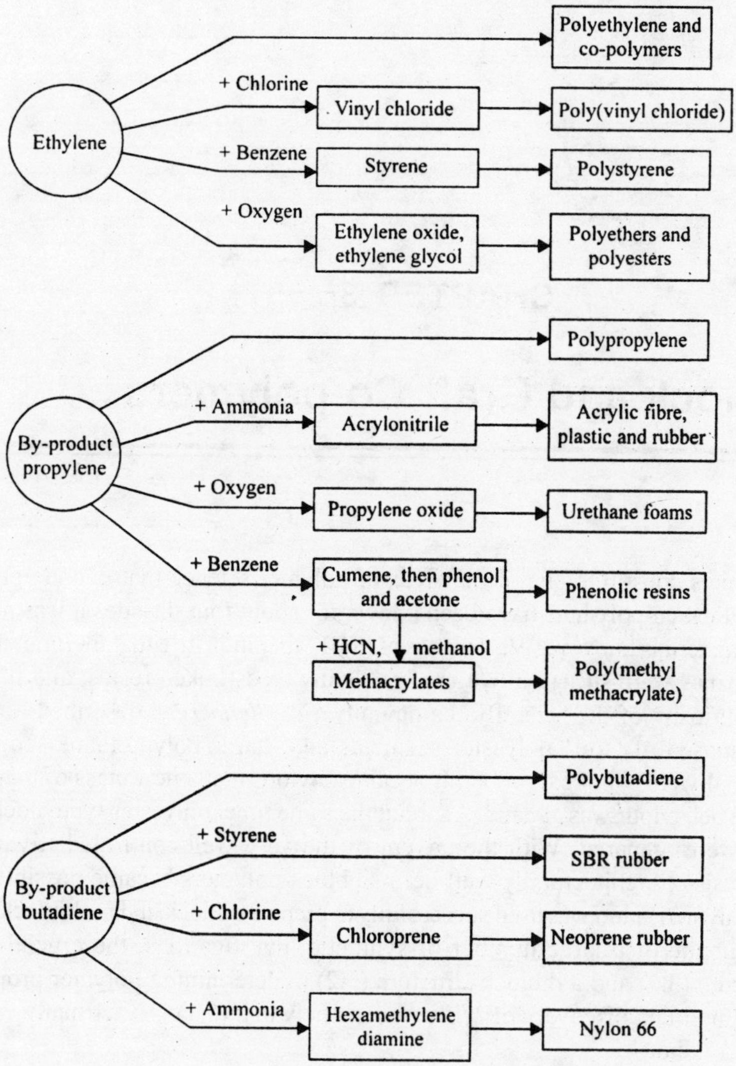

Fig. 2.9. Major polymeric products derived from ethylene and its by-products. Not all reactants or products are shown.

The product distribution can differ for various reasons. High severity operation may involve higher temperatures and longer residence times during cracking, which increases conversion per pass. Moderate severity operation usually produces more by-products but lessens the amount of feed wasted in less salable materials such as coke.

With any feed, the hydrocarbon source mixed with steam is pyrolised at 850°C or higher by passage through a tube set in a cracking furnace. The steam is added to decrease formation of coke on the tube walls. The hot effluent gases are quenched with water or oil to stop the reaction. Separations using distillation and absorption are used to obtain pure products as well as to recycle some materials.

CHAPTER 3

Block and Graft Co-polymers

INTRODUCTION

Block and graft co-polymers constitute one of the areas of polymer science that is undergoing a very rapid growth. Although block co-polymers have been known for about four decades it was not until the discovery of the class of materials known as Vulcolanes at I.G. Farbenindustrie that their potential began to be appreciated. Vulcolanes were structurally complex, lightly cross-linked resins, and thus, did not arouse the appropriate curiosity of the scientific community. *Schollenberger* described a completely linear polyurethane composed of 'soft' polyester segments and 'hard' polyurethane sequences and which behaved as a 'virtually cross-linked' rubber. It became obvious that a new class of materials with some unusual and unique behaviour was at hand. At about the same time, polyester-type block polymers with similar properties were prepared. With the advent of the very well-controlled 'living' anionic polymerisation the synthesis of architecturally well-defined block polymers became possible.

The 'living' anionic polymerisation was used successfully to prepare diblock and multiblock structures. It is noteworthy that, in spite of a large number of synthetic investigations, the crucial difference between a triblock structure (3.1) and a diblock structure (3.2) in determining polymer properties was not recognised until the important discovery of *Milkovich*, namely that (3.1) is a 'virtually cross-linked elastomer', whereas (3.2) is not.

Polystyrene-polybutadiene-polystyrene
Structure 3.1

Polystyrene-polybutadiene
Structure 3.2

This discovery led to a series of new classical investigations from which broad novel principles were evolved. These, in turn, led to the study of several new systems which demonstrated the generality of the initial concepts and produced unique, high-performance materials.

Graft co-polymers are structurally similar to block co-polymers [see structures (3.3) and (3.4)] except that the co-monomer blocks are grafted as side chains to a backbone. In the example (structures 3.3 and 3.4) the A segments are hard and the B segments are soft.

As can be seen, a triblock is equivalent to a grafted product with two grafted chains situated at the terminals of the backbone. In spite of this similarity, rigorous structure-property correlations in the area of graft co-polymers have been seriously hindered because of the difficulty of preparing well-defined structures. Indeed, the synthesis of a homopolymer-free grafted product in which the number, length, spacing, and distribution of the side chains are known is still an unresolved challenge. Therefore, although graft co-polymers have attained enormous industrial importance, as typified by high-impact polystyrene and acrylonitrile-butadiene-styrene co-polymer, one can only assume that the broad principles developed for block polymers are valid here as well. In fact, recent work with non-radical initiators appears to indicate the correctness of the foregoing assumption. Thus, the main thrust of the present discussion is from a structure-property viewpoint, emphasises block co-polymers.

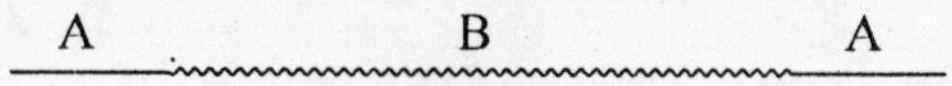

Structure (3.3)
(A triblock material)

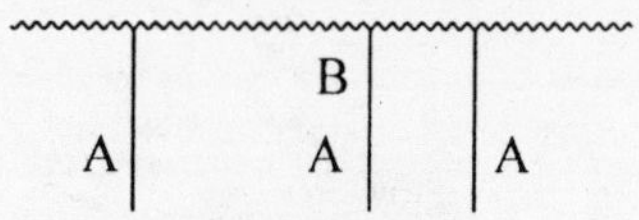

Structure (3.4)
(A grafted material)

CHEMISTRY OF BLOCK CO-POLYMER FORMATION

Linear Styrene-Diene Systems by Anionic Polymerisation

It will suffice to indicate here that investigation of this class of co-polymers was extremely important in developing an understanding of the principles of block and graft co-polymers. It was the spark which was necessary to re-kindle the interest of the scientific and technological community in these very important materials. As a result, many novel, unique, high-performance resins were synthesised and several have become commercial products.

Hydrogenated Styrene-Diene Systems

The first systems that will be discussed are the hydrogenated versions of the above styrene-diene co-polymers. Hydrogenation studies have been conducted in several laboratories. Catalyst systems employed include nickel acetylacetonate-triisobutylaluminium, nickel salicylate-butyllithium, cobalt hexanoate-butyllithium, etc. Most of these were homogeneous and yielded high conversions.

In general, the central diene segment is reduced under relatively mild conditions (e.g., 50°C, 50 psi). The reduction of the polystyrene end blocks, on the other hand, requires more stringent reaction parameters (e.g., 300°C and 1000 psi). The hydrogenation is generally conducted in the solvent used for co-polymerisation, e.g., cyclohexane. There are several advantages for these modified block co-polymers in comparison to their precursors. The most obvious advantage resides in the fact that one has an easily obtainable product devoid of the oxidatively unstable butadiene or isoprene units. Schematically, this desirable saturated and stable product is the result of the reaction shown in equation 3.1.

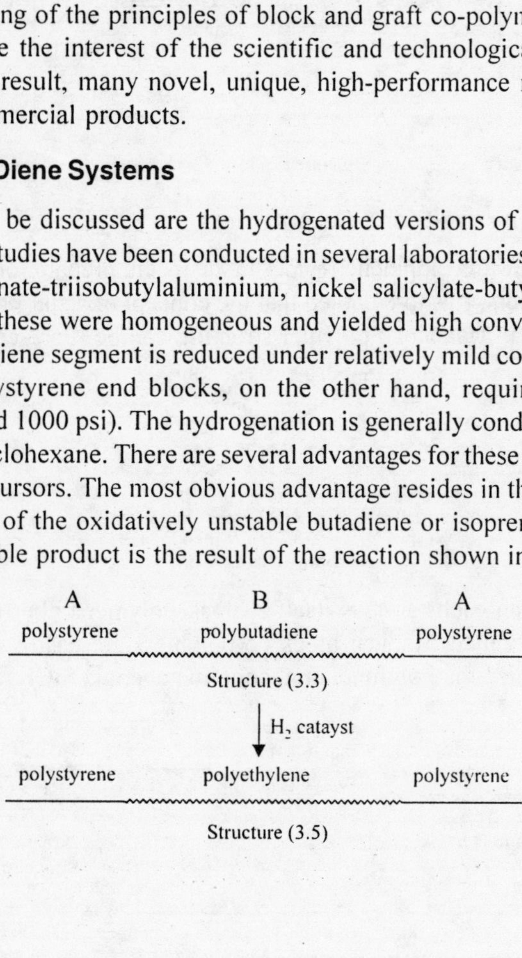

... (3.1)

Structure (3.5)

If (3.3) were to be composed of 1,4-micro-structure only, a crystalline linear polyethylene segment would be formed in structure (3.5). As a result, while intrinsically interesting structure (3.5), would not have thermoplastic elastomeric characteristics. However, by properly modifying the polymerisation reaction conditions leading to, (3.3), e.g., using an ether co-solvent, one can obtain a block co-polymer (3.6) in which the central polydiene segment contains both 1,4 and 1,2 configurations:

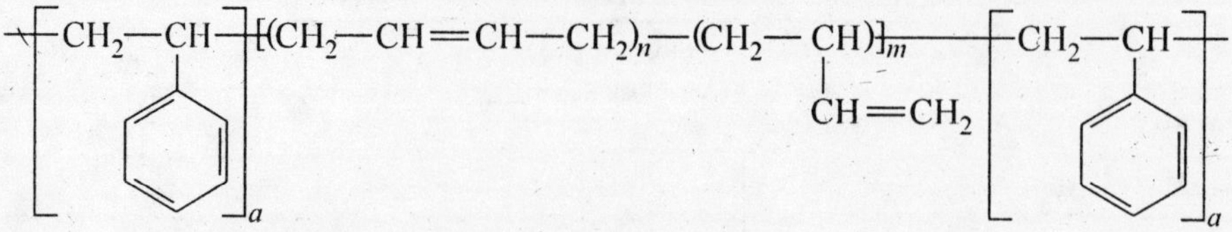

Structure (3.6)

Reduction of the central segment of (3.6) yields what can be considered to be an amorphous poly(ethylene-co-butene) segment. As a result, the elastomeric behaviour of the precursor (3.6) is maintained in the hydrogenated product. A material of this type was introduced commercially by the Shell Chemical Company under the trade name of Kraton G. In essence, Kraton G can be represented by formula (3.7). The sequences of n and m are, of course, quite short so as to eliminate cyrstallinity in the centre block.

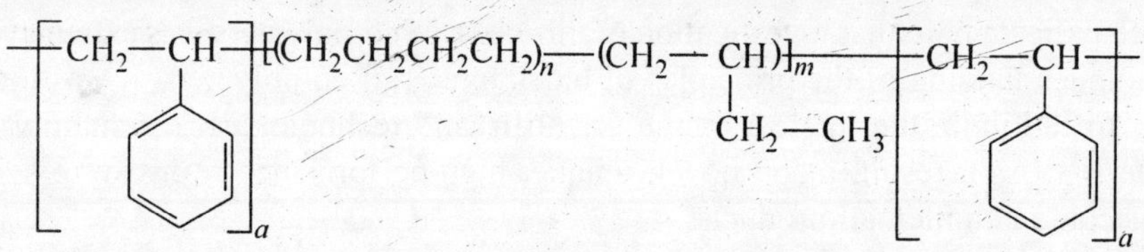

Structure (3.7)

Star-Shaped (or Radial) Styrene-Diene Co-polymers

A very interesting, equally new development, resides in the recent preparation and study of the so-called 'radial' or 'star' block polymers. It is to be noted that the concept of a star polymer has been known for more than a decade. A '5-arm' star block polymer structure can be represented by (3.8).

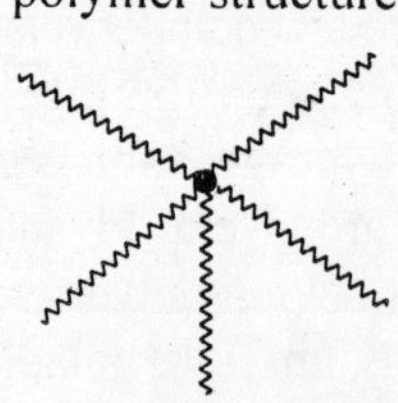

Structure (3.8)

Similar structures in which each 'arm' is itself a block polymer rather than a homopolymer were subsequently prepared and studied. Radical block co-polymers are easily synthesised by the method shown in equation 3.2, in which the polyfunctional coupling agent is $SiCl_4$, divinylbenzene, etc.

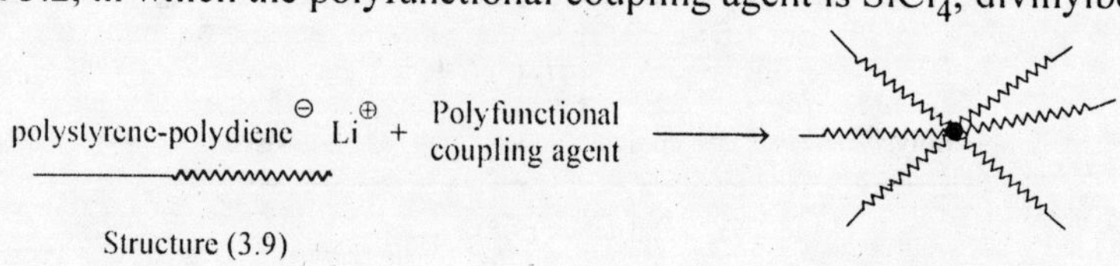

Structure (3.10) ... (3.2)

The technique has been refined by *Fetters* who has optimised the synthesis to the point that materials with up to 29 or more arms were obtained under very well-controlled conditions. An extremely interesting feature of these radial resins are their low melt and solution viscosities, even at very high-molecular weights.

However, it is important to note that in addition to elastomers, rigid radial co-polymers have been developed as well. An important class of tough styrene-rich resins has been commercialised by the Phillips Petroleum Company under the trade name of K Resins. The main attribute of these materials resides in the combination of transparency (due to the small domain size) and satisfactory impact resistance.

These properties make the resin useful for many packaging applications. For some as yet unexplained reason (possibly molecular weight) their linear counterparts do not display the same satisfactory high degree of ductility.

Styrene-Siloxane Materials

Other block co-polymers that have been studied extensively in recent years are the siloxane-styrene (or α-methylstyrene) systems. They present a significant advantage over the diene-containing materials since the soft segment (e.g., polybutadiene) is replaced by the more oxidatively stable poly(dimethylsiloxane).

Low-temperature performance is essentially equivalent. However, the tensile strengths of the diene-based co-polymers are higher than those of their siloxane analogs. This phenomenon is not completely understood, but may be the reflection of variations in molecular entanglements. The block co-polymers can be prepared by the following sequence.

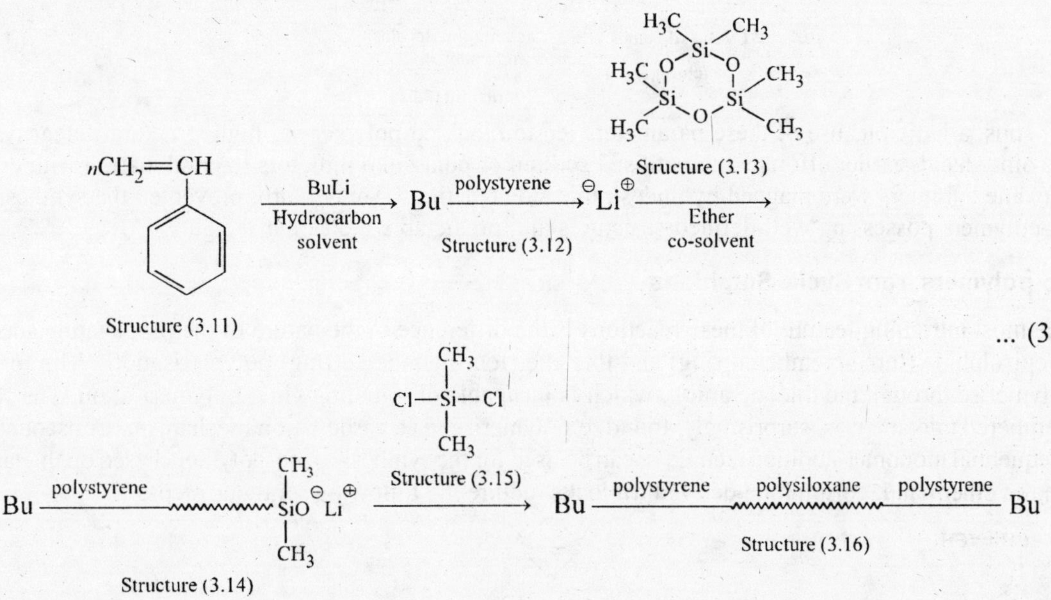

... (3.3)

The coupling step is required since the silanolate anion is not sufficiently basic to re-initiate styrene polymerisation. The scheme shown in equation 3.3 represents a way of preparing an ABA triblock co-polymers.

Multiblocks of the type $-[A-B]_n-$ have been prepared by a similar route utilising a difunctional anionic polystyrene initiator:

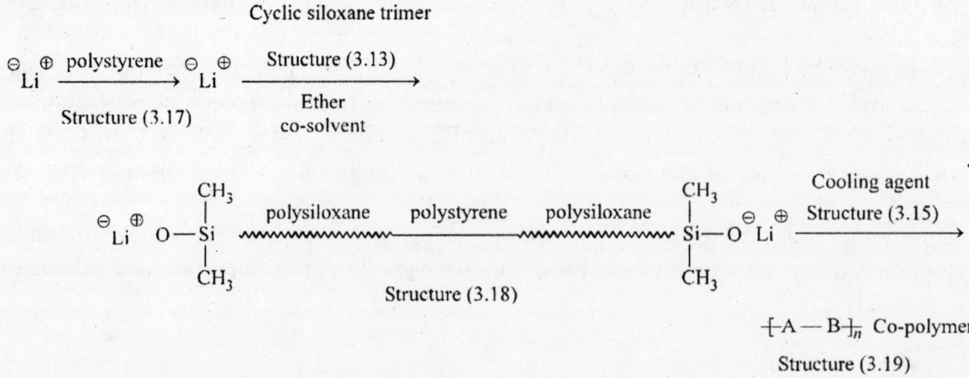

... (3.4)

The synthetic schemes 3.3 and 3.4 are based on the fundamental research conducted by *Bostick* and *Davies*. These investigators found that cyclic siloxane trimers such as (3.13) are readily polymerised to linear structures when a lithium initiator and a mildly 'promoting' solvent (an ether) are utilised. This discovery was rationalised in the following manner: (i) internal strain encourages ring opening to form linear structures; (ii) lithium-based systems do not promote fast equilibrations and redistributions of the type shown in equation 3.5; and (iii) the ether enhances the efficiency of the lithium initiator without producing the reactions (eq. 3.5).

$$\underset{\text{Structure (3.20)}}{\text{Cyclic siloxanes}} \rightleftharpoons \underset{\text{Structure (3.21)}}{\text{Linear oligomeric and high-molecular weight polysiloxanes}} \quad \text{... (3.5)}$$

Thus, a judicious use of these parameters led to block co-polymers of high structural integrity. On the other hand, earlier efforts which utilised sodium or potassium initiators in combination with cyclic siloxane tetramers were plagued by undesirable side reactions. This in turn, prevented the synthesis of co-polymers possessing well-defined structures and predictable molecular weights.

Co-polymers from Cyclic Sulphides

The most intriguing feature of these reactions is the difference in the nature of the propagating species in episulphide (three-membered ring) and thietane (four-membered ring) polymerisations. The former polymerise through the thiolate anion, which is incapable of initiating vinyl polymerisation. The four-membered thietane was, surprisingly, found to polymerise via a carbonation mechanism. Consequently, a sequential monomer-addition technique can be used for the synthesis of co-polymers based on thietanes. On the other hand, episulphide-derived triblocks require the following synthetic steps:

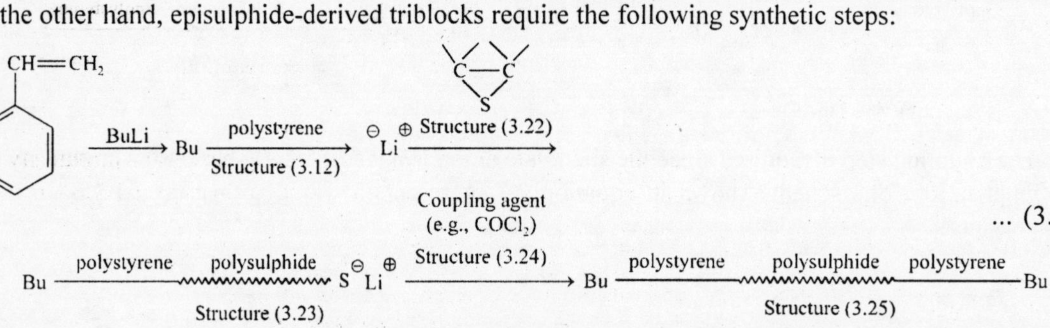

... (3.6)

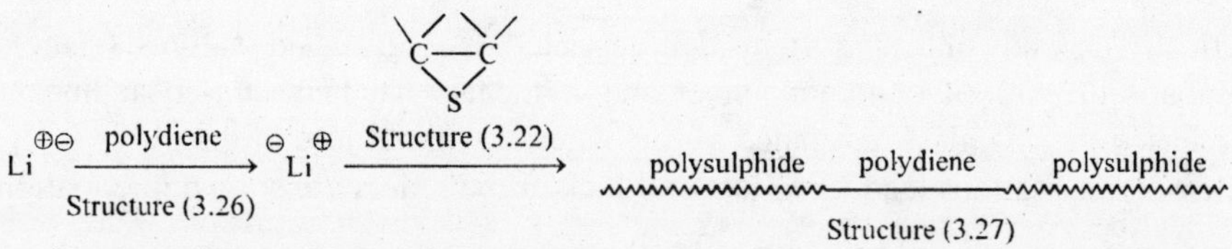

... (3.7)

Synthetic sequence 3.6 is applicable to elastomeric block co-polymers in which the episulphide (e.g., propylene sulphide) leads to the 'soft' sequence. Approach 3.7 is useful for systems in which the sulphide segment (derived for instance from ethylene sulphide) is crystalline and functions as the 'hard' block.

Lactam-Based Co-polymers

Block co-polymers based on polyamide segments constitute another class of materials wherein both elastomeric and rigid resins have been prepared. Although none has achieved a significant commercial status, it is felt that these polymers do possess potential for the future.

Lactam block co-polymers can be prepared by two methods. The first involves the ring opening polymerisation of a cyclic amide monomer initiated by an appropriately terminated oligomer. Anionic, hydrolytic, and cationic polymerisations are, in principle, feasible approaches. Note, however, that experimental conditions utilised in the hydrolytic and cationic approaches are inherently such that the probability of concurrent homopolymer formation is approximately the same as that of block co-polymer formation. On the other hand, the anionic route is believed to be the method of choice for the formation of segmented structures. It is true that several side reactions can take place in the anionic initiated polymerisation of lactams. However, the most important feature of this polymerisation is that it occurs at relatively low temperatures (100–150°C) when conducted in the presence of an initiator. Therefore, for all practical purposes, homopolymer formation is low and can be further minimised by proper control of experimental conditions.

The second method of preparing polyamide-containing block co-polymers is based on the use of functionally (amine, etc.) terminated polyamides. These are used for coupling with other suitable terminated oligomers or for initiating the polymerisation of another cyclic monomer, e.g., lactones or alkylene oxides. The anionic polymerisation of lactams to yield block co-polymers is illustrated by the following examples. It was used successfully for the synthesis of poly(arylene ether)–nylon-6 block co-polymers. The details are shown in equations 3.8 and 3.9 for the particullar poly(aryl ether) known under the trade name of Union Carbide Polysulphone.

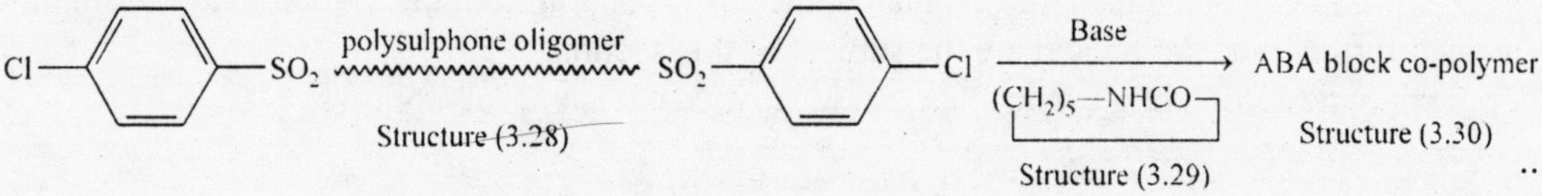

... (3.8)

Polysulphone P-1700 $\xrightarrow{\text{As above}}$ Mixtures of AB and ABA block co-polymers

Structure (3.31) Structure (3.32)

... (3.9)

Chain-scission by the lactam anion of the ether linkage in polysulphone (3.31) yields the initiating species for the reaction described in equation 3.9.

Both approaches take advantage of lactam initiation by the activated aromatic halide. Obviously, in the second method (eq. 3.9) the block integrity of the product is low. However, the method is more simple and practical. The study of these systems demonstrated two significant features: (i) their excellent 'mechanical' compatibility with either homopolymer, and (ii) the fact that their semi-crystalline morphology dramatically increased the chemical resistance of the material without loss of ductility.

Other anionic polymerisations leading to lactam block co-polymers are shown in equation 3.10.

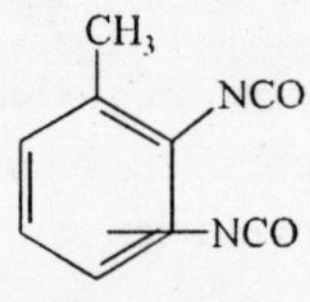

Structure (3.34)

$$\underset{\text{Structure (3.33)}}{\overset{\text{poly(oxypropylene) glycol}}{HO\sim\sim\sim\sim\sim OH}} \xrightarrow{\text{Structure (3.34)}}$$

$$\underset{\text{Structure (3.35)}}{OCN\sim\sim\sim\sim\sim NCO} \xrightarrow{\text{Lactam, basic catalyst}} \underset{\text{Structure (3.36)}}{\text{ABA elastomeric co-polymer}} \quad \ldots (3.10)$$

Similar work was also performed with poly(ε-caprolactone). Several patents describe structurally related products based on poly(oxyethylene) glycol that are useful as textile additives. Another example in which a hydrocarbon polymer is combined with nylon-6 is shown below.

$$\text{(H)X}\frac{\text{polystyrene or}}{\text{poly(}\alpha\text{-methylstyrene)}}\text{X} \xrightarrow[\text{base}]{\text{Lactam}} \underset{\text{Structure (3.38)}}{\text{AB or ABA rigid co-polymer}} \quad \ldots (3.11)$$

$X = COOC_2H_5$

$X = OCOCl$

Structure (3.37)

Since both monofunctional and difunctional initiators (eq. 3.11) were used, AB and ABA structures resulted from the reaction.

Linear Polyethers

These polymerisations may, depending on the monomer structure, be initiated by either anionic or cationic methods. The intriguing feature of some of these reactions is their 'living' character. While these characteristics have been observed in a number of instances, the true 'living' nature was unequivocally demonstrated in only a few cases. The 'living' anionic polymerisation of some cyclic ethers has been well documented, on the other hand, the corresponding cationic (oxonium ion) examples definitely need further elucidation and are the subject of this section.

A cationic polymerisation of a cyclic ether can ideally, be envisioned as follows:

$$\underset{\text{Structure (3.39)}}{\widehat{R\ O}} \xrightarrow[\text{Initiation}]{M^{\oplus}X} \underset{\text{Structure (3.40)}}{MRO^{\oplus}X^{\ominus}} \xrightarrow{n\ \text{Structure (3.39)}} \underset{\substack{\text{'Living' polymer}\\ \text{Structure (3.41)}}}{MR(\widehat{O\ R})_nO^{\oplus}X^{\ominus}} \quad \ldots (3.12)$$

If no side reactions were to occur, then structure 3.41 could serve as a macromolecular initiator for a second cationically polymerisable monomer. As a result, well-defined block co-polymers should be obtained. In reality, the situation can be summarised as follows: Small rings (e.g., ethylene oxide, propylene oxide) are very reactive; the oxonium ion intermediate, however, leads to several secondary reactions (eliminations, chain transfer, etc). For the larger cyclic monomers, e.g., hexene oxide, etc., ring opening polymerisation is not favoured thermodynamically. Thus, the correct balance of thermodynamics, reactivity, and oxonium ion stability has only been realised with four and five-membered rings, e.g., oxetanes and tetrahydrofuran. Note, that even in these preferred cases, ring-chain equilibrium demands that very careful control (e.g., temperature, monomer concentration, etc.) be maintained throughout the reaction.

Utilising the concepts summarised above, it is possible to prepare interesting elastomeric block co-polymers possessing the following structures:

A(AB)A

Structure (3.42)

where A is a polymeric segment derived from 3,3-bis(chloromethyl) oxetane and AB is a random co-polymer centre block consisting of units derived from 3,3-bis(chloromethyl) oxetane and tetrahydrofuran.

Linear Isobutylene-Styrene Systems

Cationic polymerisation of vinyl monomers to form blocks has been reported by *Kennedy* and *Melby*, who prepared styrene-isobutylene block co-polymers. The overall reaction scheme is shown below in equation 3.13.

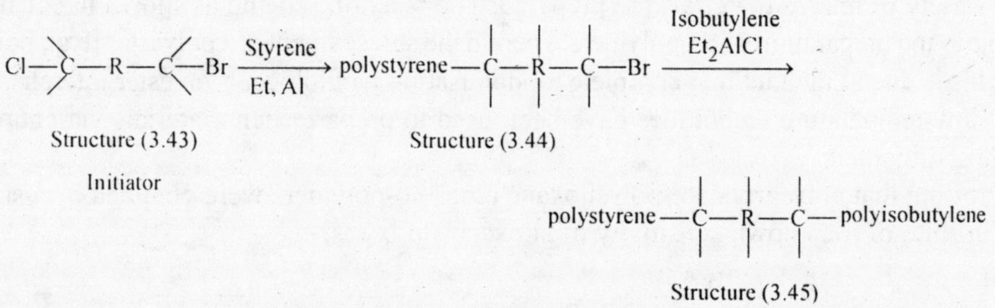

... (3.13)

A significant difference in reactivity exists between tertiary chlorides and bromides. This fact, combined with monomers that display different reactivities and with the concurrent utilisation of different organoaluminium species, form the basis for the synthetic scheme outlined above.

Formation of Random Block Polymers by Step-Reaction Polymerisation

The synthesis of linear polyurethanes via the scheme utilised by *Schollenberger* is a classical example of step-reaction polymerisation that results in a co-polymer with random block distribution. The method is illustrated below in equation 3.14:

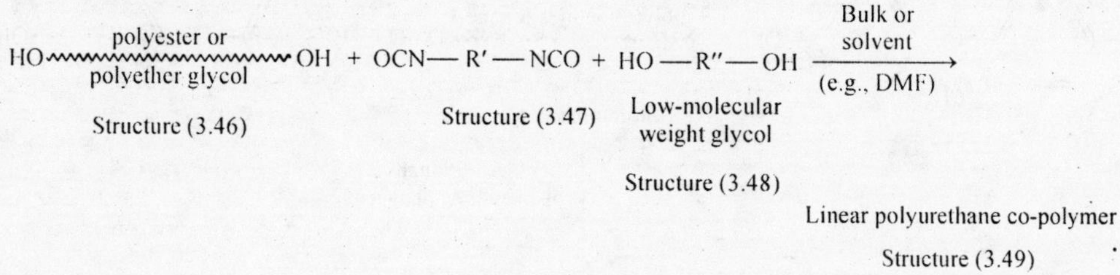

... (3.14)

Examination of the reactions involved in equation 3.14 explains the reasons for the observed random placement. Indeed, the formation of the hard block by reactions of 3.47 and 3.48 takes place simultaneously with two other processes. These are (i) the linking of the hard and soft blocks; and (ii) the self-condensation of the soft segment, e.g., reaction of 3.46 and 3.47. It is, in principle, possible to avoid the randomness by first forming the hard segment and subsequently reacting it with the polymeric glycol. However, solubility problems often render this approach impractical. Two other polymeric systems have been prepared in recent years via routes generally similar to those described above. Both are now commercial materials. The systems are cyrstalline polyester-polyether elastomers, polycarbonate-polysiloxane block co-polymers.

The synthetic approaches utilised for their preparation are outlined in equations 3.15 and 3.16.

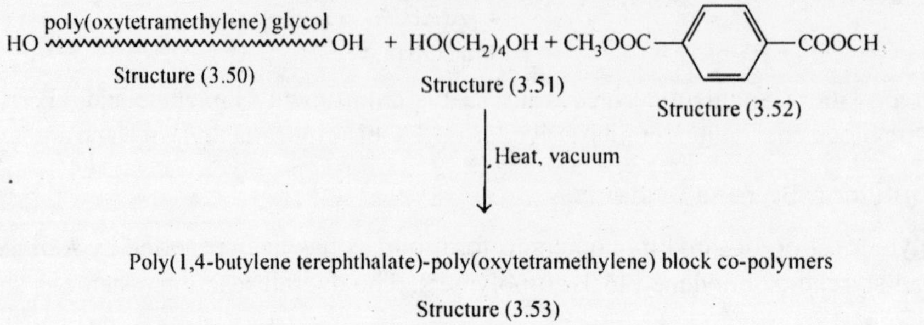

... (3.15)

A wide variety of related materials are prepared. The reaction scheme as shown in equation 3.15 is not amenable to the preparation of co-polymers wherein the soft segment is a polyester [e.g., poly(butylene adipate)]. This is due to the fact that complete randomisation would occur via ester interchange. On the other hand, low-temperature techniques have been used to prepare such materials via coupling of the corresponding oligomers.

It is interesting that polycarbonate-polysiloxane block co-polymers were obtained by a similar route (eq. 3.16), in spite of the known sensitivity of siloxanes to bases.

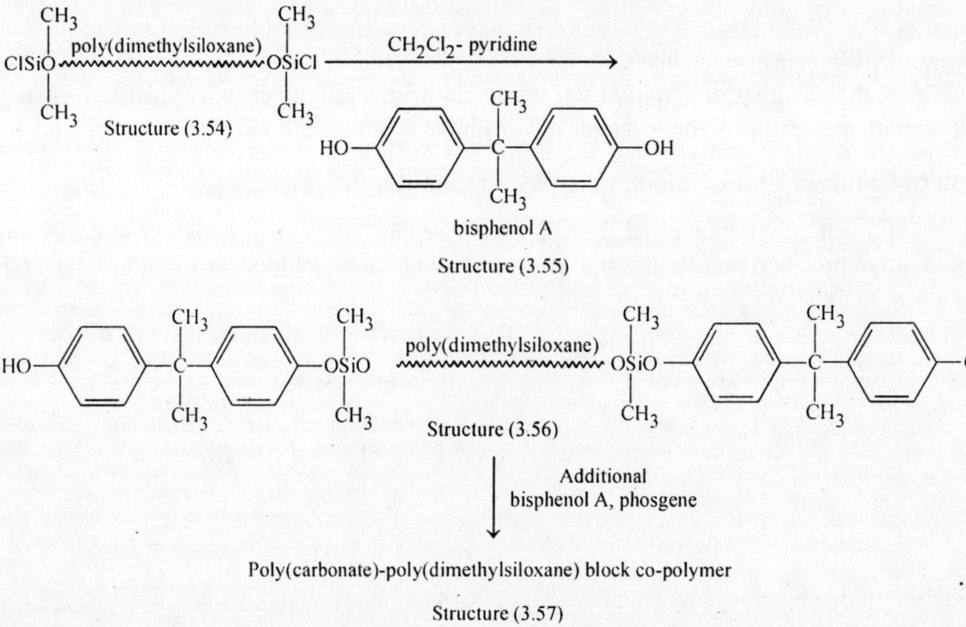

... (3.16)

Interfacial techniques have also been claimed to be successful.

Co-polymers Possessing Well-Defined $(AB)_n$ Alternating Structures

In all of the examples cited in the previous section, one preformed oligomer was reacted with the monomers that would eventually form the second block and simultaneously attach it to the preformed oligomer. A different route was chosen by *Noshay* and *Matzner* for the synthesis of a series of organosiloxane block co-polymers. It is based on the utilisation of two preformed oligomers possessing mutually reactive end groups but incapable of reacting with themselves. Thus, oligomer A can only react with oligomer B and vice verse. As a result, well-defined $(AB)_n$ alternating block co-polymers are obtained. Obviously, the polydispersity present in the starting oligomer also exists in the co-polymer products. Moreover, this feature is compounded by the polydispersity introduced during the step-reaction polymerisation itself.

The preparation of siloxane-based co-polymers takes advantage of the well-known hydroxyl-silylamine reaction as shown in equation 3.17.

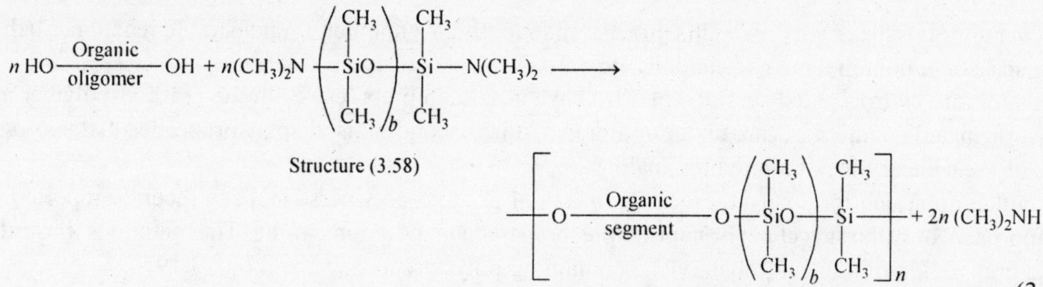

Structure (3.58)

... (3.17)

Structure (3.59)

The reaction rate depends upon the nature of the hydroxyl end groups. Phenolic hydroxyl (e.g., derived from a polysulphone) give an extremely fast reaction at 70°C; on the other hand, the process is rather slow, even at 180°C when aliphatic hydroxyl groups (e.g., derived from hydroxyl-terminated poly(α-methylstyrene) are used. This method has proved very general and permitted the preparation of a wide variety of block co-polymers.

The organic segments that were used in conjunction with poly(dimethylsiloxane) are: UCC polysulphone; the polycarbonate of 1,1,3,3,-tetramethylcyclobutane-1,3-diol; the polyterephthalate derived from bisphenol A; the polyisophthalate derived from bisphenol A; poly (α-methylstyrene) with hydroxyl termination polyamide oligomers; and poly(oxyphenylene) oligomers.

CHEMISTRY OF GRAFT CO-POLYMER FORMATION

Anionic Polymerisation

Anionic polymerisation techniques have been applied to the synthesis of backbones, side chains, and the graft co-polymers themselves. These methods have been employed with both hydrocarbon and amide-based co-polymers. Examples of some of the recently reported systems are discussed below.

Styrene-diene and related systems

It is, of course, well established that the styrene-diene linear block co-polymer syntheses are based on anionic techniques. A logical extension of this concept consists in preparing, by an analogous route, the corresponding styrene-diene graft co-polymers.

A typical synthesis is illustrated in equation 3.18.

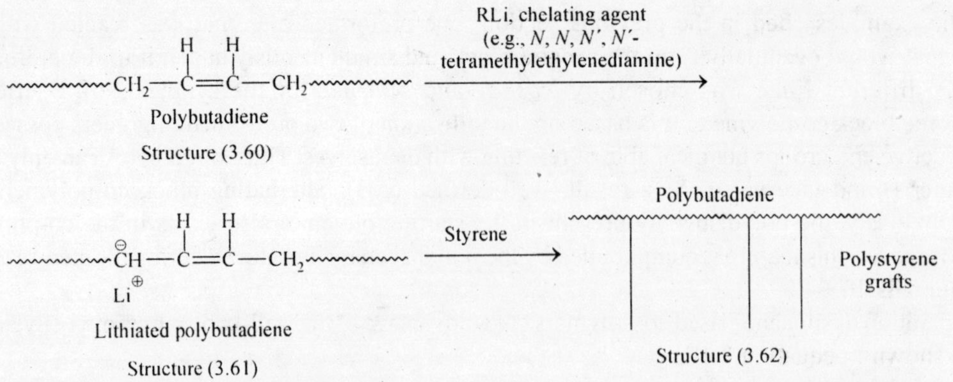

... (3.18)

Several groups have worked on this process and identified a number of undesirable reactions such as chain scissions, homopolymer formation, etc.

In addition, hydrogenated versions of 3.61 have also been prepared. Both the graft co-polymer and its hydrogenated counterpart can, under optimum synthetic conditions, display properties that approach those of their linear block co-polymer analogs.

Another approach for the synthesis of a variety of graft co-polymers has been recently reported. In this approach, only the side chain is, in principle, prepared via the anionic route. The method is illustrated by the preparation of a linear polyethylene containing polystyrene side chain (eq. 3.19).

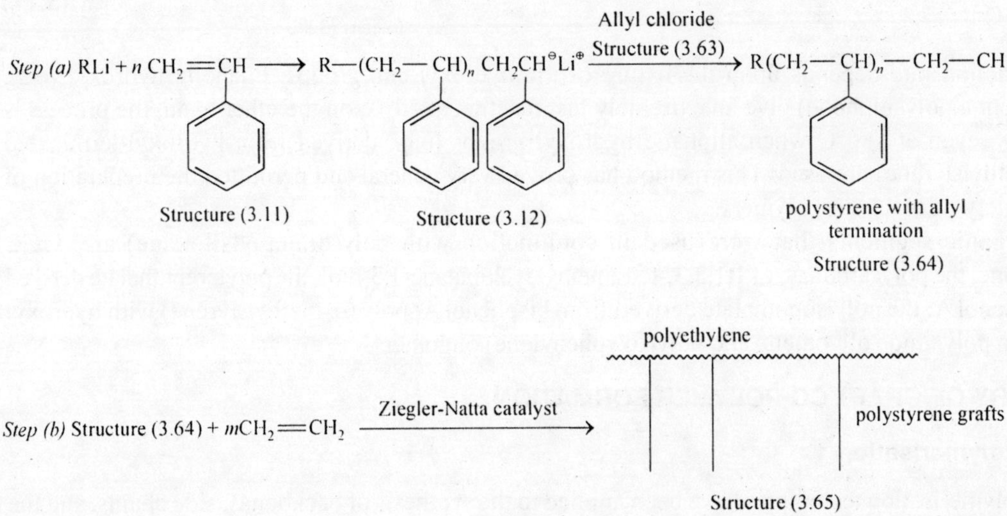

... (3.19)

Step (a) offers a general synthetic method to prepare a variety of 'macro-molecular monomers'. The latter are then co-polymerised, via known polymerisation mechanisms, to produce the backbone.

Lactam-based co-polymers

Anionic polymerisation leading to graft co-polymers containing polyamide side chains can be envisioned as follows:

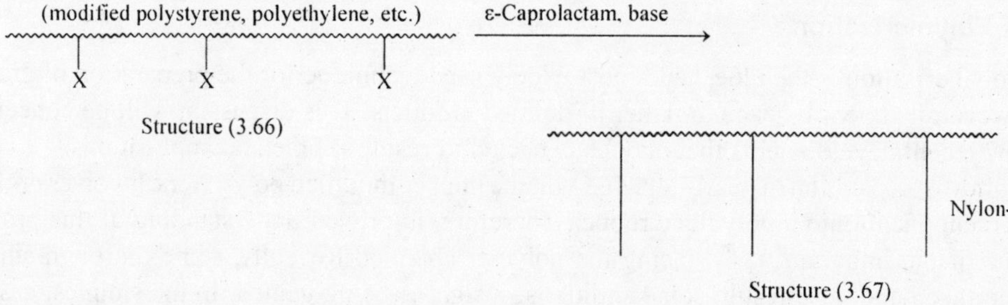

... (3.20)

X is a group capable of initiating the anionic ring-opening polymerisation of a lactam. In most of the cases studied, X was either an ester (COOR), acid halide (COCl), or *N*-carbamoyllactam.

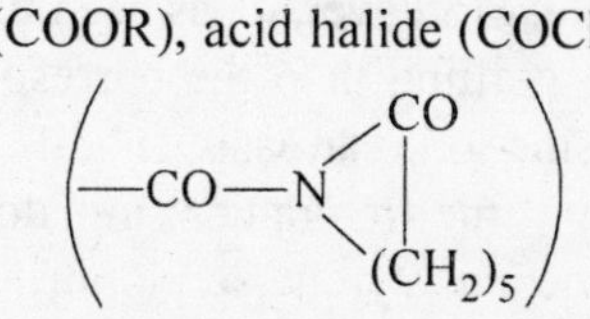

The preparation of the backbones is possible by conventional means. The graft polymerisation step is conducted under conditions typical of an anionic nylon-6 polymerisation. It is to be noted that the polyamide grafts can be prepared via hydrolytic and cationic processes as well.

Cationic Polymerisation

One of the classical methods for generating organic cations consists in forming a complex between a Lewis acid and an organic halide containing, preferably, a labile carbon-halide bond.

$$\underset{\substack{\text{Lewis acid}\\ \text{Structure (3.68)}}}{R_3Al} + \underset{\substack{\text{Organic halide}\\ \text{Structure (3.69)}}}{R'X} \longrightarrow \underset{\text{Structure (3.70)}}{\overset{\oplus}{R'}\ \overset{\ominus}{[AlR_3X]}} \quad \text{... (3.21)}$$

This well-known principle was utilised successfully for the preparation of graft polymers with a poly(vinyl chloride) backbone. Indeed, poly(vinyl chloride) contains labile allylic chloride sites and is, therefore, amenable to the scheme shown in equation 3.21. The cation formed via that route serves as the polymerisation macro-initiator. Two systems were prepared and extensively studied: rigid grafts based on polystyrene and flexible grafts based on polyisobutylene. Extraction studies and mechanical properties indicated that true grafting had occurred. Two aspects of this work are of particular interest: (i) the grafting process, *per se*, removes chlorine sites that are inherent sources of instability; as a result the material displays enhanced thermal resistance; and (ii) at higher polyisobutylene contents one may have a permanently plasticised poly(vinyl chloride) composition.

The same principles were also applied to a flexible backbone, e.g., ethylene-propylene-diene terpolymers (EPDM). The presence of unsaturation in these materials allows for the introduction of labile-halide sites, analogous to those in poly(vinyl chloride). The co-polymers prepared with rigid polystyrene side chains has mechanical behaviour typical of a vulcanised rubber. Recently, an EPDM backbone containing both chlorine and bromine functionality was used to synthesise 'bigraft' co-polymers in which grafts of styrene and α-methylstyrene coexist side-by-side. This is an interesting system; since polystyrene and poly(α-methyl-styrene) segments have been shown to be miscible in block co-polymers even at very high-block-molecular weights.

Free-Radical Polymerisation

Free-radical polymerisation is the oldest and most widely used technique for the preparation of grafted materials. However, it generally leads to rather ill-defined products. This discussion will be concerned only with a few recent developments that may be expected to result in practical applications.

As is well known, preparation of nearly all large-volume impact-modified polystyrene involves grafting of the styrene component onto a polydiene rubber. Therefore, improved understanding of this process is certainly one of the important tasks facing the polymer chemist. Recently, some thorough studies have been undertaken and, as a result, some additional insight is being gained. In the simplest case of thermal grafting, one must rely nearly exclusively on chain transfer of the growing polymeric radical to the polydiene to generate the grafting site. Moreover, it has been found that the 1,2 microstructure of the butadiene units is more responsive to grafting than the corresponding 1,4 structures. The practical importance of these finding is obvious: since one can easily modify the anionic polymerisation conditions to favour the 1,2 structure. The nature of the free-radical initiator and its effect on chemical grafting have also been studied both with styrene and styrene-acrylonitrile systems.

A product that is generating considerable industrial interest is the TPR thermoplastic rubber. It combines the elastomeric behaviour of ethylene-propylene-diene terpolymer with the high melting point of polypropylene. Although the structure is not known, it is believed that the polymer is obtained via free-radical grafting of ethylene-propylene-diene terpolymer onto polypropylene.

Structure-Property Characteristics

Block and graft co-polymers differ significantly from random or alternating co-polymers. The most interesting feature is their ability to yield two-phase systems in which the two phases are linked by a chemical bond. The latter ensures a fine dispersion or intermixing of the phases thus yielding some unique properties. Rigid-rigid, rigid-soft, and soft-soft block and graft co-polymers have been prepared and studied.

Although several well-defined block structures have been prepared, graft co-polymers, in most cases, are rather ill-defined with respect to side-chain spacing and distribution. Therefore, rigorous property-structure correlation has thus far only been achieved for the former. One would expect the findings to hold true for graft co-polymers as well, and the available data seem to support this general contention.

Phase Separation and Morphology

Homopolymer blends displaying even slight structural difference generally exhibit discrete two-phase morphology. In order to achieve complete miscibility, basic thermodynamic principles (eq. 3.22) require a negative value of the *Gibbs* free energy

$$\Delta G = \Delta H - T\Delta S \qquad \text{... (3.22)}$$

where ΔH = enthalpy of mixing;

ΔS = entropy of mixing;

T = absolute temperature.

The recent important studies related to polymer-polymer solubility such as the two-component solubility parameter approach by *Shaw*, the adaptation of *Flory's* equation of state thermodynamics by *McMaster*, and the solvent probe techniques discussed by *Kwei* and *Olabisi* are considered major advances and should be consulted by the interested reader. In a typical polymer mixture, the morphology consistently varies as a function of the composition. Generally, in the 0.30–0.70 volume fraction range, both components can exhibit some degree of phase continuity. Geometric considerations require that components present at less than 0.30 volume fraction exist as dispersed phases.

The existence of covalent bonds clearly imposes a restriction to phase separation due to a decrease in the entropy caused by configurational considerations. Two basic theories of micro-phase separation in block co-polymers have been developed by *Meier* and *Krause*. Both theories rely heaviiy on *Flory's* interaction parameter χ_{AB} and, hence, are somewhat limited to non-polar polymer systems where only dispersive interactions are important.

By using the simplest case of an AB block co-polymer, *Meier* established the thermodynamic criteria for domain formation. He calculated that higher block molecular weights were required for phase separation in a co-polymer than in simple homopolymer blends. This theory predicted that in the case of a styrene-butadiene AB block co-polymer, phase separation would occur when the polybutadiene block mw is above 50,000 and the polystyrene block mw is above 5,000–10,000. The most stable domain structure was also considered as a function of block molecular weight (or composition). It was shown that the predicted spherical domains existed as the preferred morphology at low A block mw (e.g., M_A/M_B less than 0.25). When the ratio of M_A/M_B was above 0.03, lamellar structure was favoured. Cylindrical domains were predicted in the intermediate region. These predictions have qualitatively been shown to be correct via electron microscopy and small-angle X-ray studies.

Note, however, that block co-polymers can display various morphologies dependent upon the method of preparation. This is particularly true in solvent-cast systems where the solvent can be preferentially absorbed in one phase at the point of phase separation. The resultant morphology was theoretically treated by *Meier* for block co-polymer–solvent systems where it was emphasised that the morphology in the solvent-free block co-polymer will probably be fixed in a non-equilibrium state.

A different approach, which is strictly thermodynamic and not concerned with morphology, was developed by *Krause*. The critical interaction parameter $(\chi_{AB})_{cr}$ was related to the number of blocks per molecule, block molecular weight, homopolymer concentration, and component block concentration. The expression is given in equation 3.23

$$(\chi_{AB})_{cr} = \frac{zV_r}{(z-2)V_B V_A{}^M n_B{}^C}\left[-\ln\left(V_A{}^M\right)^{V_A{}^C}\left(V_B{}^M\right)^{V_B{}^C} - \frac{N_{HA}}{N_C}\ln V_A{}^M + 2(m-1)\frac{S_{dis}}{R} - \ln(m-1)\right] \quad \text{... (3.23)}$$

where V_B = volume of B repeat units; $n_B{}^C$ = number of B units in each block co-polymer molecule; $\Delta S_{\text{dis}}/R$ = entropy lost when one segment of molecule is immobilised (assumed = 1.0); $V_A{}^C$ and $V_B{}^C$ are volume fractions of monomers A and B in block co-polymer molecule; $V_A{}^M$ and $V_B{}^M$ are volume fractions of monomers A and B in total mixture; z = lattice coordination number (assumed = 8); V_r = volume of a lattice site; m = number of blocks per molecule; N_{HA}, N_C = number of homopolymer molecules and total number of molecules.

Experimental data on poly(α-methylstyrene)-polystyrene AB block co-polymer [$(\chi_{AB})_{cr}$ = 0.002] and the corresponding homopolymer blends have shown that phase separation occurs in homopolymer blends, but that one-phase behaviour is observed in a block co-polymer where the block molecular weights were greater than those of the homopolymer. *Krause's* theoretical treatment also considers the important case where cyrstallisation is a mode of phase separation.

Transitions

Amorphous transitions

Above certain critical block molecular weights, block and graft co-polymers exhibit micro-phase separation. Exceptions can occur for blocks with similar polar and dispersive forces and/or specific interactions. The modulus-temperature behaviour of single-phase versus two-phase block and graft co-polymers is contrasted in Fig. 3.1. The determination of phase behaviour is commonly accomplished via modulus-temperature measurements, mechanical loss data and/or thermal analysis (i.e., differential scanning calorimetry).

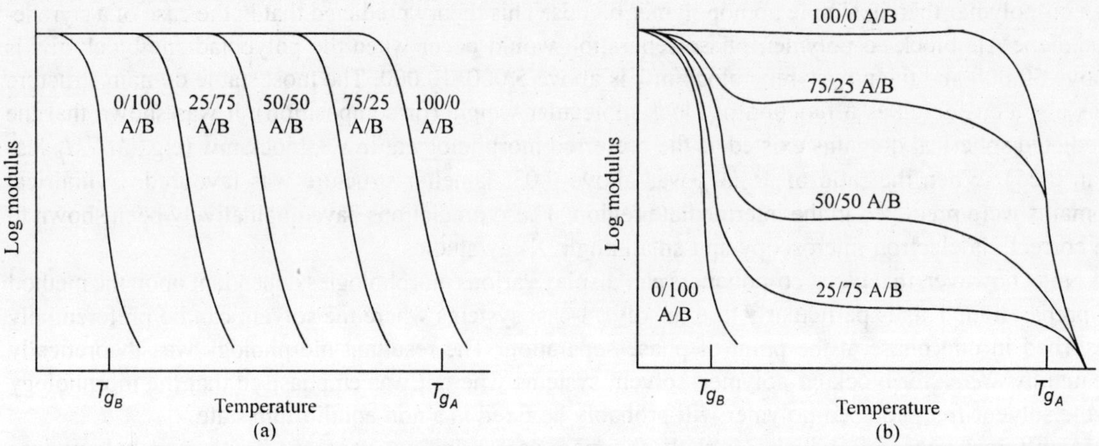

Fig. 3.1. Generalised modulus-temperature behaviour of block or graft co-polymers. (a) Random co-polymers or soluble block or graft co-polymers; and (b) phase-separated block or graft co-polymers.

Examples of phase-separated systems include bisphenol A-derived polycarbonate-polysiloxane, polysulphone-polysiloxane, and polystyrene-polydiene block co-polymers. Single amorphous phase block co-polymers (as determined by the existence of a single T_g intermediate between the constituents) include polystyrene-poly(α-methylstyrene) and polysulphone-bisphenol A-derived polycarbonate.

The magnitude of the respective transitions has been shown to be influenced by the method of preparation. Casting from different solvents leads to morphological changes and, hence, continuous structure variation. Examples of this dependence have been shown for poly(methyl methacrylate) grafted on natural rubber and polysulphone-poly(dimethylsiloxane) block co-polymer. Mechanical loss-temperature data are shown in Fig. 3.2. Mechanical deformation has also been observed to yield changes in transitional characteristics. *Canter* observed that pre-stressing styrene-diene block co-polymers sharpened the polystyrene transition with a resultant lowering of modulus between the constituent glass-transition temperatures. In cases of limited solubility, sharp micro-phase separation may not occur, and the characteristic T_g may be slightly shifted and, indeed, new intermediate transitions may emerge. This was observed by *Beecher* for styrene-butadiene block co-polymers cast from different solvent systems. It is important to point out that in well-phase-separated block co-polymers, T_g will show a predictable dependence on block molecular weight. Such behaviour was observed for polysiloxane block co-polymers for which the hard block were either polysulphone or bisphenol A derived polycarbonate. Similar dependence of T_g on molecular weight was reported for both systems (with appropriate correction for T_g at infinite molecular weight).

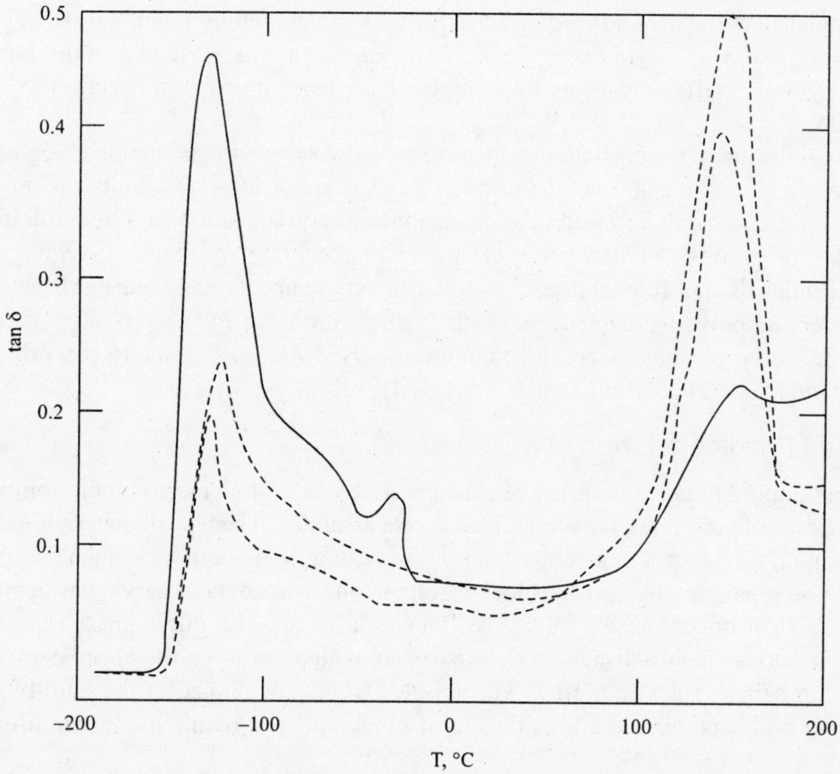

Fig. 3.2. Mechanical loss curves for the 5000–5000 $\overline{M}_n$—(AB)$_n$—polysulphone-poly(dimethylsiloxane) block co-polymer prepared under different conditions. - - -, cast from tetrahydrofuran; —. . . .—. . . .— , cast from xylene; —, moulded at 320°C.

In the case of single-phase block co-polymers, T_g values intermediate between the infinite molecular weight T_g values are observed. This behaviour was displayed by the 5000–5000 block $\overline{M}_n$ bisphenol A-derived polycarbonate-polysulphone block co-polymer.

Crystalline transitions

Crystallinity provides another means for promoting phase separation in block and grafts co-polymers. Thermoplastic polyurethanes are the earliest example of block co-polymers in which phase separation occurs via development of crystallinity. Important differences in crystallinity and melting point are noted for block vs random co-polymers. The principle was illustrated in polyester block co-polymers investigated by *Charch* and *Shivers*. In random co-polymers a rapid decrease of the melting point (i.e., crystallinity) is observed with co-monomer incorporation. This is much less pronounced in block systems because of the long sequences. Similar results have been reported for polyamide structures.

Other examples of co-polymers containing crystalline segments include polysulphone-nylon-6, polyethylene-polypropylene, and poly(butylene terephthalate)-poly(oxytetramethylene) blocks as well as grafts of nylon-6 with polystyrene and with polyethylene.

In the polysulphone-nylon-6 block co-polymer, the degree of crystallinity and melting point of the polyamide block are virtually the same as those of the nylon homopolymer. However, grafting of nylon-

6 onto poly(ethylene-co-acrylic acid) produced reductions in the melting point, degree of crystallinity, and crystallisation rate of the nylon component. This behaviour was attributed to the low molecular weight of the nylon-6 grafts as well as to a single-phase behaviour of the fraction existing in the amorphous state.

One additional feature of the crystalline transition behaviour is worthy of mention. Segments that are essentially amorphous in the isotropic undeformed state may develop crystallinity upon deformation. This phenomenon is commonly referred to as stress-induced crystallisation, and its result is to produce excellent ultimate properties. Maintenance of elastomeric behaviour (e.g., recovery) requires that stress-induced crystallisation disappear upon stress removal. Rubbery segments exhibiting this behaviour include certain polyethzer and polyester sequences [i.e., polytetrahydrofuran and poly(ε-caprolactone) present in thermoplastic polyurethanes. A recently commercialised material utilising the principle is the poly(butylene terephthalate)-poly(tetrahydrofuran) $(AB)_n$ block co-polymer.

Mechanical and Rheological Properties

One of the most important characteristics of elastomeric block co-polymers is their ability to display rubbery properties while still retaining thermoplastic processability. The type of materials exhibiting this behaviour is comprised of hard segments (high T_g and amorphous) and soft segments (low T_g and amorphous). Phase separation of the hard block results in physical domains serving as tie points which are similar to covalent bonds present in a cross-linked elastomer. The micro-phase separation of the hard segments results in highly dispersed rigid particles which reinforce the elastomeric matrix in a manner similar to carbon-black-reinforced vulcanisates. Judicious choice of the hard-block content, block co-polymer architecture, and a high degree of block purity is required to attain useful ultimate properties.

AB and BAB elastomeric block co-polymers (A = hard block) have been shown to have inferior tensile-strength and permanent-set properties compared to ABA and $(AB)_n$ ($n > 1$) systems. A new class of block co-polymers, radial block co-polymers, combine excellent ultimate properties with easier processing than their linear counterparts.

Elastomeric block co-polymers, based on polyurethanes or on polyester-polyether blocks, exhibit high tensile strength (above 5000 psi) and excellent wear and abrasion resistance. These properties are attributed to the ability of the soft block of these materials to stress crystallise. The phenomenon can even be visually observed. High extension results in opacity which disappears upon removal of stress. The styrene-diene ABA block co-polymers exhibit strengths characteristic of carbon black filled, cross-linked elastomers (ca 2000–3000 psi). Hydrogenated versions of styrene-butadiene have tensile strengths approaching those of the thermoplastic polyurethanes (ca 5000 psi). This improvement has been suggested to be due to the better phase separation expected with an ethylene-butylene block compared to a polybutadiene block. Moreover, the presence of ethylene sequences in the centre block may contribute to stress-induced crystallisation.

Block co-polymers exhibit a marked dependence of mechanical properties on the method of sample preparation. Variations in casting solvent (from an A block preferred solvent to a B block preferred solvent) bring about significant variations in the stress-strain and dynamic mechanical behaviour. Typical results are shown in Table 3.1 for the polysulphone-poly(dimethylsiloxane) $(AB)_n$ block co-polymer. The effect is most prominent in the initial stage of the stress-strain curve, i.e., in the tensile modulus. The behaviour can be rationalised by considering the phase separation process and the resultant morphological changes. Phase separation occurs after a certain concentration of solvent has evaporated. At that point, the morphology is determined by the distribution of the solvent between the respective phases. If the solvent is preferentially dissolved in the soft block phase, lower-modulus, more elastomeric materials are obtained. It must be pointed out that the resultant morphology of solvent-cast samples will generally be in a non-equilibrium state, as has been theoretically discussed by *Meier*. This behaviour is not unique to block co-polymers. In fact, the initial experiments showing this effect were conducted with a poly(methyl methacrylate) natural rubber graft co-polymer.

Table 3.1. Variation of mechanical properties by method of film preparation[a].

Method of sample preparation	*1% secant modulus, psi*	*100% modulus, psi*	*Tensile strength, psi*	*Elongation %*
Cast from tetrahydrofuran	35,500	1,760	3,310	410
Cast from chlorobenzene	33,040	1,700	3,020	355
Cast from chloroform	34,490	1,775	2,970	355
Cast from benzene	22,350	1,570	3,010	517
Cast from toluene	16,760	1,460	2,260	310
Cast from xylene	9,500	1,375	1,870	350
Cast from ethyl acetate	8,850	1,250	1,890	250
Moulded at 320°C	8,200	–	1,720	70

[a] (5000/5000 $\overline{M}_n$ polysulphone-poly(dimethylsiloxane) block co-polymer.

The modulus-composition behaviour for polymer-polymer blends is generally quite similar to that of their block or graft co-polymer counterparts. However, the ultimate properties of the block or graft co-polymers are generally superior due to the excellent interfacial adhesion obtained by the imposition of a covalent bond at the phase boundary. Examples where block co-polymers exhibited ultimate properties superior to this counterpart homopolymer blends include polysulphone–nylon-6, polystyrene-polysiloxane, and of course, the polystyrene-polydiene ABA block co-polymers.

Modulus-composition data can be used to assess the continuous phase of block co-polymers. *Kerner's* analysis for a two-phase system has been used to determine the continuous-phase contribution of the respective components of the polysulphone-poly(diemthyl siloxane) block co-polymers. It was shown that the predicted continuous structure based on modulus data agreed well with a similar analysis derived from permeability data. *Faucher* recently suggested that the use of *Kerner's* polyagglomerate analysis could adequately predict the modulus-composition relationship for styrene-diene block co-polymers. However, this approach should clearly not be applied in the case of films cast from a solvent preferential to one block. An interesting study of the behaviour of styrene-butadiene ABA block co-polymers was presented by *Matsuo* and *Sagaye*. Samples prepared by compression moulding gave superior strengths above 50 wt per cent styrene, whereas samples prepared by casting from cyclohexane (a poor polystyrene solvent) were stronger below 50 wt per cent styrene. Morphological characterisation of the continuous-phase structure agreed with strength data.

A characteristic feature of block co-polymers is that the melt viscosity is higher than that which would be expected from the homopolymer constituents. This has been attributed to the fact that the network structure exists above the major transitions of the constituent blocks. *Kraus* and *Gruver* noted that even at low shear rates non-Newtonian behaviour was observed and a temperature-shear rate super-position of the data was not possible. At high frequencies, *Arnold* and *Meier* found that typical thermoplastic behaviour was observed, indicating that the network domains can be broken down. *Matzner* hypothesised that the melt viscosity was related to the difference in solubility parameter between the constituent blocks. In a series of poly(dimethylsiloxane)-based block co-polymers, it was shown that processability improved (melt viscosity decreased) as the hard block solubility parameter approached that of soft block. Recently, radial or star-shaped block co-polymers have been shown to display lower melt viscosity while at the same time maintaining equal or superior mechanical behaviour.

Transport Behaviour

Whereas mechanical properties of block co-polymers have been investigated in detail, other physical properties (i.e., electrical, thermal conduction, and permeability) have received less attention. The permeability of polysiloxane block co-polymers is of particular interest (i.e., as membranes for blood oxygenators) since high permeability, as well as penetrant selectivity, can be attained while permitting enhanced film properties (stiffness and strength) over those of cross-linked silicone rubber. An example of this behaviour is the permeability of polysulphone-poly(di-methylsiloxane) block co-polymers studied by *Robeson*. Fig. 3.3 illustrates the oxygen permeability as a function of composition.

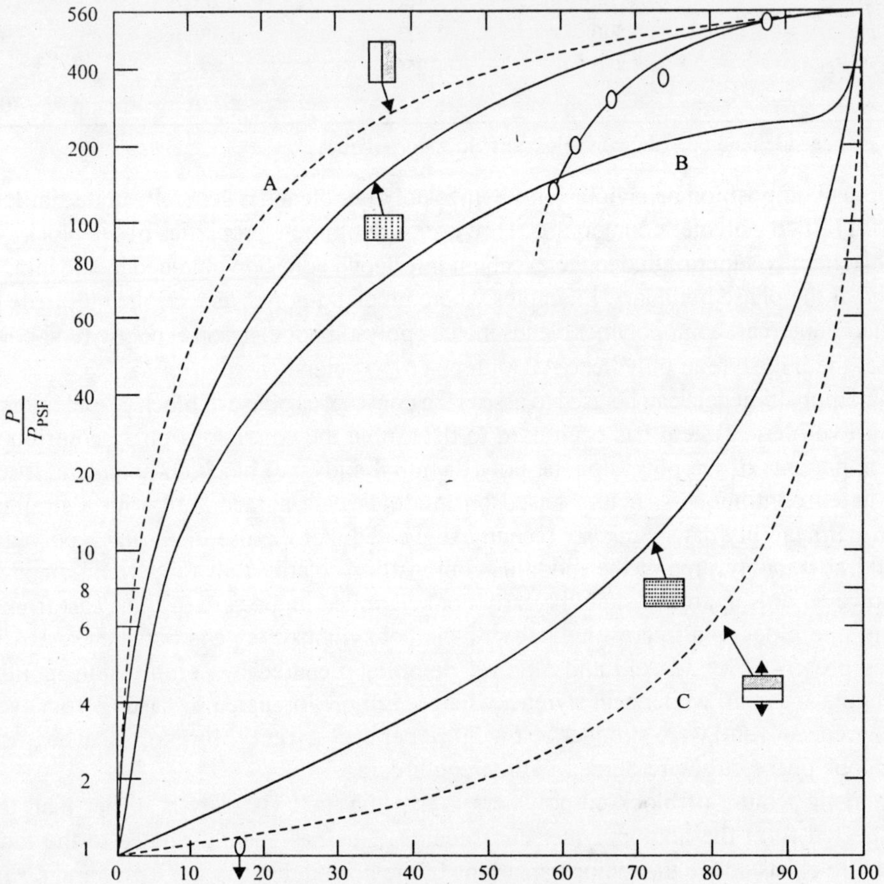

Fig. 3.3. Analysis of the phase continuity of the —$(AB)_n$—polysulphone-poly(dimethylsiloxane) block co-polymers based on *Maxwell's* equations for oxygen permeability. A, poly(dimethylsiloxane) continuous; B, both polysulphone and poly(dimethylsiloxane) equally continuous; and C, polysulphone continuous.

Maxwell's equation was utilised to determine the continuous-phase structure from oxygen permeability data. Data for the latter agreed well with modulus data determinations, thus lending credence to both approaches. It was hypothesised that once the continuous-phase structure was determined (i.e., from modulus or permeability data) other transport properties could then be calculated (i.e., thermal conductivity or electrical conductivity) from, values for the homopolymer constituents.

A considerable amount of data has been reported on graft co-polymers as potential candidates for reverse osmosis membranes. Examples of these include poly(sodium acrylate) grafted on polyethylene and polystyrene grafted on cellulose acetate.

The generalised relationship of permeability vs composition for random co-polymers, one-phase block or graft co-polymers, and two-phase block or graft co-polymers is illustrated in Fig. 3.4. Oxygen values for the nylon-6 ethylene-acrylic acid graft co-polymer indicated one-phase behaviour which was in full agreement with transition results also showing a single T_g. Oxygen permeability for the styrene-methacrylonitrile block co-polymers (Fig. 3.5) showed the expected two-phase behaviour, with the block co-polymer exhibiting significantly better oxygen barrier properties (at high methacrylonitrile content) that the random co-polymer of equal composition.

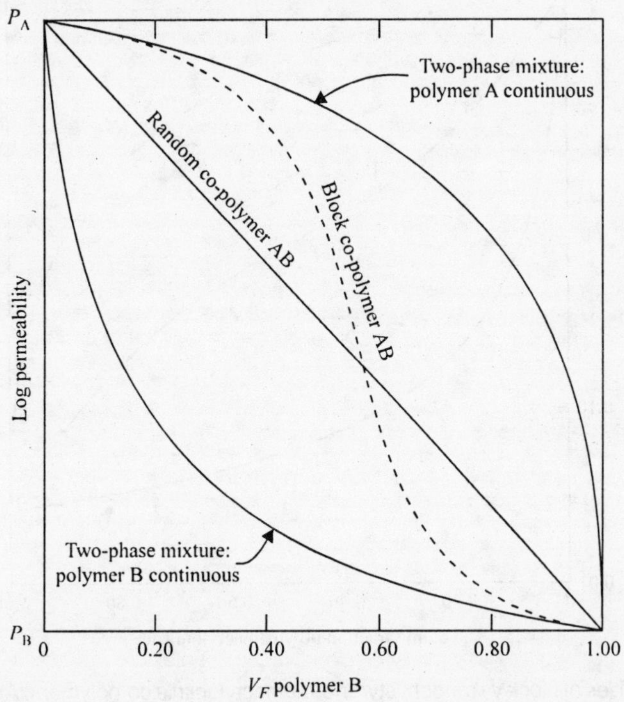

Fig. 3.4. Permeability relationships for block and random co-polymers.

Blending Phenomena

The principal applicability of graft co-polymers lies in the *in situ* polymerisation of styrene and styrene-acrylonitrile mixtures with unsaturated elastomers yielding impact modified polystyrene and acrylonitrile-butadiene-styrene co-polymer. During this process a significant amount of grafted matrix polymer on a rubbery backbone is formed. The variables inherent in this process have been systematically investigated including the importance of agitation during phase separation, occluded polystyrene to improve rubber efficiency, and the importance of rubber-phase cross-linking.

Whereas graft co-polymer-homopolymer blends are commercially utilised in large volumes (e.g., high-impact polystyrene, acrylonitrile-butadiene-styrene co-polymers, impact modifiers for vinyl plastics), block co-polymer–homopolymer systems have not yet reached a similar stage of development. However, there are some interesting studies which have been reported and which deserve mention. *Inoue* studied the addition of homopolymer constituents to an AB styrene-isoprene block co-polymer and concluded that if the homopolymer molecular weight was greater than that of the constituent block, the homopolymer would form a discrete phase in a blend with the block co-polymer. In studies of polysulphone–nylon-6 block co-polymers, a desirable average of mechanical properties was obtained over a broad composition range of blends comprising either polysulphone or nylon-6 and the block co-polymer. However, when both homopolymers were added to the block co-polymer, brittle materials were obtained. The brittle behaviour was believed to be the result of poor interfacial adhesion between the constituent homopolymers, thus yielding a flaw in the composite.

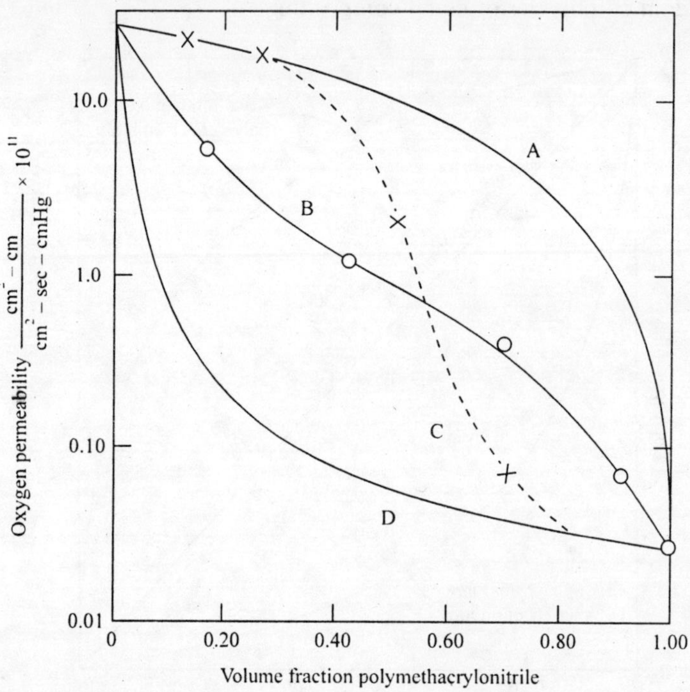

Fig. 3.5. Oxygen permeabilities of block vs random styrene-methacrylonitrile co-polymers. A, polystyrene continuous phase, *Maxwell's* equation; B, styrene-methacrylonitrile random co-polymers; C, styrene-methacrylonitrile block co-polymers; and D, polymethacrylonitrile continuous phase *Maxwell's* equation.

Most blending studies have dealt with phase-separated block and graft co-polymers. However, an investigation of blends of homopolymer constituents with the one-phase α-methylstyrene-styrene AB block co-polymers has been reported and is of considerable interest. At similar molecular weights, poly(α-methylstyrene) addition to the block co-polymer produced a single-phase system, whereas polystyrene addition resulted in a two-phase system (polystyrene homopolymer and block co-polymer phases). A ternary blend of the block co-polymer and the homopolymer constituents resulted in a polystyrene-rich phase and a poly(α-methylstyrene)-block co-polymer phase.

Blends of block and graft co-polymer with rigid homopolymer have been cited in many examples where the impact strength of the rigid polymer is improved. The incorporation of a little or 5 wt. per cent of a polysulphone-poly(dimethylsiloxane) $(AB)_n$ block co-polymer in polysulphone was found to enhance the notched Izod impact strength by a factor of ten. Polycaprolactone-polyol blocks incorporated into a cross-linked epoxy network resulted in both improved impact resistance and dramatically improved thermal shock resistance.

APPLICATIONS

Commercial acceptance of block co-polymers has centred upon their utility as thermoplastic elastomers. The primary advantages of the thermoplastic systems vis-a-vis their chemically cross-linked analogs include fast-cycle injection mouldability, re-use of scrap, transparency, greater colour selections possible (i.e., no carbon black required for reinforcement), fabrication via solvent casting, etc. However, such large-volume markets as automotive tyres are excluded owing to property limitations, such as high-temperature compression set and poor solvent and oil resistance. Therefore, most of the market penetration has been directed toward mechanical goods and speciality products. These areas have shown dramatic growth in the last decade and this trend has been projected to continue in the future.

Thermoplastic polyurethanes were the first commercial thermoplastic elastomers, and they are used in applications where excellent strength, wear, and oil resistance are required. Typical end uses include footwear, automotive bumpers and facia, adhesives and coatings, etc. Novel applications such as snowmobile treads and horseshoes have also been developed.

Polyester-polyether block co-polymers based on poly(butylene terephthalate) and poly(oxytetramethylene) are now commercially available under the trade name Hytrel. This material has characteristics that are similar to those of thermoplastic polyurethanes. It has been reputed to display better processing stability and low-temperature properties.

The styrene-diene-styrene ABA block co-polymers have shown promising growth in a variety of end-uses. They have found applications in footwear, adhesives, sporting equipment, and general-purpose rubber goods (e.g., toys). Second-generation ABA materials wherein the centre polybutadiene block has been hydrogenated to yield a saturated ethylene-butylene structure have now been commercialised under the trade name of Kraton G. Obviously, this transformation results in vastly improved resistance to thermal oxidation and to weathering. Such improvements should extend the end-use markets into areas where the precursor materials were of marginal utility or even unacceptable (e.g., automotive exterior parts).

Styrene-diene AB block co-polymers have been commercial materials for many years; they have been shown to impart improved low-temperature flexibility, increased hardness, and improved processability in blends with other conventional elastomers. The styrene-diene elastomers are also useful impact modifying additives for rigid thermoplastic polymers such as polystyrene. Rigid, high styrene-butadiene radial block co-polymers have been recently introduced under the trade name of K-Resins. These materials are transparent, relatively tough, and especially useful in clear packaging applications.

Siloxane block co-polymers wherein poly(dimethylsiloxane) soft segments are combined with either polystyrene or polycarbonate hard segments have been proposed in membrane applications, such as blood oxygenators. Polysiloxane containing materials such as these offer the possibility of achieving high oxygen and carbon dioxide permeabilities while at the same time displaying superior mechanical properties. Moreover, since external reinforcing fillers and cross-linking additives are not required, fabrication of thin pinhole-free films is facilitated. Elastomeric block co-polymers based on polysulphone-poly(dimethylsiloxane) have been shown to be efficient impact modifiers for polysulphone. Thermosetting epoxy resins can also be toughened through the use of co-reacting carboxy-terminated butadiene-acrylonitrile oligomers or poly (ε-caprolactone) blocks.

One of the oldest, but still important, applications for block co-polymers is in non-ionic surfactants. For example, polyurethane-foam stabilisers are based on alkylene oxide-silicone block or graft co-polymers. The polyether segment is soluble in the polyurethane. The silicone block is phase separated and is believed to reside at the gas polyurethane interface where it is of major importance in production of uniform foam cell structures.

SECTION II

Types of Polymerisation

CHAPTER 4

Step Reaction (Condensation) Polymerisation

INTRODUCTION

Step-growth polymerisation is defined as one that involves a random reaction of two molecules that may be any combination of a monomer, oligomer, or a longer-chain molecule. High molecular-weight polymer is formed only near the end of the polymerisation when most of the monomer has been depleted. In chain-growth polymerisation, the only chain-extension reaction is that of attachment of a monomer with an 'active' chain. The active end may be a free-radical or ionic site (i.e., anion or cation). In contrast to step-growth polymerisation, some high-molecular-weight polymer is formed in the early stages of the chain-growth polymerisation.

MECHANISM OF STEPWISE POLYMERISATION

Types of Condensation Polymers

Table 4.1. lists some representative products of stepwise polymerisation. Of these, the most important in terms of the size of the commercial market are the aliphatic polyamides or nylons (e.g., nylon-6, nylon-6,6 and nylon-6,10) and polyester [e.g., poly(ethylene terephthalate)]. Speciality or engineering-grade step-growth polymers include polycarbonate, aromatic polyamides, polyimides, polysulphones, polyurethanes, and poly(2,6,-dimethyl-1,4-phenylene oxide).

The type of products formed in a condensation reaction is determined by the functionality of the monomers, that is, by the average number of reactive functional groups per monomer molecule. Monofunctional monomers give only low-molecular-weight products and bifunctional monomers give linear polymers. Polyfunctional monomers, with more than two functional groups per molecule, give branched or cross-linked (three-dimensional) polymers. The properties of the linear and the three-dimensional polymers differ widely.

Some examples of commercially-important step-growth polymerisations are illustrated in Fig. 4.1. Most step-growth polymerisations involve a classical condensation reaction such as esterification (Fig. 4.1a), ester interchange (Fig. 4.1b), or amidation (Fig. 4.1c). Note that two routes exist for the preparation of the aromatic polyester, poly(ethylene terephthalate) (PET)—polyesterification of terephthalic acid and ethylene glycol (Fig. 4.1a) and an ester interchange involving dimethyl terephthalate and ethylene glycol (Fig. 4.1b). These polymerisations and the preparation of the aliphatic polyamide, nylon-6,6 (Fig. 4.1c), are examples of A–A/B–B step-growth condensation-polymerisations. Each of the two monomers is bifunctional and contains the same functionality at each end (i.e., A or B functional group). For example, PET may be formed (Fig. 4.1a) by the polycondensation of a dicarboxylic acid (terephthalic acid) and a diol (ethylene glycol). Alternately, an aliphatic polyester, polycaprolactone, can be formed by the self-condensation of ω-hydroxycaproic acid (Fig. 4.1d). Since the functional end groups of this acid are different (i.e., a carboxylic acid and a hydroxyl group at opposite ends), this polyesterification is an example of an A–B step-growth polycondensation.

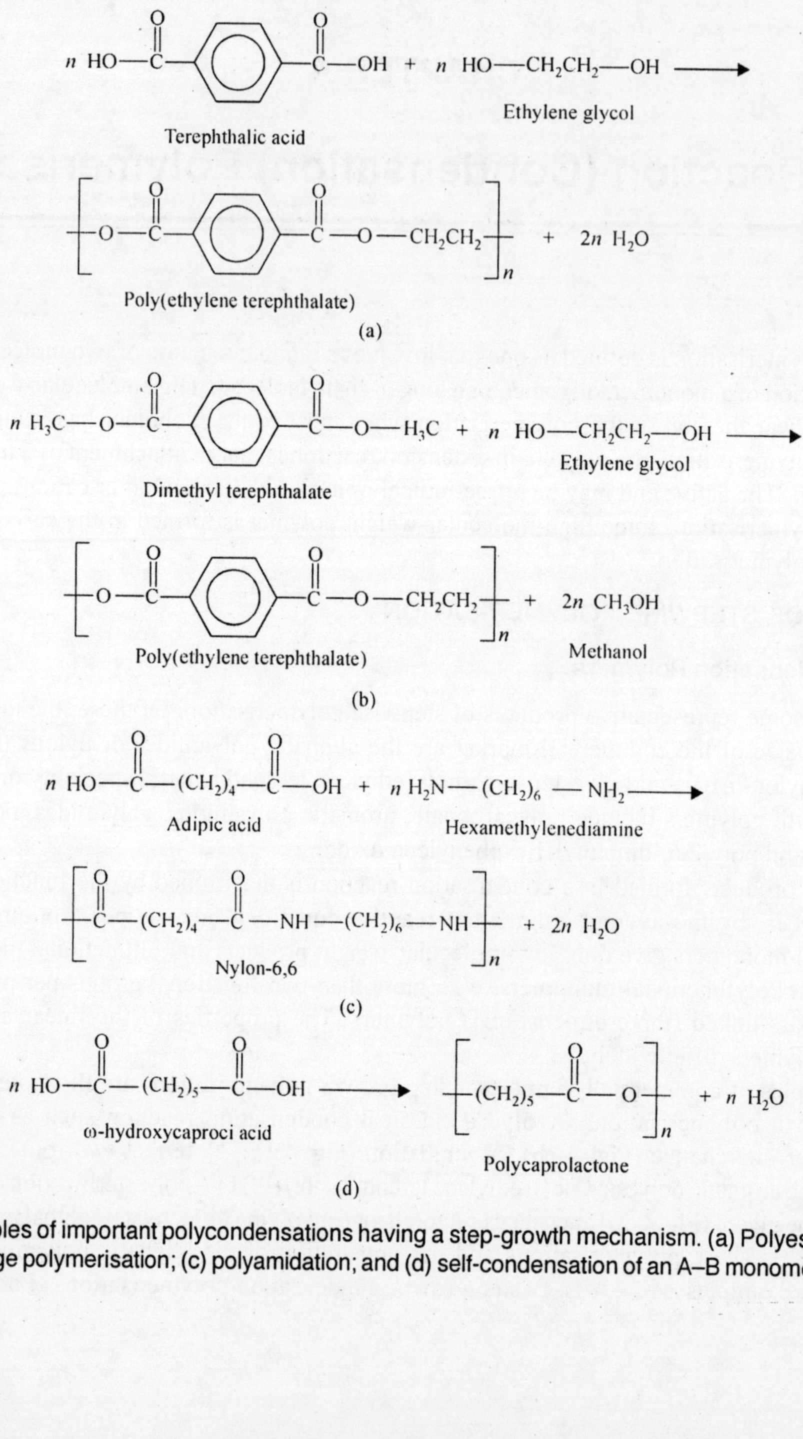

Fig. 4.1. Examples of important polycondensations having a step-growth mechanism. (a) Polyesterification; (b) ester-interchange polymerisation; (c) polyamidation; and (d) self-condensation of an A–B monomer.

Table 4.1. Typical step-reaction polymers.

Type	*Interunit linkage*	*Examples*
Polyester	$-\overset{O}{\overset{\|\|}{C}}-O-$	$HO(CH_2)_xCOOH \longrightarrow HO[-(CH_2)_xCOO-]_yH + H_2O$
		$HO(CH_2)_xOH + HOOC(CH_2)_{x'}COOH \longrightarrow HO\left[-(CH_2)_{x'}O\overset{O}{\overset{\|\|}{C}}(CH_2)_{x'}\overset{O}{\overset{\|\|}{C}}O\right]_y H + H_2O$
		$CH_2OH-CHOH-CH_2OH + HOOC(CH_2)_xCOOH \longrightarrow$ three-dimensional network $+ H_2O$
Polyanhydride	$-\overset{O}{\overset{\|\|}{C}}-O-\overset{O}{\overset{\|\|}{C}}-$	$HOOC(CH_2)_xCOOH \longrightarrow HO[-CO(CH_2)_xCOO-]_yH + H_2O$
Polyacetal	$-O-\underset{R}{\overset{H}{C}}-O-$	$HO(CH_2)_xOH + CH_2(OR)_2 \longrightarrow HO[-(CH_2)_xOCH_2O-]_y(CH_2)_xOH + ROH$
Polyamide	$-\overset{O}{\overset{\|\|}{C}}-NH-$	$NH_2(CH_2)_xCOOH \longrightarrow H[-NH(CH_2)_xCO-]_yOH + H_2O$
		$NH_2(CH_2)_xNH_2 + HOOC(CH_2)_{x'}COOH \longrightarrow H[-NH(CH_2)_xNHCO(CH_2)_{x'}CO-]_yOH + H_2O$
Polyurethane	$-O-\overset{O}{\overset{\|\|}{C}}-NH-$	$HO(CH_2)_xOH + OCN(CH_2)_{x'}CNO \longrightarrow [-O(CH_2)_xOCONH(CH_2)_{x'}NHCO-]_y$
Polyurea	$-NH-\overset{O}{\overset{\|\|}{C}}-NH-$	$NH_2(CH_2)_xNH_2 + OCN(CH_2)_{x'}CNO \longrightarrow [-NH(CH_2)_xNHCONH(CH_2)_{x'}NHCO-]_y$
Silk fibroin	$-\overset{O}{\overset{\|\|}{C}}-NH-$	$NH_2CH_2COOH + NH_2CHRCOOH \longleftarrow H[-NHCH_2CONHCHRCO-]_yOH + H_2O$
Cellulose	$-C-O-C-$	$C_6H_{12}O_6 \longleftarrow [C_6H_{10}O_4]-O-[C_6H_{10}O_4]- + H_2O$

(Contd ...)

Type	*Interunit linkage*	*Examples*
Phenol-aldehyde	$—CH_2—$ $—CH_2OCH_2—$	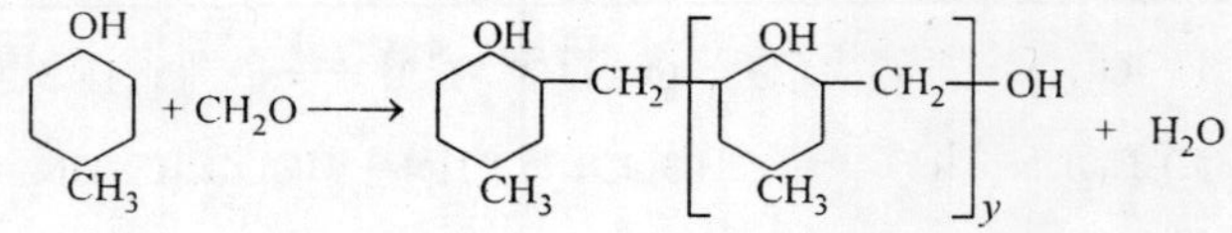
		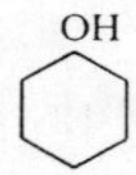 $+ CH_2O \longrightarrow H_2O$ + Three-dimensional network with $—CH_2—$ and $—CH_2OCH_2—$ bridges between positions on rings *o* and *p* to hydroxyls
Urea-aldehyde	$—NH—CHR—NH—$ $—CHR—N(—CHR—)—CHR—$	$NH_2CONH_2 + CH_2O$ (1:1 ratio) $\longrightarrow —NHCONH—CH_2—NHCONH—CH_2— + H_2O$ With excess CH_2O, three-dimensional network + H_2O
Polysulphide	$—S—$	$Cl(CH_2)_xCl + Na_2S \longrightarrow Cl[—(CH_2)_xS—]_yNa + NaCl$
	$—S—S—$	$HS(CH_2)_xSH$ + oxidation $\longrightarrow HS[—(CH_2)_xS(=S)S—]_y(CH_2)_xSH + H_2O$ or H_2S
	$—S(=S)—S(=S)—$	$Cl(CH_2)_xCl + Na_2S_4 \longrightarrow Cl[—(CH_2)_x S—S—]_y Na + NaCl$
Polysiloxane	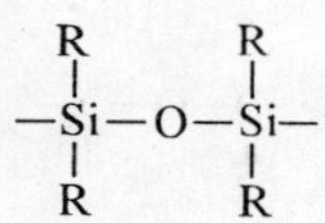	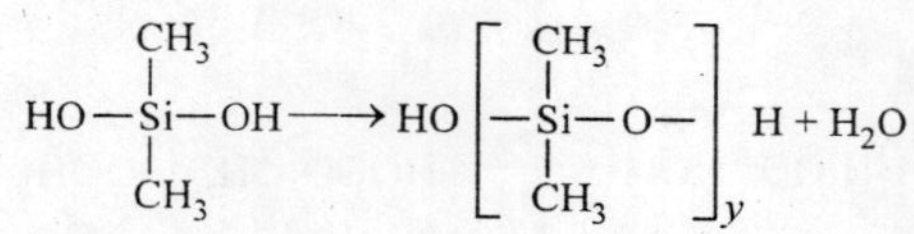
	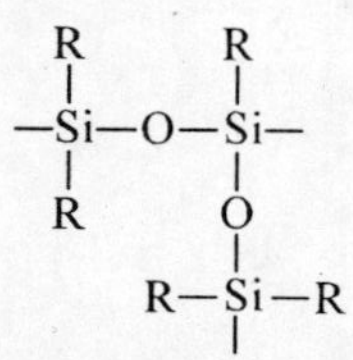	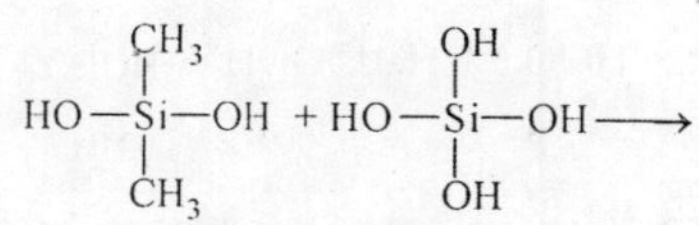 three-dimensional network + H_2O

Two non-condensation-type step-growth polymerisations are illustrated in Fig. 4.2. Fig. 4.2a shows the polymerisation of a polyurethane prepared by the ionic addition of a diol (1,4-butanediol) to a diisocyanate (1,6-hexane diisocyanate). Note that, unlike a condensation polymerisation, liberation of a small molecule is not a by-product of this polymerisation. Shown in Fig. 4.2b is the polymerisation of a high-temperature thermoplastic, poly(2,6-dimethyl-1,4-phenylene oxide), by the oxidative-coupling polymerisation of 2,6-xylenol. The mechanism of this polymerisation is free radical, but the kinetics of this and the previous polymerisation are distinctly step-growth, which means that high-molecular-weight polymer is obtained only at the end of the polymerisation. Normally, the kinetics of a free-radical polymerisation are chain growth.

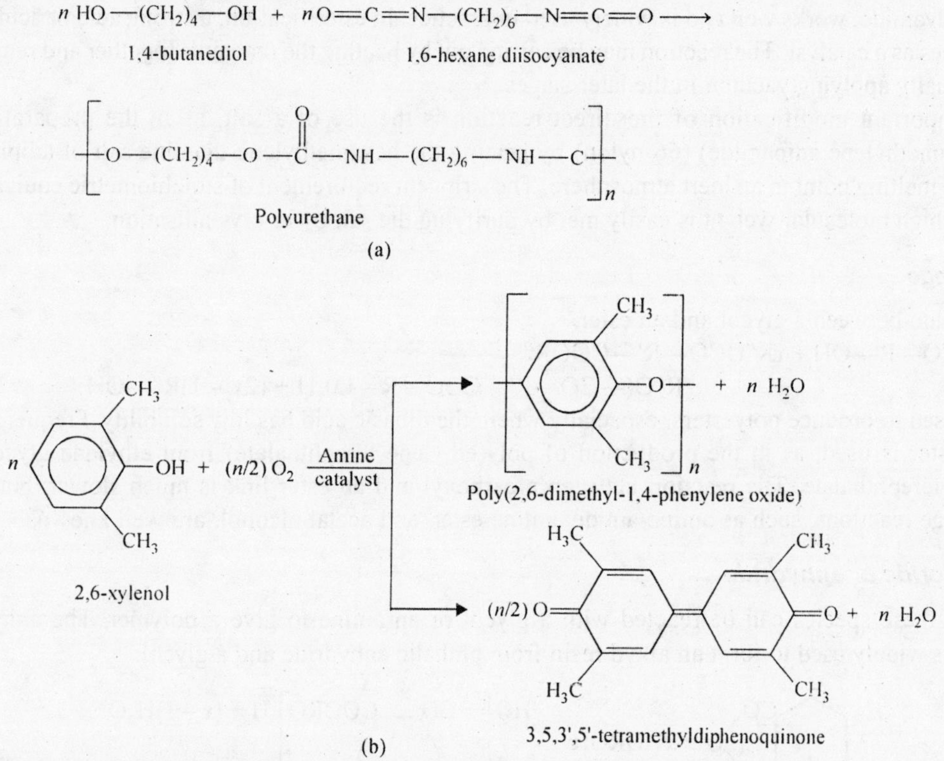

Fig. 4.2. Two non-condensation step-growth polymerisations. (a) Addition polymerisation of a polyurethane; and (b) oxidative-coupling polymerisation of 2,6-xylenol to yield a high-molecular-weight polymer or a low-molecular-weight quinone as a by-product.

Carbonyl Addition-Elimination Mechanism

The most important reaction that has been used for the preparation of condensation polymers is that of addition and elimination at the carbonyl double bond of carboxylic acids and their derivatives. The generalised reaction is

$$R-\overset{\overset{\displaystyle O}{\|}}{C}-X + Y{:} \longrightarrow \left[R-\overset{\overset{\displaystyle O{:}}{|}}{\underset{\underset{\displaystyle X}{|}}{C}}-Y \right] \longrightarrow R-\overset{\overset{\displaystyle O}{\|}}{C}-Y + X{:}$$

where R and R′, (below) may be alkyl or aryl groups, X may be OH, OR′, NH_2, NHR′, $O\overset{O}{\overset{\|}{C}}R'$, or Cl; and Y may be $R'O^-$, R′OH, $R'NH_2$, or $R'COO^-$. The species in the bracket is considered to be a metastable intermediate, which can either return to the original state by eliminating Y or proceed to the final state by eliminating X. The following paragraphs provides some typical examples of this reaction.

Direct reaction

The direct reaction of a dibasic acid and a glycol to form a polyester, or a dibasic acid and a diamine to form a polyamide, works well and is widely used in practice. In esterification, a strong acid or acidic salt often serves as a catalyst. The reaction may be carried out by heating the reactants together and removing water, usually applying vacuum in the later stages.

An important modification of the direct reaction is the use of a salt, as in the preparation of poly(hexamethylene adipamide) (66-nylon) by heating the hexamethylene diamine salt of adipic acid above the melting point in an inert atmosphere. The stringent requirement of stoichiometric equivalence to obtain high molecular weight is easily met by purifying the salt by re-crystallisation.

Interchange

The reaction between a glycol and an ester,

$$x\text{HO—R—OH} + x\text{R}''\text{OCO—R}'\text{—COOR}'' \longrightarrow \text{R}''\text{O(—CO—R}'\text{—COO—R—O)}_x\text{H} + (2x-1)\text{R}''\text{—OH}$$

is often used to produce polyesters, especially where the dibasic acid has low solubility. Frequently the methyl ester is used, as in the production of poly(ethylene terephthalate) from ethylene glycol and dimethyl terephthalate. The reaction between a carboxyl and an ester link is much slower, but other interchange reactions, such as amine-amide, amine-ester, and acetal-alcohol, are well known.

Acid chloride or anhydride

Either of these species can be reacted with a glycol or an amine to give a polymer. The anhydride reaction is widely used to form an alkyd resin from phthalic anhydride and a glycol:

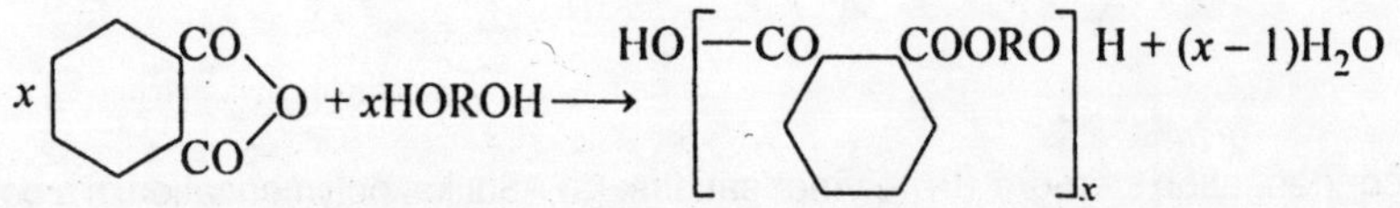

The condensation in bulk of an acid chloride with a glycol is not useful because of side reactions leading to low-molecular-weight products, but the reaction of an acid chloride with a diamine is a valuable means of preparing polyamides.

Interfacial condensation

The reaction of an acid halide with a glycol or a diamine proceeds rapidly to high-molecular-weight polymer if carried out at the interface between two liquid phases, each containing one of the reactants. Very high-molecular-weight polymer can be formed. Typically, an aqueous phase containing the diamine or glycol and an acid acceptor is layered at room temperature over an organic phase containing the acid chloride. The polymer formed at the interface can be pulled off as a continuous film or filament. The method has been applied to the formation of polyamides, polyurethanes, polyureas, polysulphonamides, and polyphenyl esters. It is particularly useful for preparing polymers that are unstable at the higher temperatures usual in step-reaction polymerisation.

Ring versus chain formation

In addition to polymer formation, bifunctional monomers may react intramolecularly to produce a cyclic product. Thus hydroxy acids may give either lactones or polymers on heating, amino acids may give lactams or linear polyamides, and so on.

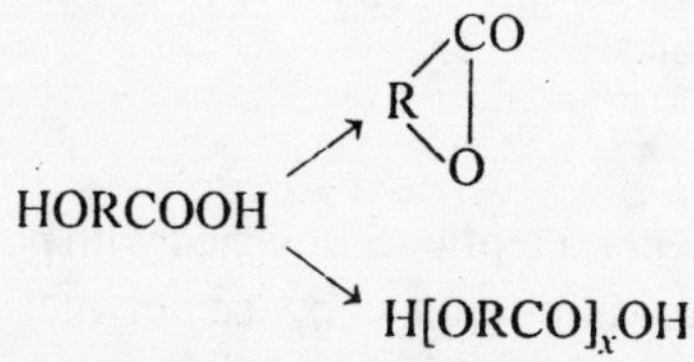

The chief factor governing the type of product is the size of the ring that can be formed. If the ring contains less than five atoms or more than seven, the product is usually linear polymer. If a ring of five atoms can form, it will do so to the exclusion of linear chains. If six- or seven-membered rings can form, either type of product can result. Larger rings can sometimes be formed under special conditions.

The lack of formation of rings with less than five atoms is explained by the strain imposed by the valence angles of the ring atoms. Five-membered rings are virtually strain free, and all larger rings can be strain free if non-planar forms are possible. As the ring size increases, the statistical probability of forming rings becomes smaller. The formation of five-membered rings to the complete exclusion of linear polymer is not thoroughly understood.

The formation of high polymer from cyclic monomers of the type just discussed is often, kinetically, an anionic chain polymerisation. This class of reaction includes the commercially important production of nylon from caprolactam.

Other Mechanisms

Carbonyl addition-substitution reactions

The reaction of aldehydes with alcohols, involving first addition and then substitution at the carbonyl groups, is of great practical and historic importance in stepwise polymerisation. The general reaction, leading to acetal formation, is

$$\mathrm{R\overset{\overset{\displaystyle O}{\|}}{C}H + R'OH \longrightarrow R\underset{\underset{\displaystyle OR'}{|}}{\overset{\overset{\displaystyle OH}{|}}{C}}H \xrightarrow{R'OH} R\underset{\underset{\displaystyle OR'}{|}}{\overset{\overset{\displaystyle OR'}{|}}{C}}H}$$

In addition to polyacetals, important polymers formed in this way include those of formaldehyde and phenol, urea, or melamine.

Nucleophilic substitution reactions

These reactions are important from the standpoint of commercial organic polymers, primarily because of their use in the polymerisation of epoxides. The most common epoxide monomer is epichlorohydrin, which reacts with a nucleophile N: as follows:

$$\mathrm{N\!:} + \mathrm{\underbrace{H_2C{-}CH}_{O}CH_2Cl} \longrightarrow \mathrm{NCH_2\underset{\underset{\displaystyle \ddot{O}\!:^-}{|}}{C}HCH_2Cl}$$

Typically, the nucleophile is a bifunctional hydroxy compound such as bisphenol A, and the reaction proceeds.

A second nucleophilic substitution reaction of historic interest in polymer chemistry is that used for the production of the early polysulphide rubbers from aliphatic dichlorides and sodium sulphide:

$$x\mathrm{ClCH_2CH_2Cl} + x\mathrm{Na_2S}_n \longrightarrow -[\mathrm{CH_2CH_2S}_n-]_x + 2x\mathrm{NaCl}$$

This type of reaction can be used for the production of semi-organic and inorganic polymers, and is probably also the basis for the formation of natural polysaccharides and polynucleotides by polymerisation in living organisms catalysed by enzymes.

Double-bond addition reactions

Although addition reactions at double bonds are often associated with polymerisation by chain mechanisms this is not necessarily the case, and many important stepwise polymerisations are based on this reaction. Of major interest among these is the ionic addition of diols to diisocyanates in the production of polyurethanes:

$$x\mathrm{HOROH} + x\mathrm{OCNR'CNO} \longrightarrow -\!\!\left[\mathrm{OR}\overset{\mathrm{O}}{\overset{\|}{\mathrm{C}}}\mathrm{NHR'NH}\overset{\mathrm{O}}{\overset{\|}{\mathrm{C}}}-\right]_x$$

The free-radical addition of dithiols to unconjugated diolefins also proceeds by a stepwise mechanism.

Free-radical coupling

These reactions lead to several polymerisation schemes of interest, including the preparation of arylene ether polymers, polymers containing acetylene units, and arylene alkylidene polymers. If [Ox] is an oxidising agent, the first of these can be written as

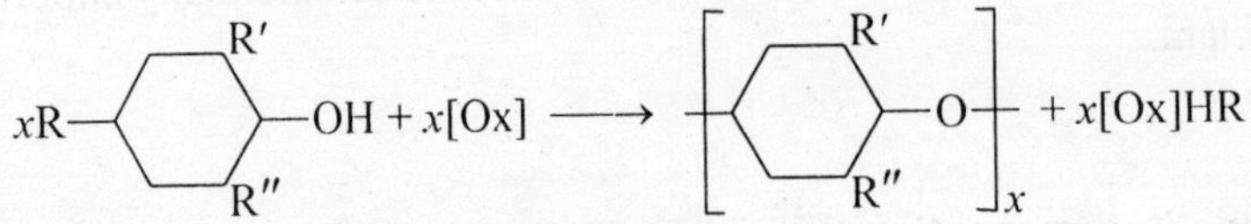

Aromatic electrophilic-substitution reactions

Reactions of this type, using standard Friedel-Crafts catalysts, produce polymers by a step-growth mechanism. A typical example is the production of poly(*p*-phenylene)

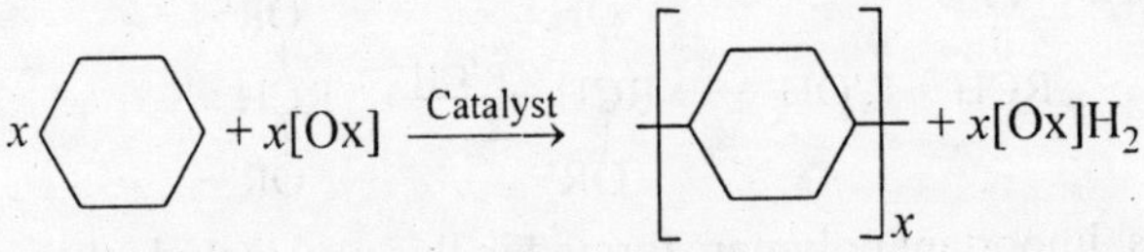

KINETICS AND STATISTICS OF LINEAR STEPWISE POLYMERISATION

Reactivity and Molecular Size

The similarity between monofunctional and polyfunctional step reactions is illustrated in the chemical equations in this chapter. Other similarities exist in the influence of temperature and catalysts on the rates of the reactions. It is tempting to substitute the simple concept of reactions between functional groups for that of the myriad separate reactions of each molecular species with all the others. With this substitution, the kinetics of stepwise polymerisation is essentially identical with that of simple condensation; without it, analysis of the kinetics would be almost hopelessly difficult.

If the simplification is to be made, the rate of the reaction of a group must be independent of the size of the molecule to which it is attached. This assumption is amply justified by experimental evidence. The rate constants of condensation reactions in a homologous series reach asymptotic values independent of chain length quite rapidly and show no tendency to drop off with increases in molecular size. The diluting effect of the large chain must, of course, be taken into account by adjusting to constant molar composition of reactive groups. In addition, the rate constants for monofunctional and bifunctional reagents are identical for sufficiently long chains separating the reactive groups of the bifunctional compound.

Kinetics of Stepwise Polymerisation

With the concept of functional group reactivity independent of molecular weight, the kinetics of stepwise polymerisation becomes quite simple. The formation of a polyester from a glycol and a dibasic acid may be taken as an example. It is well known that this reaction is catalysed by acids. In the absence of added strong acid, a second molecule of the acid being esterified acts as a catalyst. The reaction is followed by measuring the rate of disappearance of carboxyl groups †.

$$-\frac{d[\mathrm{COOH}]}{dt} = k[\mathrm{COOH}]^2[\mathrm{OH}] \qquad \text{...(4.1)}$$

If the concentrations c of carboxyl and hydroxyl groups are equal,

$$\frac{-dc}{dt} = kc^3 \qquad \text{...(4.2)}$$

and

$$2kt = \frac{1}{c^2} - \text{const.}$$

It is convenient to introduce the extent of reaction p, defined as the fraction of the functional groups that has reacted at time t. Then

$$c = c_0(1-p) \qquad \text{...(4.3)}$$

and

$$2{c_0}^2kt = \frac{1}{(1-p)^2} + \text{const.} \qquad \text{...(4.4)}$$

or a plot of $1/(1-p)^2$ should be linear in time. Typical experimental data (Fig. 4.3) bear this out over a wide range of reaction times, but in fact the linear portion corresponds to only a relatively narrow range of conversion, roughly 80–93 per cent.

† Stepwise polymerisation typically involves equilibrium reactions of the type.

$$\mathrm{A + B} \underset{k_f}{\overset{k_f}{\rightleftharpoons}} \mathrm{C + D}$$

where the rates of the forward and reverse reactions are k_f[A][B] and k_r[C][D], respectively. At equilibrium these rates are equal, whence $K = k_r/k_f$ = [A][B]/[C][D]. If the system is far from equilibrium, as in the initial stages of polymerisation, the reverse reaction is negligibly slow, and changes in the concentrations of the reactants may be considered to result from the forward reaction alone, as in Eq. 4.1 and those following.

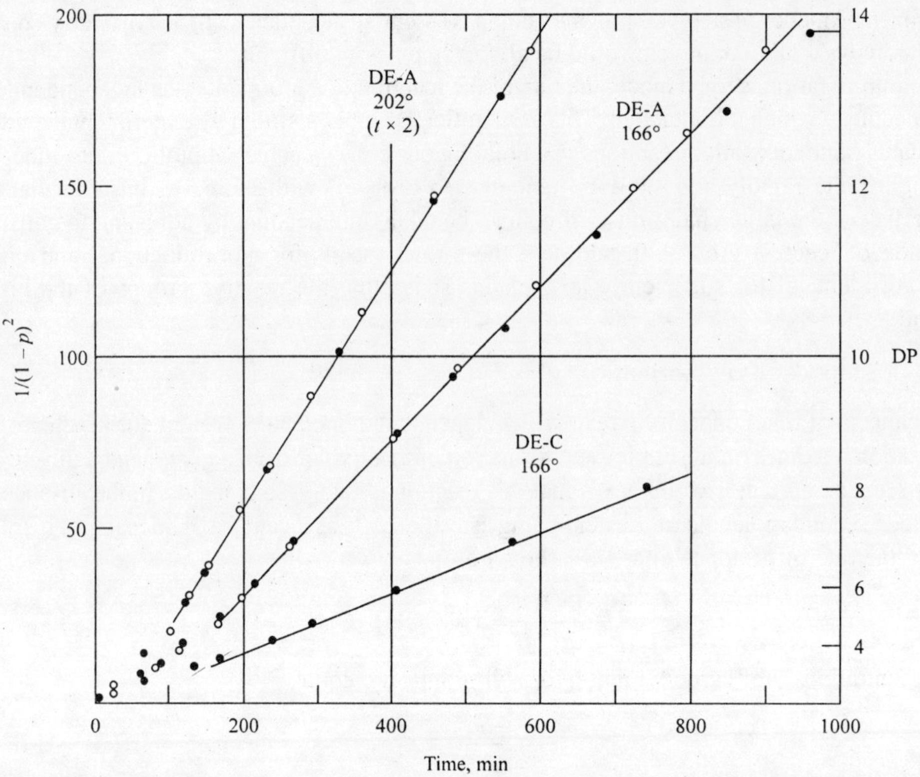

Fig. 4.3. Reactions of (di)ethylene glycol (DE) with adipic acid (A) and caproic acid (C). Time values at 202°C have been multiplied by 2.

Deviations at lower conversions are attributed to changes in the polarity of the reaction medium as the monomers are converted to polymer, and at higher conversions to an increase in the rate of the reverse reaction.

If only bifunctional reactants are present and no side reactions occur, the number of unreacted carboxyl groups equals the total number of molecules N in the system. If acid or hydroxyl groups separately (not in pairs) are defined as structural units, the initial number of carboxyls present is equal to the total number N_0 of structural units present. The number-average degree of polymerisation $\bar{x}_n$ is simply

$$\bar{x}_n = \frac{N_0}{N} = \frac{c_0}{c} = \frac{1}{1-p} \qquad ...(4.5)$$

The scale of DP in Fig. 4.3 shows that uncatalysed esterifications require quite long times to reach high degrees of polymerisation. Greater success is achieved by adding to the system a small amount of catalyst, whose concentration is constant throughout the reaction. In this case the concentration of the catalyst may be included in the rate constant:

$$-\frac{d[\mathrm{COOH}]}{dt} = k'[\mathrm{COOH}][\mathrm{OH}]$$

$$-\frac{dc}{dt} = k'c^2 \qquad \text{...(4.6)}$$

$$c_0 k't = \frac{1}{1-p} + \text{const.}$$

Hence $\bar{x}_n$ increases linearly with reaction time (Fig. 4.4). Similar reactions have been shown to be linear up to at least $\bar{x}_n = 90$, corresponding to molecular weights of 10,000.

Statistics of Linear Step-Reaction Polymerisation

Rate equations similar to Eq. 4.6 may be written for each molecular species in the reaction mixture. From these it is possible to derive the distribution of molecular weights in a step-reaction polymer. The same relation is, however, more readily derived from statistical considerations.

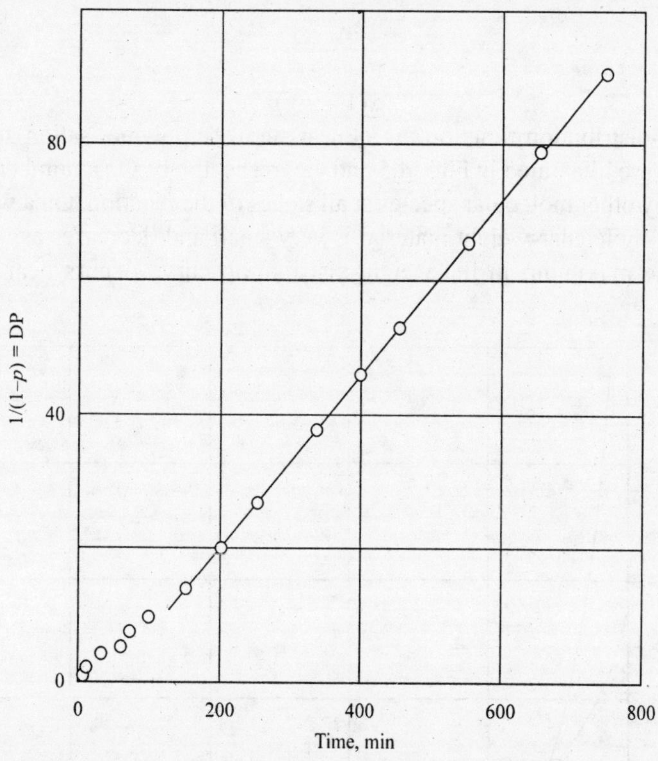

Fig. 4.4. Reaction of ethylene glycol with adipic acid, catalysed by 0.4 mol. per cent of *p*-toluenesulphonic acid.

The statistical analysis of stepwise polymerisation yields results that are equivalent to and in some cases more easily obtained than those of kinetic considerations. The method assumes independence of reaction rate and molecular size. The extent of reaction p is defined as the probability that a functional group has reacted at time t. This definition is entirely equivalent to the previous definition of p as the fraction of the functional groups that has reacted. It follows that the probability of finding a functional group unreacted is $1 - p$.

It is now necessary to find the probability that a given molecule selected at random is an x-mer, that is, contains x repeating units. Such a molecule contains $x - 1$ reacted functional groups (e.g., carboxyls) and, on the end, one unreacted group of this type. The probability of finding a single reacted carboxyl group in the molecule is p, and that of finding $x - 1$ of them in the same molecule is p^{x-1}. The presence of one unreacted group has a probability of $1 - p$. Hence the probability of finding the complete molecule is $p^{x-1}(1 - p)$. This is also equal to the fraction of all the molecules that are x-mers. If there are N molecules in all, the total number of x-mers is

$$N_x = Np^{x-1}(1 - p) \quad \text{...(4.7)}$$

If the total number of units present is N_0, $N = N_0(1 - p)$, and

$$N_x = N_0(1 - p)^2 p^{x-1} \quad \text{...(4.8)}$$

This is the number-distribution function for a linear stepwise polymerisation at extent of reaction p.

The weight fraction w_x of x-mers is given by

$$w_x = \frac{xN_x}{N_0}$$

$$= x(1 - p)^2 p^{x-1} \quad \text{...(4.9)}$$

This is the weight-distribution function for a linear stepwise polymerisation at extent of reaction p. Equations 4.8 and 4.9 are illustrated in Figs. 4.5 and 4.6, respectively. On a number basis, monomers are more plentiful than any other molecular species at all stages of the reaction. On a weight basis, however, the proportion of low-molecular-weight material is very small and decreases as the average molecular weight increases. The maximum in the weight-distribution curve occurs near the number-average molecular weight.

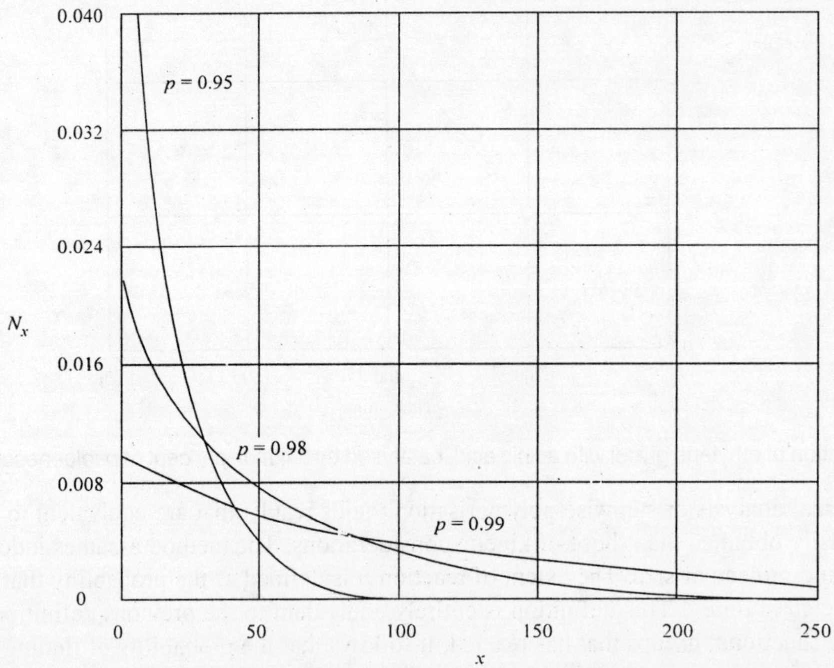

Fig. 4.5. Number- or mole-fraction distribution of chain molecules in a linear step-reaction polymer for several extents of reaction p.

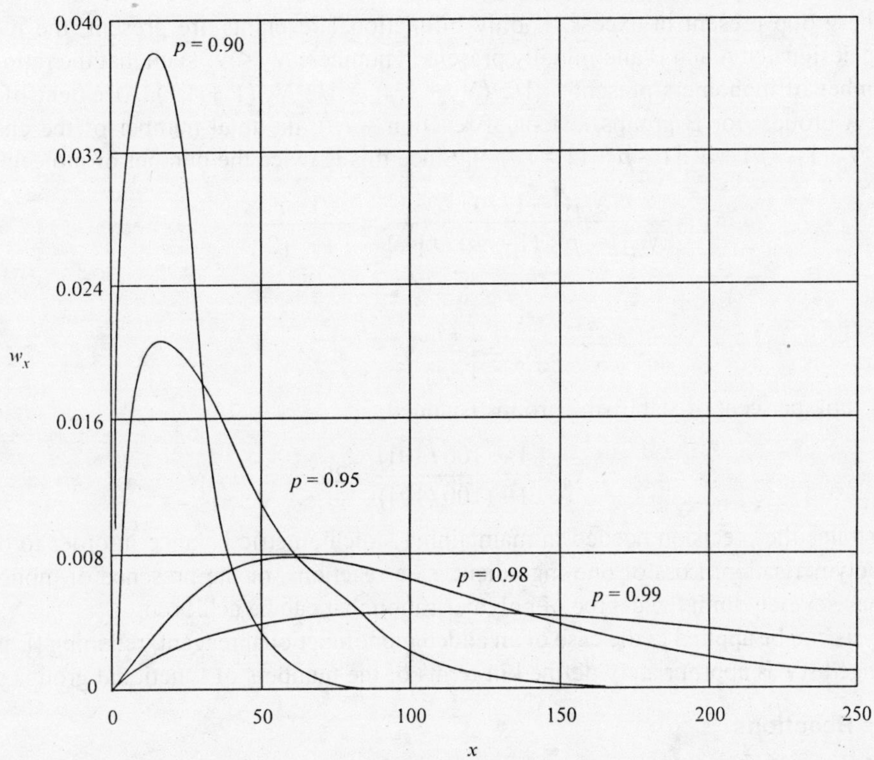

Fig. 4.6. Weight-fraction distribution of chain molecules in a linear step-reaction polymer for several extents of reaction *p*.

Various average degrees of polymerisation may be derived from the weight or number-distribution functions according to the definitions of the corresponding molecular-weight averages. The number-average degree of polymerisation $\bar{x}_n$ calculated in this way is that given in Eq. 4.5. The weight-average degree of polymerisation is

$$\bar{x}_w = \frac{1+p}{1-p} \qquad \text{...(4.10)}$$

The breadth of the molecular-weight distribution curve $\bar{x}_w / \bar{x}_n = 1 + p$; at large extents of reaction $\bar{x}_w / \bar{x}_n \rightarrow 2.0$. This has been confirmed experimentally for nylon.

Molecular-Weight Control

Although initial molecular-weight control in a stepwise polymerisation can be achieved by stopping the reaction (e.g., by cooling) at the desired point, the stability of the product to subsequent heating may not be adequate with respect to changes in molecular weight.

The easiest way to avoid this situation is to adjust the composition of the reaction mixture slightly away from stoichiometric equivalence, by adding either a slight excess of one bifunctional reactant or (as in molecular-weight stabilisation of nylons by acetic acid) a small amount of a monofunctional reagent. Eventually the functional group deficient in amount is completely used up, and all chain ends consist of the group present in excess. If only bifunctional reactants are present, the two types of groups being designated A and B and initially present in numbers $N_A < N_B$ such that the ratio $r = N_A/N_B$, the total number of monomers present is 1/2 $(N_A + N_B) = 1/2\ N_A\ (1 + 1/r)$. At extent of reaction p (defined for A groups; for B groups, extent of reaction = rp), the total number of the chain ends is $N_A(1 - p) + N_B\ (1 - rp) = N_A[1 - p + (1 - rp)/r]$. Since this is twice the number of molecules present,

$$\bar{x}_n = \frac{N_A(1+1/r)/2}{N_A[1-p+(1-rp)/r]/2} = \frac{1+r}{1+r-2rp} \qquad \text{...(4.11)}$$

As $p \to 1$,

$$\bar{x}_n = \frac{1+r}{1-r} \qquad \text{...(4.12)}$$

Thus if 1 mol. per cent of stabilising groups is added,

$$\bar{x}_n = \frac{1+(100/101)}{1-(100/101)} = 201$$

This illustrates the precision needed in maintaining stoichiometric balance in order to obtain high degrees of polymerisation. Loss of one ingredient, side reactions, or the presence of monofunctional impurities may severely limit the degree of polymerisation that can be achieved.

The analysis can be applied to the case of an added monofunctional reagent, retaining (Eqs. 4.11 and 4.12) unchanged if r is appropriately defined in terms of the numbers of functional groups present.

Interchange Reactions

In interchange reactions, which may occur freely at elevated temperature, as in polymer melts, the molecular weights of the reacting molecules can change, since, for example, two molecules of average length may react to give one longer and one shorter than average. However, the total number of molecules and hence $\bar{x}_n$ do not change.

It can be shown that if free interchange takes place, the final molecular-weight distribution is always the most probable distribution defined by Eqs. 4.8 and 4.9. The change can be demonstrated by mixing two polymers of different $\bar{x}_n$ and following a weight-average property (such as melt viscosity) as a function of time at elevated temperature.

Multichain Polymer

Another type of step-reaction polymer of interest is produced by polymerising a bifunctional monomer of type A–B with a small amount of monomer of functionality f of the type R–A_f. If the reaction is carried nearly to completion, the resulting polymer consists of f chains growing out from a central unit R. Network structures cannot occur, since no units of the type B–B are present. The case $f = 1$ corresponds to the most probable distribution of Eq. 4.9. The polymer formed when $f = 2$ is also linear; those with higher functionalities are branched.

Since the length of each of the branches is statistically determined independent of all the others, the probability of having a molecule with several branches all longer or shorter than average becomes smaller as the number of branches increases.

The molecular-weight distribution thus becomes narrower with increasing functionality, as shown in Fig. 4.7.

The distribution breadth is given by:

$$\frac{\bar{x}_w}{\bar{x}_n} = 1 + \frac{1}{f} \qquad ...(4.13)$$

For $f = 2$, $\bar{x}_w / \bar{x}_n = 1.5$.

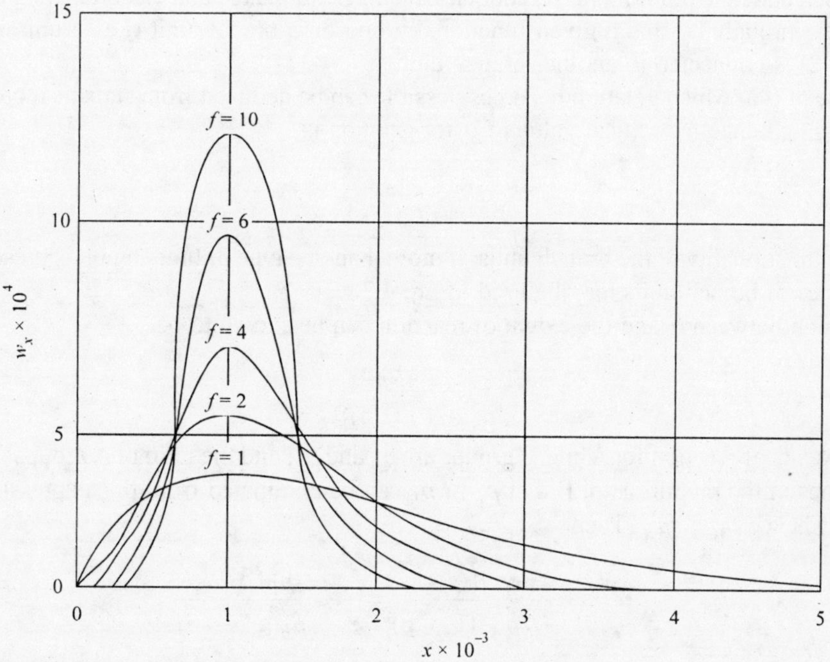

Fig. 4.7. Molecular-weight distribution for multichain polymers with functionality f as shown.

POLYFUNCTIONAL STEP-REACTION POLYMERISATION

Three-dimensional step-reaction polymers are produced from the polymerisation of reactants with more than two functional groups per molecule. The structures of these polymers are more complex than those of linear step-reaction polymers. Three-dimensional polymerisation is complicated experimentally by the occurrence of gelation, or the formation of essentially infinitely large polymer networks in the reaction mixture.

The sudden onset of gelation marks the division of the mixture into two parts: the gel, which is insoluble in all non-degrading solvents, and the solution, which remains soluble and can be extracted from the gel. As the polymerisation proceeds beyond the gel point, the amount of gel increases at the expense of sol, and the mixture rapidly transforms from a viscous liquid to an elastic material of infinite viscosity. An important feature of the onset of gelation is the low number-average molecular weight of the mixture at the gel point, where its weight-average molecular weight becomes infinite.

In the statistical consideration of three-dimensional step polymerisation, it is assumed that all functional groups are equally reactive, independent of molecular weight or viscosity, and that all the reactions occur between functional groups on different molecules. Errors resulting from these assumptions affect only the numerical agreement between theory and experiment, and not the overall considerations of the theory.

Gelation

Prediction of the Gel Point

In order to calculate the point in the reaction at which gelation takes place, a branching coefficient α is defined as the probability that a given functional group on a branch unit (i.e., a unit of functionality greater than 2) is connected to another branch unit.

The value of α at which gelation becomes possible can be deduced from statistical considerations to be just $1/(f-1)$; hence the critical value of α for gelation is

$$\alpha_c = \frac{1}{f-1} \qquad \text{...(4.14)}$$

Here f is the functionality of the branch units; if more than one type of branch unit is present, an average f over all types of branch units may be used in Eq. 4.7.

The relation between α and the extent of reaction can be shown to be

$$\alpha = \frac{p_A p_B \rho}{p_A p_B (1-\rho)} \qquad \text{...(4.15)}$$

where the extents of reaction for A and B groups are p_A and p_B, and the ratio of A groups on branch units to all A groups in the mixture is ρ. Either p_A or p_B can be eliminated from this expression by defining $r = N_A/N_B$, whence $p_B = rp_A$. Then

$$\alpha = \frac{rp_A^2\rho}{1 - rp_A^2(1-\rho)} = \frac{p_B^2\rho}{r - p_B^2(1-\rho)} \qquad \text{...(4.16)}$$

Simpler relations can be derived for several cases. When equal numbers of A and B groups are present, $r = 1$ and $p_A = p_B = p$:

$$\alpha = \frac{p^2\rho}{1-p^2(1-\rho)} \qquad \text{...(4.17)}$$

When there are no A–A units, $\rho = 1$ and

$$\alpha = rp_A^2 = \frac{p_B^2}{r} \qquad \text{...(4.18)}$$

If both conditions apply, that is, $r = \rho = 1$,

$$\alpha = p^2 \qquad \text{...(4.19)}$$

Finally, with only branch units present the probability that a functional group on a branch unit leads to another branch unit is just the probability that it has reacted:

$$\text{(branch units only)} \qquad \alpha = p \qquad \text{...(4.20)}$$

The equations hold for all branch-unit functionalities with the definitions given for r and ρ.

Experimental observations of the gel point

The gel point can be observed precisely as the time when the polymerising mixture suddenly loses fluidity, for example, when bubbles no longer rise in it. If the extent of reaction has been followed as a function of time, say, by removing aliquots of the solution and titrating for the number of functional groups present, the value of p at the gel point can be determined. In several cases investigated by *Flory*, observed values of α_c were always slightly higher than those theoretically required. This discrepancy is attributed to the reaction of some functional groups to form intramolecular links, which do not contribute to network structures. The reactions must therefore be carried slightly further to reach the critical point.

Molecular-Weight Distributions in Three-Dimensional Step-Reaction Polymers

The distribution functions for three-dimensional polymers are derived with somewhat more difficulty than those for the linear case, Eqs. 4.1 and 4.2. They depend upon the functionality and relative amounts of all the units involved. Only one example is discussed, for simplicity that of the reaction of equivalent quantities of two trifunctional monomers, all three of the functional groups of each monomer being equally reactive. In this case the weight-distribution function is

$$w_x = \left[\frac{(fx-x)!f}{(x-1)!(fx-2x+2)!}\right] p^{x-1}(1-p)^{fx-2x+2} \qquad \text{...(4.21)}$$

This equation is analogous to Eq. 4.9. The first factor (in brackets) arises because of the numerous geometric isomers of a polyfunctional x-mer. (In the linear case this factor is just x.) The second term is the probability of finding x–1 links in an x-mer, and the last term is the probability of finding $(f-2)x+2 = fx-2x+2$ unreacted links or ends.

Equation 4.21 is illustrated in Figs. 4.8 to 4.10. In comparison with the linear case, the weight distributions of branched step-reaction polymers of increasing functionality are progressively broader at equivalent extents of reaction (Fig. 4.8). Fig. 4.9 shows that the distributions broaden out with increasing extent of reaction p (or, alternatively, α, in the trifunctional case).

In Fig. 4.10 the weight fraction of the various molecular species is plotted against $\alpha = p$. In contrast to the linear case (Figs. 4.9 and 4.10 versus Fig. 4.6), the weight fraction of monomer is always greater than the amount of any one of the other species, the weight fractions of higher species being successively lower. The extent of reaction at which the weight fraction of any species reaches its maximum shifts continuously to higher values for higher molecular weights. In no case, however, does the maximum occur beyond the gel point ($\alpha = 1/2$).

Up to the gel point the sum of the weight fractions of all the species present must equal unity; beyond this point the sum of the weight fractions of all finite species drops below unity as the weight fraction of gel increases.

The number-average degree of polymerisation is given by

$$\bar{x}_n = \frac{1}{1-fp/2} \qquad \text{...(4.22)}$$

and the weight-average degree by

$$\bar{x}_w = xw_x = \frac{1+p}{1-(f-1)p} \qquad \text{...(4.23)}$$

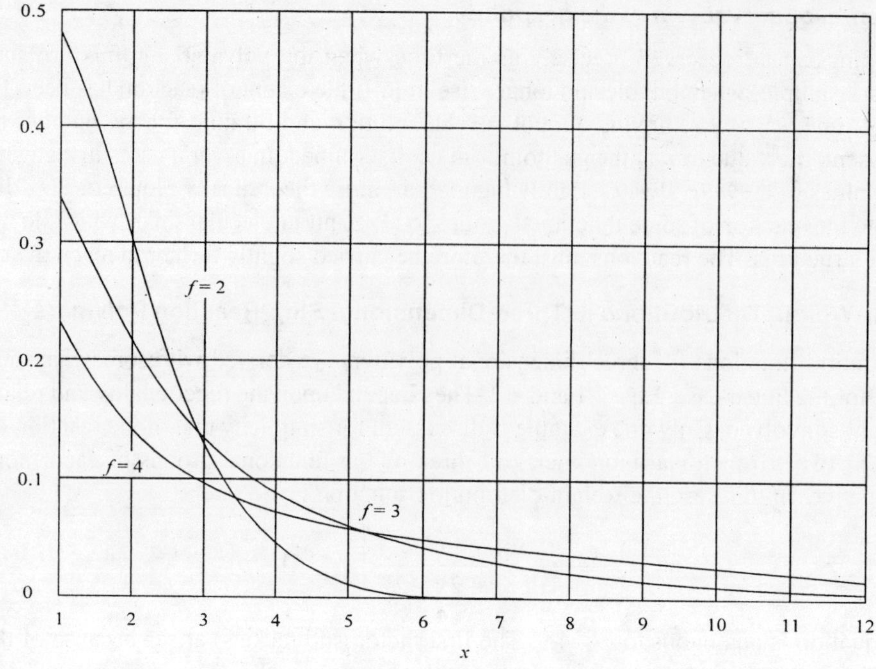

Fig. 4.8. Molecular-weight distribution in stepwise polymerisation for monomers with functionality f, at $p = 0.3$.

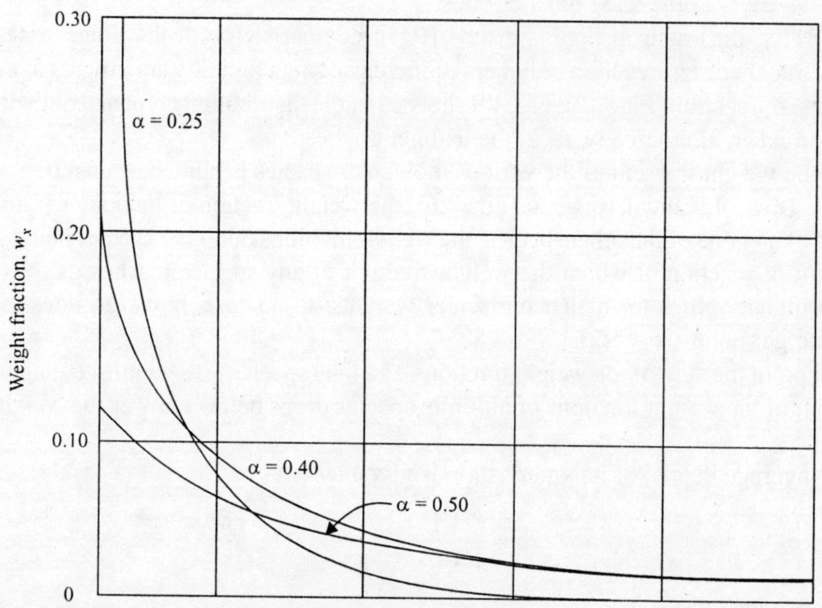

Fig. 4.9. Molecular-weight distribution in step-reaction polymer formed from trifunctional units at various stages in the reaction, denoted by $\alpha = p$.

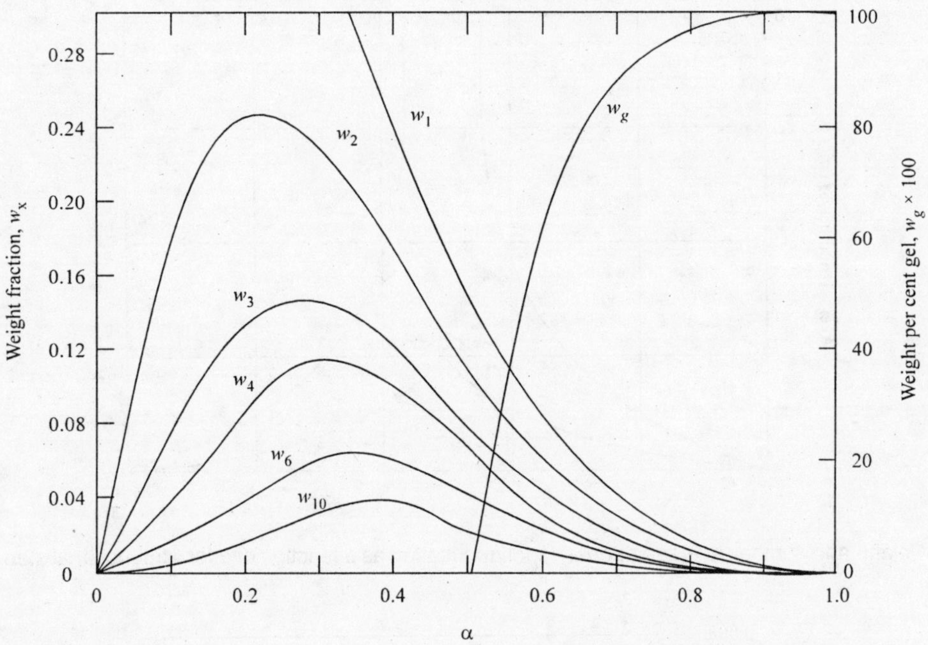

Fig. 4.10. Weight fractions of various molecular species in a trifunctional stepwise polymerisation as a function of $\alpha = p$. The weight fractions of the finite species are calculated from Eq. 4.21, and that of the gel from Eq. 4.24.

At the gel point the weight-average degree of polymerisation becomes infinite. As may be seen in Fig. 4.11, where both averages are plotted against $\alpha = p$, $\bar{x}_n$ attains only a value of 4 at this point. The very large values of $\bar{x}_w / \bar{x}_n$ near the gel point illustrate the extreme breadth of the distributions. (The portions of the curves in Fig. 4.11 above the gel point refer only to the solution fraction).

The increase in distribution breadth with increasing degree of branching is illustrated in Fig. 4.12. These curves represent the calculated distributions for bifunctional step reactions in which the branching factor α was varied by varying the small amount of added trifunctional units. The curves were calculated for $\bar{x}_n = 50$ in each case. The curves are drawn so that the total area under each curve extended to infinite x is the same.

Equation 4.21 is valid for finite species beyond the gel point. If it is summed over all values of x and subtracted from unity, the weight fraction of gel results. For trifunctional units this is

$$w_g = 1 \frac{(1-p)^3}{p^3} \qquad \text{...(4.24)}$$

The distribution functions for $\alpha > 1/2$ are identical with those at the corresponding lower value $1 - \alpha$, except that they are reduced by a factor $w_s = 1 - w_g$, the total weight fraction of sol present at that value of α.

Thus the complexity of the distribution of the sol fraction above the gel point is presumed to become less until, at $\alpha = 1$, only monomer is left in a vanishingly small amount. The reduction of $\bar{x}_w / \bar{x}_n$ shown for values of α increasing above 1/2 in Fig. 4.9 is another way of expressing the same phenomenon.

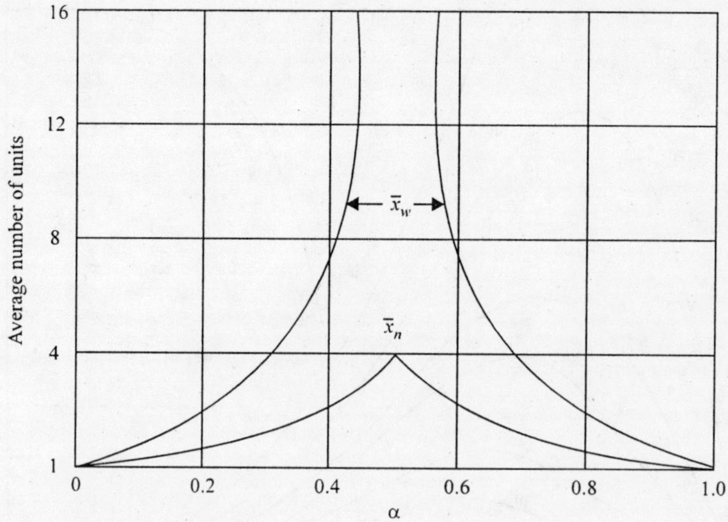

Fig. 4.11. Weight- and number-average degree of polymerisation as a function of α for a trifunctional step-reaction polymer.

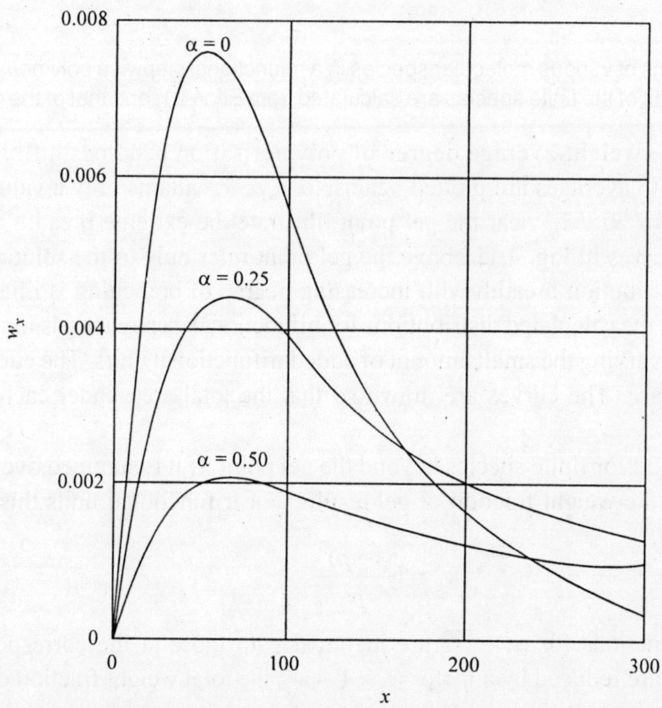

Fig. 4.12. Weight distribution for step-reaction polymers made to $\bar{x}_n = 50$ with various degrees of branching α induced by the addition of trifunctional units: α = 0 indicates no branching, and α = 0.50 is the critical point for gelation.

Although the discussion was confined to the case of equivalent quantities of two trifunctional monomers, and Figs. 4.8 to 4.12 refer specifically to this case, Eqs. 4.21 to 4.23 are more broadly applicable if p is replaced with α, which is then determined in the case of interest by r, p, and ρ, as indicated in Eq. 4.16.

CHAPTER 5

Radical Chain (Addition) Polymerisation

INTRODUCTION

Chain polymerisation is characterised by a self-addition of the monomer molecules, to each other, very rapidly through a chain reaction. No by-product is formed; the product has the same elemental composition as that of the monomer. The bifunctionality is provided by the double bonds present in the monomer. Compounds containing reactive double bonds can, therefore, undergo a chain polymerisation reaction. Typical examples are vinyl compounds ($CH_2{=}CHX$), allyl compounds ($CH_2{=}CH.CH_2X$), olefins ($CH_2{=}CHR$) and dienes ($CH_2{=}CR{-}CH{=}CH_2$). Since a majority of these monomers fall under the 'vinyl' category, chain polymerisation is also customarily termed 'vinyl polymerisation'. Table 5.1 shows some typical monomers capable of undergoing chain polymerisation.

MECHANISM OF VINYL POLYMERISATION

The polymerisation of unsaturated monomers typically involves a chain reaction. It can be initiated by methods typical for simple gas-phase chain reactions, including the action of ultraviolet light. It is susceptible to retardation and inhibition. In a typical chain polymerisation, one act of initiation may lead to the polymerisation of thousands of monomer molecules.

The characteristics of chain polymerisation suggest that the active centre responsible for the growth of the chain is associated with a single polymer molecule through the addition of many monomer units. Thus polymer molecules are formed from the beginning, and almost no species intermediate between monomer and high-molecular-weight polymer are found. Of several postulated types of active centre, three have been found experimentally: cation, anion, and free radical. Free-radical polymerisation is discussed in this chapter.

The concept of vinyl polymerisation as a chain mechanism is not new, dating back to *Staudinger's* work in 1920. In 1937 *Flory* showed conclusively that radical polymerisation proceeds by and requires the steps of initiation, propagation, and termination typical of chain reactions in low-molecular-weight species.

The carbon-carbon double bond is, because of its relatively low stability, particularly susceptible to attack by a free radical. The reaction of the double bond with a radical proceeds well for compounds of the type $CH_2{=}CHX$ and $CH_2{=}CXY$, called vinyl monomers. (Monomers in which fluorine is substituted for hydrogen may be included in this class).

Table 5.1. Monomers capable of undergoing chain polymerisation.

Nomenclature	*Formula*
Vinyl compounds	
Acrylamide	$CH_2{=}CHCONH_2$
Acrylic acid	$CH_2{=}CHCOOH$
Acrylonitrile	$CH_2{=}CHCN$
Bromotrifluoroethylene	$BrFC{=}CF_2$
Butyl acrylate	$CH_2{=}CHCOOC_4H_9$
α-Cyanomethyl acrylate	$CH_2{=}C(CN)COOCH_3$
Methacrylic acid	$CH_2{=}C(CH_3)COOH$
Methacrylonitrile	$CH_2{=}C(CH_3)CN$
Methylmethacrylate	$CH_2{=}C(CH_3)COOCH_3$
α-methyl styrene*	$CH_2{=}C(CH_3)C_6H_5$
Styrene	$CH_2{=}CHC_6H_5$
Nitro ethylene	$CH_2{=}CHNO_2$
p-Nitro styrene	$CH_2{=}CHC_6H_4NO_2$
Tetrafluoro ethylene	$F_2C{=}CF_2$
Vinyl acetate	$CH_2{=}CHOCOCH_3$
Vinyl carbazole	$CH_2{=}CH(C_{12}H_8N)$
Vinyl chloride	$CH_2{=}CHCl$
Vinyl ethyl ether	$CH_2{=}CHOC_2H_5$
Vinyl ethyl ketone	$CH_2{=}CHCOC_2H_5$
2-Vinyl pyridine	$CH_2{=}CH(C_5H_4N)$
Vinylidene chloride	$CH_2{=}CCl_2$
Allyl compounds	
Allyl acetate*	$CH_2{=}CHCH_2OCOCH_3$
Allyl alcohol*	$CH_2{=}CHCH_2OH$
Allyl chloride*	$CH_2{=}CHCH_2Cl$
Olefins	
Ethylene	$CH_2{=}CH_2$
Propylene	$CH_2{=}CHCH_3$
Isobutylene	$CH_2{=}C(CH_3)_2$
Dienes	
1,3 Butadiene	$CH_2{=}CH{-}CH{=}CH_2$
Chloroprene	$CH_2{=}CH{-}CCl{=}CH_2$
Isoprene	$CH_2{=}CCH_3{-}CH{=}CH_2$

*These monomers are difficult to polymerise.

Not all vinyl monomers yield high polymer as a result of radical polymerisation. Aliphatic hydrocarbons other than ethylene polymerise only to oils, and 1,2-di-substituted ethylenes not at all. Among compounds of the type $CH_2=CXY$, those in which both groups are larger than CH_3 polymerise slowly, if at all, by radical mechanisms.

Chain polymerisation is characteristic of compounds with multiple bonds [e.g., ethylene $CH_2=CH_2$, isobutylene $(CH_3)_2C=CH_2$, and vinyl chloride $CH_2=CHCl$], or of unstable rings containing hetero-atoms (e.g., ethylene oxide CH_2-CH_2 bridged by O).

The specific feature of chain polymerisation reactions is that development of the kinetic chain is accompanied by growth of the molecular chain.

The chain polymerisation reaction consists of three main steps:

1. Activation, or initiation, of the monomer molecule:

$$M \longrightarrow M^* \text{ (excited monomer molecule or active centre)}$$

2. Chain propagation:

$$M^* + M \longrightarrow M_1^*$$
$$M_1^* + M \longrightarrow M_2^*$$
$$M_2^* + M \longrightarrow M_3^*$$
$$M_{n-1}^* + M \longrightarrow M_n^* \text{ (activated growing polymer molecule)}$$

3. Chain termination:

$$M_n^* \longrightarrow M_n \text{ (unactivated polymer molecule)}$$

The active centre in chain polymerisation reactions may be a free radical or an ion, and the reaction is one of radical or ionic polymerisation, respectively.

RADICAL POLYMERISATION

In radical polymerisation physical or chemical effects give rise to a free radical:

$$M \text{——} R\bullet$$

This primary free radical reacts with the double bond of an unexcited monomer molecule and adds to its, forming a new radical capable of further interaction with the initial monomers (chain propagation):

$$R\bullet + M \longrightarrow R\text{–}M\bullet$$
$$R\text{–}M\bullet + M \longrightarrow R\text{–}M\text{–}M\bullet$$
$$\cdots\cdots\cdots\cdots\cdots\cdots\cdots\cdots$$
$$R\text{–}M\text{–}M\bullet + nM \longrightarrow R\text{–}M_{n+1}\text{–}M\bullet$$

The resulting macroradical may change into an inactive polymer molecule (chain termination) in the following ways:

1. Recombination of primary radicals.
   $R\bullet + R\bullet \longrightarrow R\text{–}R$
2. Combination of macroradicals.
   $R_{n-1}\text{–}M\bullet + R_{m-1}\text{—}M\bullet \longrightarrow P_{n+m}$
3. Combination of a macroradical with a primary radical.
   $R_{n-1}\text{ –}M\bullet + R_m\bullet \longrightarrow P_{n+m}$
4. Disproportionation.

METHODS OF INITIATION

Activation of a monomer to transform it into a primary radical requires the application of energy. This process can be brought about by the action of heat, light, or an ionising radiation (α-, β-, or γ-rays) or by adding to the system extraneous free radicals or substances (initiators) that break down readily into free radicals. Depending on the method of formation of free radicals the polymerisation is said to be thermally, photochemically, radiation, or chemically initiated.

Thermally Initiated Polymerisation

Thermally initiated polymerisation is a polymerisation wherein the monomer molecule is activated by heat. This is a very important method because in practice polymerisation very often involves heating. However, in the pure form thermal polymerisation is not very probable; it is usually accompanied by polymerisation under the action of catalysts or initiators.

Photochemical Polymerisation

Photochemical polymerisation is a polymerisation in which the molecule is activated as a result of the absorption of luminous energy. Photochemical polymerisation may occur at temperatures whereat polymerisation initiated by other methods does not proceed.

Radiation Polymerisation

In radiation polymerisation the molecule is activated (free radicals are formed) by the action of ionising radiations. The process proceeds subsequently by a radical mechanism.

Chemically Initiated Polymerisation

Chemically initiated polymerisation, one of the most widely used methods, consists in activating the monomer molecules by substances capable of decomposing into free radicals when heated. Such substances include inorganic and organic peroxides (hydrogen peroxide, benzoyl peroxide), hydroperoxides, and diazo compounds. The widely used initiator benzoyl peroxide decomposes readily when heated into two free radicals:

$$C_6H_5COO{-}O{-}COC_6H_5 \longrightarrow C_6H_5\bullet + C_6H_5COO\bullet + CO_2$$

These radicals serve as active centres; they cause chain propagation by combining with monomers:

$$C_6H_5COO\bullet + CH_2{=}\underset{\displaystyle X}{\underset{|}{CH}} \longrightarrow C_6H_5COO{-}CH_2{-}\underset{\displaystyle X}{\underset{|}{CH}}\bullet$$

$$C_6H_5{-}COOCH_2{-}\underset{\displaystyle X}{\underset{|}{CH}}\bullet + CH_2{=}\underset{\displaystyle X}{\underset{|}{CH}} \longrightarrow C_6H_5COOCH_2{-}\underset{\displaystyle X}{\underset{|}{CH}}{-}CH_2{-}\underset{\displaystyle X}{\underset{|}{CH}}\bullet$$

In this case the initiator molecule residues form part of the growing polymer chain.

Oxidation-Reduction Initiation

Oxidation-reduction initiation is often used in polymerisation. Here free radicals formed as a result of an oxidation-reduction reaction initiate the polymerisation, e.g.,

$$ROOH + Fe^{2+} \longrightarrow RO\bullet + OH^- + Fe^{3+}$$

Oxidation-reduction reactions require much less activation energy than the thermal decomposition of peroxides to free radicals (approx. 10 kcal/mole as compared to 30–35 kcal/mole). This makes it possible to carry out the polymerisation at lower temperatures.

CHAIN PROPAGATION

Chain propagation and termination reactions do not depend on the method of activation of monomer. Polymer chain propagation is accomplished by addition of monomer molecules to free radicals (formation of macro-radicals). The chain propagation reaction determines the rate of the polymerisation, the molecular mass of polymer, the structure of the polymer chain, i.e., the mode of monomer addition, ('head-to-tail' or 'head-to-head'), the degree of branching, etc.

Chain polymerisation reactions may be accompanied by chain transfer and by reactions of macro-radicals with double bonds in the polymer chain. Chain transfer consists in interaction of a growing polymer radical with a saturated molecule AB, resulting in termination of the molecular chain

$$R{-}CH_2{-}CHX\bullet + AB \longrightarrow R{-}CH_2{-}CHX{-}A + B\bullet$$

with the simultaneous formation of a new radical B•, which starts a new chain.

Chain transfer reactions usually involve the interaction of active centres with monomer, polymer, or solvent molecules. The new active centre may form in the middle of the chain and cause branching:

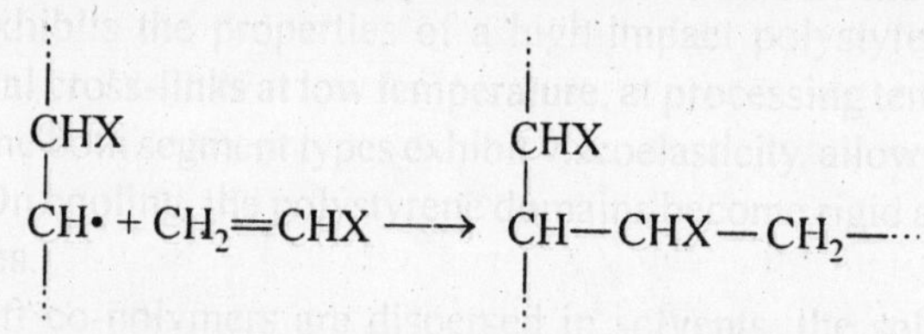

Chain transfer reactions account, for instance, for the formation of branched products upon radical polymerisation of ethylene.

Of great importance in the polymerisation of diene compounds is interaction of the polymeric radical with a double bond:

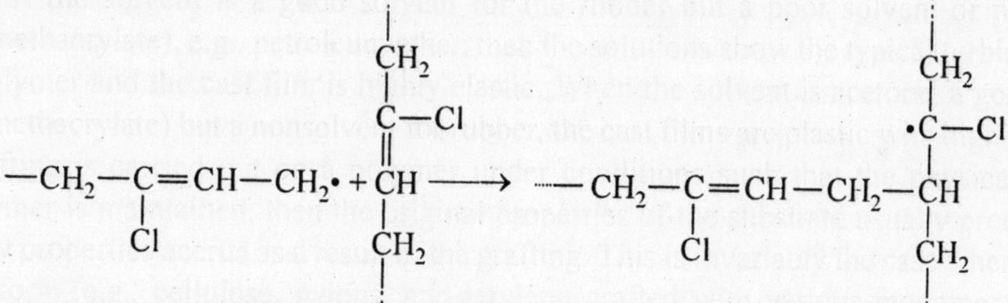

This reaction results in highly branched polymers, and their numerous repetitions will give network structures.

CHAIN TERMINATION

Chain termination results from saturation (deactivation) of the macro-radical and may occur by combination of free radicals, disproportionation or chain transfer.

1. Combination consists in the mutual saturation of two macro-radicals or of a macro-radical and a low-molecular free radical:

$$R{-}CH_2{-}CHX\cdots CH_2{-}CHX\bullet + \bullet CHX{-}CH_2\cdots CHX{-}CH_2{-}R' \longrightarrow$$
$$\longrightarrow R{-}CH_2{-}CHX\cdots CH_2{-}CHX{-}CHX{-}CH_2\cdots CHX{-}CH_2{-}R'$$

2. Disproportionation is the transfer of a hydrogen atom of one macro-radical to another to form two macromolecules, one of them with a double bond at its end:

$$\mathrm{R{-}CH_2{\cdots}CHX{\cdots}CH_2{\cdots}CHX\bullet + \bullet CHX{-}CH_2{\cdots}CHX{-}CH_2{-}R'} \longrightarrow$$
$$\longrightarrow \mathrm{R{-}CH_2{-}CHX{\cdots}CH{=}CHX + CH_2X{-}CH_2{\cdots}CHX{-}CH_2{-}R'}$$

Chemical substances capable of reacting with free radicals to terminate the reaction chain are called polymerisation inhibitors (examples are hydroquinone and trinitrobenzene). Chemical compounds which are only chain-transfer agents, i.e., do not affect the rate of polymerisation but determine the molecular mass of polymer, are called polymerisation regulators, or modifiers. These substances are added to the reaction mixture in strictly rated amounts.

Purity of initial monomer

The rates of radical polymerisation reactions and the properties of the resulting products are very sensitive to impurities in the monomer, which may act as initiators or inhibitors. Therefore to obtain a homogeneous high-quality polymer very pure monomers should be taken for the polymerisation.

Inhibition and retardation

A retarder is defined as a substance that can react with a radical to form products incapable of adding monomer. If the retarder is very effective, no polymer may be formed; this condition is sometimes called inhibition and the substance an inhibitor. The distinction is merely one of degree. The two phenomena are illustrated in Fig. 5.1, where the rate of polymerisation is displayed in idealised fashion. The action of a retarder is two-fold: It both reduces the concentration of radicals and shortens their average lifetime and thus the length of the polymer chains.

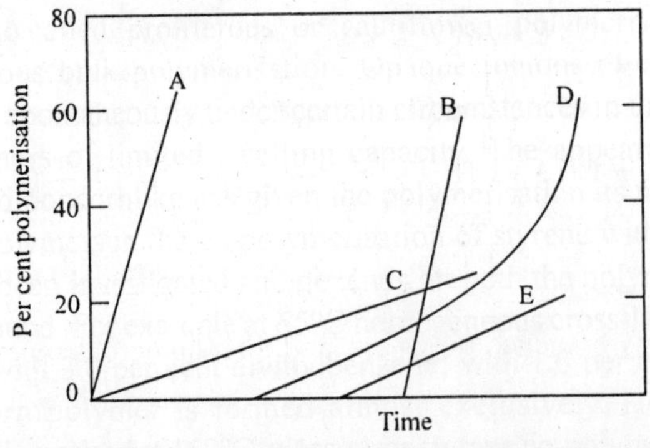

Fig. 5.1. Idealised time-conversion curves for typical inhibition and retardation effects: curve A, normal polymerisation; curve B, simple inhibition; curve C, simple retardation; curve D, non-ideal inhibition; and curve E, inhibition followed by retardation.

In the simplest case, the retarder may be a free radical, such as triphenylmethyl or diphenyl-picryl-hydrazyl, which is too unreactive to initiate a polymer chain. The mechanism of retardation is, simply the combination or disproportionation of radicals. If the retarder is a molecule, the chemistry of retardation is more complex, but the product of the reaction must be a radical too unreactive to initiate a chain. Inhibitors are useful in determining initiation rates, since their reaction with radicals is so rapid that the decomposition of inhibitor is independent of its concentration but gives directly the rate of generation of radicals. As a result, the length of the induction period before polymerisation starts is directly proportional to the number of inhibitor molecules initially present. This number then represents the number of radicals produced during the time of the induction period. For this analysis to be valid it must be known that one molecule of inhibitor is exactly equivalent to one radical. There are a number of reasons why this may not be so, and each case must be investigated separately.

Configuration of Monomer Units in Vinyl Polymer Chains

The addition of a vinyl monomer to a free radical can take place in either of two ways:

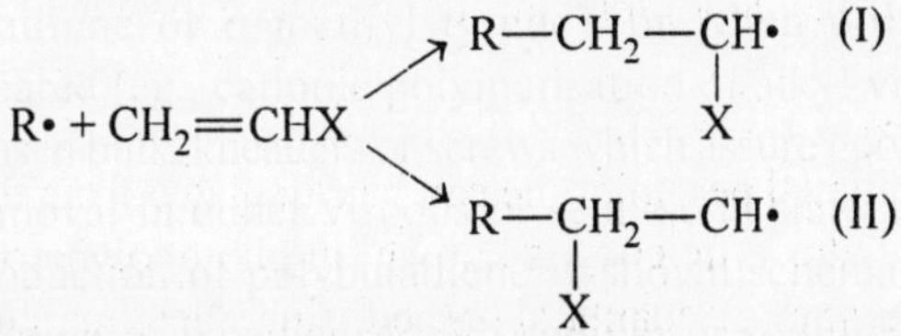

the reaction leading to the more stable product being favoured. Since the unpaired electron can participate in resonance with the substituent X in structure I but not in II, reaction I is favoured. Steric factors also favour reaction I.

The occurrence of reaction I exclusively would lead to a head-to-tail configuration in which the substituents occur on alternative carbon atoms:

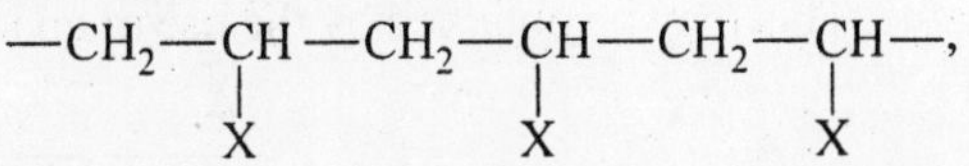

etc.

Alternative possibilities are a head-to-head, tail-to-tail configuration:

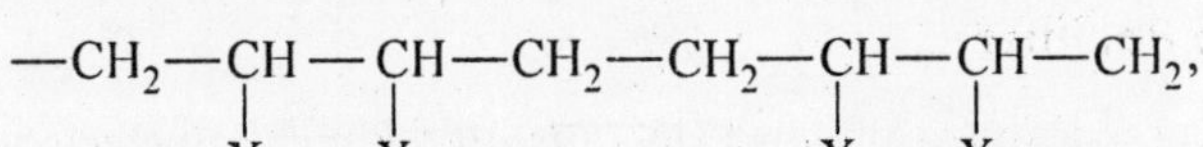

etc.

or a random structure containing both arrangements. The possibility of obtaining a regular head-to-head, tail-to-tail configuration exclusively by chain polymerisation is remote, and it appears that only an occasional monomer unit enters the chain in a reverse manner to provide an isolated head-to-head, tail-to-tail linkage.

Laboratory Methods in Vinyl Polymerisation

Small-scale laboratory polymerisations, carried to low conversion with pure monomer and minor amounts of initiator and other agents, are of great importance in obtaining kinetic data for substantiating the kinetics and mechanisms of polymerisations. Laboratory manuals are available describing the methods and experiments that are most useful. The experimental data of interest include the overall rate of polymerisation, the rate of initiation (and transfer where present), and the molecular weight of the polymer.

Rate of polymerisation

A chemical method or isolation and weighing of polymers can be used to measure the rate of polymerisation, but it is customary to work in an evacuated and sealed system to avoid extraneous effects due to the presence of oxygen.

Polymerisation rate is then followed through changes in a physical property of the system, such as density, refractive index, or ultraviolet or infrared absorption. Measurement of density is usually the most sensitive and convenient technique. The reaction is carried out in a dilatometer, a vessel equipped with a capillary tube in which the liquid level can be measured precisely. The decrease in volume on polymerisation is relatively large, being 21 per cent for methyl methacrylate and 27 per cent for vinyl acetate. In practice, a few hundredths per cent polymerisation can be detected. The experiments are usually carried out in the region below 5 per cent polymer so as to avoid deviations from constant polymerisation rate due to depletion of reactants and other factors. The dilatometer is placed in a constant-temperature bath to ensure isothermal conditions throughout the experiment.

Rate of initiation

The number of polymer chains initiated per unit time can be calculated from the number of free radicals produced in the system, if the initiator efficiency is known. In photopolymerisation the number of free radicals produced is twice the number of quanta absorbed, since one quantum of light decomposes a single initiator molecule into two free radicals. The number of quanta absorbed in the reaction vessel can be measured in a separate experiment.

The initiation rate obtained in this way can be checked by computing the number of polymer molecules formed per unit time from the overall rate of polymerisation and $\bar{x}_n$, provided that chain transfer is absent and that the mechanism of termination is known. Since this is usually not the case, an independent check on the rate of initiation is desirable. This may be obtained by the use of an inhibitor or retarder for the polymerisation.

KINETICS OF VINYL RADICAL POLYMERISATION

Conversion of Monomer to Polymer

The chemical equations discussed above under *Mechanism of Vinyl Polymerisation* for initiation, propagation, and termination of vinyl radical polymerisation can be generalised in the following mathematical scheme.

Initiation in the presence of an initiator I may be considered in two steps: first, the rate-determining decomposition of the initiator into free radicals R•,

$$\mathrm{I} \xrightarrow{k_d} 2\mathrm{R}\bullet \qquad \text{... (5.1)}$$

and, second, the addition of a monomer unit to form a chain radical $M_1^\bullet$,

$$\mathrm{R}\bullet + \mathrm{M} \xrightarrow{k_d} \mathrm{M}_1^\bullet \qquad \text{... (5.2)}$$

where the k's in these and subsequent equations are rate constants, with subscripts designating the reactions to which they refer.

The successive steps in propagation,

$$\begin{aligned} \mathrm{M}_1^\bullet + \mathrm{M} &\xrightarrow{k_p} \mathrm{M}_2^\bullet \\ \mathrm{M}_2^\bullet + \mathrm{M} &\xrightarrow{k_p} \mathrm{M}_3^\bullet \\ &\vdots \\ \mathrm{M}_x^\bullet + \mathrm{M} &\xrightarrow{k_p} \mathrm{M}_{x+1}^\bullet \end{aligned} \qquad \text{... (5.3)}$$

are assumed all to have the same rate constant k_p, since radical reactivity is presumed to be independent of chain length.

The termination step involves combination

$$M_x^{\bullet} + M_y^{\bullet} \xrightarrow{k_{tc}} M_{x+y} \qquad \text{... (5.4)}$$

or disproportionation

$$M_x^{\bullet} + M_y^{\bullet} \xrightarrow{k_{td}} M_x + M_y \qquad \text{... (5.5)}$$

Except where it is necessary to distinguish between the two mechanisms, the termination rate constant is denoted k_t.

The rates of the three steps may be written in terms of the concentrations (in brackets) of the species involved and the rate constants. The rate of initiation is

$$\upsilon_i = \left(\frac{d[\text{M}\bullet]}{dt}\right)_i = 2fk_d[\text{I}] \qquad \text{... (5.6)}$$

where the factor f represents the fraction of the radicals formed by Eq. 5.1 that is successful in initiating chains by Eq. 5.2. The rate of termination is

$$\upsilon_t = -\left(\frac{d[\text{M}\bullet]}{dt}\right)_t = 2k_t[\text{M}\bullet]^2 \qquad \text{... (5.7)}$$

For many cases of interest the concentration of free radicals [M•] becomes essentially constant very early in the reaction, as radicals are formed and destroyed at identical rates. In this steady-state condition $\upsilon_i = \upsilon_t$ and Eqs. 5.6 and 5.7 may be equated to solve for [M•]

$$[\text{M}\bullet] = \left(\frac{fk_d[\text{I}]}{k_t}\right)^{1/2} \qquad \text{... (5.8)}$$

The rate of propagation is essentially the same as the overall rate of disappearance of monomer, since the number of monomers used in Eq. 5.2 must be small compared to the used in Eq. 5.3 if polymer is obtained. Then

$$\upsilon_p = -\frac{d[\text{M}]}{dt} = k_p[\text{M}][\text{M}\bullet] \qquad \text{... (5.9)}$$

or, with substitution from Eq. 5.8,

$$\upsilon_p = k_p\left(\frac{fk_d[\text{I}]}{k_t}\right)^{1/2}[\text{M}] \qquad \text{... (5.10)}$$

Thus the overall rate of polymerisation should, in the early stages of the reaction, be proportional to the square root of the initiator concentration and, if f is independent of [M], to the first power of the monomer, concentration. This is true if the initiator efficiency is high. With very low efficiency, f may be proportional to [M], making υ_p proportional to $[\text{M}]^{3/2}$.

The proportionality of the overall rate to the square root of initiator concentration has been confirmed experimentally in a large number of cases. Typical experimental data are shown in Fig. 5.2. The straight lines are drawn with the theoretical slope of 1/2.

The deviation for polystyrene at low initiator concentration (Fig. 5.2) is due to the presence of some thermal initiation. This phenomenon, peculiar to very pure styrene and a few of its derivatives, without any initiator present, involves the thermal dimerisation of styrene followed by a reaction with still another styrene to produce a radical capable of initiating polymerisation.

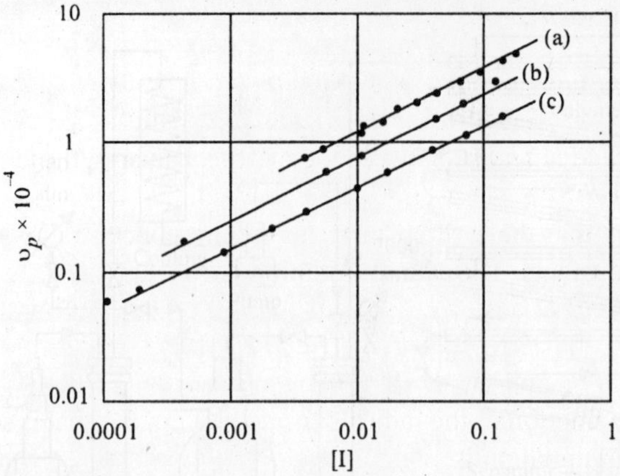

Fig. 5.2. Dependence of rate of polymerisation υ_p on concentration of initiator [I] for (a) methyl methacrylate with azobisisobutyronitirile; (b) methyl methacrylate with benzoyl peroxide; and (c) styrene with benzoyl peroxide.

Evaluation of rate constants

If the rate constant for initiator decomposition and the initiator efficiency are known, the ratio of rate constants, k_p^2/k_t can be evaluated from the overall polymerisation rate: From Eqs. 5.6 and 5.10 we have,

$$\frac{k_p^2}{k_t} = \frac{2\upsilon_p^2}{\upsilon_i[M]^2} \quad \text{... (5.11)}$$

These rate constants can be separated only by non-steady-state methods, described below. Measurements of k_p^2/k_t as a function of temperature lead to useful information about the thermochemistry of polymerisation.

Overall rate as a function of conversion

If the initiator concentration does not vary much during the course of polymerisation and the initiator efficiency is independent of monomer concentration, polymerisation proceeds by first-order kinetics; that is, the polymerisation rate is proportional to monomer concentrations. In some systems, such as the benzyol peroxide-initiated polymerisation of styrene, the reaction is accurately first order up to quite high conversions (Fig. 5.3).

The polymerisation of certain monomers undiluted or in concentrated solution is accompanied by a marked deviation from first-order kinetics in the direction of an increase in reaction rate and molecular weight termed autoacceleration or the gel effect.

The effect is particularly pronounced with methyl methacrylate, methyl acrylate, or acrylic acid (Fig. 5.4).

It is independent of initiator and is due to a decrease in the rate at which the polymer molecules diffuse through the viscous medium, thus lowering the ability of two long-chain radicals to come together and terminate.

Because termination is in fact diffusion controlled for most liquid-phase polymerisations, even at low conversion, the dependence of diffusion rate on the viscosity of the medium leads to the gel effect at high polymer concentrations, with high-molecular-weight polymers, or in the presence of inert solutes increasing the viscosity of the medium.

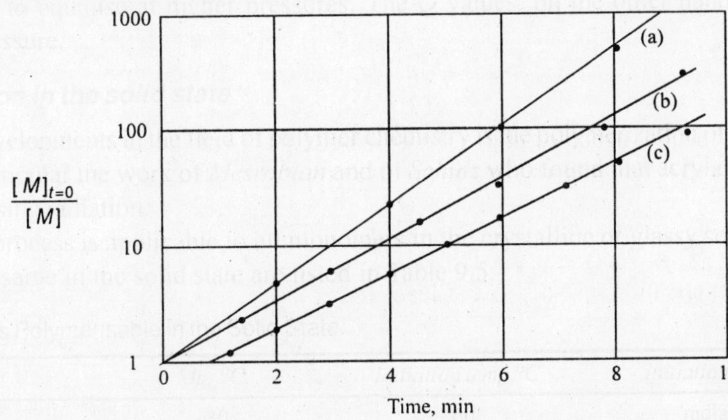

Fig. 5.3. Polymerisation of vinyl phenylbutyrate in dioxane with benzoyl peroxide initiator at 60°C plotted as a first-order reaction. At $t = 0$, monomer concentration was (a) 2.4 g/100 ml; (b) 6.0 g/100 ml; and (c) 7.3 g/100 ml.

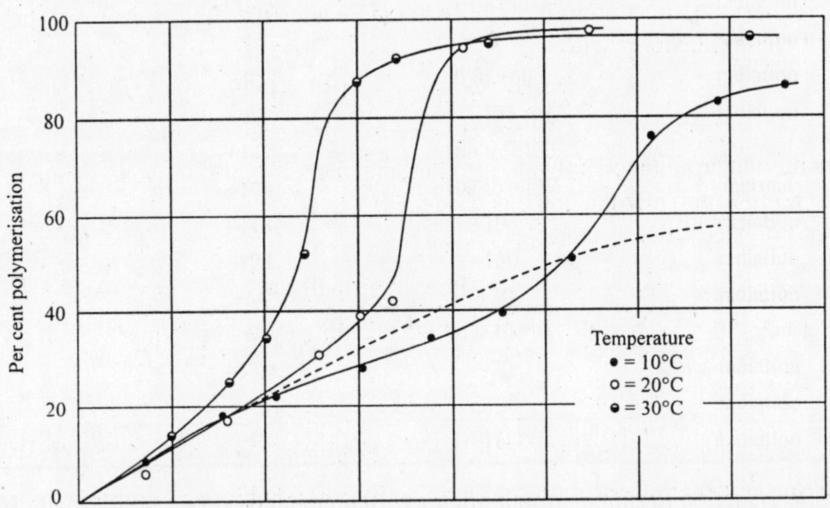

Fig. 5.4. Autoacceleration of polymerisation rate in poly (methyl methacrylate).

The decrease in termination rate leads to an increase in overall polymerisation rate and in molecular weight, since the lifetime of the growing chains is increased. At quite high conversions (70–90 per cent) the rate of polymerisation drops to a very low value as the system becomes glassy and monomer can no longer be supplied to the growing chain ends.

Degree of Polymerisation and Chain Transfer

Kinetic chain length

The kinetic chain length ν is defined as the number of monomer units consumed per active centre. It is therefore given by $\upsilon_p/\upsilon_i = \upsilon_p/\upsilon_t$:

$$\nu = \frac{k_p}{2k_t}\frac{[\mathrm{M}]}{[\mathrm{M}\bullet]} \quad \text{... (5.12)}$$

Eliminating the radical concentration by means of Eq. 5.9,

$$\nu = \frac{k_p^2}{2k_t}\frac{[\mathrm{M}]^2}{\upsilon_p} \quad \text{... (5.13)}$$

For initiated polymerisation, use of Eq. 5.8 leads to

$$\nu = \frac{k_p}{2(fk_d k_t)^{1/2}}\frac{[\mathrm{M}]}{[\mathrm{I}]^{1/2}} \quad \text{... (5.14)}$$

If no reactions take place other than those already discussed, the kinetic chain length should be related to $\bar{x}_n$, the degree of polymerisation: For termination by combination $\bar{x}_n = 2\nu$, and for disproportionation $\bar{x}_n = \nu$. This is found to be precisely true for some system, but for others wide deviations are noted in the direction of more polymer molecules than active centres. These deviations are the results of chain-transfer reactions:

$$\mathrm{M}_x\bullet + \mathrm{XP} \longrightarrow \mathrm{M}_x\mathrm{M} + \mathrm{P}\bullet$$

$$\mathrm{P}\bullet + \mathrm{M} \longrightarrow \mathrm{PM}\bullet, \qquad \text{etc.}$$

where P may be monomer, initiator, solvent, or other added chain-transfer agent. (Transfer to polymer is omitted, since no new polymer molecule is produced).

The degree of polymerisation is therefore

$$\bar{x}_n = \frac{\text{Rate of growth}}{\sum \text{Rates of all reactions leading to dead polymer}}$$

$$= \frac{\upsilon_p}{fk_d[\mathrm{I}] + k_{tr,\mathrm{M}}[\mathrm{M}][\mathrm{M}\bullet] + k_{tr,\mathrm{S}}[\mathrm{S}][\mathrm{M}\bullet] + k_{tr,\mathrm{I}}[\mathrm{I}][\mathrm{M}\bullet]} \quad \text{... (5.15)}$$

where the terms in the denominator represent termination by combination and transfer to monomer, solvent, and initiator, respectively. If termination is by disproportionation, the first term becomes, $2fk_d[\mathrm{I}]$. If transfer constants are defined as

$$C_M = \frac{k_{tr,\mathrm{M}}}{k_p}, \qquad C_S = \frac{k_{tr,\mathrm{S}}}{k_p}, \qquad C_I = \frac{k_{tr,\mathrm{I}}}{k_p} \quad \text{... (5.16)}$$

then (assuming termination by combination)

$$\frac{1}{\bar{x}_n} = \frac{k_t}{k_p^2}\frac{\upsilon_p}{[\mathrm{M}]^2} + C_M + C_S\frac{[\mathrm{S}]}{[\mathrm{M}]} + C_I\frac{k_t}{k_p^2 f k_d}\frac{\upsilon_p^2}{[\mathrm{M}]^3} \quad \text{... (5.17)}$$

The above analysis assumes that the radical formed in the transfer process is approximately as reactive as the original chain radical, otherwise retardation or inhibition results.

Transfer to solvent

In the presence of a solvent, and by properly choosing conditions to keep other types of chain transfer to a minimum, Eq. 5.17 reduces to

$$\frac{1}{\bar{x}_n} = \left(\frac{1}{\bar{x}_n}\right)_0 + C_s \frac{[S]}{[M]} \qquad \text{... (5.18)}$$

where $(1/\bar{x}_n)_0$ combines the polymerisation and transfer to monomer terms. The linear dependence of $1/\bar{x}_n$ on [S]/[M] is illustrated in Fig. 5.5. In experiments of this type the ratio $\upsilon_p/[M]^2$ must be held constant throughout if $(\bar{x}_n)_0$ is to be identified with the observed degree of polymerisation in the absence of solvent.

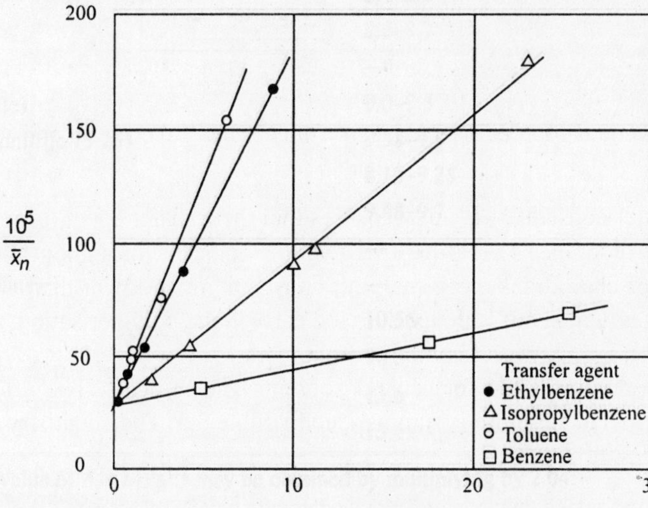

Fig. 5.5. Effect of chain transfer to solvent on the degree of polymerisation of polystyrene.

Transfer to monomer and initiator

If only the terms corresponding to polymerisation, transfer to monomer, and transfer to initiator are kept in Eq. 5.17, the resulting equation for $\bar{x}_n$ is quadratic in υ_p. This behaviour has been observed, for example, in the benzoyl peroxide-initiated polymerisation of styrene.

The contribution of transfer to monomer is constant and independent of polymerisation rate, and that of transfer to initiator increase rapidly with increasing rate, since high rate requires high initiator concentration (Fig. 5.6).

Control of molecular weight by transfer

Chain-transfer agents with transfer constants near unity are quite useful in depressing molecular weight in polymerisation reactions.

This is often of great commercial importance, for example, in the polymerisation of dienes to synthetic rubbers, where chain length must be controlled for ease of processing.

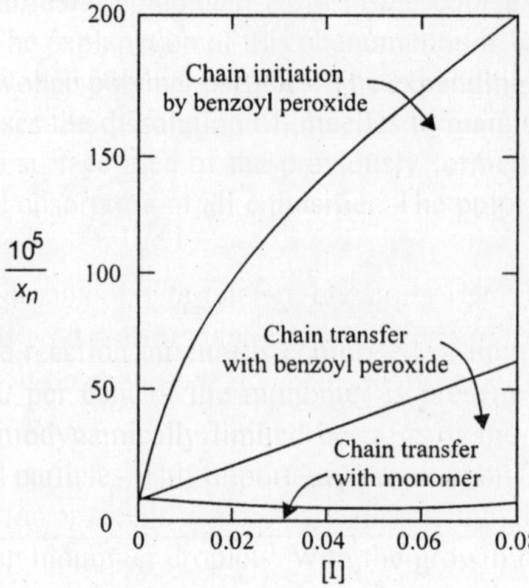

Fig. 5.6. Contribution of various sources of termination in the polymerisation of styrene with benzoyl peroxide at 60°C.

The choice of transfer constant near unity ensures that the transfer agent, or regulator, is consumed at the same rate as the monomer so that [S]/[M] remains constant throughout the reaction. Too large quantities are needed of chain-transfer agents with constants much lower than unity, and agents with transfer constants greater than about 5 are used up too early in the polymerisation. Aliphatic mercaptans are suitable transfer agents for several common monomers.

Table 5.2 lists transfer constants for several solvents and radicals. The effect of chain-transfer reactions on chain length and structure is summarised in Table 5.3. The chemistry and kinetics of chain transfer have been reviewed.

Table 5.2. Chain transfer constants for various solvents and radicals at 60°C.

	$C_s \times 10^4$		
Solvent	*Styrene*	*Methyl methacrylate*	*Vinyl acetate*
Benzene	0.023	0.040	1.2
Cyclohexane	0.031	0.10 (80° C)	7.0
Toluene	0.125	0.20	22
Chloroform	0.5	1.77	150
Ethylbenzene	0.67	1.35 (80° C)	5.5
Triethylamine	7.1	8.3	370
Tetrachloroethane	18 (80°C)	0.155	107
Carbon tetrachloride	90	2.40	9600
Carbon tetrabromide	22,000	2700	28,700 (70° C)

Table 5.3. Effect of chain-transfer reactions on chain length and structure.

Reaction	Effect on υ_p	$\overline{M}_n$	$\overline{M}_w$	Structure
To small molecule, giving active radical	None	Decreases	Decreases	None
To small molecule, retardation or inhibition	Decreases	May decrease or increase	May decrease or increase	None
To polymer, intermolecular	None	None	Increases	Produces long branches
To polymer, intramolecular	None	None	Increases	Produces short branches

Inhibition and Retardation

The kinetics of chain polymerisation in the presence of added inhibitor or retarder can be described by adding to the scheme of Eqs. 5.1 to 5.5 the reaction

$$M_x\bullet + Z \xrightarrow{k_z} M_x + Z\bullet \quad \text{... (5.19)}$$

where Z is the inhibitor. It is assumed that the radical Z· does not initiate polymerisation, and is terminated without regenerating Z. Application of the steady-state assumption leads, in direct analogy with Eq. 5.17, to the relation (again for termination by combination)

$$\frac{1}{\bar{x}_n} = \frac{k_t}{k_p^2}\frac{\upsilon_p}{[M]^2} + C_z\frac{[Z]}{[M]} \quad \text{... (5.20)}$$

where $C_z = k_z/k_p$.

Table 5.4 lists inhibition constants for some common monomers and inhibitors.

Table 5.4. Inhibitor constants for various inhibitors and monomers at 50°C.

Inhibitor	C_z Styrene	Methyl acrylate	Methyl methacrylate	Vinyl acetate
Nitrobenzene	0.326	0.00464	–	11.2
Trinitrobenzene	64.2	0.204	–	404
p-Benzoquinone	518	–	5.5 (44°C)	–
DPPH	–	–	2000 (44°C)	–
Oxygen	14,600	–	33,000	–

Determination of Individual Rate Constants

It was shown above that the rate constants for propagation and termination occur together as the ratio k_p^2/k_t in the equations for both overall polymerisation rate and kinetic chain length. Thus these rate constants cannot be evaluated separately through steady-state measurements. Recourse must be had to the study of transient phenomena, such as the rate of polymerisation before the free-radical concentration has reached its steady-state value. The duration of this transient region depends on the length of time a free radical exists from its formation in the initiation step to its demise in termination.

Sector method

If a photopolymerisation is begun by turning on light of intensity I, the concentration of free radicals grows gradually to its steady-state value over a time interval that is a function of the lifetime of the radicals. When the light is turned off, the radical concentration decays in a similar manner. At the steady-state the rate of polymerisation is proportional to the square root of the light intensity: $\upsilon_p \simeq I^{1/2}$. On the other hand, if the flashes are so rapid that the radical concentration changes very little during a single flash, the effect is the same as that produced by reducing the intensity by a constant factor but leaving it on all the time: If the light and dark periods are equal , $\upsilon_p \simeq (\frac{1}{2} I)^{1/2}$. The rates for long and short flashes differ, in this case, by $\sqrt{2}$ (Fig. 5.7).

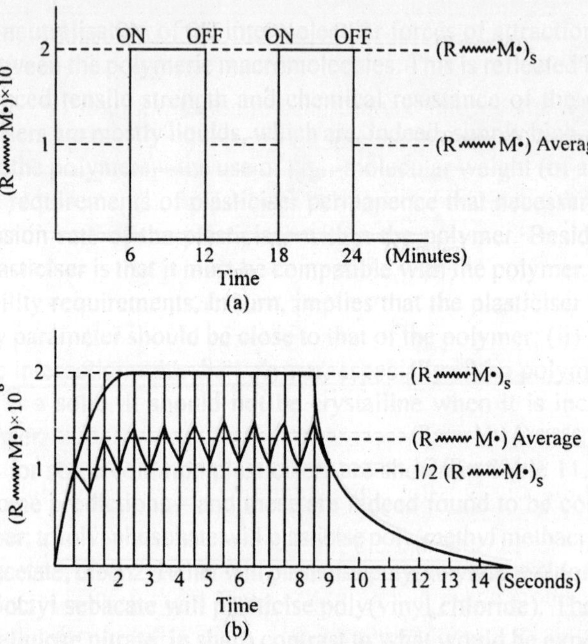

Fig. 5.7. Radical concentration versus time in the sector method: (a) long flashes compared to τ, for which the maximum radical concentration is, for example, 2×10^{-8} mole/litre and its average is 1×10^{-8}; and (b) short flashes, for which the average radical concentration is approximately $\sqrt{2} \times 10^{-8}$.

The rate of flashing at which the transition occurs can be related to the mean lifetime τ_s of the free radicals. This is defined as

$$\tau_s = \frac{\text{Number of radicals}}{\text{Number disappearing per unit time}} = \frac{[\text{M}\bullet]}{2k_t[\text{M}\bullet]^2} = \frac{1}{2k_t[\text{M}\bullet]} \quad \text{... (5.21)}$$

or, if the radical concentration is substituted from Eq. 5.9,

$$\tau_s = \frac{k_p}{2k_t}\frac{[\text{M}]}{\upsilon_p} \quad \text{... (5.22)}$$

Thus τ_s, [M], and the overall polymerisation rate υ_p suffice to calculate k_p/k_t. With the ratio k_p^2/k_t obtained from other data, the individual rate constants may be evaluated.

The relation between τ_s and the flashing rate is complex; in practice, the experimental curves of polymerisation rate (or radical concentration) against flashing rate are compared to the theoretical relation (Fig. 5.8). In Fig. 5.8 the ratio t/τ_s, where t is the time light is on, is plotted against $\bar{\upsilon}_p/(\upsilon_p)_s$, where $\bar{\upsilon}_p$ is the rate with flashing illumination and $(\upsilon_p)_s$ is the rate with steady illumination, for several values of the ratio r of dark to light intervals.

In practice, the light is usually flashed by placing a motor-driven rotating sector between the light source and the reaction vessel.

In the case of vinyl acetate, the radical lifetime is of the order of 4 sec, depending, of course, upon polymerisation rate, according to Eq. 5.22. The rate constants show that the growing chain adds about 10^4 monomers per second; the kinetic chain length is the product of this figure and τ_s, or about 4×10^4. The value of $\bar{x}_n$ is about 10 times less than this because of chain transfer to monomer; thus about 10 transfers must occur during the life of a growing chain. This represents a relatively large amount of transfer to monomer.

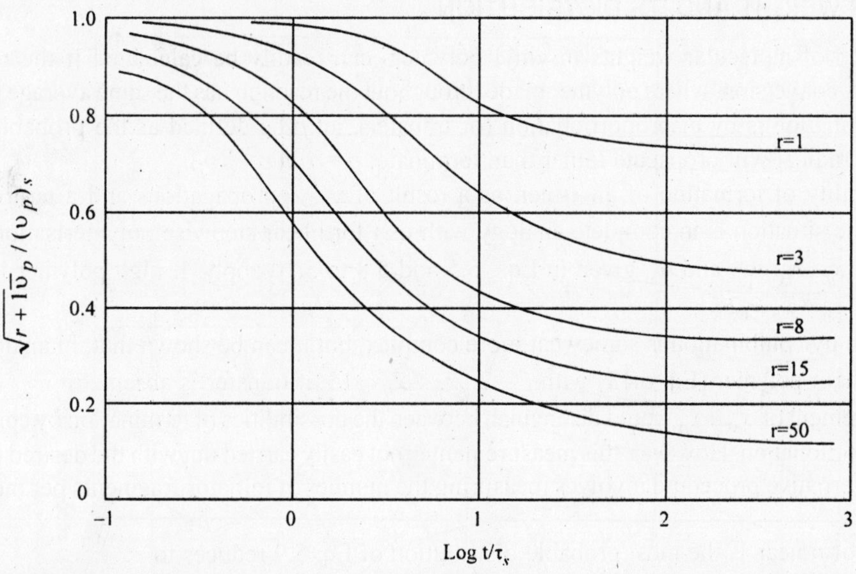

Fig. 5.8. Ratio of polymerisation rates for flashing and steady illumination plotted against $\log(t/\tau_s)$ for several values of the ratio r of dark to light periods.

Other methods

The kinetic scheme for emulsion polymerisation, allows separation of the individual rate constants, since the growth of a single chain effectively takes place in an isolated system.

In a viscosity method, now little used, the polymerisation is carried out in a viscometer, measurement of molecular weight by viscosity being substituted for measurement of rate. The change in viscosity after photoinitiation is stopped provides data obtained under transient conditions.

Dead-end polymerisation, which refers to polymerisation with a limited amount of initiator, allows the determination of the initiation rate constant k_d and initiator efficiency f. As the initiator is depleted, the reaction stops short of completion at the point where the half-life of the propagating chain is comparable to that of the initiator.

Values of the propagation and termination constants for several monomers are tabulated in Table 5.5. The rate constants vary over wide ranges. Since they depend upon the reactivity of both the radical and the monomer, it is difficult to draw conclusions about the effect of structure on reactivity from these data alone. Co-polymerisation experiments are more useful in this respect.

Table 5.5. Rate constants for propagation and termination at 50°C.

Monomer	k_p *(litre/mole-sec)*	k_t *(litre/mole-sec* $\times 10^{-7}$*)*
Styrene	209	115
Methyl methacrylate	410	24
Methyl acrylate	1000	3.5
Vinyl acetate	2640	117
Vinyl chloride	11000	2100

MOLECULAR WEIGHT AND ITS DISTRIBUTION

The distribution of molecular weights in vinyl polymers can readily be calculated if the reaction is restricted to low conversion, where polymer made throughout the reaction has the same average molecular weight. If termination is by disproportionation (or transfer), let p be defined as the probability that a growing chain radical will propagate rather than terminate: $p = \upsilon_p/(\upsilon_p + \upsilon_t)$.

The probability of formation of an x-mer, as a result of $x - 1$ propagations and a termination, is $p^{x-1}(1-p)$. The situation is in complete analogy with that for linear stepwise polymerisation, and the expression for $\bar{x}_n, N_x, w_x$, and $\bar{x}_w$ given in Eqs. 5.5 and 5.8 to 5.10 apply. If high polymer is formed, that is, $p \to 1$, $\bar{x}_w / \bar{x}_n = 2$.

Termination by combination is somewhat more complex, but it can be shown that, in analogy to the case of multichain polymer (Eq. 5.13) with $f = 2$, $\bar{x}_w / \bar{x}_n = 1.5$ if transfer is absent.

The measurement of $\bar{x}_w / \bar{x}_n$ would distinguish between the possibilities of termination by combination and by disproportionation. However, this measurement is not easily carried out with the desired precision; a preferable alternative procedure involves measuring the number of initiator fragments per molecule of polymer.

For values of p near 1, the most probable distribution of Eq. 5.9 reduces to.

$$w_x = xp^x(\ln p)^2$$

$$\bar{x}_n = \frac{-1}{\ln p} \qquad \text{... (5.23)}$$

$$\bar{x}_w = \frac{-2}{\ln p}$$

It can be shown that both this distribution, which describes low-conversion chain polymers terminating by disproportionation, and that describing termination by combination are special cases of a more general distribution function:

$$w_x = \left(\frac{y^{z+1}}{z!}\right)x^z e^{-xy}$$

$$\bar{x}_n = \frac{z}{y} \qquad \text{... (5.24)}$$

$$\bar{x}_w = \frac{z+1}{y}$$

For $z = 1$, Eq. 5.24 reduces to Eq. 5.23 with $y = -\ln p$. For $z = 2$ it describes the case of termination by combination. Higher values of z correspond to narrower distributions such as might be found in fractionated samples.

In contrast to the most probable distribution, the number distribution corresponding to Eq. 5.24 does not indicate a large number of very small molecules; hence it is preferred for describing polymers from which the small molecules have been removed, for example, by precipitation of the polymer from solution.

As polymerisation proceeds, the degree of polymerisation of the polymer being formed changes according to Eq. 5.14. If the rate of initiation is constant as is usual, $\bar{x}_n$ decreases throughout the reaction as [M] decreases, approaching zero at complete conversion. At the same time $\bar{x}_w / \bar{x}_n$ increases with increasing conversion, approaching infinity as a limit. Typical values of $\bar{x}_w / \bar{x}_n$ for high-conversion vinyl polymers seldom exceed 5, however, except in special cases, such as autoacceleration in the polymerisation of methyl methacrylate, where values of $\bar{x}_w / \bar{x}_n$ as high as 10 have been found. The above distribution functions are not well suited to the description of these broader distributions. Among empirical functions suggested for these cases is the following form, which has a high-molecular-weight 'tail' when the exponent ρ is positive and less than 1 (here Γ is the gamma function):

$$w(x) = \left[\rho \Big/ \Gamma\left(\frac{z+1}{\rho}\right)\right] y^{(z+1)/\rho} x^z e^{-yx^\rho}$$

$$\bar{x}_n = \Gamma\left(\frac{z+1}{\rho}\right) \Big/ y^{1/\rho} \qquad \text{... (5.25)}$$

$$\bar{x}_w = \Gamma\left(\frac{z+2}{\rho}\right) \Big/ y^{1/\rho}\, \Gamma\left(\frac{z+1}{\rho}\right)$$

Where chain transfer to polymer is involved, extremely broad molecular weight distributions can results. The abnormal breadth arises from the fact that each transfer reaction adds a branch point to the molecule; after the branch has grown, the molecule has increased in molecular weight. But the probability of branching is proportional (approximately) to weight-average molecular weight; hence branched molecules tend to become more highly branched and still larger. The result is that the distribution of molecular weights has a long, high-molecular-weight 'tail'.

Among the empirical distribution functions suitable for describing fractions from which low-molecular-weight species have been removed is the so-called logarithmic normal distribution:

$$w(x) = \frac{e^{-\sigma^2/2}}{\sigma\sqrt{\pi}} \frac{1}{x_0} e^{-(\ln x - \ln x_0)^2/2\sigma^2}$$

$$\bar{x}_n = x_0 e^{\sigma^2/2} \qquad \text{... (5.26)}$$

$$\bar{x}_w = x_0 e^{3\sigma^2/2}$$

In this distribution the logarithm of the molecular weight follows a normal distribution function. The ratio $\bar{x}_w / \bar{x}_n$ is identical to $\bar{x}_z / \bar{x}_w$.

Other empirical equations have been proposed for cases of unusually broad distributions by *Wesslau, Tung, Gordon* and *Roe*.

Polymers with extremely narrow molecular-weight distribution are produced by certain types of anionic polymerisation. Their molecular-weight distribution is described by the Poisson distribution

$$w(x) = \frac{\mu}{\mu+1} \frac{x e^{-\mu} \mu^{x-2}}{(x-1)!}$$

$$\bar{x}_n = \mu = 1 \qquad \text{... (5.27)}$$

$$\frac{\bar{x}_w}{\bar{x}_n} = 1 + \frac{\mu}{(\mu+1)^2}$$

EFFECTS OF TEMPERATURE AND PRESSURE ON CHAIN POLYMERISATION

Thermochemical and thermodynamic data gives valuable insight into the mechanisms of vinyl polymerisation reactions. By the usual Arrhenius treatment, the temperature dependence of the rates of the various steps in radical chain polymerisation can be separated into an energy of activation representing the amount of energy that the reactant molecules must have to be able to react on collision, and a frequency factor which allows estimation of the fraction of the collisions that lead to reaction. Enthalpies of polymerisation are readily measured and may also be calculated from bond dissociation energies. Differences between the two values give an indication of the amount of steric hindrance in the polymer. Free energies of polymerisation reactions can be calculated for some simple cases which gives information on the thermodynamic equilibria involved in polymerisation.

Enthalpies and Free Energies of Polymerisation

Enthalpies of polymerisation

These may be obtained from the heats of combustion of monomer and polymer or directly by a calorimetric experiment; they may also be calculated from bond dissociation energies. When the two results are compared, it is seen that the observed enthalpy of polymerisation is equal to or somewhat lower than the calculated heat.

For vinyl compounds with two substituents on the same carbon atom the difference is 3–8 kJ/mole. This is attributed to steric repulsion between the substituent groups in the polymer.

For monosubstituted vinyl derivatives the difference is much lower; only for styrene is it definitely greater than 4 kJ/mole.

Enthalpies of polymerisation (Table 5.6) vary over a wide range, reflecting differences in resonance stabilisation and conjugation, steric strain in monomer and polymer, and (if present) dipole interactions and hydrogen bonding.

Entropies and free energies of polymerisation

The entropy of polymerisation is largely determined by the loss of the translational entropy of the monomer on polymerisation. Values of ΔS are therefore relatively insensitive to the monomer type (Table 5.6). Both ΔH and ΔS are negative, but the exothermic nature of the polymerisation reaction outweighs the influence of the entropy term in

$$\Delta G = \Delta H - T\Delta S$$

so that the free energies are negative at usual temperatures. Experimental methods for obtaining these quantities have been reviewed.

Table 5.6. Enthalpies and entropies of polymerisation at 25°C.

Monomer	$-\Delta H$ *(kJ/mole)*	$-\Delta S$ *(J/mol-deg)*
Tetrafluoroethylene	156	112
Vinyl chloride	96	–
Ethylene	95	100
Vinyl acetate	88	110
Propylene	86	116
Acrylonitrile	77	109
Isoprene	74	101
Butadiene	73	86
Styrene	70	105
Methyl methacrylate	56	117
Isobutylene	52	120
α-Methylstyrene	35	104

At elevated temperatures the entropy term increases, and when $\Delta H = T\Delta S$, $\Delta G = 0$ and the polymerisation reaction is in equilibrium with the reverse reaction of depolymerisation. This temperature is known as the ceiling temperature for the polymer in question. Some ceiling temperatures for pure monomers are 61°C for α-methylstyrene, 220°C for methyl methacrylate, and 310°C for styrene.

The calculation of the enthalpies and entropies of reaction of free radicals has been carried out for hydrocarbons and fluorocarbons and applied to the synthesis equations for polyethylene and polytetrafluoroethylene. Values of enthalpy and free energy for the addition of monomer to the radical become less negative as the length of the chain radical increases in the first few propagation steps. After the growing free radical reaches a length of about eight carbon atoms, all successive steps have the same enthalpy and free energy of reaction, the same as those for the overall polymerisation step.

Activation Energies and Collision Factors

Overall polymerisation

The ratio of rate constants k_p^2 / k_t or $k_p / \sqrt{k_t}$ is obtained from the overall polymerisation rate and the rate of initiation (Eq. 5.11).

Expressing a rate constant by an Arrhenius-type equation gives, for example,

$$k_p = A_p e^{-E_p/RT} \quad \text{... (5.28)}$$

where A_p is the collision frequency factor for propagation and E_p the energy of activation for propagation. Application to the rate-constant ratio gives

$$\frac{k_p}{\sqrt{k_t}} = \left(\frac{A_p}{\sqrt{A_t}}\right) e^{-(E_p - E_t/2)/RT} \quad \text{... (5.29)}$$

By plotting log $(k_p / \sqrt{k_t})$ against $1/T$, $E_p - 1/2E_t$ and $A_p / \sqrt{A_t}$ can be evaluated. For many polymerisation reactions, $E_p - 1/2E_t$ is about 20–25 kJ/mole. This is equivalent to an increase in $k_p / \sqrt{k_t}$ of about 30–35 per cent for every 10°C near room temperature. Equation 5.29 refers only to the ratio of rate constants $k_p / \sqrt{k_t}$ and does not include the effect of temperature on the rate of initiation. In photopolymerisation, where the rate of initiation is independent of temperature, Eq. 5.29 represents the complete activation energy.

In an initiated polymerisation, υ_p is proportional to $k_p / \sqrt{k_d / k_t}$, and the observed activation energy is $1/2E_d + (E_p - 1/2E_t)$. The activation energy E_d for spontaneous decomposition of the initiator is about 125 kJ/mole for benzoyl peroxide or azobisisobutyronitrile. The activation energy for initiated polymerisation is therefore slightly greater than 80 kJ/mole. This corresponds to a two to threefold increase in rate for a 10°C temperature change.

The activation energy for thermal polymerisation is of the same order of magnitude, yet this reaction is much slower than normal because of the extremely low collision factor for thermal initiation. The normal value of A for a bimolecular reaction is $10^{11} - 10^{13}$; for the thermal initiation of polystyrene it is about $10^4 - 10^6$.

Degree of polymerisation

If transfer reactions are negligible, the temperature coefficient of molecular weight depends on the initiation process. If an initiator is used, it follows from Eq. 5.14, by virtue of the differential form $d \ln k/dT = E/RT^2$ of the Arrhenius equation, that

$$\frac{d \ln \bar{x}_n}{dT} = \frac{E_p - E_t/2 - E_d/2}{RT^2} \quad \text{... (5.30)}$$

Since $E_p - 1/2E_t$ is 20–25 kJ and E_d is about 125 kJ, this quantity is negative and the molecular weight decrease with increasing temperature. The same is true in thermal polymerisation.

In photopolymerisation in the absence of transfer, $d \ln \bar{x}_n / dT = (E_p - 1/2E_t)/RT^2$, which is positive. This is the only case in which molecular weight increases with temperature.

If transfer reactions are controlling, it is found that molecular weight decreases with increasing temperature. By plotting log C_s or $-\log \bar{x}_n$ against $1/T$, $E_{tr,S} - E_p$ and $A_{tr,S}/A_p$ may be evaluated. The quantity $E_{tr,S}$ usually exceeds E_p by 20–60 kJ/mole, higher activation energies going with poorer transfer agents.

Since the frequency factor for transfer usually exceeds that for propagation, the higher activation energy is responsible for the slower rate of transfer compared to monomer addition.

Absolute rates

Activation charges and frequency factors of propagation and termination can be separated if the individual rate constants are known. For most monomers E_p is near 30 kJ/mole, and E_t varies from 12 to 20 kJ/mole. The frequency factors vary rather widely, suggesting that steric effects are more important than activation energies in setting the values of the rate constants. For example, A_p is much lower for methyl methacrylate with two substituents on the same ethylenic carbon atom that it is for less hindered monomers. All the values of A_p are considerably lower than normal, because it is likely of low probability of the transition leading to unpairing of the double-bond electrons in the monomer.

Polymerisation-depolymerisation equilibrium

Some polymers decompose by stepwise loss of monomer units in a reaction that is essentially the reverse of polymerisation. The depolymerisation reaction can be incorporated into the kinetic scheme for chain polymerisation. At ordinary temperatures the scheme is not changed because the rate constant for depolymerisation is small. The activation energy of depolymerisation is quite high (40–110 kJ/mole) compared to that of propagation, however, and at some elevated temperature the rates of polymerisation and depolymerisation become equal and the ceiling temperature is reached, above which polymer free radicals, in the presence of monomer at 1-atm pressure, depolymerise rather than grow.

Effects of Pressure

The effects of high pressure (up to, say, several thousand atmospheres) on polymerisation go far beyond the direct effect of pressure on the volume of the system. Both equilibrium constants (recall that $\Delta G = -RT \ln K$) and rate constants are affected, the former through the thermodynamic relation

$$\left(\frac{\partial RT \ln K}{\partial P}\right)_\tau = -\Delta V \qquad \text{... (5.31)}$$

and the latter through the Eyring rate theory by replacing K with k and ΔV with the volume change between the reactants and the transition state.

Since the volume change on polymerisation is usually negative, an increase in pressure favours the formation of polymer and is accompanied by an increase in the ceiling temperature, often 15–20°C per 1000 atm.

The polymerisation of styrene at high pressures has been studied in more detail than that of other monomers. The rate of dissociation of initiator is reduced, but the rate of polymerisation is sufficiently increased to outweigh this, the overall polymerisation rate being seven or eight times higher at 3000 atm than at ordinary pressures. The rate of termination, being diffusion controlled, is decreased. All these factors are in the direction of higher molecular weights, but the full extent of the predicted increase is not observed, since the rate of transfer to monomer also increases with pressure and controls the molecular weight at 3000 atm and above.

CHAPTER 6

Ionic and Co-ordination Chain (Addition) Polymerisation

INTRODUCTION

Ionic polymerisations follow the same basic steps as free radical chain growth polymerisation (i.e., initiation, propagation, and termination), however, there are some important differences which are also discussed in this chapter. Either a carbanion or carbonium ionic site can be formed in the initation process. Polymerisation of vinyl monomers with an electron withdrawing group can proceed by an anionic pathway, while monomers with an electron donating group (e.g., methyl) can polymerise by a cationic mechanism. In ionic polymerisation the active centres initiating the chain reactions are ions. Ionic polymerisation proceeds in the presence of catalysts and is called catalytic polymerisation. Depending on the charge of the ion formed, the polymerisation may be cationic or anionic.

SIMILARITIES AND CONTRASTS IN IONIC POLYMERISATION

The mechanisms of these polymerisations are less thoroughly understood than that of radical polymerisation, for several reasons. Reaction systems are often heterogeneous, involving inorganic catalysts and organic monomers. Usually large effects may be produced by a third component (co-catalyst) present in very low concentration. Polymerisation often leads to very-high-molecular-weight polymer at an extremely high rate, increasing enormously the difficulty in obtaining kinetic data or even reproducible results.

As indicted in Table 6.1, the type of monomer polymerising best by each mechanism is related to the polarity of the monomer and the acid-base strength of the ion formed. Monomers with electron-donating groups attached to the double-bonded carbons form stable carbenium ions and polymerise best with cationic catalysts. Conversely, monomers with electron-withdrawing substituents form stable anions and require anionic catalysts. Free-radical polymerisation may be considered an intermediate case that is favoured by moderate electron withdrawal from the double bond plus conjugation in the monomer. Many monomers can polymerise by more than one mechanism, including co-ordination polymerisation.

Initiation of ionic polymerisation usually involves the transfer of an ion or an electron to or from the monomer, with the formation of an ion pair. The counterion of this pair stays close to the growing chain end throughout its lifetime, particularly in media of low dielectric constant. Several possible structures, including bound or contact ion pairs, solvated or solvent-separated ion pairs, and free ions can sometimes be distinguished experimentally or may be in equilibrium.

In contrast to radical polymerisation, termination in ionic polymerisation never involves the reaction between two growing chains, but usually involves the unimolecular reaction of a chain with its counterion or a transfer reaction leaving a species too weak to propagate. An agent that retards radical polymerisation, such as oxygen, often has little influence on ionic polymerisation; but impurities that neutralise the catalysts prevent the reaction, and many aromatic, heterocyclic, olefinic, and acetylenic compounds retard or terminate ionic polymerisation. In addition to these differences, ionic polymerisation can usually be distinguished from free-radical polymerisation by its very high rate and low (often below 0°C) reaction temperature.

Table 6.1. Types of chain polymerisation suitable for common monomers.

	Polymerisation mechanism[a]			
Monomer type	*Radical*	*Cationic*	*Anionic*	*Co-ordination*
Ethylene	+	+	–	+
Propylene and α-olefins	–	–	–	+
Isobutylene	–	+	–	–
Dienes	+	–	+	+
Styrene	+	+	+	+
Vinyl chloride	+	–	–	+
Vinylidene chloride	+	–	+	–
Vinyl fluoride	+	–	–	–
Tetrafluoroethylene	+	–	–	+
Vinyl ethers	–	+	–	+
Vinyl esters	+	–	–	–
Acrylic and methacrylic esters	+	–	+	+
Acrylonitrile	+	–	+	+

[a]+ = high polymer formed; – = no reaction or oligomers only.

Kinetics

It is convenient to consider some common features of cationic and anionic polymerisation.

Formation of initiator

In many instances the species that initiates ionic polymerisation is produced by ionisation of the added initiator, for example, H^+ from $BF_3 . H_2O$ or $C_5H_{11}^-$ from potassium amyl. Sometimes the added substance reacts with the monomer or with the solvent to produce the initiating species, as in the reaction of a strong base with dimethyl sulphoxide to form the initiating anion $CH_3SOCH_2^-$.

Initiation

In the step of initiation the initiating species can react with the monomer in one of two ways: transfer of one electron without the formation of a bond between the two and transfer of two electrons with bond formation.

The transfer of a single electron from initiator to monomer always produces a radical ion, two of which generally dimerise to produce a diion. Three cases can be differentiated. The first is electron transfer from an electron donor to the monomer, for example, from the naphthalide ion (formed by the ionisation of sodium naphthalide) to styrene:

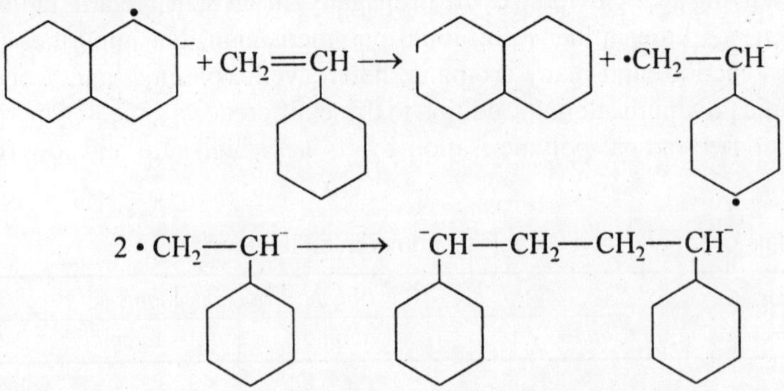

The second and third cases of single-electron transfer are, respectively, electron transfer from a monomer to an electron acceptor and the formation of a charge-transfer complex between a monomer with electron-donor or electron-acceptor properties and another molecule.

When two electrons are transferred, the monomer and the initiating species are bonded together. Again three cases can be distinguished:

1. A cation can add to a monomer in a nucleophilic reaction to form a monomer cation, for example, in the initiation of isobutylene polymerisation by boron trifluoride:

$$BF_2OH^-H^+ + CH_2 = C(CH_3)_2 \longrightarrow BF_3OH^-C^+\,(CH_3)_3$$

2. An anion can add to a monomer in an electrophilic reaction to form a monomer anion, as in the intiation of styrene polymerisation by potassium amyl:

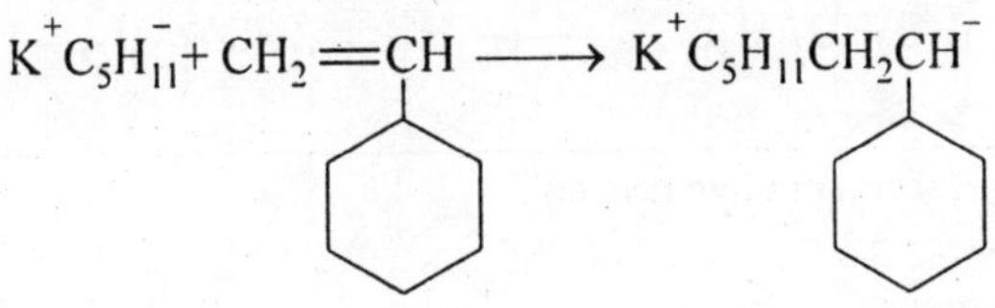

3. A neutral molecule and a monomer can add to produce a zwitterion, as in the case of initiation of polymerisation of a lactone with a tertiary amine:

$$R_3N + \overline{OCH_2CH_2CO} \longrightarrow R_3N^+CH_2CH_2COO^-$$

Propagation and termination

Whether the initiation step has led to a monoion or a diion, propagation takes place by the further addition of monomers to the growing chain, accompanied by its counterion. Termination involves the destruction of the growing anion or cation, possibly by recombination with the counterion or reaction with another species, but in many instances such a reaction does not take place and there is no termination. These cases, known as 'living polymers'.

CATIONIC, OR CARBONIUM POLYMERISATION

This polymerisation involves the formation of a carbonium ion, this being a polar compound with tricovalent carbon atom carrying a positive charge, $R-\overset{+}{C}H-R'$ Catalyst of carbonium polymerisation are compounds with pronounced electron-acceptor properties (aluminium chloride, stannic chloride, titanium tetrachloride, boron fluoride, etc.), the polymerising monomer being an electron donor (e.g., styrene in the presence of $SnCl_4$).

A carbonium ion interacts with a monomer molecule, the reaction of chain growth being accompanied with the communication of a positive charge along the chain. Consequently, the growing chain itself is a cation with a molecular mass increasing in the course of polymerisation. The pattern of the growing chain of polystyrene, being polymerised in the presence of $SnCl_4$, is given below:

$$[SnCl_4]^- CH_2-\underset{C_6H_5}{\underset{|}{CH}}-CH_2-\underset{C_6H_5}{\underset{|}{CH}}-\cdots-CH_2-\underset{C_6H_5}{\underset{|}{\overset{+}{C}H}}$$

Chain termination apparently occurs as a result of the mutual collision of the ends of a growing ion; in addition the catalyst is split off. Hence, in cationic polymerisation, the catalyst is not in the macromolecule. High rate of polymerisation at low temperatures is a characteristic of ionic polymerisations. It is often difficult to establish uniform reaction conditions before the reactants are consumed. The polymerisation of isobutylene by $AlCl_3$ or BF_3 takes place within a few seconds at –100°C, producing polymer of molecular weight up to several million. Both rate and molecular weight are much lower at room temperature. These considerations limit the industrial application of cationic polymerisation to the above-mentioned example.

Mechanism

As already discussed the most satisfactory theory of cationic polymerisation involves the carbenium ion as the chain carrier. For example, in the polymerisation of isobutylene with boron trifluoride catalyst, the first step is the reaction of the catalyst and co-catalyst, for example, water, to form a catalyst–co-catalyst complex that donates a proton to an isobutylene molecule to give a carbenium ion, $(CH_3)_3C^+$. This ion then reacts with monomer with the reformation of carbenium ion at the end of each step. The 'head-to-tail' addition of monomer to ion is the only one possible for energetic reasons. Since the reaction is in general carried out in a hydrocarbon medium of low dielectric constant, the anion and the growing cationic end form an ion pair. The termination reaction can take place by the rearrangement of the ion pair to yield a polymer molecule with terminal unsaturation, plus the original complex,

$$CH_3[CH_3)_2CCH_2]_xC(CH_3)_2^{+}(BF_3OH)^{-}$$
$$\downarrow$$
$$H^{+}(BF_3OH)^{-} + CH_3[(CH_3)_2CCH_2]_xC\begin{matrix} \nearrow CH_2 \\ \searrow CH_3 \end{matrix}$$

Here the catalyst–co-catalyst complex is regenerated, and many kinetic chains can be produced from each catalyst–co-catalyst species. An alternative termination mechanism, in which the catalyst or co-catalyst combines with growing chain, is unlikely in this example, but can take place when a covalent bond is formed, as in the polymerisation of styrene with trifluoroacetic acid:

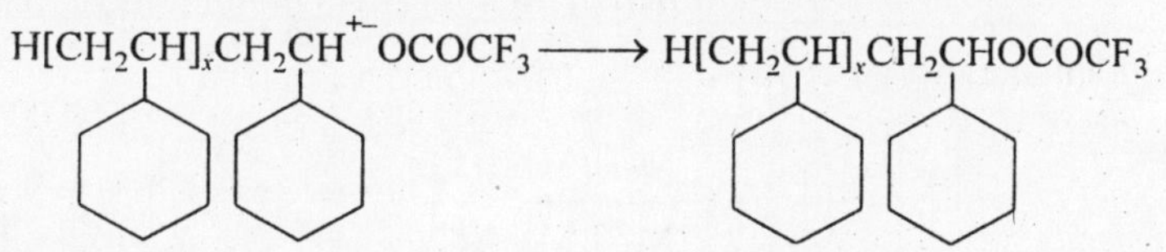

Chain transfer to monomer can also take place, and chain transfer to polymer is known and leads to branched polymers. The efficiency of the catalyst is dependent on the acid strength of the catalyst–co-catalyst complex. Also, the efficiency of the catalyst as a terminator should be related to the base strength of the complex anion. Thus the more active a molecule is as a catalyst, the less active it is as a terminator. Although the requirements of the isobutylene-BF_3 system are fully satisfied by the mechanism just given, generalisation of the scheme should be applied with caution. The experimental data on other systems, for example, do not conclusively prove the necessity of a co-catalyst.

Kinetics

Although many cationic polymerisations proceed so rapidly that it is difficult to establish the steady-state, the following kinetic scheme appears to be valid. Let the catalyst be designated by A and the co-catalyst by RH. Initiation, propagation, termination, and transfer may be represented as follows:

$$\begin{aligned} A + RH &\overset{k}{\rightleftharpoons} H^+{}^-AR \\ H^+{}^-AR + M &\xrightarrow{k_i} HM^+{}^-AR \\ HM_x{}^+{}^-AR + M &\xrightarrow{k_p} HM_{x+1}{}^+{}^-AR \\ HM_x{}^+{}^-AR &\xrightarrow{k_t} M_x + H^+{}^-AR \\ HM_x{}^+{}^-AR + M &\xrightarrow{k_{tr}} M_x + HM^+{}^-AR \end{aligned} \quad \text{... (6.1)}$$

The rate of initiation is

$$\upsilon_i = Kk_i\,[A][RH][M] \quad \text{... (6.2)}$$

where [A] is the catalyst concentration. (If the formation of $H^+{}^-AR$ is the rate-determining step, v_i may be independent of [M] and the kinetic scheme should be appropriately modified).

Since, in strong contrast to radical polymerisation, termination is first order,

$$\upsilon_t = k_t[M^+] \quad \text{... (6.3)}$$

where $[M^+]$ is written as an abbreviation for $[HM^+{}^-AR]$. By the steady-state assumption (which appears to hold in most cases, despite uncertainty as to whether such a state is actually achieved),

$$[M^+] = \frac{Kk_i}{k_t}[A][RH][M] \quad \text{... (6.4)}$$

The overall polymerisation rate is

$$\upsilon_p = k_p[M][M^+] = K\frac{k_i k_p}{k_t}[A][RH][M]^2 \quad \text{... (6.5)}$$

If termination predominates over transfer,

$$\bar{x}_n = \frac{\upsilon_p}{\upsilon_t} = \frac{k_p[M^+][M]}{k_t[M^+]} = \frac{k_p}{k_t}[M] \quad \text{... (6.6)}$$

whereas if transfer predominates,

$$\bar{x}_n = \frac{\upsilon_p}{\upsilon_{tr}} = \frac{k_p[M^+][M]}{k_{tr}[M^+][M]} = \frac{k_p}{k_{tr}} \quad \text{... (6.7)}$$

Since υ_p is proportional to $k_i k_p/k_t$, and, if termination predominates, $\bar{x}_n$ is proportional to k_p/k_t, the overall rate of polymerisation increases with decreasing temperature if $E_t > E_i + E_p$, and the molecular weight increases as the temperature of reaction decreases, if $E_t > E_p$. Since the propagation step involves the approach of an ion to a neutral molecule in a medium of low dielectric constant, no activation energy is necessary. The termination step, on the other hand, requires the rearrangement of an ion pair and thus involves an appreciable activation energy. These conditions are exactly the reverse of those found in free-radical polymerisation.

The dielectric constant of the medium has a significant effect on the nature of the ion pair, which can range between the extremes of complete covalency and completely free ions. It is often convenient to consider two types of propagating species, an ion pair and a free ion, in equilibrium. The nature of the solvent and the counterion can alter the course of the polymerisation greatly by shifting this equilibrium. Large increases in υ_p and $\bar{x}_n$ are usually observed as the solvating power of the medium is increased: The concentration of the free ion, which propagates more rapidly, is increased at the same time. Similarly, the larger and the less tightly bound the counterion, the more rapid is the propagation.

ANIONIC, OR CARBANION POLYMERISATION

This polymerisation involves the formation of a carbanion, a compound with a trivalent carbon atom carrying a negative charge. Anionic polymerisation occurs in the presence of catalysts which yield electrons readily, i.e., electron donors; examples are sodium or potassium amide, triphenylmethylsodium, alkali metals, alkylalkalies, etc. Polymerisations occur by the carbanion mechanism in the case of monomers containing electronegative substituents at one of the carbon atoms connected by a double bond; such monomers are acrylonitrile, methyl methacrylate, etc.

Potassium amide dissociated in a liquid ammonia medium:

$$KNH_2 \rightleftharpoons K^+ + NH_2^-$$

and the anion NH_2^- is added to a monomer molecule. The chain growth reaction is always accompanied with a transfer of a negative charge along the chain; consequently, the growing chain is always an anion of growing size. Chain termination occurs as a result of the collision of a growing ion with a molecule of the medium, such as an ammonia molecule:

$$(NH_2)-\left(CH_2-\underset{R}{\underset{|}{CH}}\right)_n-CH_2-\bar{C}HR + NH_3 \longrightarrow NH_2^- +$$

$$+ NH_2-\left(CH_2-\underset{R}{\underset{|}{CH}}\right)_n-CH_2-CH_2-R$$

Thus, every growing chain contains the NH_2 group.

Ionic polymerisation may occur in the presence of more than one catalyst (co-catalysts), e.g., in the presence of alkyl metals and chlorides of metals of variable valency ($TiCl_3$, $TiCl_4$).

Anionic polymerisation has great advantages, since spontaneous chain termination does not occur. Monomolecular polymers of a high molecular mass may be produced upon polymerisation in the presence of metalorganic compounds and alkali metals and in the absence of impurities which are capable of causing chain termination. Anionic polymerisation is an up-to-date method of producing monomolecular polymers. In anionic polymerisation, the end groups of a growing macromolecule possesses high activity and great stability. Therefore, the polymers produced by this method retain active centres at the end of the chain; later, these centres are capable of initiating the polymerisation of monomers. Such polymers are known as 'living polymers'. The process of polymerisation occurs when a new portion of monomer is added to such a living polymer. Such a method is used to produce, for example, block co-polymers.

Mechanism

The conventional method of initiation of ionic chain involves the addition of a negative ion to the monomer, with the opening of a bond or ring and growth at one end:

$$NH_2^- + CH_2{=}CHX \longrightarrow H_2NCH_2CHX^-$$

$$CH_3O^- + \underbrace{CH_2CH_2}_{O} \longrightarrow CH_3OCH_2CH_2O^-$$

(In the equations of this section the counterions are omitted; however, their role is essentially the same as in cationic polymerisation). Simultaneous growth from more than one centre can be obtained from polyvalent ions such as those derived from

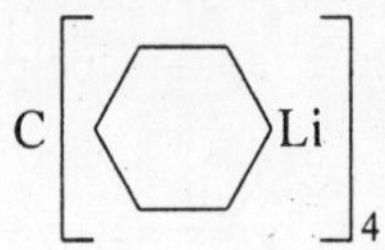 or $N(CH_2CH_2O^-Na^+)_3$

The more basic the ion, the better it serves to initiate chains. Thus OH^- will not initiate the anionic polymerisation of styrene, NH_2^- initiates fairly well, but ⬡ CH_2^- is powerful. Similarly, more acid monomers require less basic ions, with the acidity of the monomer depending on the strength of the X–M⁻ bond formed in initiation and the stability of the resulting ion.

Initiation can also occur by the transfer of an electron to a monomer of high electron affinity. If D or D^- is an electron donor,

$$D + M \longrightarrow D^+ + M^-$$

or

$$D^- + M \longrightarrow D + M^-$$

Presumably, M^- is less reactive than a true carbanion or a free radical, but the addition of a monomer to M^- gives a species that contains one radical end and one anion end:

$$CH_2 = \bar{C}HX + CH_2 = CHX \longrightarrow \underset{\text{anion end}}{:\bar{C}H_2CHX}\underset{\text{radical end}}{CH_2\dot{C}HX}$$

This species can add monomer from the two ends by different mechanisms. Two radical ends may dimerise, however, leaving a divalent anion to propagate.

Propagation in anionic polymerisation may be conventional or may be more complex, as in the elimination of CO_2 *N*-carboxyanhydrides. In contrast to radical polymerisation, the β (unsubstituted) carbon atom at the end of the growing chain is the site of addition of the next monomer.

As in cationic polymerisation, termination is always unimolecular, usually by transfer. The recombination of a chain with its counterion or the transfer of a hydrogen to give terminal unsaturation, frequent in cationic polymerisation, is unlikely in anionic mechanisms, as may be recognised by considering the small likelihood of transferring H^- when the counterion is Na^+. Thus in anionic polymerisation termination usually involves transfer, and the kinetic chain is broken only if the new species is too weak to propagate.

This leads to the unique situation in which, by careful purification to eliminate all species to which transfer can occur, the termination step is effectively eliminated and the growing chains remain active indefinitely. This case is described below under 'Living' polymers.

Kinetics

The kinetics of anionic polymerisation may be illustrated by the polymerisation of styrene with potassium amide in liquid ammonia.

$$KNH_2 \xrightarrow{k} K^+ + NH_2^-$$

$$NH_2^- + M \xrightarrow{k_i} NH_2M^- \qquad \text{... (6.8)}$$

$$NH_2M_x^- + M \xrightarrow{k_p} NH_2M_{x+1}^-$$

$$NH_2M_x^- + NH_3 \xrightarrow{k_{tr}} NH_2M_x + NH_2^-$$

The usual kinetic analysis leads for high degree of polymerisation to

$$\upsilon_p = \frac{Kk_pk_i}{k_{tr}}\frac{[NH_2^-]}{[NH_3]}[M]^2 \qquad \text{... (6.9)}$$

and

$$\bar{x}_n = \frac{k_p[M]}{k_{tr}[NH_3]} \qquad \text{... (6.10)}$$

Living Polymers

Since the termination step usually involves transfer to some species not essential to the reaction, anionic polymerisation with carefully purified reagents may lead to systems in which termination is lacking. The resulting species, called 'living' polymers, can be prepared, for example, by polymerising styrene with sodium naphthalene.

Kinetic analysis shows that the polymer can have an extremely narrow distribution of molecular weight and for all practical purposes be essentially monodisperse; if initiation is rapid compared to propagation, the molecular-weight distribution is the *Poisson* function for which $\bar{x}_w / \bar{x}_n \simeq 1 + 1/\bar{x}_n$. The polymer can be 'killed' by addition of a terminating agent, for example, water, at the end of the reaction.

The living polymer technique provides an unique opportunity for the preparation of block co-polymers of precisely defined composition.

CO-ORDINATION POLYMERISATION

Co-ordination polymerisation is far from simple in terms of mechanism, kinetics, or application. The most important catalysts are solids, leading to heterogeneous polymerisation systems, but soluble catalyst systems are known. Additional complications to the study of co-ordination system arise because the same catalysts, under other conditions, can initiate polymerisation by cationic, anionic, or even free-radical mechanisms.

Quite often co-ordination polymerisation is carried out using a catalyst in the form of a slurry of small solid particles in an inert medium (a fluid-bed process) or a supported solid catalyst. Heterogeneous polymerisation is not essential for co-ordination polymerisation and the development of stereospecificity, however, for sufficiently polar monomers.

Polymerisation reactions, especially of olefins and dienes, catalysed by organometallic compounds, fall under the category of co-ordination polymerisation.

The first step in the polymerisation process is the formation of a monomer-catalyst complex between the organometallic compound and the monomer:

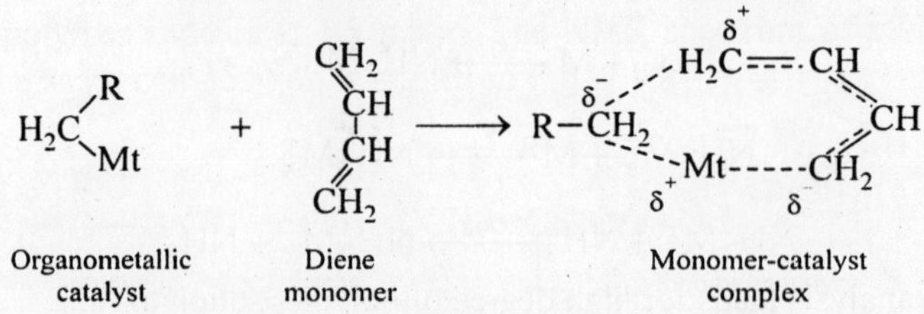

where Mt represents transition metals such as Ti, Mo, Cr, V, Ni or Rh.

In the formation of this monomer-catalyst complex, a co-ordination bond is involved between a carbon atom of the monomer and the metal of the catalyst. Hence, the polymerisation effected by such catalyst systems is called 'co-ordination polymerisation'.

The co-ordinated metal-carbon bond formed in the monomer-catalyst complex acts as the active centre from where propagation starts.

The incoming monomer is incorporated at the active centre of the metal-carbon bond, from where the chain growth starts:

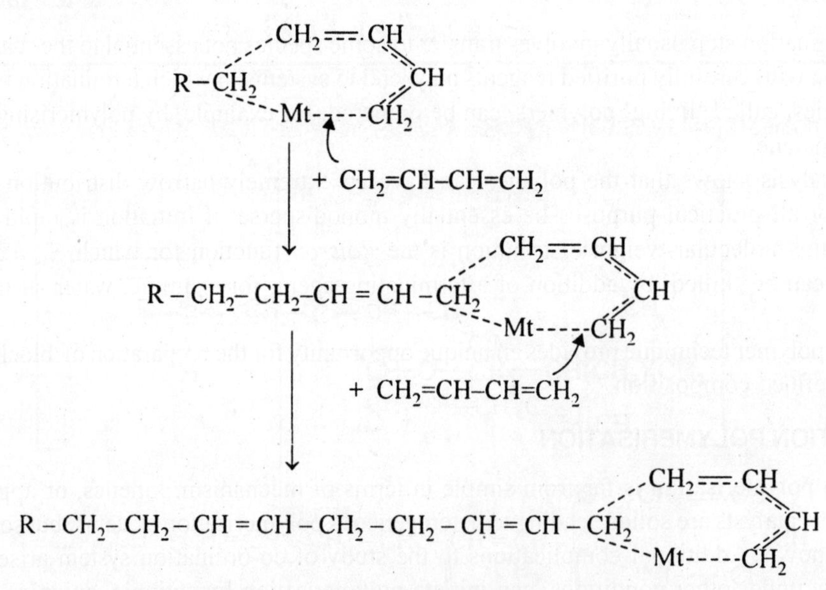

and so on . . . .

In co-ordination polymerisation, the catalyst-monomer complex is a heterogeneous system, with the metal ion in the solid phase and the carbanion of the alkyl group in the solvent phase. The monomer is inserted in between the metal ion and the carbanion with the result that the polymer chain formed is pushed out from the solid catalyst surface. For this reason, co-ordination polymerisation is also known as insertion polymerisation.

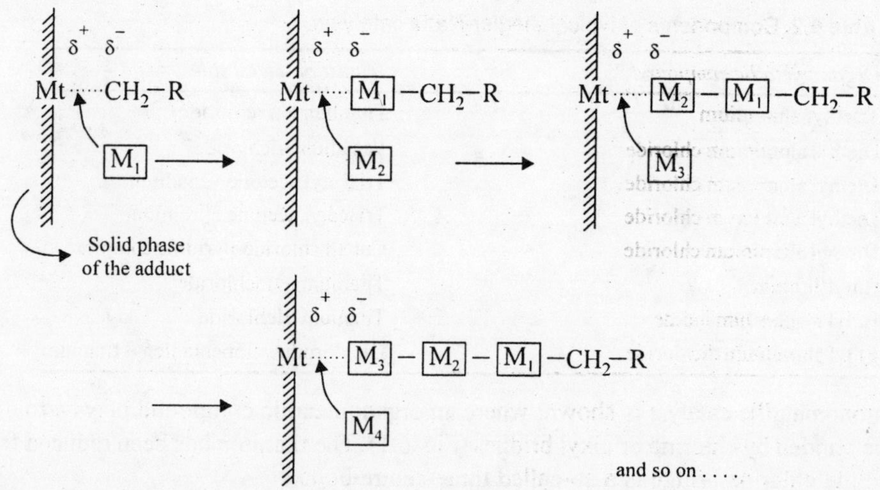

Here, M_1, M_2, M_3, and so on are the first, second, third (and so on) monomer units added to the growing chain. Another interesting feature of co-ordination polymerisation is that, depending on the polarity of the metal-carbon bond and on that of the solvent medium, the metal counter ion is placed in a particular spatial arrangement with respect to the anion.

In certain catalyst systems, therefore, the spatial arrangement has a tremendous effect on the spatial orientation of the incoming monomer and also on the manner in which the monomer is inserted into the growing chain. The specific spatial arrangement of the monomeric unit inserted into the growing chain imparts stereo-regularity to the polymer formed. By choosing a proper catalyst/solvent system, the co-ordination mechanism can be used to formulate highly stereo-regular polymers. The well-known Ziegler-Natta catalysts belong to this category of the polymerisation system.

Ziegler-Natta Catalysts

The Ziegler-Natta or, often, Ziegler catalysts have the remarkable property of polymerising a wide variety of monomers to linear and stereoregular polymers. Ethylene is polymerised to a highly linear chain, in contrast to the products of radical polymerisation. Polypropylene may be made in either the isotactic or syndiotactic form, but higher α-olefins yield only isotactic polymers. Dienes such as butadiene and isoprene may be polymerised to products that are almost exclusively *cis*-1,4, *trans*-1,2, isotactic-1,2, or syndiotactic-1,2, depending on choice of catalyst and conditions. Many other examples exist.

Ziegler catalysts are complexes formed by the interaction of alkyls of metals of groups I–III in the periodic table with halides and other derivatives of transition metals of groups IV–VIII. Some typical compositions are listed in Table 6.2.

There are literally hundreds of combinations that are active in polymerisation. Current thinking is that the transition metal, with its *d* orbitals available for both σ and π bonding, is the catalyst centre. A co-ordination complex is formed with the organometallic part of the system supplying some of the ligands. At least one empty co-ordination site accommodates the growing polymer chain; this is indicated here by the square box:

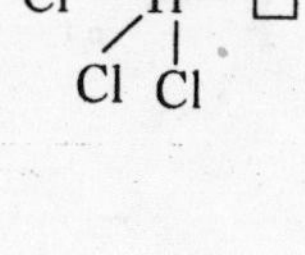

Table 6.2. Components of typical Ziegler-Natta catalysts.

Organometallic compound	*Transition-metal salt*
Triethyl aluminium	Titanium tetrachloride
Diethyl aluminium chloride	Vanadium trichloride
Diethyl aluminium chloride	Triacetyl acetone vanadium
Diethyl aluminium chloride	Triacetyl acetone chromium
Diethyl aluminium chloride	Cobalt chloride-pyridine complex
Butyl lithium	Titanium tetrachloride
Butyl magnesium iodide	Titanium trichloride
Ethyl aluminium dichloride	Dichlorodicyclopentadienyl titanium

Here a monometallic catalyst is shown; where an organometallic compound plays a role, the metal atoms may be bonded by chlorine or alkyl bridges (Fig. 6.1). The titanium has been reduced from Ti(IV) to Ti(III) and the chlorine bridge is a so-called three-centre bond.

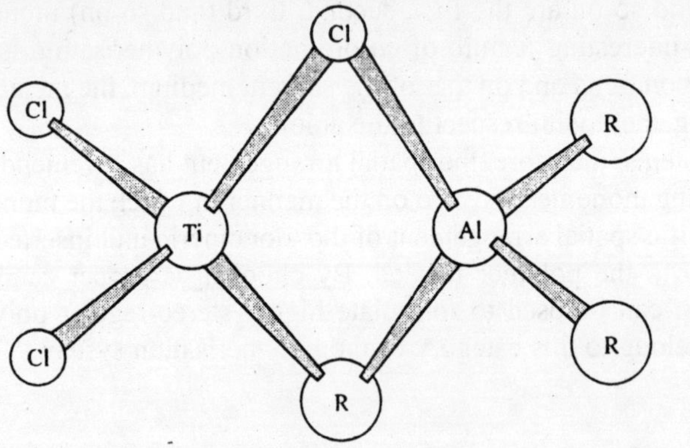

Fig. 6.1. Structure postulated for the metal halide-metal alkyl complex of a typical co-ordination catalyst.

Mechanisms

Because mechanistic studies are so much easier when carried out in a homogeneous system, much information has been obtained from experiments with soluble Ziegler-Natta catalysts. *Henrici-Olive* studied the system *bis*-cyclopentadienyl titanium (IV) dichloride (Cp_2TiCl_2), and an aluminium alkyl, usually dichloroethyl aluminium.

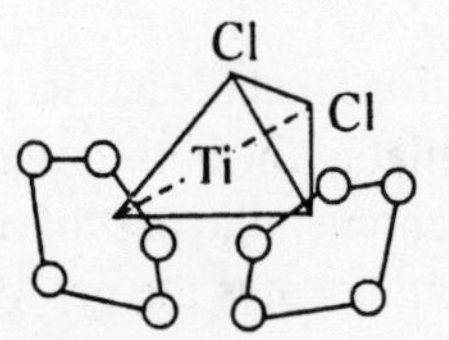

The active catalyst species was identified as

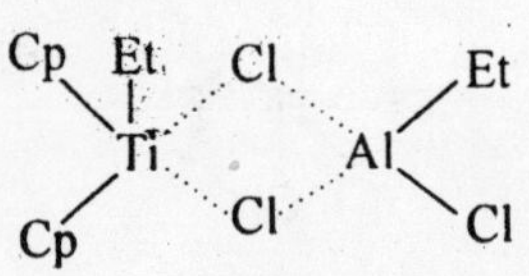

The transition metal was identified as the active site. The growth step in the polymerisation of ethylene by catalysts of this general type is presumed by most investigators to be that shown in Fig. 6.2. The ethylene is co-ordinated to the free site at the transition metal and inserted between the metal and the alkyl group R or, later, the growing chain. Both the course and the rate of the reaction can be influenced by the nature of the other ligands present.

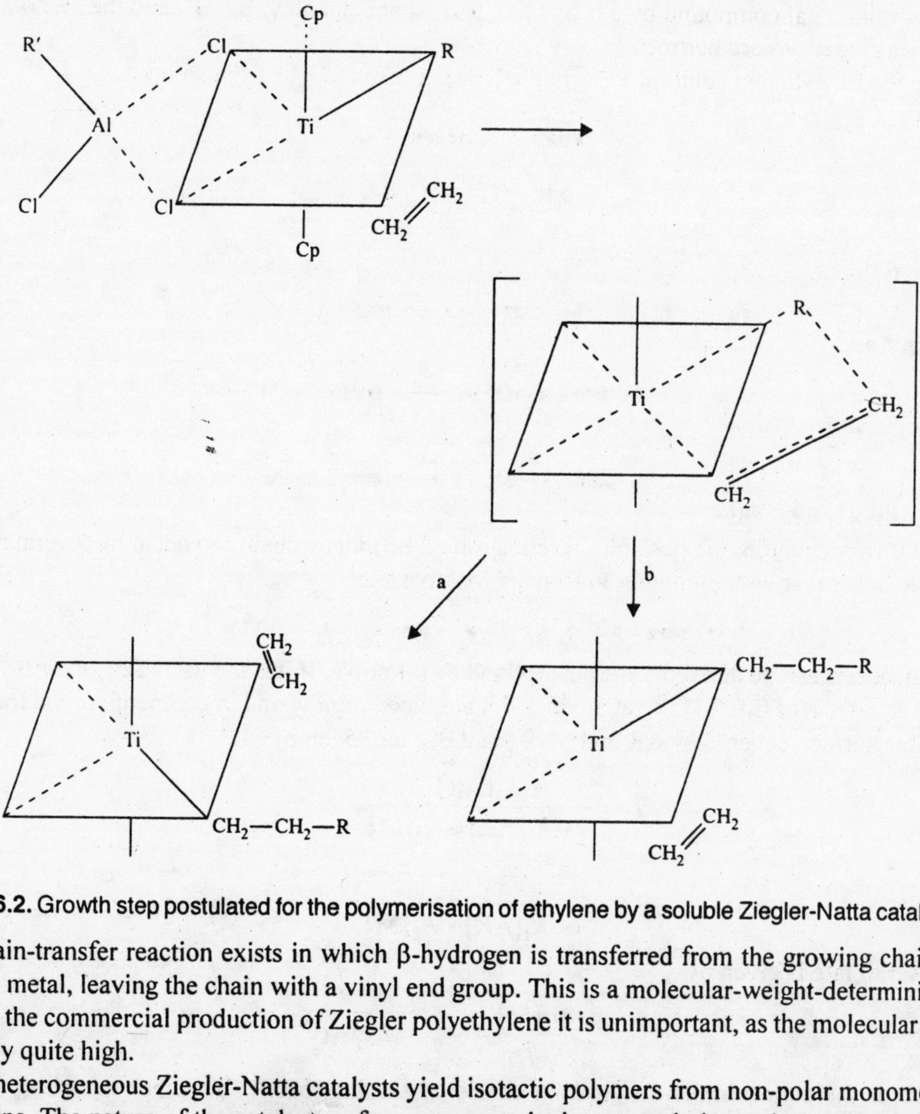

Fig. 6.2. Growth step postulated for the polymerisation of ethylene by a soluble Ziegler-Natta catalyst.

A chain-transfer reaction exists in which β-hydrogen is transferred from the growing chain to the transition metal, leaving the chain with a vinyl end group. This is a molecular-weight-determining step, though in the commercial production of Ziegler polyethylene it is unimportant, as the molecular weights are usually quite high.

Only heterogeneous Ziegler-Natta catalysts yield isotactic polymers from non-polar monomers such as α-olefins. The nature of the catalyst surface appears to be important in imparting stereoregularity in these cases. Propagation probably takes place at active sites on the crystal surface of the transition-metal compound, $TiCl_3$, for example. The sites are formed by the reaction of the group I–III soluble metalorganic compound, for example, $Al(C_2H_5)_3$ or $Al(C_2H_5)_2Cl$, with the crystal surface, yielding species of the type shown in Fig. 6.1. The chemical and crystal structures of the catalyst determine the orientation of the monomer as it is added to the growing chain. The driving force for isotactic propagation thus results from interaction between the monomer and the ligands of the transition metal at the active site. Several other models have been proposed, but none appear to be as widely applicable.

Kinetics

Considering here the important but complex case of heterogeneous polymerisation, we denote the surface of the transition-metal compound by a heavy bar, $\rule{1.5em}{0.6ex}$, the metal alkyl by AR, and the monomer by M. The following reactions are pertinent:

Adsorption of AR from solution to form the active site, and of M from solution,

$$\rule{1.5em}{0.6ex} + \mathrm{AR} \underset{}{\overset{k_1}{\rightleftharpoons}} \rule{1.5em}{0.6ex}\!-\!\mathrm{AR}$$

$$\rule{1.5em}{0.6ex} + \mathrm{M} \overset{k_2}{\rightleftharpoons} \mathrm{M}\!-\!\rule{1.5em}{0.6ex} \qquad \text{... (6.11)}$$

Initiation,

$$\mathrm{M}\!-\!\rule{1.5em}{0.6ex}\!-\!\mathrm{AR} \xrightarrow{k_i} \rule{1.5em}{0.6ex}\!-\!\mathrm{A}\!-\!\mathrm{M}\!-\!\mathrm{R} \qquad \text{... (6.12)}$$

Propagation,

$$\mathrm{M}\!-\!\rule{1.5em}{0.6ex}\!-\!\mathrm{A}\!-\!\mathrm{M}_x\!-\!\mathrm{R} \xrightarrow{k_p} \rule{1.5em}{0.6ex}\!-\!\mathrm{A}\!-\!\mathrm{M}_{x+1}\!-\!\mathrm{R} \qquad \text{... (6.13)}$$

Transfer with monomer,

$$\mathrm{M}\!-\!\rule{1.5em}{0.6ex}\!-\!\mathrm{A}\!-\!\mathrm{M}_x\!-\!\mathrm{R} \xrightarrow{k_{tr}} \rule{1.5em}{0.6ex}\!-\!\mathrm{A}\!-\!\mathrm{M}\!-\!\mathrm{R} + \mathrm{M}_x \qquad \text{... (6.14)}$$

where M_x is polymer.

Other transfer reactions are possible. Termination of the kinetic chain can occur by several reactions, an example being that with monomer to form an inactive site:

$$\mathrm{M}\!-\!\rule{1.5em}{0.6ex}\!-\!\mathrm{A}\!-\!\mathrm{M}_x\!-\!\mathrm{R} \xrightarrow{k_t} -\rule{1.5em}{0.6ex}\!-\!\mathrm{A}\!-\!\mathrm{M}\!- + \mathrm{M}_x\mathrm{R} \qquad \text{... (6.15)}$$

Rate equations can be derived if the adsorption law is known. If it is the usual *Langmuir* isotherm, in which the equilibria of Eq. 6.11 are attained and maintained rapidly and in competition, the fractions of the available surface covered by AR and M, θ_A and θ_M, are given by

$$\theta_A = \frac{K_1[\mathrm{AR}]}{1 + K_1[\mathrm{AR}] + K_2[\mathrm{M}]} \qquad \text{... (6.16)}$$

$$\theta_M = \frac{K_2[\mathrm{M}]}{1 + K_1[\mathrm{AR}] + K_2[\mathrm{M}]} \qquad \text{... (6.17)}$$

The overall rate is given by

$$\frac{-d[\mathrm{M}]}{dt} = (k_p + k_{tr})\theta_M[\mathrm{C}^*] \qquad \text{... (6.18)}$$

where [C*] is the concentration of growing chains. Assuming that [C*] reaches the steady-state, we can equate the equations for initiation and termination:

$$k_i\theta_A\theta_M = k_t[\mathrm{C}^*]\theta_M \qquad \text{... (6.19)}$$

from which, by elimination of [C*], we obtain

$$-\frac{d[\mathrm{M}]}{dt} = \frac{(k_p + k_{tr})k_i}{k_t} \frac{K_1K_2[\mathrm{AR}][\mathrm{M}]}{(1 + K_1[\mathrm{AR}] + K_2[\mathrm{M}])^2} \qquad \text{... (6.20)}$$

Similarly, an expression for $\bar{x}_n$ can be written. These equations are oversimplified for clarity by assuming only one transfer and one termination reaction. *Tait* has extended and confirmed these kinetics for the polymerisation of 4-methylpentane-1 by vanadium trichloride and triisobutyl aluminium.

Other Co-ordination Polymerisations

Polymerisation with supported metal oxide catalysts

Catalysts consisting of metals or metal oxides adsorbed on or complexed with the surface of a solid carrier are of commercial interest for the low-pressure polymerisation of olefins. Although often used in fixed-bed processes, the catalysts can be slurried and employed in fluid-bed operation like the Ziegler-Natta catalysts. Common catalyst compositions include oxides of chromium or molybdenum, or cobalt and nickel metals, supported on silica, alumina, titania, zirconia, or activated carbon.

The mechanisms and kinetics of the reactions are still largely unknown, although many proposals like those made for Ziegler-Natta polymerisation have been presented.

Alfin polymerisation

The 'Alfin' catalyst consists of a suspension, in an inert solvent like pentane, of a mixture of an alkylenyl sodium compound (such as allyl sodium), an alkoxide of secondary alcohol (such as sodium isopropoxide), and an alkali halide (such as sodium chloride). The catalyst is highly specific for the polymerisation of dienes into the 1,4-forms.

RING-OPENING POLYMERISATION

Polymers may be prepared by routes involving ring opening, which must be classed stoichiometrically as addition, since no small molecule is split off in the reaction. Some typical examples are given in Table 6.3. The polymerisation of these compounds has some aspects of both chain and step polymerisation as far as kinetics and mechanism are concerned. It resembles chain polymerisation in that it proceeds by the addition of monomer (but never of larger units) to growing chain molecules.

However, the chain-initiating and subsequent addition steps may be similar and proceed at similar rates; if so, these are not chain reactions in the kinetic sense. As in stepwise polymerisation, the polymer molecules continue to increase in molecular weight throughout the reaction.

Table 6.3. Typical ring-scission polymers.

Type	*Example*
Lactone	$\underbrace{O(CH_2)_x\,CO}$ $\longrightarrow$ $-[O(CH_2)_x\,CO-]_y$
Lactam	$\underbrace{HN(CH_2)_x\,CO}$ $\longrightarrow$ $-[HN\,(CH_2)_x\,CO-]_y$
Cyclic ether	$\underbrace{(CH_2)_xO}$ $\longrightarrow$ $-[(CH_2)_x\,O-]_y$
Cyclic anhydride	$\underbrace{CO(CH_2)_x\,CO}_{O}$ $\longrightarrow$ $-[CO(CH_2)_x\,COO-]_y$
N-Carboxyanhydride	$\underbrace{COCHRNHCO}_{O}$ $\longrightarrow$ $-[COCHRNH-]_y + CO_2$

Most of these cyclic compounds polymerise by ionic chain mechanisms in the presence of strong acids or bases when water and alcohols are excluded. These polymerisations are often very rapid. They are of commercial interest in the polymerisation of caprolactam, ethylene oxide, and 3,3-bis-(chloromethyl)oxetane. These are true chain reactions.

CHAPTER 7

Co-Polymerisation

INTRODUCTION

Co-polymerisation is a polymerisation involving two or more monomers of different structures. The properties of polymers synthesised by co-polymerisation can be varied over a wide range by adjusting the monomer ratio. Besides, some unsaturated compounds, which are incapable of separate polymerisation, polymerise readily together with other unsaturated compounds.

Co-polymerisation can be used to obtain polymers of linear, branched, and cross-linked structures. Thus, cross-linked polymers are formed if the molecule of one of the monomers contains two double bonds, even if the amount of monomer added is very small. Such monomers are called cross-linking agents. For example, if an insignificant amount of divinylbenzene

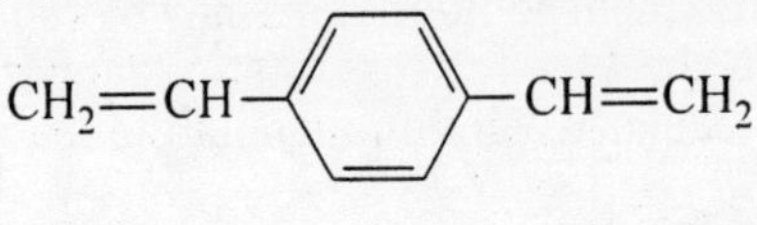

is added to styrene, they will co-polymerise to form an insoluble, non-melting co-polymer of cross-linked structure:

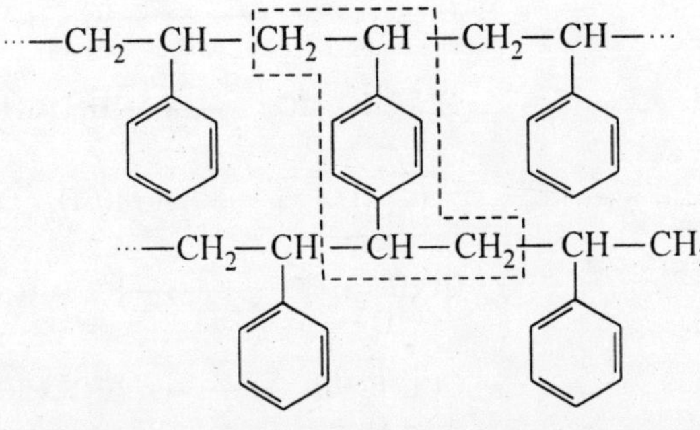

This material is used in practice for making ion exchange resins. However, the formation of cross-linking polymers during polymerisation may be undesirable. For instance, when preparing monovinylacetylene for chloroprene synthesis from acetylene, divinylacetylene $CH_2{=}CH{-}C{\equiv}C{-}CH{=}CH_2$ forms as a by-product. Its presence in chloroprene polymerisation causes the formation of a polychloroprene of cross-linked structure, which makes the polymer difficult to fabricate.

Therefore, to obtain high-quality polychloroprene the monovinylacetylene must be thoroughly freed from impurities. More recently, the ability to produce polymer containing long sequence of two or more different monomers (block and graft co-polymers) has led to new products with unique and valuable properties. The kinetic scheme for chain reaction co-polymerisation is developed for radical reactions; extension to ionic systems is straight-forward.

KINETICS OF CO-POLYMERISATION

Co-polymerisation involves the simultaneous chain-growth polymerisation of a mixture of two or more monomers. Aside from the general kinetic considerations which govern these chain reactions, as described earlier, there is imposed an additional feature, i.e., the relative participation of the different monomers during the growth of the chain. This new parameter is most important because it controls the composition of the co-polymer. Systems involving more than two monomers are difficult to resolve in this respect, but it has been found possible to treat the case of a pair of monomers with relative ease.

In the chain addition polymerisation of two monomers, regardless of the mechanism involved, the growing chain always must make a choice of reacting with one of the two monomers. Furthermore, there are two kinds of growing chains, depending on which type of monomer unit occupies the growing end. Thus, four types of propagation steps can be written as follows for any chain co-polymerisation of two monomers, assuming that the reactivity of the chain end depends only on the chain end monomer unit (terminal model).

$$M_1^* + M_1 \xrightarrow{k_{11}} M_1^* \qquad \text{... (7.1)}$$

$$M_1^* + M_2 \xrightarrow{k_{12}} M_2^* \qquad \text{... (7.2)}$$

$$M_2^* + M_2 \xrightarrow{k_{22}} M_2^* \qquad \text{... (7.3)}$$

$$M_2^* + M_1 \xrightarrow{k_{21}} M_1^* \qquad \text{... (7.4)}$$

where M_i^* and M_i refer to the growing chain and the monomer, respectively, as before, and the subscripts refer to the two kinds of monomer in the mixture. It can be seen that these four propagation reactions lead to four propagation rate constants, as shown. Hence the rate of consumption of each monomer may be expressed by the equations:

$$d[M_1]/dt = k_{11}[M_1^*][M_1] + k_{21}[M_2^*][M_1] \qquad \text{... (7.5)}$$

$$d[M_2]/dt = k_{12}[M_1^*][M_2] + k_{22}[M_2^*][M_2] \qquad \text{... (7.6)}$$

Because it is the relative rate of consumption of the two monomers which will decide the composition of the chain, it can be expressed by dividing Eq. 7.5 by Eq. 7.6:

$$\frac{d[M_1]}{d[M_2]} = \frac{k_{11}[M_1^*][M_1] + k_{21}[M_2^*][M_1]}{k_{12}[M_1^*][M_2] + k_{22}[M_2^*][M_2]} \qquad \text{... (7.7)}$$

It is obvious at once that Eq. 7.4 is quite intractable for direct use; however, it is possible to simplify it considerably by using the 'steady-state' treatment, analogous to the one previously described. This is done by assuming the rate of Eq. 7.2 to be equal to that of Eq. 7.4, and this leads to the equivalence

$$[M_2^*] = \left(\frac{k_{12}}{k_{21}}\right)[M_1^*]\frac{[M_2]}{[M_1]}$$

which, when inserted into Eq. 7.7 and after appropriate re-arrangements, yields

$$\frac{d[M_1]}{d[M_2]} = \frac{[M_1]}{[M_2]}\left[\frac{r_1[M_1]+[M_2]}{r_2[M_2]+[M_1]}\right] \quad \text{... (7.8)}$$

where $r_1 = k_{11}/k_{12}$ and $r_2 = k_{22}/k_{21}$. The parameters r_1 and r_2 are known as the monomer reactivity ratios, as they express the relative reactivity of each of the two kinds of growing chain ends with their 'own' monomer as compared with the 'other' monomer. They may in fact be considered as expressing the 'homo-polymerisation' tendency of each type of monomer relative to crossover with the co-monomer.

Equation 7.8, which relates the instantaneous composition of the co-polymer ($d[M_1]/d[M_2]$) to the prevailing monomer concentrations, can be used to determine the values of r_1 and r_2. Many such values have been recorded. Typical values of these parameters for styrene co-polymerisations are shown in Table 7.1, which illustrates the wide variations that prevail. The relative reactivity actually expresses the relative reactivity of each of the monomers shown toward the styrene radical compared with the reaction with styrene monomer. Thus, the r_1 and r_2 values permit some conclusions about the expected composition of the co-polymer obtained at any given monomer ratio.

For example, it can be deduced from Table 7.1 that a co-polymer of styrene and maleic anhydride ($r_1r_2 \Rightarrow 0$) would be strongly 'alternating' as it would be improbable to have a sequence of two maleic anhydride units. Also, it would obviously be extremely difficult to prepare a co-polymer of styrene and vinyl acetate, because the latter monomer would be virtually excluded from the styrene polymerisation.

Table 7.1. Monomer reactivity ratios for free radical co-polymerisations with styrene (M_1):

Monomer (M_2)	r_1	r_2	*Relative reactivity* ($1/r_1$)
Maleic anhydride	0.097	0.001	10.3
2,5-Dichlorostyrene	0.268	0.810	3.73
Methyl methacrylate	0.585	0.478	1.71
Methyl acrylate	0.871	0.148	1.15
Vinylidene chloride	1.839	0.087	0.54
Diethyl maleate	6.07	0.01	0.16
Vinyl acetate	18.8	0.02	0.05

It is also obvious from Eq. 7.8 that the co-polymer composition would not necessarily correspond to the co-monomer charge, depending on the values of r_1 and r_2. A desirable system would, of course, be one in which this were the case, i.e., where the co-monomers enter into the co-polymer in the ratio of their concentrations:

$$\frac{d[M_1]}{d[M_2]} = \frac{[M_1]}{[M_2]} \quad \text{... (7.9)}$$

This is defined as an 'azeotropic' co-polymerisation, by analogy to the distillation of two miscible liquids. Equation 7.9, would apply under the conditions where

$$\frac{r_1[M_1]+[M_2]}{r_2[M_2]+[M_1]} = 1 \quad \text{... (7.10)}$$

and this would be valid, for example, where $r_1 = r_2 = 1$.

In that case, Eq. 7.9 could apply for all charge ratios. That is, the two types of growing chains show no particular preference for either of the two monomers; this is described as a random co-polymerisation. Also Eq. 7.10 would be valid when $r_1 = r_2$ and $[M_1] = [M_2]$; i.e., an azeotropic co-polymerisation would result only at equimolar charge ratios. In general, Eq. 7.10 is valid when

$$\frac{[M_1]}{[M_2]} = \frac{(r_2 - 1)}{(r_1 - 1)} \qquad \text{... (7.11)}$$

This means that any co-polymerisation will be of an azeotropic type at the particular co-monomer charge ratio indicated by Eq. 7.11; however, it also means that an azeotropic co-polymerisation is possible only when both r_1 and r_2 are either greater than 1 or less than 1.

It is important to emphasise that this kinetic treatment is valid for any chain polymerisation mechanisms, i.e., free radical, cationic, anionic, and co-ordination. In the case of the ionic mechanisms, however, the type of initiator used and the nature of the solvent medium may influence the r_1 and r_2 values. This is due to the fact that the growing chain end in ionic systems is generally associated with a counterion, so that the structure and reactivity of such chain ends can be expected to be affected by initiator and the solvent.

The kinetic scheme for chain-reaction co-polymerisation is developed for radical reactions; extension to ionic systems is straightforward.

Co-polymer Equation

Dostal made the first attack on the mechanism of co-polymerisation by assuming that the rate of addition of monomer to a growing free radical depends only on the nature of the end group on the radical chain. Thus monomers M_1 and M_2 lead to radicals of types $M_1\bullet$ and $M_2\bullet$. There are four possible ways in which monomer can add:

Reaction	Rate
$M_1. + M_1 \longrightarrow M_1^\bullet$	$k_{11}[M_1^\bullet][M_1]$
$M_1\bullet + M_2 \longrightarrow M_2^\bullet$	$k_{12}[M_1^\bullet][M_2]$
$M_2\bullet + M_1 \longrightarrow M_1^\bullet$	$k_{21}[M_2^\bullet][M_1]$
$M_2^\bullet + M_2 \longrightarrow M_2^\bullet$	$k_{22}[M_2^\bullet][M_2]$ ... (7.12)

The kinetics of co-polymerisation was more fully elucidated by *Alfrey*, *Mayo*, *Simha*, and *Wall*. To Dostal's reaction scheme they added the assumption of the steady-state applied to each radical type separately, that is, the concentrations of $M_1\bullet$ and $M_2\bullet$ must each remain constant. It follows that the rate of conversion of $M_1\bullet$ to $M_2\bullet$ must equal that of conversion of $M_2\bullet$ to $M_1\bullet$:

$$k_{21}[M_2\bullet][M_1] = k_{12}[M_1\bullet][M_2] \qquad \text{... (7.13)}$$

The rates of disappearance of the two types of monomer are given by

$$-\frac{d[M_1]}{dt} = k_{11}[M_1\bullet][M_1] + k_{21}[M_2\bullet][M_1] \qquad \text{... (7.14)}$$

$$-\frac{d[M_2]}{dt} = k_{12}[M_1\bullet][M_2] + k_{22}[M_2\bullet][M_2]$$

By defining $r_1 = k_{11}/k_{12}$ and $r_2 = k_{22}/k_{21}$ and combining Eqs. 7.13 and 7.14, it can be shown that the composition of co-polymer being formed at any instant is given by

$$\frac{d[M_1]}{d[M_2]}=\frac{[M_1]}{[M_2]}\frac{r_1[M_1]+[M_2]}{[M_1]+r_2[M_2]} \qquad \text{... (7.15)}$$

This is known as the co-polymer equation; it has been verified by many experimental investigations of co-polymer composition.

Monomer Reactivity Ratios

The monomer reactivity ratios r_1 and r_2 are the ratios of the rate constant for a given radical adding its own monomer to the rate constant for its adding the other monomer. Thus $r_1 > 1$ means that the radical $M_1\bullet$ prefers to add M_1; $r_1 < 1$ means that it prefers to add M_2. In the system styrene (M_1)–methyl methacrylate (M_2), for example, $r_1 = 0.52$ and $r_2 = 0.46$; each radical adds the other monomer about twice as fast as its own.

Since the rate constants for initiation and termination do not appear in Eq. 7.15, the composition of the co-polymer is independent of overall reaction rate and initiator concentration. The reactivity ratios are unaffected in most cases by the presence of inhibitors, chain transfer agents, or solvents. Even in heterogeneous systems they remain unchanged, unless the availability of the monomers is altered by their distribution between phases. A change from a free radical to an ionic mechanism, however, changes r_1 and r_2 markedly.

A few typical values of monomer reactivity ratios are given in Table 7.2.

Table 7.2. Typical monomer reactivity ratios.

Monomer 1	*Monomer 2*	r_1	r_2	*T(°C)*
Acrylonitrile	1,3-Butadiene	0.02	0.3	40
	Methyl methacrylate	0.15	1.22	80
	Styrene	0.04	0.40	60
	Vinyl acetate	4.2	0.05	50
	Vinyl chloride	2.7	0.04	60
1,3-Butadiene	Methyl methacrylate	0.75	0.25	90
	Styrene	1.35	0.58	50
	Vinyl chloride	8.8	0.035	50
Methyl methacrylate	Styrene	0.46	0.52	60
	Vinyl acetate	20	0.015	60
	Vinyl chloride	10	0.1	68
Styrene	Vinyl acetate	55	0.01	60
	Vinyl chloride	17	0.02	60
Vinyl acetate	Vinyl chloride	0.23	1.68	60

Types of Co-polymerisation

Ideal

A co-polymer system is said to be ideal when the two radicals shows the same preference for adding one of the monomers over the other: $k_{11}/k_{12} = k_{21}/k_{22}$, or $r_1 = 1/r_2$, or $r_1r_2 = 1$. In this case the end group on the growing chain has no influence on the rate of addition, and the two types of units are arranged at random along the chain in relative amounts determined by the composition of the feed and the relative reactivities of the two monomers. The co-polymer equation reduces to $d[M_1]/d[M_2] = r_1[M_1]/[M_2]$.

Alternating

Here each radical prefers to react exclusively with the other monomer: $r_1 = r_2 = 0$. The monomers alternate regularly along the chain, regardless of the composition of the monomer feed. The co-polymer equation simplifies to $d[M_1]/d[M_2] = 1$.

Most actual cases lie between the ideal and the alternating systems: $0 < r_1r_2 < 1$. A third possibility, with both r_1 and r_2 greater than unity, corresponds to the tendency to form block co-polymers.

INSTANTANEOUS COMPOSITION OF FEED AND POLYMER

Let F_1 and F_2 be the mole fractions of monomers 1 and 2 in the polymer being formed at any instant:

$$F_1 = 1 - F_2 = \frac{d[M_1]}{d([M_1]+[M_2])} \quad \text{... (7.16)}$$

If f_1 and f_2 similarly represent mole fractions in the monomer feed,

$$f_1 = 1 - f_2 = \frac{[M_1]}{[M_1]+[M_2]} \quad \text{... (7.17)}$$

the co-polymer equation can be written as

$$F_1 = \frac{r_1 f_1^2 + f_1 f_2}{r_1 f_1^2 + 2 f_1 f_2 + r_2 f_2^2} \quad \text{... (7.18)}$$

It is apparent that in general F_1 does not equal f_1, and both f_1 and F_1 change as the polymerisation proceeds. The polymer obtained over a finite range of conversion consists of a continuous distribution of varying composition.

Equation 7.18 may be used to calculate curves of feed versus instantaneous polymer composition for various monomer reactivity ratios. Such curves for a series of ideal co-polymerisation ($r_1r_2 = 1$) are shown in Fig. 7.1. Except for pairs of monomers having very similar reactivities, only a small range of feeds gives co-polymer containing appreciable amounts of both components.

Several curves for non-ideal cases are shown in Fig. 7.2. These curves show the effect of increasing tendency toward alternation. As alternation increases, more and more feeds yield a co-polymer containing a good deal of each component. This tendency makes practical the preparation of many important co-polymers.

If one of the monomers is very much more reactive than the other, the first polymer formed contains mostly the more reactive monomer. Later in the polymerisation this monomer is largely used up, and the last polymer formed consists mainly of the less reactive monomer. Styrene and vinyl acetate form such a system.

For cases in which both r_1 and r_2 are less (or greater) than unity, the curves of Fig. 7.2 cross the line representing $F_1 = f_1$. At the point of intersection polymerisation proceeds without a change in the composition of feed or polymer. This is known as azeotropic co-polymerisation.

Solution of Eq. 7.14 with $d[M_1]/d[M_2] = [M_1]/[M_2]$ gives the critical composition for the azeotrope:

$$\frac{[M_1]}{[M_2]} = \frac{1-r_2}{1-r_1} \quad \text{... (7.19a)}$$

or, from Eq. 7.18,

$$(f_1)_c = \frac{1-r_2}{2-r_1-r_2} \quad \text{... (7.19b)}$$

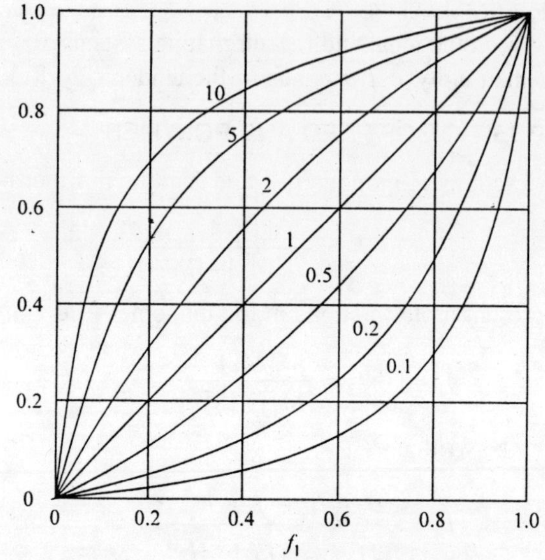

Fig. 7.1. Instantaneous composition of co-polymer (mole fraction F_1) as a function of monomer composition (mole fraction f_1) for ideal co-polymers with the values of $r_1 = 1/r_2$ indicated.

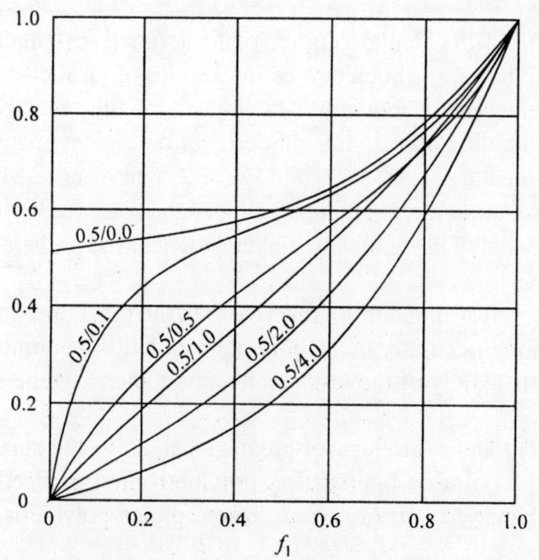

Fig. 7.2. Instantaneous composition of co-polymer F_1 as a function of monomer composition f_1 for the values of the reactivity ratios r_1/r_2 indicated.

Evaluation of Monomer Reactivity Ratios

The usual experimental determination of r_1 and r_2 involves polymerising to low conversion for a variety of feed compositions. The polymers are isolated and their compositions measured. One of several methods of analysis of the data may be used:

1. Direct curve fitting on polymer-monomer composition plots. This is a poor method, since the composition curve is rather insensitive to small changes in r_1 and r_2.
2. Plots of r_1 versus r_2. The co-polymer equation may be solved for one of the reactivity ratios:

$$r_2 = \frac{[M_1]}{[M_2]}\left[\frac{d[M_2]}{d[M_1]}\left(1+\frac{[M_1]}{[M_2]}r_1\right)-1\right] \quad \text{... (7.20)}$$

Each experiment with a given feed gives a straight line; the intersection of several of these allows the evaluation of r_1 and r_2 (Fig. 7.3). If the experimental errors are high, the lines may not intersect in a single point; the region within which the intersections occur give information about the precision of the experimental results.

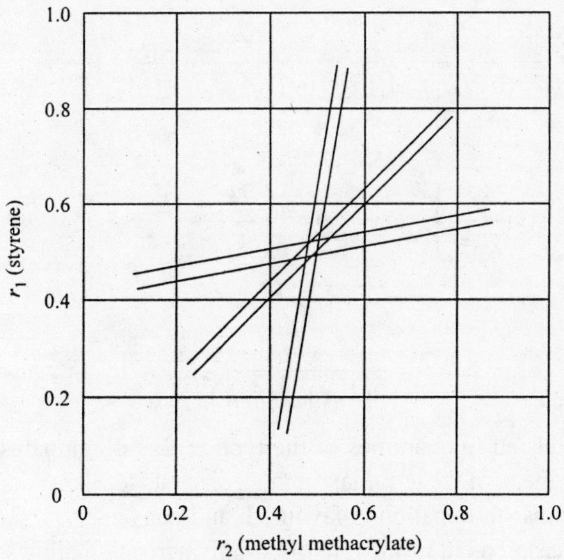

Fig. 7.3. Evaluation of monomer reactivity ratios by graphic solution of the co-polymer equation.

3. Plots of F_1 versus f_1. Equation 7.18 can be re-arranged to

$$\frac{f_1(1-2F_1)}{(1-f_1)F_1} = r_2 + \frac{f_1^2(F_1-1)}{(1-f_1)^2 F_1}r_1 \quad \text{... (7.21)}$$

   This is the equation of a straight line with slope r_1 and intercept r_2. Each experiment gives one point on the line. The least-squares treatment of a series of such points gives the best values of r_1 and r_2 in a straightforward way.
4. Analysis of data giving the co-polymer composition as a function of conversion, using the integrated form of the co-polymer equation.

Rate of Co-polymerisation

The discussion so far has considered only the relative rates of the four possible propagation steps in a binary system. The overall rate of co-polymerisation depends also on the rates of initiation and termination.

If the initiator releases radicals that have a high efficiency of combination with each monomer, the two types of initiation steps need not be considered separately. Steady-state conditions apply both to the total radical concentration and to the separate concentrations of the two radicals. Three types of termination must be considered, involving all possible pairs of radical types. The steady-state condition for all the radicals requires that the initiation and termination rates be equal:

$$\upsilon_i = 2k_{t_{11}}[M_1\bullet]^2 + 2k_{t_{12}}[M_1\bullet][M_2\bullet] + 2k_{t_{22}}[M_2\bullet]^2 \quad \text{... (7.22)}$$

Here $k_{t_{11}}$ and $k_{t_{12}}$ are the termination rate constants for pairs of like radicals, and $k_{t_{12}}$ is the constant for the cross-reaction. If this condition is applied to the equation for the overall polymerisation rate,

$$-\frac{d([M_1]+[M_2])}{dt} = \upsilon_p$$

$$= \frac{(r_1[M_1]^2 + 2[M_1][M_2] + r_2[M_2]^2)(\upsilon_i^{1/2} / \delta_1)}{\{r_1^2[M_1]^2 + 2(\phi r_1 r_2 \delta_2 / \delta_1)[M_1][M_2] + (r_2\delta_2/\delta_1)^2[M_2]^2\}^{1/2}} \quad \text{... (7.23)}$$

where

$$\delta_1 = \left(\frac{2k_{t_{11}}}{k_{11}^{\ 2}}\right)^{1/2}, \qquad \delta_2 = \left(\frac{2k_{t_{22}}}{k_{22}^{\ 2}}\right)^{1/2} \quad \text{... (7.24)}$$

and

$$\phi = \frac{k_{t_{12}}}{2(k_{t_{11}})^{1/2}(k_{t_{22}})^{1/2}} \quad \text{... (7.25)}$$

The δ values are simply related to quantities of the type $k_p^2/2k_t$; ϕ compares the cross-termination rate constant to the geometric mean of the termination rate constants for like pairs of radicals. A value of ϕ greater than 1 means that cross-termination is favoured, and conversely.

If r_1 and r_2 are known, measurement of the rate of co-polymerisation allows ϕ to be determined. A typical result is shown in Fig. 7.4. The value of ϕ greater than unity is typical of most co-polymerisations; indeed, cross-termination is markedly favoured, except for nearly ideal co-polymerisations. It is likely that differences in polarity are responsible for the preference for cross-termination, as well as for deviations from ideal co-polymerisation. As a result, the co-polymerisation of pairs of monomers of quite different polarities is likely to be slower than the mean of the rates for the separate homo-polymerisations. The above treatment assumed that the termination process is controlled by the chemical nature of the growing-chain radical ends. It is known, however, that termination in radical polymerisation is generally diffusion controlled. It is often useful to express the rate equation in terms of a single termination constant k_t.

$$\upsilon_p = \frac{(r_1[M_1]^2 + 2[M_1][M_2] + r_2[M_2]^2)\upsilon_i^{1/2}}{k_t^{1/2}(r_1[M_1]/k_{11} + r_2[M_2]/k_{22})} \quad \text{... (7.26)}$$

Ideally, k_t might be expected to be the mole-fraction-weighted average of the termination rate constants for the two homopolymerisations, but this appears to be an oversimplification, possibly due to penultimate effects.

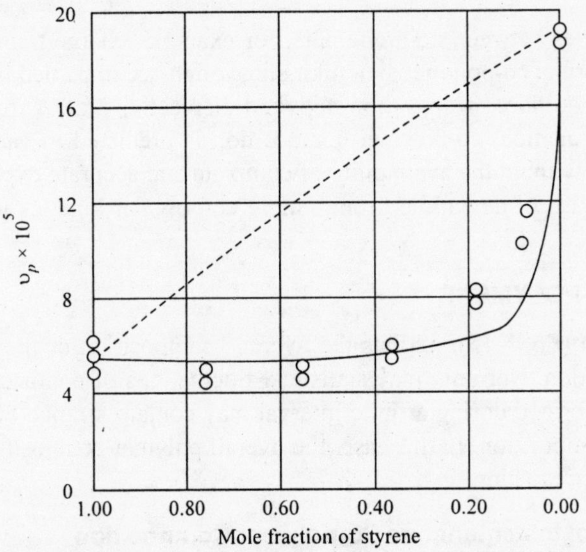

Fig. 7.4. Co-polymerisation rate data for the system styrene-methyl methacrylate at 60°C. The broken line assumes $\phi = 1$; the solid line is calculated for $\phi = 13$.

Extensions of Co-polymerisation Kinetics

Multicomponent co-polymerisation

The mathematics of the co-polymer equation has been extended to the cases of three or more monomers, but the quantitative treatment is quite complex; for example, three monomers require the specification of nine propagation and six termination reactions, and six reactivity ratios. Some simplification results from rather liberal assumptions about the steady-state and the probabilities of finding certain sequences of monomers in the resulting polymer. The predicted results are in good agreement with experiment.

Penultimate and remote-unit effects

Ham have developed the early suggestion of *Merz*, that units before the last may affect the reactivity of a chain radical. This is often the case, the effect being particularly pronounced with monomers of type CHX=CHX, where X is a polar substituent. It has been detected in co-polymers of styrene with maleic anhydride and fumaronitrile.

When the penultimate monomer unit does have some influence on the subsequent addition of monomer, four reactivity ratios, corresponding to the four possible types of radicals, must be considered. Effects of units still farther back along the chain have been detected experimentally in some cases.

Depropagation during co-polymerisation

If one or both of the monomers has a tendency to depropagate during a co-polymerisation, an additional complication is introduced. This deviation from the simple treatment can be detected by the dependence of the co-polymer composition on the absolute concentration of the monomers, even though their relative concentration is invariant.

Complex participation

The presence of complexes between co-monomers, for example, charge-transfer complexes also can cause deviations from normal co-polymerisation kinetics, often accompanied by an enhanced tendency toward alternation. Four additional propagation steps and six reactivity ratios are required in the analysis. As in the case of depropagation, co-polymer composition is predicted to vary with temperature and monomer concentration. Without the availability of complete and accurate experimental data it is often difficult to assign the cause of deviations from simple co-polymer kinetics to one or another of the above models.

COMPOSITION OF CO-POLYMERS

The co-polymer equation (Eq. 7.15) predicts the average composition of the polymer formed at any instant in the polymerisation. Not only may statistical fluctuations of composition about this average occur, but also polymer formed during a finite interval may contain a range of compositions resulting from variation of feed composition. In this case, the overall polymer composition can be calculated by integrating the co-polymer equation.

Statistical Fluctuation of Instantaneous Co-polymer Composition

Stockmayer showed that the spread in co-polymer composition due to statistical fluctuations is small in practical cases. Alternating co-polymers give the minimum spread in composition; even in ideal co-polymers, where the spread is greater and is a function of the degree of polymerisation, the effect is not large. For a co-polymer containing on the average 50 per cent of each monomer, only 12 per cent of the polymer molecules contain less than 43 per cent of either monomer at $\bar{x}_n = 100$, while only 12 per cent of the molecules contain less than 49.3 per cent of either monomer at $\bar{x}_n = 10{,}000$. The distribution of compositions follows a *Gaussian* curve. This small spread in composition is neglected in the discussion to follow.

The probability of occurrence of sequences of *n* similar monomer units can be calculated statistically for an ideal co-polymer. The number $N(n)$ of such sequences is

$$N(n) = p^{n-1}(1-p) \qquad \text{... (7.27)}$$

where p is the probability of addition of that monomer,

$$p = \frac{r_1[\mathrm{M}_1]}{r_1[\mathrm{M}_1]+[\mathrm{M}_2]} \qquad \text{... (7.28)}$$

Except for a monomer present in large excess, the sequences are always short. Any tendency toward alternation clearly leads to shorter sequences than those in an ideal co-polymer.

Integration of the Co-polymer Equation

The direct integration of the co-polymer equation gives a results that is not convenient for calculations. The most convenient method for calculating co-polymer composition and distribution involves the use of Eq. 7.18. For a system in which $F_1 > f_1$, when dM moles of monomer have polymerised, the polymer contains $F_1\ dM$ moles of monomer 1, and feed contains $(f_1 - df_1)(M - dM)$ moles of monomer 1. For a material balance,

$$Mf_1 - (M - dM)(f_1 - df_1) = F_1\, dM \qquad \text{... (7.29)}$$

Combining this with Eq. 7.18 gives

$$\frac{dM}{M} = \frac{df_1}{F_1 - f_1} \qquad \text{... (7.30)}$$

which can easily be integrated by numerical or graphic means to give the desired composition of polymer as a function of conversion. One convenient closed-form result of this integration is given below:

$$\log\left(\frac{M}{M_0}\right) = \frac{r_2}{1-r_2}\log\left(\frac{f_1}{(f_1)_0}\right) + \frac{r_1}{1-r_1}\log\left(\frac{f_2}{(f_2)_0}\right) + \frac{1-r_1r_2}{(1-r_1)(1-r_2)}\log\left(\frac{(f_1)_0 - \varepsilon}{f_1 - \varepsilon}\right) \qquad \text{... (7.31)}$$

where $\varepsilon = (1 - r_2)/(2 - r_1 + r_2)$.

Variation of Co-polymer Composition With Conversion

Ideal system

The system styrene (M_1)–2-vinylthiophene (M_2) is nearly ideal, with r_1=0.35 and r_2=3.10. The variation of the composition of polymer formed at any instant in this system with conversion and initial feed composition is shown in Fig. 7.5.

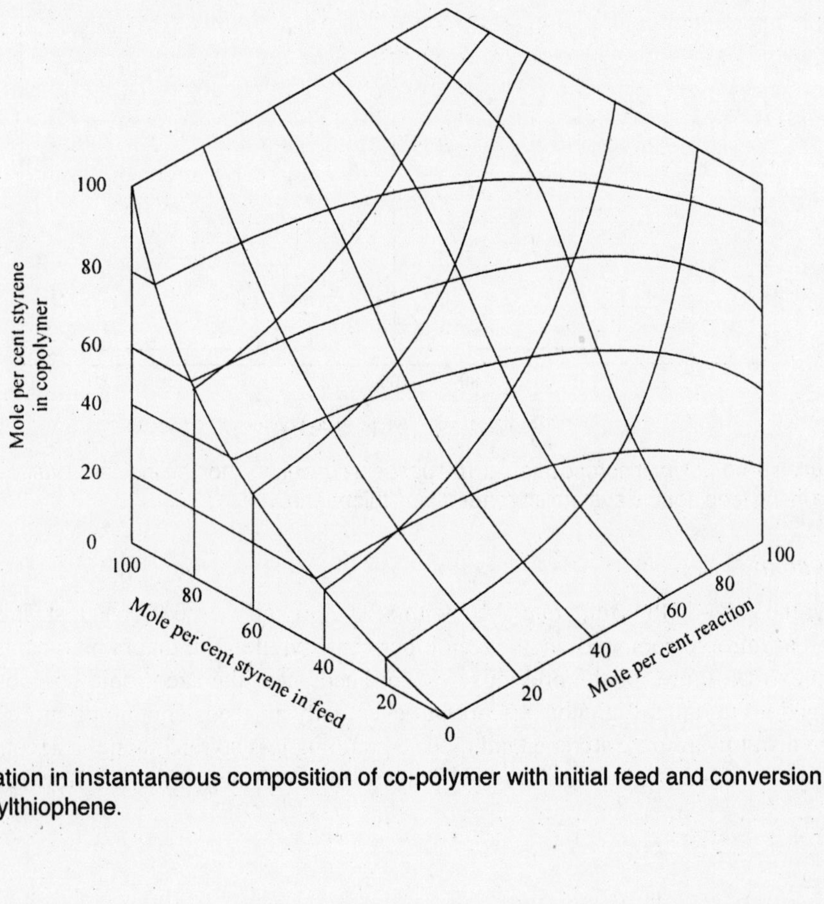

Fig. 7.5. Variation in instantaneous composition of co-polymer with initial feed and conversion for the system styrene–2-vinylthiophene.

The greater reactivity of the vinylthiophene causes it to be depleted until at the last only styrene remains, and pure polystyrene is the final product formed. The overall polymer composition varies less with conversion and ultimately approaches the initial feed composition as required for a material balance.

The polymer composition distribution for 100 per cent conversion for several initial feeds is shown in the bar graphs in Fig. 7.6. Each block represents a 5 per cent interval in co-polymer composition. The U-shaped distribution curves arise whenever the monomer reactivity ratios are less than 0.5 and greater than 2, respectively. They indicate that appreciable quantities of two distinct compositions predominate, with little material of intermediate composition.

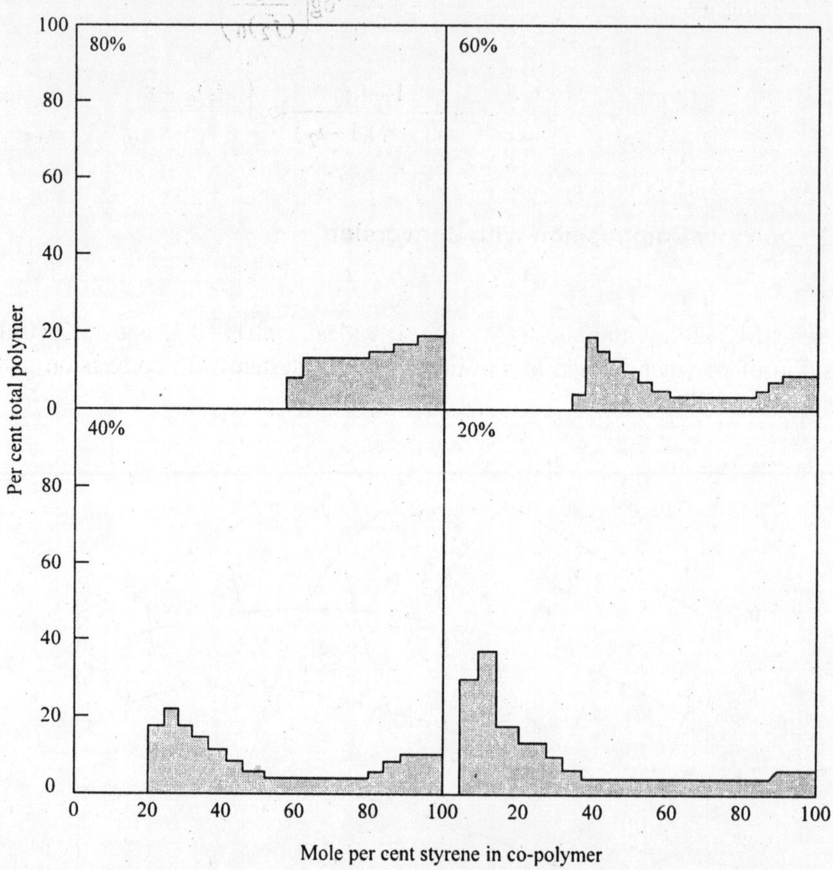

Fig. 7.6. Distribution of co-polymer composition at 100 per cent conversion, for the indicated values of initial mole per cent styrene in the feed, for the system styrene–2-vinylthiophene.

Alternating system

The system styrene (M_1)-diethyl fumarate (M_2) with $r_1 = 0.30$ and $r_2 = 0.07$ has a strong tendency to alternate and an azeotropic composition at 57 mol. per cent styrene. The diagrams for this co-polymer are more complex than those for the previous system. Feeds near the azeotropic composition remain almost unchanged up to rather high conversion, but eventually all feeds containing more than 57 mol. per cent styrene drift toward pure styrene, and those containing less styrene than the azeotropic amount yield pure diethyl fumarate. Thus (Fig. 7.7) pure polystyrene is formed at the last from all initial feeds higher in styrene than the azeotropic, and pure poly(diethyl fumarate) from all initial feeds containing less than 57 mol. per cent styrene. The overall polymer composition shows markedly less change with conversion for all feeds and is of course invariant at the azeotropic composition.

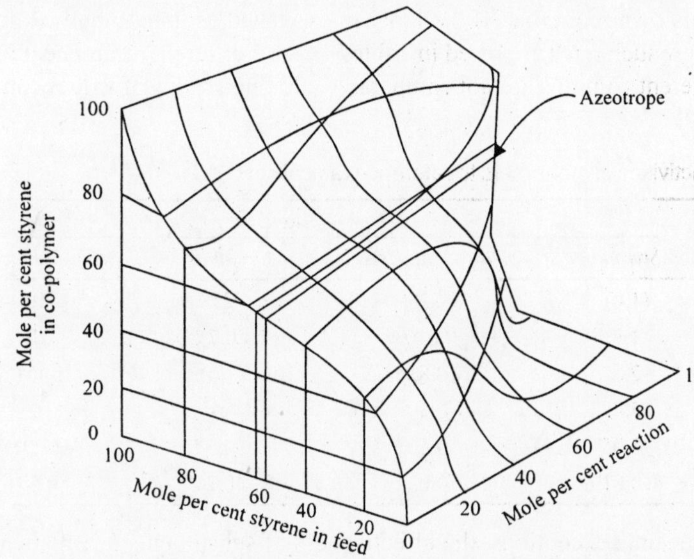

Fig. 7.7. Variation in instantaneous composition of co-polymer with initial feed and conversion for the system styrene-diethyl fumarate.

Attainment of homogeneity in co-polymers

It is often desirable to produce a polymer that is homogeneous in composition rather than one having a broad distribution of compositions. The foregoing discussion suggests two ways of doing this: (i) the reaction can be stopped short of complete conversion. For example, the styrene-diethyl fumarate co-polymer from feed containing 40 per cent styrene, when stopped at 75 per cent conversion, contains only polymer having between 44 and 52 per cent styrene; and (ii) the polymerisation can be started with a feed giving the desired polymer and the feed composition maintained by adding an increment of more reactive monomer as the polymerisation proceeds. Both methods find commercial use.

MECHANISMS OF CO-POLYMERISATION

Free-Radical Co-polymerisation

The order of reactivity of monomers toward free radicals not only is a function of the reactivity of the monomers, but also depends on the nature of the attacking radical. This is illustrated by the tendency of many monomers to alternate in a co-polymer chain. The two factors, general reactivity and alternating tendency, are predominant in determining the behaviour of monomers in co-polymerisation.

The reactivity of monomers and radicals in co-polymerisation is determined by the nature of the substituents on the double bond of the monomer. These substituents influence reactivity in three ways: They may activate the double bond, making the monomer more reactive; they may stabilise the resulting radical by resonance; or they may provide steric hindrance at the reaction site.

Reactivity of monomers

The relative reactivities of monomers to a reference radical can be derived from the monomer reactivity ratio. The inverse of this ratio is the rate of reaction of the reference radical with another monomer, relative to that with its own monomer. If the latter rate is taken as unity, relative monomer reactivities can be examined. A few such rates are listed in Table 7.3. As a different reference point is taken for each radical, values in different columns are not comparable, although ratios of values can be compared from column to column.

Table 7.3. Relative reactivities of monomers to reference radicals at 60°C.

	Reference radical				
Monomer	*Styrene*	*Methyl methacrylate*	*Acrylonitrile*	*Vinyl chloride*	*Vinyl acetate*
Styrene	(1.0)	2.2	25	50	100
Methyl methacrylate	1.9	(1.0)	6.7	10	67
Acrylonitrile	2.5	0.82	(1.0)	25	20
Vinylidene chloride	5.4	0.39	1.1	5	10
Vinyl chloride	0.059	0.10	0.37	(1.0)	4.4
Vinyl acetate	0.019	0.05	0.24	0.59	(1.0)

Although there are some exceptions, the effectiveness of substituents in enhancing the reactivity of the monomer lies in about the order

$$—C_6H_5 > —CH{=}CH_2 > —COCH_3 > —CN > —COOR > —Cl > CH_2Y > —OCOCH_3 > —OR$$

The effect of a second substituent on the same carbon atom is usually additive.

This order of reactivity corresponds to the resonance stabilisation of the radical formed after the addition. In the case of styrene the radical can resonate among forms of the type:

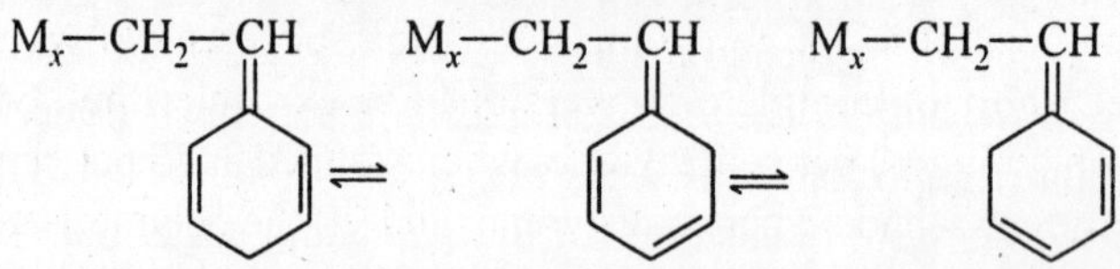

The radical is stabilised with a resonance energy of about 80 kJ/mole. At the other extreme, substituents that have no double bonds conjugated with the ethylenic double bond give radicals having very low resonance energies (5–15 kJ/mole), since only polar or unbonded forms can contribute to the resonance.

Substituents that stabilise the product radical tend also to stabilise the monomer, but the amount of stabilisation is much smaller than for the radical; in styrene the monomer is stabilised to the extent of about 12 kJ/mole.

Thus the stabilising effect of a substituent on the product radical is compensated for only to a limited extent by stabilisation of the monomer.

Resonance stabilisation

Resonance stabilisation of monomers and radicals can be considered in terms of a potential energy diagram of the type shown in Fig. 7.8. The energy changes shown represent situations before and after each of four possible reactions in which a resonance-stabilised (M_s) or unstabilised (M) monomer adds to a stabilised (R_s•) or unstabilised (R•) radical.

The set of four curves labeled with the four reaction possibilities represents the combined energy of the growing chain radical and the monomer before reaction, while the two curves labeled 'new radical' represent the potential energy of the new bond formed by their combination. Activation energies are represented by solid arrows, and heats of reaction by broken arrows. If the entropies of activation are all about the same (a good assumption in the absence of steric hindrance), the order of reaction rates is predicted to be the opposite of that of the activation energies:

$$R\bullet + M_s > R\bullet + M > R_s\bullet + M_s > R_s\bullet + M$$

This series summarises much of the rate information available for both co-polymerisations and homopolymerisations.

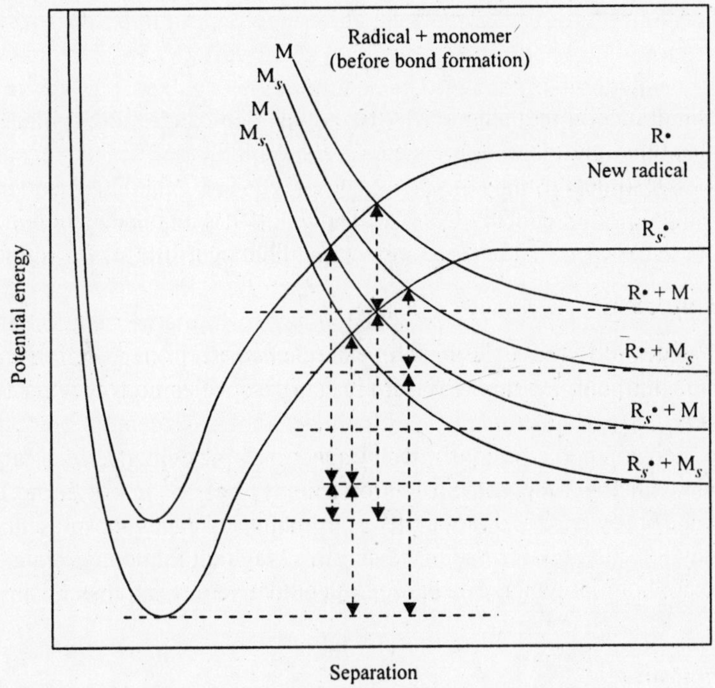

Fig. 7.8. Potential energy of a radical-monomer pair as a function of reaction co-ordinate.

Reactivity of radicals

The reactivity of radicals to a reference monomer can be derived by combining the absolute propagation rate constants k_{11} with reactivity ratios to obtain the rate constants k_{12} for radical 1 adding monomer 2. Typical data are listed in Table 7.4.

As expected, resonance stabilisation depresses the reactivity of radicals so that their order of reactivity is the reverse of that of their monomers; styrene is one of the least reactive radicals and vinyl acetate one of the most reactive. The effect of a substituent in depressing the activity of a radical is much greater than its effect in enhancing the activity of the monomer.

For example, the styrene radical is about 1000 times less reactive than the vinyl acetate radical to a given monomer if alternation effects are disregarded, but styrene monomer is only about 50 times more reactive than vinyl acetate to a given radical.

Table 7.4. Radical-monomer propagation rate constants at 60°C (litre/mole-sec).

	Radical					
Monomer	*Butadiene*	*Styrene*	*Methyl methacrylate*	*Methyl acrylate*	*Vinyl chloride*	*Vinyl acetate*
Butadiene	100	250	2,820	42,000	350,000	–
Styrene	74	145	1,520	14,000	600,000	230,000
Methyl methacrylate	134	278	705	4,100	123,000	150,000
Methyl acrylate	132	206	370	2,090	200,000	23,000
Vinyl chloride	11	8	70	230	12,300	10,000
Vinyl acetate	—	2.6	35	520	7,300	2,300

Steric effects

The effect of steric hindrance in reducing reactivity may be demostrated by comparing the reactivities of 1,1- and 1,2-disubstituted olefins to reference radicals. As indicated in the preceding discussion, the addition of a second 1-substituent usually increases reactivity 3- to 10-fold; however, the same substituent in the 2-position usually decreases reactivity 2- to 20-fold. The extent of reduction in reactivity depends on energy differences between *cis* and *trans* forms and on the possibilities for resonance.

Charge-transfer complexes

Two monomers with widely different polarities, for example, an electron donor and an electron acceptor, can form a charge-transfer complex that may act as a single species in polymerisation, leading to 1:1 alternating polymers.

The formation of such complexes is confirmed by spectroscopic evidence. Typical electron acceptors are methyl methacrylate, methyl acrylate, and acrylonitrile; typical donors for these complexes are vinyl chloride, vinyl acetate, propylene, and isobutylene. In many instances the presence of a Lewis acid enhances the tendency for complex formation. It is in this way that monomers such as isobutylene and propylene, which do not homopolymerise because of degradative chain transfer, can participate in free-radical polymerisation.

Polarity and alternation

By examining the tendency for alternation as given by the product r_1r_2, it is possible to tabulate monomers in a series arranged so that two monomers farther apart in the series have a greater tendency to alternate. Such an arrangement is shown in Table 7.5. Values of r_1r_2 generally decrease as monomers farther apart in the table are compared. There are some exceptions, suggesting specific interactions, probably steric in nature.

For example, vinyl acetate alternates more than styrene with vinylidene chloride, and less than styrene with acrylonitrile.

There is little doubt that alternation is primarily due to the polarity of the double bond in the monomer: The order of monomers in Table 7.5 closely parallels the order of the tendency of substituents around the double bond to donate electrons to the bond (hydrocarbon, acetoxy) or to withdraw them (carbonyl, cyano).

As might be expected, the tendency toward alternation (if not too great) parallels the tendency toward decreased reactivity induced by substitution of electron-withdrawing groups at the double bond.

Table 7.5. Product of reactivity ratios with monomers arranged in order of alternating tendency.

Vinyl acetate							
–	Butadiene						
0.55	0.98	Styrene					
0.39	0.31	0.34	Vinyl chloride				
0.30	0.19	0.24	1.0	Methyl methacrylate			
0.6	< 0.1	0.16	0.96	0.61	Vinylidene chloride		
0.21	0.0006	0.016	0.11	0.18	0.35	Acrylonitrile	
0.0049	—	0.021	0.056	—	0.55	—	Diethyl fumarate

Alfrey-Price treatment

Alfrey and *Price* attempted to express the factors of general reactivity and polarity quantitatively. They wrote the rate constants for co-polymerisation in the form

$$k_{12} = P_1 Q_2 \exp(-e_1 e_2) \qquad \text{... (7.31)}$$

where exp denotes the exponential, P_1 is the general reactivity of the radical $M_1\bullet$, Q_2 is the reactivity of the monomer M_2, and e_1 and e_2 are proportional to the electrostatic interaction of the permanent charges on the substituents in polarising the double bond. This form is analogous to *Hammett's* equation for the effects of nuclear substituents on the reactivity of aromatic compounds. From the definitions of the monomer reactivity ratios it follows that

$$r_1 = \frac{Q_1}{Q_2} \exp[-e_1(e_1 - e_2)]$$

$$r_2 = \frac{Q_2}{Q_1} \exp[-e_2(e_2 - e_1)] \qquad \text{... (7.32)}$$

and

$$r_1 r_2 = \exp[-(e_1 - e_2)^2] \qquad \text{... (7.33)}$$

The *Alfrey-Price* equation assumes that the same value of *e* applies to both monomer and radical. This assumption is justified only by the success of the *Q-e* scheme in interpreting relative reactivities.

It should be possible to assign *Q* and *e* values to a series of monomers from the data on a few co-polymerisation experiments and compute values of r_1 and r_2 for all the possible arrangements of the monomers in pairs. Unfortunately, this practical application of the scheme is hampered by large uncertainties in the values of *Q* and *e* that can be assigned with the present experimental error. The scheme is not entirely sound from the theoretical standpoint and is best regarded as semi-empirical. In a sense, it represents the transcription of the reactivity and polarity series of Tables 7.3 and 7.5 to equation form. *Young* provides an extensive tabulation of values of *Q* and *e*.

Ionic Co-polymerisation

The order of monomer reactivity in cationic co-polymerisation is quite different from that in free-radical polymerisation and is distinguished by the high reactivities of vinyl ethers and isobutylene. However, the reactivities correspond to the anticipated effect of substituents upon the reactivity of double bonds toward electrophilic reagents. The differences of reactivity arise from the change in electron availability in the double bond and the resonance stabilisation of the resulting carbenium ion. Compared with radical-induced co-polymerisation, the carbenium-ion reactions show a much wider range of reactivity among the common monomers, and no tendency toward alternation in the co-polymer. These characteristics mean that relatively few monomer pairs easily yield co-polymers containing large proportions of both monomers.

The data on anionic co-polymerisation suggest that the order of monomer reactivity is much different from both free-radical and carbenium-ion polymerisation. As anticipated, the order of reactivity in the anionic process is determined by the ability of substituents to withdraw electrons from the double bond and to stabilise the carbanion formed. Anionic and co-ordination polymerisations involving monomers of like electronegativity are usually ideal, since the unsubstituted carbon atom at the end of the growing chain has little influence on the course of the reaction. When the electronegativities of the monomers are widely different, however, the rates of the cross-propagation reactions may be quite different also. In extreme cases, only one monomer can add at all.

Both cationic and anionic reactivity ratios can be profoundly affected by solvent and counterion effects. These are both complicated and variable, but it appears that if the catalyst and its counterion form a tightly bound pair, solvent effects can be large, and conversely.

The contrast among the three methods of polymerisation is brought out very clearly by the results of co-polymerisation studies using the three modes of initiation. Fig. 7.9 shows the compositions of the initial polymer from a feed of styrene and methyl methacrylate using different initiators. These results verify the different polymerisation mechanisms and illustrate the extreme variations of reactivities that monomers exhibit with different type initiators. Again, the results reflect the particular selection of catalyst, solvent, and counterion almost as much as the difference between mechanisms.

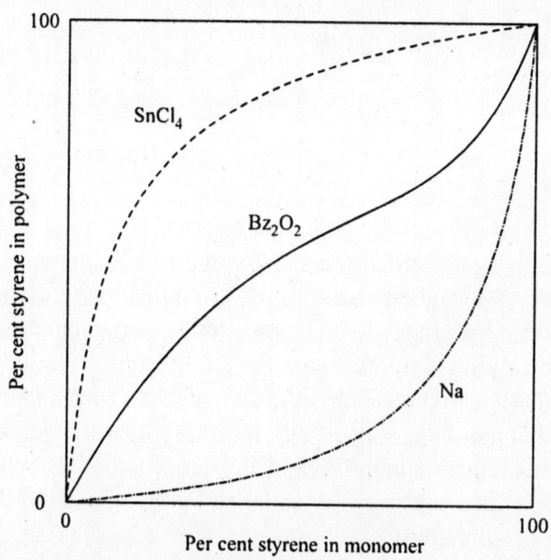

Fig. 7.9. Instantaneous composition of co-polymer as a function of monomer composition for the system styrene-methyl methacrylate polymerised by cationic ($SnCl_4$), free-radical (benzoyl peroxide, Bz_2O_2), and anionic (Na) mechanisms.

Step-Reaction Co-polymerisation

Since the reactivities of all functional groups in simple bifunctional stepwise polymerisation are essentially identical, irrespective of the length of the molecule to which they are attached, co-monomers in such systems are randomly distributed along the chain in amounts proportional to their concentrations in the feed. Thus the 'reactivity ratios' for stepwise polymerisation are unity.

Departures from this generalisation are occasionally encountered, however; for example, the esterification rate of succinic acid HOOC $(CH_2)_2COOH$, differs from that of adipic acid, $HOOC(CH_2)_4COOH$, because of the proximity of the carboxyl groups in the former compound. For similar reasons, ethylene glycol and 1,4-butanediol have different esterification rates. Also, 1,3-butanediol differs in reactivity from 1,4-butanediol because one of the hydroxyl groups of the former is secondary, whereas both of the latter are primary. Where such reactivity differences exist, the tendency is for the more reactive species to enter into the polymer first.

BLOCK AND GRAFT CO-POLYMERS

It is shown that the probability of finding long sequences of one monomer in an ordinary random co-polymer is very small, except in the trivial case where one monomer is present in large excess. Methods of synthesis of polymers containing such long sequences are of interest, however, since they may lead to polymers with properties widely different from those of either homopolymers or random co-polymers. Polymers with long sequences of two monomers can have two arrangements of chains: In block co-polymers the sequences follow one another along the main polymer chain,

—AABB—BBAA—AABB—, etc.

whereas in graft co-polymer sequences of one monomer are 'grafted' onto a 'backbone' of the second monomer type,

```
—AAA—AAA—AAA—,      etc.
  B    B    B
  B    B    B
  |    |    |
```

The ultimate aim in preparing either type is to obtain the desired co-polymer free from homopolymer or other unwanted species, on a scale that allows evaluation of physical properties. Few of the techniques described satisfy these requirements. The industrial importance of both block and graft co-polymers has increased markedly in recent years.

Block Co-polymer

Polymers with labile end groups

The usual method of preparing block co-polymers utilises a polymer with end groups that can be made to react under different conditions. In stepwise polymerisation the method is trivially simple: Two polyesters, for example of different types are separately prepared, mixed, and allowed to polymerise further. Ring-scission polymerisation can be used effectively.

By means of free-radical techniques, block co-polymers can be made in several ways. Labile end groups (as, for example, bromine incorporated through termination by transfer to CBr_4) can be activated by thermal or ultraviolet-light scission of the C—Br bond. Difunctional initiators can be used to produce block co-polymers if the two initiator groups can be decomposed independently, or polymer radicals can be formed by milling or masticating the polymer. In each case, of course, the polymer radical must be produced in the presence of a second monomer to yield the co-polymer.

By far the most important synthesis of block co-polymers is that of *Szwarc* for producing 'living' polymers by unterminated anionic polymerisation. Addition of a second monomer to the still active polymer leads to block co-polymer uncontaminated with homopolymer and with blocks of accurately known and controlled length. The process can be repeated, with care, to produce multiblock polymers. Several commercial block co-polymers are produced in this way and are already discussed in Chapter 3.

Graft co-polymers

Graft co-polymerisation results from the formation of an active site at a point on a polymer molecule other than its end, and exposure to a second monomer. Most graft co-polymers are formed by radical polymerisation. The major activation reaction is chain transfer to polymer. In many instances the transfer reaction involves abstraction of a hydrogen atom. An important commercial application is the grafting of polystyrene or styrene-acrylonitrile co-polymer onto butadiene or acrylonitrile-butadiene co-polymer rubber in the production of ABS resins. Ultraviolet or ionising radiation, or redox initiation, among other methods, can also be used to produce the polymer radicals leading to graft co-polymers.

IR ANALYSIS OF VINYL ACETATE CO-POLYMER

A rapid and accurate method for analysing poly(methyl acrylate)-poly(vinyl acetate) co-polymers has been developed. Poly(vinyl acetate) and poly-isomeric monomers, have relatively similar solubility parameters. In connection with work at the laboratory on properties of co-polymers, a rapid and accurate method of analysis was needed to determine the extent of co-polymerisation of vinyl acetate and methyl acrylate. The investigation of poly(vinyl acetate) and poly(methyl acrylate) polymers referred to as PVA and PMA by infrared spectroscopy revealed a considerable difference in the spectra in the region of ester absorption. The infrared spectrum of PMA contained maxima at 7.9, 8.35, and 8.55 microns, while the spectrum of PVA had one very strong absorption maximum at 8.0 microns, the application of simultaneous equations at 8.0 and 8.55 microns made feasible the analysis for both components in the presence of one another.

Several standard techniques have been described for quantitative analyses in the infrared. *Panda* has described the 'cell in-cell out' method where the spectrophotometer is set manually at the desired wave length, and a radiant intensity, I_0^a, is measured through air and an I^a measured through the least absorbing component. Next, as I_0^b is again obtained through air and I^b found for the unknown mixture. The log of the ratio, I_0^b/I^b divided by I_0^a/I_0^a gives the absorbance value. *Zinkel* has described the 'base line density' method where the reference I_0 is obtained by drawing a base line on both sides of the maximum to be measured. In the method employed, the absorbances log I_0^a/I^o were obtained by first scanning the infrated spectrum of the solvent to determine the I_0^a values, then to determine the I_a values from the spectrum of the solutions contained in the same cell.

Operating Conditions and Reagents

The Perkin-Elmer double beam infrared spectrophotometer with sodium chloride optics was used in the procedure. The infrared data on the samples were obtained in a 0.1-mm. sodium chloride cell with a 0.5-inch sodium chloride block in the reference beam. The following instrument settings were optimum: selector switch, manual; slit width, 180 microns; gain, 5; response, 1; response time, 3 seconds; suppression, 0; and scanning seconds; suppression, 0; and scanning speed, 1.15 microns per minute.

The acetonitrile used as solvent was reagent grade. Standard 2 per cent (w./v.) poly(vinyl acetate) and 2 per cent (w./v.) poly(methyl acrylate) in acetonitrile were prepared. Each solution was prepared by adding 2 grams of pure polymer to a 100 ml. volumetric flask and diluting to volume with acetonitirile.

Determination of Simultaneous Equations

The infrared instrument was adjusted to the described settings. The 0.1 mm. cell was filled with acetonitrile and the spectrum scanned from 7.7 to 9.5 microns. This spectrum (C in Fig. 7.10) provided the I_o values at 8.0 and 8.55 microns. The acetonitrile was flushed out of the cell with dry nitrogen and replaced with 2 per cent PVA solution. The spectrum was obtained from 7.7 to 9.5 microns (A in Fig. 7.10). The cell was flushed successively with acetonitrile and dry nitrogen. The cell was then filled with 2 per cent PMA solution and the region of 7.7 to 9.5 microns again scanned (B in Fig. 7.10).

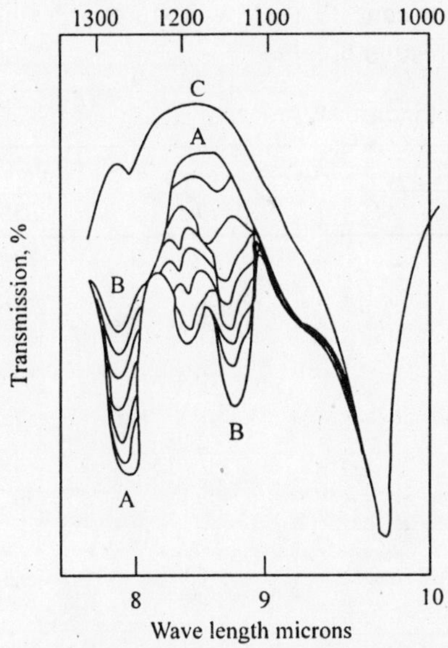

Fig. 7.10. Spectra of standard PVA-PMA mixture. A-100% PVA; B-100% PMA; C-Solvent, acetonitrile.

As shown in Fig. 7.10, the ester linkage of PVA absorbed most strongly at 8.0 microns while that of PMA absorbed strongest at 8.55 microns. The absorbance values were determined by $A = I_0/I$, where I_o was the per cent transmittance of acetonitrile and I was the per cent transmittance of the 2 per cent standard solutions at 8.0 and 8.55 microns. From these data a specific absorptivity was calculated, based on the 2 per cent (w./v.) standards. Assuming the absorbances to be additive for the two pure components, the following equations are valid:

$$A_{8.00\mu} = a_{IPMA}\, bc_{PMA} + a_{2PVA}\, bc_{PVA} \qquad \dots (7.34)$$

$$A_{8.55\mu} = a'_{IPMA}\, bc_{PMA} + a'_{2PVA}\, bc_{PVA} \qquad \dots (7.35)$$

Where a_1 and a_2 are the specific absorptivities of the indicated polymer at 8.00 microns, a'_1 and a'_2 are those at 8.55 microns, and C_{PMA} is the concentration of PMA, and C_{PVA} is the concentration of PVA. The term *b* is the cell thickness which remains constant, and cancels out when the equations are solvent simultaneously. Combining the equations gives equations 7.34 and 7.35, which give the per cent composition of PVA and PMA mixtures and/or co-polymers.

Since Equation 7.36 was derived from data obtained from 2 per cent (w./v.) solutions, it is limited to 2 per cent (w.v.) polymer solutions.

$$\% \text{ PVA} = 70.27\,(1.92\,A_{8.00\mu} - A_{8.55\mu}) \quad \text{... (7.36)}$$

$$\% \text{ PMA} = 100 - \% \text{ PVA} \quad \text{... (7.37)}$$

Procedure

Accurately weigh the dry co-polymer of PVA-PMA mixture. Dissolve it in acetonitrile and dilute to sufficient volume to make a 2 per cent (w./v.) solution. Treat the sample solution in the same manner as is used to determine the absorbances of the standard solutions. Use Equation 7.36 to determine the PVA concentration, and Equation 7.37, the percent PMA. The method was tested by preparing accurate standards containing appropriate amounts of 2 per cent PVA and 2 per cent PMA to obtain nine different standards varying from 10 to 90 per cent PVA-PMA mixtures. The curves for the standards are shown in Fig. 7.10. Table 7.6 shows the results obtained.

Table 7.6. Analysis of standard PVA/PMA mixtures.

Known %		*Found %*		
PVA	*PMA*	*PVA*	*PMA*	*Error*
10	90	10.5	89.5	0.5
20	80	20.5	79.5	0.5
30	70	31.0	69.0	1.0
40	60	40.4	59.6	0.4
50	50	48.8	51.2	1.2
60	40	57.8	42.2	2.2
70	30	69.8	30.2	0.2
80	20	80.7	19.3	0.3
90	10	90.9	9.1	0.9
				Av. dev. = 0.80%
				Std. dev. = 0.99%

Co-polymers of PMA and PVA were prepared by the usual solution polymerisation technique using reagent grade benzene as reflux solvent, and 2-azobsisisobutyronitrile as catalyst. These samples were analysed by the infrared method and the compositions were calculated using Equation 7.36.Theoretical values of co-polymer composition at a specific degree of conversion were calculated from the monomer reactivity ratios. Table 7.7 compares some values obtained theoretically with those shown in Fig. 7.11, which were determined by infrared analysis.

Table 7.7. Comparison of PVA/PMA co-polymers compositions.

		% PMA	
Sample	*% Conversion*	*Experimental Infrared*	*Theoretical*
1	4.8	44.3	45.2
2	15.6	34.7	31.0
3	23.3	24.7	21.0

From these results it is concluded the PVA/PMA co-polymers composition can be analysed rapidly and accurately by infrared measurement at the selected frequencies.

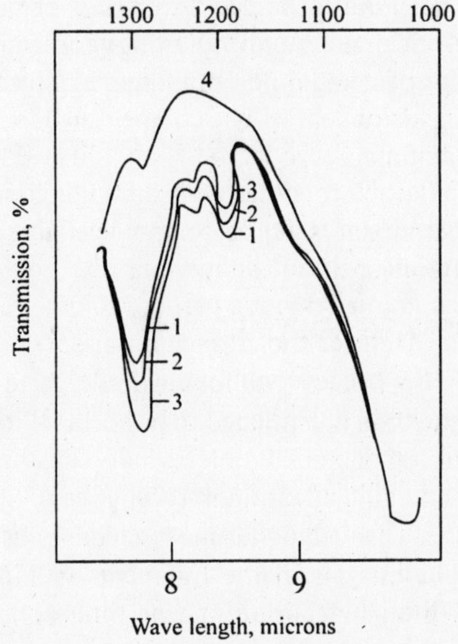

Fig. 7.11. Spectra of PVA-PMA co-polymers. Mole fraction of poly(vinyl acetate): 1. 0.557, 2. 0.653, 3. 0.753, 4. Solvent, acetonitrile.

Discussion

The method is based on the fact that methyl esters give different infrared absorption spectra between 7.7 and 9.0 microns than esters derived from large alcoholic groups. The following structures are depicted for the isomeric PVA and PMA polymers.

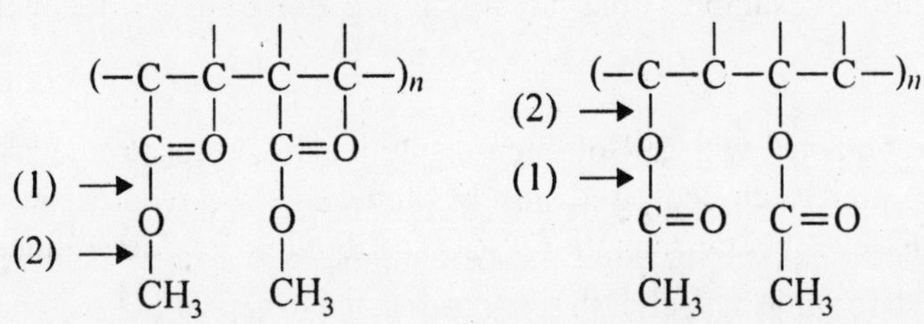

PMA PVA

It has been investigated that the ester linkage may give rise to different vibrations depending upon the nature of the groups attached to the oxygen in the 1 and 2 positions. The alcoholic groups in PVA is polymeric and sufficiently long to have a dampening influence that could give rise to one strong resonance frequency. This absorption band was located at 8.0 microns. In PMA on the other hand, the alcoholic group is a lighter methyl groups that might more easily give rise to more than one frequency from the 1 and 2 bond vibrations. Fig. 7.10 shows that methyl ester give rise to three absorption bands (8.0, 8.35, and 8.55 microns) in this region. The general nature of three bands originating from methyl esters of long chain acids has proved observing three absorption bands in the spectra of methyl esters of long chain fatty acids.

The spectra of co-polymers showed slight shifts to higher wave lengths. The 8.0-micron band shifted to 8.05 microns and 8.55-micron band shifted to 8.57 microns. Compensation for this shift was made by taking the I_0 values of acetonitrile at the new wave length maxima. The method has the advantage that it is rapid and fairly accurate. As many as eight determinations may be completed in an hour. Since the reference intensity of acetonitile is determined in the fixed cell with each batch of samples analysed, any change in the instrument will be compensated for by changing the I_0 for acetonitrile as well as the I of the sample. Maximum reproducibility for slit width was obtained by setting the slit manually to a fixed positions of 180 microns before scanning. It would seem possible to apply the method to the analysis of other ester mixtures whose components give rise to different absorption bands in the C—O—C region of the infrared spetrum.

NMR ANALYSIS OF VINYL ACETATE POLYMER

A high resolution NMR method has been developed for the quantitative determination of the composition of vinyl acetate co-polymers. The method is estimated to be accurate to within 2 moles per cent absolute in the range of 0 to 20 mole per cent comonomer. Problems involving the determination of the composition of copolymers have received considerable attention in recent years. Several analytical techniques have been utilised for studies of this type. The techniques most generally used involve either pyrolysis of the polymer with gas chromatographic analysis of the fragments or infrared absorption analysis of the polymer. These techniques are often inaccurate and/or time consuming. With the pyrolysis gas chromatography scheme, the pyrolysis mode and proper operating conditions must be determined; with infrared analysis, the preparation of satisfactory thin films is often difficult.

Quantitative high resolution NMR spectroscopy offers a convenient alternative approach to the analysis of co-polymers. Two main requirements must be satisfied in order for NMR analysis to be feasible; the polymer must have sufficient solubility in an appropriate solvent and it must produce NMR spectral peaks narrow enough to permit differentiation between comonomers. These requirements place limitations on the versatility of the technique for different polymeric materials. When applicable, however, NMR is a rapid nondestructive method for the analysis of co-polymers. The vinyl acetate co-polymers studied in this work represent additional examples of successful application of the NMR method.

Experimental

The NMR spectra were obtained with HR-60 high-resolution NMR spectrometer equipped with a V-3521 integrator. Spectra were recorded on a Mosely Model 3-S XY recorder at a sweep rate of 7.3 e.p.s./ second at an R.F. level below saturation. Scale calibration was determined by the method of modulation side bands.

Co-polymer samples for use as standards were prepared as latex emulsions. As used, this technique is expected to give negligible cross-linking and a random distribution of vinyl acetate units in the polymer. The emulsions were cast on glass and allowed to dry. Samples for NMR analysis were obtained by dissolving sufficient co-polymer in carbon tetrachloride to give 15 to 20 per cent (wt./vol.) solutions. Samples of vinyl acetate homopolymer for NMR analysis were obtained by dissolving sufficient polymer in benzene to give 5 to 10 per cent (wt./vol.) solutions.

Results and Discussion

The NMR spectrum of a vinyl acetate homopolymer is a shown in Fig. 7.12. It contains a distinct peak for the α-CH protons (CO_2CH) at 5.1 p.p.m. The remaining protons produce the large signal at 1.9

p.p.m. *Zinkel* report a spectrum for vinyl acetate homopolymer obtained in carbon tetrachloride at 110°C in which the signals of the β-protons and the methyl protons are resolved.

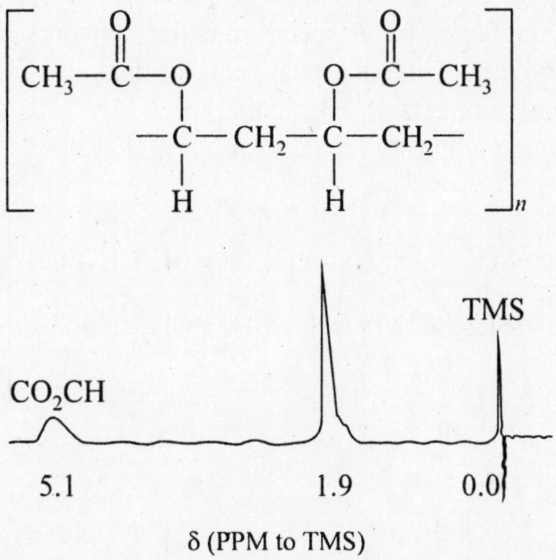

Fig. 7.12. NMR spectrum of a vinyl acetate homopolymer in benzene (solvent peaks not shown).

The NMR spectrum of a vinyl acetate-dibutyl maleate co-polymer (80:20 wt./wt.) is shown in Fig. 7.13 (the values cited in this report for the composition of these co-polymers are derived from the relative weights of the comonomers used to prepare each co-polymer). Besides the peaks in the vinyl acetate portions of the polymer (4.7 and 1.9 p.p.m.), there are two distinct peaks resulting from the comonomer. These are the peaks at 4.0 p.p.m. resulting from the α-CH_2 protons (CO_2CH_2) and at 1.0 p.p.m. resulting from the alkyl CH_2 protons.

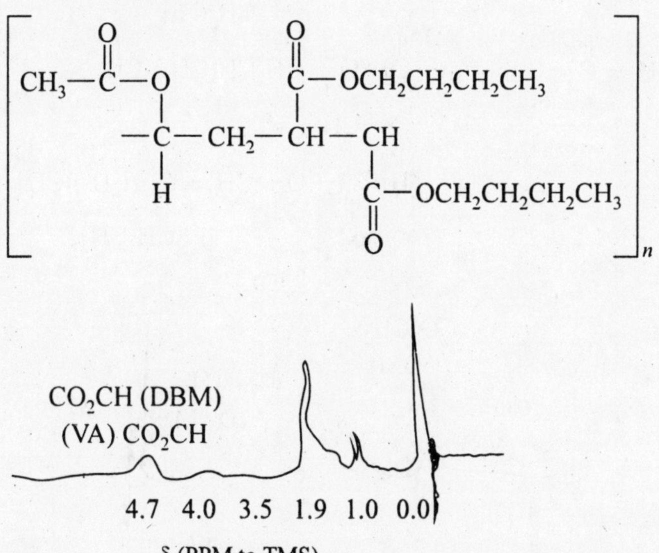

Fig. 7.13. NMR spectrum of a vinyl acetate-dibutyl maleate co-polymer (80:20 wt/wt.) in carbon tetrachloride.

The small peak at 3.5 p.p.m. is caused by residual water not removed in drying the co-polymer films. A vinyl acetate-dibutyl fumarate co-polymer of the same composition gave an identical NMR spectrum. The NMR spectrum of a vinylacetate-1-ethylhexyl acrylate co-polymer (80:20 wt./wt.) is shown in Fig. 7.14. The assignment of the peaks in this spectrum is the same as that given in Fig. 7.13. Because of the longer eight carbon alkyl group, an additional peak for the alkyl CH_2 protons in the 2-ethylhexyl acrylate portions of the polymer appears at 1.3 p.p.m. The NMR spectrum of a vinyl acetate-dioctyl maleate co-polymer of the same composition is shown in Fig. 7.15.

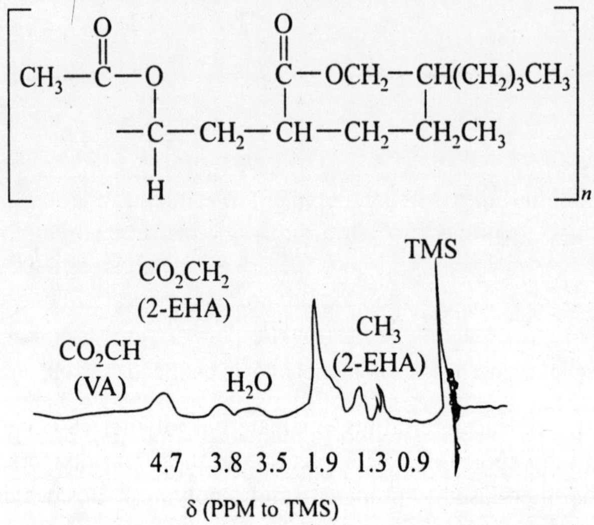

Fig. 7.14. NMR spectrum of a vinyl acetate-2-ethylhexyl acrylate co-polymer (80:20 wt./wt.) in carbon tetrachloride.

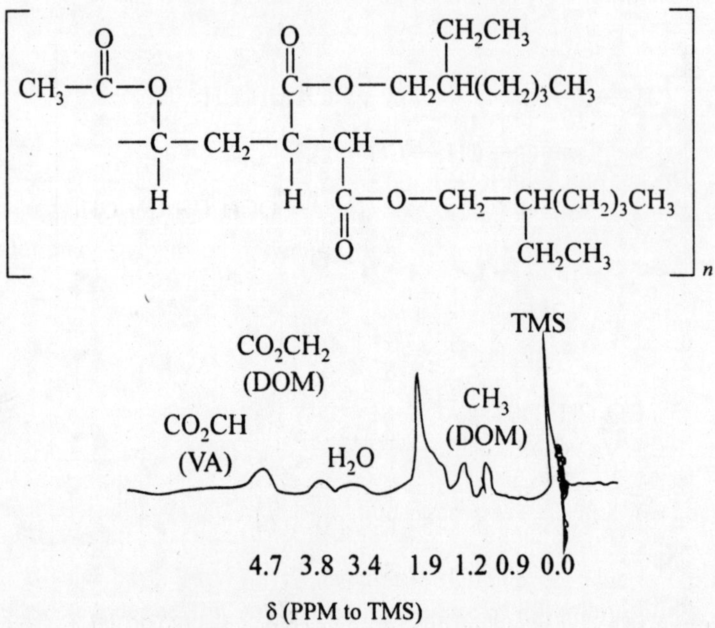

Fig. 7.15. NMR spectrum of a vinyl acetate-dioctyl maleate co-polymer (80:20 wt./wt.) in carbon tetrachloride.

The spectra of these two co-polymers differ only in the relative intensities of the various peaks. This is not unexpected since the two polymers contain the same basic structural units, differing only in the relative numbers of these units.

These polymers contain peaks specific for each comonomer the α-CH peak in the vinyl acetate portions of the polymer, and the α-CH_2 and alkyl CH_2 peaks in the comonomer portions of the polymer. A direct analysis based on integration of the α-CH and α-CH_2 peaks was attempted using the varian integrator. Although reasonably good agreement was obtained between the areas of these peaks and the areas predicated on the basis of the relative numbers of protons of these types, the precision of these measurements was low. The values obtained were quite dependent upon spectrometer operating conditions and sample viscosity. Presumably this is caused by a small overlap of the broad peaks, the degree of overlap varying with sample composition broadening which will contribute to the overlap. In the effort to minimise the problem of peak overlap and improve the accuracy and speed of analysis, peak height measurements were investigated as a quantitative procedure. Since peak height are dependent on operating variables such as viscosity, temperature, and traces of paramagnetic materials, which do not affect peaks areas, special care was taken to establish operating conditions to minimise or eliminate these factors.

Reproducible peak height ratios were obtained within sweep rate range of 4.0 to 10.0 c.p.s./ second and a sample concentration range of 5 to 25 per cent (wt./vol.). All measurements were made at the essentially constant probe temperature, which exists under normal operating conditions. Since there was no broadening of the TMS reference signal, in samples with low viscosity, it was concluded that no paramagnetic impurities were present.

Peak height measurements of different peaks can be directly related to reactant concentrations when differences in peak multiplicity are taken into account. However, the peak multiplicity could not be determined experimentally for the co-polymers reported in this study. Also, a range of chemical shifts rather than a single value for the designated proton types would be predicted because of the existence of these protons in sequences of varying length. These factors make it necessary to use working curves for the analysis rather than to calculate directly the percent comonomer from peak height or area measurements. Such calibration curves obtained from a set of vinyl acetate dibutyl maleate co-polymers are shown in Fig. 7.16. The values used for per cent comonomers are those obtained from the relative weights of the comonomers used to prepare the copolymer. Each point on the calibration curves represents the average figure from 5 NMR spectra of one or two samples. Two curves are plotted—the solids curve represents data from the alkyl CH_3 peak resulting from dibutyl maleate; the dashed curve represents data from the CH_3 peak resulting from dibutyl maleate. Two curves are included to provide a check on the analysis. A disagreement greater than 2 mole per cent between these figures suggests that an interfering component may be in the sample. Samples of vinyl acetate-dibutyl fumarate co-polymer of known composition also give points which fall on these curves. Similar calibration curves obtained from a series of vinyl acetate-2-ethylhexyl acrylate co-polymers are shown in Fig. 7.17. The two curves again represent data obtained from the α-CH_2 and alkyl CH_3 peaks of the comonomer. Fig. 7.18 shows the calibration curves obtained from a series of vinyl acetate-dioctyl maleate co-polymers.

In using these curves for analysis values for the mole per cent (or weight per cent) comonomer are obtained from each of the calibration curves of the particular co-polymer. The per cent comonomer is reported as the average of the figures from the two working curves. The method is accurate to within 2 mole per cent absolute in the ranges 0 to 20 mole per cent for the dibutyl maleate or furmarate co-polymers and 0 to 10 mole per cent for the 2-ethylhexyl acrylate or dioctyl maleate co-polymers.

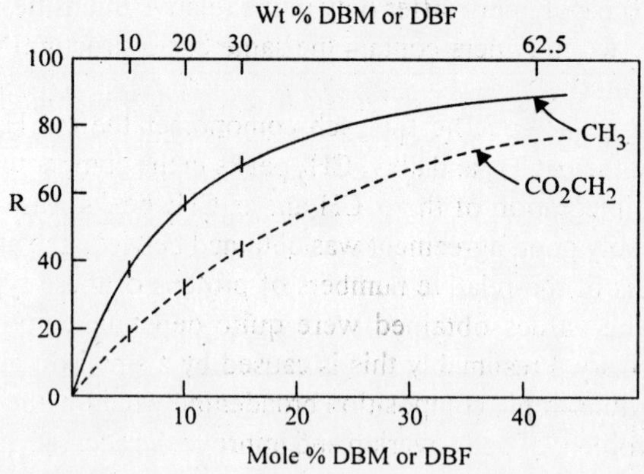

Fig. 7.16. Calibration curves for vinyl acetate-dibutyl maleate or vinyl acetate-dibutyl fumarate co-polymers.

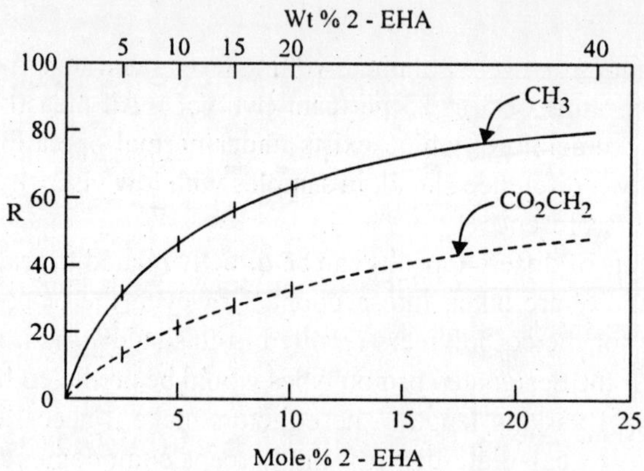

Fig. 7.17. Calibration curves for vinyl acetate-2 ethylhexyl acrylate co-polymers.

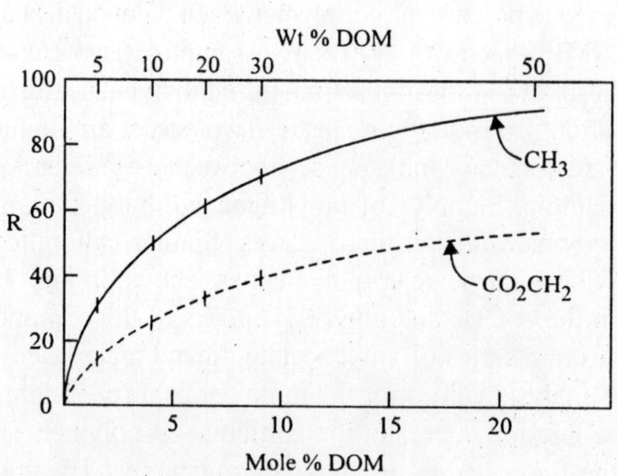

Fig. 7.18. Calibration curves for vinyl acetate-dioctyl maleate co-polymers.

At higher comonomer levels the analysis is less accurate because of the decreasing dependence of the peak height ratios (R) on per cent comonomer. An alternative method of plotting peak height data is available which will give linear calibration curves. These linear plots are obtained by including terms to take into account the differing numbers of protons of a given type per mole of reactant and the differing peak multiplicity. When these factors are included, linear calibration curves are obtained. The slopes of these curves, however, do not show good agreement with predicted values and the peak height ratio R has a value greater then zero at zero percent comonomer. This deviation presumably arises from the approximations necessary to correct for peak multiplicity as well as the chemical shift spread expected for a particular proton type. Since linear curves do not provide more accurate readings and because additional efforts is needed to prepare these curves, the nonlinear plots included in this report are used for analysis.

CHAPTER 8

Co-polymers of Ethylene and Propylene Oxides

INTRODUCTION

The co-polymers of ethylene oxide and propylene oxide are indicated as poly(propylene oxide-block-ethylene oxide). These block co-polymers are represented as follows:

$$H\text{–}(O\text{–}CH_2\text{–}CH_2)_m\text{–}O\text{–}\underset{\displaystyle CH_3}{\underset{|}{CH}}\text{–}CH_2\text{–}O\text{–}(CH_2\text{–}\underset{\displaystyle CH_3}{\underset{|}{CH}}\text{–}O)_n\text{–}O\text{–}(CH_2\text{–}CH_2\text{–}O)_m\text{–}H$$

where, m=150–80, and n=2–60.

Ethylene oxide-propylene oxide co-polymers are divided into three broad classes depending on structure and use. These classes overlap, so division is often arbitrary. Polyols are polyethers used to build up the polyurethane structure.

Important factors in this application are functionality, concentration, and reactivity of the hydroxyl end groups.

The name polyols indicates the presence of hydroxyl groups. Block co-polymers are made up of long segment of propyloxy groups and segments of ethyloxy groups. The propyloxy groups act as hydrophobes, the ethyloxy groups as hydrophiles. With the proper balance, non-ionic surfactants can be prepared.

The so-called 'functional fluids' are a third class of polyethers. This class includes the polyethylene glycols, alcohol-based polypropylene glycols, and the random co-polymers of ethylene oxide and propylene oxide. The functional fluids are used in a large number of applications, including hydraulic fluids, quenchants, lubricants, and surfactants.

The important properties for these applications are defined by the nature of the polyether backbone; the end groups are relatively unimportant.

CLASSIFICATION OF POLYALKYLENE OXIDE BLOCK CO-POLYMERS

For convenience in describing the polyalkylene oxide block co-polymers, several classifications have been arbitrarily established. First, the number of reactive hydrogen atoms present per mole of initiator has been used as a basis to distinguish among non-ionic surfactants derived from 1-butanol (Class I), propylene glycol (Class II), glycerol (Class III), ethylenediamine (Class IV), etc. Within each of the above arbitrarily defined classes, three further distinctions are set up, as follows.

All Block Non-ionics

This group (sub-class A) of non-ionic surfactants is formed by the sequential addition of one alkylene oxide to an initiator, which may be organic or inorganic, followed by the sequential addition of another alkylene oxide. This order of oxide addition results in the formation of compounds which contain a block of oxyalkylene groups in a chain, which is connected to a block of different oxyalkylene groups in a chain. Wherever a third, fourth etc., sequential oxyalkylation step occurs, whether the oxide was previously used or not, the product is still considered to fall within this class.

Block-Heteric or Heteric-Block Non-ionics

The block-heteric group of non-ionic surfactants (sub-class B) embraces those polyols that are formed by the addition of an alkylene oxide to an initiator, followed by the addition of a mixture of two or more alkylene oxides. The heteric-block group includes those polyols that are formed by the reverse order of addition to the initiator—the heteric oxyalkylene chain is formed first, followed by the addition of only one alkylene oxide. Either order of addition results in the formation of compounds which contain a block of oxyalkylene groups in a chain, connected to a heteric oxyalkylene chain, in which the different oxyalkylene groups are distributed randomly throughout the chain. The word 'heteric' is used to describe surfactants which contain oxyalkylene groups derived from alkylene oxides that are mixed prior to reaction with a compound containing one or more reactive hydrogen atoms. This order of addition produces a random distribution of oxide residues and results in polyols which have properties entirely different from the polyols derived from sequential additions of the same oxides.

All-Heteric Non-ionics

This group of surfactants (sub-class C) encompasses those polyols which are formed by the addition of a mixture of two or more alkylene oxides in a given ratio, to an initiator, followed either by the addition of the same oxide mixture in a different ratio, or by a different alkylene oxide mixture in a given ratio. This results in the formation of compounds that contain a heteric oxyalkylene chain connected to a different heteric oxyalkylene chain. The different oxyalkylene groups are randomly distributed throughout each oxylakylene chain. Obviously, it is possible to have countless combinations and variations of the three above categories, but for simplification they have been established as described. In this section various classes of EO-PO and BO co-polymers are described along with their respective applications. All these compounds may not be available commercially.

MONOFUNCTIONAL INITIATORS

All Block Non-ionics

RPE non-ionics

Surfactants of the formula $Y(\!-\!\overset{R_1}{\underset{R_3}{\overset{|}{\underset{|}{C}}}}\!-\!\overset{R_2}{\underset{R_4}{\overset{|}{\underset{|}{C}}}}\!-\!O)_n X$

were prepared, wherein Y is the residue of an organic compound containing a single hydrogen atom capable of reacting with a 1,2-alkylene oxide, R_1, R_2, R_3 and R_4 are selected from the group consisting of hydrogen, aliphatic radicals, and aromatic radicals, at least one such substituent being a radical other than hydrogen; n is greater than 6.4, and preferably greater than 8, as determined by hydroxyl number; and X is a water-solubilising group, usually polyoxyethylene, although including polybutadiene oxide, comprising 25 to 95 per cent, and preferably 40 to 75 per cent by weight of the final product. For the sake of brevity, these compositions are referred to as RPE non-ionics, where R denotes a monofunctional initiator, P a polyoxypropylene chain, and E a polyoxyethylene chain.

In this group, whole range of products can be prepared. The monofunctional initiator can be selected from a wide variety of compounds, including monofunctional alcohols, acids, mercaptans, secondary amines, and *N*-substituted amides. Although polyoxypropylene is preferred as the hydrophobe, other chains that can be utilised include polyoxybutylene, polyoxystyrene, and polyoxycyclohexene. These surfactants are said to be suitable as wetting and dispersing agents, and as lubricants and plasticisers, for the textile, leather, paper, and rubber industries. Because they possess good thermal stability, they can be used as heat transfer media and hydraulic fluids. They are claimed to be especially effective as detergent. They may be formulated with quaternary ammonium compounds to prepare excellent detergent-sanitiser compositions. These non-ionics do not haemolyse red blood cells and have no effect upon the lipoproteins present in whole blood. This suggests their use in cleaning equipment employed in blood banks, etc. The anionic sulphated esters of these non-ionics are reported to be valuable for light duty liquid detergents.

REP non-ionics

Compounds prepared by first reacting ethylene oxide, and then propylene oxide, with a monofunctional initiator, in the reverse order from that previously described, were found to have completely different surface-active properties, especially lower cloud points and foam heights. For the sake of brevity, these are referred to as REP non-ionics. The structure of the REP surfactants corresponds to the formula:

$$R(C_2H_4O)_n(C_3H_6O)_mH$$

in which R is the nucleus of the reactive hydrogen compound, n has a value such that the weight of the oxyethylene groups constitutes about 20 to 90 weight per cent of the total oxyethylene-oxypropylene composition, and m has a value such that the molecular weight of the oxypropylene is at least about 371 and up to about 25000.

Block-Heteric Non-ionics

Glycols, which contain both a tertiary and either a secondary or a primary hydroxyl groups, have been converted into non-ionic surfactants. Several oxyalkylation steps are involved to produce the desired surfactant. The glycols, such as 3-methyl-1,3-butanediol or 2-methyl-2,4-pentanediol, and similar diols with up to 22 carbon atoms, are oxyalkylated with ethylene oxide and propylene oxide in different steps. The products formed are reported to be useful as wetting, detergent, and levelling agents in the laundry, textile, and dyeing industries; as wetting agents in the acid washing of bricks and building stone, or for applying asphalt in road building; as floatation agents; as emulsifying agents for cosmetics, water-repellent textile finishes, etc.; as well as for breaking crude oil-in-water emulsions. These surfactants may be combined with a variety of reactants as chemical intermediates to produce other surfactants (cationic and anionic). Other surfactants have also been produced by utilising a mixture of oxides, to produce a block-heteric non-ionic. Some all-block non-ionic (IA) are also formed by the sequential addition of two or three alkylene oxides. In this group of compositions, styrene oxide is added to a monofunctional initiator, followed by an alkylene oxide, or a mixture of alkylene oxides, each containing from two to four carbon atoms, in the range of 5 to 50 moles per mole of styrene glycol monoether.

The formula for these surfactants is

$$ROCHCH_2O\,(\text{alkylene-O})_nH$$

where R is an unsubstituted alkyl radical containing 2 to 18 carbon atoms, or a phenyl, alkylphenyl, or chlorinated phenyl radical, and *n* is 5 to 50. These compounds are useful as detergents, emulsifying, and wetting agents.

Tergitol non-ionics

Representatives of the heteric-block non-ionic class of surfactants that are commercially available are Union Carbide's Tergitol non-ionics XD and XH. The hydrophobe of the Tergitol non-ionic groups has a minimum molecular weight of 1000 and consists of an aliphatic monohydric alcohol containing from one to eight carbon atoms to which is attached a heteric chain of oxyethylene and oxypropylene. The weight ratio of oxypropylene to oxyethylene covers the range of 95:5 to 85:15. Attached to this is the hydrophilic polyoxyethylene chain, which is from 44.4 to 54.6 per cent of the total molecular weight.

Because they exhibit good emulsifying properties with aromatic polar solvents, the Tergitols are used in the preparation of agricultural emulsifiable concentrates and as emulsifiers for silicone oils and emulsion polymerisation of acrylic esters. They also find application as dispersion agents for oil and wax formulations used in leather treating, as stabilisers for polyvinyl acetate latices, and in the formation of base latices into paints. In the cosmetic industry they impart improvement of rinse properties of shampoos and are used in the preparation of emulsion hair conditioners. They are also recommended for low-foaming automatic dish washing and dry-cleaning detergents and household detergents in general.

BIFUNCTIONAL INITIATORS

All-Block Non-ionics

Pluronic polyols

The widely known series of commercial products, Pluronic polyols made by the sequential addition of propylene oxide and then ethylene oxide to a low-molecular-weight reactive hydrogen compound, falls within this class. The compounds take advantage of the fact that polyoxypropylene glycol, with a molecular weight of 900 or more, is water insoluble and functions as a hydrophobic compounds. The addition of ethylene oxide, to constitute from approximately 20 to 90 per cent of the total weight, yields water-soluble polyols which have surface-active properties and the structure

$$HO(C_2H_4O)_a(C_3H_6O)_b(C_2H_4O)_cH$$

where b is at least 15, and $(C_2H_4O)_{a+c}$ is 20 to 90 per cent of the total weight of the compound.

Block co-polymers are produced by BASF Wyandotte under the trade name Pluronic with molecular weight of 1,100 to over 14,000. A polypropylene glycol is prepared first from propylene oxide and propylene glycol. Ethylene oxide reacts at each end of the molecule to give a block structure which contain 10–80 per cent ethyloxy groups. The polypropylene glycol block acts as a hydrophobe, whereas the ethylene oxide ends are hydrophilic. By varying the length and proportions of these constituents, polymers are produced that range from mobile liquids to solids hard enough to be flaked. These compounds are designated in such a way that the structure is identified. A letter is used to indicate the physical form: L for liquid, P for paste, and F for flakeable solid. The letter is followed by two or three digits. The first digit or digits indicates the typical moleculer weight of the polypropylene glycol unit. The final digit multiplied by 10 gives the weight per cent of the portion of the molecule derived from ethylene oxide. For example, L-31 is a liquid, 10 per cent of which is derived from ethylene oxide; its molecular weight is 950. Surface-active properties are controlled by the sizes of the hydrophobic and hydrophilic blocks.

Applications of Pluronic polyols

The widespread industrial use of the Pluronic polyols is due in part to their low-foam level and to their unusually low order of toxicity. Pluronic polyols L61 is used extensively as a defoamer; L81 and L101 find similar applications, in a dry-cleaning detergent, for antifreeze, and for starch solutions. Thus for deaerating anionic detergent slurries a combination of soap and Pluronic (L61, L64, or F68) is found useful. The same combination of soap, anionic and Pluronic L62, L64, or F68 is useful in producing a low-sudsing synthetic detergent of high cleaning power. For defoaming fermentation processes, L61, L64, and L81 are valuable, whereas Pluronic L61 is used for defoaming stable formaldehyde solutions. For inhibiting foam formation in steam-generating systems, L44, L62, L64, and F68 are recommended.

Pluronic F68 has been used as the defoaming agent in a mechanical heart-lung apparatus for circulating blood to prevent air from being drawn into the patient's bloodstream. It reduces bubble size and increases oxygenation. L61 is used for foam control by the paper industry for make-up and application of starch-sizing solutions, for cooking and dilution of starches in paper coating, for spent cooling liquors from kraft and sulphite pulping, in adhesives such as glue for box board, etc. Other Pluronic polyols used in the paper industry include L31, L62LF, and L62 for felt washing, whereas L62 and L62LF find application for wetting talc pigment, as an emulsifying agent in resin paper coatings, as preventative for self-sizing and curl, as pulp and paper drainage aids, and to increase water pick-up of absorbent papers.

L61 and L62 are used to stabilise butadiene latices in coating colours, L62 and L64 are used to control the viscosity of paper coating colours, L62 as a latex stabiliser with starch adhesives, L62LF or L64 to improve cooking liquor penetrations into wood chips, L62LF or F68LF for deinking of waste papers, and F68LF as a dispersant for pitch.

In automatic dishwashers, Pluronic polyols L62, L62D, L62LF and F68 find widespread use as a rinse additive in the final rinse cycle or as the low-foaming surfactant in the detergent. Block co-polymers of propylene and ethylene oxides that have been capped with propylene oxide, or with propylene oxide and then benzyl chloride, are also recommended for similar uses. Pluronic polyols are useful in soap detergent bars with F68 particularly effective in reducing soap-scum formation in dilute solution of a combination soap detergent bar.

A liquid-dishwashing-detergent composition uses Pluronic polyols L61 and L62. The light-duty is converted into a heavy-duty product by the addition of suitable alkaline builders. Pluronic polyols L61, L62 and L64 are used in heavy-duty liquid products, and L62 and other ethylene and propylene oxide block co-polymers are utilised in concentrated liquid washing and bleaching compositions. The Pluronic polyols can also be used to stabilise hydrogen peroxide solution during storage. Pluronic L62 is used as the wetting agent in an acid etching bath for photographic plates. The bath consists of 30 per cent hydrochloric acid, 2 per cent hydrofluoric acid, 0.1 per cent Pluronic, iron chloride, and copper chloride. In acid pickling baths, the Pluronic polyols are used effectively with 15 per cent sulphuric acid, 30 per cent phosphoric acid, and 50 per cent muriatic acid solutions. An aerosol oven cleaner utilises Pluronic L62 and an acid dairy cleaner uses Pluronic F68. A phosphating cleaning composition is based on Pluronic polyols and F68 is used in an abrasive composition. In pharmaceutical industry, Pluronic F68 finds diversified applications, because of its relatively non-toxic nature, although other Pluronic polyols are also utilised. Thus, F68 or P75 is incorporated in fecal stool-softeners. An antibacterial mouthwash utilises F68; the same polyol is used in a vaginal suppository and as the preferred secondary emulsifier in an injectable anaesthetic.

The Pluronic polyols are useful as pharmaceutical excipients and in dentifrice compositions. Pluronic P75 is recommended for use in poultry-feed compositions; L62 and other polyoxypropylene-polyoxyethylene block polymers are used as a drench for bloat control of cattle. Pluronic polyols are also useful in nasal sprays, eye and ear drops, and shampoos, as an emulsifier in anhydrous depilatory sticks, and in controlling the viscosity of shampoos. Pluronic polyol F68 is used in shampoo compositions which posses good lubricity and hair-conditioning properties.

Pluronic F68 is recommended as the most satisfactory wetting agent for calamine lotion. Pluronic F68 is effective in shake cultures as a means of growing fibroblast in suspension, using a rotary shaker, has also been used in basal medium for studying suspension cultures of cells, and is utilised as a diluent for dissolving a medicament for parenteral injections.

The germicidal surface-active properties of iodine-Pluronic complexes have been extensively explored and utilised. F68 is used in a free-flowing water-soluble germicidal composition as well as in an acid cleaning composition with germicidal properties. A phosphoric acid iodine detergent-germicidal sanitiser uses L62 and F68. Other germicidal products based on Pluronic and iodine or iodine and bromide use L44, L62, L64, P65, F68 and P85. In a multipurpose cleaner-germicide deodoriser, the use of Pluronic L62, L64 or F68 not only gives detergency and good suds control but it also solubilises the chlorinated bisphenol.

Pluronic polyols are being used extensively as wetting agents. Thus, L44 is used in a corrosion inhibitor, and L61 is used in a starch slurry as a drainage aid. Pluronic F68 has found use in the final extraction process in sugar mills to increase the penetration of water and thereby yield more sugar from the stalks. Block co-polymers of ethylene and propylene oxides containing at least 60 per cent ethylene oxide are claimed to improve the process for the purification of crude hydroxymethyl furfural by vacuum distillation. Pluronic L62 improves the adherence of aggregate coating compositions to wet mineral aggregates; L64 is useful in releasing microscopic bubbles of gas in protein films and coatings, thus eliminating fish eyes, pinholes, and surface roughness.

The addition of a polyoxypropylene polyoxyethylene glycol to the developing bath is reported to improve the sharpness of prints made with offset lithographic plates having a photosensitive colloid layer containing suitable diazo compounds. The Pluronic polyols find use as dye-leveling agents in the dyeing of fabrics, as wetting agents in descalant compositions, in acid cadmium plating baths, and in other areas.

The Pluronic polyols are useful as antistatic agents. Pluronic L44 gives durable static protection to acrilan fabrics after five washings, whereas L62 and L64 give good temporary static protection to acrilan. Pluronic polyols find applications as dispersing agents. F68 is useful for dispersing talc pigments by preventing the agglomeration of pigment particles on addition to aqueous systems. It performs a similar function in latex paints. Pluronic L44 is preferred for dispersing resins prepared from steam-cracked petroleum fractions. The emulsifying properties of Pluronic polyols are utilised in varied industrial applications. Thus, L62 and L64, among others, are used for agricultural chemicals, and F68 is widely used as the emulsifier for oral and intravenous therapeutic fat products, as well as for oil-in-water emulsions of vitamins. F38 and P66 are also suitable for similar emulsions.

Pluronic polyols L44, L62 and L64 are used as coolant and lubricant in metal-cutting and grinding operations. Oil-in-water emulsions useful as cutting fluids incorporate Pluronic polyols as emulsifying agents, whereas water-in-oil emulsions of dyestuffs require Pluronic L44, L62, and L64. In the preparation of emulsion latices from hydrocarbon polymers, Pluronic F68 is found useful. Polyvinyl acetate latex emulsions and similar coating compositions utilise any of several Pluronic polyols, such as L61, L62, L64, or F68. The Pluronic polyols have excellent plasticising action on a wide range of thermoplastic and thermosetting resins. This has led to their use in the manufacture of resinoid grinding wheels.

Various resin compositions use a Pluronic polyol such as L31, L44, L61, L62, L64, F68, L72 and P75, as a plasticiser or as a bonding agent. The manufacture of polyurethane products, prepared from several Pluronic polyols, includes L42, L44, L61, L62, and P75. Many uses for Pluronic polyols in the polyurethane industry have been developed in the past few years. Pluronic L35 improves the rate of cure of resilient polyether urethane foams because of its ability to emulsify water and amine catalysts into the pre-polymer in the foaming process. Pluronic polyols are used in the vulcanisation of rubber, for masking compositions for coating paint-spray booths, and for various lubricating compositions. Pluronic L62 is one component of a stable gas generating composition. One of the more interesting applications for some Pluronic polyols, especially P75 and P84, although P66 and L81 can also be utilised, is as substrates for gas chromatography of polar compounds. Diatomaceous earth coated with Pluronic F68 is used as the column packing in studying volatile flavour compounds by gas-liquid-partition chromatography.

A major use of Pluronic polyols is in the petroleum industry. One of the first applications reported was a recommendation to use Pluronic polyols or their esters as petroleum demulsifiers to break water-in-oil emulsions. Pluronic polyols are also used in secondary oil recovery. Pluronic polyols L62, L64, and F6 are recommended for oil-well acidising with a dilute hydrochloric acid solution. The use of Pluronic L62 is suggested as a defoaming agent for corrosion inhibitors and for other surfactants. L64 and especially F68, are recommended to prevent scale formation in oil-field pipes, owing to their excellent dispersing properties, on calcium and magnesium ions. Several others corroborate the suitability of block co-polymers of ethylene and propylene oxides and their derivatives, to function as demulsifiers.

A well-washing fluid, containing a Pluronic polyol, eliminates the swelling of clays, such as that caused by the use of alkylphenol ethers, which interferes with the production of oil from a well. A solution of Pluronic L64 is useful to recover oil from underground formations by thermal means. Pluronic polyols are also useful as foaming agents for the removal of aqueous liquids and solids in gas wells which produce both gas and water. Boiler-water-treating compounds utilise Pluronic polyol F6 because of its ability to disperse and complex calcium and magnesium ions. Block co-polymers of propylene and ethylene oxides are usefuls ingredients in the prevention and removal of paraffin from flow lines and other surfaces in oil fields.

Miscellaneous commercial polyols

Other surfactant block co-polymers of propylene and ethylene oxides are those offered commercially in Germany by Farbwerke Hoechst. A. G. under the Genapol (PF10, PF20, PF40, and PF80) and Dissolvan (4400, 4411, 4422 and 4433) trade names. The latter are 65 per cent solutions of polyoxypropylene-polyoxyethylene block co-polymers, containing between 50 and 43 weight per cent ethylene oxide, in methanol, and are designed for use as petroleum demulsifiers. In England, Lankro Products offer polyoxyethylene-polyoxypropylene condensates under the Ethylan (PB, PC) trade name. Additional surfactants derived from ethylene and propylene oxides, but of undisclosed structure, are reported to be commercially available in the USA from Process Chemicals Co., Retzloff Chemical Co., and NALCO Chemical Co., Visco Products Co. Another commercially available product, of undisclosed structure from the Wyandotte Chemicals Corp. is Pluronic RA-10. Described as a polyoxyalkylene condensate, it is a 100 per cent active low-foaming liquid surfactant with excellent detergency properties. It is reported not to exhibit gel formation at any concentration in water and is infinitely soluble in water, isopropyl alcohol, and propylene glycol. It is recommended for use as a rinse aid or surfactant in machine-dishwashing formulations, in heavy-duty liquid and granular detergents, and in all-purpose cleaners.

Polyoxybutylene-polyoxyethylene block polyols

Other series of block co-polymers (Class IIA) are those derived from butylene and ethylene oxides. The use of a hydrophobe consisting solely of a higher α,β-alkylene oxide condensed on an organic compound having a plurality of reactive hydrogen atoms was first disclosed by the Wyandotte Chemicals Corp. and latter by the Dow Chemical Co. A higher alkylene oxide is defined as one which contains four or more carbon atoms in its chemical structure. Although 2-butylene oxide is preferred, the higher alkylene oxides such as amylene oxide, cyclohexene oxide, styrene oxide, and others may be used. To the hydrophobe, which has a minimum molecular weight of 500, is added sufficient ethylene oxide so that the hydrophile comprises 20 to 90 per cent of weight of the surfactant. For detergency purpose, the oxyethylene should constitute 40 to 85 per cent of the surfactant.

The surfactants have formula:

$$\mathrm{Y}[(\overset{\overset{\mathrm{R_1}}{|}}{\underset{\underset{\mathrm{R_3}}{|}}{\mathrm{C}}}-\overset{\overset{\mathrm{R_2}}{|}}{\underset{\underset{\mathrm{R_4}}{|}}{\mathrm{C}}}-\mathrm{O})_n\,(\mathrm{CH_2CH_2O})_n\mathrm{H}]_x$$

where Y is the residue of an organic compound containing x active hydrogen atoms; R_1, R_2, R_3 and R_4 are selected from the group consisting of hydrogen, aliphatic radicals, and aromatic radicals, the substituents being selected so that the R's contain a total of at least two carbon atoms; n has a value such that the molecular weight of the compound, exclusive of the oxyethylene chain, is at least 500; and z has a value such that the oxyethylene groups constitute 20 to 90 per cent by weight of the compound and x is a plural number.

A Dow patent covers polyoxybutylene glycols with a molecular weight range of 1000 to 2000. The hydrophilic portion contributes from 20 to 90 per cent by weight of the compound. Block co-polymers derived from polyoxypropylene, and polyoxyethylene glycols of an average molecular weight of less than about 3000, that are substantially water insoluble, are used in a composition for spinning synthetic fibre and textile fibres of high acrylonitrile polymers.

PEP polyols

Still another series of surfactants in class IIA is that formed by the sequential addition of first ethylene oxide and then propylene oxide to a reactive hydrogen-containing compound. It was found that physical properties differ markedly from the Pluronic type of polyols formed from the same initiators, by reversing the order of addition of ethylene oxide and propylene oxide. Initiators have from two to six reactive hydrogen atoms per molecule and maximum of six carbon atoms. The molecular weight of each oxypropylene chain is at least 900. These polyols are made in the standard manner for making alkylene oxide polymers.

Heteric-Block Non-ionics

Heteric-block polyols

Surfactants are derived from a hydrophobe consisting of a mixture of higher α,β-alkylene oxides condensed on an organic compound having two or more reactive hydrogen atoms. Another series of heteric-block compounds (Class IIB) is prepared from a wide variety of low-molecular-weight water-soluble compounds which contain from two to six reactive hydrogen atoms, condensing these with a mixture of propylene and ethylene oxides, containing 5 to 40 weight per cent of ethylene oxide, followed by oxyethylation.

The ethylene oxide-propylene oxide mixture used can be varied widely as long as a definite minimum molecular weight of about 1000 is used. Effective surfactants are obtained when the molecular weight of the hydrophobe intermediate ranges up to 25,000 and higher. The hydrophile can be varied to constitute from 5 to 90 per cent of the total molecular weight. Depending upon the number of reactive hydrogen atoms in the molecule, two to six, surfactants fall into class IIB to VIB, but for the sake of clarity, all are considered together.

All-Heteric Non-ionics

Heteric polyols

A class of surfactants is possible with a wide variety of properties, from the use of a hydrophobe which can consist of a mixture of alkylene oxides, followed by a hydrophile which comprises either a mixture of the same oxides, but in a different proportion, or a mixture of different oxide. The inititiator may consist of an organic compounds having from one to six reactive hydrogen atoms, selected from a group comprising alkanols, alkyl ethers of alkanols, alkylamines, alkane amides, carboxylic acid, etc., each having up to six carbon atoms per molecule. This means that these compounds could fall into classes IC, IIC, IIIC, IVC, etc., depending upon the initiator. To avoid confusion, all are considered together here. The hydrophobic polyoxyalkylene chain attached to the nucleus has an average oxygen to carbon ratio of not more than 0.4. It can consist of polymers of oxypropylene, oxybutylene, oxyamylene, and oxyethylene alone or with each other, and with oxyethylene. The molecular weight of the initiator plus the hydrophobic polyoxyalkylene chain is at least 400 to 900 and up to 25,000. The heteric hydrophilic chain attached to the other end of the hydrophobe consists essentially of 95 to 40 weight per cent of oxyethylene groups and 5 to 60 weight per cent of higher oxyalkylene groups selected from oxypropylene, oxybutylene, etc., and mixtures thereof. The heteric hydrophilic chain has an average oxygen/carbon ratio greater than 0.4 and constitutes from 5 to 90 weight per cent of the mixture of compounds.

One of the compound is commercially available from the Wyandotte Chemicals Corp. as Plurodat HA 430. The properties of this non-ionic suggest use as a rinse aid, as the low foaming surfactant in egg-washing formulations or in machine dishwashing, as the surfactant in a low-sudsing heavy-duty detergent or as a foam-control agent for the adhesive portion of pigmented paper coatings. In this latter application the liquid non-ionic may be added to the starch or other binder solution prior to mixing with the pigment slurry.

Phenyl heteric polyols

Complex mixtures of linear polyglycols in which the polyoxyalkylene chains contain both the phenyl-substituted oxyethylene group ($–C_6H_5C_2H_3O–$) and one or more lower oxyalkylene groups having the general formula ($C_nH_{2n}O$) have been reported. Ethylene and propylene oxides are the preferred lower alkylene oxides although 1,2-butylene or isobutylene oxide may also be used. Any glycol, or aliphatic diol, or mixture of two or more such glycols, may be used as the initiator, but ethylene or 1,2-propylene or styrene glycol is preferred. Alternatively, a minor amount of water may be employed to react with the styrene or alkylene oxide to form the corresponding amount of glycol starting material.

The initiator is usually employed in amounts corresponding from one to ten per cent, preferably one to five, based on the sum of weight of the oxides used as reactants. The polymeric composition has an average molecular weight of from 480 to 2000. The styrene oxide and the alkylene oxide are combined in amounts corresponding to from 0.1 to 9 in molecular proportions of styrene oxide per molecular equivalent proportion of the lower alkylene oxide.

TRIFUNCTIONAL INITIATORS

All-Block Non-ionics

Glycerol-initiated polyols

Block co-polymers, using glycerol as the initiator, are described in several patents. In some patents assigned to the Dow Chemical Co. the polyoxypropylene hydrophobe attached to the glycerol initiator consists of from 48 to 66 mole per cent (about 55 to 72 weight per cent of the surfactant) per mole of glycerol.

The number of propylene oxide groups ranges from 63 to 126 per glycol nucleus. The hydrophile consists of from 33 to 135 moles of oxyethylene per mole of glycerol, provided it is present in an amount between 40 and 80 weight per cent of the polyoxypropylene-glycerol polymer portion. These surfactants are useful as textile chemicals and as wetting, dispersing, and emulsifying agents, especially for agricultural insecticides, particularly those which are chlorinated hydrocarbons. The emulsifying power of these non-ionic polyols is unusual. They form stable oil-in-water emulsions extremely rapidly and with very little mixing. They are also used for dispersing solids, such as pigments, in liquids and in blending and thickening agents. They have low oral toxicity and such low skin absorptivity as to be valuable ingredients in cosmetic preparations and dishwashing detergents.

Polyglycol 112 polyols

A similar group of surfactants also derived from glycerol, to which propylene oxide and then ethylene oxide are sequentially added, is claimed to be effective for defoaming. In this series, the ethylene oxide content ranges from about 9 to 28 per cent, with the balance essentially propylene oxide. Molecular weights are at least about 2400 and higher. The Dow Chemical Co. offers them commercially as the polyglycol 112 (trademark of Dow Chemical Co.) line. Two members, 112-2 and 112-3, are available; only the former is designed in the surfactant industry. The applications of this surfactant are primarily as a demulsifying agent in petroleum industry and an antifoaming agent for both aqueous and non-aqueous systems.

Triol esters

The formation of partial esters of polyoxyalkylene triols has been noted. Preparation of a polyoxyethylene-polyoxypropylene glycerol having a molecular weight of between 4000 and 4500, and containing 37.5 per cent oxyethylene, is described.

Heteric-Block Non-ioncs

Heteric-block polyols

Heteric block surfactants derived from a trifunctional initiator, including glycerol or trimethylpropane have a heteric hydrophobe utilising only propylene and ethylene oxides, and an oxyethylene hydrophile. The polyols, seems to be satisfactory for use in a rinse aid, in combination with other non-ionic surfactants.

Silicone Containing Non-ionics

Silicone surfactants

Silicone surfactants are generally co-polymers of a polysiloxane and one or more oxyalkylene polymer chains. A typical general chemical structure may be represented as

$$R'[R_2SiO)_y]_a[(C_nH_{2n}O)_xR'']_b$$

in which y denotes the number of siloxane units and has a value of at least two or three, n denotes the number of carbon atoms in the oxyalkylene group and is an integer from two to four, x denotes the length of the oxyalkylene chain and is at least five, b is an integer whose value may be from one to six and denotes the number of polyoxyalkylene chains, a is an integer of one or two, and R, R′, and R″ represent hydrocarbon radicals.

The polyoxyalkylene chain may be an all-block non-ionic, a heteric-block non-ionic, or an all-heteric non-ionic and may be derived from oxyethylene, oxy-1,2-propylene, oxy-1,3-propylene, oxybutylene, and combinations thereof. In Union Carbide's L-520 or in General Electric's Silicone SF-1034 for example, which has a structure of

$$C_2H_5Si[O(CH_3)_2O)_2(C_2H_4O)_x(C_3H_6O)_yC_4H_9)]_3$$

the organic portion comprises a 50 per cent ethylene oxide and 50 per cent propylene oxide heteric co-polymer terminated with a butoxy group. The silicon portion has a molecular weight of approximately 1500, and each of the three branch chains of the organic portion also has an average molecular weight of 1500. Since many silicone surfactants appear to have three organic groups, they are arbitrarily included in class III.

Without a surfactant, the formation of flexible urethane foams would not be feasible. It has been suggested that the function of a silicone surfactant in a flexible urethane foam system is three-fold:

Emulsification

The water-soluble co-polymer acts as an emulsifying agent resulting in a smooth reaction among the three compounds, water, polyol and isocyanate.

Bubble nucleation

The surfactant depresses the surface tension of the system, helping to speed the release of the bubbles of gas as they are generated. This results in finer cells.

Bubble stability

A high surface viscosity, resulting from a surfactant concentration at the bubble surface, promotes film resilience and increases stability of the bubble. Polysiloxane-oxyalkylene co-polymers, together with an organo-tin compound, are reported useful of preparing curing compositions particularly adapted for use in manufacturing stable flexible urethane foams. Sillicone surfactants are also extremely useful in rigid urethane foams. Here the surfactant contributes to maximum foam rise, resulting in a lower density and a better insulator. Without the surfactant, the foam often cannot negotiate bends, so that more pouring sites are required, with added labour cost. Second, the silicone surfactant, by slightly altering the surface chemistry or the rising foam, produces a finer cell structure with a higher percentage of closed cells. This results in better insulation and lower moisture permeability. Third, a foam with greater uniformity of cell structure is obtained. This eliminates the formation of voids and variation in density throughout the foam, and results in a uniform load factor, thereby contributing to structural soundness.

Organosilicone surfactants are produced commercially in the USA by the General Electric Co., the Dow Corning Corpn., and the Union Carbide Corpn.

TETRAFUNCTIONAL INITIATORS

All-Block Non-ionics

Amine-initiated polyols

Non-ionic surfactants have been derived from reactive compounds containing at least one nitrogen atom and four reactive hydrogen atoms. An important characteristic of this class of compounds, like the Pluronic polyols, is that the nucleus must be lower-molecular-weight compound which is itself not hydrophobic. The initiator can have up to six carbon atoms per molecule, and must contain one or more nitrogen atoms, and may contain from two to six reactive hydrogen atoms.

Aliphatic primary diamines are preferred, since they give products with the highest detergency values, although aromatic amines, heterocyclic compounds, alkanolamines, etc., can be used. The addition of oxypropylene chains, having a total molecular weight of atleast 900 and up to about 2500, provides the hydrophobe. The oxyethylene chains are connected to these and are present in the amount of from 20 to 90 weight per cent of the mixture. Although the use of pure propylene oxides is preferred, trace amounts of ethylene oxide can be included to prepare the hydrophobe, without significantly altering the detergency properties of the final composition.

Similarly, the ethylene oxide, which is then condensed with the hydrophobe, can contain trace amounts of propylene oxide without significant alteration of the detergent properties of these compositions.

Tetronic polyols

Surfactants of the above type can have from one through six reactive hydrogen atoms and therefore fall into classes IA through VIA. However, they have been placed in IVA, tetrafunctional initiator, because only those derived from ethylenediamine are commercially available, under the Tetronic trade name, from the Wyandotte Chemical Corpn. They have the structure

$$\begin{array}{ccc} H(C_2H_4O)_y(C_3H_6O)_x\diagdown & & \diagup(C_3H_6O)_x(C_2H_4O)_yH \\ & NCH_2CH_2N & \\ H(C_2H_4O)_y(C_3H_6O)_x\diagup & & \diagdown(C_3H_6O)_x(C_2H_4O)_yH \end{array}$$

where the molecular weight of each polyoxypropylene group ranges from at least 225 to about 6250 and y is 20 to 90 per cent of the total weight of the product.

The tetronic polyols are identified by three-digit numbers, which indicate the composition of the particular surfactant. The first digit identifies the average molecular weight of the hydrophobe. The second is always zero and the third digit represents 1/10 of the approximate weight per cent of the hydrophile group in the molecule.

Applications of tetronic polyols

In addition to their value as demulsifiers in the petroleum industry the tetronic polyols of the seven to nine series reported to be useful for incorporation in drilling fluids to inhibit hydration of clay or water-swellable formations. The tetronic polyols with a low oxyethylene content are used as defoamers in many areas. The deaeration of anionic detergent compositions is reported with the use of tetronic polyols 704 and 707. The property of the tetronic polyols to sequester lime-soap curd is taken advantage of in a combination soap detergent bar composition. In a patent on soap detergent bars, the tetronics are utilised to improve cold water solubility and to replace coconut oil soap in a coco-tallow soap bar, without loss of lather properties. Other detergent compositions also include tetronic polyols. Tetronic 908 is one of the preferred surfactants of a solid rinse-aid bar used in automatic dishwashing machines. The tetronic polyols can be used as dispersing or wetting agents in a kit designed to allow the 'do-it-yourself' housewife to dye her own fabrics at home. The dispersing property of tetronic 908 is recommended for a talc pigment composition.

The antistatic properties of the tetronic polyols are reported satisfactory for use in polyethylene as well as for synthetic resin moulding powders. Phenolic-formaldehyde resin compositions utilise the tetronic polyols because of their excellent plasticising action and resistance to flaking. Tetronic 908 is recommended as a co-polymer in the preparation of vinyl acetate graft co-polymers. The addition of a block co-polymer oxypropylene-oxyethylene adduct of ethylene diamine to viscose in extruding results in cross-circular filaments with a non-crenulated surface and a thick skin. Improvement in the process for producing alkali viscose solutions from wood pulp, utilising tetronic polyols (304, 504, 701, 702, 704, 707 and 908) has been disclosed. The tetronic polyols are finding extensive application in the preparation of both rigid and flexible polyurethane foams. Patents in the urethane field reveal the advantages of using the ethylene-diamine-derived block co-polymers of ethylene and propylene oxides. A patent has also been granted for the use of tetronic polyols in the vulcanisation of rubber. Block co-polymer oxypropylene-oxyethylene adducts of ethylenediamine are useful as thickening agents for aqueous-based hydraulic-fluid compositions. In the treatment of metal surfaces to inhibit the corrosive action of liquids, tetronic polyols 702, 904, and 908 are of value. Tetronic 908 is used as one of the emulsifiers in a latex-base paint. One patent discloses the use of tetronic polyols, such as 304, 504 and 908, for stabilising 3 to 90 weight per cent hydrogen peroxide solutions over the pH range of 1 to 9. The polyols concentration in the finished product can range from 2 to 30 weight per cent.

It is reported that low levels of tetronic 908 exert a synergistic effect on the antibacterial effectiveness of 3,4,4′-trichlorocarbanilide in soap, using *S. aureus* as the test organism. Higher concentration levels, however, act as neutralising agents.

Heteric-Block Non-ionics

Pentaerythritol-initiated polyols

In addition to the pentaerythritol-derived heteric-block polyols covered previously, other oxyalkylated derivatives of pentaerythritol are claimed in a DeGroote patent. Their chief use is in the breaking of water-in-oil emulsions, but they are also claimed to be valuable as wetting, detergent, and dye-levelling agents, among other applications. The preferred compositions, which are used as demulsifying agents in refinery continuous desalting operations, contain about twice as much ethylene oxide by weight as propylene oxide, and the weight ratio of the ester to the total oxide is 1:2.

PENTAFUNCTIONAL INITIATORS

All-Block Non-ionics

Polyamine-initiated polyols

Block co-polymers of diethylenetriamine, a pentafunctional initiator, are the products made by the sequential addition of propylene and ethylene oxides, by the normal oxyalkylation procedure. Products made by the reverse order of addition, namely, ethylene oxide added to the initiator first, followed by propylene oxide, are also described. The minimum initiator content is 1.75 per cent, with a maximum of 10 per cent of the final product. This indicates that the theoretical molecular weight of the products formed ranges from about 1030 to 5890. The propylene oxide content varies from 50 to 88 per cent and the ethylene oxide content varies from 10 to 48 per cent of the final product. The products are found useful for resolving petroleum emulsions of the water-in-oil type. A similar series of block co-polymers, derived from a different pentafunction initiator, is that based on 3,3′-iminobispro-pylamine.

These surfactants are made by the sequential addition of propylene and ethylene oxides, or vice versa. The standard oxyalkylation procedure is utilised. The 3,3′-iminobispropylamine in the finished product comprises between 1.75 and 10 per cent, corresponding to a theoretical molecular weight range of from 1310 to 7500. The oxypropylene chain comprises 50 to 88 per cent, and the oxyethylene content ranges from 10 to 48 per cent of the finished product. These products are also used for breaking water-in-oil emulsions.

HEXAFUNCTIONAL INITIATORS

All-Block Non-ionics

Hexitol-initiated polyols

Block co-polymers that are derived from a hexafunctional initiator have been prepared from hexitols, such as sorbitol or mannitol. They may be represented by the formula:

$$R([O(CH_2)_x]_m[O(CH_2)_y]_nOH)_6$$

where R is a residue of hexitol, x and y represent 2 or 3 and are of different values; m and n each represent an integer above 6 and preferably below about 35, and the ratio of m to n within the limits of from 3:1 to 1:3. The polymers are used for petroleum-demulsifying agents, as humectants, and as thickening agents.

Triethylenetetramine-initiated polyols

Another series of surfactants derived from a hexafunctional initiator is based on triethylenetetramine. The theoretical molecular-weight range of these surfactants is about 1460 to 8350. The range of oxypropylene and oxyethylene contents is 50 to 88 per cent, and 10 to 48 per cent, respectively. Similar homologous-block co-polymers are derived from tetraethylenepentamine (Class VIIA) and higher polyethylene polyamines. The surfactants derived from tetraethylenepentamine have a theoretical molecular weight range of about 1890 to 10,800. All the products are recommended as useful for breaking water-oil emulsions.

OCTAFUNCTIONAL INITIATORS

All-Block Non-ionics

Sucrose-initiated polyols

Block co-polymeric surfactants derived from an octafunctional initiator are those obtained from sucrose. By oxyalkylation of the sucrose with propylene or butylene oxide, for example, the hydrophilic nature of the sucrose is radically altered to convert it to a xylene-soluble compound. This is then followed by oxyethylation or by reaction with ethylene oxide and then glycide.

Alternatively, the oxyethylation may precede the oxypropylation. The per cent of sucrose in the final product is less than 10 and preferably 2 to 7 per cent. The ethylene oxide content is preferably 23 to 70 per cent and the propylene oxide content is from 23 to 70 per cent, with the total of all three being 100 per cent.

Thus for the classification of various EO-PO co-polymers and their respective applications have been described in details. This should given an idea about the wide range of products available and also the spectrum of their applications. The products are said to be useful as emulsifying agents, as components in the preparation of micellar solutions and for resolving petroleum emulsions of the water-in-oil type.

Functional Fluids

Monohydroxy polymers of propylene oxide

The commercial series of monohydroxy polymers of propylene oxide are linear and based on an alcohol. They are characterised by an alkoxy group at one end and a hydroxy group on the other. Polymers of this type are available in a wide range of molecular weights. The principal manufacturers are Union Carbide Corporation (Ucon LB series) and Olin Corporation (Poly F WI series).

Random co-polymers of propylene oxide and ethylene oxide

The two commercial classes of random co-polymers of ethylene oxide and propylene oxide are linear monols and diols. The monohydroxy polymers have an alkoxy group on one end and a hydroxy group on the other, whereas the dihydroxy polymers have a hydroxy groups on each end. These polymers are available in a wide range of molecular weights (Table 8.1).

Table 8.1. Random EO-PO co-polymers.

Polyether type	*Starter*	*Ethyloxy groups %*	*Molecular weight range*	*Trade name*	*Manufacturer*
Monol	Alcohol	50	270–4,000	Ucon 50-HB	Union Carbide Corp.
			270–4,000	Poly G WS	Olin Corp.
			1,800–4,000	Jeffox WL	Texaco Chemical Company
			730–4,000	Pluracol W	Wyandotte Chemicals Corp.
Diol	Glycol	75	1,050–22,000	Ucon 75-H	Union Carbide Corporation
			2,500–12,000	Poly G WT	Olin Corporation
			2,500	Jeffox WL	Texaco Chemical Company
			7,000–22,000	Pluracol V	Wyandotte Chemicals Corp.

Handling and storage

Polypropylene glycols, random co-polymers of propylene oxide and ethylene oxide, and urethane polyols are all liquids and can be stored in carbon-steel tanks. High viscosity fluids are stored in heated tanks. Urethane polymers must be protected from atmospheric moisture. Centrifugal pumps can be used for fluids of up to about 500 sq. mm (=cSt), but higher viscosity fluids may require a gear or rotary pump.

Health and safety

The alcohol or diol-based polymers of propylene oxide and co-polymers of propylene oxide and ethylene oxide are relatively non-toxic. The lower molecular weight species may be slightly toxic when taken orally. Alcohol-based propylene oxide polymers and random polymers with viscosities above 500 SUs at 37.8°C exhibit low-order oral toxicity. Fluids below 500 SUs may cause slight-to-moderate eye injury. The product below 300 SUs may cause transient reddening of the skin. None of the higher molecular weight fluids cause eye or skin irritation. Owing to very low vapour pressures, no respiratory problems are expected. However, exposure to mists or vapours generated at high temperature may result in inhalation of harmful material. Adequate ventilation should be provided to avoid prolonged or extensive inhalation of the vapours or oxidation products. A variety of these products have been patch-tested on humans with no indication of sensitivity. Animal lifetime feeding and skin-painting studies give no evidence that long-term repeated exposure to these polymers can cause adverse health effects.

As expected, the block co-polymers are also very low in toxicity. Any toxicity or irritation decreases with increasing molecular weight. The urethane polyols are also of very low toxicity. The lower molecular amine and phenolic compounds have an increased toxicity. All of these polymer types burn, but they have low vapour pressures and high flash points.

Applications

Co-polymers of propylene oxide and ethylene oxide have achieved commercial importance in such diverse areas as fire-resistant hydraulic fluids, lubricants, quenchers and polyether polyols for urethanes. These and other applications have grown to a point where the preparation of propylene oxide consumes a significant portion of the total propylene production. Most of propylene oxide production is used for polymer formation.

The properties and uses of the co-polymers of ethylene oxide and propylene oxide are related to the properties and use of their respective homopolymers. The properties of co-polymers of ethylene oxide and propylene oxide are intermediate between those of the polyethylene glycols and polypropylene glycols. Water solubility depends on the proportion of ethylene oxide. Higher proportions of ethylene oxide increase the resemblance to a polyethylene glycol. Mechanistic conclusions as to the reactions occurring in the preparation of these co-polymers are generally drawn by analogy from studies of the relevant homopolymer. The homopolymers are usually the models for the properties of the co-polymer. Commercially, polypropylene glycols are often classed with the co-polymers because of the similarities of their applications and properties. Inclusion of polypropylene glycols widens the property range of a product line of alkylene oxide-derived polyethers. The general structure of homopolymers and co-polymers of ethylene oxide and propylene oxide is shown below:

$$(CH_2-\overset{\overset{\large R_1}{|}}{C}H-O)_x-(CH_2-\overset{\overset{\large R_2}{|}}{C}H-O)_y$$

where, $R_1 = R_2 = H$ ethylene oxide homopolymer
$R_1 = R_2 = CH_3$ propylene oxide homopolymer
$R_1 = H, R_2 = CH_3$ co-polymer

The presence of the ether linkages is responsible for the unique properties of these polymers. The nomenclature used by Chemical Abstracts for ethylene oxide polymers is oxirane polymers. Propylene oxide polymers are identified as 'methyloxirane polymers', and ethylene-oxide-propylene oxide co-polymers are identified as either 'oxirane, polymer with methyloxirane' or 'oxirane, methylpolymer with oxirane'. In general usage the ethylene oxide-derived polymers are called polyethylene glycols or polyethylene oxide. Similar names exist for the propylene oxide analogues. The co-polymers are usually identified by trade names or simply as ethylene oxide-propylene oxide co-polymers. These should follow the nomenclature polyethylene oxide-co-propylene oxide or polypropylene oxide-co-ethylene oxide, the first for polymers with a higher proportion of ethyloxy groups, the second for polymers with a higher proportion of propyloxy groups. The infix-co is for co-polymers of random or unidentified structure. For block structure the -co is replaced by -b-.

Surfactants

Propylene oxide is homopolymerised or co-polymerised with ethylene oxide—water, alcohols, glycols, polyhydroxy alcohols, amines, and acids are used as initiators to give surface-active agents. The hydrophobic/hydrophilic nature of the co-polymer surfactants can be altered by changing the ratio of propylene oxide to ethylene oxide. An increase in this ratio results in a more hydrophobic surfactant. These surfactants are used in a variety of ways, e.g., in detergents, textiles defoamers, hair dressings, and lubricants.

The surface characteristics of block co-polymers make them useful as dispersants, emulsifiers, wetting agents, foam stabilisers, flocculants, demulsifiers, etc., in many industrial and pharmaceutical preparations. The most widely used block co-polymeric surfactants are those of ethylene oxide (EO) and propylene oxide (PO). Such block co-polymers behave like typical hydrophilic-hydrophobic agents where the surface activity may be tailored just by adjusting the molecular characteristics. For this reason and because of their non-toxicity, EO-PO block co-polymers are used widely in cosmetics (e.g., dentifrices, mouthwashses, aerosol antiperspirant preparations) in the paper industry (e.g., in transfer coatings), in the textile industry (e.g, as antistatic agents), in agricultural formulations (e.g., in the dispersion of pesticides), as detergents, in the petroleum industry and in water treatment. More recent applications of EO-PO block co-polymers as emulsifiers and demulsifiers for various industrial preparations, as dispersants for hydrophobic particles, e.g., pesticides in aqueous media, have also been described.

Other block co-polymer with ethylene oxide as the hydrophilic sequence are being tested for use as surfactants. Recent fundamental studies on alpha-butylene oxide-ethylene oxide block co-polymers of the type AB, ABA, and BAB, and on styrene-ethylene oxide block co-polymers have shown their excellent surface activity. The principal criteria for judging surfactant suitability apart from performance are toxicological and ecological characteristics. Cationic and non-ionic surfactants have come to play an increasingly important role along with their anionic counterparts. Despite the wide choice of possibilities, however, only a very few surfactants account for the major share of the market, partly as a result of economic factors.

Approximately 14 per cent of the ethylene oxide consumed in the USA is used in the manufacture of non-ionic surfactants. These are derived by addition of ethylene oxide to fatty alcohols, alkyl phenols, tall oil alkyl mercaptans and various polyols, such as, polypropylene glycol, sorbitol, mannitol, and cellulose.

They are used in household detergents, industrial surfactant applications, in emulsion polymerisation, textiles, paper manufacture, and recycling, etc.

Defoamers/antifoaming agents

Ethylene oxide-propylene oxide co-polymers are surfactants and are used as defoamers. These product are used in starch solution tanks, evaporation processes, and certain distillation procedures. They must be used in relatively high temperature operations because they may become insoluble in water at elevated temperatures by the cloud point mechanism function.

Polyalkylene oxide co-polymers are frequently encountered in liquid phase antifoam components. For example, alternating co-polymers of polyethylene oxide and polypropylene oxide reduce the foam in the acid gas scrubbing process. Sometimes the polyether is mixed with other liquids, e.g., polypropylene oxide is mixed with polydimethylsiloxane. Alternatively, the polyether is co-polymerised with a silicone. Polyvinyl ethers are also used and are particularly effective in combating foam in mineral oil lubricants.

Dentifrices

EO-PO block co-polymers have been found to be useful as a gelling agent in dentifrices. They have been used in alcohol-based mouthwash. The block co-polymer with ethylene oxide content of >40 per cent prevents the formation of cloudy appearance.

Cosmetics

In the fields of aerosol antiperspirants, it has been reported that certain block co-polymers prevent the staining of clothing after repeated use of the antiperspirant formulation.

Medical applications

Many applications of the EO-PO block co-polymers have been reported. The use of a block co-polymer with at least 50 per cent ethylene oxide content has been reported in a process for preparation of a stable and concentrated antiserum from human or animal plasma and serum, by fractional precipitation. Several block co-polymer amines and their tetraesters have been found to be useful as hypocholesterolaemic agents in animals and man. One EO-PO block co-polymer has been used as an emulsifying agent in radiological studies, and in artificial blood. An antiseptic skin cleaning formulation based upon chlorhexidine gluconate has been developed containing 25 per cent of a block co-polymer. A method for enhancing drug or antibiotic levels in the blood has been reported by oral administration of a capsule containing the drug and a block co-polymer.

Paper

In the paper industry, the preparation of a single transfer coating for paper utilises a block co-polymer on production scale. It has been used as a wetting and dispersing agent to apply coating on a backing surface of the paper sheet. It has been reported that the moisture level in a sheet of cellulose, such as paper, can be stabilised by using a polyalkylene oxide as a stabilising agent and a block co-polymer to enhance the rate of absorption of the polyglycol by the sheet material.

Textiles

The textile industry uses block co-polymers as antistatic agents for fibres. An effective antistatic agents giving improved performance to nylon was obtained by chain extending a block co-polymer amine with a diepoxide or a diisocyanate. Even better antistatic effectiveness is achieved by further reaction with a sulphuric acid derivative, such as sodium paratoluene sulphonate. A novel method for softening laundry was reported by tumbling it in a damp state with coated polystyrene foam spheres. By dip-coating the spheres in a blend of a block co-polymer, sodium tallow alcohol sulphate slurry, and ethyl alcohol, the softener was readily transferred to the laundry while tumbling in a dryer.

Lubricants

Improved lubricating oil compositions containing conventional additives may be obtained by incorporating relatively small amounts, as little as 0.01 per cent, of a block co-polymer amine.

Chemical description and applications of various block polymers are given in Table 8.2.

Table 8.2. Chemical description and applications of block polymers.

Block polymers *Supplier/Trade name*	*Chemical description*	*Functionality application*
Akcros Chemicals		
Monolan 2000/E12; 8000/E80 flake; P222	Condensate of EO with PO, having two terminal hydroxyl groups	Defoaming agent; low foam; wetting agent agrochemicals; emulsion polymerisation; resins; latexes; textile industry
Monolan 2500/E30	Condensate of EO with PO, having two terminal hydroxyl groups	–
Monolan 8000/E80	Block polymer	Co-dispersant aquisition pigment systems
Monolan P 165	Complex of EO-PO co-polymer having three terminal hydroxyl groups	Hydraulic fluids
Akzo Nobel		
Berol 370	Block polymer	Low foam rinse aids
Berol 374	Block polymer	Low foam dishwashing agents
Pluronic PE 3100, 4300 6100, 6200, 6400, 6800, 8100, 9200, 9400	PO-EO block co-polymer	Low foam
Pluronic RPE 2520, 3110	EO-PO co-polymer block	–
Dr. Th. Boehme		
Filipan 7521	EO-PO block polymer	Spin finish textile industry; fibre production
Chemax		
Chemal BP-261, 262 262LF, BP 2101	Difunctional block polymer ending in primary hydroxyl group	Defoaming agent metal working; cosmetics; paper industry; textile industry; dishwashing agents; lubricants
Henkel		
Texadril 2010	EO/PO block co-polymer	Low foam; wetting agent; co-emulsifier plastics industry; coatings
Hoechst		
Genapol PF10, PF20, PF40, PF80	Block polymer	Low foam; dispersant detergents; dishwashing agents
Genapol PN-73	Diamine-based block polymer	–
Huls		
Marlox LM 75/30	Polyethylene-polypropylene glycol ether	Foam regulator low foam industrial cleaners
ICI		
ATLAS G-3888	Capped EO/PO block polymer	Household and industrial applications; agrochemicals; textile industry

(Contd ...)

Block polymers *Supplier/Trade name*	*Chemical description*	*Functionality application*
Hypermer B246, B259, B261	Non-ionic block co-polymer surfactant	Emulsifier polymer industry
Synperonic PE/25R2	POE/POP block co-polymers	Emulsifier; foam control; wetting agent agrochemicals; household and industrial applications
Ukanil 2239	POE/POP co-polymers	Defoaming agent; wetting agent agrochemicals; textile industry and institutional cleaners
Kao Corporation		
DANOX LF-102	EO-PO block co-polymers	–
FINDET SE-1851	EO-PO block co-polymers	–
FINDET SE-2411	EO-PO block co-polymers	–
Dr. W. Kolb		
IMBENTINPAP/3100	EO/PO co-polymer	Personal care products; household detergents; industrial and institutional cleaning; pharmaceuticals
Olin Chemicals		
Poly-Tergent E-17A	EO/PO block polymer	Low foam
PPG		
MACOL 1	Polymers of PO and EO arranged in blocks	Lubricant; wetting agent; w/o emulsifier; defoaming agent; metal working
Protex		
PROX-ONIC EP 1090-1	Difunctional block polymer ending in primary hydroxyl group	Defoaming agents dishwashing agents; metal working; cosmetics; paper industry; textile industry
Surfachem		
SURFAC HT 10	Block co-polymer of PO and EO	–
Union Carbide		
TERGITOL XD	High molecular weight co-polymer of EO and PO	Emulsifier; dispersant insecticides; herbicides, silicone oils; leather industry; emulsion polymerisation, paint industry
Witco		
REWOPAL PO	EO/PO block polymer	Wetting agent non-foaming cleaners
Synthionic D 9500	PO/EO condensate	Detergent; wetting agent; low foam dishwashing agents

CHAPTER 9

Polymerisation Techniques

INTRODUCTION

So far the chemical aspects of polymerisation reactions, the terminology included terms such as polymerisation system and catalyst systems have been discussed. These refer to the physical aspects of polymerisation reactions which decided whether the monomer is polymerised in its condensed gaseous state, whether it is polymerised as such or along with other inert components such as solvents or non-solvents. Factors such as the nature of the monomer, the type of polymerisation mechanism chosen, the required physical form of the polymer and the viability of the process for industrial production are some of those dictating the physical conditions under which polymerisation is to be carried out. These varying physical conditions lead to different polymerisation techniques which are discussed below.

BULK POLYMERISATION

Polymerisation in bulk, that is, from undiluted low-molecular-weight starting materials, is the simplest and oldest method for the synthesis of macromolecules. It is important not only for preparations and kinetic studies in the laboratory but also in industrial production. In comparison with other methods of polymerisation, it has the following advantages: (i) the equipment or apparatus required is relatively simple; (ii) the reaction is relatively rapid and the yield good; (iii) polymers of high purity are formed directly from the reaction; and (iv) the polymers are obtained in a form immediately processable. The technique of polymerisation in bulk is applicable to both addition and condensation polymerisations. Fundamental differences exist, however, in the reaction mechanisms of the two processes. Polycondensation in bulk, the principal use of which is in the preparation of linear polyesters and polyamides, is, however, not considered to be a true bulk polymerisation.

Method and Theory

Unsaturated compounds can be polymerised, in bulk in solution, in emulsion, or in suspension. The choice of method depends upon, among other things, the chemical properties of the monomers and polymers, the equipment necessary, and the end use of the polymers. The most obvious method for the preparation of polymers from unsaturated compounds is, at least for liquid monomers, bulk polymerisation. Bulk polymerisation is the polymerisation of pure monomers, during which the only additives present are such aids to polymerisation as initiators, chain regulators, or activators.

Bulk polymerisation is termed homogeneous or heterogeneous according to whether the product remains dissolved in the monomer or precipitates during the reaction. In Table 9.1 the various polymerisation methods mentioned above are compared and contrasted.

In principle, all polymerisable unsaturated or cyclic compounds can be polymerised in bulk. The procedure is particularly suited to liquid or liquifiable monomers, although it is applicable to gaseous or solid monomers. The importance of bulk polymerisation lies, above all, in its universal applicability in the laboratory for polymer preparation and for studies of polymerisation kinetics and of co-polymerisation behaviour.

Table 9.1. Comparison of polymerisation methods.

Method	*Advantages*	*Disadvantages*
	Homogeneous	
Bulk (batch)	Minimum contamination; simple equipment for making castings.	Strongly exothermic; broadened molecular-weight distribution at high conversion; complex if small particles required
Bulk (continuous)	Lower conversion per pass leading to better heat control and narrower molecular-weight distribution	Requires agitation, material transfer, separation, and recycling
Solution	Ready control of heat of polymerisation; polymer solution may be directly usable	Not useful for dry polymer because of difficulty of complete solvent removal
	Heterogeneous	
Suspension	Ready control of heat of polymerisation; suspension or resulting granular polymer may by directly usable.	Continuous agitation required; contamination by stabiliser possible; washing, drying possibly compacting required.
Emulsion	Rapid polymerisation to high molecular weight and narrow distribution, with ready heat control; emulsion may be directly usable.	Contamination with emulsifier, etc., almost inevitably leading to poor colour and colour stability; washing, drying and compacting may be required

Heat of polymerisation

Since the polymerisation being considered here takes place by an exothermic chain reaction, problems of heat removal frequently arise during polymerisation of large amounts of material, especially when done on an industrial scale. The propagation step in the polymerisation of unsaturated compounds results in the conversion of one double bond into two single bonds, and it is highly exothermic. Although the overall heat of reaction is the sum of the contributions of the different reaction steps (chain initiation, propagation, and termination), the major portion arises from the propagation reaction. The heat of polymerisation of monomers, determined either directly through calorimetric measurements or through comparison of heats of combustion of monomers and polymers, varies between 10 and 30 kcal/mole. It has, consequently, a considerable influence on the course of the reaction and is one of the factors that must be taken into consideration in bulk polymerisation (Table 9.2).

The rate of polymerisation of unsaturated compounds increases with increasing temperature. Moreover, since the thermal conductivity of polymers is low and the viscosity of the solution of polymer in monomer increases rapidly, careful temperature control is indispensable in bulk polymerisation.

Table 9.2. Heat of polymerisation.

Polymers		*Heat of polymerisation kcal/mole*	*Remarks*
		Homopolymerisation	
Vinyl compounds			
Ethylene		22.3–24.-2	
Styrene		16.1–17.5	
Acrylic acid		18.5	In water
Acrylonitrile		17.3	
Acrylamide		19.8	In water
Vinyl acetate		21.3	
Vinyl chloride		14.4–16.7	
Butyl vinyl ether		14.4	
Methacrylic acid		15.8	
Methyl methacrylate		13.0–13.9	In water
α-Methylstyrene		8.0–8.4	
Tetrafluoroethylene		39.0	
Dienes			
1, 3-butadiene		17.4	1, 2 addition
1, 3-butadiene		18.4	1, 4 addition
Isoprene		16.9–17.9	
Chloroprene		16.2	
Formaldehyde and ring compounds			
Formaldehyde		12.2–13.0	
Ethylene oxide		22.6–24.9	
Trimethylene oxide		19.3	
Tetrahydrofuran		2.9–5.0	
Propylene oxide		18.0	
3, 3-di(chloromethyl)- 1-oxacyclobutane		20.0	
		Co-polymerisation	
Monomer A	*Monomer B*		
1-butene	Sulphur dioxide	10.6	Excess B
Vinyl acetate	Maleic anhydride	20.2	Excess A
Styrene	Methyl methacrylate	13.2–16.6	
Styrene	Butadiene	17.1–17.7	
Styrene	Vinyl acetate	17.2–18.1	

The effect of temperature is particularly marked if the polymerisation proceeds quite rapidly or if it takes place in large reaction vessels so that the surface-to-volume ratio is small. If heat removal is not sufficiently rapid, the spontaneous rise in temperature can, in extreme cases, lead to explosive polymerisation. The polymers resulting from uncontrolled reaction have undersirably low molecular weights and a wide molecular-weight distribution. Also, discolourations due to side reactions often present.

In contrast to addition polymerisations, problems of heat removal do not ordinarily arise with bulk condensation polymerisations, and this method is therefore commonly used for the industrial preparation of linear condensation polymers. Condensation reactions are only mildly exothermic and proceed not by a chain mechanism, but by a step mechanism in which each step requires the same activation energy. Therefore, linear condensation polymers frequently have low melt viscosities, which facilitate heat removal and escape of gas bubbles. Consequently, even with large quantities reacting without stirring at temperatures over 200°C there is no danger of spontaneous overheating.

Initiation

Two further factors that must be considered in addition polymerisation in bulk, which have considerable influence on the course of the reaction both for radical and for ionic polymerisations, are the initiation process and the gel effect. In principle, bulk polymerisation can be applied for radically and ionically initiated polymerisations. However, apart from research and a few ionic polymerisations used industrially, nearly all bulk polymerisations are by radical mechanisms initiated either by peroxides and azo compound or by light. The uncatalysed polymerisation produced by heating a liquid monomer in an inert atmosphere is also termed thermal polymerisation.

The dependence of the rate of polymerisation and of the molecular-weight average upon concentrations of monomer and initiator has been determined for a number of monomers. Only the beginning of polymerisation has been taken into consideration in order to avoid complications that may appear later in the course of the reaction (branching, gel effect, interference of decomposition products of the catalyst, etc.). For radical polymerisation of vinyl compounds, almost without exception, a direct proportionality between the rate of polymerisation R_p and the square root of the initiator concentration $\sqrt{[I]}$ has been established (Fig. 9.1). In analogy to this, a linear relationship holds between 1/DP and $\sqrt{[I]}$ (Fig. 9.2). Depending upon the temperature range of the polymerisation, one has a choice among various radical formers. Several of the more important of these initiators and the temperature ranges in which they decompose at a useful rate are given as follows: potassium peroxydisulphate, 40-80°C; benzoyl peroxide, 60–90°C; acetyl peroxide, 40–50°C; cumyl hydroperoxide, 50–100°C; di-*t*-butyl peroxide, 80–150°C; 2,2′-azobisisobutyronitrile, 30–100°C; and *t*-butyl perbenzoate, 90–100°C. The rate of the initiation reaction is dependent upon the rate of decomposition of the initiator.

Radical polymerisations have a high activation energy (about 30 kcal/mole for the common initiators) and are highly dependent on temperature. In contrast to this, the initiation reaction of ionic polymerisations has only a small activation energy (4–10 kcal/mole), so that the rate of polymerisation is little affected by temperature. In many cases ionic polymerisations take place with explosive violence even at temperatures below –50°C (e.g., the anionic polymerisation of styrene in tetrahydrofuran at –70°C, or the cationic polymerisation of isobutylene in liquid ethylene at –100°C).

Accordingly, cationic or anionic bulk polymerisaions of vinyl monomers are more difficult to control than the corresponding radical polymerisations exceptions are the preparation of liquid oligomers where temperature control by constant stirring is possible, e.g., the polymerisation of isobutylene and of alkyl vinyl ethers to viscose liquid polymers.

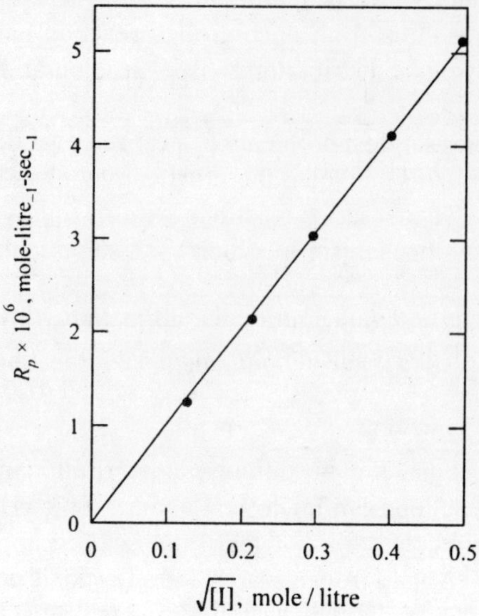

Fig. 9.1. Dependence of rate of polymerisation, R_p, on initiator concentrationl [I], in the bulk polymerisation of styrene with benzoyl peroxide at 27°C.

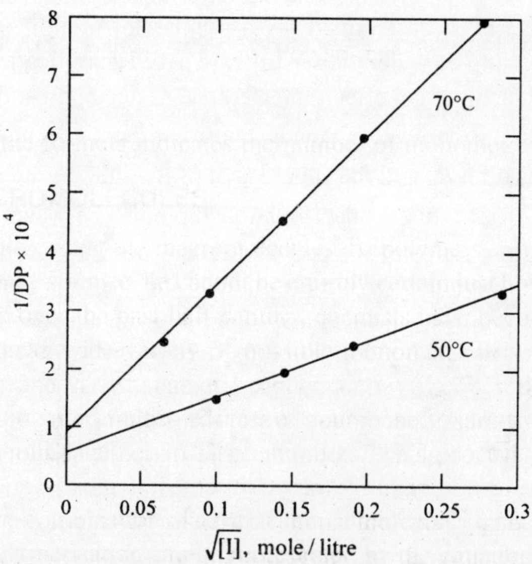

Fig. 9.2. Dependence of degree of polymerisation, DP, on initiator concentration, [I], in the bulk polymerisation of methyl methacrylate with benzoyl peroxide at 50 and 70°C.

Gel effect

Normally the polymerisation of unsaturated compounds follows first-order kinetics with respect to monomer concentration. This clearly holds true in the region of low conversions. However, if the course of polymerisation is followed to higher conversions, such as are reached in bulk polymerisation, sharp deviations from first-order kinetics nearly always occur. With some monomers, at conversions over 30 per cent, a phase of accelerated polymerisation rate and of increased molecular weight appears. The phase of autocatalysis, which is of great importance in bulk polymerisation, is called the get effect or Trommosdorff effect. It introduces further methacrylic acid derivatives (Fig. 9.3).

The gel effect is independent of initiator. It is based on a severe curtailment of the mobility of polymer chains in viscous systems, leading to a drastic reduction of chain terminations by re-combination of two polymer radicals. Thus, the normal steady-state equilibrium between radicals being formed through decomposition of initiator and those disappearing through chain termination is disturbed. More radicals appear per unit time, and the overall rate of polymerisation is thus increased.

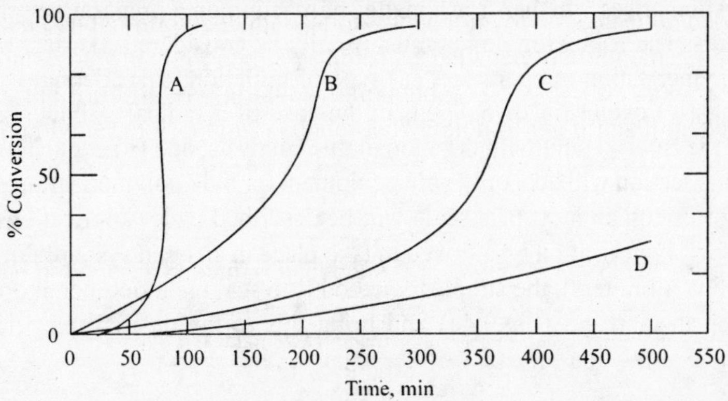

Fig. 9.3. Bulk polymerisation of various acrylic monomers at 60°C. (A) Acrylonitrile with 0.1 per cent benzoyl peroxide. (B) Butyl methacrylate with 1.0 per cent benzoyl peroxide. (C) Methyl methacrylate with 0.1 per cent benzoyl peroxide. (D) Butyl methacrylate with 0.1 per cent benzoyl peroxide.

At the same time the molecular weight of the resulting polymer increases, since the average lifetime of the growing chain is longer. This accounts for the great heterogeneity with respect to molecular weight in bulk polymerisation. If it is not possible to remove the heat of polymerisation liberated by the gel effect, a 'runaway' reaction takes place.

The extent of the gel effect varies greatly. Whereas it is scarcely apparent in the bulk polymerisation of styrene, which is homogeneous, in heterogeneous polymerisations, i.e., of acrylonitrile or of vinyl chloride, the polymerisation rate increases from the instant when the polymer begins to separate out in colloidal form. Here the increased in rate is so rapid that the reaction becomes truly explosive. In polymerisations at temperatures over 100–120°C the gel effect does not ordinarily appear because of increased mobility of chains. In some instances, however, the gel effect is desirable. Polymer is then dissolved in the monomer before the beginning of the reaction to diminish the induction period and to attain a rapid polymerisation and a high molecular weight.

The occurrence of the gel effect is not limited to polymerisation in bulk. It occurs also in precipitative polymerisations and is used, without its undesirable aspects, in emulsion and suspension polymerisations to obtain a better time-yield and a product of higher molecular weight.

Chain transfer

Chain transfer also plays a part in bulk polymerisations, especially with monomers having reactive hydrogen atoms, eg., vinyl acetate.

At high conversions and high polymerisation temperatures there is a tendency to form branched and even cross-linked polymers through chain transfer with polymer molecules already existing. The extent of branching or cross-linking at various conversions may be determined through measurements of viscosity and solubility. To avoid these side reactions, regulators such as mercaptans and other compounds with reactive hydrogen atoms may be added. Such transfer reactions are deliberately used in the preparation of graft co-polymers.

Equipment and Apparatus

The experimental procedures for bulk polymerisation are manifold, since they must be adapted to the specific properties of the monomer and to special polymerisation conditions. Moreover, in laboratory procedures the question arises whether it is a matter of reaction on a preparative scale or of reaction for kinetic measurements. The following points must usually be considered: (i) state of aggregation of the monomer at the polymerisation temperature; (ii) type of initiation; (iii) adequacy of heat removal and temperature control; (iv) exclusion of oxygen, in the case of a radical polymerisation where oxygen itself is not used as the catalyst, and with organometallic catalysts; and (v) exclusion of water from ionic polymerisations. This section will deal only with equipment for bulk polymerisation on a small laboratory scale. Equipment for use on an industrial scale will be described under applications, below.

In no case should polymerisation be allowed to take place in an open system. For liquid monomers in small quantities (a few milliliters), the simplest method is to seal the monomer and initiator in an ampul or a test tube under an inert gas (Fig.9.4a) and bring this to the desired temperature in a constant-temperature bath.

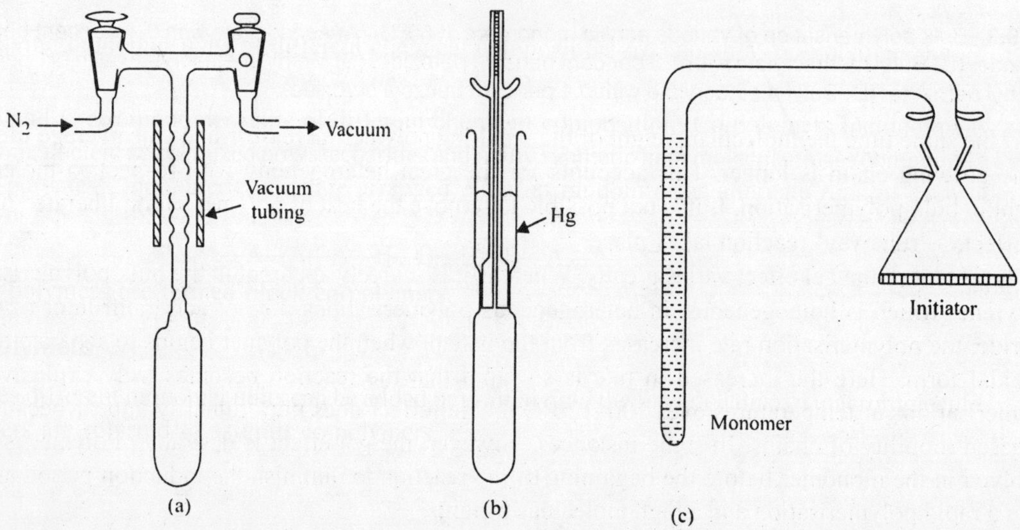

Fig. 9.4. Laboratory apparatus useful in bulk polymerisation. (a) Simple device for melting monomer samples under nitrogen or under vacuum. (b) Dilatometer for measuring volume contraction during polymerisation. (c) Apparatus for polymerisation in the gas phase.

For kinetic measurements, all preparatory steps are best carried out under high vacuum, before the reaction vessel is sealed. With larger quantities (up to 2 litres) multinecked roundbottomed flasks provided with stirrer, reflux condenser, thermometer, and inlet and outlet for an inert gas may be used. It is often useful to also include an arrangement for sampling. The course of the polymerisation can then be followed either by precipitating and weighing the polymer or by determining the remaining monomer content by titration of the sample.

If the polymerisation is to be followed precisely, it is best to work in dilatometers for kinetic measurements. In these, the measure of the volume contraction occurring during polymerisation may be followed. In the simplest, the full polymerisation chamber is closed with a graduated capillary with which the liquid level can be read precisely (Fig.9.4b).

Gaseous monomers can be polymerised under normal, reduced, or super atmospheric pressures. The apparatus for polymerisation at atmospheric pressures is the most convenient. Boiling isoprene can be polymerised in a flask with a reflux condenser in which metallic sodium or potassium initiator is suspended in the monomer vapour. The polymer is deposited as a rubbery precipitate on the metal. In another simple apparatus (Fig. 9.4c), the gas-phase polymerisation of isoprene catalysed by aluminum chloride may be measured. If the monomer is to be polymerised under increased pressure to improve the space time yield, sealed tubes, pressure vessels, or autoclaves may be used. In autoclaves, the polymerisation may be followed by measuring the decrease in pressure. The best-known industrial process of this type, the high-pressure polymerisation of ethylene, will be described in detail under Bulk polymerisation of gaseous monomers.

No special apparatus has yet been developed for polymerisation in the solid state. Single crystals of monomer or powdered crystals are sealed in ampuls, usually under high vacuum. The polymerisation can then be initiated by radiation or heat, or, in the case of some heterocycles, by Lewis acids.

Applications

The great advantage of bulk polymerisation, both for the laboratory and for industrial processes, is the special purity of the polymer formed, occasioned by the absence of many of the additives necessary in other procedures. The most important fields of application of the process can be summarised as follows: Laboratory research, kinetic measurements, co-polymerisation studies; preparation of viscous polymers of relatively low degrees of polymerisation, e.g., of plasticisers, adhesives, alkyd resins, etc.; raw materials for injection moulding, i.e., thermoplastic polymers; production of sheets, blocks, rods, and other shapes where polymerisation takes place directly in the form desired; monomer polymer mixtures as adhesives, plastic cements, and prosthetic materials used, for example, in dentistry; reinforced and unreinforced casting resins; and new developments in the field of block and graft polymerisation.

Nevertheless, bulk polymerisation has not attained the universal industrial importance of solution, emulsion, and suspension polymerisations. The various methods of bulk polymerisation of vinyl monomers will be described in the following sections.

Bulk polymerisation of liquid monomers

In the bulk polymerisation of liquid monomers, the monomer itself serves as reaction midium at the beginning of the polymerisation; the end of the reaction takes place in a highly viscous solution or in the solid polymer. (Heterogeneous polymerisation, in which the polymer is insoluble in the monomer and immediately separates, will be treated separately).

Laboratory polymerisation

As mentioned above, the chief importance of bulk polymerisation of unsaturated compounds lies in its universal usefulness in the laboratory. It is used here as a means of determining the polymerisation tendency of monomers free from inhibitor, the activity of initiators, the influence of contaminants, reactivity ratios in co-polymerisation, etc. Purity of the monomer is always indispensable, since the chain-reaction polymerisation is highly sensitive to impurities. Impurities of much less than one part per thousand can severely disturb the course of the reaction. The most important contaminants are inhibitors, oxygen, and by-products from the preparation of the monomers and the decomposition and oxidation products of the monomers. Removal of inhibitors is of prime importance in purification. Inhibitors are usually chosen from among compounds that can be easily and quantitatively separated, eg., phenolic materials which can be extracted with bases. Any distillations, crystallisations, or sublimations necessary to purification are best carried out under a protective layer of a gas such as nitrogen or argon. To obtain extremely pure monomers such as are necessary for kinetic studies, polymerisation of a pre-purified sample is initiated. After 5-15 per cent conversion all interfering impurities are consumed and the purified monomer can be distilled off. Along with infrared and ultraviolet analyses and gas chromatography, the polymerisation behaviour itself can be used as a test for the absence of impurities. The absence of an induction period in a catalysed polymerisation and the precise reproducibility of reaction rate under standard conditions have proved to be the most rigorous tests of the purity of the monomer.

Determination of conversion

In order to follow exactly the course of polymerisation, a continuous determination of conversion is necessary. The most obvious method is the separation and weighing of the polymer or the titration of the monomer present. However, since the course of the reaction is interrupted, both methods are troublesome and time-consuming. In the conversion of monomer to polymer, the physical properties of the reaction mixture are altered. The values of these properties can be followed, by appropriate procedures, without interference with the course of the reaction, so that any number of points on the time-conversion curve can be taken in a single polymerisation run. Besides dilatometry, measurements of viscosity, refractive index, ultraviolet and infrared absorption, dielectric constant, and vapour pressure are most often used to follow the course of the reaction. The various methods have been compared but by far best suited to determination of conversion is the observation of the contraction in volume (dilatometry) resulting from the increasing difference in density between monomer and polymer during polymerisation (Table 9.3).

Table 9.3. Density of unsaturated monomers and their polymers.

	Density, d_4^{20}	
Compound	*Monomer*	*Polymer*
Butadiene	0.6276	0.906
Isoprene	0.6805	0.906
Isobutylene	0.6005	0.912
Methyl acrylate	0.9556	1.223
Methyl methacrylate	0.9438	1.213
Styrene	0.9063	1.059
Vinyl acetate	0.9359	1.191
Vinyl chloride	0.9109	1.406

The volume contraction corresponding to 100 per cent conversion amounts to 16.19 per cent for styrene at 25°C, 22.06 per cent for vinyl acetate, 22.50 per cent for methyl methacrylate, and 25.0 per cent for isoprene; it can, according to all existing evidence, be linearly interpolated. Even conversions of less than 1 per cent can be accurately determined. The great advantage of the use of dilatometry, besides its high sensitivity, is the fact that the density is not dependent on small differences in structure or, to any noteworthy extent, upon degree of polymerisation. (It is in just this way that the equally sensitive viscometric methods are limited for accurate determination of conversion.)

Industrial polymerisation

From the industrial standpoint, the bulk polymerisation of liquid monomers can be divided into continuous processes and batch polymerisations (the latter also sometimes known as cast polymerisation or *in situ* polymerisation).

Continuous bulk polymerisation

The best known example of continuous bulk polymerisation on an industrial scale is the production of polystyrene. Styrene is one of the few monomers for which bulk polymerisation is the most commonly used manufacturing process. Initiation is by a radical mechanism, and is either purely thermal or else is catalysed by azo compounds or peroxides.

This continuous bulk polymerisation of styrene is carried out in two steps: a pre-and a post-polymersation. The first step consists of the partial polymerisation of styrene. This step either can be carried out in a single stationary reactor, or it can be tied in directly with the other units, as shown in Fig. 9.5. The pre-polymerisation takes place under nitrogen in two aluminium kettles of 2 m^3 capacity each. These are heated, with stirring, to 80°C by a hot-water jacket or hot-water coils, and the monomer is continuously added at about 25 kg/hr per vessel. The course of polymerisation is followed by measurement of, for example, refractive index. After a holding time of some 55 hr, up to one-third of the monomer is polymerised. The polymerisation mixture, which is highly viscous and water-clear, is fed into the upper end of the vertical column reactor in which the post-polymerisation is carried out. The reactor has a capacity of about 15,000 kg of polymerisation mixture and is divided into six zones. The zones, which are independent of each other, are heated with superheated steam to temperatures of 100–180°C, except for the last zone, which is heated electrically to 220°C. At these temperatures the polymerisation proceeds smoothly and with considerable speed.

The high temperature at the end of the column guarantees an essentially complete polymerisation. Mechanical mixing between the zones is prevented by the increasing viscosity of the polymer solution. After an average residence time of 28–30 hr the molten polymer (at 220°C) leaves the column through an extruder. It is conveyed over an air and water-cooled steel belt to comminuting equipment, where it is chopped and ground.

Although the system operates under greatly varying conditions, particularly temperature, the properties of the polystyrene produced are remarkably constant, once the dynamic equilibrium of the system has been properly adjusted. The operation runs continuously under completely automatic control. The product obtained still contains a small amount of styrene monomer, which, however, is not troublesome in most applications (eg., moulding powders). All volatile components may be removed by sintering the polymer at 120–130°C under a vacuum of 1–10 mm Hg. Moreover, polystyrene is extruded industrially under vacuum. Heating with infrared radiation or degassing in towers by the application of vacuum have also been used for removing residual monomer.

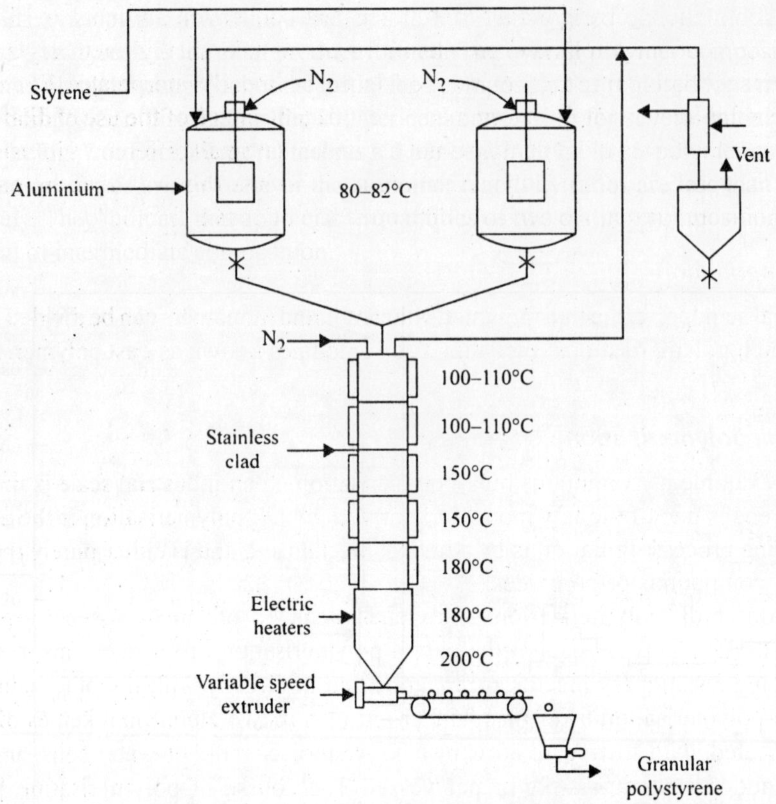

Fig. 9.5 Vertical column reactor for the continuous bulk polymerisation of styrene.

Another process for the continuous bulk polymerisation of styrene is shown in Fig. 9.6. The monomer is pre-polymerised in a polymerisation kettle (A) at 80–200°C to 30–50 per cent conversion. The viscous polymer mixture then goes to a screw conveyor (B) in which the temperature is slowly raised until the greater part of the monomer has reacted. At the end of this step the product, now at a temperature of about 200°C, is introduced into a vaporisation column (C), in which fine filaments of molten polymer fall through a vacuum chamber. Under these conditions the fraction of residual monomer and other volatile components is reduced to a very low value.

Polystyrene with somewhat improved heat resistance can be obtained if it is possible to avoid the formation of styrene oligomers such as occurs at polymerisation temperatures over 180°C. In order to take this into account in a bulk polymerisation, instead of completing the polymerisation process at a higher temperature, the residual monomer can be distilled off. This is done in the manufacture of polystyrene a syrup polymerised up to 40 per cent conversion is allowed to come into contact in a vacuum chamber (160 mm Hg pressure) with steam heated rollers. The monomer thus removed from the polystyrene is condensed and returned to the polymerisation mixture.

The bulk polymerisation of vinyl acetate is also of commercial interest. However, owing to its high chain-transfer constant, linear poly (vinyl acetate) can be obtained only at low temperatures or through the addition of chain regulators such as aliphatic aldehydes and chlorinated hydrocarbons (carbon tetrachloride). The polymerisation is usually carried out at the boiling temperature of the monomer

(72–73°C) with organic peroxides such as diacetyl peroxide, benzoyl peroxide, or peroxides of higher aliphatic acids (eg., dilauryl peroxide) as initiators. An incremental addition of cold monomer facilitates temperature control. Polymerisation is usually carried out only to conversions of 30–50 per cent and then the monomer is distilled from the viscous mass. Continuous polymerisation in a column reactor can be used only for poly (vinyl acetate) with a softening point below the decomposition temperature of the polymer, i.e., only poly (vinyl acetate) of intermediate degrees of polymerisation. However, the most important method of preparing this polymer is actually by emulsion polymerisation.

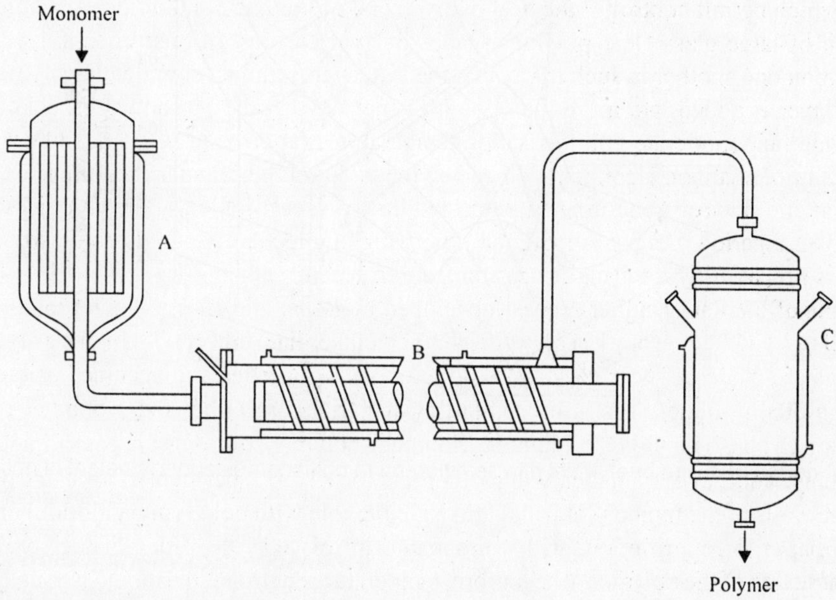

Fig. 9.6 Schematic representation of equipment for continuous polymerisation. (A) pre-polymerisation reactor (30–50 per cent conversion); (B) polymerisation screw (90–95 per cent conversion); (C) vapouriser for removal of residual monomer.

Batch polymerisation

For reasons of economy, industrial procedures tend to be continuous processes. Industrial batch-type processes, therefore, are usually used only when the polymer can be prepared directly in the desired form. Batch or *in situ* polymerisation of methyl methacrylate in particular has been developed industrially. Here flawless blocks, sheets, bars, rods, and tubes are polymerised directly from monomer or from monomer polymer mixture. The special position of methacrylic esters in this process lies in their ability to polymerise completely and in the combination of optical purity and excellent resistance to ageing and weathering of the products. An important factor in the economics of methyl methacrylate plastics is, moreover, the fact that the monomer can be recovered by pyrolysis of scrap. There are several requisites for the successful batch polymerisation of methacrylic esters. These are, in particular, design of the provision for heating and cooling and purity of the monomer. Note should be taken here of the pronounced gel effect (Trommsdorff effect) of acrylic and methacrylic acid derivatives (Fig. 9.3.)

In the simplest version of this process, polymers are prepared for use as raw materials for lacquer or for injection-moulding compounds. Polymerisation is carried out in flat chambers that facilitate removal of the heat of reaction. Especially low-molecular-weight resinous polymers, such as are used as components of special alkyd resins, may be prepared in such reactors equipped with stirrers. To avoid the sudden development of heat caused by the gel effect in the polymerisation of large quantities, the monomer is added to the polymerisation mixture at such a rate that the monomer-polymer composition is not altered during the process. After the last addition, the reaction mixture is heated until conversion is complete, and the polymer is withdrawn at the elevated temperature as the melt.

In order to obtain flawless, homogeneous poly(methyl methacrylate) by direct polymerisation, moulds must be used which permit control of the heat evolved and of volume contraction. A simple method for the preparation of large glass-clear polymer shapes is layer casting. In this process, layers a few cm thick are laid upon one another in such a way that the last layer is almost completely polymerised before the following layer is added. No more than one layer per day can be put down, especially with large blocks of polymer and moderate polymerisation temperatures (approx. 60°C). Rods of polymer can be produced in a similar manner. Here a monomer-polymer solution held under pressure is polymerised in a water-bath at the desired reaction temperature; the surface of the water-bath rises slowly and continuously over a period of days. A material of very high molecular weight is obtained, which is thus superior to bars prepared by the usual extrusion from the melt.

The principle of the flat-chamber process developed by *Rohm* and *Haas*, which promotes good heat removal, is still used in the preparation of poly (methyl methacrylate) sheet with high clarity and flawless surfaces. Forms are made of sheet glass, and, if the optical properties and accuracy of the surface are important, high-quality plate glass is used. Starting materials are either the monomer or monomer-polymer solution that also contain the initiator. Monomer-polymer solutions are easy to pour, cause few leakage problems, and have a smaller volume contraction (10–15 per cent). To take up the volume contraction, the walls of the moulds (or cells) are made flexible with poly (vinyl chloride) or with rubber wrapped with paper, or, more often, with cellophane. In this way the sides of the form can follow exactly the contraction of the polymer. The uniform temperature control of the mould is usually maintained by the use of warm air. If the removal of heat is not uniform, the degree of conversion within the sheet varies and stresses result, giving rise to optical disturbances in the sheet. Depending on the initiator used, the polymerisation temperature chosen lies between 20 and 50°C. Polymerisation times range from 1–2 days to 1–2 weeks or more for thicker sheets. The final polymerisation, that is, the polymerisation of the last 5–10 per cent of the monomer, must take place in the vicinity of, or even slightly over, the softening temperature in order to end the polymerisation in a reasonable length of time.

This procedure can also be applied to other monomers and is particularly suited to polyester-styrene mixtures. Addition of bifunctional cross-linking agents such as divinylbenzene or ethylene glycol dimethacrylate increases the danger of flaws in the sheet, but results in products with heigher glass-transition temperatures and greater strength, as well as better temperature and solvent resistance. Polymerisation in moulds is of considerable significance in the electronics industry. Electronic components are embedded in monomer-polymer systems that are then polymerised. Besides polyester resins, *N*-vinylcarbazole polymers, distinguished by their high softening temperatures and low loss factors, are sometimes used for this process.

Polymerisation of monomer polymer solutions and slurries

The problems of heat transfer and volume contraction, which arise particularly in batch polymerisation, can be considerably lessened by the use of polymer-monomer solutions and slurries. For embedding anatomical and biological specimens, for example, coloured solutions of poly(methyl methacrylate) in the monomer are used. The solutions are injected into body cavities (eg., veins or arteries) and polymerised by immersion in warm water. After mechanical or hydrolytic removal of the coverings, accurate copies of the cavities are obtained. When glasslike clarity of the product is not important, the use of catalyst-containing monomer-polymer slurries has proved suitable. From 40 to 70 per cent monomer is mixed into 60–30 per cent powdered polymer (or, better, fine beads). The plastic paste resulting from swelling and partial solution of the polymer beads is pressed into the desired form and polymerised.

This process is much used in dentistry for dental plates, tooth fillings, and dentures; the rubber and cellulose nitrate preparations used in this field before 1940 have been completely displaced by methacrylate polymers. To obtain natural-coloured artificial teeth, the dentist mixes (a few minutes before use) about one part of polymer beads, containing the initiator (benzoyl peroxide) and the pigment, with about three parts of monomer stabilised with inhibitor. The resulting plastic paste is pressed into the plaster cast and polymerised in a water-bath at 60–90°C, frequently by raising the temperature in increments. Although methyl methacrylate is used almost exclusively as monomer, vinyl chloride-vinyl acetate co-polymer is also coming into use. Redox-initiated polymerisation (at 25°C) of polymer monomer slurries also has numerous applications in the dental field. In this case the polymerisation mixture contains peroxide (eg., 0.5–1.0 per cent benzoyl peroxide) and, as activators, 0.5–7.0 per cent of sulphinic acids or their salts, mercaptans, or tertiary amines such as *N, N*-diethylaniline or *N, N*-dimethyltoluidine. To increase the strength and reduce swelling of the artificial teeth, up to 5 per cent of a bisacrylate or allyl acrylate may be added as cross-linking agent.

Such monomer-polymer slurries are also widely used outside the dental field. For example, the following uses may be mentioned: soft polymethacrylate and polyacrylate as obturators for facial or jaw restoration by plastic surgery; as moulds for casting and for other forms; as moulding materials, calking materials, and cements; and as materials for embedding specimens and for the preparation of thin sections for microscopic study.

A wide variation of properties of the end product is possible through use of different polymers and co-polymers. For example, co-polymers containing acrylic acid or maleic acid give resins with high adhesion, particularly to metals. A rapidly developing field for poly (methyl methacrylate) solution in monomer is in reinforced as well as in unreinforced casting resins.

Heterogeneous bulk polymerisation

The polymerisation kinetics of vinyl monomers in reaction media (in this case, monomer) in which the polymer is insoluble deviate sharply from those of ordinary polymerisations. The polymerisations proceed throughout with higher reaction rates and the polymers formed have higher molecular weights than in homogeneous polymerisations under comparable conditions. Acrylonitrile, vinylidence cyanide, nitroethylene, and halogen-substituted ethylenes behave in this way.

Detailed investigations of the heterogeneous bulk polymerisation of vinyl chloride and acrylonitrile have been carried out. With vinyl chloride the rate of polymerisation is accelerated up to 40 per cent conversion by the precipitating polymer. The molecular weight is almost independent of catalyst concentration and of conversion. This phenomenon can be explained by assuming chain transfer of growing radicals with terminated ('dead') polymer. Immobile free radicals with long lifetimes are thus produced on the surface of the polymer particles. However, the radicals do not become entrapped in the polymer but, because of the high chain-transfer tendency of vinyl chloride, are transferred from the polymer surface back to the monomer. In keeping with this theory, the acceleration of polymerisation by the precipitating polymer is proportional to the surface of the particles. In contrast to this, radical transfer by the monomer is very slow in the otherwise analogous course of bulk polymerisation of acrylonitrile; thus there is permanent trapping of the radicals in the polymer coils. The lifetime of these radicals at room temperature is several hours; they can be detected by chemical methods and by electron paramagnetic reasonance. If such radical-containing polymers are heated in the presence of vinyl compounds, an extremely rapid polymerisation takes place with the formation of graft or block co-polymers.

Although very pure polymers result from bulk polymerisation, since no solvents, emulsifiers, or suspension stabilisers are added, the process is seldom used industrially. Most monomers polymerise violently in bulk and thus complicate control of the reaction. Usually in these cases precipitative polymerisation, i.e., in solvents in which the monomer but not the polymer is soluble, is more easily carried out.

Use of heterogeneous bulk polymerisation is advantageous only if the polymerisation increases the reaction rate as well as the degree of polymerisation of relatively slowly reacting monomers. An example of this is the industrial bulk polymerisation of chlorotrifluoroethylene, which usually reacts extremely slowly. Continuous low-temperature polymerisation can be carried out using trichloroacetyl peroxide catalyst. The precipitated polymer is filtered off and the monomer returned to the reaction. The French company uses a bulk polymerisation process for vinyl chloride in which horizontally mounted reactors with a capacity of 12,000 litres are rotated about their axes, thus making possible the uniform removal of the heat of reaction. Howevere, the reaction cannot be allowed to run to completion, because the danger of explosive polymerisation exists. Heterogeneous bulk co-polymerisation of maleic acid with methyl vinyl ether may be mentioned as a further example of the industrial use of this process. These two monomers give alternating co-polymers. The vinyl ether itself may serve as the solvent for maleic anhydride, since a deficiency of one component is without influence on the polymer composition.

Popcorn polymerisation

'Popcorn' polymerisation also called 'proliferous' or 'cauliflower' polymerisation, may be considered a modification of heterogeneous bulk polymerisation. Opaque regions of characteristic macroscopic appearance can be produced spontaneously under certain circumstances in unsaturated systems capable of forming insoluble polymers of limited swelling capacity. The appearance of these precipitates cauliflower-like, crusted, and popcornlike has given the polymerisation its name.

The formation of popcorn polymers in the co-polymerisation of styrene with divinylbenzene and other bifunctional monomers has been investigated and depends on both the polymerisation temperature and the content of divinyl compound. For example at 65°C homogeneous cross-linked polystyrene is formed with either 0.3 per cent or with 3.0 per cent divinylbenzene; with 1.0 per cent divinylbenzene, on the other hand, opaque popcorn polymer is formed almost exclusively. However, if the 1 per cent divinylbenzene mixture is polymerised at 150°C, glass-clear styrene co-polymer is again obtained. Methyl acrylate even tends to form popcorn polymers without the addition of a divinyl compound; here the insolubility may be attributed to cross-linking through chain transfer.

These polymers have been quite troublesome in the industrial synthesis of styrene-butadiene co-polymers, since the proliferous growth leads to obstruction of pipes in the plant equipment. Sodium nitrite, iodine, peroxides, and nitric acid may be used to eliminate the formation of such polymers. The active chain ends of popcorn polymers remain reactive for a long time. If popcorn polymers are introduced into their own or another monomer, they initiate polymerisation again with the formation of grafted popcorn polymers. The proliferous polymers will retain their ability to grow even if heated at 260°C for 20 hr before being allowed to come in contact with fresh monomer.

Ionic bulk polymerisation of vinyl compounds

The ionic polymerisation of vinyl compounds differs from radical polymerisation particularly in the low activation energy of the initiation reaction. Consequently, the rate of polymerisation has only a slight dependence on temperature. Chain termination ordinarily does not take place spontaneously, but only when impurities or deliberately added terminating agents are present. Accordingly, ionic bulk polymerisations are even more difficult to control than radical reactions. In both anionic and cationic polymerisations, the temperature must be kept low; this presents no problem in the laboratory.

Industrially, ionic bulk polymerisations are usually carried out with slow reactions (e.g., the anionic polymerisation of 1,3,-butadiene or α-methylstyrene), or when polymers with low degrees of polymerisation are to be prepared (eg., cationic polymerisation of alkyl vinyl ethers or isobutylene).

For anionic polymerisations in bulk, kneaders or screws which assure good mixing of the polymerisation charge and adequate heat removal in either viscous or semi-solid state are suitable. A polymerisation screw for the continuous production of polybutadiene is shown schematically in Fig. 9.7. The pilot-plant or industrial polymerisation of butadience with sodium or sodium-potassium catalyst to the so-called Buna rubbers, Buna 32, Buna 85, Buna 115, is carried out in equipment of this type. (The attached number is an indication of viscosity.)

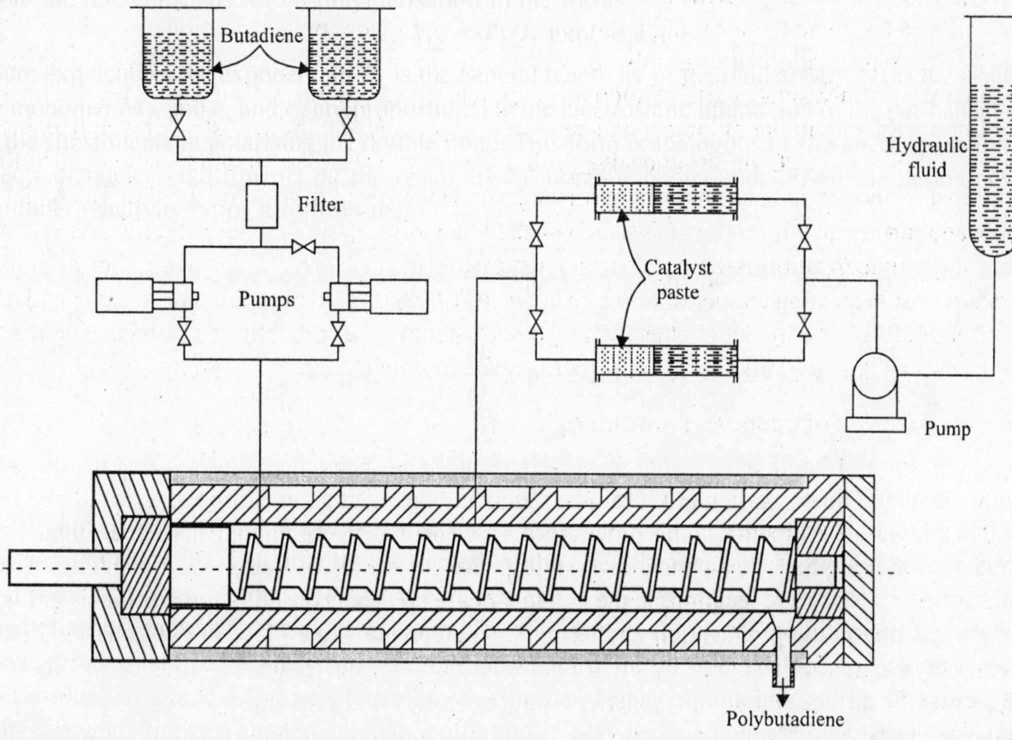

Fig. 9.7. Schematic representation of a polymerisation screw for the synthesis of polybutadience or of butadiene-styrene co-polymer.

The butadience or butadiene-modifier mixture is fed into the reactor by pumps under a pressure of about 1 atmosphere. A paste of 10–20 per cent sodium or potassium in paraffin is used as catalyst. The metal paste is added by means of a hydraulic system after the first cooling stage to avoid plugging the safety valve. The polymer leaves the apparatus in the form of red-to-violet coloured strands. This colour disappears (deactivation of the catalyst) on milling with 1–2 per cent fatty acid. The consumption of alkali metal lies between 0.1 and 0.3 per cent based on monomer; the reaction temperature is 50–70°C.

Adjustment to a particular molecular weight can be brought about by modifiers added during polymerisation. Aldehydes, ketones, acetals, amines, nitriles, thioethers, halogenated hydrocarbons, etc., have been so used. By increasing the amount of modifer the molecular weight can be decreased to such an extent that low-molecular-weight liquid polymers or oligomers are obtained.

The activity of the alkali metal catalyst increases from lithium to cesium. There are no significant differences in the structures of the diene polymers prepared with sodium, potassium, rubidium, or cesium. Lithium catalysts, on the other hand, show a pronounced structure specificity toward formation of cis-1,4 polymers.

The cationic polymerisation of vinyl ethers is also of industrial importance. In contrast to radical initiators, cationic polymerisation initiators effect a rapid and vigorous polymerisation of the electron-rich double bond of vinyl ethers. The growing polymer cation is stabilised through resonance with the alkoxy or aryloxy group. Besides the not too useful mineral acids, Lewis acids such as aluminium chloride, zinc chloride, stannic chloride, ferric chloride, boron trifluoride, and boron trifluoride dihydrate are used. Because of its easy purification, convenient handling, and good solubility, boron trifluoride-ethyl etherate has become the most important initiator of vinyl ether polymerisation.

Industrial polymerisations are usually carried out in pressure vessels. In order to maintain better control frequently only small parts of monomer and catalyst solution are mixed. After the onset of polymerisation, the remaining monomer and catalyst are then added continuously. For example, methyl vinyl ether is mixed at 5°C in an autocalve with a 3 per cent catalyst solution of boron trifluoride dihydrate in dioxane. As soon as polymerisation begins, the temperature rises to 10-55°C. The remaining monomer and further quantities of catalyst are added to the polymerising mixture at reflux. After the last addition, the polymerisation is completed under pressure in the closed autoclave either at constant temperature or at increasing temperatures up to about 100°C. A drop in pressure indicates the end of the polymerisation. Poly (methyl vinyl ether) with the consistency of a soft resin is obtained. The reaction temperatures for the higher alkyl vinyl ethers are somewhat higher, eg., 50–80°C.

Bulk polymerization of gaseous monomers

The polymerisation of undiluted monomers in the gas phase has been much investigated. It was at first mistakenly assumed that a particularly useful insight into the exact course of polymerisation could be obtained in this way. The kinetics of this process are, however, relatively complicated, since immediately after initiation the polymer precipitates as a solid phase. It is thus difficult to differentiate between reactions in the gas phase and reactions in the polymer phase. Whereas the initiation reaction is ordinarily only in the gas phase, the propagation and termination steps may take place in both phases. In some instances, polymerisation is started by the initiation at polymer nuclei on the walls of the container. Reactions may be carried out at atmospheric, reduced, or elevated pressures.

Initiation can be brought about by alkyl or acyl radicals. The usual method is direct or sensitised photopolymerisation. Many monomers photopolymerise in the gas phase only on addition of sensitisers such as acetone and diacetyl. Styrene, acrylonitrile, and vinyl chloride belong to this class. An example of anionic gas-phase polymerisation was mentioned in (Fig. 9.4c) A simple apparatus for the boron trifluoride-catalysed polymerisation of isobutylene in the gas phase has been described by *Evans* and *Meadows*. The onset of polymerisation can usually be recognised visually by the formation of a cloud in the gas.

Experimental procedures have been described, in particular by *Melville* for the investigation of photopolymerisation under reduced pressure (1–200 mm Hg.). Vapourised monomer (methyl methacrylate, vinyl acetate, methyl vinyl ketone, or vinyl chloride, etc) with or without a photosenitiser, is exposed to ultraviolet radiation of wavelengths under 3000Å. The reaction is followed by measuring the drop in pressure.

Although gas-phase polymerisation under normal pressures has often been investigated, it has no industrial importance. Unsaturated hydrocarbons, especially ethylene and acetylene, form only low-molecular-weight linear or cyclic products. They also act as inhibitors when added as a co-monomers in other gas-phase polymerisations. The lack of reactivity is due to the high activation energy of the reaction between the free radicals and the unsaturated hydrocarbons. Butadiene, initiated either by photolysis with accetone as photosensitiser or by acetyl radicals, attains a degree of polymerisation of only about 10–15. Vinyl compounds with polar groups, however, polymerise rapidly under the same conditions with the formation of high-molecular-weight products. This reaction is much more dependent upon the nature of the monomer than is polymerisation in solution.

It was first observed with chloroprene that after illumination of the monomer in the gas-phase polymerisation continued for up to several days in the dark, if sufficient monomer was present. The same observation has also been made for other 1,1-disubstituted vinyl monomer such as methyl methacrylate. The photopolymerisation of vinyl acetate in the gas phase indicates that in some cases active sites are formed to which a purley radical character cannot be ascribed. In this case polymerisaiton is inhibited by inert gases such as argon and helium. These gases function not by stopping growing chains but by deactivating by collision the photoexcited vinyl acetate molecule.

In order to obtain better space–time yields gas-phase polymerisations are usually carried out under increased pressure. This is the case, for example, in the most important industrial high-pressure polymerisation, that of ethylene. The process is classified as gas-phase polymerisation, although strictly speaking polymerisation in the presence of a gas phase takes place in the highly swollen liquid polymer phase. Because of the use of pressures of over 500 atm and the difficulty of removing the heat of polymerisation about 800 cal/g. complex equipment is required.

Although batch polymerisation is simpler to operate, it is not economical since it must be carried out slowly if it is to be reproducible and controllable. At high conversions branching increases sharply, which results in adverse effects on the physical properties of the polyethylene. As a result, ethylene is now polymerised entirely by continuous process. Pressures of 1000–3000 atm, or even higher, and temperatures of 100–250°C are used. Traces of oxygen initiate polymerisation very easily. Impurities such as hydrogen and acetylene act as chain-transfer agents and must be carefully removed if high-molecular-weight compounds are to be obtained. Besides oxygen, peroxides, hydroperoxides, or azo compounds may be used as initiators. A plant for the continuous polymerisation of polyethylene is shown in Fig. 9.8. Highly purified ethylene with an oxygen content of 0.05–0.1 per cent is compressed in several steps to a pressure of 1500 atm and forced into the polymerisation tube. Polymerisation is begun at 190–220°C and the temperature at the end of tube allowed to drop to just over the melting point of polyethylene, i.e., to about 130°C. The residence time in the reactor is about 2-6 min; the conversion in a single pass lies between 10 and 20 per cent. The space–time yield is about 600 to 900 g/ (litre) (hr). After the oxygen-containing by-products (especially formaldehyde) have been washed out the excess ethylene can be returned to the compressor. The molecular weight of the polyethylene thus prepared decreases with increasing temperature and decreasing pressure. Polymersation at a pressure of 1,500 atm with 0.01 per cent oxygen produces polyethylene with a molecular weight of 18,000; with 0.07 per cent oxygen, 10,000 molecular weight; with 0.13 per cent oxygen, 6,000 molecular weight. Still higher oxygen contents in the ethylene lead to explosive reactions.

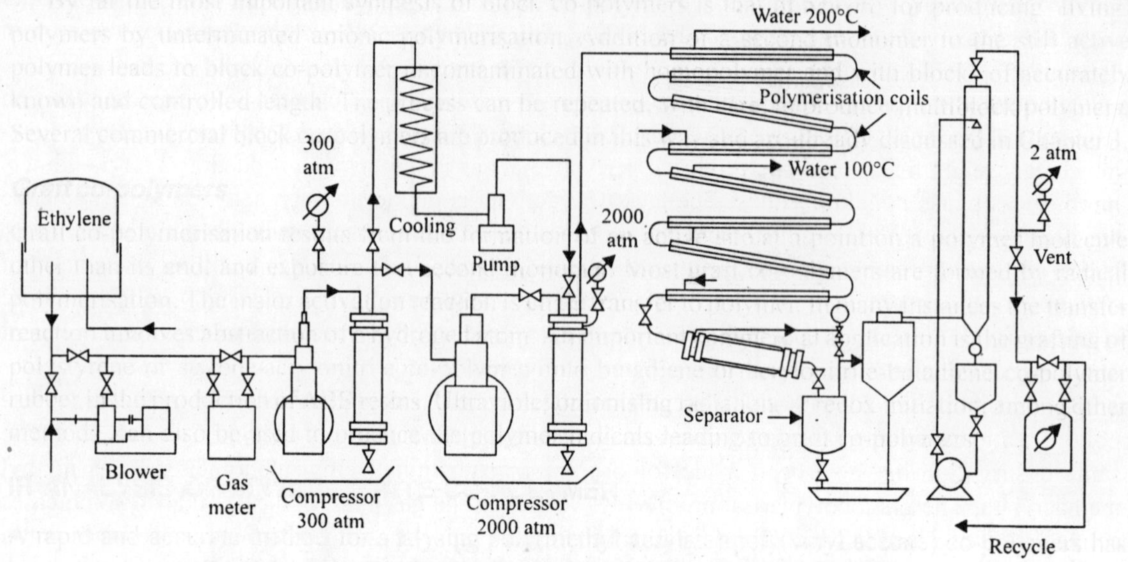

Fig. 9.8. Flow diagram for the continuous production of polyethylene.

Although the high-pressure polymerisation of ethylene was long the uncontested process for the manufacture of polyethylene, the low-pressure process (1–150 atm) for the synthesis of linear polyethylene and of other polyolefins is now of greater importance. However, these methods are solution and precipitative polymerisations.

Bulk polymerisation under pressure

Free-radical bulk polymerisation can be considerably accelerated, with simultaneous increase in molecular weight, through the use of increased pressure. This has been demonstrated, for example, for polystyrene by *Gillham* (Table 9.4) and also holds for the bead polymerisation (suspension polymerisation) of styrene and the emulsion polymerisation of butadiene.

Table 9.4. Effect of pressure on polymerisation of styrene at 100°C

Pressure, atm	*Max polymerisation rate, % conversion/hr*	*% Conversion*	η_{sp}/c^a
1	3.40	44.9	1.50
1000	5.06	88.0	1.96
2000	15.9	60.6	2.09
3000	34.3	71.6	2.32
4000	50.5	49.9	2.52

[a] 0.1% Chloroform solution at 20°C.

The rate of polymerisations under pressure is generally constant up to about 30 per cent conversion and then increases extremely rapidly owing to the gel effect. The dependence of the rate of polymerisation on initiator concentration has the usual square-root relationship at low pressures. At greatly increased pressures the exponent becomes smaller. Investigation of the influence of pressure on the reactivity ratios of the co-monomers in the systems styrene-acrylonitrile and methyl methacrylate acrylonitrile shows that they increase in both cases. The *e* values for these two monomer pairs are pressure-dependent and tend to equalise at higher pressures. The *Q* values, on the other hand, remain largely uninfluenced by pressure.

Bulk polymerisation in the solid state

One of the newer developments in the field of polymer chemistry is the polymerization of solid monomers. This followed in particular the work of *Mesrobian* and of *Schulz* who found that acrylamide was rapidly polymerised by ionising radiation.

In principle the process is applicable to all monomers in the crystalline or glassy solid state. Several monomers polymerisable in the solid state are listed in Table 9.5.

Table 9.5. Monomers Polymerisable in the Solid State.

Monomer	*Mp, °C*	*Initiation temp, °C*	*Initiation method*
Styrene	–30	–60	radiation
Vinyl acetate	–139	–196	radiation
N-vinylcarbazole	+64	+20	radiation
N-vinylpyrrolidinone	+13.5	+7	radiation
Methacrylic acid	+16	0	radiation
Acrylamide	+85	0 to +60	radiation
Acrylonitrile	–83	–135	radiation
p-acetamidostyrene	+147	+100 to +135	thermal
p-benzamidostyrene	+178	+107 to +157	thermal
Formaldehyde	–118	–196	radiation
Acetaldehyde	–123	–150	radiation
Trioxane	+64	+30	radiation
Trioxane	+64	+30	BF_3, $SnCl_4$
β-propiolactone	–33.4	–70	radiation
Diketene	–6.5	–130	radiation
3, 3-bis(chloromethyl) oxetane	+17	+16.4	radiation

In most cases lattice defects and crystal boundaries are the starting points of polymerisation. Electron microcopy indicates that the polymer chains grow in the interphase between crystalline and amorphous regions. If monomers with high lattice energies such as salts of acrylic acid are used, or if the polymerisation is carried out at temperatures well under the melting point of the monomer, it is clearly a reaction in the solid state. On the other hand, if polymerisation is carried out only a few degrees under the melting point, in many cases local melting or a disorientation of the neighboring monomer units caused by the heat of polymerisation cannot be excluded. As would be expected, in the polymerisation of crystalline monomers the reactivity of the monomer, the manner of growth of the chain, and the properties of the resulting polymer are determined or influenced by the geometry of the crystal lattice of the monomer. It has been shown that in the radiation-induced polymerisation of trioxane, β-propiolactone, diketene, and other heterocycles, the monomer lattice induces an orientation into the crystalline polymer. *Morawetz* demonstrated a chain growth controlled by the monomer lattice for vinyl compounds such as acylated *p*-aminostyrene and salts of acrylic acid.

In reactions initiated with boron trifluoride (cationic) or with ultraviolet light or thermally (radical) the mechanism of polymerisation in the solid state has been unequivocally determined. On the other hand, polymerisation is more often initiated through ionising radiation, and here a clear differentiation between ionic and radical mechanisms is frequently not possible. The usual criteria, such as co-polymerisation characteristics and inhibitor effects, applicable to the polymerisation of liquid monomers are not valid here, since foreign molecules are not normally built into the crystal lattice. In some cases an overlapping of ionic and radical mechanisms must be assumed. A phenomenon typical of radiation-induced polymerisation is postpolymerisation (aftereffect), which occurs with some monomers (eg., acrylamide, β-propiolactone) and which extends in some cases to as long as several months after removal of the radiation source.

Co-polymerisation in bulk

The polymerisation of a mixture of two monomers usually produces co-polymers, ie., polymers built up from units of both monomers. The tendency of monomers to form such co-polymers differs and does not correspond to the polymerisability of the individual monomers with themselves.

Experimentally, co-polymerisation investigations have largely been carried out as bulk polymerisations by the usual methods. To prevent a depletion of the more rapidly polymerising component, laboratory co-polymerisation reactions are interrupted at less than 10 per cent conversion. Industrially this is not economically possible. In these cases the more rapidly polymerising component is added continuously. The polymerisation equipment does not differ basically from that used for homopolymerisation.

Some diene rubbers, as well as several types of impact-resistant polystyrene and of polyester casting resins, are among the more important commercial examples of co-polymerisation in bulk. Under the heading of impact-resistant polystyrene are included co-polymers of styrene with acrylonitrile, butadiene and the corresponding co-polymers with preformed polymers.

Graft and block co-polymers are also sometimes prepared in bulk.

Unsaturated polyester resins

If polyesters containing carbon-to-carbon double bonds along the backbone chain are dissolved in vinyl monomers, unsaturated polyester resins can be prepared by co-polymerisation of these two components. This is therefore a special form of graft co-polymerisation in bulk. Maleic anhydride or fumaric acid, which combine the advantages of low cost and good properties, are usually used for the introduction of double bonds in the backbone. The greater the content of carbon-to-carbon double bonds in the backbone chain, the more closely cross-linked, i.e., the more rigid, the cured resin.

Styrene in particular is an important inexpensive comonomer, useful both for the good properties that it imparts and for its ready co-polymerisation tendency. Vinyl toluene, triallyl cyanurate, methyl methacrylate (especially in conjunction with styrene to improve weatherability), and diallyl phthalate may also be used. The latter co-polymerises casily but reaction is relatively slow. The starting mixtures for the preparation of unsaturated polyester resins usually consist of 0.6–0.7 mole of a polyester of molecular weight 2000–6000 diluted with 0.3–0.4 mole of the desired monomer to a pourable liquid. Polymerisation inhibitors, eg., hydroquinone, are added to stabilise the system. The cure, that is, the bulk polymerisation of this mixture is initiated with peroxides and is nearly always carried out in moulds.

To obtain a rapid cure at room temperatures, promotes or activators are added to accelerate peroxide decomposition, e.g., tertiary amines for diacyl peroxides. For hydroperoxide catalysts, soluble cobalt salts such as cobalt naphthenate and cobalt octanoate are used. The gel effect is in action from the beginning in these high-viscosity systems (curing takes place very rapidly and must be taken into consideration particularly with large quantities and high reaction temperatures.

Bulk polymerisation of heterocyclic monomers

Numerous heterocyclic compounds such as lactams, cyclic ethers, cyclic amines, cyclic acetals, and cyclic anhydrides can be polymerised by a mechanism involving ring opening. Initiation can be either catalytic or through the addition of a compound forming end groups. The former results in an ionic chain reaction, often extremely rapid, brought about by strong acids or bases. In reactions of the latter type, e.g., the water-initiated polymerisation of ε–caprolactam, the initiation and the individual propagation steps show no difference in principle; that is, they proceed with similar activation energies. They are, therefore, not chain reactions in the kinetic sense, but stepwise polymerisations. Problems of heat removal thus do not arise.

Bulk polymerisation of cyclic compounds is a frequently used laboratory and industrial method for the preparation of polyesters, polyacetals, polyamides, and polyesters. Bulk polymerisation of caprolactam has long been carried out on an industrial scale, and that of other heterocycles, such as ethylene oxide and 3,3-bis (chloromethyl) oxetane, is achieving increasing industrial importance.

EMULSION POLYMERISATION

Emulsion polymerisation is a heterogeneous reaction process in which unsaturated monomers or monomer solutions are dispersed in a continuous phase with the aid of emulsifier system and polymerised with free radical initiators. In emulsion polymerisation the physical degree of sub-division of reaction mixture in the colloidal size range has pronounced influence on the course of the polymerisation, the characteristics of the product and its performance. The diversity of application for emulsion polymerisation is evident in a wide range of products, which include synthetic rubbers, cement and concrete additives, and drug delivery system.

Theory of emulsion polymerisation

Harkins theory

The Harkins model of emulsion polymerisation, which is presented here, is based on monomers such as styrene and butadiene, water-soluble initiators such as persulphate ion, and anionic surfactants, which form micelles in water. The course of batch reaction is described in terms of three intervals. Polymerisation within small polymer particles is a central concept in the Harkins mechanism.

Interval I

The reaction mixture of the beginning of interval I consist of a continuous water phase and a dispersion of monomer droplets 1–10 μm.

The water phase contains the initiator, buffer, small quantities of dissolved monomer and minor quantities of emulsifier. Most of the monomer is located in the large droplets, but some is solubilised within the micelles and some dissolved in the water. It was proposed that the polymerisation reaction starts when water-borne free radicals migrate into the monomer-swollen micelles. Polymerisation within these micelles depletes the monomer and establishes a thermodynamic driving force, which cause mass transfer from the monomer droplets through the water phase to the newly formed and growing polymer particles.

Particle nucleation is normally completed early in the course of a batch reaction; conversions of 2–10 per cent are typical. The explanation of this phenomenon is based on the new surface area formed by the growing monomer swollen polymer particles. The expanding particles provide a new hydrophobic surface, which, in turn causes the dissolution of micelles to maintain the micelle-solution equilibrium. Nucleation stops when the surface area of the previously formed particles plus that of the monomer droplets is adequate for the adsorption of all emulsifier. The point in the reaction sequence constitutes the end of interval I.

Interval II

At the end of interval I, the reaction mixture is comprised of monomer-swollen polymer particles and monomer droplets; i.e., 90 per cent of the monomer is present in the droplets. The monomer and polymer particles are thermodynamically limited because of the free energy required to expand the surface area of the colloidal particles. This important phenomenon is due to the degree of sub-division of the reaction sites. The reaction proceeds by polymerisation within the particles. Monomer is supplied by diffusion from the shrinking monomer droplets. With the growth of the particles during interval 2, the total interfacial area increases, usually to values higher than can be saturated with the emulsifier used in the initial charge. Interval 2 ends when the monomer is depleted from the droplets. At this point in the reaction, the degree of saturation of the particle surfaces by emulsifier is at its lowest value.

Interval III

Overall monomer conversion at the beginning of this interval varies considerably among different monomer systems; for styrene, it is typically 40 per cent. Further polymerisation reduces the monomer concentration in the particles. For some latex products, very high conversion is desired in order to reduce problems with residual monomer later.

In this regime, polymerisation takes place with higher polymer concentrations at the reaction loci, leading to more branching and cross-linking. This is particularly important with some synthetic elastomers. Lower polymerisation temperatures, chain-transfer agents, and controlled conversion govern product quality in such systems. In the last case, polymerisation is stopped at the desired conversion, and the unreacted monomer is recovered and recycled.

This model represents a useful basis for examining the influence of various reaction and reactor parameter on the course of the polymerisation and the characteristics of the latex.

Mechanisms of particle formation

Three major mechanisms have been proposed for particle formation in emulsion polymerisation namely, micellar, homegeneous and droplet nucleation.

Contributions of different nucleation mechanisms to particle formation

In conventional emulsion polymerisations, all three-particle formation mechanisms in principle are operating simultaneously. However, their relative contributions to particle formation can vary considerably. Typically, one mechanism dominates particle formation depending on the sufractant concentration, the monomer solubility in the aqueous phase, and the level of sub-division of the monomer solubility droplets. Micellar nucleation is considered to be the primary nucleation mechanism for monomers with relatively low water solubilities ($[M]_{aq}$<15 mmol dm^{-3}). Homogeneous nucleation, on the other hand, is considered to be the primary mechanism of particle formation for monomers with relatively high water solubilities ($[M]_{aq}$<170 mmol dm^{-3}). Homogeneous nucleation is also proposed as the primary mechanism for particle formation in systems where the surfactant concentration is below its C.M.C. and in surfactant-free emulsion polymerisation systems.

In the latter, the stabilisation of the primary particles and growing particles is attributed to the presence of the ionic groups (such as sulphate) from the initiator fragments on the surface of the particles, which leads to electrostatic stabilisation of these particles. Particle formation by the droplet nucleation mechanism is considered to be insignificant in conventional emulsion polymerisations. However, when the droplets are in the sub micron size range, such as, in miniemulsion or microemulsion systems, they become the primary source of particle formation. Droplet nucleation dominates in miniemulsion systems comprising monomers or macro monomers with extremely low water solubilities (<1 mmol dm^{-3}).

Rate of polymerisation (Quantitative aspect)

An expression for the rate of polymerisation can be obtained by considering first the rate in single polymer particle in which propagation is occuring (i.e., a particle containing a radical) and then the number of such particles. The rate of polymerisation R_p at any instant is given by the product of the concentration of active particles [*P*.] and the rate of propagation in a particle R_p.

$$R_p = k_p[M.][P.] \quad \text{... (9.1)}$$

$$R_p = \frac{10^3 N\tilde{n} k_p [M.]}{N_A} \quad \text{... (9.2)}$$

For the polymerisation rate equation (9.2) applies to Intervals II and III where only polymer particles exist (no micelles). It is during Intervals II and III that the overwhelming per cent of monomer conversion to polymer takes place.

The value of $\tilde{n}$ during Intervals II and III is of critical importance in determining R_p and has been the subject of much theoretical and experimental work. Three cases 1, 2 and 3 can be distinguished based upon the work of *Smith* and *Edward.*

The major differences between the three cases are the occurrence of radical diffusion out of the polymer particles (desorption), the particle size, modes of termination, and the rates of the initiation and termination processes relative to each other and to the other reaction parameters. The quantitative interplay of these factors leading to case 1,2 or 3 behaviour has been discussed.

Case 2 : $\tilde{n}$ = 0.5

This is the case with most of the emulsion polymerisations. It occurs when desorption of radicals does not occur or is negligible compared to the rate of radicals entering particles (absorption) and the particle size is too small, relative to the bimolecular termination rate constant, to accommodate more than one radical. Under these conditions, a radical entering a polymer particle is trapped within that particle and undergoes propagation until another radical enters, at which point there is essentially instantaneous termination. Any polymer particle will be active half of the time and dormant the other half of the time. In other words, at any given moment half of the polymer particles contain one radical and are growing while the other half are dormant. The number of radicals per particle $\tilde{n}$ averaged over all the particles is 0.5. Case 2 behavior also requires the initiation rate not be excessively low and negligible termination of radicals in the aqueous phase. This is often typical of styrene polymerisation.

Case 1 : $\tilde{n}$ < 0.5

The average number of radicals per particle can drop below 0.5 if radical desorption from particles and termination in the aqueous phase are not negligible. The decrease in $\tilde{n}$ is larger for small particles sizes and low initiation rates. This is often typical for monomers with water solubility.

Case 3: $\tilde{n}$ > 0.5

Some fraction of the polymer particles must contain two or more radicals per particle in order for $\tilde{n}$ to be larger than 0.5 since there will always be a fraction (a very significant fraction) which has zero radical per particle. This occurs if the aqueous phase and desorption are not important and the initiation rate is not too low.

Simultaneous increase in polymerisation rate and molar mass

One of the most important outcomes of the heterogeneity of the emulsion polymerisation process is that the number of polymerisation sites is within order of magnitude of the number of radicals generated over a wide range of temperature, and initiator concentrations is responsible for the ability to simultaneously increase both the polymerisation rate and the polymer molar mass.

This is the key feature of emulsion polymerisation, which distinguishes it from other polymerisation. In the latter processes, usually any attempt to increase the rate typically leads to a reduction in the molar mass. This illustrated by comparing the following general expressions for the instantaneous rate polymerisation, R_p, and the number-average of polymerisation, x_n, for emulsion polymerisation versus bulk, solution or suspension polymerisations. For emulsion polymerisation, these quantities can be expressed as

$$R_p = \frac{k_p[\mathrm{M}]\tilde{n}N}{\mathrm{N}_A} \quad \text{... (9.3)}$$

$$x_n = \frac{k_p[\mathrm{M}]\tilde{n}N}{R_i} \quad \text{... (9.4)}$$

where k_p is the rate coefficient for propagation, $[M]$ the monomer concentration in the particles, $\tilde{n}$ the average number of radicals per particle, N the number of latex particles per unit volume, R_i the rate of radical generation (or total rate of radical absorption), and N_A the Avogadro's constant.

For bulk, solution or suspension polymerisations:

$$R_p = k_p \cdot [M.] \left(R_i / 2k_t \right)^{1/2} \quad \text{... (9.5)}$$

$$x_n = k_p [M.] \left(2 / R_i K_t \right)^{1/2} \quad \text{... (9.6)}$$

in which k_t is the rate coefficient for termination Equations (9.3) and (9.4) demonstrate that both the polymerisation rate and molar can be simultaneously increased in an emulsion polymerisation by increasing the number of polymer particles, N, at a constant initiation rate. This is typically accomplished by increasing the surfactant concentration. Increasing the rate of initiation in bulk, solution or suspension polymerisations may also increase the polymerisation rate; however, this leads to a reduction in the molar mass as shown by equation (9.6).

Another important feature of emulsion polymerisation is the average number of radicals per particle, $\tilde{n}$, in equations (9.3) and (9.4), which is a key parameter in determining both the rate and molar mass profiles during the course of the polymerisation. This number is a function of the rate of radical generation, the number of polymer particles, radical entry into particles, radical exit (i.e. desorption after entry) and termination reactions. The average number of radicals per particle is not constant during the entire polymerisation process, and its value is determined by the above factors, which in turn is affected to a great extent by the water solubility of the monomer. The following equation is derived based on above consideration,

$$N = k\left(p/_{\mu} \right)^{(1-z)} (a_s S)^z \qquad \ldots (9.7)$$

Where $0.6 < z < 1.0$ and depends on the chain transfer constant and water solubility of the monomer. The sum of the exponents for the dependence of the number of particles with respect to emulsifier (S) and rate of initiator decomposition (R_i) is always equal to unity. An increase in the order with respect to S will be accompanied by a decrease in order with respect to p. This is usually true for monomers with water solubilities higher than styrene, such as vinyl acetate, methyl methacrylate and vinyl chloride.

Kinetics of Emulsion Polymerisation

Investigations of the kinetics of emulsion polymerisation typically involve the determination of the polymerisation rate and its dependence on the polymerisation parameters such as the monomer, emulsifier and initiator concentrations, and the temperature. The experimental approach typically starts with determination of the conversion-time curves by varying the above parameters one at a time, followed by calculation of the polymerisation rate by taking the first derivative of the conversion with respect to time over certain time intervals.

The rate profiles are usually plotted versus time or versus conversion. In order to complete the kinetic analysis, one needs to know the number of particles and the monomer concentration in the particles as a function of conversion, which allows for calculation of the average number of radicals per particle, n. The monomer concentration in a particle can be estimated from thermodynamic analysis and knowledge of the unreacted amount of monomer, the particle number of particle size.

Effect of various parameters on the kinetics of emulsion polymerisation

Initiator

A source of free radicals is needed for water borne emulsion polymerisation. The free radicals can be produced by thermal decomposition of peroxy compounds like persulphate, or by redox reactions like the persulphate/bisulphite couple or by γ-radiation. The free-radical initiators can be either water or oil soluble, determining the preferred phase in which the free radicals will be produced.

Initiator concentration

Initiator concentration and type affect kinetics of polymerisation and thereby properties of latex. Increasing the initiator concentration, the rate of polymerisation increases and the molar mass of the polymer formed decreases.

Type of initiator

Water–soluble initiators

The most frequently used initiators are the salts of peroxydisulphate, i.e. per sulphates. The thermal decomposition yeilds sulphate radical anions, which contribute to the charged character of the latex particle. Sodium, potassium and ammonium salts are generally interchangeable and are used in the temperature range of 50–90°C. Water-soluble initiators are thought to form oligomeric radicals of varying length in the aqueous phase which are then adsorbed or captured by micelles or existing particles. Hydro-peroxides are soluble in both the aqueous hydroxyl and in the monomer-swollen particles. The decomposition generates a hydroxyl and another oxygenated radical which could partition in either phase dependent on the type of hydro-peroxide used. Cumene hydro-peroxide in styrene polymerisation has been postulated to decompose in the particle interphase region, between monomer-swollen particle and water, and the hydroxyl radical enters the water phase and the non-polar radical diffuses into the particles.

Oil-soluble initiators

Oil-soluble initiators such as benzoyl peroxide, 2,2-azobis (isobutyronitrile) are sometimes used. They partition preferentially into the monomer-swollen particle and generate the free radical in that confined space. The competitive growth rates were the same as in the presence of water-soluble initiators.

Redox initiators

The combination of certain oxidising and reducing agents will produce free radicals even at low temperature. This can be particularly useful when high molar mass polymers are sought with low levels of branching. Common redox systems are: persulphate-bisulphite and persulphate-hydroperoxide. Sodium formaldehyde sulphoxylate has been used with a wide number of oxidising agents like cumene hydroperoxide. The persulphate/iron (II) redox pair has been long known and in practically all of the above redox systems iron (II) is used as a 'co-catalyst'.

A fast dissociative redox initiator decreases the induction period and prevents homogeneous nucleation. Therefore, the conversion curves do not contain zero-order period in mini-emulsion polymerisation. Co-polymerisation can be carried out successfully at lower temperature (45°C). The induction period was decreased and the polymerisation was quicker.

Emulsifier

Emulsifiers are a key formulation variable in emulsion polymerisation. The emulsifying agent or soap has a multifold function to play in the emulsion polymerisation process.

1. Act to solubilise the monomer droplets in an emulsion form.
2. Serve to solubilise monomer within emulsifier micelles.
3. Stabilise the formed latex particles as well as the particles, which continue to grow during polymerisation.
4. Act to solubilise polymer.
5. Serve as a site for nucleation of particles.
6. Act as chain transfer agents or retarders.

A successful emulsion polymerisation process requires an acceptable emulsification system. An acceptable emulsification results from a combination of optimum agitation and the correct amount of the proper emulsifying agent or soap. Emulsifiers are categorised into four major classes: anionic, cationic, non-anionic, and zwitterionic.

Anionic emulsifier

The negatively charged hydrophilic head group of the anionic surfactants may comprise sulphate, sulphonate, sulphosuccinate or phosphate groups attached to an extended hydrophobic backbone. Anion emulsifier often used includes sodium lauryl (dodecyl) sulphate, sodium dodecylbenzene sulphonate (SDBS), sodium dodecyl allyl sulphosuccinate phosphate ester surfactants. The anionic soap systems seem to be less subject to disturbances by minor changes in the other process variables. Thus they are more reliable in the sense of confidence of trouble-free operations.

Cationic emulsifier

The cationic surfactants are used infrequently in emulsion polymerisation applications since they are not compatible with the anionic surfactants and with the (typically) negatively charged latex particles. However, in special situation, such as when an anionic system would be repelled by a surface to which it is to be applied, cationic systems are useful. These surfactants are usually of the following types: salts of longs chain amines, polyamins and their salts, quatemary ammonium cetyltrimethyl ammoniumbromide is probably the most commonly used cationic emulsifier. Cationic emulsifiers are used for making cationically charged latex particles for special application in paper coatings and asphalt additives.

Non-ionic emulsifier

One special advantage of this type of emulsifier is that the latex is insensitive to a wide range of pH. The structure of these emulsifiers commonly include a long-chain alkyl group attached through an oxygen or nitrogen atom to a chain of ethylene or propylene oxide groups. Non-ionic emulsifier may be divided into the following classes: polyoxyethylenated alkylphenols, polyoxyethylenated straight-chain alcohols, polyoxyethylenated polyoxypropylene glycols, are the most commonly utilised for emulsion polymerisation formulations. Examples of the polyoxyethylenated alkylphenol type of emulsifier include the Triton® series (Union Carbide) in which octylphenol (Triton X series) or nonylphenol (Triton N series in phenol portion of the molecule.

Non-ionic surfactants are used for controlling latex particle morphology and for enhancing the post-polymerisation colloidal stability against mechanical shear freezing and added electrolytes. Emulsion co-polymerisations of styrene and butyl acrylate were conducted at 70°C using varying amounts of Triton X-405 (octyl phenoxy polyethoxy ethanol) as emulsifier. The kinetic behaviour and particle-size distributions were found to vary widely depending on the emulsifier concentration.

Zwitterionic (amphoteric) emulsifier

These types of emulsifier exhibit anionic properties at high pH and cationic properties at low pH and may be categorised as N alkylamino propionic acids N-alkyl α iminodi propionic acids imidazoline amines oxides. The sulphobetaines are amphoteric amines oxides. The sulphobetaines are amphoteric at all values of pH. These surfactants are not commonly used in emulsion polymerisation formulations.

Mixed emulsifiers

The use of anionic and non-ionic mixed emulsifier systems may be a powerful tool to ensure the control on the particle number, size and distributions.

The effect of the emulsifier on the reaction kinetics is controlled by the total amount of emulsifier, but not by the anionic/non-ionic ratio. The use of mixed surfactant systems, compared to the use of anionic emulsifier alone led to a competitive effect between decrease in particle number due to the addition of non-ionic emulsifier prevailing at high anionic emulsifier concentration and increase in particle number due to the increase of the total concentration of emulsifier prevailing at low anionic emulsifier concentration. Moreover, narrower particle-size distributions (PSDs) with mixed surfactant systems than those obtained with single anionic systems. High total concentration of emulsifier led to broad PSDs.

The seeded emulsion polymerisastion was used to study the effect of the concentration and type of the emulsifier system on the growth of particles by starting from the same initial particle number. The use of a non-ionic emulsifier, with or without an anionic one, to stabilise the growing seed in semi-continuous seeded emulsion polymerisation caused the colloidal destabilisation of the system and led to a broader particle size distribution with larger particles sizes compared to those predicted theoretically.

Polymerisable surfactant (surfmers)

Surfactants can have negative effects on the properties of the products of emulsion polymer latexes. These negative effects are caused by the fact that the surfactants when unbound to the polymer material desorb from the latex particle surface, causing destabilisation, for e.g., under high shear. The unbound surfactants can migrate through the film, forming aggregates that increase the water sensitivity of the product.

A solution to these problems can be the use of reactive surfactants. The use of reactive surfactants ensures that the surfactant moiety is bound covalently to the polymer material so that desorption from the latex particle surface or migration in the polymer film is impeded. The more promising possibility lies in the combination of a surfactant and a monomer, referred to as surfmers (polymerisable surfactants), e.g. include anionic surfmers with sulphate or sulphonate head groups, cationic surfmers, and non-ionic surfmers. Also a wide range of reactive groups can be found, for example, acrylics, acryl amides (meth) acrylates, styrenics, and maleates.

Uses of surfmers bring about improvements in properties eg., in mechanical stability and electrolyte stability of the latex (due to decreased surfactant desorption from the latex surface) and control of surface charge density.

In films cast from latexes prepared with surfmers, a decrease of surfactant migration has been reported as well as an improvement of water resistance and adhesivity.

Optimum surfmer behavior in emulsion polymerisation

To define the conditions for the optimum use of surfmers, two anionic maleates surfmers were applied in various emulsion co-polymerisation systems including St/BA/AA; VAc/VEOVA 10/AA; VAc/BA/AA and MMA/BA/VA. This led to the definition of the optimum surfmer behaviour: low conversion (surfmer not reactive) in the beginning and high conversion (surfmer reactive) toward the end of the reaction.

Effect of functional monomers

Small amounts of water-soluble organic compounds such as acrylic acids or sulphonated and cationic monomers are often used to stabilise lattices and impart desirable application properties. The distribution of such monomers is likely to be almost exclusively in the aqueous phase, where initiation and propagation reactions are extremely important. They are likely to be distributed as water-soluble polymer, on the surface of the latex particle as part of a co-polymer with the hydrophobic monomers and buried under the particle surfaces as a co-polymer, at the end of the polymerisation.

Incorporation of a small amount of functional monomers such as acrylic acid into the polymer increased the latex stability significantly. The ionised carboxyl group that is chemically incorporated into the polymer can increase the particle surface charge density and, therefore, increase the repulsive force among the interactive particles. The type of carboxylic monomers should play an important role in the particle nucleation and growth processes.

The kinetics of emulsion terpolymerisation of VAc, BA and MMA in the presence of small amounts of AA enhances the latex stability, improve the wetting properties, and exhibit an increase of the adhesion after ageing.

Monomer ratio and water solubility of monomers

Co-polymerisation rate and conversion increases with increasing the content of monomer with high water solubility. Homogeneous nucleation takes place in considerable amount due to high water solubility of monomers and hence number of particles increases.

Study on batch emulsion co-polymerisation of S/BA indicated, increase in S/BA ratio also decreases molar-mass of polymer. This is because the monomers of low S/BA ratio are easily propagated to obtain the polymer of high molar-mass. The difference between propagation rate constant of S and BA is less, but the termination rate constant of S is much higher than BA.

Heat removal and temperature

Removal of heat is often rate limiting factor in commercial system. Batch polymerisations are carried out at temperature between 40 to 55°C because large heat generation becomes rate-limiting factor. Redox initiators are used in low temperature process. With proper heat removal system, the overall rate of polymerisation increases with an increase in temperature. Temperature increases the rate by increasing both polymerisation rate constant K_p and N, the number of particles. The increase in N is due to the increased rate of radical generation at higher temperatures. Semi-continuous and continuous polymerisations are carried out at high temperatures (60–80°C), because continuous addition of monomer facilitates heat removal and temperature is controlled easily at constant levels.

Effect of O_2

Emulsion polymerisation is a free radical polymerisation whose kinetics is severely affected by the presence of small amounts of inhibitors and retarders. Oxygen is a common and powerful polymerisation suppressor that is often considered to be an ideal inhibitor, i.e. it has to disappear completely from the reaction mixture to start the polymerisation. Presence of oxygen decreases rate of polymerisation and increase induction periods. Solutions to this problem are:

1. High initiator concentration decreases induction period,
2. System and its ingredients should be purged with nitrogen.

Nomura observed decrease in polymerisation rate on increase of agitation rate due to increased oxygen transport rate from the vapour space to the aqueous phase.

It was also observed that the lower the oxygen levels in the reactor, the higher the reaction rate and the lower the inhibition time. Even after the inhibition period, a continuous flow of oxygen from the vapour space retards the polymerisation rate.

Role of mixing

In emulsion polymerisation, mixing can play a significant role in determining the kinetics of the polymerisation. In addition to maintaining a uniform temperature throughout the reactor, it controls the degree of dispersion of the immiscible oil (monomer) phase in the continuous aqueous medium. This is important throughout the first half of the reaction, during which monomer droplets serve as reservoirs supplying the monomer by diffusion to the growing (polymerising) sub-micron polymer particles. Too little agitation leads to large droplets and phase separation, which may result in diffusion limitations in systems in which the monomer is sparingly water-soluble. On the other hand, too much agitation can lead to either reduced nucleation of particles or nucleation in the monomer droplets. The stability of the particles as manifested by coagulum formation can also be dependent on the shear to which the latex is subjected during various phases of the reaction.

The effect of agitation speed and impeller type on the kinetics of emulsion co-polymerisation of styrene and *n*-butyl acrylate at 30 per cent solids was studied.

With increase in impeller speed from 400 to 1000 rpm, the initial number of micelles and, thus, the initial rate of polymerisation, decreased. This is due to the creation of a greater number of monomer droplets with correspondingly smaller sizes at the higher impeller speed, leading to the adsorption of more surfactant on the increased surface area of the monomer droplets with less surfactant being available in the aqueous phase for creating the initial particles. However, the impeller type and speed did not have any significant effect on the final number of particles and the overall rate of polymerisation. At 50 per cent solids, some differences were observed when two different impellers (Ruston radial flow and A310 fluid foil axial flow impeller were used).

High Solids Content Emulsion Polymerisations

High solid content latexes (45–50 wt per cent) are used for coating and adhesives. These latexes offer some advantages like low shipping cost, reduced drying time and better film forming properties. One of the major difficulties encountered in the preparation of this latex is the increase in viscosity once the solid content increases beyond 30 per cent, depending on the system. To overcome this problem, stable emulsions having a higher volume fraction can be prepared under certain conditions: (i) polydispersed emulsions; (ii) under the influence of centrifugal field; and (iii) the deformation of the droplets into polyhedral structures of particular types.

Polymer latex with a multimodal particle size distribution (PSD) was prepared by introducing additional surfactants during the semi-continuous emulsion polymerisation of styrene, BA and MAA. The results showed that bimodal and trimodal PSD polymer lattices can be obtained by this method and that the secondary generation of particles is greatly influenced by the nature and the amount of additional surfactants, as well as the moment when they are introduced.

High solid content through mini-emulsion polymerisation

Mini-emulsion polymerisation seems to be a good approach to produce latexes having a broad PSD for the production of high-solids-content. The viscosity of the reaction mixture limits the maximum solids content of the latex. The viscosity of a latex decreases when the particle size and the broadness of the particle size distribution increases. Owing to the different nucleation mechanism, the polymer particles produced in the mini-emulsion process are larger in size and with a broader particle size distribution than those obtained in the conventional emulsion polymerisation process. An important consequence of this behaviour is that high-solids-content latexes of low viscosity may be produced by mini-emulsion (two step and/or) polymerisation.

Types of Emulsion Polymerisation

Oil-in-water dispersions are typically classified as possessing the characteristics of one of three types of emulsions; macro-emulsions, mini-emulsions or micro-emulsion. As the names imply, a major distinction between these three categories is the size of the droplets (i.e. the discontinuous or dispersed phase) present in these stability of the emulsion.

Macro-emulsions

Macro-emulsions are the coarsest of the three in that the droplet size is relatively large (1–100 μm) and the stability is limited to minutes. Phase separation is rapid unless the system is well mixed. Droplets continuously collide and coalesce, and are broken by the shear exerted on the system. The droplet size is dependent on the system components (oil stabiliser, phase ratio) and the mixing characteristics (impeller type and speed).

Mini-emulsion

Mini-emulsion polymerisation could be defined as the polymerisation of all the monomer droplets present in the initial emulsion, where the final particle size distribution is reflected in the initial droplets size distribution (i.e. there is a one-to-one correspondence between the droplets and particles.

Mini-emulsions are considered to lie in between macro-emulsions and micro-emulsions in terms of droplets size and emulsion stability. Droplets generally range in size from 50 to 500 μm and the emulsion can be stable for as little as days and as long as months. The droplets are usually stabilised by the combination of an ionic surfactants (e.g., SLS) and a co-surfactant. The latter usually consists of either a long chain alkane (e.g., hexadecane) or alcohol (e.g., hexadecanol). The function of the co-surfactant can be two-fold; it can act in combination with the surfactant to create a barrier to droplet/droplet coalescence by arranging at the oil-water interface and/or it can limit the diffusion of the oil phase from smaller to larger droplets by virtue of its low water solubility which enables a pseudo-equlibrium to be maintained are between the droplets of different size and composition. Both mechanisms are considered to be operative for the case of the long-chain alcohols while only the latter can explain the increased stability brought about by use of the alkanes.

Mini-emulsions are typically formed by subjecting the oil/water/surfactant co-surfactant system of a high shear field created by devices such as an ultra sonifier, the Manton Gaulin homogeniser and the Microfluidiser. These rely on mechanical shear and/or cavitations to break the oil phase into sub-micron size droplets.

One of the key advantage of mini-emulsion polymerisations over conventional emulsion polymerisation was seen to be the direct control over the resulting number of particles which could be achieved by controlled in a reproducible number of monomer droplets.

In conventional emulsion polymerisation, the number of particles is chiefly controlled by the concentration of surfactant and initiator.

A key to the successful implementation of miniemulsion polymerisation lies in the ability to nucleate all of the droplets created.

Micro-emulsion

Micro-emulsions on the other hand are thermodynamically stable (i.e. indefinite stable) with droplet sizes varying from 10 to 100 nm. Relatively large quantities of mixed emulsifiers, typically consisting of an ionic surfactant (e.g,. sodium dodecyl sulphate, SLS) and a short chain alcohol (e.g. pentanol), are usually used to prepare these emulsions.

Inverse emulsion polymerisation

Inverse emulsion polymerisations is an important process for the commercial production of water-soluble polymers. To be most effective, these polymers must be of very high molar mass and their manufacturing requires a process of which overcomes the high exothermicity of the polymerisation.

Therefore, the industrial production of these polymers is most often carried out in water-in-oil dispersions. In water-in- oil polymerisation, a hydrophilic monomer (usually in aqueous solution) is dispersed in a continuous organic phase using a water in-oil soluble or a water-soluble initiator to yield inverse latex, i.e., a colloidal dispersion of water-swollen polymer particles in oil. Such a process is well suited to the preparation of high molar mass polymers at rapid reaction rates, due to the high local monomer concentration within the particles and to the fact that free radicals grow in separate particles preventing their mutual termination.

The achievement of low-viscosity media at high solids contents, easier handling, storage and dissolution are other advantages of water-in-oil polymerisation over the more conventional bulk or solution processes. For application, the polymer dissolves rapidly contrary to solid-powder polymer, which forms gels or aggregate when added to water.

Synthetic water-solution polymers are an important class of materials owing to their fundamental interest as well as for their numerous applications. For example, polymers based on acrylamide (AM) and its derivatives are widely used for their thickening properties in paints and as pushing fluids in tertiary oil recovery, as drag reduction agents and drillings fluids. Co-polymers with anionic or cationic charges are effective flocculants for many substracts, including sewage treatment, cellulosic fibres and fines for retention in paper-making, coal tailings and in general applications whenever aqueous solid-liquid separations are required.

MICRO-EMULSIONS—A NOVEL MEDIUM FOR POLYMERISATION REACTIONS

Microemulsion was discovered by *Hoar* and *Schulman* way back in 1943. Since then, its applications have extended to various technological areas like oil recovery, cosmetics, pharmaceuticals, lubrication, food technology, agricultural sprays, coatings, environmental remediation, cleaning, chemical synthesis, polymer synthesis, micro-process, media synthesis, enhanced reaction kinetics and many other applications. Today it is as versatile as conventional emulsion technology. Micro-emulsion is defined as a thermodynamical stable, 'isotropically clear dispersion of two immiscible liquids consisting of microdomain of one or both liquids stabilised by an interfacial film of surface active molecules. Micro-emulsions have distinct advantages over conventional emulsions. Like an emulsion, a micro-emulsions is generally considered to be droplets of one type of fluid encased in a surfactant shell and dispersed through a second immiscible fluid. However, micro-emulsions are thermodynamically stable phased, and as such they can be formed with defined properties that depend only on thermodynamic conditions rather than on the kinetics of formation.

Further the micro-emulsion formed is generally a homogenous and well characterised system. For example it is possible to create micro-emulsion consisting of a fairly monodispersed system of spherical droplets, in contrast to the generally broad distribution of droplets size found within an emulsion. As the name implies, micro-emulsion droplets are 50–1000 Å in dia. The equilibrium nature of micro-emulsion enable them to remain stable as long as conditions around them are unchanged and thus they are unaffected by destabilisation mechanism e.g., creaming, flocculation and coalescence etc. that so strongly affect emulsion stability. On the other hand, by changing temperature or composition to values at which the micro-emulsions is no longer energetically favourable, one may destabilise the system in a controlled fashion.

However, these advantages of micro-emulsions are costly indeed. The thermodynamic stability of these phases means that they can be formed only in specific range of temperature, pressure and composition, which does not allow one to have complete freedom in designing a system. Around 1980, a new interest developed for polymerisation in micro-emulsion as a consequence of the extensive studies performed on these systems after the 1970 oil crisis. A primary goal of the investigation was to produce thermodynamically stable latexes in the mamosite range (<50 nm) not attainable with the classical emulsion polymerisation process. Microlatexes with such characterisation are desirable in several applications, e.g., in drug delivery or micro-capsulation. It is also of a particular relevance to prepare stable and uniform inverse latexes via a micro-emulsion polymerisation process, since the classical inverse emulsion technique is know to produce unstable latexes with a broad particle size distribution.

Another appealing aspect of micro-emulsions polymerisation was the possibility to fix permanently the labile structure of these systems and to secure more tightly solubilised molecules.

Conditions for Formation of Micro-emulsions

The standard free energy of micro-emulsion is as follows:

$$\Delta G_{\text{micro e}} = \Delta H_{\text{micro e}} - T\Delta S_{\text{micro e}} \quad \text{... (9.8)}$$

To be a favourable system $\Delta G_{\text{micro e}}$ should be negative. Usually $\Delta H_{\text{micro e}}$ is positive and is much smaller then $\Delta S_{\text{micro e}}$ which give $\Delta G_{\text{micro e}}$ negative value and thus forming micro-emulsions. Thus the micro-emulsion process in governed by the entropy gain associated with it and the driving force (the interfacial tension) must have a measurable negative value which would give a negative free energy. In order to form micro-emulsion, it is necessary that the concentration of the surfactants must be greater than that required to reduce the interfacial tension between the dispersed and continuous phase to zero and to cover the total interfacial area of all dispersed droplets.

The total interfacial tension can be defined as

$$\gamma_i - \gamma_{ow} - \pi \quad \text{... (9.9)}$$

γ_i = Oil/water interfacial tension in the absence of surfactants.

π = Spreading pressure of the added surfactants at the oil/water interface.

If $\pi > \gamma_{ow}$, $\gamma_i < 0$ micro-emulsion is formed

If $\pi < \gamma_{ow}$; $\gamma i > 0$ micro-emulsion is formed

A microemulsion is thermodynamically stable because interfacial tension is low enough than the increase in interfacial energy accompanying dispersion of one phase in the other is outweighed by the free energy decrease associated with the entropy of dispersion. Further, the free energy decrease accompanying adsorption of surfactant molecules for a bulk phase favours the existence of a large interfacial area and hence plays a major role in stabilising the micro-emulsion.

Structure of Micro-emulsions

Emulsions and micro-emulsions are both composed of a dispersion of two immiscible liquids stabilised by one or several surface active agents. The micro-structures of micro-emulsions are known to be oil in water (o/w) like emulsion. In the o/w and water in oil (w/o) like emulsions. In the o/w micro-emulsions, oil droplets are dispersed in water and are stabilised by surfactants and/or co-surfactants which are adsorbed at the interface between an oil droplet and water; in the w/o micro-emulsion, water droplets are dispersed in oil and are stabilised by surfactants and/or co-surfactants which are adsorbed at the interface between water droplet and oil (Fig. 9.9).

However, sharp distinction exists between these two types of system. Emulsion are thermodynamically unstable and turbid systems whose average droplets size (1–10μm) tends to grow continuously with time. Emulsification is a non-spontaneous process and an input of energy through vigorous stirring is needed because of the large increase in interfacial area during the formation of the emulsion. The stability of an emulsion is kinetic in origin and is determined by the potential barrier generated between two particles due to the competition between the Van der walls interaction and repulsive double layer forces. In contrast, the formation of micro-emulsion is a spontaneous process which produces an optically transparent and thermodynamically stable system with time dependent properties. Although a microemulsion can be formed from a single surfactant. It is more common to use a mixture of classical amphiphile (soap) and a weak amphiphile like a short chain alcohol, referred to as co-surfactants.

The role of a short chain alcohol is to: (i) decrease interfacial viscosity; (ii) provide additional interfacial area; (iii) reduce electrical repulsion between droplets by decreasing surface charge density; and (iv) induce appropriate curvature changes.

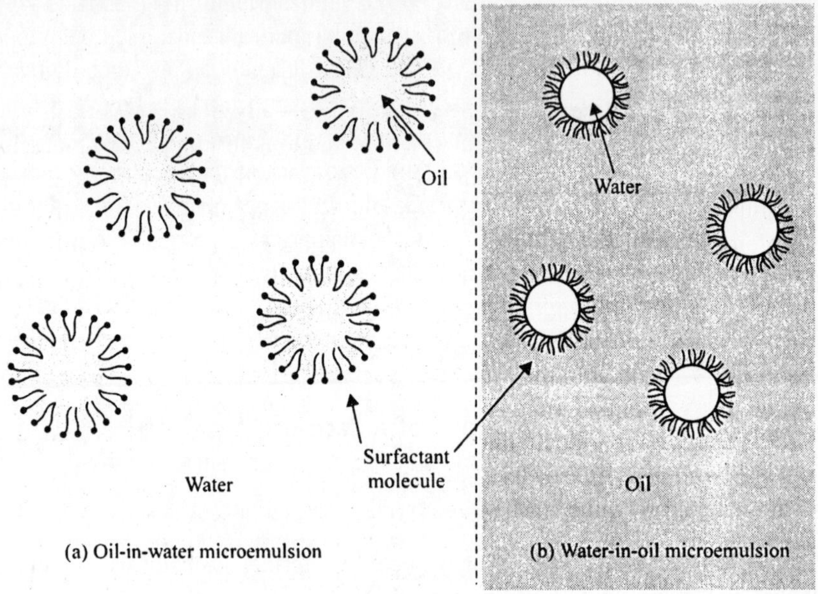

Fig. 9.9. Types of micro-emulsion.

Phase Diagram of Micro-emulsion

The phase diagram of micro-emulsion are very important to predict their phase behaviour for applications. The pseudo ternary phase diagram are usually constructed by considering surfactant plus co-surfactant and water plus salt as pseudo components. Depending on the types of micro-emulsion, there are three ideal phase diagrams as shown in (Fig. 9.10).

Two methods are known for the same.

1. Titration method.
2. Grid method.

Titration method

In this method, several samples having the various compositions on one of the three sides of the diagram are prepared and the samples are titrated by the known amount of one remaining component. After each titration, equilibrium is checked and the composition of each phases is analysed in addition to its volume. Usually sample contains surfactant, alcohols and oil (or water) and are titrated by water (or oil).

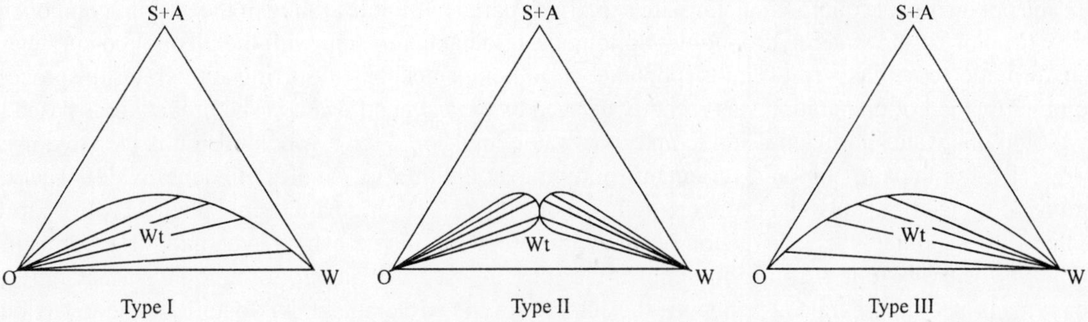

Fig. 9.10. Phase diagram of micro-emulsion.

Grid point method

In this method, a lot of samples having various compositions of the three-pseudo components are prepared at once and their equlibria are checked and composition are analysed.

Formulation and Structure of Polymerisable Micro-emulsion

A thorough research on appropriate formulation rules alongwith a good characterisation of the phase diagram of the systems (containing monomer) should precede any study on micro-emulsion polymerisation because after polymerisation a microlatex is formed which has the same characteristics as those of the starting micro-emulsion, i.e., stability, fluidity and optical transparency. Furthermore, the use of these material for application implies to find inexpensive formulation characterised by a high ratio of monomer to surfactants concentration. These problems are not trivial; and inappropriate choice of the nature and/ or composition of the components can lead to phase separation or gel formation during polymerisation process.

Among the experimental methods available to determine the structures, the scattering techniques—static and dynamic light scattering, and small angle neutron scattering are certainly the most powerful. However, these techniques require an extrapolation to zero concentration in order to eliminate the interparticle interaction. Ternary system (monomer/surfactant/water) can easily be diluted by addition or the water continuous phase. The problem is more complicated in systems with alcohol because the co-surfactant is usually distributed in the surfactant film and in the oil and water phase. However, these dilution procedure in some case is inoperative, e.g., if the micro-emulsions are bicontinuous.

Attempts have also been made to use microscopy experiments, but they have been generally unsuccessful owing to the fast molecular exchanges between the micelles which prevents a correct visualisation, free of artifacts.

Polymerisation in o/w Micro-emulsions

Various studies has been carried out on different systems which has led after polymerisation to stable and fluid microlatexes. The pioneering work in this field is that of *Schanber* and *Riess* who prepared cross-linked and non-cross-linked stable microlatex particles with 5–10 per cent solid contents of polymer.

It has been found that the microlatexes are bluish, less transparent than the original micro-emulsion and show some Tyndall effect due to an increase in the particle size during the process and the higher refractive index of polymer compared to that of monomer.

The polymerisable micro-emulsions are usually prepared by simple mixing of the various components or by titration of the water/hydrophobic monomers/surfactant emulsion with the alcohol co-surfactant. The formation of a micro-emulsion is spontaneous, although mechanical stirring accelerates the process. Another method of preparation was recently derived by *Larpent* and *Tadros*. The procedure is based on inversion of a water-in-oil emulsion to an oil-in-water emulsion. First a w/o emulsion is prepared using a low HLB surfactant, nonionic surfactant (nonyl phenol with 5 to 7 moles ethylene oxide). The w/o emulsion is then titrated with an aqueous solution of a high HLB surfactant (nonyl phenol) with 15 to 17 moles ethylene oxide. Phase inversion occurs with formation of a gel which eventually transforms into a low viscosity o/w micro-emulsion upon further titration. A small amount of anionic surfactant (0.05 per cent) is also incorporated to enhance the stability of the w/o emulsion and facilitate the inversion.

Polymerisation in w/o Micro-emulsion

Several studies have been carried out dealing with the polymerisation of water soluble monomers in globular w/o microemulsion. First it should be noted that there is a distinct difference in the formulation of these system as compared to that of w/o micro-emulsion. In the case of o/w, the hydrophobic monomer like styrene, MMA used as oil in the recipes are in liquid form. Therefore the resulting micro latexes consists of particles where the polymer is in the bulk state. In contrast, in w/o systems the water soluble monomers are generally powder and must be first dissolved in water usually (1:1 mass ratio). Thus the inverse microlatexes are formed of polymer particles highly swollen by water and dispersed in the organic continuous phase. This means that the basic w/o micro-emulsion are formed at least of four components-water/monomers/oil/surfactants, even in the absence of co-surfactants.

Polymerisation in Biocontinuous Micro-emulsions

The formulations improves significantly when the globular w/o or o/w microemulsions is replaced by a micro-emulsion with discontinuous structure. In the later case, the amount of monomer can reach 25 per cent and that of surfactant is reduced to about 10 per cent. The final microlatexes are perfectly transparent, stable and fluid even at high volume fraction approx. 60 per cent. So far only water soluble monomers have been co-polymerised inside the aqueous domain of nonionic bicontinuous micro-emulsion, neutral monomers. Acrylamide anionic monomer, e.g., sodium acrylate and sodium -2-acrylamide-2-methyl propane-sulphonate or cationic monomer e.g., methacryloxythyltrimethyl ammonium chloride have been investigated.

HYDROXYETHYLCELLULOSE AS A PROTECTIVE COLLOID IN EMULSIONS POLYMERISATION

This section illustrates the use of hydroxyethylcellulose (HEC) in the following emulsion polymerisation systems:

1. Vinyl acetate, homopolymers and co-polymers for coatings.
2. Polyvinyl acetate for adhesive.
3. Acrylic emulsions for coatings.

Chemistry of HEC

HEC is the hydroxyethyl ether of cellulose whose polymer molecule is composed of anhydroglucose units, each of which has three hydroxyls capable of reactions. By treating the cellulose with sodium hydroxide and reacting with ethylene oxide, hydroxyethyl substituent groups are added to the cellulose chain through an ether-oxygen linkage. Ethylene oxide may also react with the hydroxyl of the hydroxyethyl substituent to give a side chain based on two or more molecules of ethylene oxide.

The average number of moles of ethylene oxide attached to each anhydoglucose unit in the cellulose is known in the molar substitution (MS) for the derivative. HEC 250 has an average hydroxyethyl MS of about 2.5. An idealised structure for this type is shown in Fig. 9.11.

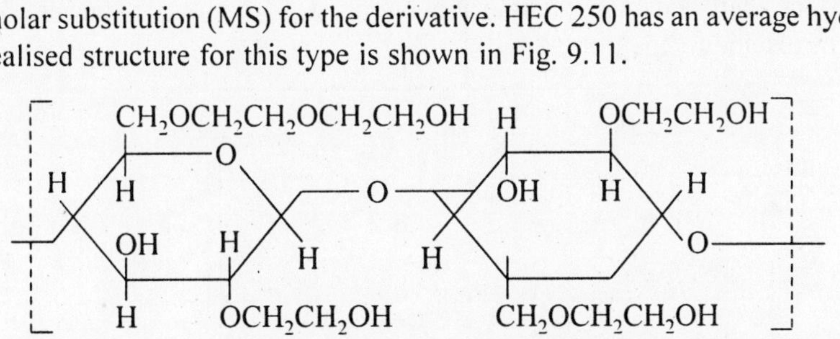

Fig. 9.11. Idealised structure of HEC.

HEC is a non-ionic cellulose ether that the been widely used for many years as a protective colloid in the production of vinyl acetate, homopolymer, and co-polymer for water-based paints. It is preferred protective colloid in this use because of its superior stabilising, thickening, and suspending properties. It also provides a number of other benefits such as batch-to-batch emulsion consistency, particle size uniformity, low grit content, and good shelf stability.

In addition, recent studies have demonstrated the utility of HEC as a protective colloid for all acrylic emulsions and for the large-particle-size polyvinyl acetate latexes for adhesives. These emulsions, stabilised with HEC have advantages over those produced with traditional stabilisers, particularly in freeze-thaw, heat, shelf-life, and mechanical stability, as well as improved compatibility with many additives. Suggested procedures for the use of HEC with these emulsions are presented in this section.

Thus a wide range of emulsion characteristics can be obtained with a variety of monomers, using HEC as a protective colloid. HEC can be used to stabilise emulsions in two ways. First, it can be postadded to thicken the emulsion and impart proper rheology; and second it can be added during the polymerisation process to prevent particle agglomeration through steric stabilisation. When used as a protective colloid, HEC also provides the following desirable benefits:

1. Control of latex particle size and viscosity by proper selection of HEC viscosity type.
2. Reduced cost and improved water resistance of the film by considerable reduction in the amount of conventional surfactant.
3. Improved freeze-thaw and storage stability, particularly in systems that contain anionic surfactants.
4. Good mechanical and heat stability, particularly during processing of the latex with various substances such as fillers and pigments.
5. Improved film integrity, due to grafting of HEC onto the latex particles during polymerisation.

Types of HEC

HEC 210 and 250 have hydroxyethyl molar substitutions that average 2.1 and 2.5 respectively (Fig. 9.12). These are the substitution types most commonly used for emulsion polymerisation. Numerous viscosity types are available for each of these substitution levels. The viscosity types available for the 250 substitution type, alongwith some typical properties of these polymers, are given in Table 9.9. The lower molecular weight viscosity types (L through G) are most commonly used for polymerisation of latexes for coating where a small particle size is desired.

All types of HEC are available in a rapid-dispersion form, designated R, that permits addition of the dry powder to water without lumping. This surface treatment facilitates solution preparation without heating, and enables addition of the dry powder to filled systems. The hydration of the R types has been inhibited, and this delay, between initial wetting and the start of hydration, can be varied. Hydration delay is also dependent of pH and temperature.

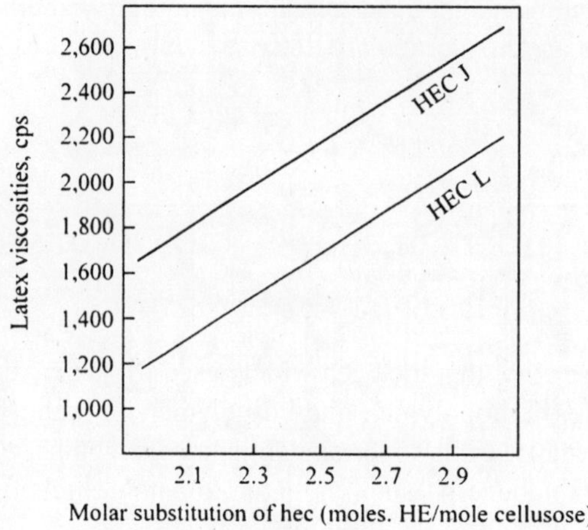

Fig. 9.12. Effect of molar substitution and molecular weight of HEC on viscosity of emulsions described in examples 1, 2 and 3 of Table 9.6.

Emulsion Polymerisation

A brief discussion of emulsion polymerisation is helpful in understanding how to utilise HEC to the best advantage in this use. Free radicals are generated by the decomposition (usually thermal) of the water-soluble initiator in the aqueous phase. These radicals are absorbed into the monomer swollen emulsifier micelles and initiate polymerisation. Polymerisation can also proceed by homogeneous nucleation. In this process, the radicals initiate polymerisation in the aqueous phase, forming an oligomer, which can then be the nucleus for a new particles. At a critical stage in the polymerisation process, the monomer-swollen latex particles become extremely viscous and are subject to agglomeration through coalescene. This constitutes the so-called 'sticky' stage. The presence of hydroxyethylcellulose in the aqueous phase provides steric stabilisation and prevents the agglomeration of these particles.

Hydroxyethylcellulose can also be grafted onto the surface of the latex particles. The grafting occurs in the aqueous phase through the abstraction of hydrogen atom from HEC by the initiator or latex radicals. The radicals formed along the HEC backbone can then initiate polymerisation or combine with growing latex radicals in solution. Thus, HEC becomes chemically bound to the particle surface and enhances the stability of the resulting latex.

After the monomer is completely reacted, stable latex particles remain whose surface is partially covered with emulsifier and HEC. Any unadsorbed HEC remaining in the aqueous phase contributes to the final latex viscosity and rheology. The use of HEC in emulsion polymerisation permits control of latex particles size and viscosity through the proper selection of both viscosity type and hydroxyethyl content. Fig. 9.12 shows the effect of these two HEC variable on the viscosity of the polyvinyl acetate and vinyl acrylic emulsions, detailed in examples 1, 2 and 3 of Table 9.6. The date in this figure show that latex viscosity is markedly increased as the hydroxyethyl molar substitution is increased from 2.1 to 2.9. Also, the latex viscosity is increased as molecular weight of the HEC is increased from the 250L to the 250J. The viscosity trends shown in Fig. 9.12 tend to be dependent on the recipe used. The relationship may be different for other recipes.

Laboratory Formulations

The formulations in Table 9.6, 9.7 and 9.8 are presented to illustrate the versatility of HEC protective colloid for emulsion polymerisation and to provide a starting point for developing a process for latex manufacture.

Polyvinyl Acetate and Vinyl-acrylic Latexes for Coatings

HEC has a long history of use in homopolymer and co-polymer vinyl acetate emulsions used for coatings. Hydroxyethylcellulose is particularly well suited for use in these latexes, as it promotes the formation of stable emulsions containing a narrow distribution of very small particles, usually desirable for coatings. The ability to control latex viscosity and particle size by varying the concentrations, molecular substitution, and molecular weight of the HEC makes it the protective colloid of choice for these systems.

Sample formulation utilising HEC 250LR for homopolymers and co-polymers of vinyl acetate are given in Table 9.6. Polymerisation procedures and properties of the emulsions are also presented. The influence of substitution and molecular weight of HEC on viscosity of the latexes of Examples 1, 2 and 3 of Table 9.6 is detailed in Fig. 9.12.

Table 9.6. Polyvinyl acetate and vinyl-acrylic emulsions for coatings.

	Parts by weight			
Ingredients	*1*	*2*	*3*	*4*
Water	433.5	462.3	433.3	438.0
HEC-250LR	12.5	12.0	7.7	6.0
Tergitol NPX	2.0	12.0	–	–
Aerosol OT (25%)	–	9.0	–	–
Borax	–	3.5	–	–
Sodium bicarbonate	1.2	–	1.1	1.0
Potassium persulphate	1.2	3.2	1.6	–
Sodium persulphate	–	–	–	1.5
Benzoyl peroxide	0.5	–	–	–
Igepal CO-997	–	–	30.7	–
Igepal CO-430	–	–	12.0	–
Igepal CO-897 (70%)	–	–	–	17.8
Igepal CO-630	–	–	–	4.2
Abex 18S (30%)	–	–	–	13.5
Vinyl acetate	549.1	448.0	436.0	440.0
2-ethylhexyl acrylate	–	50.0	–	78.0
Butyl acrylate	–	77.6	–	–
	1,000.0	1,000.0	1,000.0	1,000.0
Initial monomer addition, % of total monomer	10[(a)]	10[(a)]	20[(a)]	Nil[(b)]

(Contd ...)

Ingredients	Parts by weight 1	2	3	4
Emulsion properties				
Viscosity, cps	1,250	1,750	2,500	1,800
Total solids, %	55.3	55.1	55.5	55.0
pH	4.5	4.7	4.5	4.5–5.0
Average particle size, μ	0.5	0.5	0.34	
Freeze/thaw stability, 5 cycles	Good	Good		
Mechanical stability	Excellent	Excellent		

Table 9.7. Polyvinyl acetate for adhesives.

Ingredients	*Parts of weight* 1	2
Water	446.1	446.1
HEC 250 HR	18.0	7.0
HEC 250 MR	–	12.9
Sodium bicarbonate, 5% solution	14.4	14.4
Potassium persulphate	1.5	1.5
Vinyl acetate	520.0	518.1
	1,000.0	1,000.0
Initial monomer addition % of total monomer	3.2	2.2
Emulsion properties		
Viscosity, cps	4.600	8.600
Total solids	54.0	54.0
pH	4.3	4.5
Average particle size, μ	1.5	1

Table 9.8. All -acrylic latex emulsion.

Ingredients	*Parts by weight*
Water	515.5
HEC 250 LR	10.0
Potassium persulphate	14.0
Alpha-methylbenzyl alcohol	20.0
Ethyl acrylate	269.0
Methyl methacrylate	176.0
Methacrylic acid, glacial	5.5
	1,000.0
Initial monomer mix addition, % of total monomer	5.0
Emulsion properties	
Viscosity, cps	2,350
Total solid, %	46.0
Average particle size, μ	0.9

Polyvinyl Acetate-based Adhesives

Recent studies have demonstrated that latexes possessing particle sizes and viscosities equivalent to polyvinyl alcohol-stabilised emulsions can be made with HEC 250. Thus, the utility of HEC 250 has been extended to include polyvinyl acetate-based adhesives. Important factors are the proper choice of viscosity type of HEC 250 and the polymerisation technique.

HEC 250 HR, alone or in combination with HEC 250 MR, produces latexes possessing particle sizes greater than 1 μ and viscosities ranging from 1,000 to 8,000 cps. Critical to the success of the procedure is the incorporation of a small amount of monomer in the initial step, as shown in the formulations provided in Table 9.7. Without this initial monomer addition step, a high-viscosity submicron-particle-size latex is produced. Polyvinyl acetate latexes prepared with HEC as the protective colloid also shown improved freeze-thaw, heat and mechanical stability over similar latexes made with polyvinyl alcohol.

Another major advantage of these HEC-stabilised polyvinyl acetate emulsions is their tolerance to borax. Blends of these emulsions with borax-modified dextrins make excellent low-cost adhesives for paper, corrugated paper, and wood applications. The resultant adhesives give wood-to-wood bonding strengths equivalent to that attained with 100 per cent polyvinyl acetate latex adhesives. These blends would not be possible with polyvinyl alcohol emulsion, which are destabilised and lose viscosity in the presence of borax.

Procedure

Water is charged to a suitable polymerisation vessel. The HEC and catalyst are dispersed in the water with stirring. The temperature is raised to 75°C and the system is purged with nitrogen. The initial small monomer addition is made and the reaction is allowed to proceed for 15 minutes, at which time the mix should be quite viscous. The buffer solution is added, then continuous monomer addition is begun and maintained at a rate that sustains a small amount of reflux.

Emulsion Polymerisation with Acrylic Monomers

Research has established the utility of HEC as a protective colloid in the emulsion polymerisation of acrylic monomers. The excessive latex structuring and irreversible coagulation that resulted from the use of HEC in standard all-acrylic formulations has been eliminated by incorporating a small amount of a chain transfer agent such as alpha-methylbenzyl or benzyl alcohol into the formulation. In fact, the stability of the all-acrylic emulsions was found to be so good using HEC in the presence of alpha-methylbenzyl alcohol that it was possible to produce these latexes without the aid of surfactants. A sample formulation for a 100 per cent acrylic emulsion utilising HEC as the sole protective colloid is given in Table 9.8. Emulsions made from this recipe exhibit outstanding stability, good solvent tolerance, and powerful adhesion to polar substrates such as glass and wood. It is possible to vary the viscosities of these latexes without affecting particle size by adjusting the concentration of HEC in the formulation.

Procedure

Water is charged to a suitable polymerisation vessel. HEC alphamethylbenzyl alcohol, potassium persulphate, and the initial increment of the monomer mix are added and the system is purged with nitrogen. Stirring is continued until the HEC is completely dissolved. The reaction temperature is raised to 75°C. After 10 minutes the addition of the remaining monomer mix is started and continued on a semicontinuous basis over a 2-hr reaction period.

1. Reaction condition for examples 1, 2 and 3 were similar except for the amount of monomer added initially (Table 9.6). The procedure was as follows: Water is charged to a suitable polymerisation vessel equipped with an agitator, a jacket for heating and cooling, a condenser, a nitrogen purge, and a reservior to permit continuous addition of the monomer (or monomer mix). The HEC, buffer, surfactant, and catalyst are dispersed in the water with stirring. The temperature is raised to 75°C, the system is purged with nitrogen to remove all of the oxygen, and the stirring is continued until the HEC is completely dissolved. The initial monomer addition is then made and the reaction is permitted to proceed for 5 to 10 minutes until there is a noticeable reduction in the reflux rate.
The remaining monomer (or monomer mix) is then added uniformly, with stirring, over a 2 to 4-hr period, with the temperature held between 75 and 80°C. After the delayed addition is complete, the reaction mixture is heated to 90 to 95°C for 30 minutes to complete the reaction and reduce the residual monomer content to less than 1per cent. At the end of this reaction period, the latex emulsion is cooled to room temperature with continuous stirring, and then removed for optimum stability, the pH should be between 4 to 5: it can be adjusted by using additional buffer.
2. The polymerisation procedure for example 4 differs from that given in footnote (a) above, in the following respects: 60 ml of the water to be used is held out to dissolve 0.8 g (53 per cent) of the sodium persulphate catalyst, which is then placed in a dropping funnel for continuous addition to the vessel. The initial charge to the reaction vessel is heated to 80°C with stirring to dissolve the HEC. The system is then cooled to 72°C, and 0.7 g (47 per cent) of dry sodium persulphate is added. Continuous addition of the monomer mix and the catalyst solution is started and continued at the rates as given in Table 9.10.

Table 9.9. Typical properties of HEC.

Polymer	
Powder	White to light tan, odourless
Moisture content (as packed), %	5.0 max
Ash content (calculated as Na_2SO_4), %	4.5
Effect of heat	
Softening range, °C	135–140
Browning range, °C	205–210
Bulk density, g/ml	0.6
Solutions	
Specific gravity, 2% solution, g/ml	1.0033
Refractive index, 2% solution	1.336
pH	7
Surface tension, dynes/cm (N/m)	
HEC 250 L at 0.1%	166.8 (0.0668)
At 0.001%	67.3 (0.0673)
HEC 180 L at 0.1%	66.7 (0.0667)
At 0.001%	69.8 (0.0698)

(Contd ...)

Interfacial tension vs. Fractol, dynes/cm (n/m)	
HEC 250 L at 0.001%	25.5 (0.0255)
180 G at 0.001%	23.7 (0.0237)
Bulking value in solution, gal/lb (L/kg)	0.1 (8.34)
Films	
Refractive index	1.51
Specific gravity at 50% relative humidity, g/ml	1.34
Moisture content (equilibrium), %	
At 73°F, 50% relative humidity	6
At 73°F, 84% relative humidity	29

Table 9.10. Catalyst solution and catalyst rate.

	g/10 min	*Minutes of addition*	*Total added, g*
Monomer mix	21.6	240	518.0
Catalyst solution	2.5	243	60.8

The reaction mass is maintained at 73 ± 1°C during the entire addition period. The temperature is then raised to 90°C for 1/2 an hour to complete the reaction, then cooled.

SOLUTION POLYMERISATION

In solution polymerisation, the monomer is dissolved in a suitable inert solvent along with the chain transfer agent, whenever used. The free-radical initiator is also dissolved in the solvent medium, while the ionic and co-ordination catalysts, can either be dissolved or suspended. The presence of the inert solvent medium helps to control viscosity increase and promote a proper heat transfer.

The major disadvantage of the solution polymerisation technique is that, however, inert the selected solvent may be, chain transfer to the solvent cannot be completely ruled out and, hence, it is difficult to get very high molecular weight products. The polymer formed will also have to be isolated from the solution either by evaporation of the solvent or by precipitation in a non-solvent, and removal of their final traces is always extremely difficult. Solution polymerisation techniques, nevertheless, can advantageously be used where the polymer is used in its solution form, as in the case of certain adhesives and coating compositions, or in systems where the polymer formed is insoluble in its monomer or solvent and precipitates out as a slurry and is amenable for easy isolation.

Coming to the practical applications, industrial production of polyacrylonitrile by free-radical polymerisation and that of polyisobutylene by cationic polymerisation, use the solution technique. Block co-polymers are also made exclusively by this technique.

This process is also known as solvent polymerisation. The monomer is dissolved in an inert solvent (e.g., a saturated hydrocarbon) together with the initiator operation or as a continuous operation.

Advantages and Disadvantages of Solution Polymerisation

Advantages

1. Heat removal and control are easier.
2. Tracking the reaction is rather easy because rates of the reaction are lower. Hence, in this case, design of reactor becomes easy through known kinetics.

3. A readymade polymer solution is directly available from the reactor. This is very useful in such applications as lacquers and adhesives.

Disadvantages

1. Use of solvent lowers monomer concentration and hence, the rate of polymerisation, also lowers the average chain length of the polymer.
2. Solvents involved bring in toxicity, flammability and give higher product cost.
3. Recovery of the used solvent requires additional technology and increase the cost of production.
4. Removal of the last traces of solvent poses problems.
5. Yield per reactor volume is less than that in the mass polymerisation.

Effects of Dilution in Solution Polymerisation

The effect of adding a solvent which is inert, is to reduce the monomer concentration. Hence in consequence, the rate of polymerisation is also decreased proportionately. This leads to lowering of the molecular weight correspondingly. Fig. 9.13 shows the effect of concentration of the monomer on the rate of polymerisation.

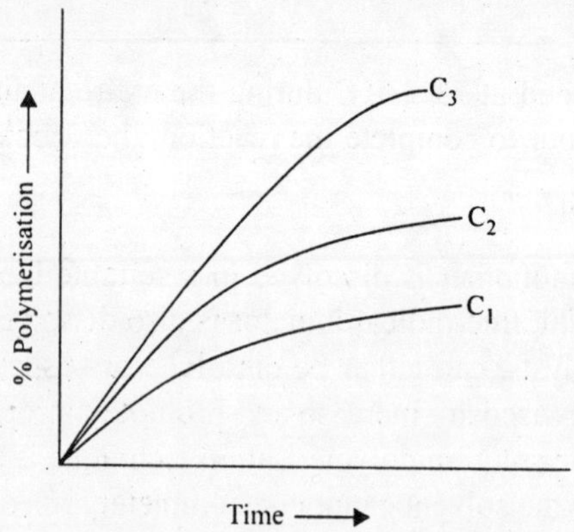

Fig. 9.13. Concentrations of polymer solution ($C_3 > C_2 > C_1$).

At higher concentrations, an acceleration of the rate of polymerisation takes place. This is called *auto acceleration* or *gel effect*. When the polymer chains grow, their movements become more dependent on the entanglement. When the concentrations are high, the entanglement becomes more, so much so that for a chain to drag itself through the entanglements becomes rather difficult. Without chain movement, active ends do not come to be terminated. Hence, propagation becomes more compared to the termination. This explains the upward trend of the plot. At still higher concentration, and at certain temperatures where the chains become completely immobile, the propagation reaction due to the movement of radicals becomes diffusion controlled. This is the reason why at very high concentration the curve tends to level off. Another aspect regarding dilution of polymer solution is that the decrease in the rate of polymerisation is proportionate to the amount of dilution but the decrease in the molecular weight is not. This can be explained if we imagine that the active radicals in addition to helping the propagation, also attack solvent molecules and extract some weakly bonded hydrogen or other atoms in the process. The small, active atoms thus extracted initiate further polymerisation. Thus more number of initiations take place and less number of propagations. This accounts for the larger fall in molecular weight.

Advantage of a diluent

The main advantage of a diluent is to take up heat of polymerisation by a rise in temperature or by vapourisation. We can sub-divide solution polymerisations by the phase present. Invariably, monomer and diluent are miscible, so the combinations are:

	1	2	3	4
Monomer and diluent	Soluble	Soluble	Soluble	Soluble
Initiator	Soluble	Insoluble	Insoluble	Soluble
Polymer	Soluble	Soluble	Insoluble	Insoluble

Some physical schemes to carry out these reactions can be illustrated with specific examples.

1. *All components soluble*. In one process ethylene containing a trace of oxygen (initiator) is compressed to 2000 atm and continuously pumped to a heated tube 5 mm ID and 20 m long (Fig. 9.14). With an average residence time of 45 s at 175°C, 22 per cent of the ethylene is converted (3 kg polymer/h). At the end of the tube a valve is opened at intervals and the pressure drops slightly as polymer and monomer are ejected in a cyclic manner. The monomer flashes off, leaving polymer to be washed, dried, and packed. In this pilot-scale example, the monomer is the diluent. The extreme pressure favours the tube geometrý. In a production unit the tube can be over a kilometre in length. Stirred autocalves operating at 2000 to 3000 atm are sometimes used in place of the tubular geometry.

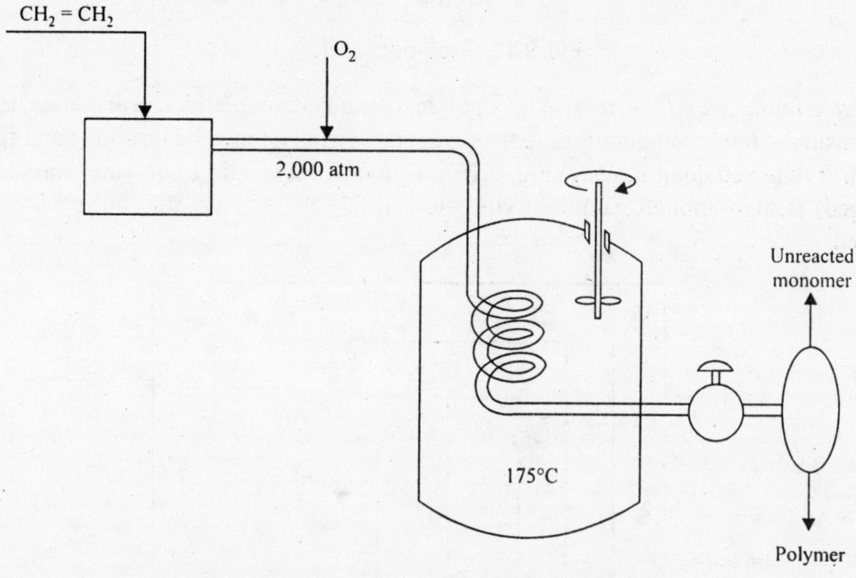

Fig. 9.14. Tubular reactor for ethylene polymerisation.

2. *Insoluble initiator*. A chromia silica alumina catalyst (an initiator that can be regenerated) will convert ethylene to linear polymer at moderate pressures. One method is to pass a dilute (2 to 4 per cent) solution of ethylene in a saturated hydrocarbon over a fixed bed of catalyst at 150 to 180°C and 300 to 700 psi (2.1 to 4.8 Mpa) (Fig. 9.15). Conversion may be high, but there is a

considerable volume of solvent to recover and recycle. Periodic reactivation of catalyst at about 50-h intervals is required.

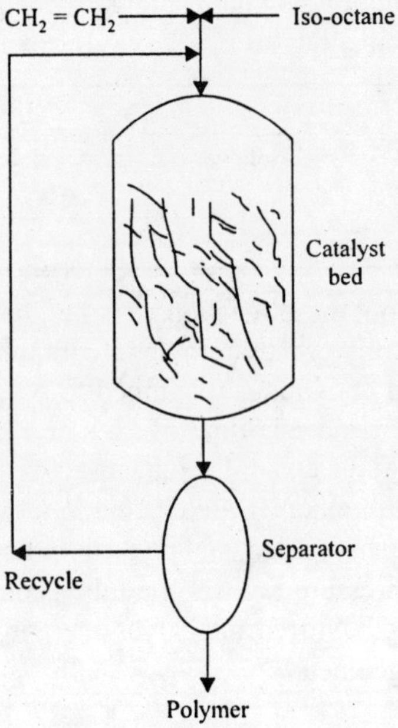

Fig. 9.15. Fixed-bed reactor.

3. *Insoluble initiator and polymer*. With a polymerising system of ethylene and an agitated catalyst suspension at lower temperatures, polyethylene and catalyst may be insoluble and form a slurry which is removed continuously (Fig. 9.16). If productivity (ratio of polymer formed to catalyst charged) is high enough, simple drying may suffice. If not, washing out of catalyst may be needed.

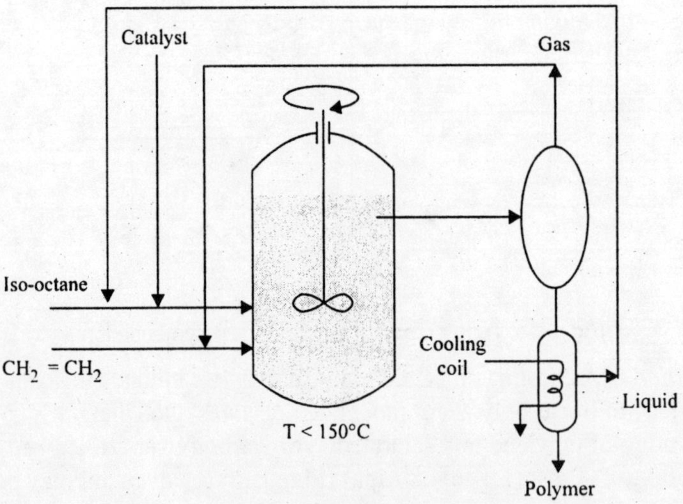

Fig. 9.16. Slurry reactor.

A fluidized-bed, gas-phase polymerisation bears some resemblance to the process just described. Fine catalyst particles are contacted with rapidly upward-moving ethylene vapours below the melting point of polyethylene. This has become an important commercial process with definite economic advantages over solvent-based systems.

4. *Insoluble polymer*. Heating an aqueous solution of acrylonitrile with persulphate as initiator at 80°C results in a stringy precipitate of polyacrylonitrile that can be filtered out and dried.

An interesting example of condensation polymerisation in this category is interfacial condensation. A diacid chloride is dissolved in a heavy, dense solvent such as tetrachloroethylene. As a stagnant layer above it is poured a solution of diamine in water. At the interface a layer of polyamide (nylon) is formed almost immediately. But the reaction stops because of the slow diffusion of reactants to each other through the polymeric interface. However, as the interface is pulled out, reactants contact each other and form more polymer (Fig. 9.17). Within limits, the rate of reaction is controlled by the rate of interface removal.

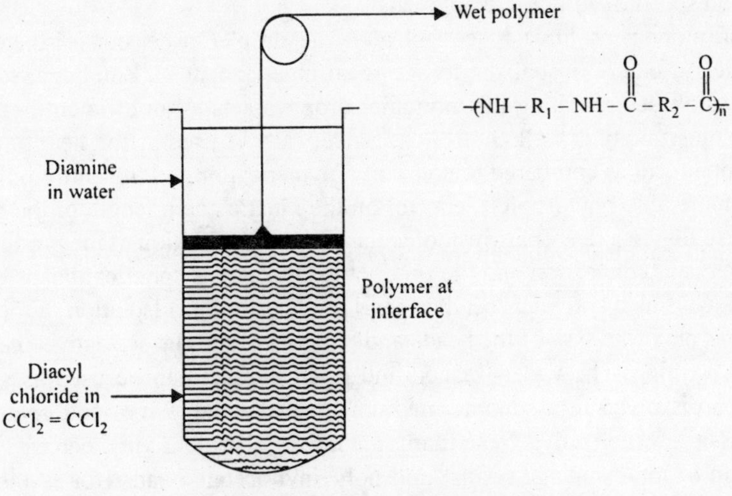

Fig. 9.17. Interfacial condensation.

With pure sebacoyl chloride and hexamethylene diamine, nylon 6–10 is formed from which fibres or films can be made.

$$\underset{\text{Hexamethylene diamine}}{H_2N-CH_2-(CH_2)_4-CH_2-NH_2} + \underset{\text{Sebacoyl chloride}}{Cl-\overset{O}{\overset{\|}{C}}-(CH_2)_8-\overset{O}{\overset{\|}{C}}-Cl} \rightarrow$$

$$\underset{\text{Nylon 6-10}}{\left[-NH-(CH_2)_6-NH-\overset{O}{\overset{\|}{C}}-(CH_2)_8-\overset{O}{\overset{\|}{C}}-\right]} + HCl$$

The demonstration based on the apparatus of Fig. 9.17 has been popularised under the name 'The Nylon Rope Trick'. It has been used at countless science fairs and at all levels of sophistication from elementary school to university. On the other hand, it is not very convenient even for laboratory preparations. A typical preparative method starts with the same ingredients as the nylon rope trick. The aqueous phase is placed in a high-speed stirrer (blender) and the organic phase added while the system is stirred at high speed. Suspending agents or surfactants may be added to stabilise the polymer-solvent dispersion which results. Plant-scale operations also use stirred tanks rather than quiescent interfaces.

For all these systems a conventional glass-lined or stainless-steel reactor is suitable for batch operations when reaction pressures and product flow characteristics are within certain limits. Recently, scraped-surface, stainless-steel reactors have been built that will handle pressures from a full vacuum to 6 atm and viscosities of 1 mPa's to 10kPa's for large batches.

SUSPENSION POLYMERISATION

Only water-insoluble monomers can be polymerised by this technique. The monomer is suspended in water, in the form of fine droplets, which are stabilised and prevented from coalescing by using suitable water-soluble protective colloids, surface active agents and by stirring. The size of the monomer droplets formed depends on the monomer-to-water ratio, the type and concentration of the stabilising agents and also on the type and speed of agitation employed.

The initiators are monomer soluble. Since each monomer droplet suspension is isolated and independent of the other droplets it can be visualised to act as an independent bulk polymerisation nucleus. The continuous aqueous phase separating the monomer droplets acts as an efficient heat transfer medium and, hence, the exothermicity is well controlled. Since water is used as the heat transfer medium, the process is also economical as compared to solution polymerisation. As the entire bulk of the monomer is divided into innumerable tiny droplets, control on the kinetic chain length of the polymer formed is also quite good, resulting in a fairly narrow molecular weight distribution of the product. Polymerisation proceeds to 100 per cent conversion and the product is obtained as spherical beads or pearls. (For this reason, this technique is also known as bead or pearl polymerisation.) Isolation of the product becomes easy as this involves only filtration of the beads and removal of the surface active agents and protective colloids by water washing. The water-washed and dried product can be used as such for moulding purposes or can be dissolved in a suitable medium for use as adhesives and coatings. Expandable polystyrene beads (from which polystyrene foams are made), styrene-divinyl benzene co-polymer beads (for the preparation of ion-exchange resins) and polyvinyl acetate beads (for further conversion into polyvinyl alcohol) are produced by the suspension technique using free-radical initiators.

This method was developed to overcome some of the disadvantages of the emulsion process, in particular the tendency to adsorb impurities due to the large surface area of the small polymer particles. Here the average particle size of the dispersed phase is between 1 and 10 microns initially (10^4 Å to 10^5Å). Unlike an emulsion, a physical suspension of polymerising drops or beads coalesce rapidly upon stopping the agitation. The agents added are an initiator and a suspension stabiliser. The usual prefered dispersion medium is water. Inert organic liquids may be used provided the monomer and the polymer are insoluble in it.

The mechanism of polymerisation in this case is analogous to mass and solution polymerisation. There being no emulsifier to establish close and intimate contact between the two phase, each bead or pearl of the monomer acts as a minute, discrete mass polymerisation system. A protective colloid is also added to increase the viscosity of the continuous phase so that it would hydrodynamically hinder the coalescence of monomer droplets. It has to be inert with regards to the polymerisation reaction. The insoluble protective colloid such as $MgCO_3$, collects at the droplet-water interface by surface tension and prevents coalescence of drops upon collision. A typical suspension process is the manufacture of polymethyl-methacrylate (PMMA).

Advantages and Disadvantages of suspension polymerisation

Advantages

1. Heat is easily removed and controlled.
2. The polymer is obtained in small granular form which is convenient and easily handled.

Disadvantages

1. Yeild per reactor volume is low.
2. It gives a polymer less pure than can be obtained by the mass polymerisation process.
3. The process cannot be easily run in a continuous manner.

A simple example of suspension polymerisation can be carried out in a glass flask with agitation.

Aqueous phase:

Water	400 ml
Poly (vinyl alcohol)	1 g

Oil Phase:

Methyl methacrylate	100g
Benzoyl peroxide	1g

Under a nitrogen atmosphere the aqueous phase is heated to 80°C. The oil phasae is added with stirring at about 150 to 300 rpm with a paddle agitator. After an hour or so a slight exotherm, 5 to 10°C, is noted. Particles of the slurry examined before this time are sticky and usually agglomerate on cessation of stirring. After the exotherm, when most conversion has taken place, the particles are hard and do not agglomerate. Initially the monomer phase is lighter than water and agitation near the surface is essential. After the exotherm, enough polymer has been formed to make the oil phase heavier than water, so that settling out occurs if stirring is inadequate. Final recovery of the beads, which are 100 to 1000 μm in diameter, is by filtration and washing.

In most cases, polymerisation is carried to high conversion and pearl-like beads of polymer result, especially if the polymer is amorphous and below its glass-transition temperature. Two important exceptions are vinyl acetate and vinyl chloride. Vinyl acetate can be polymerised to give a stable suspension of about 50 per cent solids content if the particle size is kept small (1 to 15 μm diameter). Plasticised with castor oil so that the polymer will be above T_g at room temperature, the suspension makes an excellent quick-setting adhesive for wood, paper, and cloth. Poly (vinyl chloride) is insoluble in its monomer. Suspension polymerisation can be carried to 70 to 90 per cent conversion under moderate pressure (6 atm) and temperature (80°C) by using an oil-soluble initiator such as lauroyl peroxide or azoisobutyronitrile. During this period, polymer precipitates in the droplets. Release of pressure and stripping of monomer by pulling a vacuum leaves the polymer portion of the droplets as porous particles that can be recovered by filtration and drying or directly by spray drying. Such porous particles are ideally suited for adsorbing plasticisers in 'dry-blending' operations.

A combination of solution suspension and network polymerisation has been used for a special-purpose material. *Moore* describes the co-polymerisation of styrene with divinyl benzene as a 20 per cent solution in toluene with peroxide initiator. All these ingredients are suspended in water. The swollen-gel droplets that result are well suited for gel permeation chromatography columns. Ion-exchange resins can be made from cross-linked suspension-polymerised droplets (without the solvent), since the physical form does not have to be changed and the subsequent reactions can be performed on the, beads. Sulphonation of styrene-divinyl benzene co-polymer beads leads to a strong-acid ion exchanger.

Some important distinctions between emulsion and suspension polymerisations are:

1. Emulsions usually are comprised of small particles, 0.05 to 5 μm, compared with suspensions, particles 10 to 1000 μm diameter.
2. Water-soluble initiators are used rather than monomer-soluble.
3. The end product usually is a stable *latex*-an emulsion of polymer in water rather than a filterable suspension.

Because of these conditions, the mechanism of polymerisation is basically different (The essentials of an emulsion polymerisation system are monomer, a surface-active agent (surfactant), initiator, and water. Initially, the surfactant is in the form of micelles, spherical or rodlike aggregates of 50 to 100 surfactant molecules with their hydrophobic 'tails' oriented inward and their hydrophilic 'heads' outward. These micelles form whenever the concentration of surfactant exceeds a rather low 'critical micelle concentration'.

The manner in which the critical micelle concentration for a given surfactant is measured is instructive for our purposes. If a dilute aqueous solution of an oil-soluble dye like eosin is titrated with a dilute soap solution, a soap concentration is reached where the colour suddenly disappears. The explanation is that the dye has dissolved in or has been extracted by the micellar interiors. This is the same point at which surface tension stops decreasing rapidly with added soap.

If monomer is added to a micellar dispersion, most of it remains as rather large droplets, but some of it dissolves in the micelles just as the dye does. Since they are smaller, the micelles present much greater surface area than do the droplets. Consequently, when free radicals are generated in the aqueous phase, the micelles capture most of them.

After a few per cent conversion, the system consists of (i) stabilised, monomer-swollen polymer particles rather than micelles; and (ii) monomer, which is still mainly in droplets, although it constantly diffuses to replenish the swollen particles where polymerisation continues. The result is, when monomer is quite water-insoluble,

$$R_p = k_p [M] [M\bullet] \qquad \text{... (9.10)}$$

Where [M] now is the concentration of monomer in the swollen polymer particles. Since this concentration, moles per litre of swollen particles, may remain constant from low conversions up to 70 or 80 per cent conversion, R_p should be independent of the total concentration of monomer, i.e., moles per litre of emulsion.

This means that the superficial reaction rate is pseudo-zero order in monomer. Now, it can be shown that a polymer particle cannot tolerate more than one growing chain at a time. Two radicals in the same particle mutually terminate rapidly. Therefore, on the average half the particles will contain radicals at any instant.

As an analogy, picture a group of cups standing in the rain. Each cup has received either an odd or an even number of raindrops. With enough cups and enough rain, the chances are that half the cups have received an even number of drops and the other half an odd number. If the raindrops are initiating radicals and the cups are polymer droplets, only the half that have received an odd number of droplets are growing, the even cases having been stopped by mutual termination.

So [M•] is simply $N_p/2$, where N_p is the number of particles per unit volume of emulsion and

$$R_p = \frac{k_p[M]N_p}{2} \qquad \text{... (9.11)}$$

Similar reasoning, disregarding chain transfer, and assuming termination by coupling, gives

$$\overline{x}_n = \frac{k_p N_p[\text{M}]}{d[\text{M}\bullet]/dt} \qquad \text{... (9.12)}$$

Where d[M•] /dt is the rate of radical formation (not the net rate) and may be proportionl to the square root of initiator concentration in the aqueous phase.

The number of particles per unit volume, N_p, can be related to the concentrations of surfactant and initiator. The derivation can be complex, but the Smith-Ewart approach introduces a number of simplifying assumptions. The entire process of establishing N_p takes place during the first part of polymerisation when the surfactant is still present in the form of micelles. Each effective radical generated from the initiator converts a micelle into a swollen polymer particle, which then grows at a constant rate ($\mu = dv/dt$). The capture of radicals by particles in competition with micelles is ignored for the moment. Every particle formed grows in area as well as in volume. Each increment of area requires a coating of surfactant, which is obtained from unreacted micelles. Thus micelles disappear for two reasons: Some are converted to polymer particles, and some supply surfactant to growing particles.

At some time t_1 all the surfactant in the system has coated the surface of particles, leaving none in the form of micelles. At that point, N_p no longer changes and the rate of polymerisation is given by Eq. (9.11). If effective radicals are being generated constantly at a rate ρ_r and each one starts a particle, then at t_1

$$\text{N}_p = \rho_r t_1 \qquad \text{... (9.13)}$$

Next, we can relate t_1 to the surface area of particles being formed.

1. A single particle is formed at time τ and then grows until time t_1. The volume of this particle (neglecting the initial volume of the micelle) at this time will be

$$\upsilon(t_1, \tau) = \mu(t_1 - \tau) \qquad \text{... (9.14)}$$

The surface area of the particle is related to the volume by geometry:

$$a(t_1, \tau) = (36\pi)^{1/3}\,[\mu(t_1 - \tau)^{2/3} \qquad \text{... (9.15)}$$

2. The number of particles generated in the time interval $d\tau$ is $\rho_r\, d\tau$. Thus the total surface area of particles at t_1, using Eq. (9.15), is:

$$\text{A}_t = \int_0^{t_1} a(t_1, \tau)\rho_r\, d\tau = (36\pi)^{1/3}\,(0.6\ \rho_r)_\mu^{2/3}\ t_1^{5/3} \qquad \text{... (9.16)}$$

3. The total area at t_1 is also given by

$$\text{A}_t = a_s[\text{S}] \qquad \text{... (9.17)}$$

where a_s is the surface area occupied by a mole of surfactant and [S] is the surfactant concentration in mol/litre.

4. Combining Eqs. (9.15), (9.16), (9.17), we can eliminate A_t and t_1:

$$N_p = 0.53\left(\frac{\rho r}{\mu}\right)^{0.4}(a_s[S])^{0.6} \qquad \text{... (9.18)}$$

The total number of particles per litre should vary with surfactant concentration to the 0.6 power and with initiator concentration (via ρ_r) to the 0.4 power. As noted earlier, at time t_1 (usually corresponding to a very small fraction of conversion) N_p no longer increases and the rate of polymerisation reverts to Eq. (9.11). This derivation can be modified to take into account radicals entering particles in competition with micelles. The only change is that the factor 0.53 becomes 0.37. More serious differences occur when particles are large, when initiator decomposition is very slow, or when the monomer in partly soluble in water or insoluble in polymer. However, the form of Eq. (9.18) often holds even though the proportionality factor must be determined experimentally.

An increase in surfactant concentration increases N and, therefore, increases R_p and $\bar{x}_n$ (Fig. 9.18). High molecular weights are possible in emulsion polymerisation because initiation is in one phase, water, and termination is in another phase, monomer-polymer. The combination of high molecular weight with high rate is one reason for the popularity of this method. *Seeded polymerisation* can be useful for making large-particles-size latexes. A completed 'seed' latex is diluted to give the desirable value of N_p particles per litre of emulsion. No additional surfactant is added, so no new particles are formed. Monomer is fed in and initiator is added. Polymerisation occurs in the previously formed particles, so that each one grows as monomer diffuses to it and is converted (Fig. 9.19). When the seed monomer and added monomer are different, graft co-polymers may be formed provided the dead polymer in the seed can react by means of residual unsaturation or chain transfer.

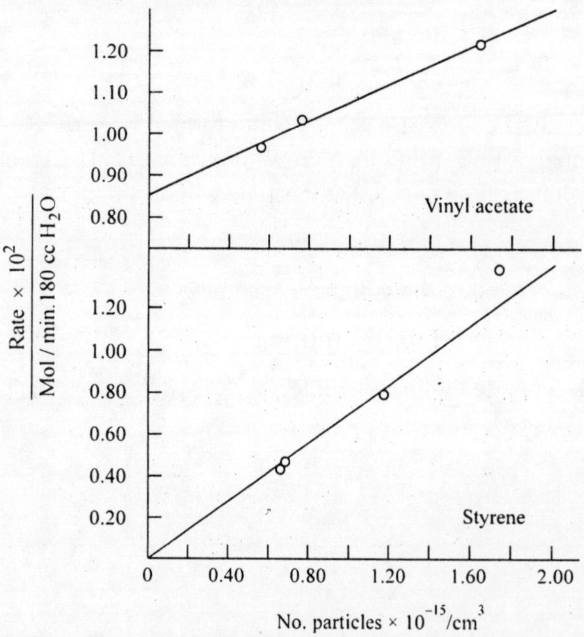

Fig. 9.18. Rate of polymerisation of styrene and vinyl acetate vs. number of particles using a seed of poly(vinyl acetate) latex.

A small amount of water solubility can alter the polymerisation characteristics. Vinyl acetate dissolves in water only to the extent of 2 per cent, but this is enough to give a combination of emulsion and solution polymerisations when emulsion polymerisation is attempted (Fig. 9.19). The rate also is proportional to the initiator concentration. Another exceptional case is vinylidene chloride, where monomer is almost insoluble in polymer but is adsorbed on the surface of polymer particles. Under these circumstances, faster stirring can accelerate the reaction (Fig. 9.20).

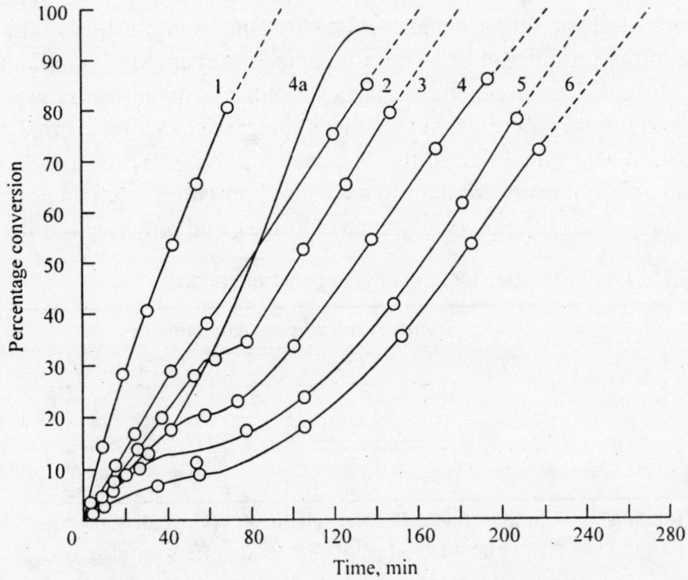

Fig. 9.19. Effect of surfactant concentration on emulsion polymerisation of vinylidene chloride at 36°C. Initiator is 0.15 g $(NH_4)_2\ S_2O_3$ and 0.15 g $Na_2S_2O_5$/100 g monomer. Sodium lauryl sulphate concentration of (i) 10.0 g; (ii) 3.0 g; (iii) 3.0 g, (4 and 4a) 2.0 g, (5) 1.0 g, (6) 0.50 g/100 g monomer. Stirring rate of 285 rev/min except for 4a, which is 756 rev/min.

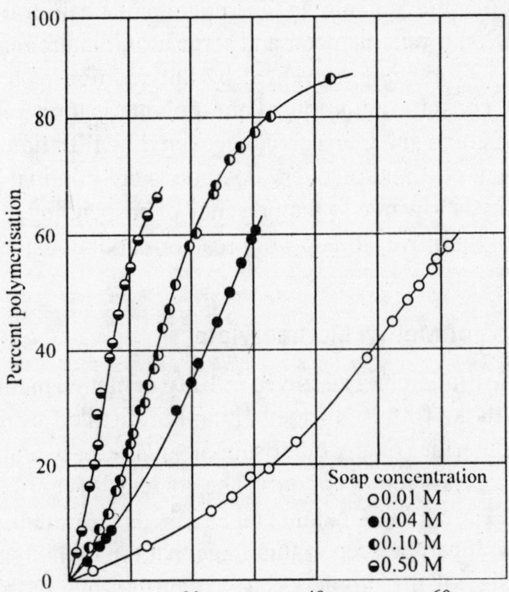

Fig. 9.20. Polymerisation of isoprene at 50°C for concentrations of emulsifier (potassium laurate).

The latex that results from emulsion polymerisation may be the desired form for the intended end use. Some paints and adhesives can use latexes directly. However, if massive polymer is desired, recovery may involve coagulation by heating, freezing, salt or acid addition, spray drying, or mechanical turbulence. Surfactants, coagulants, and initiator fragments often remain as impurities in the final product under these circumstances. In the production of styrene-butadiene rubber (SBR), sodium soaps of fatty acids are used as emulsifiers. Acidification of the final latex simultaneously destroys the surfactant and deposits free fatty acid in the coagulum, where it is a valuable adjunct to the sulphur cross-linking reaction used later on for vulcanisation. A Typical recipe for suspension polymerisation is given in Table 9.11. The method is used commercially to prepare hard, glassy vinyl polymers such as polystyrene, poly(methyl methacrylate), poly(vinyl chloride), poly(vinylidene chloride), and polyacrylonitile.

Table 9.11. Typical recipes for suspension polymerisation.

	Parts per 100 parts of monomer	
Component	*Methyl methacrylate*	*Vinyl chloride*
Peroxide initiator	~0.5	0.1–0.5
Water	~350	150–350
Stabilisers[a]	0.01–1	0.01–1

[a] Typical stabilisers include methyl cellulose, gelatin, poly(vinyl alcohol), and sodium polyacrylate. Minor amounts of emulsifiers and buffers are also used.

Improved heat transfer can be obtained by using the excellent thermal properties of water through either suspension or emulsion polymerisation. In suspension ('bead' or 'pearl') polymerisation, a batch reactor fitted with a mechanical agitator is charged with a water-insoluble monomer and initiator. Sometimes, a chain-transfer agent may be added to control molecular weight in a free-radical polymerisation. Droplets of monomer containing the initiator and chain-transfer agent are formed. These are typically between 50 and 200 μm in diameter and serve as miniature reactors for the polymerisation. Coalescence of these 'sticky' droplets is prevented by the addition of a protective colloid, typically poly(vinyl alcohol), and by constant agitation of the polymerisation mixture. Near the end of the polymerisation, the particles harden and then can be recovered by filtration, which is followed by a final washing step. Although solvent cost and recovery operations are minimal in comparison with solution polymerisation, polymer purity is low due to the presence of suspending and other stabilising additives that are difficult to completely remove. In addition, reactor capital costs are typically higher than for solution polymerisation.

Suspension Polymerisation of Methyl Methacrylate

Here, 0.1 g of benzoyl peroxide (initiator) is dissolved in 100 g of methyl methacrylate monomer contained in a 500 ml round bottomed flask. To this is added 180 ml of distilled water containing 0.02 g of cetyl dimethyl benzyl ammonium chloride (or any other suitable quarternary ammonium salt type emulsifier) and 0.05 g of polyvinyl alcohol (protective colloid). The contents of the flask are flushed with nitrogen and placed in a constant temperature bath maintained at 75 ± 1°C. A medium-speed stirrer is fitted and the contents stirred uniformly. It can be seen at this stage that the monomer gets dispersed in the water phase as tiny droplets. The size of the droplets can be adjusted by controlling the stirrer speed. The experiment is continued for 8 hours at the end of which fine spherical beads of polymethyl methacrylate settle at the bottom of the flask. The beads are filtered and washed several times with warm water to free them from the emulsifier. Salient features of different polymerisation techniques are given in Table 9.12.

Table 9.12. Salient features of different polymerisation techniques.

Bulk polymerisation (1)	*Solution Polymerisation (2)*	*Suspension polymerisation (3)*	*Emulsion polymerisation (4)*	*Melt polycondensation (5)*	*Solution polycondensation (6)*
Homogeneous system In certain cases where the polymer is insoluble in the monomer, the polymer precipitates out as polymerisation sets in, and the system becomes, heterogeneous.	Homogeneous system In certain cases, where the polymer is insoluble in the monomer/solvent, the polymer precipitates out as the polymerisation sets in and the system becomes heterogeneous.	Heterogeneous system Proper control on agitation and stabiliser needed, to keep the suspension stable.	Heterogeneous system. Agitation and proper control on the emulsifier/stabiliser needed to keep the emulsion stable.	Homogeneous system	Homogeneous system If the polymer is insoluble in the solvent, the system becomes heterogeneous.
Viscosity of the medium increases very rapidly and at high conversions, proper mass transfer becomes difficult.	Viscosity of the medium increases slowly, but not difficult to maintain proper mass transfer even at high conversions.	Viscosity of the medium does not increase appreciably even at high conversions.	Viscosity of the medium does not increase through out the polymerisation.	Viscosity of the medium increases very rapidly and at high conversions, proper mass transfer becomes difficult. Removal of by product also becomes difficult at high conversions.	Viscosity of the medium increases slowly. Removal of byproduct is comparatively easier if a suitable solvent is chosen as entrainer.
Exotherm dissipation difficult. May lead to autoacceleration.	Exotherm dissipation easy.	Exotherm dissipation easy.	Exotherm dissipation the best.	No exotherm. But proper heat transfer within the medium becomes difficult at high conversions.	No exotherm. Heat transfer is uniform.
Requires longer duration for very high conversions.	Requires longer duration for very high conversions.	Requires longer durations for very high conversions.	High conversions (almost 100%) achieved in shorter duration.	Requires longer duration for very high conversions.	Requires longer duration for very high conversions.
Suitable for producing medium to high molecular weight products.	Suitable for producing low molecular weight products.	Suitable for producing medium to high molecular weight products.	Suitable for producing very high molecular weight products.	Suitable for producing medium to high molecular weight products.	Suitable for producing medium to high molecular weight products.

(Contd ...)

Bulk polymerisation (1)	*Solution Polymerisation (2)*	*Suspension polymerisation (3)*	*Emulsion polymerisation (4)*	*Melt polycondensation (5)*	*Solution polycondensation (6)*
Most widely used in industrial processes involving free-radical, polycondensation and polyaddition polymerisation.	Suitable for free-radical, cationic, and coordination polymerisations.	Suitable for free-radical and polycondensation polymerisations.	Suitable for free-radical and polycondensation polymerisations.	Suitable for polycondensation polymerisation.	Suitable for polycondensation polymerisation.
Product is minimum contaminates; easy to isolate and can be directly used for moulding purposes.	Isolation becomes difficult. Product contaminated with solvent and solvent-related impurities.	Isolation is easy as the product is obtained as powder/beads and can be directly used for moulding purposes. Product contaminated with stabiliser.	Emulsion can be directly used for paints, adhesives, and so on. Solid product can be isolated by breaking the emulsion. Product contaminated with emulsifier, stabiliser and so on.	Product minimum contaminated. Freeing from monomers and by-product difficult since the polymer solidifies at ambient temperature.	Product usually contaminated with solvent.

CHAPTER 10

Chemical Modification of Polymers

INTRODUCTION

The terms rubber and elastomer embrace those polymers which have useful rubberlike, highly elastic properties at ambient temperatures; however, many polymers which are nonrubbers by themselves can be chemically modified to a relatively small extent to give products with very useful viscoelastic properties. For example, the introduction of a few chlorine atoms and sulphoxide groups into polyethylene changes the macromolecules so that they no longer tend to form crystalline regions, but become elastomers over a wide temperature range. These groups also allow vulcanisation, so that reversibly deforming, solvent-resistant products can be formulated. On the other hand, starting with a conventional rubber, it is possible, by means of chemical reactions, to convert the macromolecular chains so that they lose their viscoelasticity and become thermoplastic at ambient temperatures (e.g., the cyclisation of polyisoprenes). Quite apart from the question of the temperature at which observations are carried out, the borderline between the viscoelastic and the plastic states is a relatively ill-defined one. The common factor of rubbers and plastics is, of course, their macromolecular nature. Now that we have a better understanding of their structure at a molecular level (not forgetting rubber-modified plastics and "plasticised" rubbers), we are able to say that, with some exceptions, a modification that can be carried out on one polymer species may, under suitable conditions, be carried out on a polymer of a different species. Whether the same type of chemical modification will give us the properties we are seeking without the loss of properties we would wish to retain is a matter, first, for conjecture and, subsequently, for experimental verification.

In the early days of polymer chemistry and technology, a new chemical modification successfully applied to one polymer was quickly evaluated with a whole range of chemically similar reagents (e.g., the esterification of cellulose), and the same type of reaction was attempted with the available range of similar material. Today, the number of homopolymers available runs into the hundreds, and the number of random co-polymers runs into the thousands; therefore, the number of possible chemical modifications, including block and graft co-polymers, runs into astronomical figures. Only a fraction of these potential systems has been evaluated (and then frequently only in a superficial way to obtain patent coverage), so that the field is wide open for research and development. Because of its breadth and depth, the field of the chemical modification of polymers can be treated only in outline in a single chapter and only the more important reactions can be described. To serve the interest of the reader, however, this broad survey is punctuated by discussions in greater depth of areas that are of interest to rubber chemists and that, in the opinion of the writer, have considerable potential for further development. The emphasis is on the principles underlying the chemical modification of polymers; specific details of reactions conditions are deliberately omitted.

ESTERIFICATION, ETHERIFICATION, AND HYDROLYSIS OF POLYMERS

The chemical modifications discussed in this section are historically and scientifically so closely linked to one polymer, cellulose, that although the latter occurs primarily as a fibre and not an elastomer, a discussion of this group of cellulose modifications seems appropriate. Apart from the fact that some cellulose derivatives, like ethyl cellulose, when plasticised, can be quite elastomeric, the effects of modification of a basic polymer are particularly well demonstrable on a substance as stiff and highly crystalline as cellulose. Moreover, in view of the expected hydrocarbon shortage, cellulose may soon gain a new role as a polymeric starting commodity. Cellulose, identified chemically as β-1, 4-glucan, is the most widely distributed natural polymer, constituting the permanent structure of plant cell walls.

Much of the early history of the chemical modification of cellulose is related to the attempts to find a solvent for it, as its macromolecular structure was not understood at that time. *Mercer* discovered and commercialised the interaction of alkali with cellulose fibres, a process still in use under the name *mercerisation*. The initial product of the reaction, alkali cellulose, is not a chemical modification but a physical form in which water and sodium ions penetrate the macromolecular structure and reduce the hydrogen bonding, with consequent swelling of the fibres. The initial product, cellulose I, is converted to cellulose II, a complex physicochemical modification, in the final washing stage. The degree and rate of swelling in this process are dependent on the source of the cellulose, and if the fibres are stretched prior to and during the reaction, optimum interaction is achieved. Many other inorganic salt solutions swell cellulose and of these, zinc chloride has found the widest application. Aqueous solutions of thiourea, resorcinol, chloral hydrate, and benzene sulphonates also lead to limited swelling of cellulose. In all cases the reduction in physical cross-linking can be followed by a study of the X-ray diffraction diagrams of the crystalline content.

The complete solubility of cellulose in cuprammonium solutions, led to the development of the rayon industry but, as in the case of alkali cellulose, the regenerated polymer is chemically the same as the precursor. Regeneration via cellulose zanthate solutions, invented by *Cross* and *Bevan*, is another process still in use; it forms the basis for the manufacture of Cellophane. The first "chemical" modification of cellulose was achieved by *Braconnot* with the production of cellulose nitrate from a wide range of cellulosic materials. The products were highly inflammable powders which could be dissolved in concentrated acetic acid to give clear tough varnishes. (Note the conversion of a fibre to a film by chemical modification). Later highly nitrated cellulose, guncotton, was discovered schutzenberger produced acetylated cellulose using hot acetic anhydride as the reaction medium. These reactions have a common mechanism, namely, the esterification of the hydroxyl groups in the basic cellulose moiety:

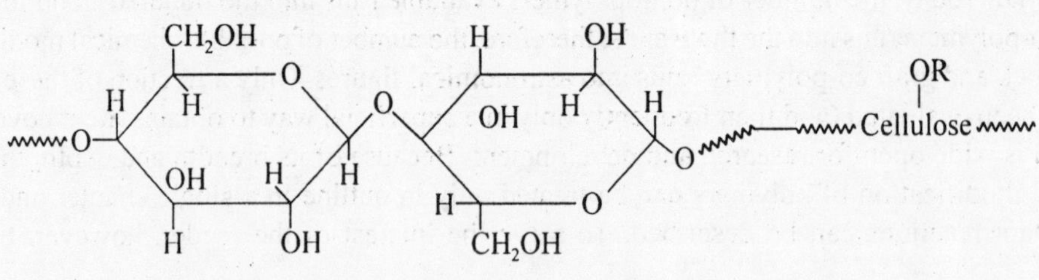

... (10.1)

Since this early work, a very large range of organic acids have been used to prepare cellulose esters, mixed esters, and ethers esters. A typical example of considerable commercial importance is the acetylation of cellulose. As in all esterifications of macromolecular materials, the accessibility of the hydroxyl groups to the esterifying acid is of prime importance. Reaction (10.1) represents complete esterification, a process which is probably never fully achieved. The identification of the esterified product is, therefore, dependent not only on the content of acetyl groups but also on the location of these groups on the macromolecular backbone. Both factors are affected by the method of preparation and the esterification conditions.

Although many esterification reactions are based on inorganic acids, for insoluble hydroxyl compounds like cellulose, xanthation is more important. Sodium hydroxide is normally used to produce the swollen alkali cellulose, which (after ageing) is reacted with carbon disulphide to form the sodium salt of cellulose xanthate:

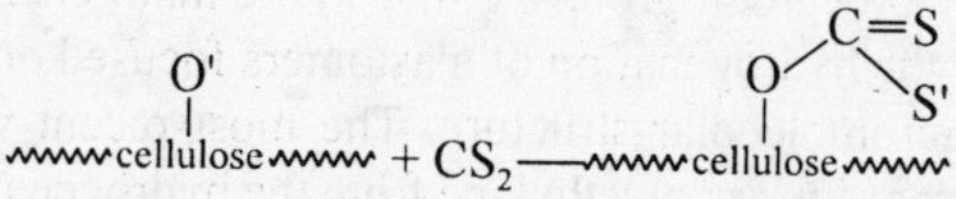
... (10.2)

The cellulose can be regenerated by spinning (or extruding as a film) into an acid bath containing salts such as sodium and zinc sulphate. During spinning or extrusion, the macromolecules are oriented in the direction of flow to give high strength to the viscose fibre or the Cellophane film. The occurrence of macromolecular orientation during spinning is very important and it is used in the chemical modification of many polymers.

The cellulose ethers constitute another important group of cellulose derivatives prepared from alkali cellulose by standard etherification reactions between the hydroxyl groups and an alkyl halide. The properties of the ethers depend on the extent of the reaction, i.e., the degree of etherification. In general the ethyl celluloses are water-insoluble thermoplastic materials, whereas methyl ether, ethyl hydroxyethyl cellulose, and carboxymethyl cellulose are soluble in cold water and are used as viscoelastic thickeners and adhesives.

For the preparation of synthetic hydroxy polymers, hydroxyl groups can be introduced by co-polymerisation of the base monomer with a hydroxy monomer. These groups can then be used for esterification or etherification, but the relatively high cost of hydroxy monomers detracts from the wide spread use of direct co-polymerisation. Instead, one introduces the groups required by the complete or partial hydrolysis of the ester groups in an appropriately hydrolysable polymer, such as poly(vinyl acetate). Complete hydrolysis yields poly(vinyl alcohol) (PVA), a water-soluble polymer with considerable utility as a stabiliser and viscosity modifier for aqueous systems:

$$\sim\!\sim CH_2-\underset{\displaystyle OH}{\underset{|}{CH}}-CH_2-\underset{\displaystyle OH}{\underset{|}{CH}}\sim\!\sim \qquad \text{... (10.3)}$$

PVA has a unique use as a strengthening fibre in conjunction with weaker materials such as merino wools in the weaving of delicate fabrics, from which it can afterward be removed by water washing. A major portion of the polymer produced is reacted with aldehydes to form the corresponding poly(vinyl formal), poly(vinyl acetal), and poly(vinyl butyral):

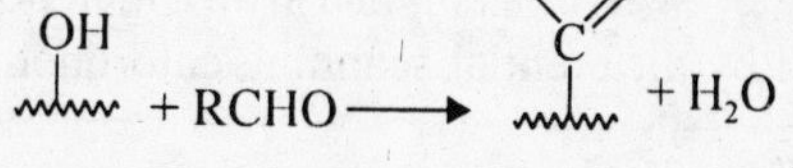
... (10.4)

There are different grades of each of these materials according to the overall molecular weight and the degree of substitution. These polymers are used as components of systems with unique adhesive properties, e.g., in the manufacture of safety glass laminates [poly(vinyl butyral) and mixed derivatives] and of metal-to-metal adhesive [poly(vinyl formal) cured with phenolics and other resins]. Reactions of poly(vinyl alcohol) with acids or anhydrides occur as normal esterifications, a route used to synthesise polymers and co-polymers which cannot be readily formed by conventional polymerisation (e.g., when the reactivity ratios of the monomers are not suitable).

Natural rubber and synthetic rubbers in general do not have hydroxyl groups in sufficient numbers for them to be used for esterification reactions. Terminal hydroxyl groups may be introduced into synthetic rubbers as terminal catalyst or initiator fragments and used for coupling or extension reactions.

HYDROGENATION OF POLYMERS

Any polymer with unsaturated hydrocarbon groups, either in the main chain or as side groups, can be hydrogenated. Early research on the hydrogenation of elastomers focused on destructive hydrogenation with consequent loss of the macromolecular structure. The most recent work in hydrogenation has produced excellent products, such as linear polyethylene from the hydrogenation of poly(1, 4-butadiene) and poly(ethylene-co-propylene) rubber from the hydrogenation of polyisoprene.

$$\text{+}CH_2{-}C{=}C{-}CH_2\text{+} \xrightarrow{H_2} \text{+}CH_2{-}CH_2CH_2{-}CH_2\text{+}_n \qquad \text{...(10.5)}$$

$$\left(\text{+}CH_2{-}\underset{\displaystyle CH_3}{\underset{|}{C}}{=}C{-}CH_2\text{+}\right) \xrightarrow{H_2} \left(\text{+}CH_2{-}\underset{\displaystyle CH_3}{\underset{|}{C}}{-}CH_2{-}CH_2\text{+}\right)_n \qquad \text{... (10.6)}$$

These reactions were carried out using a heterogeneous catalyst.

Homogeneous soluble transition metal catalysts for hydrogenation have been used to create novel polymers. Homogeneous hydrogenation catalysts are usually generated from Ziegler-type catalysts based on nickel or cobalt organic salts reduced in the presence of organoaluminium or organolithium compounds. These catalysts are used to form saturated elastomers by hydrogenating unsatured elastomers. The resulting polymers have vastly different viscoelastic properties than their unsaturated parent polymers. For example, the hydrogenation of a 99% poly(1,2-butadiene) has resulted in the formation of polybutene which has a lower glass transition temperature than its parent elastomer. It is interesting to note that hydrogenation does not affect the polymer molecular weight or backbone architecture.

The ease of hydrogenation and the resultant degree of saturation achieved reflect the microstructure of the polymer. Hydrogenation of unsatured elastomers usually proceeds in a blocky way. This is due to the different reactivities of the various double bonds. In general, double bonds of 1,2 structure are four times more reactive than double bonds of 1,4 structure. *cis*-1,4 units are more reactive than *trans*-1,4 units. Chamberline have shown that hydrogenation of 1,2 units is statistically random, whereas the hydrogenation of 1,4 units is not.

The complete hydrogenation of poly(1,4-butadiene), *cis* or *trans* structure, forms a polyethylene with a low melting point of about 115°C. It is believed that this linear polyethylene is a low-density material. It has been found that on the partial hydrogenation of a 99% *cis* poly(1,4-butadiene) showed that the hydrogenation proceeded in a blocky fashion. The 40–50% hydrogenation of poly(1,4-butadiene) (*cis* content 98%) made by nickel catalyst gives a polymer with a melt point of +98°C and a crystallisation temperature of –13°C as measured by differential scanning calorimetry (DSC). This confirms the fact that hydrogenation of *cis*-poly-1,4-butadiene proceeds in a blocky fashion to produce a block of polyethylene and a block of poly(*cis*-1,4-butadiene).

Polybutadienes made by anionic catalysts in the presence of polar modifiers contain a mixed microstructure of *cis*-1,4, *trans*-1,4, and 1,2 units. Hydrogenation of these polymers leads to interesting products. As mentioned previously, hydrogenation favours the 1,2 units over the 1,4 units by a 3 (or 44)-to-1 ratio. Because of this mismatch in reactivity, hydrogenation of a polybutadiene containing 40–50% 1,2 units produces a polymer containing a polyethylene portion with a T_m of 85–95°C and a rubbery portion with a T_g of –62°C.

Block co-polymers can also be hydrogenated to produce unique products. Hydrogenated triblock co-polymers of poly(styrene-co-butadiene-co-styrene) (SBS) are commercially available from various companies. The middle block is usually a mixed microstructure of poly(1,2-butadiene) and poly(1,4-butadiene) units. The resulting products is a hydrogenated unsaturated polymer which exhibits greater thermal and oxidative properties than the parent SBS triblock.

Similar procedures have been used to hydrogenate poly(1,4-butadiene-co-1,2-butadiene) diblocks and poly(1,4-butadiene-co-1,4-isoprene-co-1,4-butadiene) triblocks. Hydrogenation of these diblock and triblock co-polymers forms thermoplastic elastomers with crystalline and amorphous segments. All these materials exhibit crystallinity, glass transition, solubility, and dynamic mechanical loss spectra different from those of their unsaturated counterparts.

Another method of preparing saturated rubbers was developed using the diimide reduction. This method can be used to produce a high degree of saturation dependent on the type of reagent used; however, side reactions can occur in this method. Generation of the diimide from *p*-toluenesulphonyl hydrides leaves an acidic fragment which may cause cyclisation in some unsaturated elastomers.

DEHALOGENATION, ELIMINATION, AND HALOGENATION REACTIONS IN POLYMERS

Dehydrochlorination of Poly(vinyl Chloride)

The dehydrochlorination of poly(vinyl chloride) has been the subject of much investigation, particularly with the view of developing greater stability in PVC polymers and co-polymers. Like many polymeric reactions, dehydrochlorination is a complex process. The vinylene groups, created by the elimination of HCl from adjacent carbon atoms in the chain, may be the result of free radical ionic, or ion-radical steps.

$$\sim\sim CH_2—CHCl—CH_2—CHCl \sim\sim \longrightarrow HCl + \sim\sim CH=CH—CH_2CHCl \sim\sim \quad \text{... (10.7)}$$

The presence of a small proportion of head-to-head, tail-to-tail, and other configurational irregularities in the backbone structure of poly(vinyl chloride) leads to more complex elimination steps by thermal degradation (alone or in the presence of catalysts such as aluminium chloride). The introduction of ring structures, a major process during dehydrochlorination, is likewise affected by the distribution of the chlorine atoms along the polymeric backbone. Hydrogen bromide can be effectively eliminated thermally from poly(vinyl bromide), as dehydrohalogenation is a universal thermal reaction the complexities of which increase from chloride to iodide.

Thermal Elimination

The thermal elimination process can be applied to most '"substituted' groups in vinyl polymers by controlled pyrolysis at 600–700°C, producing polyvinylene compounds, e.g., by the splitting off of acetic acid from poly(vinyl acetate). By careful temperature control one can achieve bifunctional reactions and/or intramolecular cyclisations. This has been developed commercially at relatively high temperatures, in the case of the polymerisation of methacrylamide above 65°C, to yield a polymer with a substantial proportion of imide groups:

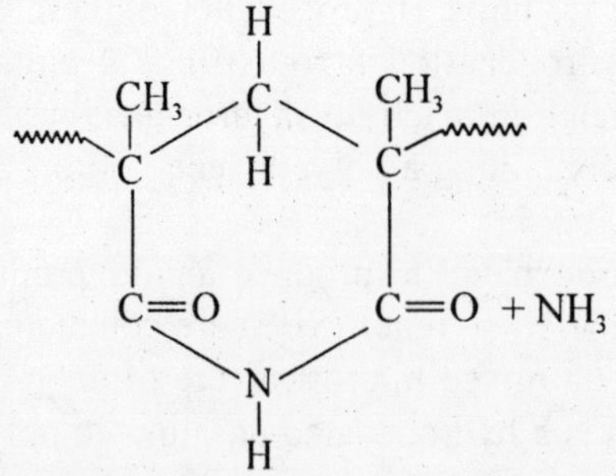

... (10.8)

Polymers and co-polymers of multifunctional vinyl monomers (using the term to cover the presence of a halogen or other reactive group in addition to the vinyl group, rather than in the sense of more than one polymerisable group), such as α-chloroacrylic acid, often undergo partial lactonisation and hydrolysis during polymerisation. Heating in alcohol solution or electrolysing alcoholic solutions, one obtains, e.g., the introduction of double lactam rings during the acid hydrolysis of poly(α-acetamineoacrylic acid)

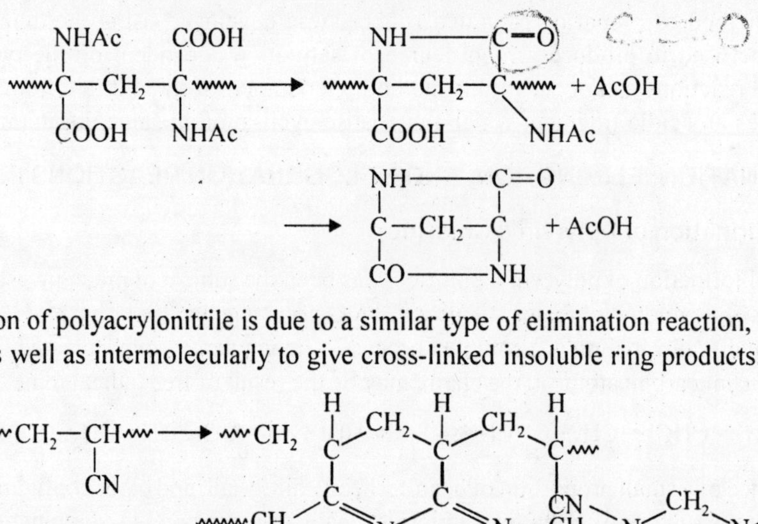

... (10.9)

The discoloaration of polyacrylonitrile is due to a similar type of elimination reaction, which in this case occurs intra- as well as intermolecularly to give cross-linked insoluble ring products:

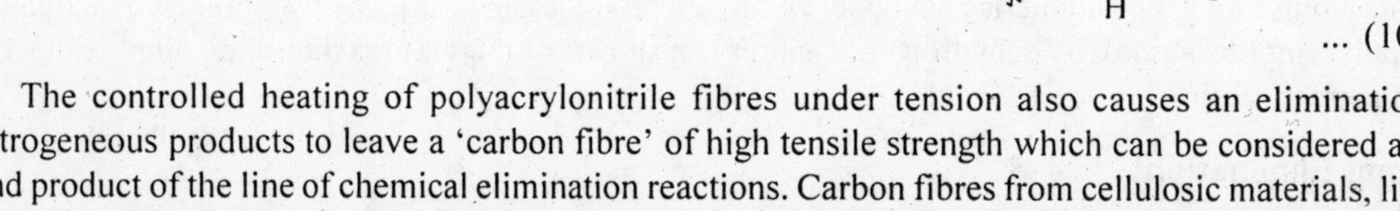

... (10.10)

The controlled heating of polyacrylonitrile fibres under tension also causes an elimination of nitrogeneous products to leave a 'carbon fibre' of high tensile strength which can be considered as the end product of the line of chemical elimination reactions. Carbon fibres from cellulosic materials, lignin, and various interpolymers and blends have been developed. The structures of these products consist largely of three-dimensional carbon networks, partially crystalline and partially graphitic or amorphous.

Halogenation of Polymers

In case of halogenation and hydrohalogenation the main problems with these reactions are the cyclisation and chain scission that occur parallel to the halogenation reaction. These introduce difficult problems in the characterisation of the resulting products. Despite these problems, several products have been prepared and commercialised. Chlorination of poly(1,4-butadiene) to prepare a product similar to poly(vinyl chloride) has been reported and this process had extensive side reactions and chain degradation. The chlorination of butyl rubber and conjugated diene-butyl rubbers gives end products that are used in the tyre industry as inner liners for air retention.

Ethylene-propylene co-polymers (EPDM) are, by their random co-polymerisation, amorphous in structure and therefore easily halogenated. EPDM has been chlorinated to improve its properties and cocurability with other rubbers. The chlorination was directed toward the termonomer dicyclopentadiene to form the allylic chloride. In this manner, EPDM was chlorinated and the resulting products had improved properties.

EPDM rubbers are modified by 1,2-addition of *N*-chlorothiosulphonamides to their olefinic sties. Such additions may be carried out in solution or without solvent in an internal mixer or extruder. The solventless reactions are facilitated by added carboxylic acids or by certain metal salts of weak acids. The modified EPDMs are of interest because of their ability to covulcanise in ozone-resistant blends with polydiene rubbers. Although less fully explored, *N*-chlorothiocarboxamides and imides also react with EPDM to produce modified products which covulcanise in blends with polydienes.

In the absence of oxygen, the chlorination of polyethylene, with or without a catalyst, can be controlled to provide products with varying chlorine content. The chlorination process is statistically random so that chlorination of polyethylene to the same chlorine content as poly(vinyl chloride) (60%) gives a product which is chemically different from PVC yet fully compatible with it. This random chlorination of polyethylene destroys its crystallinity. At a degree of chlorination corresponding to the loss of all its crystallinity, the chlorinated product becomes soluble at room temperature. The *p*-bromination of polyethylene follows a similar course to yield a rubberlike polymer at 55% bromine content.

Both chlorination and bromination of polypropylene and isobutylene lead to degradation of the main chain, with the loss of many useful properties. Degradation during chlorination can, however, be avoided at low temperatures by limiting the reaction to a maximum of about 2%. This procedure forms a useful commercial product.

The addition of hydrogen chloride to unsaturated elastomers has also received considerable attention. Extensive work has been done on the hydrochlorination of *Hevea* [poly(*cis*-1,4-isoprene)] and *Balata* [(poly(*trans*-1,4-isoprene)] rubbers. Both *cis*-1,4 and *trans*-1,4-polyisoprenes readily add hydrogen chloride following *Markovnikov's* rules with only a small amount of cyclisation.

Cyclisation of Polymers

Cationic cyclisation of unsaturated elastomers such as poly(*cis*-1,4-isoprene), poly(3,4-isoprene), poly(1,2-butadiene), and poly(1,4-butadiene) usually leads to the formation of cyclised resinous products of no commercial value.

Cyclisation of unsaturated elastomers, such as polyisoprene can be carried out in the solid state, in solution, or even in the latex. The process involves the transformation of linear macromolecular chains into much shorter ones consisting of mono-, di-, tri-, and tetrapolycyclic groups distributed randomly along the chain. Cyclisation of polyisoprene increases its glass transition temperature by 20–30°C. The mechanism of cyclisation of elastomers depends on the catalyst employed in the process.

OTHER ADDITION REACTIONS TO DOUBLE BONDS

Ethylene Derivatives

Besides the addition of halogens and hydrohalogens 'across the double bond' there are many other reagents which will react similarly with unsaturated polymers by free radical, ionic, or radical-ion mechanisms. Of prime importance is the addition of ethylene derivatives to polydienes. One of the earliest reactions of natural rubber to be studied in detail was the combination with maleic anhydride. Depending on the reaction conditions and the presence or absence of free radical initiators, one or more of four basic reactions may take place, with the products shown (the arrows indicate where the addition has taken place and the new bonds formed).

1. Intramolecular addition to the double bond within polyisoprene chains:

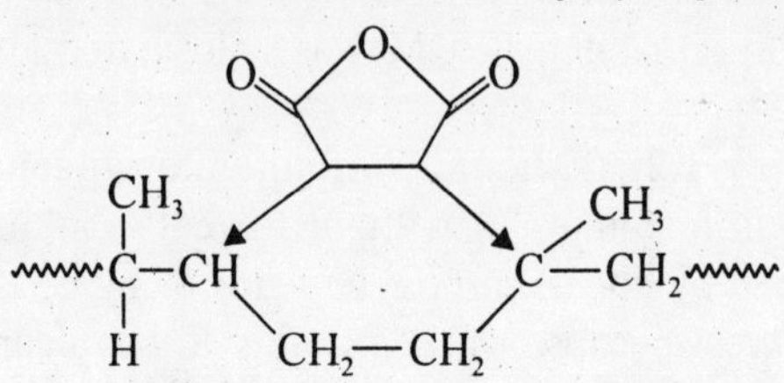

... (10.11)

2. Intermolecular addition to double bonds in different polymer chains in this group should be included the statistically possible reaction between widely separated double bonds within the same molecule:

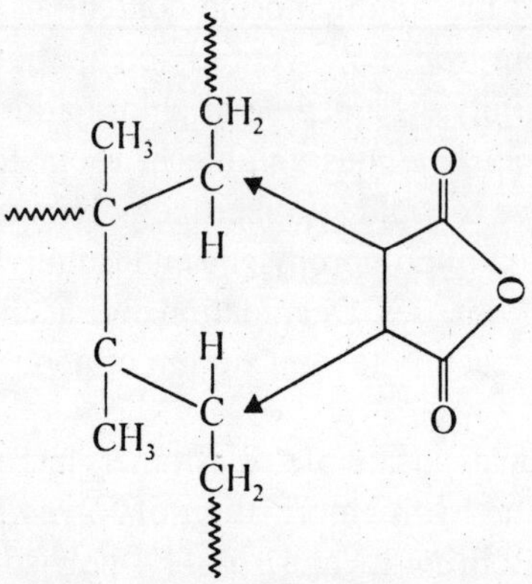

... (10.12)

3. Addition to α-methylenic carbon atoms of a polyisoprene chain:

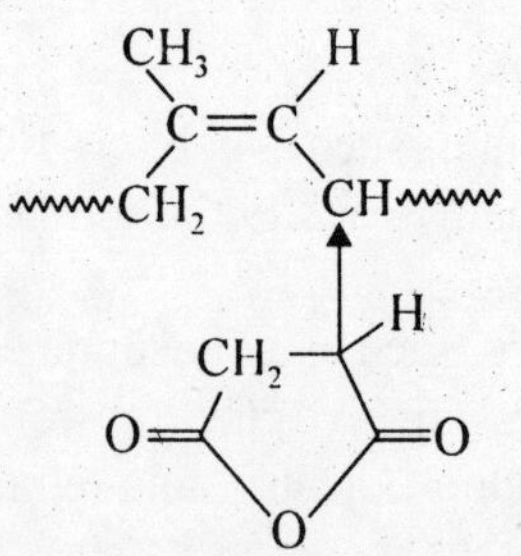

... (10.13)

4. Intermolecular addition to α-methylenic carbon atoms in adjacent chains (or widely spaced α-methylenic carbon atoms in the same macromolecule):

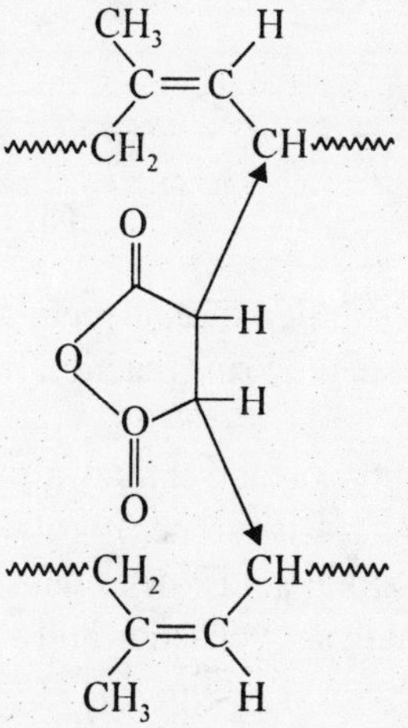

... (10.14)

In general, the overall reaction rates increase with rising temperature and in the presence of oxygen or free radical initiators, but these same conditions promote intermolecular reactions leading to gel formation. Similar reactions take place with *gutta-percha*, synthetic poly(*cis*-1,4isoprene), and poly(*cis*-1,4-butadiene). Many scientists used two-roll mills and other mastication techniques as convenient ways of blending the maleic anhydride with rubbers at elevated temperatures, but where these techniques have been used mechanochemical reactions have complicated the overall process. Reaction products of natural rubber containing 5–10% combined maleic anhydride can be vulcanised by conventional sulphur cures: of greater interest is the possibility of creating cross-linking by the use of oxides of calcium, magnesium, and zinc. Other compounds reacting similarly via activated double bonds (excluding here block or graft co-polymerisation) include maleic acid, *N*-methyl-maleimide, chloromaleic anhydride, fumaric acid, γ-crotonolactone, *p*-benzoquinone, and acrylonitrile. Other polymers with unsaturated backbones, such as polybutadiene, co-polymers of butadiene with styrene and with acrylonitrile, and butyl rubber, react in similar ways, but the recorded reaction with poly(vinyl chloride) is largely mechanochemical in nature.

Prins Reaction

Another addition to polymers with main-chain unsaturation is the Prins reaction between ethylenic hydrocarbons and compounds containing aldehydic carbonyl groups. *Kirchof*, described the reaction of natural rubber in benzene solution with aqueous formaldehyde in the presence of concentrated sulphuric acid. The general reaction of an aldehyde, RCHO, with a polyisoprene in the presence of an inorganic or organic acid or an anhydrous metal salt is represented by

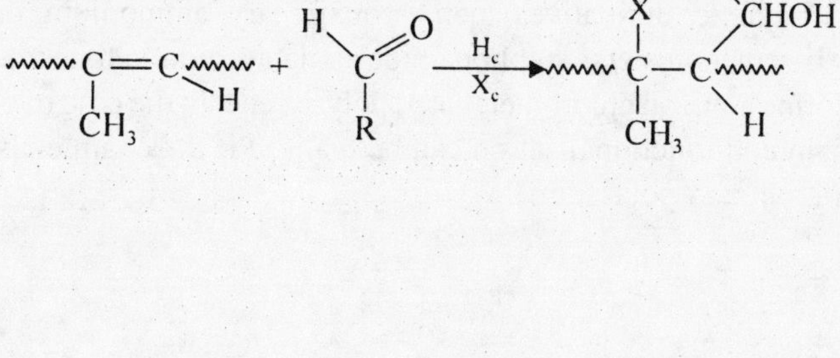

... (10.15)

In the absence of such catalysts, the reaction leads to a shift in the double bond rather than its elimination:

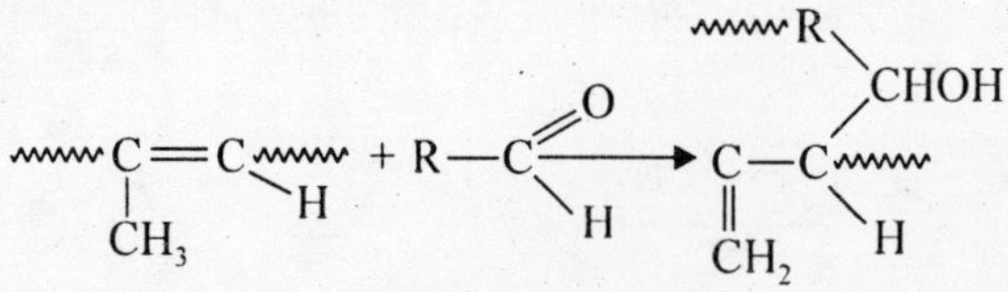

... (10.16)

These reactions can be carried out in solution or in dispersion or by reaction in the solid phase; in the last case it is again difficult to differentiate the Prins reaction from mechanochemical reactions initiated by chain rupture during mastication.

Other aldehydic compounds, such as glyoxal and chloral, also react in a similar way with polyisoprenes and unsaturated rubbers [e.g., poly(*cis*-1,4-isoprenes), poly(*cis*-1,4-butadiene) and co-polymers of isobutylene and isoprene]. The use of strong acids, or Lewis acids, causes complications, as the acids themselves, under suitable conditions, catalyse cyclisation and *cis-trans* isomerisation, and these reactions may occur simultaneously with the addition reactions.

OXIDATION REACTIONS OF POLYMERS

Uncontrolled oxidation of rubber is detrimental to its physical properties. Oxidation reactions takes place readily at unsaturated groups in polymers and are often referred to collectively as *epoxidation*; however, oxidation under controlled conditions can lead to useful products such as the epoxidised natural rubber introduced by the Malaysian Rubber Producers Association. Natural rubber in the latex form is treated with hydrogen peroxide dissolved in acetic acid. This gives 50% epoxidised natural rubber. This rubber shows very interesting physical properties and excellent carbon black dispersion. Similarly, nonaqueous epoxidisations of synthetic polyisoprene can be achieved using either hydrogen peroxide or hypochloride in t-butanol. The controlled degree of epoxidation usually leads to some interesting products. For example, the 25% epoxidised synthetic polyisoprene is an elastomer with viscoelastic properties similar to those of the unepoxidised material, but has better carbon black dispersion. This gives high modulus and tensile strength, however, higher degrees of epoxidation (60–75%) produce a resinous material which is not rubbery.

FUNCTIONALISATION OF POLYMERS

Polymers with stable backbones such as polystyrene, polyethylene, and polypropylene can be functionalised. Functionalisation of polystyrene has received considerable attention, because it is a unique polymer with aromatic rings which are capable of undergoing many nucleophilic as well as electrophilic reactions. Resin based on sulphonated polystyrene are used in ion-exchange material and catalyst binding materials.

Electrophilic substitution on polystyrene through a chlorometallation reaction yields chlorine functionality. This has opened up the possibilities of making many derivatives of polystyrene. Starting with chlorometallated polystyrene, derivatives such as quaternary, ammonium, or phosphonium salts have been made. Similarly, ethers, esters, sulphonamides, silanes, and ketones derivatives have been made by replacing the chlorine atom on chlorometallated polystyrene. In the case of polystyrene, however, it was discovered that chain end functionalisation can be realised if the chain ends were terminated by group I metals such as lithium and potassium.

MISCELLANEOUS CHEMICAL REACTIONS OF POLYMERS

Direct replacement of the hydrogen atoms of aromatic rings such as styrene or the allylic hydrogen in poly(1,4-butadiene) or poly(1,4-isoprene) can be carried out via metallation with organometallic compounds of group I such as lithium, sodium, and potassium. Usually, the yield tends to be low and the product is insoluble; however, the use of chelating diamines with organolithium compounds has increased the yield and the products are soluble in cyclohexane. For example, polystyrene has been metallated in high yields to give polylithiated polystyrene in which several functional groups have been successfully introduced. Similarly, polyisoprene and polybutadiene have been successfully metallated with either *s*-butyllithium or *t*-butyllithium in the presence of tetramethylethylenediamine (TMEDA) as 50°C. In the case of the polyisoprene, chain scission and reduction in molecular weight resulted at longer metallation temperatures and times. In many cases, these lithiated polymers have been used to prepare graft and block co-polymers.

BLOCK AND GRAFT CO-POLYMERISATION

Effects on Structure and Properties of Polymers

Some of the most significant changes in structure and properties of polymers can be brought about by either block or graft co-polymerisation. The term *block co-polymer* is applied to macromolecules made up of sequences with different chemical (or physical, i.e., tactic) structures, usually represented by A, B, and C. The sequences are of a molecular weight which would give them polymeric features even if separated. The manner in which these sequences are arranged defines the type of block co-polymer prepared. A diblock co-polymer is represented by AB, indicating that a segment with chemical composition A is connected to a segment with composition B. Other possible types of block co-polymers include triblocks ABA and ABC, for example. They may be linear or branched; the linear structures are called block co-polymers,

$$\text{AAAAAA BBBB} \cdots \text{BBBBB AAAAA} \qquad \text{... (10.17)}$$

and the branched structures graft co-polymers.

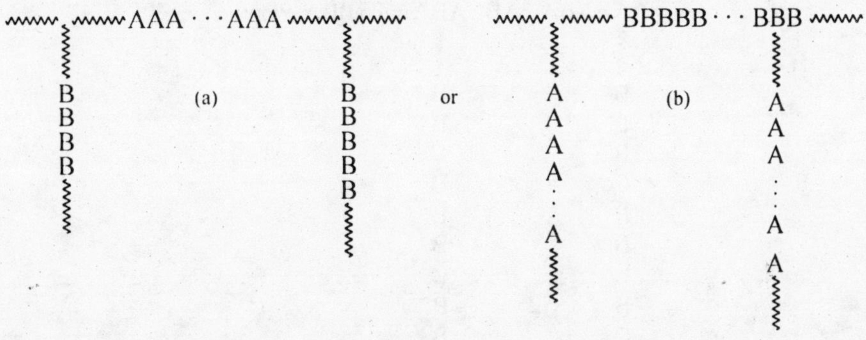

... (10.18)

Thus the same polymeric sequences may be put together as block or as graft co-polymers, with differing properties, though the major differences between the properties of block and graft co-polymers of the same constituent polymers are pronounced only in solution or in the melt. For example, natural rubber may be block co-polymerised with poly(methyl methacrylate), or methyl methacrylate monomer may be grafted onto natural rubber. In an attempt to distinguish by nomenclature one structure from the other, insertion of the letters *b* and *g*, for block and graft, respectively, between the names of the specific sequences was introduced, e.g., natural rubber-*b*-poly(methyl methacrylate) and natural rubber-*g*-poly(methyl methacrylate) in the case of the examples cited. The structures of these two macromolecules represented in equations (10.17) and (10.18(a), where the ··· AAAA ··· sequences represent natural rubber and the ··· BBBBB ··· sequences represent pply(methyl methacrylate).

The number and order of sequences may be more complicated. Block co-polymers are usually made by free radical or living polymerisations. These processes can produce polymers that consist of a pure A block connected to a pure B block, with no interphase zone of mixed A and B structure. The preparation of block co-polymers is not limited to monomers A and B, but can also encompass segments of random co-polymers.

For example, a block of a random co-polymer AB can be connected to a block of polymer A or B. Moreover, the point of attachment of the blocks can be either at the end or the middle of the polymer chain. Several examples of the various types of block co-polymers possible follow:

```
~ AAAAAA BBBB AAAAA ~     or     ~ AAAAAAAAAAAA ~
                                     B      B
                                     B      B
                                     B      B
                                     B      B          ... (10.19)
```

```
~ ABABAB   AAAAAA ~     or   ~ ABABBABB   AAAAAAAA ~
                                  A            B
                                  A            B
                                  A            B
~ ABABA    BBBBBBB ~              A            B        ... (10.20)
```

When the sequences making up the segments are random co-polymers, the prefix *co* may be introduced, with the major component monomer preceding the minor constituent. A backbone polymer of butadiene-styrene rubber grafted with styrene containing a small percentage of acrylic acid would be described as poly[(butadiene-co-styrene)–(styrene-co-acrylic acid)] and could be schematically represented as:

```
~~~~AABABAAABBABAAAABBA~~~~
    §          §
    B          B
    B          B
    B          C
    C          B
    B          B
    B          B
    B          C
    B          B
    C          B
    B          B
    §          B
               §                                  ... (10.21)
```

where A represents butadiene, B styrene, and C acrylic acid.

Block Co-polymer Synthesis

Several methods can be used to synthesise block co-polymers. Using living polymerisation, monomer A is homopolymerised to form a block of A; then monomer B is added and reacts with the active chain end of segment A to form a block of B. With careful control of the reaction conditions, this technique can produce a variety of well-defined block co-polymers. This ionic technique is discussed in more detail in a later section. Mechanicochemical degradation provides a very useful and simple way to produce polymeric free radicals. When a rubber is mechanically sheared, as during mastication, a reduction in molecular weight occurs as a result of the physical pulling apart of macromolecules. This chain rupture forms radicals of A and B which then recombine to form a block co-polymer. This is not a preferred method because it usually leads to a mixture of poorly defined block co-polymers.

These block co-polymers have found many uses in the shoe sole and adhesive industries. The physical properties were dependent on the macrostructure and microstructure of these block co-polymers.

Examples

As major examples, let us consider the three monomers butadiene, styrene, and acrylonitrile, and see how they can be block co-polymerised together by mechanochemical means. From the larger number of theoretical possibilities, 11 have been selected for discussion; these may be prepared by mastication of the following:

1. A butadiene–styrene co-polymer rubber with acrylonitrile monomer.
2. Polyacrylonitrile (plasticised) with a mixture of butadiene and styrene monomers.
3. A butadiene–acrylonitrile co-polymer rubber with styrene monomer.
4. Polystyrene with a mixture of butadiene and acrylonitrile monomers.
5. A styrene-acrylonitrile resin with a mixture of styrene and butadiene monomers.
6. Polybutadiene with a mixture of styrene and acrylonitrile monomers.
7. A butadiene–styrene rubber with polyacrylonitrile (best plasticised).
8. A styrene–acrylonitrile resin with polybutadiene.
9. A butadiene–acrylonitrile rubber with polystyrene.
10. A high styrene–butadiene resin with acrylonitrile monomer.
11. A high styrene–butadiene resin with polyacrylonitrile (plasticised).

In each example, the products would be chemically and physically different in terms of the makeup of the structural sequences, and all properties would also depend on the relative proportions of the initial components. In all mechanochemical reactions, some of the starting polymer or co-polymer remains unchanged, mainly the low-molecular-weight fraction, which is not effectively sheared, and some homopolymer may be formed from the polymerising monomers by chain transfer reactions. Varying the mastication conditions greatly influences the yield and rate of reaction; the chemical nature of the products is less affected, except that the presence of butadiene as one of the constituents (either polymer or monomer) will cause increasing gel contents with continued mastication. Processes 3, 5, 6, 8, and 9 are known to give products in which a rubber phase is dispersed in a resinous matrix; i.e., they are alternative methods for producing an A–B–S-type co-polymer (It has been found in practice that a number of monomers that normally do not polymerise by free radical processes in the temperature range 10–50°C can be block co-polymerised by cold mastication techniques, indicating ionic initiation via heterolytic scission.) The presence of a proportion of block or graft co-polymer in the system assists in stabilising the dispersion of the rubber phase in the resin matrix by acting at the phase boundary as a 'soap', i.e., a compatibilising agent at the phase boundary.

Other Methods of Effecting Mechanicochemical Reactions

Mechanicochemical degradation is the term used in describing chain scission of polymer backbones through the application of shear during a processing operation. It was previously believed that this type of process led to carbon–carbon chain scission, which usually causes a dramatic change in rheological properties. It has been found by Watson that, in the absence of oxygen, radicals produced by mechanical shear can be used to initiate the polymerisation of vinyl monomers to form block co-polymers. For example, vibromilling of natural rubber below its glass transition temperature has enabled block co-polymerisation of natural rubber with methyl methacrylate to be carried out on a small scale, with conversions as high as 86%. Similar results were achieved with styrene and with acrylonitrile.

This type of approach has also been used to attach antioxidants to unsaturated polymers. The novel approach of *Scott* using the technique employed in *Watson* enabled the attack of substituted allyl mercaptans and disulphides to olefinic double bonds employing the Kharasch reactions.

Mechanico-chemical reactions can occur during processing, when the polymer is converted to a finished product. The chain scission can occur in both saturated and unsaturated polymer backbones. For example, during processing in a screw extrúder, backbone scission in polypropylene produces long-tail free radicals which can form macroalkyl radical peroxides. These peroxides are responsible for the observed decrease in melt viscosity. In the absence of oxygen, these monoallyl macroradicals can be used to graft new monomers or such polymers as polypropylene and polyethylene to the backbone. In this manner, maleic adducts of polypropylene have been prepared, giving improved dyeability, hydrophilicity, and adhesion.

Ionic Mechanisms

Ionic mechanisms for the preparation of block co-polymers are a very important tool of the synthetic polymer chemist. A feature of many homogeneous anionic polymerisations in solution is that termination can be avoided by careful control of experimental conditions. In fact, an infinite life of the active chain end is theoretically possible, and this has led to the term *living polymers*. Polymer carbanions can resume growth after the further addition of monomer.

By changing the monomer composition, block co-polymerisation is readily initiated, and this process can be repeated. A major advantage of this type of synthesis over most free radical processes is the ability to control the chain length of the sequences by adjusting the concentrations of initiating sites and of monomer at each stage of the block co-polymerisation.

Anionic block co-polymerisation employs organolithium initiators, which have wider use because of their extended range of solubility, which includes hydrocarbons. Organolithium compounds act as initiators by direct attack of the organic anion on the monomer species, again a fast reaction and, in the absence of compound with active hydrogen atoms, without transfer or termination steps. If carefully executed, the reaction permits one to have precise control over molecular weight and (the narrow) molecular weight distribution.

The convenience of this technique has led to the development of many commercial products, including thermoplastic elastomers based on triblocks of styrene, butadiene, and isoprene. The initiator used in these systems is based on hydrocarbon-soluble organolithium initiators. In some cases, a hydrocarbon-soluble dilithio initiator has been employed in the preparation of multiblock co-polymers. Several techniques are used to prepare thermoplastic elastomers of the ABA type. A short summary of these techniques is given here.

Three-stage process with monofunctional initiators

In this technique, used, e.g., for the synthesis of block co-polymers of poly(styrene-*b*-butadiene-*b*-styrene) (SBS), a polystyrene block is formed by employing *n*-butyllithium as the initiator in an aromatic solvent. Butadiene monomer is then added to react with the polystyrene-lithium chain end to form the poly(butadiene) block. If the reaction was terminated at this stage, a poly(styrene-*b*-butadiene) co-polymer would result, which has no thermoplastic properties. Therefore, styrene monomer is added to produce the triblock SBS. The process for the preparation of SBS is very carefully controlled to avoid the formation of a diblock, as the presence of any appreciable amount of SB dramatically reduces the thermoplastic properties of SBS.

Two-stage process with difunctional initiators

Several commercial processes using difunctional initiators based on soluble organolithium compounds have been developed. These compounds can polymerise at both ends. Difunctional initiators are useful in the cases of ABA block co-polymers where B can initiate A but A cannot initiate B. These difunctional initiators are useful in the preparation of SBS. The elastomeric butadiene block is polymerised with hexane as a solvent. The added styrene monomer is also soluble in hexane. This method is also useful in preparing triblocks with hydrocarbon middle blocks and polar end blocks such as poly(methacrylonitrile-b-isoprene-*b*-methacrylonitrile).

Monofunctional initiation and coupling

In this two-stage process B is sequentially polymerised onto A, and then the two chains are coupled to yield an ABBA block co-polymer. Triblocks of SBS have been prepared using this method, with methylene dichloride as the coupling agent. The disadvantage is the formation of radical anions which can lead to contamination of the triblock with multiblock species.

Tapered block co-polymers

This method is used to form a block co-polymer which consists of two segments of essentially homopolymeric structure separated by a block of a "tapered" segment of random co-polymer composition. These are usually prepared by taking advantage of the differences in reaction rates of the component monomers. When polymerised individually in hexane, butadiene reacts six times more slowly than styrene; however, when styrene and butadiene are co-polymerised in a hydrocarbon solvent such as hexane, the reaction rates reverse and the butadiene becomes six times faster than the styrene. This leads to a tapering of the styrene in a co-polymerisation reaction.

Graft Co-polymer Synthesis

The synthesis of graft co-polymers is much more diverse, but can nevertheless be divided into groups of related processes; (i) polymer transfer; (ii) co-polymerisation via unsaturated groups; (iii) redox polymerisation; (iv) high-energy radiation techniques; (v) photochemical synthesis; and most importantly; (vi) metallation using activated organolithium with chelating diamines.

Polymer transfer

In a free radical polymerisation, chain transfer, is an important reaction. Chain transfer to a monomer, solvent, mercaptan, or other growing chain can take place. When a chain transfer reaction to another chain takes place, it creates a radical which acts as a site for further chain growth and grafting

$$P + P \rightarrow PH + P\cdot$$
$$R\cdot + P \rightarrow RH + P\cdot$$
$$P\cdot + P\cdot \rightarrow PH + P\cdot \qquad \text{...(10.22)}$$

The reaction proceeds by the transfer of a hydrogen or halogen (in the case of halogenated polymers) atom from a macromolecule P to the growing chains P· (or to an excess initiator free radical R·, thereby 'terminating' them). The reactivity is now located on the transfer molecule, which in turn initiates co-polymerisation, i.e., the growth of a grafted side chain of a newly introduced second monomer. A measure of grafting occurs with most monomer–polymer systems, especially those initiated by benzoyl peroxide, if the concentrations of polymer and initiator are high.

The simplest technique is to dissolve the polymer in the appropriate solvent; add the peroxide initiator, which abstracts a hydrogen radical and generates a radical on the polymer chain; and then add fresh monomer for grafting onto this site. This technique has been employed in grafting methylacrylate onto natural rubber and synthetic polyisoprene. In this manner, several commercially useful products such as ABS resins have been prepared; however, tyre elastomers are not made in this manner because of the generation of micro and macro gel particles, which are detrimental to physical properties. In many cases when latex grafting has been used, the product has usually been targeted toward thermoplastic applications rather than rubber applications.

Co-polymerisation via unsaturated groups

Other methods (e.g., most car body repairs) are based on the polyester-styrene co-polymerisation process (reinforced with various types of inert mesh or glass fabric), a graft co-polymerisation of styrene onto backbone unsaturated polymer of relatively low molecular weight. In general, for high grafting yields, a reasonably high concentration of pendant vinyl groups is required on the backbone polymer. For glass-reinforced plastics, the polyester resins are selected with this in view. In natural rubber, a few such groups per molecule are always present and these undoubtedly participate during normal grafting. The content of pendant vinyl groups can be increased by mastication of unsaturated rubbers under nitrogen, because the resonance structures recombine as

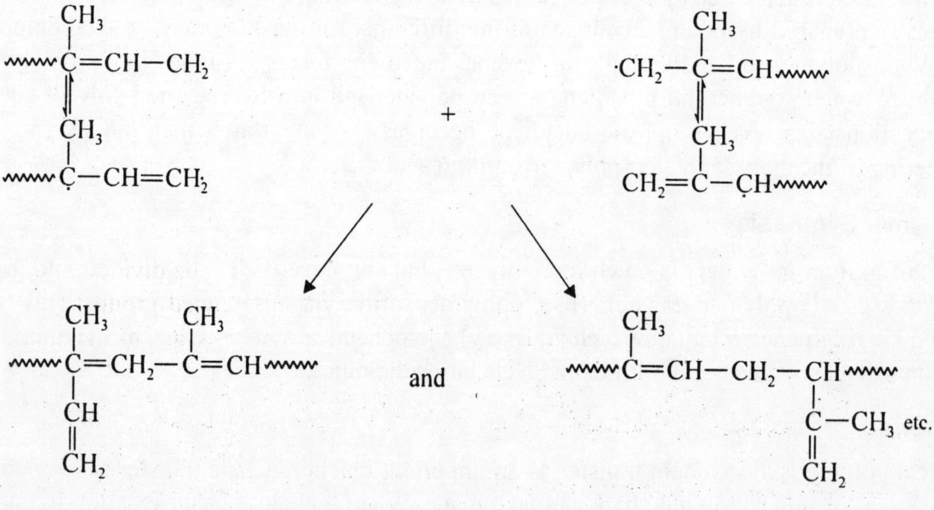

... (10.23)

The direct introduction of peroxide groups into the backbone of polymers, such as poly(methyl methacrylate), has been used to produce macromolecular initiators for the synthesis of block co-polymers, e.g., poly(methyl methacrylate-*b*-acrylonitrile) and poly(methyl methacrylate-*b*-styrene). Ozonisation can also be used, with careful control of the degree of ozonolysis, to introduce epoxy ring structures into natural rubber:

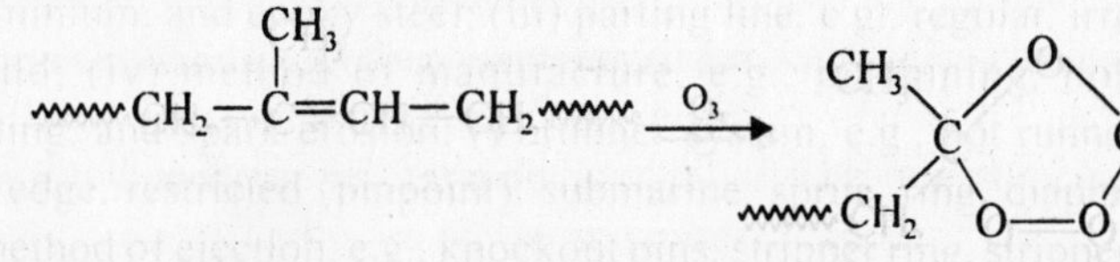

... (10.24)

By carrying out the reaction to about 4% of the available double bonds in a solvent such as toluene at a low temperature followed by a nitrogen purge, grafting can be effected by addition under nitrogen of methyl methacrylate (MMA) monomer (reacting at 80°C in sealed ampules) and formation of two MMA chains attached to the oxygens of the opened —O—O— bridge. This technique should be applicable to isoprene and butadiene co-polymers.

Redox polymerisation

Redox polymerisations are among the most popular techniques for grafting reactions, and of the possible initiator systems, ferrous ion oxidation and those based on ceric ion reduction are widely used. In a redox polymerisation, a hydroperoxide or similar group is reduced to a free radical plus an anion, while the metal ion is oxidised to a higher valency state, and at the same time a monomer is added. When the reducible group is attached to a polymeric chain, the free radical grafting sites thus formed on the macromolecular backbone act as initiators for graft co-polymerisation

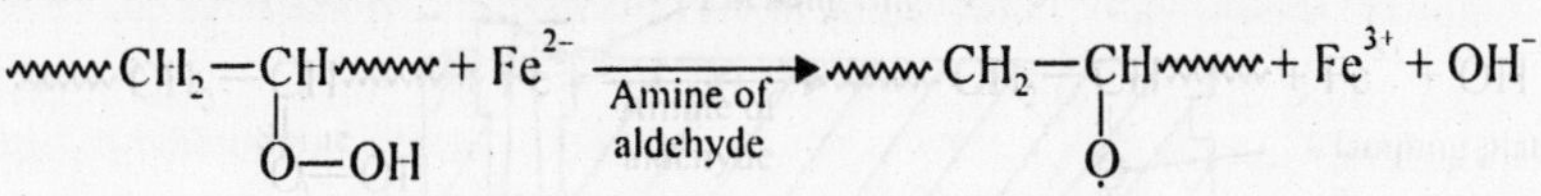

... (10.25)

This method has been used to graft methyl methacrylate to natural rubber latex. (Actually, fresh latex contains a few hydroperoxide groups per macromolecule, which can take part in grafting reactions). Recentrifuged latex concentrate is mixed with methyl methacrylate and a solution of tetraethylene pentamine is added, followed by a small quantity of ferrous sulphate solution. The homogenised blend is allowed to stand, often overnight. The graft co-polymer is isolated by coagulation. As practically all free radical sites are formed on the rubber backbone, there is very little free poly(methyl methacrylate) in the grafted system; on the other hand, some rubber chains are without grafts, as not all chains have hydroperoxy groups. Higher yields of graft co-polymer are obtained by allowing the monomer to dissolve in, and equilibrate with, the latex particles before adding the amine and ferrous ion initiator. It has been claimed that passing oxygen (air) through the latex for several hours reduces the free rubber content of the polymerisation product, but nitrogen purging is then necessary to prevent dissolved oxygen from acting as a polymerisation inhibitor.

Hydroxy polymers can be grafted by redox polymerisation by using a water-soluble peroxide, such as hydrogen peroxide in conjunction with ferrous ions. The OH radicals thus produced abstract H atoms from the hydroxy groups in the polymer, giving free radical grafting sites on the backbone. This method has been used with starch and cellulose derivatives, but considerable quantities of homopolymer are formed from the initial hydroxyl radicals in parallel with the H abstraction. By introducing a few hydroxyl groups into a copolymeric synthetic rubber, grafting can be effected, provided the presence of homopolymer can be tolerated. Mixtures of ferrous ammonium sulphate and ascorbic acid are suitable redox initiation systems. Many patents claim the preferred use of ceric ions, which easily oxidise hydroxyl groups by a radical-ion reaction:

$$R - OH + Ce^{3+} \longrightarrow R - \overset{+}{\underset{\bullet}{O}}H + Ce^{2\bullet}$$
$$R - \overset{+}{\underset{\bullet}{O}}H + OH \longrightarrow R - \dot{O} + H_2O \qquad \text{... (10.26)}$$

The advantage of this reaction lies in the fact that only hydroxyls on the polymer are converted into R—O free radicals, so that no homopolymer can be produced and pure graft is obtained.

High-energy radiation techniques

During high-energy irradiation in vacuo, e.g., from a Co source, some main-chain degradation of natural rubber and other polyisoprenes occurs:

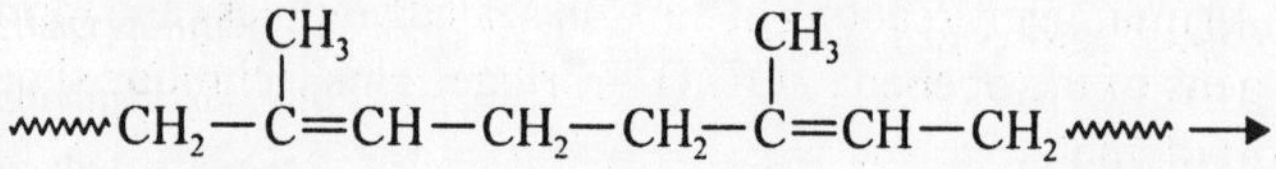

... (10.27)

Much of the irradiation energy is also adsorbed by the removal of hydrogen atoms:

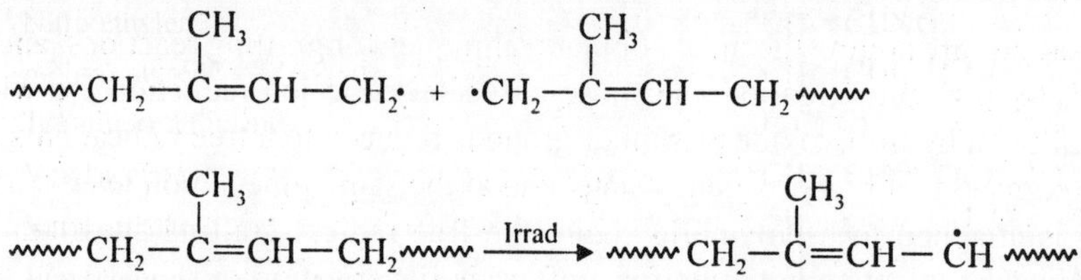

... (10.28)

The irradiation of natural rubber in the presence of a vinyl monomer thus leads primarily to a synthesis of graft co-polymers, but some block co-polymer is certainly always present. Irradiation syntheses may be carried out in solution, either in contact with liquid monomer (with or without a diluent) or in contact with monomer in the vapour phase, or in emulsion or suspension. The rubber may be preirradiated in the absence of air to produce free radicals for later monomer addition, but the life of these radicals is short as a result of mobility within the rubber matrix. Irradiation at very low temperatures makes it possible to use the trapped radicals technique for a variety of natural and synthetic rubbers. Plastics and polymers with a crystalline phase are more readily preirradiated to initiate later grafting by trapped radicals. Irradiation may also be carried out in air to introduce peroxide groupings: These groups can then be used to initiate grafting by any of the methods already discussed. Latex phase grafting is generally favoured for its simplicity; natural rubber grafts with methyl methacrylate styrene, acrylonitrile, and vinyl chloride have been made in this way.

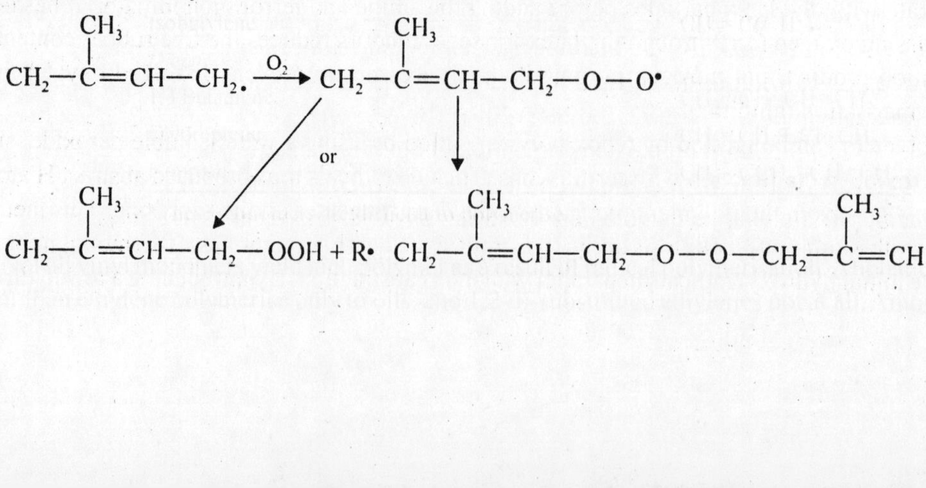

... (10.29)

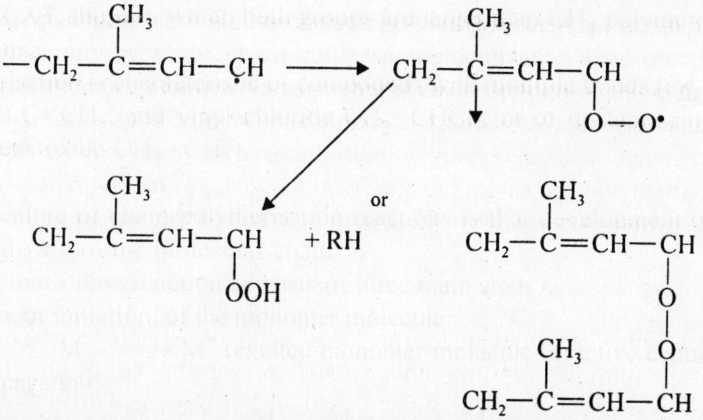

... (10.30)

The irradiation of mixed lattices for subsequent combination of the ruptured chains is another approach; it has been carried out with natural rubber and poly(vinyl chloride) lattices to prepare graft (and block) co-polymers in fairly high yields without the problem of monomer recovery. The same method has been used to graft polychloroprene onto synthetic polyisoprene dispersions and onto polybutadiene lattices of various compositions.

Photochemical synthesis

Macromolecules containing photosensitive groups which absorb energy from ultraviolet frequencies often degrade by free radical processes. The degradative process as a rule is fairly slow, but by the addition of photosensitisers, such as xanthone, benzyl, benzoin, and 1-chloroanthraquinone, the rate can be speeded up to enable graft co-polymerisation to take place in the presence of methyl methacrylate or other monomers. This can be done in the case of natural rubber in the latex phase with reasonably high yields of graft co-polymer. Natural rubber-*y*-polystyrene and poly(butadiene-*y*-styrene) have both been prepared by ultraviolet irradiation of sensitised latex–monomer dispersions. A combination of photochemical synthesis and redox-type initiation can also be carried out—a process known as one-electron oxidation—to achieve grafting with minimal homopolymer formation.

Bromine atoms on the backbone of a polymer can be liberated readily by ultraviolet irradiation to give free radical sites for grafting reactions. The bromination can be photochemically induced

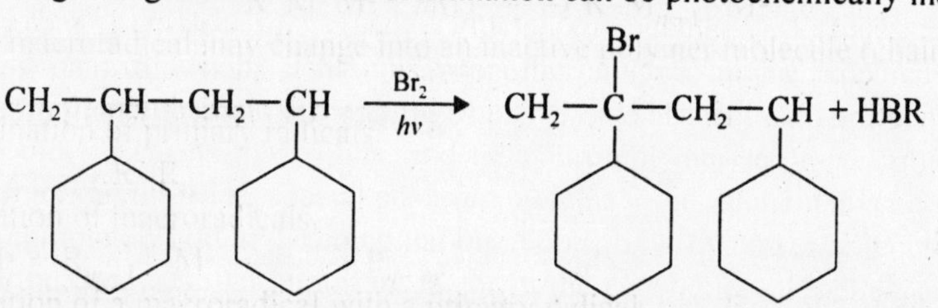

... (10.31)

or a chain transfer agent such as carbon tetrabromide may be used in the polymerisation step to introduce the labile groups,

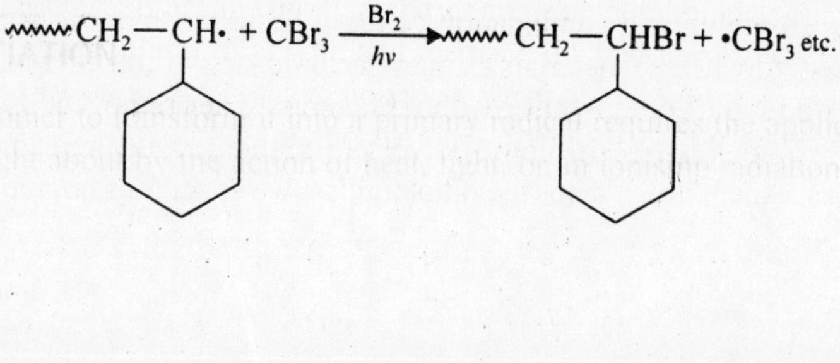

... (10.32)

With the aid of suitable sensitisers, polymers such as brominated butyl rubber, valuable because of their flame retardancy, may act as backbone polymers for a variety of grafting reactions.

An early synthesis of block co-polymers was based on the ultraviolet irradiation of poly(methyl vinyl ketone) in the presence of acrylonitrile. The initial degradative step is.

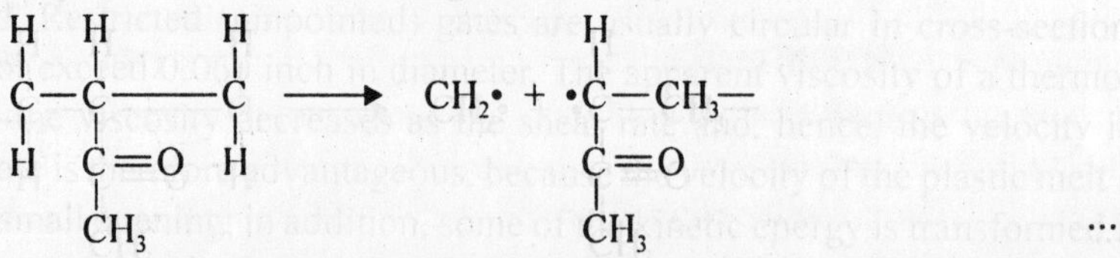

... (10.33)

This degradation reaction, supplemented by various subsequent oxidation steps, has found renewed interest in the form of the introduction of photodegradable plastics as part of the campaign to reduce plastic litter from throwaway packaging. Although as yet there has been no demand for photodegradable rubbers, the incorporation of a small percentage of a vinyl ketone into a rubber co-polymer or homopolymer would open the way to a useful synthesis of block co-polymers.

Many other syntheses of block and graft co-polymers have been reported, but enough has been said to indicate the scope of these reactions and to indicate a potential that has still to be thoroughly explored. Many grafting and block co-polymerisation systems have only been evaluated for plastic materials, but are capable of extension to rubbers.

Metallation using activated organolithium with chelating diamines

Unsaturated elastomers can be readily metallated with activated organolithium compounds in the presence of chelating diamines or alkoxides of potassium or sodium. For example, polyisoprene, polybutadiene, styrene-butadiene co-polymers, and styrene-isoprene co-polymers, can be metallated with *n*-butyllithium. TMEDA complexes (1/1 or 1/2 ratio) to form allylic or benzylic anions. The resulting allylic anion can be employed as an initiator site to grow certain branched or comb polymer species. These polymers can include polystyrene, which would form hard domains, or polybutadiene, which forms soft domains.

Research in this area has resulted in the preparation of several comb polymers. The metallation technique is a useful and versatile method as it can be used with any polymeric material which contains a few double bonds. For example, ethylene-propylene was successfully grafted with norbornene. Similar reactions were performed on polymeric materials which contain aromatic rings, such as polystyrene, poly-α-methyl styrene, and polyphenylene oxide (PPO). In general, polymeric materials that contain either side groups or mainchain allylic or acidic hydrogens can be metallated with organolithium compounds in the presence of chelating diamines. They can also be grafted with ionically polymerisable monomers to produce comblike materials.

Base Polymer Properties

Some properties of the co-polymer particularly germane to this discussion are briefly mentioned here. The properties of, say, natural rubber grafted with poly(methyl methacrylate) cannot be evaluated unless the co-polymer is isolated from either homopolymer species. The methods used are based on fractional precipitation, selective solution, or a combination of these basic techniques. In many cases, though, technologists are concerned with the materials as manufactured, so we consider in this context also the properties of the block and graft co-polymers without homopolymer removed.

The presence of two chemically different polymeric sequences in the same chain causes that macromolecule to act as a soap; i.e., it helps to compatibilise two species of homopolymer in a blend by accumulating at their interface, assisting a more gradual transition from one phase to the other, and thus reducing the interfacial energy. Microphase separation, of course, still occurs, the predominant case in practice, but macrophase separation is thereby usually prevented. In high-impact polystyrene and in A–B–S co-polymers prepared by grafting reactions, the dispersed rubber phase in the glassy matrix and the dispersed glassy phase within the rubbery particles are both prevented from forming separate phases by the graft co-polymer chains, which on a molecular scale have their rubbery segments associated with the rubber particles and their plastic segments with the glass phase. In this respect there is little difference in properties between a graft and a linear block co-polymer—the essential feature is the presence of the two types of sequences in the same macromolecule.

The block co-polymeric thermoplastic polymers owe their properties to this very structure, whereby the polystyrene end blocks (along with any homopolymeric polystyrene) form the microphase, which is dispersed within a continuous phase of polybutadiene formed from the polybutadiene segments of the central sequences in S–B–S-type block co-polymers. (For this to happen, the total volume of the polybutadiene segments must exceed the total volume of the polystyrene segments. When the reverse is the case, the product exhibits the properties of a high-impact polystyrene. Although the polystyrene 'structures' act as physical cross-links at low temperature, at processing temperatures above the softening temperature of polystyrene both segment types exhibit viscoelasticity, allowing the material to be extruded, injection moulded, etc. On cooling, the polystyrene domains become rigid again and assert their influence on the material properties.

When block and graft co-polymers are dispersed in solvents, the solutions have properties which depend on whether or not the co-polymer is eventually fully solvated. If the solvent is a 'good' solvent for both sequences, [e.g., chloroform in the case of natural rubber graft co-polymerised with poly(methyl methacrylate)], then both segment types are expanded and films cast from dilute solutions will usually be intermediate in properties to the two homopolymers (in this example the properties of a reinforced rubber film). If the solvent is a good solvent for the rubber but a poor solvent or nonsolvent for poly(methyl methacrylate), e.g., petroleum ether, then the solutions show the typical turbidity of a block or graft co-polymer and the cast film is highly elastic. When the solvent is acetone, a good solvent for poly(methyl methacrylate) but a nonsolvent for rubber, the cast films are plastic with high tear strengths.

When grafting is carried out on a polymer under conditions such that the physical form of the substrate polymer is maintained, then the original properties of the substrate usually predominate, and supplementary properties accrue as a result of the grafting. This is invariably the case when the substrate is in fibrous form, e.g., cellulose, nylons, and terylene grafted with various monomer systems. The nature of the grafting reaction to these fibres is usually such as to form a surface coating over the substrate polymer; the surface characteristics, such as dyeing, are therefore usually those of the grafting system. It is very doubtful that any blends of two polymers, or of chemically different co-polymers, can from a thermodynamic point of view ever be fully compatible. Even most block or graft co-polymers systems therefore show microphase separation which will be typical for the properties of a given system. Chemical modification, as discussed in this chapter, will in general lead to the formation of polymeric single phases, provided the reaction has been carried out homogeneously. The choice need not be restricted, however, to just these two approaches, as chemical modification can be carried out after block or graft co-polymerisation or vice versa. Very little has been published on such a consecutive use of these two physically different ways of modifying polymers chemically, so that there is considerable scope for developing new modifications of long-established rubbers, as well as generally for changing old into new polymers.

2. Disproportionation is the transfer of a hydrogen atom of one macro-radical to another to form two macromolecules, one of them with a double bond at its end:

$$R-CH_2-CHX-CH_2-CHX\bullet + \bullet CHX-CH_2-CHX-CH_2-R \rightarrow R-CH_2-CHX-CH{=}CHX + CH_2X-CH_2-CHX-CH_2-R$$

Chemical substances capable of reacting with free radicals to terminate the reaction chain are called polymerisation inhibitors (examples are hydroquinone and trinitrobenzene). Chemical compounds which are only chain-transfer agents, i.e., do not affect the rate of polymerisation but determine the molecular mass of polymer, are called polymerisation regulators, or modifiers. These substances are added to the reaction mixture in strictly rated amounts.

Purity of initial monomer

The rates of radical polymerisation reactions and the properties of the resulting products are very sensitive to impurities in the monomer, which may act as initiators or inhibitors. Therefore to obtain homogeneous high-quality polymer very pure monomers should be taken for the polymerisation.

Inhibition and retardation

A retarder is defined as a substance that can react with a radical to form products incapable of adding monomer. If the retarder is very effective, no polymer may be formed; this condition is sometimes called inhibition and the substance an inhibitor. The distinction is merely one of degree. The two phenomena are illustrated in Fig. 5.1, where the rate of polymerisation is displayed in idealised fashion. The action of a retarder is two-fold: it both reduces the concentration of radicals and shortens their average lifetime and thus the length of the polymer chains.

Fig. 5.1. Idealised time-conversion curves for typical inhibition and retardation effects: curve A, normal polymerisation; curve B, simple inhibition; curve C, simple retardation; curve D, nonideal inhibition; and curve E, inhibition followed by retardation.

In the simplest case, the retarder may be a free radical, such as triphenylmethyl or diphenylpicrylhydrazyl, which is too unreactive to initiate a polymer chain. The mechanism of retardation is simply the combination or disproportionation of radicals. If the retarder is a molecule, the chemistry of retardation is more complex, but the product of the reaction must be a radical too unreactive to initiate a chain. Inhibitors are useful in determining initiation rates, since their reaction with radicals is so rapid that the decomposition of inhibitor is independent of its concentration but gives directly the rate of generation of

SECTION III

Characterisation of Polymers

Moulded-in inserts are commonly used with thermosetting materials. However, since the screw-injection process is automatic, it is desirable to use post-assembled inserts rather than moulded-in inserts because moulded-in inserts require that the mould be held open each cycle to place the insets. A delay in the manual placement disrupts an automatic cyclic operations, affecting both the production rate and the product quality.

Tolerances of parts made by injection moulding of thermosetting materials are comparable to those produced by the compression and transfer methods described earlier. Tolerances achieved are as low as ± 0.001 inch/inch, although ordinarily tolerances of ± 0.003 to 0.005 in./in. are economically practical.

Thermosetting materials used in screw-injection moulding are modified from conventional thermosetting compounds. These modifications are necessary to provide the working time-temperature relationship required for screw plasticating. The most commonly used injection-moulding thermosetting materials are the phenolics. Other thermosetting materials often moulded by the screw-injection process include melamine, urea, polyester, alkyd, and diallyl phthalate (DAP).

Since the mid-1970s the injection moulding of glass-fibre-reinforced thermosetting polyesters gained increasing importance as better materials (e.g., low shrinkage resins, pelletized forms of polyester glass, etc.), equipment, and tooling became available. Injection-moulded reinforced thermoset plastics have thus made inroads in such markets as switch housing, fuse blocks, distributor caps, power-tool housings, office machines, etc. Bulk moulding compounds (BMC), which are puttylike FRP (fibrous glass-reinforced plastic) materials, are injection moulded to make substitutes of various metal die castings.

For injection moulding, FRP should have some specific characteristics. For example, it must flow easily at lower-than-mould temperatures without curing and without separating into resin, glass, and filler components, and it should cure rapidly when in place at mould temperature. A traditional FRP material shrinks about 0.003 inch/inch during moulding, but low-shrink FRP materials used for injection moulding shrink as little as 0.000 to 0.0005 inch/inch. Combined with proper tooling, these materials thus permit production of pieces with dimensional tolerances of ± 0.0005 inch/inch.

Proper design of parts for injection moulding requires an understanding of the flow characteristics of material within the mould. In this respect, injection-moulded parts of thermosets are more like transfer-moulded parts than to compression moulded parts. Wall-section uniformity is an important consideration in part design. Cross-sections should be as uniform as possible, within the dictates of part requirements, since moulding cycles, and therefore costs, depend on the cure time of the thickest section. (For thermoplastics, however, it is the cooling time that is critical). A rule of thumb for estimating cycle times for a 1/4-inch wall section is 30 sec for injection-moulded thermosets (compared to 45 sec for thermoplastics). As a guideline for part design, a good working average for wall thickness is 1/8 to 3/16 inch, with a minimum of 1/16 inch.

STANDARDS FOR TOLERANCES ON MOULDED ARTICLES

Moulding materials shrink after they are taken out of the mould. The extent of shrinkage depends on the character of the material and the final mould temperature. A significantly large dimensional change takes place in cold-mould articles, because the cure or hardening is accomplished outside of the mould. Moreover, many of the conventional thermosetting and thermoplastic materials continue to shrink for many months after moulding. Such data on normal material shrinkage after moulding are commonly supplied by material suppliers.

The mould shrinkage described should not, however, be confused with tolerance. Tolerance is the variation in mould shrinkage rather than the shrinkage itself. Dimensional tolerances in a moulded article

CHAPTER 11

Polymer Solutions

INTRODUCTION

Dissolving a polymer is a slow process that occurs in two stages. First, solvent molecules slowly diffuse into the polymer to produce a swollen gel. This may be all that happens—if, for example, the polymer-polymer intermolecular forces are high because of cross-linking, crystallinity, or strong hydrogen bonding. But if these forces can be overcome by the introduction of strong polymer-solvent interactions, the second stage of solution can take place. Here the gel gradually disintegrates into a true solution. Only this stage can be materially speeded by agitation. Even so, the solution process can be quite slow (days or weeks) for materials of very high molecular weight.

Knowledge of the solubility of various polymers in different solvents is important in assessing their chemical resistance and their application potentialities in the fields of paints, spinning fibres, and casting films. Important also is the knowledge of the solubility of various materials, such as plasticisers and extenders in the polymer, especially since this has an important bearing on plastics formulation.

Low molecular weight substances can readily undergo dissolution, polymers take considerable time to do so. Even at fairly low concentrations, polymer solutions are highly viscous, compared to solutions of low molecular weight substances, which are far less viscous even at high concentrations.

For example, add sodium chloride (a low molecular weight substance) to water in a beaker and stir. The salt crystals reduce in size and finally disappear. Obviously, sodium chloride molecules leave the crystal lattice and disperse in water, forming a sodium chloride solution. (Actually, sodium chloride is ionised into Na^+ and Cl^- ions which are uniformly dispersed in water but, for simplicity, sodium chloride is described as being dispersed in water as molecules). As soon as the solute comes in contact with the solvent, the solute molecules start diffusing into the solvent phase; there is thus a progressive reduction in the volume of the solute phase, which ultimately merges with the solvent phase and forms the solution phase. Low molecular weight solids in solvents dissolve faster when stirred or agitated. Now, add solid polyvinyl alcohol to water and stir. For quite some time, nothing seems to happen to the solid fragments despite vigorous stirring. Slowly, however, the fragments swell and increase in size. As against the fast dissolution undergone by the low molecular weight solutes, polymer fragments take time before undergoing any visible change. In the former case the individual molecules (and, in many cases, the ions) are discrete and separate from one another and are held together within the solid matrix only by van der Waal's forces or electrostatic forces of attraction. As soon as the solvent and the solute are brought together, the solvent molecules surround the solute molecules at the surfaces, establish solvent-solute interaction and break the solute-solute attraction. As this happens, the solute molecules, which are discrete, are isolated from the solid phase and, as their size is comparable to that of the solvent molecules, they diffuse fast into the solvent phase. Dissolution here is almost instantaneous.

Polymer molecules, on the other hand, are giant molecules as compared to solvent molecules and are made up of hundreds of chain segments. These long molecules, moreover, are not in the form of extended straight chains, but are in the form of tightly folded random coils. Individual molecular coils are also not discrete and separate but are interpenetrating and entangled with one another. There are also varying degrees of cohesive and attractive forces between different segments of the same molecular coil as well as neighbouring coils. Forces such as dispersion, induction, dipole-dipole interaction and hydrogen bonding (both intramolecular and intermolecular) hold the molecular coils and their segments together tightly. Because of the large size and coiled nature of the polymer molecules and also because of strong forces of attraction between them, solvent molecules take time to establish interactions with polymer molecules, to overcome the forces of attraction, to release individual molecules out of the chain entanglement and get them out of the polymer phase. This explains the difference in this dissolution behaviour of low and high molecular weight substances.

Because of the size and shape of the polymer molecules and other factors, the solubility relations in polymer systems are complex in comparison to those among low-molecular-weight compounds. Some empirical solubility rules have, nevertheless, been derived for applying to polymer systems, and it is also possible to make certain predictions about solubility characteristics of such systems.

The underlying reason that one material can act as a solvent for another is the compatibility of the materials—i.e., the ability of the molecules of the two materials to coexist without tending to separate. If we denote the force of attraction between the molecules of the one material A by F_{AA}, that between the molecules of another material B by F_{BB}, and represent that between one A and one B molecule as F_{AB}, then the system will be compatible and a solution will result if $F_{AB} > F_{BB}$ and $F_{AB} > F_{AA}$. On the other hand, if F_{AA} or $F_{BB} > F_{AB}$, the system will be incompatible and the molecules will separate, forming two phases. In the absence of any specific interaction (e.g., hydrogen bonding) between solvent and solute, we can reasonably assume the intermolecular attraction forces between the dissimilar molecules to be approximately given by the geometric mean of the attraction forces of the corresponding pairs of similar molecules; that is, $F_{AB} = (F_{AA}F_{BB})^{1/2}$. Consequently, if F_{AA} and F_{BB} are equal, F_{AB} will also be similar and the materials should be soluble.

SOLUBILITY PARAMETER

A measure of the intermolecular attraction forces in a material is provided by the cohesive energy, which approximately equals the heat of vapourisation (for liquids) or sublimation (for solids) per mole and so can be estimated from thermodynamic data. The cohesive energy per unit volume is called cohesive energy density, and the square root of this cohesive energy density is known as the solubility parameter (δ). The value for a solvent can thus be calculated from its latent heat of vapourisation. However, since polymers, in general, cannot be vapourised without decomposition, this method of determination of the solubility parameter is obviously inapplicable to polymers. Consequently, other methods must be used. If a solvent can be found in which a given polymer will dissolve without producing any specific reaction or association, without any heat effect and volume change, then the solubility parameter of the polymer may be taken to be equal to that of the solvent. The value of δ for a given polymer can thus be estimated by testing a large number of solvents of known δ to find out which one conforms most closely to the aforesaid requirements.

Cross-linkage, if present, makes a polymer insoluble; the δ value in such a case can be obtained by finding the solvent that produces the maximum equilibrium swelling.

In contrast to this experimental method, which is labourious and time consuming, the additive method of *Small* for calculating the solubility parameter from a set of additive constants, F, called molar attraction constants, is very simple and accurate to the first decimal place, provided that hydrogen bonding between polymer molecules is insignificant. Values of molar attraction constants for the most common groups in organic molecules have been estimated by *Small* from the vapour pressure and heat of vapourisation data for a number of simple molecules. These values of molar attraction constants, referring to 25°C, are shown in Table 11.1. The solubility parameter δ is given by the relationship,

$$\delta = \frac{\rho}{M}\sum F$$

where $\sum F$ is the molar attraction constants assumed over the groups present in the given compound; ρ and M are the density and the molecular weight of the compound. The method of computing the solubility parameter by using *Small's* table is illustrated in the following examples. [*Small's* method and constants yield values of δ in units of $(cal/cm^3)^{1/2}$. The SI value in $MPa^{1/2}$ may be obtained by multiplying by 2.04].

Table 11.1. Molar attraction constants, *F*, at 25°C.

Group	$F\left[\frac{(cal/cm^3)^{1/2}}{mole}\right]$	*Group*	$F\left[\frac{(cal/cm^3)^{1/2}}{mole}\right]$
$-CH_3$	214	H	80–100
$-CH_2-$ (single bonded)	133	O (ethers)	70
$-CH<$	28	CO(ketones)	275
$-\overset{\vert}{\underset{\vert}{C}}-$	−93	COO (esters)	310
$CH_2=$	190	CN	410
$-CH=$ (double bonded)	111	Cl single	270
$>C=$	19	Cl twinned as in $>CCl_2$	260
$CH\equiv C-$	285		
$-C\equiv C-$	222	Cl triple as in $-CCl_3$	250
Phenyl	735	Br single	340
Phenylene (*o, m, p*)	658	I single	425
Naphthyl	1146	CF_2 } in fluorocarbons	150
Ring (5-membered)	105–115	CF_3 } only	274
Ring (6-membered)	95–105	S (sulphides)	225
Conjugation	20–30	SH (thiols)	315
		ONO_2 (nitrates)	~440
		NO_2 (aliphatic)	~440
		PO_4 (organic)	~500

Example 11.1. *Dibutyl phthalate.*

Formula: $C_6H_4(COOCH_2CH_2CH_2CH_3)_2$

$M = 278$

$\rho = 1.04\ g/cm^3$

1 phenylene	at 658 = 658
2 CH_3	at 214 = 428
6 CH_2	at 133 = 798
2 COO	at 310 = 620
ΣF	= 2504

$\delta = 1.04(2504/278) = 9.4$

Example 11.2. *Polystyrene.*

Formula (for repeating unit): $-CH(C_6H_5)-CH_2-$

$$—CH—CH_2— \atop \quad | \atop \quad C_6H_5$$

M (for repeating unit) = 104

ρ (for polymer) = 1.05 g/cm³

1 phenyl	at 735 = 735
1 CH_2	at 133 = 133
1 CH	at 28 = 28
ΣF	= 896

$\Sigma F/M$ for the repeating unit = 896/104

The value of $\Sigma F/M$ for the whole polymer will evidently be the same as for the repeating unit.

Hence δ (for polymer) = $1.05(896/104) = 9.05\ (cal/cm^3)^{1/2}$.

Example 11.3. *Poly(methyl methacrylate).*

Formula (for repeating unit): $-CH_2-C(COOCH_3)(CH_3)-$

M(for repeating unit) = 100

ρ (for polymer) = 1.19 gm/cm³

2 CH_3	at 214 = 428
1 CH_2	at 133 = 133
1 COO_2	at 310 = 310
1 —C— (quaternary)	at –93 = –93
ΣF	= 778

δ (for polymer) = $1.19\ (778/100) = 9.26\ (ca./cm^3)^{1/2}$

The calculated values of Table 11.2 and all values of Table 11.3 are those derived according to *Small's* method. In applying this method to crystalline polymers, it is desirable to use a density value corresponding to the amorphous state. This 'amorphous density' can be obtained from the data above the melting point by extrapolation, or it can be obtained from other sources. Further, this method is not applicable where hydrogen bonding is significant and hence cannot be used for calculating solubility parameters of alcohols, amines, carboxylic acids, or other strongly hydrogen-bonded compounds.

Table 11.2. Solubility parameter values for polymers.

	$\delta[(cal/cm^3)]^{1/2}$	
Polymer	*Observed*	*Calculated*
Polytetrafluoroethylene	—	6.2
Polyethylene	7.9	8.1
Polypropylene	—	7.9
Polyisobutylene	8.05	7.7
Styrene-butadiene rubber (with 25% styrene)	8.09–8.6	8.5
Polyisoprene (natural rubber)	7.9–8.35	8.2
Polybutadiene	8.4–8.6	8.4
Polystyrene	8.5–9.7	9.1
Poly(vinyl acetate)	—	9.4
Poly(methyl methacrylate)	9.0–9.5	9.3
Buna *N* (butadiene/acrylonitrile 75:25)	9.38–9.5	9.2
Neoprene GN	8.18–9.25	9.4
Poly(vinyl chloride)	9.48–9.7	9.6
Bisphenol A polycarbonate	—	9.5
Poly(ethylene terephthalate)	—	10.7
Cellulose (di) nitrate	10.56	10.5
Cellulose (di) acetate	10.6	11.4
Nylon-6,6	13.6	—
Polyacrylonitrile	15.4	12.8

Conversion of units: SI value of δ in $MPa^{1/2}$ may be obtained by multiplying by 2.04.

Table 11.3. Solubility parameter values for common solvents at 25°C.

Solvents	$\delta[(cal/cm^3)]^{1/2}$
n-Hexane	7.3
n-Heptane	7.4
Diethyl ether	7.4
n-Octane	7.6
n-Decane	7.7
Methylcyclohexane	7.8
Turpentine	8.1
Camphor	8.5
2,2-Dichloropropane	8.2
n-Butyl acetate	8.3
Amyl acetate	8.5

(Contd...)

Solvents	$\delta[(cal/cm^3)]^{1/2}$
Methyl isopropyl ketone	8.5
Carbon tetrachloride	8.6
Xylene	8.8
Toluene	8.9
1,2-Dichloropropane	9.0
Ethylacetate	9.0
Benzene	9.2
Diacetone alcohol	9.2
Chloroform	9.3
Tetrahydrofuran	9.3
Trichloroethane	9.3
Tetrachloroethane	9.4
Tetraline	9.5
Methyl chloride	9.7
Methylene chloride	9.7
Ethylene dichloride	9.8
Cyclohexanone	9.9
Cellosolve	9.9
Dioxane	9.9
Carbon disulphide	10.0
Acetone	10.0
sec-Butanol	10.8
Pyridine	10.9
n-Butanol	11.4
Cyclohexanol	11.4
Isopropanol	11.5
n-Propanol	11.9
Dimethylformamide	12.1
Ethanol	12.7
Cresol	13.3
Formic acid	13.5
Methanol	14.5
Phenol	14.5
Glycerol	16.5
Water	23.4

SI value of δ in $MPa^{1/2}$ may be obtained by multiplying by 2.04.

Prediction of Solubility

In practice, it is observed that when there is no specific interaction (such as hydrogen bonding) between dissimilar molecules and also no tendency on the part of one species to crystallise, two compounds will be soluble, if their solubility parameter values, in units of (cal/cm^3), differ by no more than unity. Coupled with this criterion the values in Tables 11.1 and 11.2 permit prediction of solubility or insolubility of polymers in different solvents. Thus, for instance, it is obvious that cellulose acetate (δ = 10.6) is soluble in acetone (δ = 10.0) but not in methanol (δ = 14.5) or toluene (δ = 8.9). Natural rubber (δ = 8.3) dissolves, as expected, in toluene and carbon tetrachloride (δ = 8.6) but not in ethanol (δ = 12.7). Poly-(methyl methacrylate), an amorphous thermoplastic having a solubility parameter of about 9.2, dissolves in several liquids, including ethyl acetate (δ = 9.0), ethylene dichloride (δ = 9.8), trichloroethylene (δ = 9.3), chloroform (δ = 9.3), and toluene. Polystyrene, a hydrocarbon polymer with δ = 9.1, dissolves in several hydrocarbons with similar solubility parameters, such as benzene (δ = 9.2) and toluene. Cellulose dinitrate having a value of 10.5 is soluble in cellosolve (δ = 9.9) or an equimolar mixture of ethanol and ethyl ether (δ = 7.4).

Specific interaction

The solubility-parameter approach is applicable only when the given polymer is amorphous and there exists no hydrogen bonding or other forms of specific interaction between the polymer and the solvent. Various modifications have been proposed to account for these cases. Thus, hydrogen bonding has been used as a second parameter, in addition to δ, and the dielectric constant or some related quantity as a third but a point of diminishing return is soon reached.

Crystallisation is accompanied by a decrease of free energy, so there exists a free-energy barrier restricting the solution of crystalline polymers. If there can be some form of interaction between a crystalline polymer and a solvent and if that interaction is strong enough to overcome the free-energy barrier, then the polymer will go into solution. On the other hand, if the polymer is highly crystalline but incapable of specific interaction, then it is unlikely to dissolve. It is thus understandable that polyethylene and polytetrafluoroethylene have no solvents at room temperature. However, even if the interaction with a solvent is weak or absent, the dissolution of a crystalline polymer may yet be brought about by raising the temperature to overcome the energy barrier. Thus, although at room temperature there are no solvents for polyethylene and polytetrafluoroethylene, they do dissolve at higher temperatures. But if a highly crystalline polymer is heated above its melting point and dissolved in a solvent with which it has no specific interaction, then the polymer will tend to crystallise on subsequent cooling.

Low-density polyethylene dissolves in benzene at about 60°C. Even though at room temperature there is no solvent for polyethylene, solvents with solubility parameters close to that of polyethylene will, however, cause swelling; such swelling is more significant in low-density polymers.

Polypropylene is very similar to high-density polyethylene and has a similar solubility parameter; consequently, the two polymers tend to be swollen by the same liquids.

Solvent Type

When specific interaction comes into play in the dissolution process, depending on the solubility parameter alone for predicting solubility relations might be quite misleading, because solvents promoting different types of interactions can act differently, even if they happen to have similar solubility parameters. For instance, since poly(vinyl chloride) and the polycarbonate derived from bisphenol A have similar solubility parameters, we might expect them to exhibit similar solubility characteristics in relation to different solvents. But they do not. Thus, tetrahydrofuran and cyclohexanone are good solvents for PVC but not for the polycarbonate, whereas chloroform and methylene chloride are good solvents for the polycarbonate but not for PVC. This difference presumably resides in the fact that PVC is a proton donor, whereas the polycarbonate is a proton acceptor (though both are weak in their respective behaviour). Consequently, solvents such as cyclohexanone and tetrahydrofuran, which are proton acceptors, can dissolve PVC much better than the polycarbonate, and the reverse is obviously true in the case of solvents such as methylene chloride and chloroform since they are proton donors.

From similar arguments we can say that a polyamide such as nylon-6,6 will dissolve in proton donors, and a polymer like polyacrylonitirle will dissolve in proton acceptors. Being highly polar and crystalline, both polyamide and polyacrylontrile can, however, be expected to dissolve only in strongly interacting solvents such as formic acid, glacial acetic acid, phenols, and cresols for the former and dimethylformamide and tetramethylene sulphone for the latter.

Plasticiser Solubility

Plasticisers cause partial neutralisation of the intermolecular forces of attraction, thus producing greater freedom of movement between the polymeric macromolecules. This is reflected in an enhanced flexibility and plasticity and in reduced tensile strength and chemical resistance of the materials. Leaving aside exceptional cases, plasticisers are mostly liquids, which are, indeed, simply high-molecular-weight (hence high-boiling) solvents for the polymers—the use of high-molecular weight (of at least 300) liquids being necessitated by the basic requirements of plasticiser permanence that necessarily demands low vapour pressure and a low diffusion rate of the plasticiser within the polymer. Besides permanence, another basic requirement of a plasticiser is that it must be compatible with the polymer. In light of our foregoing discussion, the compatibility requirements, in turn, implies that the plasticiser must conform to certain demands: (i) its solubility parameter should be close to that of the polymer; (ii) it should have the ability to undergo some specific interaction with the polymer, especially if the polymer itself has tendency to crystallise; and (iii) if it is a solid, it should not be crystalline when it is incapable of some specific interaction with the polymer.

Solubility parameters for some common plasticisers are shown in Table 11.4, which on comparison with Table 11.2 allows some predictions—and these are indeed found to be correct—that aromatic oils will plasticise natural rubber; tritolyl phosphate will plasticise poly(methyl methacrylate); dimethyl phthalate will plasticise cellulose diacetate; dibenzyl ether will plasticise poly(vinylidene chloride); and octyl phthalates, tritolyl phosphate, and dioctyl sebacate will plasticise poly(vinyl chloride). The fact that camphor is an effective plasticiser for cellulose nitrate, in sharp contrast to what would be expected from consideration of the solubility parameters alone, can be explained by the ability of the carbonyl group in the camphor to undergo specific interaction with some group in the cellulose nitrate molecule. Celluloid, the first synthetic plastics material, is the product of cellulose nitrate plasticised with camphor.

In PVC formulations the plasticiser is sometimes partially replaced by some other material called an extender (such as chlorinated waxes and refinery oils), which, in itself, is not a plasticiser but can be added, up to a certain concentration, to a polymer-plasticiser combination. To make a three-component mixture compatible, the solubility parameter of the mixture of plasticiser and extender should be within unity of that of the polymer. The solubility parameter δ_{mix} of a mixture of two liquids of solubility parameters δ_1 and δ_2, respectively, can be calculated from

$$\delta_{mix} = x_1\delta_1 + x_2\delta_2$$

where x_1 and x_2 are the fraction concentrations of the two liquids. Thus, if a polymer is plasticised with a liquid of relatively high solubility parameter, then the blend would evidently tolerate a relatively higher proportion of an extender with low solubility parameter. As an example, we may cite the PVC-tritolyl phosphate formulation, which, as is to be expected, can take up more of a low-solubility-parameter extender than can the PVC-dioctyl sebacate formation.

Table 11.4. Solubility parameter values[a] for some plasticisers at 25°C.

Plasticiser	$\delta[(cal/cm^3)]^{1/2}$
Paraffinic oils	7.5
Aromatic oils	8.0
Camphor	8.5
Diisooctyl adipate	8.7
Dioctyl sebacate	8.7
Diisodecyl phthalate	8.8
Dibutyl sebacate	8.9
Diethylhexyl phthalate	8.9
Diisooctyl phthalate	8.9
Di-2-butoxyethyl phthalate	9.3
Dibutyl phthalate	9.4
Triphenyl phosphate	9.8
Tritolyl phosphate	9.8
Trixylyl phosphate	9.9
Dibenzyl ether	10.0
Triacetin	10.0
Dimethyl phthalate	10.5

[a] Values computed by *Small's* method.
SI value of δ in $MPa^{1/2}$ may be obtained by multiplying by 2.04.

Rate of dissolution

Note that the foregoing considerations of the thermodynamics of polymer dissolution only give the criteria of what may be called thermodynamically good solvents. But thermodynamically good solvents may not necessarily be kinetically good as well, since in contrast to the aforesaid thermodynamic considerations the rate of polymer dissolution depends primarily on the rapidity of diffusion of the polymer and the solvent into one another. Solvent diffusion through a polymer involves solvent molecules passing through voids and other gaps between polymer molecules. The rate of diffusion therefore largely depends on the size of molecules and the size of gaps. Solvents that promote rapid solubility are usually small compact molecules. Mixtures of a kinetically good solvent and a thermodynamicaly good solvent are often very powerful than rapid polymer solvents.

POLYMER DISSOLUTION

Let us now take up, the process of polymer dissolution in detail and, for the sake of simplicity, let us consider only polymers comprising linear flexible chains. When a polymer fragment is added to a solvent, forces of attraction or dispersion start acting between them, depending on their chemical nature, polarity and solubility parameter. When this solvent-solute interaction becomes more powerful than the solute-solute attraction, the forces holding the polymer segments together become weak and the solvent molecules force their way between the segments, break the segment-segment contacts, surround individual segments and establish contact with them. In this process, the segments unfold or loosen out from their tightly coiled conformation. In other words, the segments which were 'aggregated' in the absence of a solvent get now 'solvated'.

The onset of the solvent-solute interaction and the resultant 'solvation' and 'unfolding' of the segments constitute a slow process. To begin with, a few segments of the coil here and there unfold and get solvated and the intersegmental space thus vacated is 'raided' by more solvent molecules. As more solvent penetrates the intersegmental space, the solvent-solute interaction increases and more of the coil segments unfold and get solvated, slowly swelling up the polymer matrix.

The first stage in polymer dissolution is, therefore, characterised by a slow penetration of the solvent molecules into the interstices of the polymer coil and forcing them to 'swell'. During this swelling stage—one might call it a pseudo-induction period—the volume of the polymer matrix increases and the solvent molecules leave the solvent phase and diffuse into the polymer matrix. The polymer molecules, on the other hand, remain within the matrix itself and do not diffuse out. The phenomenon of swelling depends purely on the forces of interaction between solvent molecules and polymer segments and is not influenced by stirring or agitation. The process can, however, be hastened by exposing more polymer area for solvent interaction. It is for this reason that in order to dissolve a polymer it is put in the solvent in as many small pieces as possible and left overnight for swelling to occur. As swelling continues, more and more segments of the polymer molecule are solvated and loosened out. When all segments are solvated, the molecule as a whole, in the form of a loose coil, separates out from the swollen polymer. The loosened polymer molecule then diffuses slowly out of the polymer phase and disperses in the solvent phase, forming the solution.

Since molecules in a solid polymer are entangled with the neighbouring ones, polymer molecules during dissolution do not diffuse out of the swollen polymer one after another, but a stream of entangled molecules simultaneously diffuse and disperse in the solvent (Fig. 11.1). The rate of diffusion of polymer molecules into the solvent phase, and therefore the rate of dissolution, can be increased by agitation when the polymer is fully swollen. This stage of dissolution can be compared with the dissolution of low molecular weight substances.

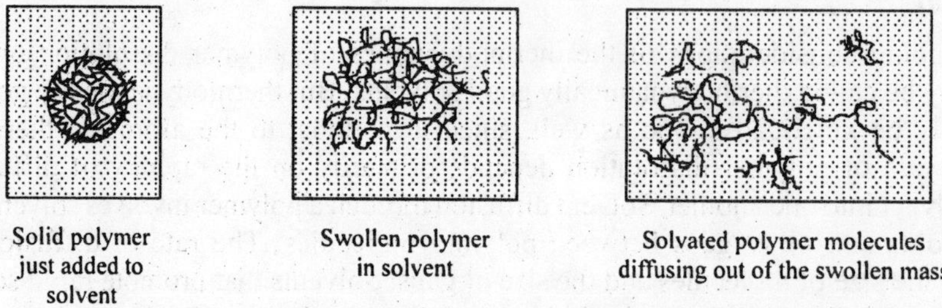

Fig. 11.1. Schematic illustration of different stages of polymer dissolution.

Even though all the chain segments of a polymer molecule in solution are unfolded and fully solvated, the molecule does not assume the shape of an extended straight chain. The coiled nature of the molecule is still retained but with a very much expanded coil volume and with the empty space between the unfolded segments of the coil being occupied by solvent molecules. We can now visualise that the polymer coil, along with solvent molecules filled within, moves as a whole in the solvent phase, forming what is known as the 'hydrodynamic' sphere or ellipsoid. The solvent held within the polymer coil can be called the 'bound' solvent and that surrounding the coil as the 'free' solvent. The 'bound' solvent, however, is in equilibrium with the 'free' solvent because solvent molecules diffuse in and out of the expanded polymer coil at a constant rate. The apparent volume occupied by the expanded coil (along with the 'bound' solvent) is referred to as the 'hydrodynamic volume' of the polymer molecule in the solution under flow.

The striking difference in the viscosities of a low molecular weight solution and a polymer solution can now be explained as follows:

In the former case, the molecular size of the solute and that of the solvent are comparable and the solute does not swell in volume in the process of dissolution; there is a free molecular mobility with respect to the solute without much increase in intermolecular friction. Hence, the viscosity of the solution is not very much different from that of the solvent. In the case of polymer solutions, however, the solute molecules, which are gigantic compared to the solvent molecules, swell considerably in volume during the process of dissolution. The giant size and the increased volume of the polymer molecule restrict its molecular mobility in the solution and increase the intermolecular friction. The polymer solution, therefore, is highly viscous.

THERMODYNAMICS OF POLYMER SOLUTIONS

The behaviour of polymers toward solvents is characteristic and different from that of low-molecular-weight substances. The size and conformations of dissolved polymer molecules require special theoretical treatment to explain their solution properties. Conversely, it is possible to obtain information about the size and shape of polymer molecules from studies of their solution properties.

The first group of pro~erties of interest includes those depending upon equilibrium between two phases, one or both of which is a solution of the polymer. In this section we discuss situations in which one phase in pure solvent, and in other section solubility phenomena in which both phases contain polymer.

Thermodynamics of Simple Liquid Mixtures

From the condition for equilibrium between two phases may be derived relations such as that for the free energy of dilution of a solution:

$$\Delta G_A = kT \ln\left(\frac{p_A}{p_A^{\,0}}\right) \qquad \text{... (11.1)}$$

where ΔG is the free energy of dilution resulting from the transfer of one molecule of liquid A from the pure liquid state with vapour pressure p_A^0 to a large amount of solution with vapour pressure p_A. This expression is written in terms of one of the colligative properties of solutions: vapour-pressure lowering, freezing-point depression, boiling point elevation, and osmotic pressure. The values of these properties cannot be related to the composition of the system by pure thermodynamic reasoning. It is necessary to know the type of variation of one property with concentration; the variation of the others may then be deduced.

Ideal solutions

In the simplest type of mixing, the molecules of components A and B have roughly the same size and shape and similar force fields. They may then form an ideal solution, defined as one in which *Raoult's* law is obeyed. This law states that the partial vapour pressure of each component in the mixture is proportional to its mole fraction. Therefore,

$$p_A = p_A^{\,0}\frac{N_A}{N_A + N_B} = p_A^{\,0} n_A \qquad \text{... (11.2)}$$

where n denotes mole fraction, and Eq. 11.1, for example, becomes

$$\Delta G_A = kT \ln n_A \quad \text{... (11.3)}$$

The total free energy of mixing is

$$\Delta G = N_A \Delta G_A + N_B \Delta G_B$$
$$= kT(N_A \ln n_A + N_B \ln n_B) \quad \text{... (11.4)}$$

The conditions for ideal mixing imply that the heat of mixing $\Delta H = 0$; that is, the components mix without change in enthalpy. Since $\Delta G = \Delta H - T\Delta S$, the entropy of mixing is given by

$$\Delta S = -k(N_A \ln n_A + N_B \ln n_B) \quad \text{... (11.5)}$$

which is positive for all compositions, so that, by the second law, spontaneous mixing occurs in all proportions.

Other types of mixing

In practice, few liquid mixtures obey *Raoult's* law.

Three types of deviations are distinguished:

1. 'Athermal' solutions, in which $\Delta H = 0$ but ΔS is no longer given by Eq. 11.5.
2. 'Regular' solutions, in which ΔS has the ideal value but ΔH is finite.
3. 'Irregular' solutions, in which both ΔH and ΔS deviate from the ideal values.

It is usually found in systems of similar-sized molecules that ΔS is nearly ideal when $\Delta H = 0$; therefore athermal solutions are nearly ideal. Many mixtures are found for which ΔH is finite, however. Such cases arise when the intermolecular force fields around the two types of molecules are different. Expressions for the heat of mixing may be derived in terms of the cohesive-energy density of these force fields.

Entropy and Heat of Mixing Polymer Solutions

Deviations from ideal behaviour

Polymer solutions invariably exhibit large deviations from *Raoult's* law, except at extreme dilutions, where ideal behaviour is approached as an asymptotic limit. At concentrations above a few per cent, deviations from ideality are so great that the ideal law is of little value for predicting or correlating the thermodynamic properties of polymer solutions. Even if the mole fraction is replaced with the volume fraction, in view of the different sizes of the polymer and solvent molecules, there is not a good correlation with the experimental results.

Entropy of mixing

Deviations from ideality in polymer solutions arise largely from small entropies of mixing. These are not abnormal, but are the natural result of the large difference in molecular size between the two components. They can be interpreted in terms of a simple molecular model. The molecules in the pure liquids and the mixture are assumed to be representable without serious error by a lattice. A two-dimensional representation of such an arrangement for non-polymer liquids is shown in Fig. 11.2a. Whereas the molecules of a pure component can be arranged in only one way on such a lattice, assuming that they cannot be distinguished from one another, the molecules of a mixture of two components can be arranged in a lattice in a large but calculable number of ways, W. By the Boltzmann relation, the entropy of mixing $\Delta S = k \ln W$. Equation 11.5 results for the case of molecules that can replace one another indiscriminately on the lattice.

It is assumed that the polymer molecules consist of a large number x of chain segments of equal length, flexibly joined together. Each link occupies one lattice site, giving the arrangement of Fig. 11.2(b). The solution is assumed to be concentrated enough that the occupied lattice sites are distributed at random rather than lying in well-separated regions of x occupied sites each. It can now be seen qualitatively why the entropy of mixing of polymer solutions is small compared to that with normal solutes. There are fewer ways in which the same number of lattice sites can be occupied by polymer segments: Fixing one segment at a lattice point severely limits the number of sites available for the adjacent segment. The approximate calculation of W for such a model is due separately to *Flory* and *Huggins*; their results (which differ only in minor details) are known as the *Flory-Huggins* theory of polymer solutions. The entropy of mixing is analogous to that given in Eq. 11.5 for simple liquids: For polymer solutions.

$$\Delta S = -k\,(N_1 \ln v_1 + N_2 \ln v_2) \qquad \text{... (11.6)}$$

where subscript 1 denotes the solvent and 2 the polymer, and v_1 and v_2 are volume fractions defined as

$$v_1 = \frac{N}{N_1 + xN_2} \qquad \text{... (11.7)}$$

$$v_2 = \frac{xN_2}{N_1 + xN_2}$$

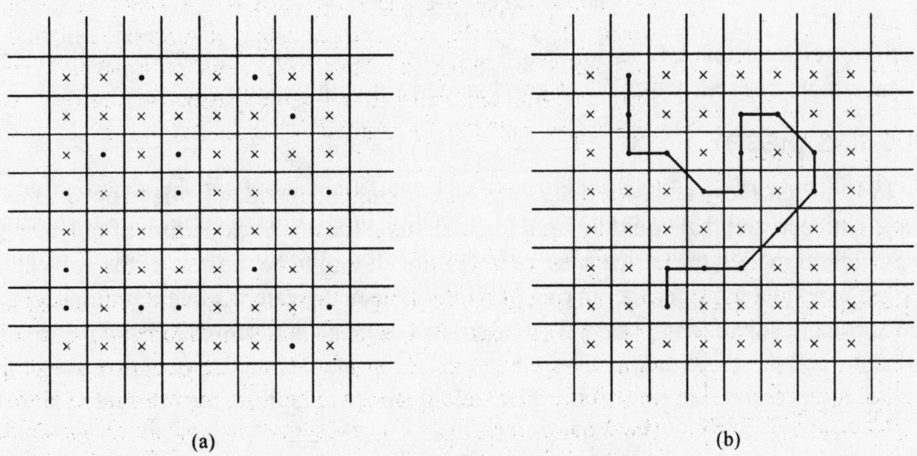

Fig. 11.2. Two-dimensional representation of (a) non-polymer liquids and (b) a polymer molecule located in the liquid lattice.

Heat and free energy of mixing

The heat of mixing of polymer solutions is analogous to that of ordinary solutions: † (†The nomenclature of *Flory* is used throughout. *Huggins* writes μ_h^0 instead of χ_1, and *Hildebrand* expresses ΔH in terms of the solubility parameter, as described earlier).

$$\Delta H = \chi_1 kTN_1 v_2 \qquad \text{... (11.8)}$$

where χ_1 characterises the interaction energy per solvent molecule divided by kT. Combining Eqs. 11.6 and 11.8 gives the Flory-Huggins expression for the free energy of mixing of a polymer solution with normal heat of mixing:

$$\Delta G = kT\,(N_1 \ln v_1 + N_2 \ln v_2 + \chi_1 N_1 v_2) \qquad \text{... (11.9)}$$

From this expression may be derived many useful relations involving experimentally obtainable quantities. For example, the partial molar free energy of mixing is

$$\Delta\overline{G}_1 = kT\left[\ln(1-v_2)+\left(1-\frac{1}{x}\right)v_2+\chi_1 v_2^2\right] \quad \text{... (11.10)}$$

and from it is obtained the osmotic pressure

$$\pi = -\frac{kT}{V_1}\left[\ln(1-v_2)+\left(1-\frac{1}{x}\right)v_2+\chi_1 v_2^2\right] \quad \text{... (11.11)}$$

where V_1 is the molecular volume of the solvent. If the logarithmic term in Eq. 11.11 is expanded and only low powers of v_2 are retained, the following equation for the osmotic pressure is obtained:

$$\pi = \frac{kT}{V_1}\left(\frac{v_2}{x}+(\tfrac{1}{2}-\chi_1)v_2^2+\ldots\right) \quad \text{... (11.12)}$$

In expansions of this type the coefficient of v_2^2 is known as the second virial coefficient A_2. Here it is defined as

$$A_2 = \frac{\overline{v}_2^2}{N_0 V_1}\left(\frac{1}{2}-\chi_1\right) \quad \text{... (11.13)}$$

where $\overline{v}_2$ is the specific volume of the polymer. This simple treatment does not describe the temperature and molecular-weight dependence of A_2 in good agreement with experiment.

FLORY-HUGGINS THEORY

In the early 1940s, *Paul Flory* and *Maurice Huggins*, working independently, developed a theory based upon a simple lattice model that could be used to understand the non-ideal nature of polymer solutions. In the *Flory-Huggins* model, the lattice sites, or holes, are chosen to be the size of the solvent molecule. As the simplest example, consider the mixing of a low-molecular-weight solvent (component 1) with a low-molecular-weight solute (component 2). The solute molecule is assumed to have the same size as a solvent molecule, and therefore only one solute or one solvent molecule can occupy a single lattice site at a given time. A representation of the lattice model for this case is illustrated in Fig. 11.3.

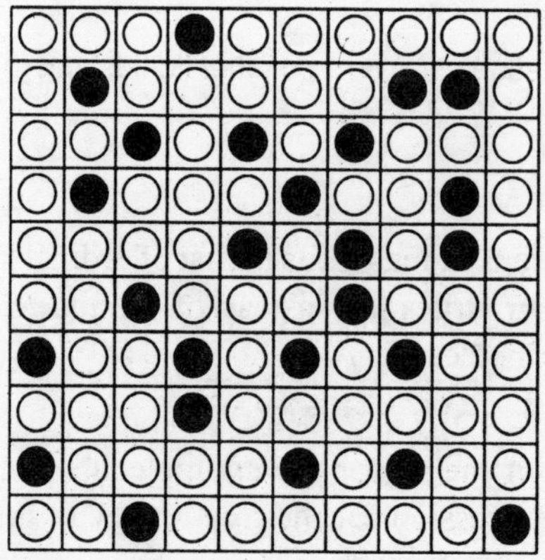

Fig. 11.3. Representation of two-dimensional *Flory-Huggins* lattice containing solvent molecules (O) and a low-molecular-weight solute (●).

The increase in entropy due to mixing of a solvent and solute, ΔS_m, may be obtained from the Boltzmann relation

$$\Delta S_m = k \ln \Omega \qquad \text{... (11.14)}$$

where k is Boltzmann's constant (1.38×10^{-23} JK^{-1}) and Ω gives the total number of ways of arranging n_1 solvent molecules and n_2 solute molecules, where $N = n_1 + n_2$ is the total number of lattice sites.

The probability function is given as

$$\Omega = \frac{N!}{n_1!n_2!} \qquad \text{... (11.15)}$$

Use of *Stirling's* approximation

$$\ln n! = n \ln n - n \qquad \text{... (11.16)}$$

leads to the expression for the entropy of mixing as

$$\Delta S_m = -k\,(n_1 \ln x_1 + n_2 \ln x_2) \qquad \text{... (11.17(a))}$$

or

$$\Delta S_m = -R\,(x_1 \ln x_1 + x_2 \ln x_2) \qquad \text{... (11.17(b))}$$

where R is the ideal gas constant † and x_1 is the mole fraction of the solvent given as

$$x_1 = \frac{n_1}{n_1 + n_2} \qquad \text{... (11.18)}$$

Equation 11.17 (a and b) is the well-known relation for the entropy change due to mixing of an ideal mixture, which can also be obtained from classical thermodynamics of an ideal solution following the Lewis-Randall law ‡.

Equations 11.17(a and b) can be written for a multicomponent system having N components as

$$\Delta S_m^{id} = -R \sum_{i=1}^{N} x_i \ln x_i \qquad \text{... (11.19)}$$

† $R = N_A k$ where N_A is Avogadro's number.

‡ The relationship between the partial-molar Gibbs free-energy and fugacity is given as

$$d\overline{G}_i \equiv d\mu_i = RTd \ln \hat{f}_i$$

Integration from the standard state to some arbitrary state gives

$$\overline{G}_i - G_i^o \equiv \overline{\Delta G}_i = RT \ln \frac{\hat{f}_i}{f_i^0}$$

where $\hat{f}_i$ is the fugacity of component i in a mixture and f_i^0 is the standard-state fugacity. Substitution of the *Lewis-Randall* law

$$\hat{f}_i^{id} = x_i f_i^0$$

gives

$$\overline{\Delta G}_i^{id} = RT \ln x_i$$

Since the thermodynamic properties of a solution are the sum of the product of the mole fraction and the partial-molar property of each of m components in the mixture, it follows that

$$\Delta G^{id} = RT\sum_{i=1}^{m} x_i \ln x_i$$

Since

$$\overline{\Delta G_i}^{id} = \overline{\Delta H_i}^{id} - T\overline{\Delta S_i}^{id}$$

and $\Delta H_i^{id} = 0$ for an ideal solution, we have

$$\overline{\Delta S_i}^{id} = -\frac{1}{T}\overline{\Delta G_i}^{id} = -R \ln x_i$$

and

$$\Delta S_i^{id} = -R\sum_{i=1}^{m} x_i \ln x_i$$

The entropy of mixing a low-molecular-weight solvent with a high-molecular-weight polymer is smaller than given by Eq. 11.19 for a low-molecular-weight mixture. This is due to the loss in conformational entropy resulting from the linkage of individual repeating units along a polymer chain compared to the less ordered case of unassociated low-molecular-weight solute-molecules dispersed in a low-molecular-weight solvent. In the development of an expression for ΔS_m for a high-molecular-weight polymer in a solvent, the lattice is established by dividing the polymer chain into r segments, each the size of a solvent molecule, where r is the ratio of polymer volume to solvent volume. For n_2 polymer molecules, the total number of lattice sites is then $N = n_1 + rn_2$. A lattice containing low-molecular-weight solvent molecules and a single polymer-chain is illustrated in Fig. 11.4.

Without going into details of the derivations,† the expression for the entropy change due to mixing obtained by *Flory* and *Huggins* is given as

$$\Delta S_m = -k\,(n_1 \ln \phi_1 + n_2 \ln \phi_2) \qquad \text{... (11.20)}$$

where ϕ_1 and ϕ_2 are the lattice volume fractions of solute (component 1) or polymer (component 2), respectively. These are given as

$$\phi_1 = \frac{n_1}{n_1 + rn_2} \qquad \text{... (11.21(a))}$$

and

$$\phi_2 = \frac{rn_2}{n_1 + rn_2} \qquad \text{... (11.21(b))}$$

For a polydisperse polymer, Eq. 11.20 may be modified as

$$\Delta S_m = -k\left(n_1 \ln \phi_1 + \sum_{i=2}^{N} n_i \ln \phi_i\right) \qquad \text{... (11.22)}$$

---

† An excellent review of the development of the lattice model is given by *Flory.*

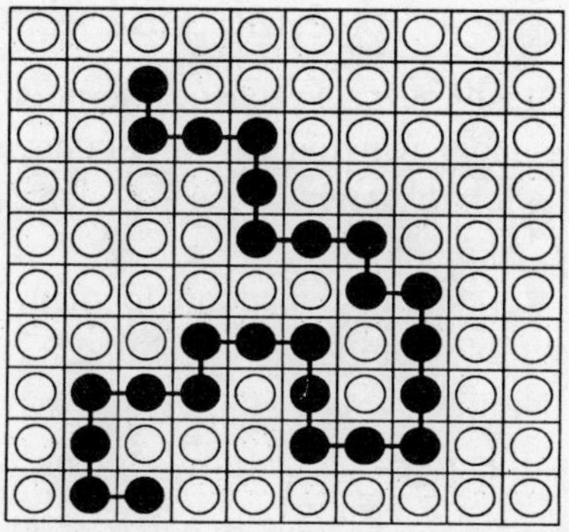

Fig. 11.4. Lattice model for a polymer chain in solution. Symbols represent solvent molecules (O) and polymer-chain segments (●).

where the summation is over all polymer chains (N) in the molecular-weight distribution. For simplicity, the most commonly used form of the entropy expression, Eq. 11.20, will be used in further discussion. Equation 11.20 provides the entropy terms in the expression for the Gibbs free energy of mixing, ΔG_m, of a polymer solution given as

$$\Delta G_m = \Delta H_m - T\Delta S_m \quad \text{.... (11.23)}$$

Once an expression for the enthalpy of mixing, ΔH_m, is known, expressions for the chemical potential and activity of the solvent can be obtained as

$$\Delta\mu_1 = \mu_1 - \mu_1^0 = \overline{\Delta G_m} = \left(\frac{\partial \Delta G_m}{\partial n_1}\right)_{T,p} \quad \text{... (11.24)}$$

where $\overline{\Delta G_m}$ is the partial-molar Gibbs free energy of mixing and the activity is related to the chemical potential as

$$\ln a_1 = \frac{\Delta\mu_1}{kT} \quad \text{... (11.25)}$$

For an ideal solutions, $\Delta H_m = 0$. Solutions for which $\Delta H_m \neq 0$ but for which ΔS_m is given by Eqs. 11.17(a) and 11.17(b) are termed regular solutions and are the subject of most thermodynamic models for polymer mixtures. The expression that *Flory* and *Huggins* gave for the enthalpy of mixing is

$$\Delta H_m = z n_1 r_1 \phi_2 \Delta\omega_{12} \quad \text{... (11.26)}$$

where z is the lattice coordination number or number of cells that are first neighbours to a given cell, r_1 represents the number of 'segments' in a solvent molecule for consideration of the most general case, and $\Delta\omega_{12}$ is the change in internal energy for formation of an unlike molecular pair (solven-polymer or 1–2) contacts given by the mean-field expression as

$$\Delta\omega_{12} = \omega_{12} - \frac{1}{2}(\omega_{11} + \omega_{22}) \quad \text{... (11.27)}$$

where ω_{ij} is the energy of i–j contacts. It is clear from Eqs. 11.26 and 11.27 that an ideal solution ($\Delta H_m = 0$) is one for which the energies of 1–1, 1–2 and 2–2 molecular interactions are equal.

Since z and ω_{12} have the character of empirical parameters, it is convenient to define a single energy parameter called the Flory interaction parameter, χ_{12}, given as

$$\chi_{12} = \frac{zr_1\Delta\omega_{12}}{kT} \qquad \text{... (11.28)}$$

The interaction parameter is a dimensionless quantity that characterises the interaction energy per solvent molecule (having r_1 segments) divided by kT. As Eq. 11.28 indicates, χ_{12} is inversely related to temperature but is independent of concentration.

The expression for the enthalpy of mixing may then be written by combining Eqs. 11.26 and 11.27 as

$$\Delta H_m = kT\chi_{12}n_1\phi_2 \qquad \text{... (11.29)}$$

Combining the expression for the entropy (Eq. 11.20) and enthalpy (Eq. 11.29) of mixing gives the well-known Flory-Huggins expression for the Gibbs free energy of mixing

$$\Delta G_m = kT(n_1 \ln \phi_1 + n_2 \ln \phi_2 + \chi_{12}n_1\phi_2) \qquad \text{... (11.30)}$$

From this relationship, the activity of the solvent (Eq. 11.25) can be obtained from Eq. 11.30.

$$\ln a_1 = \ln (1-\phi_2) + \left(1 - \frac{1}{r}\right)\phi_2 + \chi_{12}\phi_2^2 \qquad \text{... (11.31)}$$

In the case of high-molecular-weight polymers for which the number of solvent-equivalent segments, r, † is large, the $1/r$ term within parentheses on the right-hand side of Eq. 11.31 can be neglected to give

$$\ln a_1 = \ln (1-\phi_2) + \phi_2 + \chi_{12}\phi_2^2 \qquad \text{... (11.32)}$$

The Flory-Huggins equation is still widely used and has been largely successful in describing the thermodynamics of polymer solutions; however, there are a number of important limitations of the original expression that should be emphasised. The most important are the following:

1. Applicability only to solutions that are sufficiently concentrated that they have uniform segment-density.
2. There is no volume change of mixing (whereas favourable interactions between polymer and solvent molecules should result in a negative volume change).
3. There are no energetically-preferred arrangements of polymer segments and solvent molecules in the lattice.

There have been a number of subsequent developments to extend the applicability of the original *Flory-Huggins* theory and to improve agreement between theoretical and experimental results. For example, *Flory* and *Krigbaum* have developed a thermodynamic theory for dilute polymer solutions. *Koningsveld* and others have improved the agreement of the original *Flory-Huggins* theory with experimental data by an empirical modification of χ_{12} to include a composition dependence and to account for polymer polydispersity. Both of these approaches are presented briefly in the following section. More recent approaches employ equation-of-state theories such as those developed by *Flory* and others for which a volume change of mixing can be incorporated.

FLORY-KRIGBAUM AND MODIFIED FLORY-HUGGINS THEORY

Flory-Krigbaum Theory

Flory and *Krigbaum* have provided a model to describe the thermodynamics of a dilute polymer solution in which individual polymer chains are isolated and surrounded by regions of solvent molecules.

† For a polymer sample with a distribution of molecular weights, r may be taken to be the number-average degree of polymerisation, $\overline{X}_n$.

In contrast to the case of a semi-dilute solution addressed by the Flory-Huggins theory, segmental density can no longer be considered to be uniform. In their development, *Flory* and *Krigbaum* viewed the dilute solution as a dispersion of clouds consisting of polymer segments surrounded by regions of pure solvent. For a dilute solution, the expression for solvent activity was given as

$$\ln a_1 = (\kappa_1 - \psi_1)\,\phi_2^2 \qquad \text{... (11.33)}$$

where κ_1 and ψ_1 are heat and entropy parameters, † respectively. They defined an 'ideal' or theta (θ) temperature as

$$\theta = \frac{\kappa_1 T}{\psi_1} \qquad \text{... (11.34)}$$

from which Eq. 11.33 can be written as

$$\ln a_1 = -\psi_1\left(1 - \frac{\theta}{T}\right)\phi_2^2 \qquad \text{... (11.35)}$$

It follows from Eq. 11.35 that solvent activity approaches unity as temperature approaches the θ temperature. At the θ temperature, the dimension of a polymer chain collapse to unperturbed dimension (i.e., in the absence of excluded-volume effects).

Modified Flory-Huggins Theory

In the original lattice theory, χ_{12} was given an inverse dependence upon temperature (Eq. 11.28) but there was no provision for a concentration dependence which experimental studies has shown to be important. *Koningsveld* and others have introduced an empirical dependence to improve the agreement with experimental data by casting the Flory-Huggins expression in the general form

$$\Delta G_m = RT\,(\phi_1 \ln \phi_1 + \phi_2 \ln \phi_2 + g\phi_1\phi_2) \qquad \text{.... (11.36)}$$

In Eq. 11.36, g is an interaction energy term for which the concentration dependence can be given as a power series in ϕ_2 as

$$g = g_0 + g_1\phi_2 + g_2\phi_2^2 + \;.... \qquad \text{.... (11.37)}$$

where each g term, g_k (k=1,2,3 ...), has a temperature dependence that can be expressed in the form

$$g_k = g_{k,1} + \frac{g_{k,2}}{T} \qquad \text{... (11.38)}$$

EQUATION-OF-STATE THEORIES

Although the Flory-Huggins theory is still useful as a starting point for describing polymer thermodynamics, there are a number of weaknesses. For example, the simple lattice model does not accommodate a volume change of mixing, which can be significant in the case of a thermodynamically-good solution. Such an inability to incorporate a volume change of mixing can lead to particular weakness in the prediction of phase equilibria. Substantial improvement in the theoretical treatment of polymer thermodynamics has been obtained by adopting a statistical-thermodynamics approach based upon an equation of state (EOS) as first proposed by *Flory*. Other successful EOS theories have been proposed by *Sanchez* and a by *Simha*. For the purpose of providing an introduction to the use of EOS theories in the treatment of polymer thermodynamics, only the Flory EOS-theory, which was the first and is still widely used, is described in this section.

† $\overline{\Delta H_1} = RT\kappa_1\phi_2^2;\;\; \overline{\Delta S_1} = R\psi_1\phi_2^2$

Flory Equation of State

Thermodynamic variables in statistical thermodynamics are obtained from a suitable partition function which can be simple or complex function depending upon the size and physical state of the molecule being considered. The simplest partition functions are obtained for monatomic and diatomic gases such as helium and nitrogen. The partition function chosen by *Flory* for the polymer was obtained from contributions by internal (i.e., intramolecular chemical-bond forces) and external (i.e., intermolecular forces) degrees of freedom. The internal contribution is dependent upon temperature, while the external contribution is dependent upon both temperature and volume. The Flory partition function can be given in reduced form as

$$Z = Z_{comb}(gv^*)^{rnc}(\tilde{v}^{1/3} - 1)^{3rnc}\exp(rnc/\tilde{v}\tilde{T}) \qquad \text{... (11.39)}$$

Here Z_{comb} is a combinatory factor, g is an inconsequential geometric factor, v^* is a characteristic (specific) volume per segment (usually called the hard-core or closed-packed volume), $\tilde{v}$ is a reduced volume per segment defined in terms of the characteristic volume, r is the mean number of segments per molecules, n is the number of molecules (or mers), c is the mean number of external degrees of freedom per segment,† and $\tilde{T}$ is a reduced temperature as defined later. The exponential term in Eq. 11.39 is related to the configurational or mean potential energy (in van der Walls form), which is inversely proportional to volume.

Statistical thermodynamics provides the following equation to obtain an EOS from the partition function:

$$p = kT\left(\frac{\partial \ln Z}{\partial V}\right)_T \qquad \text{... (11.40)}$$

The resulting EOS obtained from the partition function given by Eq. 11.39 is given in reduced form as

$$\frac{\tilde{p}\tilde{v}}{\tilde{T}} = \frac{\tilde{v}^{1/3}}{\tilde{v}^{1/3} - 1} - \frac{1}{\tilde{T}\tilde{v}} \qquad \text{... (11.41)}$$

where $\tilde{p}$ is the reduced pressure. The reduced parameters are defined in terms of the characteristic parameters (three EOS parameters, v^*, T^*, and p^*, for each of the pure components) as

$$\tilde{v} = \frac{v}{v^*} \qquad \text{... (11.42)}$$

$$\tilde{T} = \frac{T}{T^*} = \frac{2v^* cRT}{s\eta} \qquad \text{... (11.43)}$$

and

$$\tilde{p} = \frac{p}{p^*} = \frac{2pv^{*2}}{s\eta} \qquad \text{... (11.44)}$$

† The total number of degrees of freedom in the system is $3rnc$.

The characteristic parameters can be obtained from experimental *PVT* data. ‡ Representative values for Flory EOS-parameters evaluated at 25°C for four low-molecular-weight organic compounds and for four polymers are given in Table 11.5. In general, v^* and T^* increase with increasing temperature while p^* decreases.

Table 11.5. Representative values of Flory equation-of-state parameters at 25°C.

Polymer	v^* $(cm^3\ g^{-1})$	T^* (K)	p^* $(J\ cm^{-3})$
Polystyrene	0.8098	7420	547
Polydimethylsiloxane	0.8395	5530	341
Natural rubber	0.9432	6775	519
Polyisobutylene	0.9493	7580	448
Toluene	0.9275	5197	547
Cyclohexane	1.0012	4721	530
Benzene	0.8860	4709	628
Methyl ethyl ketone	0.9561	4555	582

‡ Differentiation of the EOS (Eq. 11.41) with respect to temperature at constant pressure yields at zero pressure the characteristic hard-core volume as

$$v^* = v\left[\frac{3+3\alpha T}{3+4\alpha T}\right]^3$$

Differentiation of the EOS with respect to temperature at constant volume yields at zero pressure the characteristic pressure

$$p^* = \gamma T \tilde{v}^2 = (\alpha / \beta) T \tilde{v}^2$$

where α is the thermal-expansion coefficient

$$\alpha = \frac{1}{V}\left(\frac{\partial V}{\partial T}\right)_p$$

β is the compressibility coefficient

$$\beta = -\frac{1}{V}\left(\frac{\partial V}{\partial p}\right)_T$$

and γ is the thermal-pressure coefficient

$$\gamma = \left(\frac{\partial p}{\partial T}\right)_V$$

Finally, the characteristic temperature is obtained from the EOS by letting $\tilde{p} = 0$:

$$T^* = \frac{T\tilde{v}^{4/3}}{\tilde{v}^{1/3} - 1}$$

Adaptation of the Flory EOS to mixtures is based upon the following two premises:

1. Core volumes of the solution components are additive.

$$v^* = \frac{n_1 v_1^* + n_2 v_2^*}{n_1 + n_2} \qquad \text{... (11.45)}$$

2. The intermolecular energy depends on the surface area of contact between molecules and/or segments.

Since a segment is an arbitrary unit, the segment size can be chosen such that $v_1^* = v_2^* = v^*$. This gives the following mixing rules for the mixture:

$$\frac{1}{r} = \frac{\phi_1}{r_1} + \frac{\phi_2}{r_2} \qquad \text{... (11.46)}$$

where ϕ_1 and ϕ_2 are segment (or core-volume) fractions:

$$\phi_1 = 1 - \phi_2 = \frac{n_1 v_1^*}{n_1 v_1^* + n_2 v_2^*} \qquad \text{... (11.47)}$$

The number of contact sites, s, for the mixture is given as

$$s = \phi_1 s_1 + \phi_2 s_2 \qquad \text{... (11.48)}$$

where s_i is the surface area of segment of component i. In a similar manner, the mean number of external degrees of freedom for the mixture can be written as

$$c = \phi_1 c_1 + \phi_2 c_2 \qquad \text{... (11.49)}$$

The site fraction is defined as

$$\theta_2 = 1 - \theta_1 = \frac{\phi_2 s_2}{s} \qquad \text{... (11.50)}$$

The characteristic pressure for the mixture is given as

$$p^* = \phi_1 p_1^* + \phi_2 p_2^* - \phi_1 \theta_2 X_{12} \qquad \text{... (11.51)}$$

where X_{12} is called the exchange interaction-parameter, defined as

$$X_{12} = \frac{s_1(\eta_{11} + \eta_{22} - 2\eta_{12})}{2v^{*2}} = \frac{s_1 \Delta\eta}{2v^{*2}} \qquad \text{... (11.52)}$$

where the η_{ij} terms are energy parameters for the i–j segment pairs and $\Delta\eta = \eta_{11} + \eta_{22} - 2\eta_{12}$. The exchange interaction-parameter is analogous to $\Delta\omega_{12}$ in the *Flory-Huggins* theory (Eq. 11.27) but has the dimensions of energy density. Finally, the characteristic temperature for the mixture is given as

$$T^* = \frac{\phi_1 p_1^* + \phi_2 p_2^* - \phi_1 \theta_2 X_{12}}{\dfrac{\phi_1 p_1^*}{T_1^*} + \dfrac{\phi_2 p_2^*}{T_2^*}} \qquad \text{... (11.53)}$$

The EOS of the mixture is given in the same form as that of the pure component (Eq. 11.41) except that the reduced parameter refer to those of the mixture. The reduced volume of the mixture, $\tilde{v}$, may be obtained from the EOS with $\tilde{p}$ set to zero for low pressures and $\tilde{T}$ defined by use of the characteristic temperature given by Eq. 11.53 for the mixture. Subsequently, other important quantities can be calculated. For example, the molar enthalpy of mixing is given as

$$\Delta H_m = \frac{\theta_2 x_1 v_1^* X_{12}}{\tilde{v}} + x_1 v_1^* p_1^* \left(\frac{1}{\tilde{v}_1} - \frac{1}{\tilde{v}} \right) + x_2 v_2^* p_2^* \left(\frac{1}{\tilde{v}_2} - \frac{1}{\tilde{v}} \right) \quad \text{... (11.54)}$$

where

$$x_1 = \frac{n_1}{n_1 + n_2} \quad \text{... (11.55)}$$

An important relationship is the Flory-EOS expression for $\Delta\mu_1$ given as

$$\Delta\mu_1 = RT\left[\ln \phi_1 + \left(1 - \frac{v_1^*}{v_2^*}\right) \phi_2 \right] + \frac{\theta_2{}^2 v_1^* X_{12}}{\tilde{v}} + v_1^* p_1^* \left[3\tilde{T}_1 \ln\left(\frac{\tilde{v}_1{}^{1/3} - 1}{\tilde{v}^{1/3} - 1} \right) + \frac{1}{\tilde{v}_1} - \frac{1}{\tilde{v}} \right] \quad \text{... (11.56)}$$

Alternately, Eq. 11.56 has been given in the form.

$$\Delta\mu_1 = RT\left[\ln \phi_1 + \frac{1 - r_1}{r_2}\phi_2 \right] + p_1^* v_1^* \left[3\tilde{T}_1 \ln\left(\frac{\tilde{v}_1{}^{1/3} - 1}{\tilde{v}^{1/3} - 1} \right) + \tilde{v}_1{}^{-1} - \tilde{v}^{-1} + \tilde{p}_1(\tilde{v} - \tilde{v}_1) \right]$$
$$+ (\chi_{12} - TQ_{12}\tilde{v}) v_1^* \frac{\theta_2^2}{\tilde{v}} \quad \text{... (11.57)}$$

The principal difference between Eqs. 11.56 and 11.57 is the appearance of Q_{12} in the last term of Eq. 11.57. This parameter is called the non-combinational entropy correction and generally is used as an adjustable parameter. Comparison of the standard *Flory-Huggins* relationship (Eq. 11.31), where $\Delta\mu_1 = RT \ln a_1$ and $\phi_1 = 1 - \phi_2$) with the *Flory* EOS (Eq. 11.56) shows that the first term within brackets in Eq. 11.56 is simply a combinatory term. Despite its cumbersome form, the Flory-EOS theory provides substantial improvement over the earlier lattice theory. For example, the theory reasonably predicts an excess volume of mixing as

$$\frac{V^{\mathrm{E}}}{V} = \frac{v^{\mathrm{E}}}{\tilde{v}} = 1 - \frac{\phi_1 \tilde{v}_1 + \phi_2 \tilde{v}_2}{\tilde{v}} \quad \text{... (11.58)}$$

Furthermore, it is capable of modeling the complete range of the observed phase behaviour of polymer solutions, as discussed in the next section.

PHASE EQUILIBRIA

Whether or not a polymer and solvent are mutually soluble, or miscible, is governed by the sign of the Gibbs free energy of mixing, ΔG_m, which is related to the enthalpy and entropy of mixing by Eq. 11.23. Three different dependencies of ΔG_m on solution composition (i.e., volume fraction of polymer) at constant temperature are illustrated in Fig. 11.5. If ΔG_m is positive over the entire composition range, as illustrated by curve I, the polymer and solvent are totally immiscible over the complete composition range and will co-exist at equilibrium as two distinct phases. Two other possibilities are those of partial and total miscibility, as illustrated by curves II and III, respectively. For total miscibility, it is necessary that both

$$\Delta G_m < 0 \quad \text{... (11.59)}$$

and that the second derivative of ΔG_m with respect to the volume fraction of solvent (component 1) or polymer (component 2) be greater than zero

$$\left(\frac{\partial^2 \Delta G_m}{\partial \phi_2^{\ 2}}\right)_{p,T} > 0 \qquad \text{... (11.60)}$$

over the entire composition range. Both conditions for miscibility are satisfied by curve III but not curve II, which exhibits two minima in ΔG_m, and therefore the derivative criterion expressed by Eq. 11.60 is not satisfied at all points along the ΔG_m-composition curve. A solution that exhibits such minima will phase-separate at equilibrium into two phases containing different composition of both components. The compositions of the two phases are given by the points of common tangent as illustrated in Fig. 11.5 where the composition of the solvent-rich phase is identified as ϕ_2^A and that of the polymer-rich phase as ϕ_2^B.

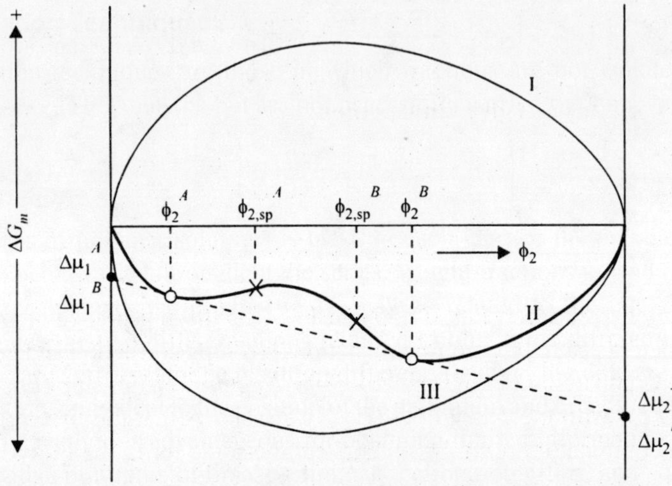

Fig. 11.5. Idealised representation of three generalised possibilities for the dependence of the Gibbs free energy of mixing, ΔG_m, of a binary mixture on composition (volume fraction of polymer, ϕ_2) at constant pressure and temperature. I. Total immiscibility; II. partial miscibility; and III. total miscibility. Curved II represents the intermediate case of partial miscibility whereby the mixture will separate into two phases whose composition (O) are marked by the volume-fraction coordinates, ϕ_2^A and ϕ_2^B, corresponding to points of common tangent to the free-energy curve. Spinodal points, compositions $\phi_{2,sp}^A$ and $\phi_{2,sp}^B$, occur at the points of inflection (×).

Phase equilibrium is strongly affected by solution temperature. In fact, any of three types of phase behaviour illustrated in Fig. 11.5 may result by a change in the temperature (or pressure) of the system. Our usual experience with solutions of low-molecular-weight compounds is that solubility increases with an increase in temperature as illustrated by the phase diagram shown in Fig. 11.6. In this example, the solution is homogeneous (i.e., the two components are totally miscible) at temperatures above the point identified as UCST which stands for the upper critical solution temperature as described below. At lower temperatures (i.e., below the UCST), phase separation may occur depending upon the overall composition of the mixture. At a given temperature below the UCST (e.g., T_1), compositions lying outside the curves are those constituting a homogeneous phase, while those lying inside the curves are thermodynamically unstable and therefore the solution will phase-separate at equilibrium. The composition of the two phases, identified as phases A and B, are given by points lying along the curve called the binodal.

The binodal is the loci of points that satisfy the conditions for thermodynamic equilibrium of binary mixture given as

$$\mu_1{}^A = \mu_1{}^B \quad \text{... (11.61(a))}$$

and

$$\mu_2{}^A = \mu_2{}^B \quad \text{... (11.61(b))}$$

As the chemical potential is given by the derivative of the Gibbs free energy with respect to compositions (Eq. 11.24), the chemical potentials are obtained graphically from the intercepts of the common tangent drawn to curve II with the free energy axes as illustrated in Fig. 11.5.

Between the binodal and the unstable region lies the metastable region, which is bounded by the spinodal. In the metastable region, the system can resist small concentration fluctuations but will eventually equilibrate to the stable two-phase state given by the binodal. Points lying along the spinodal correspond to the points of inflection identified in curve II of the free-energy diagram (Fig. 11.5) and satisfy the relationship

$$\left(\frac{\partial^2 \Delta G_m}{\partial \phi_2{}^2}\right)_{p,T} = 0 \quad \text{... (11.62)}$$

The binodal and spinodal coincide at the critical point, which satisfies the following equality for the third derivative of the Gibbs free energy with respect to composition:

$$\left(\frac{\partial^3 \Delta G_m}{\partial \phi_2{}^3}\right)_{p,T} = 0 \quad \text{... (11.63)}$$

In the case of the upper critical solution temperature (UCST), the critical point lies at the top of the phase diagram as shown in Fig. 11.6.

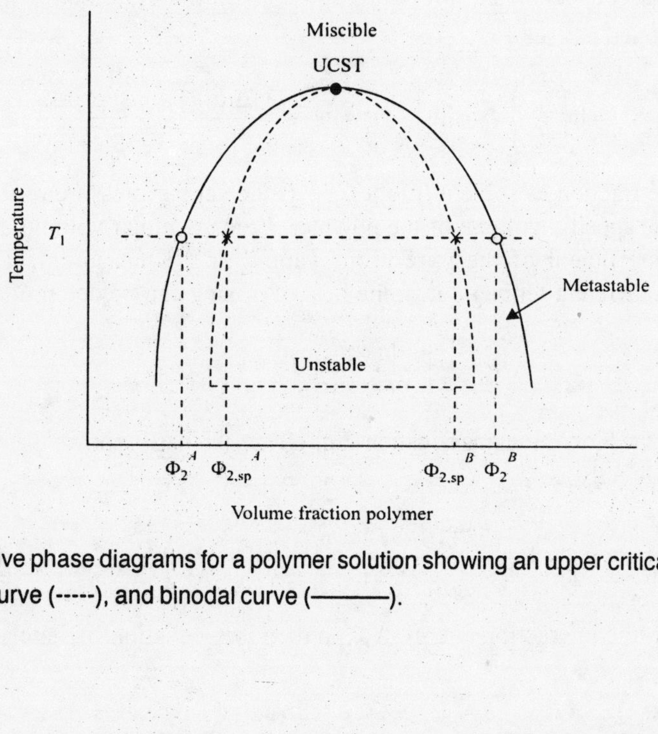

Fig. 11.6. Representative phase diagrams for a polymer solution showing an upper critical solution temperature (UCST) (●), spinodal curve (-----), and binodal curve (———).

Although the UCST behaviour of dilute polymer solutions had been observed for many years, it was not until 1961 that phase separation of polymer solutions was first reported to occur with an increase in temperature. In this case, the binodal and spinodal curves coincide at a temperature and composition called the lower critical solution temperature or LCST. One serious limitation of the *Flory-Huggins* theory is that it fails to predict LCST behaviour. The more recent equation-of-state theories are much more successful in predicting the entire range of phase behaviour.

DETERMINATION OF THE INTERACTION PARAMETER

Experimentally, χ_{12} as well as the exchange interaction parameter, X_{12}, in the Flory-EOS theory can be determined by a variety of techniques, including several scattering method and from the melting-point depression of semi-crystalline polymers. By far the most commonly used method to determine polymer solvent as well as polymer-polymer interaction parameters has been inverse gas-chromatography or IGC.

Inverse Gas-Chromatography

The term 'inverse' is used to indicate that the substance being characterised constitutes the stationary phase (i.e., the bed packing) rather than the mobile phase, as is the case in traditional gas chromatography. The stationary phase is prepared by coating a thin layer of a polymer or polymer blend from a dilute solution onto a commercial packing material in the form of small beads. A fluidised bed sometimes is used in the coating process. The coated packing is vacuum dried to remove all residual solvent and then packed into a GC column which is heated to approximately 50°C above the glass-transition temperature (T_g). A solvent probe is then injected into the carrier gas, and the time for the probe to be eluted from the column is measured.

During its passage, the probe is free to be sorbed into the liquid polymer coating of the packed bed. The extent of solubility (i.e., activity) is directly related to the retention time from which a specific retention volume V_g, can be calculated. From this value, the infinite-dilution volume-fraction activity coefficient is then obtained as

$$\ln\gamma_1^{\infty} = \ln\left(\frac{a_1}{\phi_1}\right)^{\infty} = \lim_{\phi_1\to 0}\left(\frac{a_1}{\phi_1}\right) = \ln\left(\frac{273.16Rv_2}{V_g p_1^{0}}\right) - \frac{p_1^{0}(B_{11}-V_1)}{RT} \quad \text{... (11.64)}$$

where ϕ_1 is the volume fraction of solvent (probe), p_1^{0} is the vapour pressure of the probe (solvent) in the carrier gas, v_2 is the specific volume of the polymer, V_1 is the molar volume of the probe, and B_{11} is the second virial coefficient of the pure probe vapour at the measurement temperature. From measurement of γ_1^{∞} at different temperatures, the heat of mixing can be determined as

$$\Delta H_m = R\left[\frac{\partial \ln \gamma_1^{\infty}}{\partial(1/T)}\right] \quad \text{.... (11.65)}$$

From Eq. 11.64 and the Flory-Huggins equation (Eq. 11.32), it is easily shown that the Flory interaction-parameter is obtained directly as

$$\chi_{12} = \ln\left(\frac{273.16Rv_2}{V_g p_1^{0} V_1}\right) - \frac{p_1^{0}(B_{11}-V_1)}{RT} - 1 \quad \text{... (11.66)}$$

In a similar manner, the interaction energy, X_{12}, in the Flory equation-of-state also can be obtained.

EFFECT OF MOLECULAR WEIGHT ON SOLUBILITY

In a polydispersed system (containing both low and high molecular weight components), low molecular weight components get fully solvated to start with and then they diffuse out, forming a solution, while higher molecular weight components are still in the mode of progressive solvation.

When ΔH is negative, the high molecular weight components also get fully solvated with time and go into solution. The higher the molecular weight, the longer the time taken for dissolution since the number of segments to be solvated is larger. On the other hand, when ΔH is positive, the point of equilibrium may be reached before the high molecular weight components also get fully solvated. At the point of equilibrium, $T.\Delta S$ becomes equal to $+\Delta H$ and, therefore, molecules with a molecular weight below a certain critical value will be in solution and those above it will be in a swollen state. Such a system, where equilibrium exists between the solution phase and the swollen solute phase, is referred to as a 'gel-sol' system ('gel' referring to the swollen polymer and 'sol' to the polymer solution) and the equilibrium as 'gel-sol' equilibrium. The critical molecular weight value, which determines the solubility (or otherwise) in such an equilibrium system, can be shifted by changing the 'solvent power' using binary solvent mixtures in varying proportions. Such binary solvent systems usually comprise a 'good' solvent (with $-\Delta H$ value) and a 'poor' solvent or a 'non-solvent' (with $+\Delta H$ values). All components of the polymer sample go into solution in a good solvent. When a non-solvent is progressively added to this, because of the $+\Delta H$ factor of the non-solvent, the equilibrium is shifted in such a way that some of the segments in the molecular coils of the highest molecular weight components are brought into aggregation (i.e., they get desolvated) and, hence, these molecules form the gel phase. This is to say that the highest molecular weight fraction of the polymer sample now precipitates out of the solution phase. When a further quantity of the non-solvent is added, the equilibrium is further shifted in such a way that the biggest molecules among the remaining molecules in solution precipitate out. As we progressively increase the amount of non-solvent, more and more polymer of comparatively lesser molecular weight precipitates out. This principle is utilised in the fractional precipitation method for separating a polymer sample into different fractions.

SOLUBILITY OF CRYSTALLINE AND AMORPHOUS POLYMERS

At this stage an interesting question arises. Is the foregoing concept of dissolution true for both amorphous as well as crystalline polymers? Indeed, the thermodynamic feasibility is true for both. That is, the dissolution of a polymer in a solvent will occur, be it crystalline or amorphous, whenever the ΔH value is negative or, if positive, its magnitude is less then $T.\Delta S$. Amorphous and crystalline polymers, however, differ in their phase state and, hence, it is but natural for them to behave differently as far as the physical process of their dissolution is concerned. In the gross sense of the term, dissolution of an amorphous polymer can be compared to the mixing of two liquids. Let us elaborate; two liquids, since both of them are in the same phase state, can freely mix in all proportions to form a single homogeneous phase, provided the mixing results in a negative ΔF. Likewise, an amorphous polymer, since it exists in the liquids phase state, can easily undergo dissolution in a solvent which is also in the liquid phase state, so long as ΔF is negative. On the other hand, dissolution of a crystalline polymer involves mixing of two substances existing in two different phase states. The crystalline polymer is in the crystalline phase state, while the solvent is in the liquid phase state. Since the two different phases cannot mix and form a single homogeneous system, mixing or dissolution is impossible unless both of them are brought into the same phase state. The first step involved in dissolving a crystalline polymer, therefore, is to bring the same into the liquid phase state, which can be done in two ways: (i) by heating the crystalline polymer to its melting point; and (ii) by bringing in conditions so as to destroy the long-range order (which is responsible for the crystalline state). The reduced solubility on account of crystallinity or increased intermolecular cohesive forces is common with respect to both polymers as well as low molecular weight solutes. However, owing to the long chain-link nature of the polymer molecular and the presence of strong interchain interactions, the phenomenon is quite significant in the case of polymer dissolution. Crystalline polymers are not, as a rule, soluble in most solvents below their melting points. They are soluble below their melting points only in a few solvents where strong interaction between the polymer and the solvent (by way of hydrogen bonding or dipole interaction) overcomes the interchain cohesive forces within the polymer and destroys the crystalline lattice structure. For example, polyethylene can be dissolved in hydrocarbon solvents such as toluene only when heated. While atactic polystyrene is soluble in many solvents at ambient temperatures, isotactic polystyrene can be dissolved only at elevated temperatures. Nylon 6, in which the crystallinity is due to intermolecular hydrogen bonding, can be dissolved at ambient temperatures only in solvents such as formic acid or phenol, which are capable of forming strong hydrogen bondings. For the same reason, the fully aromatic polyamides can be dissolved only in concentrated sulphuric acid.

The free energy equation governing the solubility of crystalline polymers near their melting point can then be expressed as:

$$\Delta F = (\Delta H_m + \Delta H_f) - T.(\Delta S_m + \Delta S_f) \quad \text{... (11.67)}$$

where the subscripts *m* and *f* denote mixing and fusion, respectively.

HEAT OF DISSOLUTION AND THE SOLUBILITY PARAMETER

The heat of mixing (ΔH_m) of two liquids, whose dissolution involves only dispersion forces (endothermic mixing), has been expressed by *Hildebrand* and *Scott* as follows:

$$\Delta H_m = V_m\left[\left(\frac{\Delta E_1}{V_1}\right)^{1/2} - \left(\frac{\Delta E_2}{V_2}\right)^{1/2}\right]^2 \phi_1\phi_2 \quad \text{... (11.68)}$$

where H_m = heat of mixing; V_m = total volume of the two liquids; ΔE_1 and ΔE_2 = energy of vapourisation of liquids 1 and 2; V_1 and V_2 = the molar volumes of the two liquids; and ϕ_1 and ϕ_2 = volume fractions of liquids 1 and 2.

The term ($\Delta E/V$) indicates the amount of energy required to vapourise a unit volume of the liquid. In the process of vapourisation, energy is spent to overcome the cohesive force holding the molecules together and to let the molecules completely separate out into the gaseous state. The term ($\Delta E/V$), therefore, is a measure of the cohesive force that holds molecules in the liquid together and has been designated as the 'cohesive energy density' (CED). *Hildebrand* and *Scott* proposed the term 'solubility parameter' (δ), for the square root of the CED:

$$\delta = \left(\frac{\Delta E}{V}\right)^{1/2} \quad \text{... (11.69)}$$

Now, ΔH_m can be related to the solubility parameters of the two liquids by the equation

$$\frac{\Delta H_m}{V_m\phi_1\phi_2} = (\delta_1 - \delta_2)^2 \quad \text{... (11.70)}$$

The square of the difference between the solubility parameters of two liquids decides the magnitude of the heat of mixing. Hence, for the positive heat of mixing to be at a minimum value so as to let mixing occur, the difference between the solubility parameters ($\delta_1 - \delta_2$) of the two liquids must be as low as possible. When δ_1 equal δ_2, ΔH_m becomes zero and mixing will occur readily because of the entropy factor.

The above theory of mixing based on solubility parameters is valid also for polymer dissolution and Eq. 11.70 can be simplified as

$$\Delta H = (\delta_p - \delta_s)^2 \, \phi_p \phi_s \qquad \text{... (11.71)}$$

where the subscripts p and s denote the polymer and solvent, respectively. Equation 11.71 predicts that polymer will dissolve only in solvents having δ values very close to that of the polymer. If δ_p and δ_s values are very much different, dissolution will not take place. This has been verified in the case of amorphous polymers. The theory, however, fails when applied to crystalline polymers as well as to solvents with strong polar groups and hydrogen bonding. As already mentioned, the solubility parameter theory is based on dispersion forces and, hence, does not apply to polymer-solvent systems with positive interaction where ΔH is negative. Dissolution can occur in such cases even when δ_p and δ_s are quite apart.

$$\pi = \left[\left(\frac{RT}{\overline{M}_n} \right) c + RT \left(\frac{\overline{v}_p^2}{\overline{V}_s} \right) \left(\tfrac{1}{2} - \chi \right) c^2 + \ldots\ldots \right] \qquad \text{... (11.72)}$$

or

$$\frac{\pi}{c} = RT \left(\frac{1}{\overline{M}_n} + Bc \right) \qquad \text{... (11.73)}$$

where B is termed as the second virial coefficient. The force of interaction between the polymer molecules and the solvent is measured by this term. Equation 11.73 demonstrates the need to extrapolate the experimental osmotic pressure data to zero concentration at which the higher power concentration terms may be neglected and the equation gets reduced to:

$$\left(\frac{\pi}{c} \right)_{c \to 0} = \frac{RT}{\overline{M}_n} \qquad \text{... (11.74)}$$

The term on the left-hand side of Eq. 11.74 indicates π/c at zero concentration or infinite dilution.

NATURE OF POLYMER MOLECULES IN SOLUTION

Staudinger's pioneering work led to the present-day knowledge of the macromolecular nature of polymers. Even prior to his work, starch and cellulose were known to be made up of glucose units. However, their abnormal behaviour as displayed, for instance, during fibre formation or due to the high viscosity of their solutions, was attributed to the formation of aggregates of oligomeric glucoside molecules (up to deca-sachharides) through intermolecular secondary valence forces. Starch solution in water was, therefore, thought of as a colloidal solution of aggregated molecules of oligomeric α 1,4,—glucosides and was compared to a soap solution where molecular aggregates (micelles) form a colloidal solution.

By some simple and interesting experiments, *Staudinger* demonstrated that starch or cellulose particles in solution are individual (and not aggregate) molecules and each molecule is made up of thousands of glucoside rings linked through primary covalent bonds to form long chain-like molecules. He took a cellulose sample and carefully measured its molecular weight in an aqueous copper ammonium hydroxide solution. The sample was then converted into cellulose nitrate by treatment with nitric acid, when the –OH groups were replaced by $-ONO_2$ groups. The product formed had completely different solubility properties in that it was insoluble in copper-ammonium hydroxide but was soluble in acetone, methyl ethyl ketone, etc. The molecular weight of the nitro-ester derivative was measured from its solution in acetone. This nitro-ester derivative was then carefully regenerated by saponification and the cellulose thus obtained was again measured for molecular weight from the copper-ammonium hydroxide solution. The molecular weight of the original cellulose before nitration and that of the cellulose after regeneration were found to be the same. *Staudinger* also found that the particle weights of cellulose in the copper-ammonium hydroxide solution and those of the cellulose nitrate in acetone solution lead to the same degree of polymerisation. Based on these findings, he argued that the cellulose molecule is made up of a number of glucosidic rings connected through primary valence linkages and that these rings and linkages remain intact during conversions.

If the solutions were made up of aggregates of molecules of oligomeric glucosides, then: (i) the degree of aggregation of cellulose molecules with —OH groups in copper-ammonium hydroxide due to secondary valence forces would be quite different from that of cellulose nitrate containing $—ONO_2$ groups in the organic solvent; and (ii) the particle weights would not lead to the same degree of polymerisation. *Staudinger* further proved this by showing that fatty acid soaps, which can form a colloidal solution in water, when esterified, form only a normal solution, and not a colloid, in organic solvents. He also pointed out that substances that can form colloidal solutions by way of molecular aggregation or micelle formation do so only under specific conditions and in specific solvents, whereas polymers exhibit a colloidal nature in whichever solvents they undergo dissolution. He thus differentiated colloidal sols of silver and gold as well as soap solutions from polymer solutions. He called the former 'micellar colloids' and the latter 'molecular colloids' (Fig. 11.7), and postulated a macromolecular structure for polymer molecules. According to *Staudinger*, the large size of solute molecules in polymer solutions (although comparable in size to colloidal particles) is not due to the formation of the molecular aggregation by secondary valence forces as in the case of micellar colloids. Instead, these are due to the enormous size of the polymer molecules resulting from the primary valence linkages between a very large number of monomeric units.

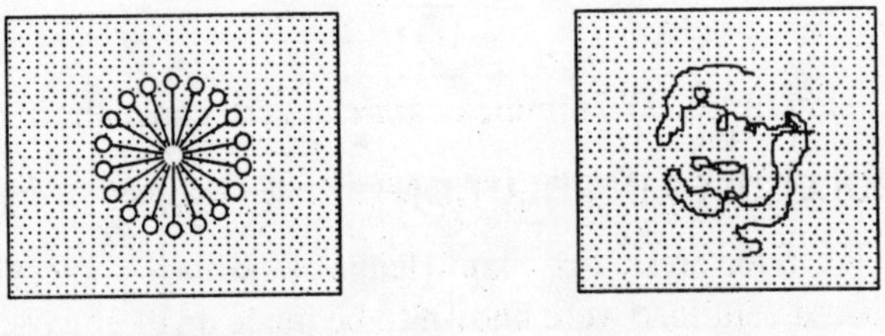

Fig. 11.7. Schematic representation of: (a) aggregates of soap molecules forming 'micellar colloid'; and (b) single polymer molecule forming 'molecular colloid'.

Although a polymer molecule can be visualised as a long chain or a thin long thread, in reality, it does not exist in solution as a stretched out chain or thread, but as a random coil. Such randomly coiled molecules; (i) can remain isolated and independent of their neighbouring molecules in the case of very dilute solutions; (ii) can be interpenetrating and entangled with the neighbouring ones in the case of moderately concentrated solutions; and (iii) can be aggregated into a uniform jelly-like mass in the case of swollen gels.

SIZE AND SHAPE OF MACROMOLECULES IN SOLUTION

In order to get a reasonably correct picture of the size and shape of a polymer molecule, we consider one which is free and not entangled with the neighbouring ones. Since polymers cannot be vapourised, a state in which the molecules can remain independent of each other, we can only visualise a system wherein a single polymer molecule is suspended within a space occupied by a large number of solvent molecules. In other words, we can consider an extremely dilute solution of a polymer in a solvent, and see what the polymer molecule looks like.

We have seen that a linear polymer molecule is made up of several segments linked to each other in a linear fashion and that each of these segments can rotate around fixed angles at the links between the successive segments.

Let us first assume that the polymer molecule suspended in a solvent medium is free from any interactive forces. Here, the absence of interaction means that there are no inter segmental attraction forces nor any other interaction forces exerted by the solvent medium on the above mentioned segments. The segmental conformations are governed purely by random flight probability, restricted only by the angle between successive segments and by a constraint on their rotation due to stearic hindrance. The molecule, over a period of time, assumes innumerable conformations. When all such conformations are statistically averaged over a period of time, the probability is that, most of the time, the two ends of the chain would be separated by a distance r_0 (Fig. 11.8) which is extremely small compared to the distance between the two ends of the molecule when it is fully stretched out.

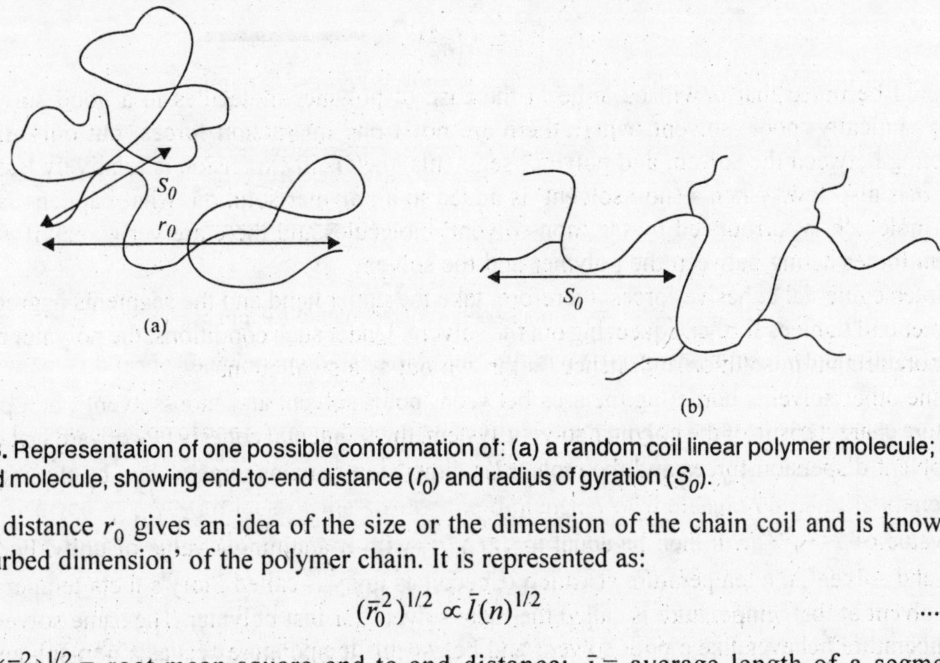

Fig. 11.8. Representation of one possible conformation of: (a) a random coil linear polymer molecule; and (b) a branched molecule, showing end-to-end distance (r_0) and radius of gyration (S_0).

This distance r_0 gives an idea of the size or the dimension of the chain coil and is known as the 'unperturbed dimension' of the polymer chain. It is represented as:

$$(\bar{r}_0^2)^{1/2} \propto \bar{l}(n)^{1/2} \quad \text{... (11.75)}$$

where $(\bar{r}_0^2)^{1/2}$ = root-mean-square end-to-end distance; $\bar{l}$ = average length of a segment; and n = number of segments in the chain.

When we deal with branched (instead of linear) polymers with the same number (n) of segments of length $\bar{l}$ the main backbone length gets reduced as some of the segments are utilised in forming the branches. The random coil conformation of a branched polymer, therefore, results in a lower dimension as compared to the linear chain of the same molecular weight. Such a coil will contain more than two chain ends, and we cannot speak in terms of the end-to-end distance while denoting the coil dimension. The term radius of gyration, $\bar{S}_0$, defined as the average distance between the molecular coil centre of gravity and the chain ends, is used for this purpose Fig. 11.8. Such a definition and concept of radius of gyration are true in the case of linear polymer molecules also and has been shown to be related with the end-to-end distance by the equation

$$\left(\bar{r}_0^2\right)^{1/2} = 6\left(\bar{S}_0^2\right)^{1/2} \qquad \text{... (11.76)}$$

In actual polymer solutions, however, different types of interaction forces do exist. Cohesive forces act between the polymer chain segments and, as a consequence, the segments try to come closer and tighten the coil. While this happens, strong interaction as well as dispersion forces act between the polymer segments and solvent molecules, pushing the segments apart and loosening the coil. The actual dimension of the polymer chain suspended in a solvent will, therefore, depend on the net result of these opposing forces and will be different from the 'unperturbed dimension'. In thermodynamically 'good' solvents, strong interaction forces are developed between the chain segment and the solvent. The polymer coil, therefore, opens up considerably increasing the coil dimension. If we can denote the end-to-end distance of a chain in a solvent as (r), the same can be related with (r_0) through an expansion factor α which describes the increase in size of the polymer coils in solution over the unperturbed dimension:

$$\left(\bar{r}^2\right)^{1/2} = \alpha\left(\bar{r}_0^2\right)^{1/2} \qquad \text{... (11.77)}$$

It should be noted that α will be large in the case of polymer molecules in a good solvent. In a thermodynamically 'poor' solvent, where there are no strong interaction forces, but only dispersion forces acting between the solvent and polymer segments, the chain expansion is relatively less and the value of α is also low. When a 'non-solvent' is added to a polymer solution, what happens is that the polymer molecule is surrounded by the 'non-solvent' molecules and there are very weak if at all any, dispersion forces acting between the polymer and the solvent.

The intersegmental cohesive forces, therefore, take the upper hand and the segments aggregate, and the polymer coil tightens further, squeezing out the solvent. Under such conditions, the polymer molecule can no more remain in solution and, hence, is thrown out as a swollen mass.

In some other solvents bordering the area between 'poor' solvent and 'non-solvent', at a particular temperature characteristic of the polymer/solvent system, the segmental cohesive forces are just balanced by the solvent dispersion forces and the molecular dimension remains unperturbed by these opposing interactions.

The value of $(\bar{r}^2)^{1/2}$ will then be equal to $(\bar{r}_0^2)^{1/2}$, with α attaining a value of unity. For a given polymer and solvent, the temperature at which α becomes unity is called Flory's theta temperature (θ) and the solvent at that temperature is called the theta solvent for that polymer. The same solvent above the θ temperature behaves like a poor solvent and below this temperature acts as a non-solvent.

The theta temperature can, therefore, be defined as the temperature at which polymer molecules are just held in solution or are at the threshold of precipitation. The expansion factor α of a polymer in a given solvent, however, varies slightly with the molecular weight—the higher the molecular weight, the higher the α value.

VISCOSITY OF DILUTE POLYMER SOLUTIONS

The viscosity of a dilute polymer solution depends on several factors, namely, the nature of the polymer and that of the solvent, their concentrations, the polymer molecular weight, the temperature and the shear rate. (The effect of the shear rate has been neglected in the following discussion for the sake of simplicity).

At the molecular level, the viscosity of a polymer solution is a direct measure of the hydrodynamic volume of the polymer molecules. Hydrodynamic volume, as we have seen is the apparent volume occupied by the expanded or swollen molecular coil along with the imbibed solvent and can be defined in terms of the expansion factor α and unperturbed end-to-end distance $(\bar{r}_0^2)^{1/2}$. We have earlier seen the dependence of α on the solvent nature and of $(\bar{r}_0^2)^{1/2}$ on the polymer molecular weight. It, therefore, follows that a polymer will exhibit a higher viscosity in a good solvent than in a poor solvent and that, in the same solvent, the viscosity will be directly proportional to the molecular weight.

Viscosity and Molecular Weight

Staudinger, used viscosity as a measure of the molecular weight of the polymer, to postulate his hypothesis about the long-chain nature of polymer molecules. He proposed the relationship:

$$\eta_{sp} = K_s cM \qquad \text{... (11.78)}$$

where K_s is a constant for a given polymer/solvent/temperature; c the concentration; M the molecular weight and η_{sp} the specific viscosity denoting the increase of viscosity of a polymer solution over that of the pure solvent according to the relation

$$\eta_{sp} = \frac{\eta - \eta_0}{\eta_0} \qquad \text{... (11.79)}$$

where η = viscosity of the polymer solution and η_0 = viscosity of the solvent.

It can be seen from Eq. 11.78 that the specific viscosity is dependent on the concentration. In order to quantify a viscosity function of a polymer in a solvent, which will be independent of the concentration, the limiting value of reduced viscosity (η_{sp}/c) or that of the inherent viscosity ($\ln \eta_r/c$) at infinite dilution is chosen and termed 'intrinsic viscosity' or 'limiting viscosity number', $[\eta]$:

$$(\eta_{sp} / c)_{c\to 0} = [\eta] = (\ln \eta_r / c)_{c\to 0} \qquad \text{... (11.80)}$$

The term $[\eta]$ has been related to the two viscosity functions through the following two equations by *Huggins* and *Kraemer*, respectively:

$$\frac{\eta_{sp}}{c} = [\eta] + k'[\eta]^2 c \qquad \text{... (11.81)}$$

$$\frac{\ln \eta_r}{c} = [\eta] - k''[\eta]^2 c \qquad \text{... (11.82)}$$

where k' and k'' are constants for a given polymer/solvent/temperature system. For many linear flexible polymer systems, k' often indicates the measure of the solvent power: the poorer the solvent, the higher the values of k'. The value of $k' + k''$ has been shown to be 0.5.

It should be noted that $[\eta]$ is dependent on polymer/solvent/temperature and can be computed by measuring the viscosity of dilute solutions of the polymer in a given solvent at various concentrations

(c less than 0.5 g/dl) and extrapolating the viscosity functions to zero concentration according to Eqs. 11.81 and 11.82. Representative plots are shown in Fig. 11.9.

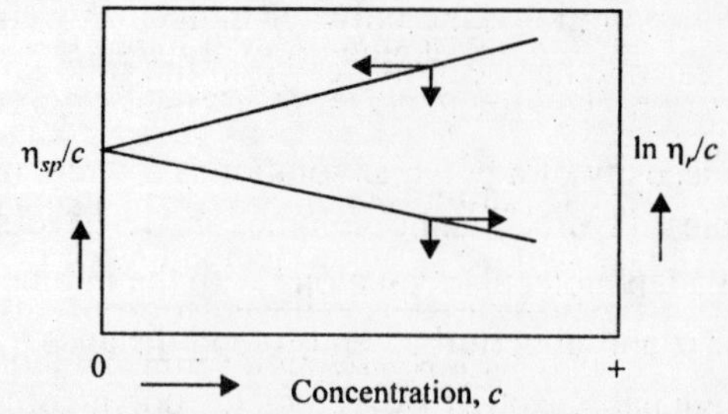

Fig. 11.9. Typical plots of viscosity data for computing intrinsic viscosity, $[\eta]$, according to Eqs. 11.81 and 11.82. The ordinate intercept gives $[\eta]$.

Staudinger's equation was subsequently replaced by the famous *Mark-Houwink* equation:

$$[\eta] = K.\overline{M}^a \quad \text{... (11.83)}$$

where K and a are constants for a given polymer/solvent/temperature system. The value of a varies from 0.6 to 0.8 and becomes 0.5 in theta solvents, while the value of K varies widely from one system to another. K and a values for some polymer/solvent systems are given in Tables 11.6 and 11.7.

Table 11.6. Viscosity-molecular weight constants for polymer/solvent systems at 30°C.

Polymer	*Solvent*	*Constants*	
		$K \times 10^3$ (ml/g)	a
Cellulose triacetate	Chloroform	4.5	0.90
	o-Cresol	6.15	0.90
Cellulose trinitrate	Ethyl acetate	2.50	1.01
Natural rubber	Benzene	18.50	0.74
Polyacrylic acid	1,4-dioxane	76.00	0.50
Polyacrylonitrile	γ-Butyrolactone	34.2	0.70
	Dimethyl farmamide	20.9	0.75
Polybutadiene–high *cis*	Benzene	33.7	0.715
	Toluene	30.5	0.725
Polybutadiene–high *trans*	Toluene	29.4	0.753
Polybutadiene–*cis, trans* mix	Toluene	39.0	0.713
Polybutyl methacrylate	Benzene	4.0	0.77
Polyethyl acrylate	Acetone	20.0	0.66
	Benzene	27.7	0.67
	Chloroform	31.4	0.68
	Ethyl acetate	26.0	0.66
Polyethylene terephthalate	Trifluoro acetic acid	43.3	0.68
	Phenol/tetrachloroethane (3/5 vol)	22.9	0.73
	Phenol/trichlorophenol (10/7 vol)	630.0	0.47

(Contd...)

Polymer	*Solvent*	*Constants*	
		K × *10*^{3} (*ml/g*)	*a*
Polyisobutylene	Benzene	61	0.56
	Carbon tetrachloride	29	0.68
	Cyclohexane	27.6	0.69
	Toluene	20.0	0.67
Polyisoprene–*cis*	Toluene	20.0	0.728
Polymethyl acrylate	Acetone	28.2	0.52
	Benzene	3.56	0.798
	Toluene	7.79	0.697
Polymethyl methacrylate–*atactic*	Acetone	7.70	0.70
	Acetonitrile	39.30	0.50
	Benzene	5.20	0.76
	Toluene	7.0	0.71
Polymethyl methacrylate–*isotactic*	Acetone	23.0	0.63
	Benzene	5.2	0.76
Polypropylene–*atactic*	Benzene	33.8	0.67
	Cyclohexane	20.9	0.76
	Toluene	21.8	0.725
Polypropylene–*syndiotactic*	Heptane	31.2	0.71
Polystyrene–*atactic*	Benzene	11.5	0.73
	Chloroform`	4.9	0.794
	Decalin (100% *trans*)	61	0.53
	Dichloro ethane	8.38	0.74
	Toluene	11	0.725
Polystyrene–*isotactic*	Benzene	10.6	0.735
	Chloroform	25.9	0.734
	Toluene	11.0	0.725
Polyvinyl acetate	Acetone	10.2	0.72
	Acetonitrile	41.5	0.62
	Benzene	56.3	0.62
	Ethyl formate	32	0.65
	Methanol	31.4	0.60
Polyvinyl alcohol	Water	45.3	0.64
Polyvinyl butyrate	Benzene	11.15	0.735
Polyvinyl chloride	Chlorobenzene	71.2	0.59
	Cyclohexanone	16.3	0.77
	Tetrahydrofuran	83.3	0.83
Polyvinyl methyl ether	Benzene	76.0	0.60
Polyvinyl pyrrolidone	Methanol	23.0	0.65
	Water	39.3	0.59
SBR-rubber	Cyclohexane	31.6	0.70
	Toluene	37.9	0.71

Table 11.7. Viscosity-molecular weight constants for polyisobutylene in toluene at different temperatures.

Polymer	*Solvent*	*Temp. (°C)*	*Constants* $K \times 10^3$ *(ml/g)*	*a*
Polyisobutylene	Toluene	0	40	0.60
Polyisobutylene	Toluene	15	24	0.65
Polyisobutylene	Toluene	30	20	0.67
Polyisobutylene	Toluene	50	20	0.68
Polyisobutylene	Toluene	60	13.5	0.71
Polyisobutylene	Toluene	90	12.6	0.72

Flory and *Fox* (based on Einstein's theory of viscosity, which state that the specific viscosity of suspension of spheres in a liquid is proportional to the volume fraction of the spheres) deduced an empirical relationship connecting the hydrodynamic volume with the intrinsic viscosity. They showed that the hydrodynamic volume is proportional to the cube of root-mean-square end-to-end distance:

$$[\eta] = \Phi \frac{\left(\overline{r^2}\right)^{3/2}}{M} \qquad \text{... (11.84)}$$

where Φ is a universal constant independent of polymer type, solvent and temperature. When $[\eta]$ is expressed in dl/g, the value of Φ is 2.84×10^{21}.

Since $\bar{r}^2 = \alpha^2 \bar{r}_0^2$ [from Eq. 11.77], equation (11.84) can be re-written as:

$$[\eta] = \Phi \left\langle \frac{\bar{r}_0^2}{M} \right\rangle^{3/2} M^{1/2} \alpha^3 \qquad \text{... (11.85)}$$

The term $\Phi \left\langle \frac{\bar{r}_0^2}{M} \right\rangle^{3/2}$ is a constant for a polymer and is independent of solvent and temperature and can be denoted by K_θ.

Now, the final *Flory-Fox* equation for molecular weight determination can be written as

$$[\eta] = K_\theta M^{1/2} \alpha^3 \qquad \text{... (11.86)}$$

and at *Flory's* theta temperature, when α becomes 1, the equation is reduced to

$$[\eta]_\theta = K_\theta M^{1/2} \qquad \text{... (11.87)}$$

where $[\eta]_\theta$ is the intrinsic viscosity at the theta temperature in a theta solvent.

From Eq. 11.83 we can see that the experimental determination of the molecular weight of a polymer from viscosity measurements requires a prior knowledge of the two constants K and a. However, if the measurements are made, in a theta solvent, the knowledge of one constant (K_θ) alone is necessary.

Theta solvent are known for limited polymers and even a very slight variation in the temperature condition makes the system shift from the theta conditions, making measurements less reliable. Attempts were, therefore, made to compute viscosity functions that would very closely resemble $[\eta]_\theta$, from measurements made in common solvents at ambient temperatures. According to *Bawn*, viscosity measurements carried out in a solvent/non-solvent mixture in which the polymer is just at the point of incipient precipitation will yield the intrinsic viscosity at the precipitation point, $[\eta]_P$ which is supposed to be identical with $[\eta]_\theta$. $[\eta]_P$ can be computed by measuring $[\eta]$ in solvent/non-solvent mixtures of varying proportions and extrapolating the data to that proportion where the polymer is at the threshold of precipitation, i.e., precipitation point. This will avoid actual measurement at the precipitation mixture and the related uncertainties associated with such measurements. A typical plot is shown in Fig. 11.10.

Cleverdone and *Smith* showed that when the η_{sp}/c versus c plots of a polymer sample in different solvents are extrapolated to negative concentrations, they converge at a point called the 'true intrinsic viscosity', $[\eta]_T$ (Fig. 11.11) and claimed that $[\eta]_T$, unlike $[\eta]$, should be independent of the solvent. They, however, found that $[\eta]_T$ was lesser than $[\eta]_P$ and $[\eta]_\theta$. Later it was argued that although a plot of η_{sp}/c versus c for a hypothetical theta solvent with a k' value of 0.5 would also converge at the same point (at $[\eta]_T$) at a negative concentration, such a plot would not be able to give any idea of the intrinsic viscosity in that solvent at zero concentration, i.e., $[\eta]_\theta$, since k' alone cannot define the slope of the curve. They, nevertheless, put forward an alternative method of extrapolating the ln η_r/c versus c plots to a positive concentration and showed that such curves do meet at a common point of intersection. They termed the intrinsic viscosity at that point as $[\eta]_R$ (Fig. 11.11). According to them the plots of a hypothetical theta solvent should meet at the same point and since the slope of that plot would be zero, $[\eta]_R$ should be identical with $[\eta]_\theta$.

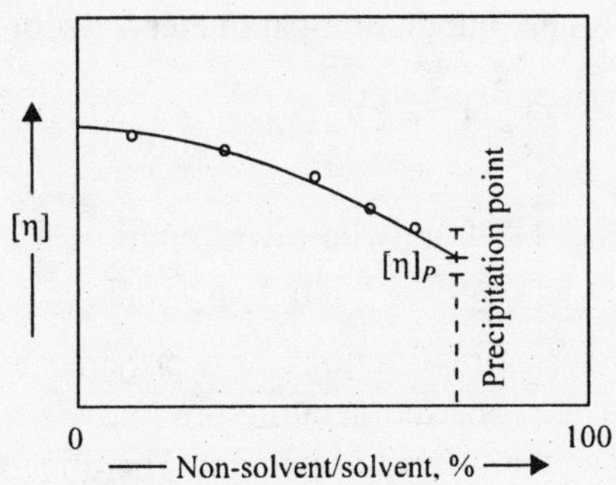

Fig. 11.10. Computation of $[\eta]_P$ from $[\eta]$ measurements in a series of solvent/non-solvent mixtures of different proportions and extrapolating to precipitation point.

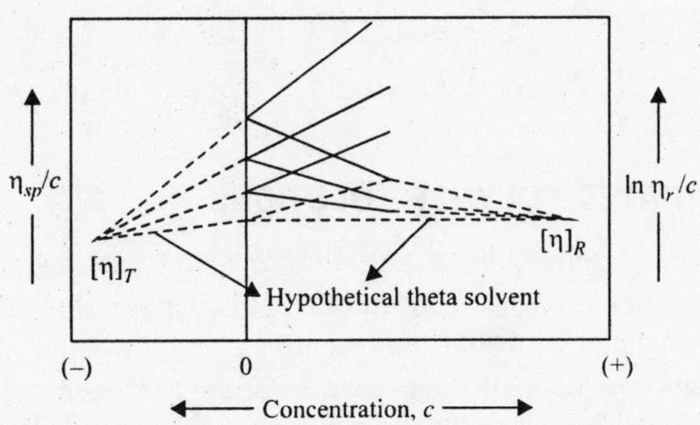

Fig. 11.11. Extrapolation of η_{sp}/c and ln η_r/c plots for computing $[\eta]_T$ and $[\eta]_R$, respectively. Note that $[\eta]_T$ is less than $[\eta]_R$

Further, $[\eta]_P$ and $[\eta]_R$ are believed to be reasonably close to $[\eta]_\theta$ and, hence, can be utilised to compute molecular weights according to the equation

$$[\eta]_R = [\eta]_P = [\eta]_\theta = K_\theta M^{1/2} \qquad \text{... (11.88)}$$

Unperturbed Dimension and Intrinsic Viscosity

In dilute solutions, we know that polymer molecules exist as expanded or swollen chain coils and that the degree of swelling depends on the nature of the solvent. The swelling is higher in a good solvent than in a poor solvent. In a theta solvent, the coil assumes the unperturbed dimension. It is possible to calculate the unperturbed dimension using the following equation by knowing the molecular weight and the value of $[\eta]_\theta$

$$[\eta]_\theta = \Phi\left\langle \frac{\bar{r}_0^2}{M} \right\rangle^{3/2} M^{1/2} \qquad \text{... (11.89)}$$

As already mentioned, the measurement of viscosities at theta conditions demands exacting experimental conditions. Moreover, theta temperatures may not lie in a convenient, experimentally realisable range. Attempts have, therefore, been made to compute the unperturbed dimension from viscosity measurements using any solvent under normal conditions. The most popular relationship is that of *Stock-mayer* and *Fixman*, which is as follows:

$$[\eta] = K_\theta M^{1/2} + 0.51\Phi BM \qquad \text{... (11.90)}$$

where B is related to the Flory-Huggins interaction parameter χ according to the equation

$$B = \bar{v}_p^2(1-2\chi)\bar{V}_s N_A \qquad \text{... (11.91)}$$

where N_A is Avogadro's number.

For easy graphical treatment, Eq. 11.70 may be rewritten as

$$[\eta]/M^{1/2} = K_\theta + 0.51\Phi BM^{1/2} \qquad \text{... (11.92)}$$

If the molecular weights of a series of fractions of a polymer and their corresponding intrinsic viscosities in any single solvent at any one temperature are known, a simple plot of $[\eta]/M^{1/2}$ against $M^{1/2}$ as per Eq. 11.92 will give K_θ as the ordinate intercept. The unperturbed root-mean-square end-to-end distance can then be calculated from the equation

$$K_\theta = \Phi\left\langle \frac{\bar{r}_0^2}{M} \right\rangle^{3/2} \qquad \text{... (11.93)}$$

The interaction parameter χ for the solvent used can also be calculated by knowing the slope of the plot and by using Eq. 11.91.

VISCOSITY OF CONCENTRATED POLYMER SOLUTIONS

While the study of the viscosity behaviour of a dilute polymer solution is of significant theoretical importance, that of the flow behaviour of a concentrated polymer solution is of practical importance. Many industrial problems associated with adhesives, coatings, cast films and spun fibres cannot be fully appreciated without a proper knowledge of the properties of concentrated polymer solutions.

In a concentrated solution (generally concentration above 5 per cent by weight), polymer molecules do not exist as isolated chain coils, but are interpenetrating and entangled with each other. Shear effects are, therefore, pronounced in the case of the viscosity behaviour of concentrated solutions. Any viscosity measurement will have to be made at different shear rates and extrapolated to zero shear so as to get the true picture of the solution behaviour.

The concentration dependence of viscosity of concentrated solutions is expressed by the equation

$$\frac{\eta_{Sp}}{c[\eta]} = \exp\left\{\frac{k'[\eta]c}{1-bc}\right\} \qquad \text{... (11.94)}$$

where b is a constant. The values of k' and $[\eta]$ for a polymer-solvent system can be found out using Eq. 11.81.

Similarly, the viscosity as a function of molecular weight and concentration can be expressed by the equation

$$\eta = K^* \, [c^{\alpha/\beta} M]^{3.4} \qquad \text{... (11.95)}$$

Here K^* is a characteristic constant for a polymer-solvent system and is temperature dependent. Its value is roughly of the order of 2.0×10^{-15} when c is expressed in g/ml and η in poise. The value of α/β ranges from 1.38 to 1.85 with an average of 1.5.

FRACTIONATION OF POLYMERS BY SOLUBILITY

In this section are discussed only the most important and widely practiced techniques for fractionating polymers by solubility differences.

Bulk Fractionation by Non-Solvent Addition

Fractional precipitation in bulk is carried out by adding the non-solvent to a dilute solution of the polymer until a slight turbidity develops at the temperature of fractionation. To ensure the establishment of equilibrium the mixture may be warmed until it is homogeneous and allowed to cool slowly back to the required temperature, which should thereafter be carefully maintained. The precipitated phase is allowed to settle to a coherent layer, and the supernatant phase is removed. A further increment of precipitant is added and the process repeated. The polymer is isolated from the precipitated phase, which may still be relatively dilute (perhaps 10 per cent polymer).

Refractionation is often used to achieve better separation. A more efficient procedure is to restrict the initial experiments to very dilute solutions, or to refine each initial fraction by a single re-precipitation at higher dilution, returning the unprecipitated polymer from this separation to the main solution before the next fraction is removed. The solvent and the precipitant should be chosen so that precipitation occurs over a relatively wide range of solvent composition yet is complete before too high a ratio of precipitant to solvent is reached. Other important considerations are the stability and volatility of the liquids and their ability to form a highly swollen, mobile gel phase. Relatively simple equipment, for example, a three-neck flask of several litres capacity is adequate for bulk precipitation fractionation.

Column Elution

Solvent-gradient elution

In elution methods, polymer is placed in contact with a series of liquids of gradually increasing solvent power. Species of lowest molecular weight, and thus highest solubility, dissolve in the first liquid, and successively higher molecular-weight fractions in subsequent liquids. To ensure rapid equilibration, the polymer must be present as a very thin film. For convenience, the polymer is often applied to a finely divided substrate such as sand or glass beads, which is subsequently used to pack a column. By the use of mixing vessels, a continuous gradient of solvent–non-solvent composition can be produced and eluted through the column.

Thermal-gradient elution

A modification of the solvent gradient method utilises a small temperature gradient from one end to the other of the column, in addition to the gradient of solvent composition. The sample is initially confined to a small zone of the packing at the warm end of the column. Each species undergoes a series of solution and precipitation steps as it progresses down the column.

Analytical Precipitation Techniques

Analytical fractionation techniques are those in which fractions are not isolated, in contrast to the methods discussed above. They may be but are not necessarily carried out on a smaller scale than their preparative counterparts.

Summative fractionation

In this method, a series of polymer solutions is used. In each, part of the polymer is precipitated, the point of fractionation being varied throughout the series. Weight fractions w_x and molecular weights M_x of the precipitates are obtained; the product w_xM_x may be related to the integral of the cumulative distribution curve. Since the two differentiations required to obtain the differential distribution would introduce intolerably large errors into the resulting differential curve, the data are properly analysed to provide only information equivalent to the position of the maximum and breadth of the molecular-weight distribution curve. The method is advantageous for obtaining limited information about the molecular-weight distribution with a minimum of time, preliminary calibration effort, and equipment expense.

Turbidimetric titration

In this technique a precipitant is added slowly to a polymer solution, and the turbidity due to precipitated polymer is measured by the decrease in intensity of a beam of transmitted light or by the increase in intensity of scattered light. Selection of an appropriate solvent–non-solvent system for a given polymer requires much preliminary study, and it is usually not possible to obtain accurate information on molecular-weight distribution from the experimental data. Nonetheless, the method is attractive for control purposes once the proper operating conditions are established.

Effect of Polymer Structure on Solubility-Based Fractionation

Effect of chemical type

It must be re-emphasised that the solubility of polymers is determined primarily by their chemical composition, and in the usual case molecular weight is only a secondary variable. If chemical composition differences exist among the polymer species in a sample, fractionation is likely to occur primarily as a result of these differences and to yield information on them rather than on the distribution of molecular weights. Turbidimetric titration has often been used to explore qualitatively the effects of chemical composition in co-polymer systems.

Effect of chain branching

On the basis of limited information, it appears that branching increases the solubility of high polymers. Thus at any given stage in a solubility-based fractionation process, the branched polymer being separated will consist of a mixture of species, some with low branching and low molecular weight, others more branched but with compensating higher molecular weight, all having the same solubility. Additional information is needed to characterise such a system in terms of either variable separately.

Effect of crystallinity

Precipitation of polymer to a crystalline, rather than an amorphous phase can lead to fractionation by molecular weight (with control of the other variables just mentioned) only if the (depressed) crystalline melting point is a strong function of molecular weight. This is the case only for polymers of rather low molecular weight (say, less than 20,000), but in these cases the technique works rather well for the production of narrow fractions in large amounts.

Treatment of Data

Information about the distribution of molecular weights in a polymer is often the goal of a fractionation experiment. Such information requires not only that the variables mentioned in the previous paragraphs be controlled, but also the measurement of the weight and molecular weight of each fraction. For convenience, intrinsic viscosity is often used for the molecular-weight measurement, but this must be limited to experiments with linear polymers. Despite this fact, the entropy of mixing, calculated from the heat and free energy, was in fair agreement with theory except for the dilute-solution region.

The properties of other systems do not in general conform as well to the predictions of the theory. For example, the anticipated independence of χ_1 of v_2 is usually not observed (Fig. 11.12) (but see the corresponding-state theories described below).

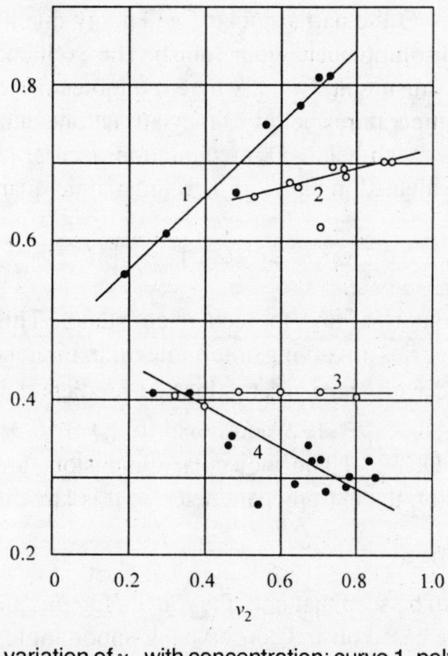

Fig. 11.12. Experimentally observed variation of χ_1 with concentration: curve 1, poly(dimethyl siloxane) in benzene; curve 2, polystyrene in benzene; curve 3, rubber in benzene; and curve 4, polystyrene in toluene.

Dilute Solutions

Flory-Krigbaum theory

The lattice model used in the Flory-Huggins treatment neglects the fact that a very dilute polymer solution must be discontinuous in structure, consisting of domains or clusters of polymer chain segments separated on the average by regions of polymer-free solvent. *Flory* and *Krigbaum* assume such a model in which each cloud of segments is approximately spherical, with a density that is maximum at the centre and decreases in an approximately *Gaussian* function with distance from the centre. Within the volume occupied by the segments of one molecule, all other molecules tend to be excluded. Within such an excluded volume occur long-range intramolecular interactions whose thermodynamic functions can be derived. The partial molar heat content, entropy, and free energy of these interactions are, respectively,

$$\Delta\overline{H}_1 = kT\kappa_1 v_2^2$$

$$\Delta\overline{S}_1 = k\psi_1 v_2^2$$

$$\overline{G}_1 = kT(\kappa_1 - \psi_1)v_2^2 \qquad \text{... (11.96)}$$

where, by comparison with Eq. 11.12,

$$\kappa_1 - \psi_1 = \chi_1 - \frac{1}{2} \qquad \text{... (11.97)}$$

It is convenient to define as a parameter the Flory temperature Θ such that $\Theta = \kappa_1 T_1/\psi_1$. It follows that

$$\psi_1 - \kappa_1 = \psi_1\left(1 - \frac{\Theta}{T}\right) = \tfrac{1}{2} - \chi_1 \qquad \text{... (11.98)}$$

and that at the temperature $T = \Theta$ the partial molar free energy due to polymer-solvent interactions is zero and deviations from ideal solution behaviour vanish. The excluded volume becomes smaller as the solvent becomes poorer and vanishes at $T = \Theta$, where the molecules interpenetrate one another freely with no net interactions. At temperatures below Θ they attract one another and the excluded volume is negative. If the temperature is much below Θ, precipitation occurs.

The parameter α can be evaluated in terms of thermodynamic quantities:

$$\alpha^5 - \alpha^3 = 2C_m\psi_1\left(1 - \frac{\Theta}{T}\right)M^{1/2} \qquad \text{... (11.99)}$$

where C_m lumps together numerical and molecular constants. This equation, which involves the assumption of validity of theories involving intermolecular interactions as well as intramolecular interactions, must be accepted with some reservations. It predicts that α increases without limit with increasing molecular weight; since $\overline{r_0^2}$ is proportional to M, $\overline{r^2}/M$ should increase with M. It also leads to the fact that, at $T = \Theta$, $\alpha = 1$ and molecular dimensions are unperturbed by intramolecular interactions. Since α depends on the entropy parameter ψ, it is large in better solvents.

Corresponding-State Theories

From what has been said, it can be seen that both the *Flory-Huggins* and *Flory-Krigbaum* theories have serious shortcomings. Both are based on and conserve the important features of the theories of 'regular' solutions of small molecules. Only the appropriate combinatorial entropy of mixing has been modified to fit the polymer case. The most important assumption that is retained is that there is no volume change on mixing.

The deviations between theory and experiment for the interaction parameter χ_1 led to its re-interpretation as a combined entropy and enthalpy parameter, as in Eqs. 11.11 and 11.12. The entropy term was considered to be a small negative correction, corresponding to the plausible idea of a small increase in order as polymer–polymer and solvent–solvent interactions were replaced with polymer-solvent interactions. Typical values of ψ_1 and κ_1 show that this is not the case, for the entropic contribution to χ_1 is positive and much larger than the enthalpic term. Still other failures of the traditional theories of the thermodynamics of polymer solutions are already discussed .

These difficulties have been overcome in corresponding-state theories first developed by *Prigogine* and extended and put into practice by Flory, and others. The major new factor in the theories is the recognition of the dissimilarity in the free volumes of the polymer and the solvent as a result of their great difference in size, the usual solvent being much more expanded than the polymer. Mixing is not unlike the condensation of a gas (the solvent) into a dense medium (the polymer). The total volume change on mixing is usually negative and is accompanied with a negative (exothermic) ΔH and a negative contribution to ΔS. Thus the total ΔH on mixing consists of the usual positive interaction term Eq. (11.8) and the new negative term; both contribute to a new interaction parameter like χ_1. Similarly, the entropy of mixing consists of the Flory-Huggins combinatorial term Eq. (11.6) plus the new negative term; only the latter contributes to the new χ. The relative magnitude of these contributions is already discussed.

It can be said, however, that a basic assumption of the free-volume theories is that all liquids, polymers when non-crystalline, and mixtures of these follow the same reduced equation of state. In *Flory's* formulation this can be written as

$$\frac{\tilde{p}\tilde{V}}{\tilde{T}} = \frac{\tilde{V}^{1/3}}{\tilde{V}^{1/3}-1} - \frac{1}{\tilde{V}\tilde{T}} \quad \text{... (11.100)}$$

which simplifies further at low pressures, including atmospheric, where $\tilde{p} \to 0$. Here $\tilde{p}$, $\tilde{V}$, $\tilde{T}$ are the ratios of the real pressure, volume and temperature to reduction parameters that can be evaluated from the thermal coefficient of expansion and the isothermal compressibility. It is possible to calculate reduced parameters of mixtures as well as for the pure components; from the resulting equations of state we can evaluate the volume change on mixing, the heat and entropy of mixing, and such details as the dependence of the new χ on concentration.

It has been found that volume changes on mixing can be predicted to within 10–15 per cent of the experimental values, and the correct variation of χ with concentration is predicted, in contrast to the result of the *Flory-Huggins* theory discussed above.

Scaling concepts

From the previous discussion we know the properties of polymer chains in both dilute and concentrated solutions. Dilute non-overlapping coils in a good solvent are swollen, with a size proportional to $x^{0.60}$. On the other hand, in very concentrated solutions or melts the chains are essentially ideal, with a size proportional to $x^{0.50}$, and they interpenetrate one another very strongly. What has happened in between? The only available theory for a long time was that of *Flory* and *Huggins*, and it is not adequate at low and intermediate concentrations. The solution has now been elucidated both experimentally, through neutron-scattering experiments and theoretically, through the applications of scaling concepts.

The two extreme concentration regions described above are separated by a region of transition in properties at a concentration $c = c^*$ where the chain coils begin to overlap (Fig. 11.13). The scaling properties of c^* are important. It is comparable to the local segment concentration within a single coil:

$$c^* \simeq x / (\bar{r}^2)^{3/2} \sim x^{-0.80} \qquad \text{... (11.101)}$$

In terms of the volume fraction, $v_2^* \sim x^{-0.80}$ and is quite small ($\sim 10^{-3}$) for large x.

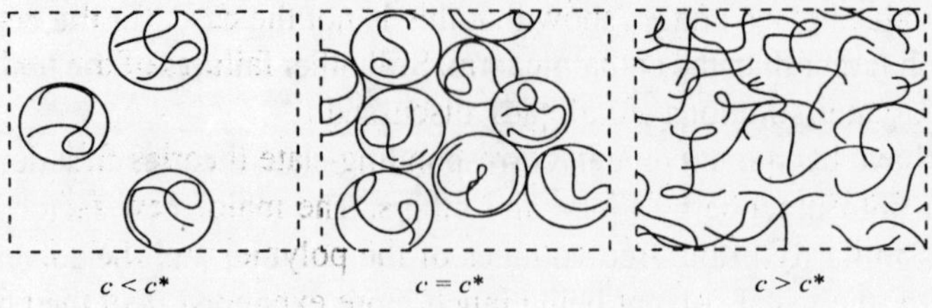

Fig. 11.13. Schematic of chain conformations in the (a) dilute; (b) transition; and (c) semi-dilute concentration regions.

When c is small compared to c^* (the dilute-solution region), *Flory* showed that the equation of state takes the form

$$\frac{\pi}{kT} \simeq \frac{c}{x} + (\text{const.}) \left(\frac{c}{x}\right)^2 (\bar{r}^2)^{3/2} + ... \qquad \text{... (11.102)}$$

which leads by scaling concepts to $A_2 \sim x^{-0.20}$, in good agreement with experiment. In the transition region, where v_2 is in the broad region between v_2^* and 1, application of scaling concepts to the osmotic pressure leads to $\pi/kT \sim v_2^{2.25}$, in contrast to the Flory-Huggins prediction of v_2^2. This has been confirmed experimentally.

CHAPTER 12

Measurement of Molecular Weight and Size of Polymers

INTRODUCTION

A molecule has a group of atoms which have strong bonds among themselves but relatively weak bonds to adjacent molecules. Examples of small molecules are water (H_2O), methanol (CH_3OH), carbon dioxide, and so on. Polymers contain thousands to millions of atoms in a molecule which is large; they are also called macromolecules. Polymers are prepared by joining a large number of small molecules called monomers. Polymers can be thought of as big buildings, and monomers as the bricks that go into them.

Monomers are generally simple organic molecules containing a double bond or a minimum of two active functional groups. The presence of the double bond or active functional groups act as the driving force to add one monomer molecule upon the other repeatedly to make a polymer molecule. This process of transformation of monomer molecules to a polymer molecule is known as polymerisation. For example, ethylene, the prototype monomer molecule, is very reactive because it has a double bond. Under the influence of heat, light, or chemical agents this bond becomes so activated that a chain reaction of self-addition of ethylene molecules is generated, resulting in the production of a high-molecular weight material, almost identical in chemical composition to ethylene, known as polyethylene, the polymer of ethylene (Fig. 12.1).

The difference in behaviour between ordinary organic compounds and polymeric materials is due mainly to the large size and shape of polymer molecules. Common organic materials such as alcohol, ether, chloroform, sugar, and so on, consist of small molecules having molecular weights usually less than 1000. The molecular weights of polymers, on the other hand, vary from 20,000 to hundreds of thousands.

The name polymer is derived from the Greek *poly* for many and *meros* for parts. A polymer molecule consists of a repetition of the unit called a *mer*. Mers are derived from monomers, which, as we have seen for ethylene, can link up or polymerise under certain conditions to form the polymer molecule. The number of mers, or more precisely the number of repetitions of the mer, in a polymer chain is called the degree of polymerisation (DP). Since the minimum length or size of the molecule is not specified, a relatively small molecule composed of only, say, three mers might also be called a polymer. However, the term polymer is generally accepted to imply a molecule of large size (macromolecule). Accordingly, the lower-molecular-weight products with low DP should preferably be called oligomers (oligo = few) to distinguish them from polymers. Often the term high polymer is also used to emphasise that the polymer under consideration is of very high molecular weight.

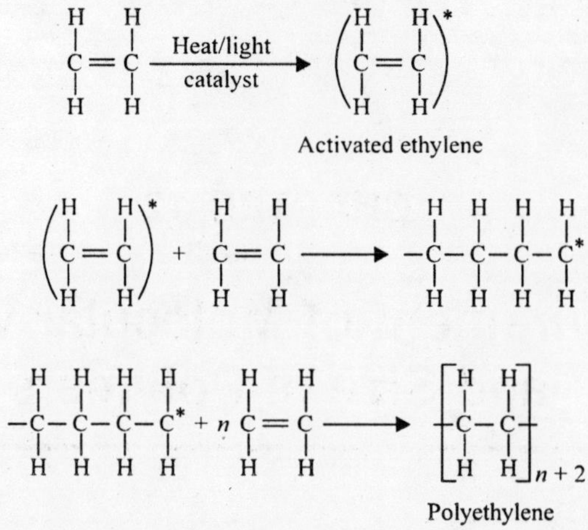

Fig. 12.1. Intermediate steps during formation of polyethylene.

Because of their large molecular size, polymers possess unique chemical and physical properties. These properties begin to appear when the polymer chain is of sufficient length—i.e., when the molecular weight exceeds a threshold value—and becomes more prominent as the size of the molecule increases. The dependence of the softening temperature of polyethylene on the degree of polymerisation is shown in Fig. 12.2a.

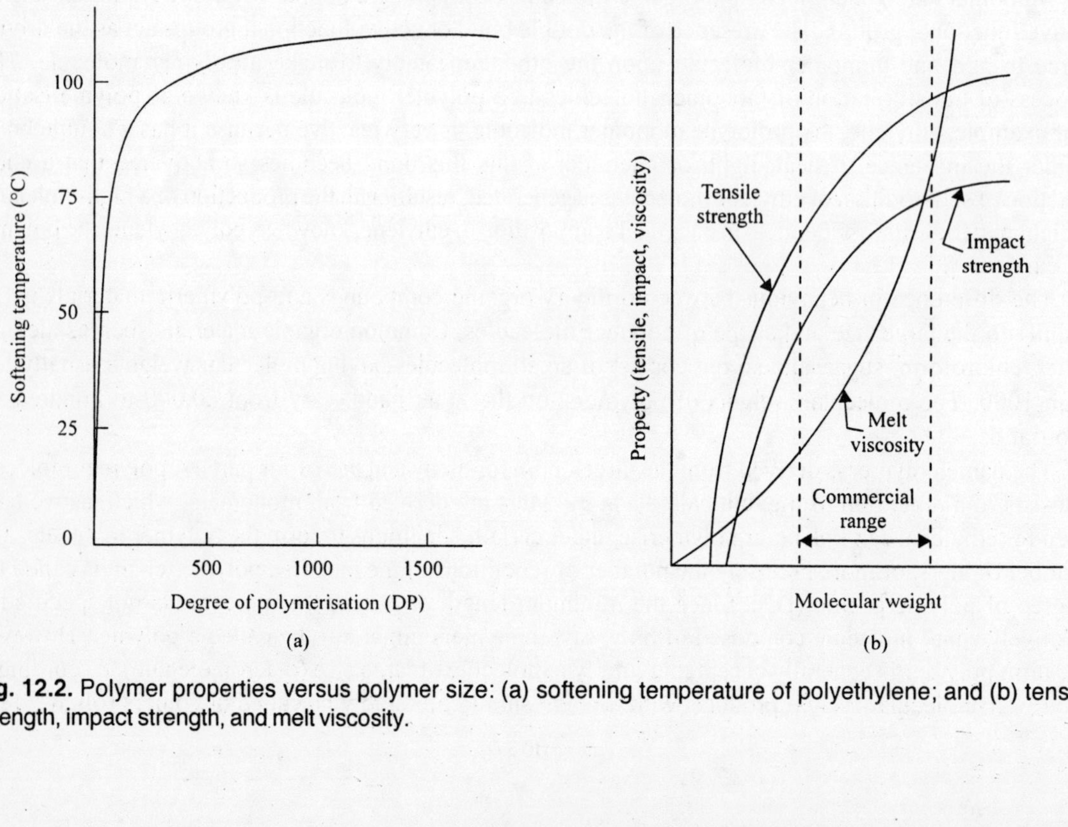

Fig. 12.2. Polymer properties versus polymer size: (a) softening temperature of polyethylene; and (b) tensile strength, impact strength, and melt viscosity.

The dimer of ethylene is a gas, but oligomers with a DP of three or more (that is, C_6 or higher paraffins) are liquids, with the liquid viscosity increasing with the chain length. Polyethylenes with DPs of about 30 are greaselike, and those with DPs around 50 are waxes. As the DP value exceeds 400 or the molecular weight exceeds about 10,000, polyethylenes become hard resins with softening points about 100°C. The increase in softening point with chain length in the higher-molecular-weight range is small. The relationship of such polymer properties as tensile strength, impact strength, and melt viscosity with molecular weight is indicated in Fig. 12.2b. Note that the strength properties increase rapidly first as the chain length increases and then level off, but the melt viscosity continues to increase rapidly. Polymers with very high molecular weights have superior mechanical properties but are difficult to process and fabricate due to their high melt viscosities. The range of molecular weights chosen for commercial polymers represents a compromise between maximum properties and processability.

WEIGHT, SIZE AND DISTRIBUTION OF MOLECULES

Because there is such diversity among polymer molecules, a number of techniques for defining and quantifying these characteristics are in use by the industry—and are also of value to processors and end-users as a determinant of polymer properties.

One such parameter relates to the size of the molecules in the polymer and is known as molecular weight (MW), which refers to the average weight of the molecules in the mixture of different sizes of molecules that make up the polymer. It is expressed either as a number average, based on the sum of the number fractions of the weight of each species or size of molecule present, or as a weight average, based on the weight fractions of each species or size of molecule present in the polymer.

The molecular weight of a polymer has a significant effect on its properties. Thus, higher molecular weight polymers tend to be tougher and more chemically resistant, whereas low molecular weight polymers tend to be weaker and more brittle. In the polyethylene family, for example, low molecular weight polyethylenes are almost waxlike in characteristics, whereas ultra-high molecular weight polyethylenes offer outstanding chemical resistance and toughness (although, conversely, the higher the molecular weight is, the more energy in the form of temperature and pressure required to process the material).

Another expression applicable to molecular weight is the degree of polymerisation (DP). This refers to the number of monomer molecules that combine to form a single polymer molecule and is estimated by dividing the number-average molecular weight of the polymer by the molecular weight of the monomer. The relationship can be expressed as:

$$\text{MW of polymer} = \text{DP} \times \text{MW of monomer}$$

And, finally, it is important to know something about the molecular weight distribution within the polymer, that is, the relative proportions of molecules of different weight. Obviously, if one could create a mono-disperse polymer, all of the molecules would be of a single size. This has not been achieved commercially, however, and so another parameter of definition used to describe polymers is the distribution of the various sizes of molecules within a poly-disperse polymer, that is, the breadth of distribution or ratio of large, medium, and small molecular chains in the resin. If the resin is made up of chains close to the average length, it is called narrow; if it is made up of chains of a wider variety of lengths, it is called broad.

In general, narrow-range resins have better mechanical properties than broad-range resins, although, as with the case of the higher molecular weight materials, they are somewhat more difficult to process.

The molecular weights of polymers and the molecular weight distribution are determined indirectly by measuring the properties of the polymers themselves or of their solutions and correlating this information with the type of molecular weight believed to correspond to the type of property measured.

For example, chemical analysis of the end-groups present in polymer molecules, studies of the boiling points and freezing points of solutions of polymers, and osmotic pressure studies on polymer solutions yield data on the number-average molecular weight of the polymer. Light scattering in polymer solutions and sedimentation methods in the ultra-centrifuge yield data related to the weight-average molecular weights of polymers. Such methods yield the overall average molecular weight of the polymer samples. To get direct data on molecular weight distributions, the polymers first must be separated by fractionation methods into rather sharp fractions of samples of relatively monodisperse molecular weight.

MOLECULAR WEIGHT OF POLYMERS

In ordinary chemical compounds such as sucrose, all molecules are of the same size and therefore have identical molecular weights (M). Such compounds are said to be monodisperse. In contrast, most polymers are polydisperse. Thus a polymer does not contain molecules of the same size and, therefore, does not have a single molecular weight. In fact, a polymer contains a large number of molecules—some big, some small. Thus there exists a variation in molecular size and weight, known as molecular-weight distribution (MWD), in every polymeric system, and this MWD determines to a certain extent the general behaviour of polymers. Since a polymer consists of molecules of different sizes and weights, it is necessary to calculate an average molecular weight ($\overline{M}$) or an average degree of polymerisation DP).

The molecular weights commonly used in the characterisation of a polydisperse polymer are the number average, weight average, and viscosity average. Polymer properties versus polymer related to-softening temperature of polyethylene and tensile strength, impact strength, and melt viscosity (schematic) are given in Fig. 12.2.

Consider a sample of a polydisperse polymer of total weight W in which N = total number of moles; N_i = number of moles of species i (comprising molecules of the same size); n_i = mole fraction of species i; W_i = weight of species i; w_i = weight fraction of species i; M_i = molecular weight of species i; and x_i = degree of polymerisation of species i.

MEASUREMENT OF MOLECULAR WEIGHT AND SIZE

Polymers are non-homogeneous with respect to molecular sizes and individual characteristics of varying lengths, etc. Hence a concept of an average molecular weight is adopted. Each approach to determine the molecular weight is unique in its concept and yields a value which very rarely coincides with those obtained by other methods.

The following are some of the important methods by which molecular weights of polymers are determined:

1. Method based on counting the number of molecules in a known mass. It gives the number-average molecular weight, $\overline{M}_n$.
2. Method based on the weights of various fractions of a polymer when subjected to distillation. It gives the weight-average molecular weight $\overline{M}_w$.
3. Method of obtaining z-average molecular weight. This gives the z-average molecular weight, $\overline{M}_z$.

4. The viscosity measurement method.
5. The ebullioscopic or the elevation of boiling point method and the cryoscopic or the depression of freezing point method.
6. The vapour-pressure method.
7. The osmotic pressure method.
8. The sedimentation velocity method.
9. The sedimentation equilibrium method.
10. The analysis of light scattering or the spectrophotometer method.
11. The gel-chromatographic analysis method.

Number-Average Molecular Weight

The number of moles in a sample of polymer is the sum of all the molecular species present in the sample. Thus the total number of molecules = ΣN_i, in the sample. Similarly, the total weight of the sample is the sum of the weight of each molecular species:

$$W = \Sigma\, w_i = \Sigma\, N_i M_i \qquad \text{... (12.1)}$$

Hence, the number-average molecular weight is defined as

$$W = \Sigma\, w_i / \Sigma\, N_i = \Sigma\, N_i M_i / \Sigma\, N_i \qquad \text{... (12.2)}$$

The accuracy of this method, depends on accuracy with which the values of N_i and M_i are determined. The number of moles present in a sample are usually obtained by the end group analysis method or the osmotic pressure method.

The high molecular weight of polymers is responsible for many of the properties that make polymers valuable as a class of materials. Although there is no sharp dividing line, we can draw an imaginary one at a molecular weight of about 2000, since that is near the limit of convenient purification by distillation or extraction. Also, differences between members of a homologous series differing in steps of 100 or so (the molecular weight of a typical monomer) become so slight as to prevent clean separations.

Some polymers such as polyethylene, poly(ethylene oxide), and dimethyl silicones are commercially available in sizes ranging almost continuously from monomer or dimer up to molecular weights in the millions.

Except for the smaller members of each series, each of these products is a mixture of different-sized molecules having an average molecular weight, which we can define and measure in several ways, and a distribution of molecular weight. Distributions usually have been rather tedious to measure and not as often specified as the average molecular weight. However, new analytical techniques are changing this situation.

If someone told you that he wanted to drop into your hand, from a height of 1 ft, a series of 1000 steel balls with an average diameter of 2.4 inch, you might agree to help him in his playful experiment. After all, a 2.4 inch diameter ball weighs only about 2 lb. However, if he said the average diameter was 23.6 inch, your attitude might be less co-operative. Both numbers could refer to the same series of balls, the difference being in the manner in which the 'average' diameter is calculated. If the population is as shown in Table 12.1, we can calculate an average diameter in several ways:

The average diameter $\overline{D}_L$, based on length (one dimension):

$$\overline{D}_L = \frac{\Sigma\, N_i D_i}{\Sigma\, N_i} = \frac{2400}{1000} = 2.4 \text{ inch} \qquad \text{... (12.3)}$$

Table 12.1. Hypothetical distribution of balls.

Number of balls N_i	*Diameter* D_i, *in*	*Length* N_iD_i	*Area* (× 1/π) $N_iD_i^2$	*Volume* (× 6/π) $N_iD_i^3$
900	1	900	900	900
50	5	250	1250	6250
50	25	1250	31250	781250
$\Sigma N_i = 1000$		$\Sigma N_iD_i = 2400$	$\Sigma = 33400$	$\Sigma = 788400$

The average diameter $\overline{D}_A$, based on area (two dimensions):

$$\overline{D}_A = \frac{\Sigma N_i D_i^2}{\Sigma N_i D_i} = \frac{33400}{2400} = 13.9 \text{ inch} \quad \text{... (12.4)}$$

The average diameter $\overline{D}_V$, based on volume (three dimensions):

$$\overline{D}_V = \frac{\Sigma N_i D_i^3}{\Sigma N_i D_i^2} = \frac{788400}{33400} = 23.6 \text{ inch} \quad \text{... (12.5)}$$

Although $\overline{D}_L$ reflects the preponderant number of small balls, the 1-inch balls represent only about 0.1 per cent of the total volume. $\overline{D}_V$ reflects the importance of the few large balls, which represent 99 per cent of the volume (and weight) of the system. Incidentally, each 25 inch ball weighs about 1 tonne.

Let us now imagine a population of polymers with molecular weights distributed as in Table 12.1, where N is now the number of molecules of molecular weight $M = D$. The quantity we called N_iD_i before is now $N_iM_i = W_i$, the weight of species with molecular weight M_i. We have used before the concept of a number-average molecular weight M_n, where

$$M_n = \frac{\text{Total weight of system}}{\text{Molecules in system}} \quad \text{... (12.6)}$$

In terms of any population, then

$$M_n = \frac{\Sigma N_i M_i}{\Sigma N_i} = \frac{\Sigma W_i}{\Sigma (W_i / M_i)} \quad \text{... (12.7)}$$

As in the case of the steel balls, M_n is very sensitive to the concentration of low-molecular-weight species. The weight-average molecular weight M_w is defined as

$$M_w = \frac{\Sigma N_i M_i^2}{\Sigma N_i M_i} = \frac{\Sigma W_i M_i}{\Sigma W_i} \quad \text{... (12.8)}$$

In correlating such important polymer properties as viscosity or toughness, M_w often is a more useful parameter than M_n. Higher averages are defined as

$$M_z = \frac{\Sigma N_i M_i^3}{\Sigma N_i M_i^2} = \frac{\Sigma W_i M_i^2}{\Sigma W_i M_i} \quad \text{... (12.9)}$$

$$M_{z+1} = \frac{\Sigma N_i M_i^4}{\Sigma N_i M_i^3} = \frac{\Sigma W_i M_i^3}{\Sigma W_i M_i^2} \quad \text{... (12.10)}$$

These are used less often than M_n and M_w. The ratio of M_w to M_n is a measure of the broadness of a distribution, since each is influenced by an opposite end of the population. The quantity M_w/M_n is the polydispersity index.

Z-Average Molecular Weight

It is also determined by the sedimentation equilibrium technique and is defined as

$$\overline{M}_z = \frac{\Sigma W_i M_i^2}{\Sigma W_i M_i} = \frac{\Sigma N_i M_i^3}{\Sigma N_i M_i^2} \quad \text{... (12.11)}$$

The accuracy of this method is dependent on the accuracy of the values of w_i, M_i, etc. For linear condensation polymers,

$$\overline{M}_n : \overline{M}_w : \overline{M}_z = 1 : (1+p) : \frac{1+4p+p^2}{1+p} \quad \text{... (12.12)}$$

where, p is called the extent of reaction. When $p \rightarrow 0$, the three molecular weights approach one another in value and when $p \rightarrow 1$ $\overline{M}_n : \overline{M}_w : \overline{M}_z \rightarrow 1:2:3$. This amply demonstrates that molecular weights of polymers should be referred specifically.

Viscosity-Average Molecular Weight

It is determined by viscosity measurements (for dilute solutions using capillary viscometers such as, Ostwald-Fenske viscometer and for concentrated and viscous solutions rotational viscometers such as Brook-Fields viscometer or a Mooney viscometer) and is to be used in the following equation, assuming that the intrinsic viscosity, and the molecular weight relationship of the sample follows the *Staudinger's* equation,

$[\eta] = KM^a$, where a is constant dependent upon the polymer-solvent system.

$$\overline{M}_v = [\Sigma W_i M_i^a]^{1/2} \left(\frac{\Sigma N_i M_i^{a+1}}{\Sigma N_i M_i} \right)^{\frac{1}{a}} \quad \text{... (12.13)}$$

When a equals unity, the viscosity-average reduces to the weight-average.

Polydispersity Index

The ratio of weight-average molecular weight to number-average molecular weight is called the dispersion or polydispersity index (I). It is a measure of the width of the molecular-weight distribution curve (Fig. 12.3) and is used as such for characterisation purposes. Normally I is between 1.5 and 2.5, but it may range to 15 or greater. The higher the value of I is, the greater is the spread of the molecular-weight distribution of the polymer. For a monodisperse system (e.g., pure chemicals), $I = 1$.

There is usually a molecular size for which a given polymer property will be optimum for a particular application. So a polymer sample containing the greatest number of molecules of that size will have the optimum property. Since samples with the same average molecular weight may possess different molecular-weight distributions, information regarding molecular-weight distribution is necessary for a proper choice of polymer for optimum performance. A variety of fractionation techniques, such as fractional precipitation, precipitation chromatography, and gel permeation chromatography (GPC), based on properties such as solubility and permeability, which vary with molecular weight, may be used for separating polymers of narrow size ranges.

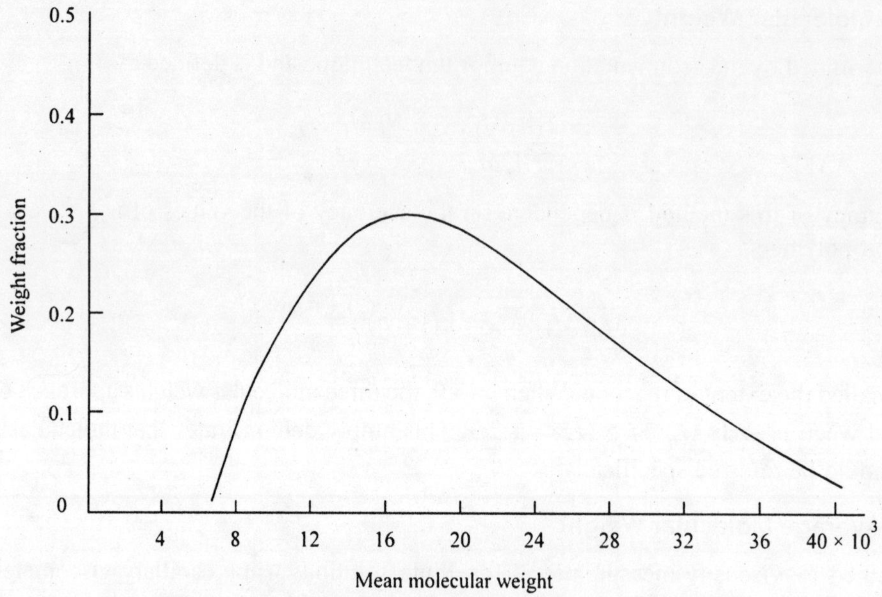

Fig. 12.3. Molecular-weight distribution of a polymer.

Example 12.1. A sample of poly(vinyl chloride) is composed according to the following fractional distribution (Fig. 12.3).

Wt. fraction	0.04	0.23	0.31	0.25	0.13	0.04
Mean mol. wt. × 10^{-3}	7	11	16	23	31	39

(a) Compute $\overline{M}_n, \overline{M}_w, \overline{DP}_n$, and $\overline{DP}_w$.

(b) How many molecules per gram are there in the polymer?

Solution: (a)

Wt. fraction (w_i)	*Mean mol. wt.* (M_i)	$w_i \times M_i$	w_i/M_i
0.04	7000	280	0.57×10^{-5}
0.23	11000	2530	2.09×10^{-5}
0.31	16000	4960	1.94×10^{-5}
0.25	23000	5750	1.09×10^{-5}
0.13	31000	4030	0.42×10^{-5}
0.04	39000	1560	0.10×10^{-5}
Σ		19110	6.21×10^{-5}

From Eq. 12.2

$$\overline{M}_n = \frac{1}{6.21 \times 10^{-5}} = 16100 \text{ g / mole}$$

From Eq. 12.5

$$\overline{M}_w = 19110 \text{ g / mole}$$

1 mer weight of vinyl chloride (C_2H_3Cl)
= (2) (12) + (3) (1) + 35.5 = 62.5 g/mer

$$\overline{DP}n = \frac{16100 \text{ g / mole}}{62.5 \text{ g / mer}} = 258 \text{ mers / mole}$$

$$\overline{DP}w = \frac{19110 \text{ g / mole}}{62.5 \text{ g / mer}} = 306 \text{ mers / mole}$$

(b) Number of molecules per gram

$$= \Sigma \frac{w_i}{M_i} \text{(Avogadro number)}$$

$$= (6.21 \times 10^{-5})\,(6.02 \times 10^{24})$$

$$= 3.74 \times 10^{19} \text{ molecules/g.}$$

Cryoscopic and Ebullioscopic Methods

These two methods are based on the fact that when a substance is completely soluble in a solvent, the lowering of the freezing point or the elevation of the boiling point of the solvent is proportional to the amount and the nature of the substance dissolved in it.

The equation for the freezing point curve or the boiling point line may be written as:

$$\Delta T/T_x T = -(R/L_x) \ln a_s \qquad \text{... (12.14)}$$

where

$\Delta T = (T_{f.s} - T_f)$ or $(T_b - T_{b,s})$, $T_{f,s}$ and T_f are the freezing points of the solvent and the solution, respectively, and T_b and $T_{b,s}$ are the boiling points of the solution and the solvent respectively.

$T_x = T_{f,s}$ or $T_{b,s}$ as the case may be.

R = the gas constant.

L_x = the molar heat of fusion or the molar heat of evaporation of the solvent, as the case may be.

a_s = non-ideality factor or the activity coefficient.

For very dilute solutions (*Raoult's* law is valid)

$$\Delta T = KC = K \frac{W_A}{W_s} \frac{1000}{M_A} \qquad \text{... (12.15)}$$

where

$$K = K_f \text{ or } K_b \qquad \text{... (12.16)}$$

the molar freezing point depression constant of the solvent, or the molar boiling point elevation constant of the solvent, respectively.

$\Delta T = \Delta T_f$ for freezing point depression method and

$= \Delta T_b$ for boiling point elevation method.

C = molality of the solution.

The value of K is calculated from the relationship,

$$K = R[T_s^2 M_s/1000L] \quad \text{... (12.17)}$$

where L = molar heat of fusion or of evaporation of the solvent, as the case may be.

These two methods have some disadvantages when they are used for low molecular weight compounds. They are as follows:

1. The vapour pressure depression is too small compared to the changes in temperatures.
2. The methods are really valid only at the freezing point or the boiling point of the solvent and therefore experiments conducted at room temperatures do not yield dependable data.

Vapour-Pressure Method

Considering the polymer solution as an ideal solution where *Raoult's* law is applicable, the number-average molecular weight $\overline{M}_n$ for component A is given as $[P_s^0 \, W_A \, \overline{M}_s / \Delta P W_s]$, where, P_s^0 = pure component vapour pressure; $\Delta P = P_s^0 - \pi$, with π as the total pressure acting on the solution.

But solutions of polymers are not ideal, and hence incorporating non-ideal aspect in the above relationship gives

$$\overline{M}_n = \frac{W_A M_S}{(1 - a_s) W_S} \quad \text{... (12.18)}$$

where

W_A, W_S = weight fractions of component A and the solvent, respectively.

a_s = factor of non-ideality, i.e., ($P_S = a_s \, P_s^0$)

Osmotic Pressure Method

This method is suitable for high molecular weight materials for two reasons, (i) high molecular weight substances have osmotic pressure in a range convenient to measure with accuracy; and (ii) preparation of semi-permeable membranes is easy for large polymer molecules.

The semi-permeable membrane permits only the solvent molecules (Fig. 12.4) to diffuse through it and not the solute molecules. Thus the polymer solution gets diluted by the solvent. The liquid level in the limb of the solution chamber rises.

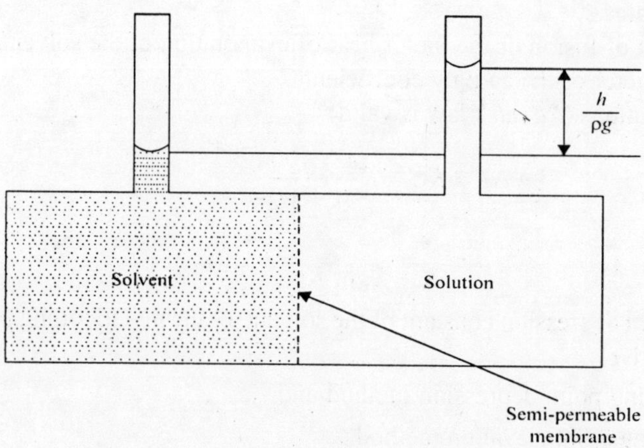

Fig. 12.4. Osmotic pressure measurement.

The increase in entropy of the system by dilution continues until it is countered by the increase in the pressure head. At this point, the chemical potential is the same in both the chambers. Hence, the pressure difference indicated by the levels of the two limbs give the osmotic pressure, π. Measurements are made for several dilutions and a plot of (π/C) versus C is drawn and the same is extrapolated for zero concentration, to give the $\overline{M}_n$ of the polymer as:

$$\lim_{C \to 0}\left(\frac{\pi}{C}\right) = \left(\frac{RT}{\overline{M}_n}\right) \qquad \text{... (12.19)}$$

where C is the polymer concentration wt/vol. The major problem of this process is finding a suitable membrane for a particular system. This method is found suitable for polymers having $\overline{M}_n$ in the range of 50,000 to 10,00,000.

Sedimentation Velocity Method

The principle of this method is *Svedberg's* postulation which states that particles suspended in a liquid, distribute themselves according to their size or mass when put in a centrifugal field. This distribution of particles then has a mathematical relationship with the velocity at which the sedimentation or segregation is taking place. Larger molecules sediment faster than smaller ones. Also the rate of sedimentation in a given field varies directly with the centrifugal force and inversely with the frictional resistance encountered by the molecules. The centrifugal force, in its turn, depends upon the molecular weight of the sample while the frictional resistance depends upon the molecular volume and also the molecular shape. The assumption is that the centrifugal force is countered by the frictional resistance and the molecular diffusion is negligible. For an element at a distance χ from the centre of rotation, moving with an angular velocity ω, the acceleration is $\omega^2\chi$ and the centrifugal force on a gram mole of solute is given as [$\omega^2\chi$ (effective mass of one g mole)] = [$\omega^2\chi M_A(1- \bar{v}_A\rho)$], where, M_A is the partial specific volume of the solute and ρ is the density of the solution. The frictional resistance is given as $f\,(dx/dt)$ where, f is the constant of friction per gram mole of solute and (dx/dt) is the sedimentation velocity.

Therefore,

$$M_A(1-\bar{v}_A\rho)\omega^2\chi = f\left(\frac{dx}{dt}\right) \qquad \text{... (12.20)}$$

This is *Svedberg's* equation where the only unknown is f. When the solute molecules are large and independent of the suspension medium, f can be determined from the relationship, $D_0 = (RT/f)$, where D_0 is the diffusivity at zero concentration of the solute. The rate of sedimentation for unit centrifugal field is given as:

$$\frac{dx/dt}{\omega^2 x} = \frac{m(1-\bar{v}_A\rho)}{\rho} = S_0(\text{say}) \qquad \text{... (12.21)}$$

Hence, M_A is given as

$$M_A = \frac{RT}{1-\bar{v}_A\rho}\frac{S_0}{D_0} \qquad \text{... (12.22)}$$

where the values of S_0 and D_0 have to be obtained separately. This method is not so very suitable for unfractionated polymers.

Sedimentation Equilibrium Method

Here the polymer solution is subjected to a moderate centrifugal field and at intervals of time the changes registered in the concentration of the solution is noted. Unlike in the earlier method, here the diffusion effects are not negligible. Hence, sedimentation and diffusion is considered to be taking place at the same time. The rate of sedimentation caused by the centrifugal force creates a concentration gradient of the solute molecules. But, there would be a natural tendency for the molecules in the higher concentration layer to diffuse back to the lower concentration layer. These two would counter each other and at equilibrium become equal so that

$$C\frac{dx}{dt} = D\frac{dc}{dx} \quad \text{... (12.23)}$$

where, C is the concentration of the solute at a distance x from the centre of rotation. Using the *Svedberg* equation and substituting for f with RT/D_0, we get

$$M_A(1-\bar{v}_A\rho)\omega^2 xC = RT\frac{dc}{dx} \quad \text{... (12.24)}$$

which on integration gives,

$$\overline{M}_A = \frac{2RT\ln(C_2/C_1)}{\omega^2(1-\bar{v}_A\rho_s)(x_2^2-x_1^2)} \quad \text{... (12.25)}$$

This method is satisfactory for unfractionated polymer samples. It gives both $\overline{M}_w$ and $\overline{M}_z$ depending upon whether the variation of concentration is determined by the light absorption method or by the refractive index method, respectively. For determining $\overline{M}_z$, the C_2/C_1 has to be replaced with Z_2X_1/Z_1X_2, where Z is the scale displacement which is proportional to the concentration gradient.

This method may be used for polymers of molecular weights ranging between 5,000 and 500,000.

Analysis of Light Scattering method

When a beam of light is passed through a non-absorbing polymer solution, the intensity of the scattered light is dependent on the size and shape of the polymer molecules.

The light lost due to scattering is related to the incident light as

$$I = I_0 e^{-\tau l} \quad \text{... (12.26)}$$

where, I_0 and I are the intensities of light entering and leaving the cell, respectively; τ is the turbidity of the solution and l is the length of the cell.

τ is related to the change in osmotic pressure with concentration in the *Debye* equation as:

$$\frac{HC}{\tau} = \frac{1}{RT}\frac{d\pi}{dc} \quad \text{... (12.27)}$$

Also

$$\frac{\pi}{c} = \frac{RT}{M} + BC \quad \text{... (12.28)}$$

where

$$B = \frac{RT}{M_S}\frac{\rho_s}{\rho_A}(0.5-\mu) \quad \text{... (12.29)}$$

Therefore,

$$\frac{HC}{\tau} = \frac{1}{M_w} + \frac{2B}{RT}C \quad \text{... (12.30)}$$

where

$$H = \frac{32\pi^3}{3\lambda^4 N_0} n^2 \left(\frac{dn}{dc}\right)^2 \quad \text{... (12.31)}$$

with,

C = concentration

n = R.I. of the solution

N_0 = Avogadro number

λ = wavelength of the light (in vacuum)

π = osmotic pressure

For a series of solutions of different concentrations, the τ and the n values are noted. The molecular weight is obtained by plotting (HC/τ) against C and extrapolating to zero concentration. It would give the value of $1/M_w$ and hence the M_w.

There are some advantages as well as some disadvantages of this method.

Advantages

1. It gives the width of the distribution curve.
2. Precision of the method increases with increase in the molecular weight of the sample.
3. It is very useful for low concentrations where polymer solutions approach ideal behaviour.

Disadvantages

1. Impurities or dust particles or coarser particles give errors to light scattering measurements.
2. The method is not very good for low molecular weight materials.
3. Colour in the polymer or in the solvent may affect the accuracy of the method.

Gel-Permeation Chromatography Method

The method (Fig. 12.5) is also called the size exclusion chromatography method. In this method, a chromatographic column is filled with the gel or the porous solid beads which are rigid and which have pore sizes similar to that of the polymer molecules. The usual materials for the gel are glass for polar molecules and highly cross-linked polystyrene for non-polar molecules.

A solvent stream is allowed to flow through the column. To this is introduced a dilute solution of the polymer. The molecules smaller than the pore size of the gel or beads will enter the beads while the larger ones will pass on. Thus, the larger molecules will have a shorter retention time and the smaller ones will have greater retention time. The retention time or volume may be plotted against the amount of eluted molecules to give a typical chromatogram.

For the gel-permeation chromatography (GPC), no theoretical relationship has been worked out relating the molecular weight and the retention time or the retention volume. Hence, the retention volumes are used for making calibration plots relating molecular sizes and hence molecular weights. Thus, from the chromatogram, knowing a particular retention volume, the corresponding molecular weight can be obtained from the calibration plot (Fig. 12.6).

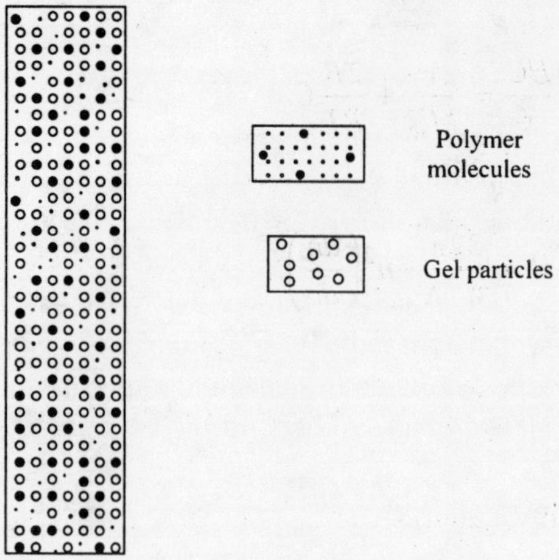

Fig. 12.5. A GPC column.

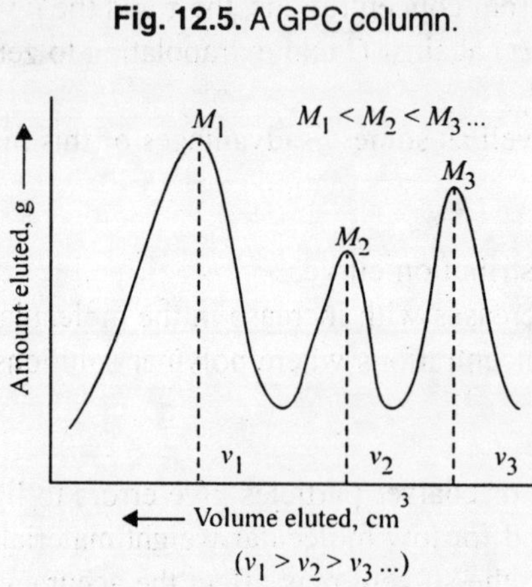

Fig. 12.6. A standard chromatogram.

A standard plot is made with the amount of polymer eluted versus the volume eluted (even retention volume or retention time may be plotted instead). The peak values for the samples of known molecular weights are noted from this plot. Next, the log M versus v plot is made, which gives the calibration plot (Fig. 12.7) from which corresponding to any experimentally determined value of v, the molecular weight of the polymer M, may be obtained.

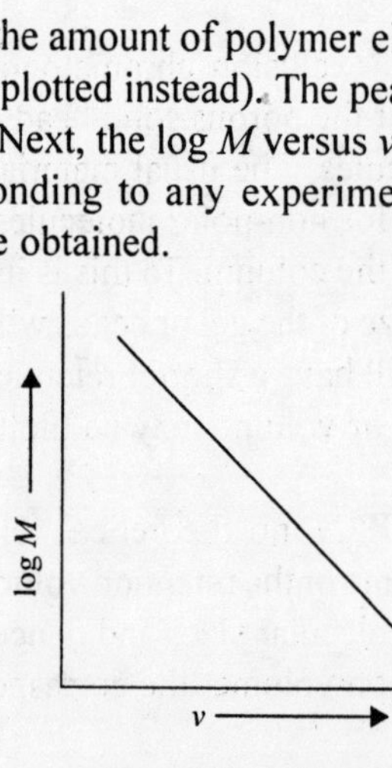

Fig. 12.7. Calibration plot.

Alternatively, a plot of cumulative weight per cent versus molecular weight may be made from which any molecular weight corresponding to a particular cumulative weight per cent may be read. A typical plot appears as shown in Fig. 12.8.

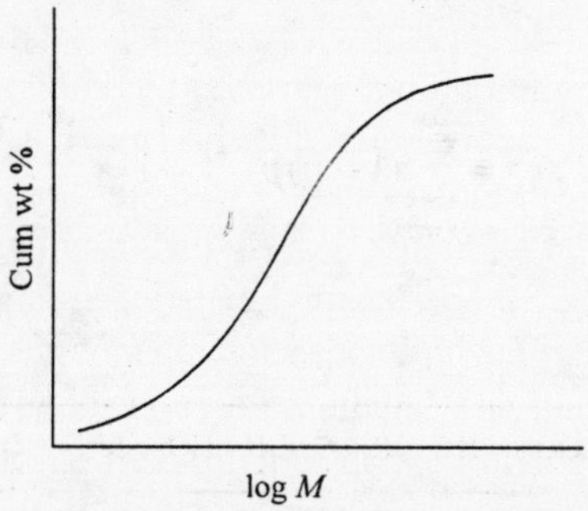

Fig. 12.8. Cumulative weight versus molecular weight.

THEORETICAL DISTRIBUTIONS

With the possible exception of some naturally occurring proteins, all polymers are mixtures of many molecular weights ('polydisperse'). This is a consequence of the random nature of polymerisation reactions. For example, take the case analysed by *Flory* in which the monomer is an ω-hydroxy acid.

$$HO{-}CH_2{-}CH_2{-}CH_2{-}CH_2{-}CH_2\overset{\overset{\large O}{\|}}{C}{-}OH$$

$$2HO{-}R{-}\overset{\overset{\large O}{\|}}{C}{-}OH \rightarrow HO{-}R{-}\overset{\overset{\large O}{\|}}{C}{-}O{-}R{-}\overset{\overset{\large O}{\|}}{C}{-}OH + H_2O \text{ etc.}$$

(Dimer)

This is a typical stepwise polymerisation in which each successively higher molecular weight, dimer, trimer, etc., is just as reactive as monomer. An important parameter is the fraction converted, p. If we have a population of such monomers, each can react to form an ester group. Each esterification eliminates one carboxyl and one hydroxyl group. With five monomers (Fig. 12.9) there would be a maximum of five esterifications ($p = 1$). Initially, p is 0. After one reaction, $p = 0.20$. Abbreviating the unreacted sites by the symbol U and the esterified sites by the symbol E, we can generate a distribution of molecular weights by random reaction of 50 out of 100 sites ($p = 0.5$). If we had alternated sites of reaction with unreacted sites, we would have generated 50 dimers and no other species. Another possible extreme would be to generate 49 monomers and 1 polymer molecule with a degree of polymerisation x of 51. We have 50 molecules in either case. With random reaction (Fig. 12.9), monomers predominate in number of molecules N_x, although not necessarily in weight xN_x. If our sample population were much larger, as it is in any macroscopic polymerisation, we could assign a probability $(P_r)_x$ that a monomer unit selected at random is a part of a polymer with degree of polymerisation x. We expect this to be a function of p, which is the probability that any particular reactive site has been esterified. The probability that monomer unit is still unreacted varies from 1 when $p = 0$ to 0 when $p = 1$. We surmise that

$$(P_r)_1 = 1 - p \qquad \text{... (12.32)}$$

The likelihood that a unit has become part of a dimer is p time $(P_r)_1$. Each additional unit is less likely, since the cumulative probability is represented by a product of the probabilities of each step. Thus we see that

$$(P_r)_x = (1 - p)p^{x-1} \qquad \text{... (12.33)}$$

Also,

$$\sum_{1}^{\infty}(P_r)_x = \sum_{x=1}^{\infty}(1-p)p^{x-1} = 1 \qquad \text{... (12.34)}$$

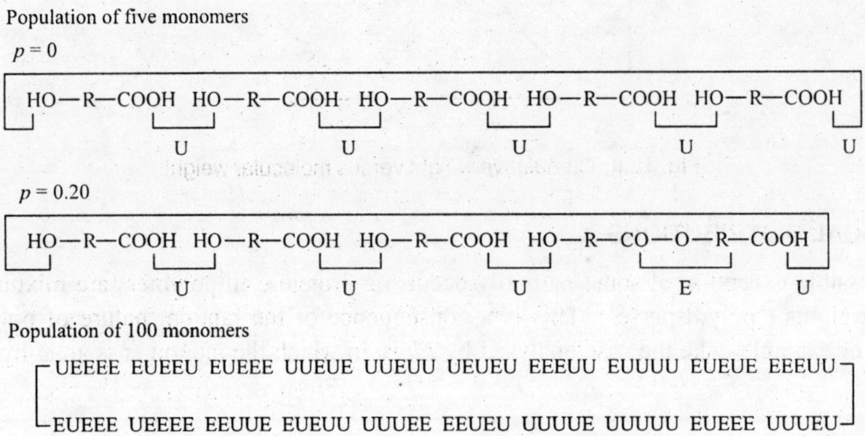

Summary:

	x	N_x	xN_x
Monomer (—UU—)	1	25	25
Dimer (—UEU—)	2	15	30
Trimer (—UEEU—)	3	2	6
Tetramer (—UEEEU—)	4	4	16
Pentamer (—UEEEEU—)	5	2	10
Hexamer (—UEEEEEU—)	6	1	6
Heptamer (—UEEEEEEU—)	7	1	7
		50	100

Fig. 12.9. Generations of molecular weight distribution by random reaction.

Now we can call $(P_r)_x$ the mole fraction of x-mer, since it is a normalised probability (i.e., the summation over all values of x is unity). According to Eq. 12.33, the degree of polymerisation with the largest mole fraction always is 1, namely, monomer. Keep in mind that we are dealing with a particular polymerisation scheme in which all species are at equilibrium. The number-average degree of polymerisation $\bar{x}_n$ for a distribution represented by Eq. 12.33 is

$$\bar{x}_n = \frac{\text{Number of monomer units originally charged}}{\text{Number of molecules in system}} = \frac{1}{1-p} \qquad \text{... (12.35)}$$

Disregarding the contribution to molecular weight made by the end groups, the molecular weight of a polymer M is related to the degree of polymerisation x by

$$M = M_0 x \qquad \text{... (12.36)}$$

where M_0 is the molecular weight of monomer unit in the chain. We can write Eq. 12.33 for x and for $\bar{x}_n$ and combine the two to give

$$\ln \frac{(P_r)_x}{(P_r)_{xn}} = (x - \bar{x}_n) \ln p \quad \text{... (12.37)}$$

But $-\ln p \cong 1 - p = 1/\bar{x}_n$, so that

$$\ln \frac{(P_r)_x}{(P_r)_{xn}} \cong 1 - \frac{x}{\bar{x}_n} \quad \text{... (12.38)}$$

This 'normalised' equation for mole fraction as a function of degree of polymerisation is shown in Fig. 12.10.

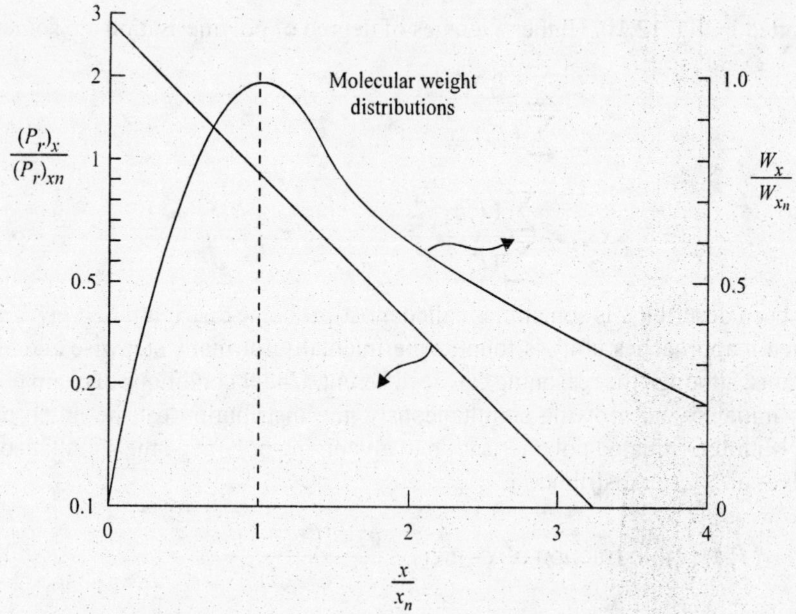

Fig. 12.10. Normalised distributions according to Eqs. 12.38 and 12.39.

A second kind of average degree of polymerisation can be defined as

$$\bar{x}_w = \frac{\sum W_x x}{\sum W_x} \qquad x = 1, 2, 3, \ldots \quad \text{... (12.39)}$$

where W_x is proportional to the weight of x-mer and $\bar{x}_w$ is the weight-average degree of polymerisation. The weight of x-mer from Eq. 12.33 is

$$W_x = x(1-p)p^{x-1} M_0 \quad \text{and} \quad \sum W_x = \frac{M_0}{1-p} \quad \text{... (12.40)}$$

Thus we have

$$\bar{x}_w = \frac{\sum (P_r)_x x^2}{\sum (P_r)_x x} = \frac{\sum x^2(1-p)p^{x-1}}{\sum x(1-p)p^{x-1}} = \frac{1+p}{1-p} \quad \text{... (12.41)}$$

The weight fraction of x-mer w_x is given by

$$w_x = \frac{W_x}{\sum W_x} = x(1-p)^2\, p^{x-1} \quad \text{... (12.42)}$$

As in Eq. 12.38, we can have a 'normalised' distribution:

$$\ln \frac{w_x}{w_{x_n}} = \ln \frac{x}{x_n} + (x - x_n)\ln p \cong \ln \frac{x}{x_n} - \frac{x}{x_n} + 1 \quad \text{... (12.43)}$$

This also is plotted in Fig. 12.10. Higher averages of degree of polymerisation are sometimes useful:

$$\bar{x}_z = \frac{\sum (P_r)_x x^3}{\sum (P_r)_x x^2} \quad \text{... (12.44)}$$

$$\bar{x}_{z+1} = \frac{\sum (P_r)_x x^4}{\sum (P_r)_x x^3} \quad \text{... (12.45)}$$

What we have been describing is sometimes called most-probable distribution. It gives $x_w/x_n = 1 + p$. The ratio of 2 when p approaches unity is found experimentally for many stepwise and also for many chain polymerisations, although the reasoning can be different. Under conditions of chain polymerisation with all polymers initiated and growing simultaneously, the distribution can be much narrower. For example, in a butyl-lithium-initiated polymerisation to give living polymers, the simultaneous growth of polymer chains gives a *Poisson* distribution.

$$(P_r)_x(\text{mole fraction of } x\text{ - mer}) = \frac{e^{1-x_n}(x_n - 1)^{x-1}}{(x-1)!} \quad \text{... (12.46)}$$

or

$$\ln (P_r)_x = 1 - x_n + (x-1)\ln (x_n - 1) - \ln [(x-1)!] \quad \text{... (12.47)}$$

$$w_x(\text{weight fraction of } x\text{ - mer}) = \frac{x(P_r)_x}{x_n} \quad \text{... (12.48)}$$

$$\frac{x_w}{x_n} = 1 + \frac{x_n - 1}{x_n^2} \quad \text{... (12.49)}$$

It can be seen that even when x_n is small, say about 10, x_w/x_n is nearly unity. Polymers can be produced on a commercial scale with

$$\frac{M_w}{M_n} = \frac{x_w}{x_n} < 1.1$$

by using butyl lithium as the initiator.

We can further illustrate the differences in the various averages by examining the sample of Fig. 12.9 (Table 12.2). In this example, monomer (x =1) has the highest mole fraction but not the highest relative weight (W_x). The average degrees of polymerisation are also calculated according to the equations presented. As mentioned before, the usefulness of each of these numbers may depend on the particular property we are looking at. The reactivity of a urethane pre-polymer may depend entirely on x_n. On the other hand, the viscosity of a polymer melt often varies more directly with x_w.

EMPIRICAL DISTRIBUTION MODELS

The distribution of molecular weights has been shown to be important in many diverse applications including flow of melts and solutions, ageing and weathering behaviour, adhesion, and flocculation. Because of the difficulties involved in measuring a distribution in detail, one shortcut has been to postulate a reasonable mathematical form or model for the distribution and then evaluate the parameters from x_n and x_w.

Table 12.2. Analysis of distribution from Fig. 12.9.

Degree of polymerisation x	*Number of moles* N_x	*Mole fraction* $P_r = N_x/\Sigma N_x$	*Relative number of monomer units in x-mer* $W_x/M_0 = x(P_r)_z$	*Weight fraction* $w_x = W_x/\Sigma W_x$	xw_x	x^2w_x
1	25	0.50	0.50	0.25	0.25	0.25
2	15	0.30	0.60	0.30	0.60	1.20
3	2	0.04	0.12	0.06	0.18	0.54
4.	4	0.08	0.32	0.16	0.64	2.56
5	2	0.04	0.20	0.10	0.50	2.50
6	1	0.02	0.12	0.06	0.36	2.16
7	1	0.02	0.14	0.07	0.49	3.43
	50	1.00	2.00	1.00	3.02	12.64

$$x_n = \frac{\sum x(P_r)_x}{\sum (P_r)_x} = \frac{2.00}{1.00} = 2.00$$

$$x_w = \frac{\sum x^2(P_r)_x}{\sum x(P_r)_x} \quad \text{or} \quad \frac{\sum xw_x}{\sum w_x} = \frac{3.02}{1.00} = 3.02$$

$$x_z = \frac{\sum x^3(P_r)_x}{\sum x^2(P_r)_x} \quad \text{or} \quad \frac{\sum x^2w_x}{\sum xw_x} = \frac{12.64}{3.02} = 4.19$$

In general, such a model would give w_x, the weight fraction of x-mer, as a function of x, x_w, x_n, and perhaps other parameters. The 'most-probable' distribution, Eq. 12.44, has only one parameter p, which can be evaluated from either x_n (Eq. 12.37) or x_w (Eq. 12.43). If the experimentally determined ratio x_w/x_n is not nearly 2 for high-molecular-weight material, the model cannot be used.

The *Schulz* distribution is a more general form of the most-probable:

$$w_x = \frac{a}{x_n \Gamma(a+1)} \left(\frac{ax}{x_n} \right)^a \exp\left(-\frac{ax}{x_n} \right) \quad \text{... (12.50)}$$

where $x_w/x_n = (a+1)/a$ and $\Gamma(a+1)$ is the gamma function of $a + 1$. When $a = 1$ and x is large, this becomes the same as the most-probable since $x_n = 1/(1-p)$ and $(x-1)\ln p \cong -x/x_n$. However, any value of x_w/x_n can be inserted. Qualitatively, the curves will resemble Fig. 12.10 except that the overall breadth will reflect the influence of the polydispersity index x_w/x_n.

It will be seen in the next section that some experimental methods of measuring distribution results in a cumulative distribution function $W(x)$, from which the differential distribution $w_x = dW/dx$ can be derived. We are speaking now of distributions in the higher ranges of x_n, where the approximation $\sum w_x = \int w_x \, dx$ is applicable. It has been observed that a plot of $W(x)$ against x on logarithmic probability paper often gives a straight line. In terms of w_x this becomes the *Wesslau* distribution:

$$w_x = \frac{1}{\beta \pi^{1/2}} \frac{1}{x} \exp\left(-\frac{1}{\beta^2} \ln^2 \frac{x}{x_0} \right) \quad \text{... (12.51)}$$

where $\beta^2 = \ln (x_w/x_n)^2$ and $x_0 = x_n \exp(\beta^2/4)$. Once again, specification of x_w and x_n defines the entire distribution. As with the *Schulz* model, only the central value of the distribution, reflected by x_n, and the breadth, measured by x_w/x_n, can be changed. On a cumulative weight $[W(x)]$ basis, the *Schulz* and *Wesslau* models give similar curves (Fig. 12.11). On a differential (w_x) basis, qualitative differences become apparent (Fig. 12.12).

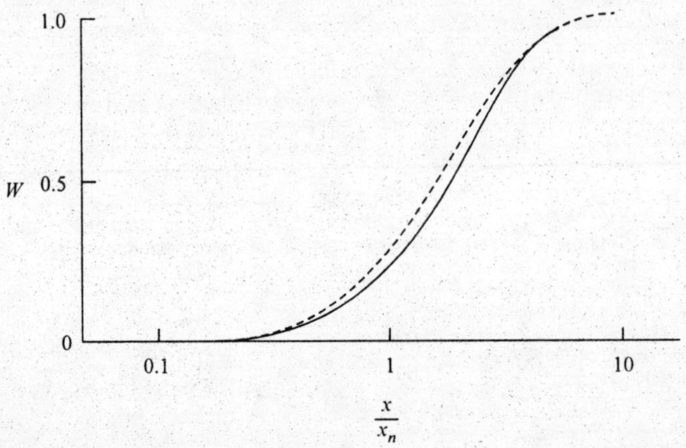

Fig. 12.11. Cumulative weight fraction *W* as a function of ratio of degree of polymerisation to number-average value for *Schulz* (solid line), and *Wesslau* (dashed line) models.

When dealing with high molecular weights ($x > 100$), it becomes convenient to treat distributions as being continuous rather than finite. As an example, let us assume that the cumulative weight fraction $W(x)$ of a polymer with degree of polymerisation less than x is given by

$$\begin{aligned} W(x) &= 0.01\,(x-10) && \text{for } 10 < x < 110 \\ W(x) &= 0 && \text{for } x < 10 \\ W(x) &= 1 && \text{for } x > 110 \end{aligned} \quad \text{... (12.52)}$$

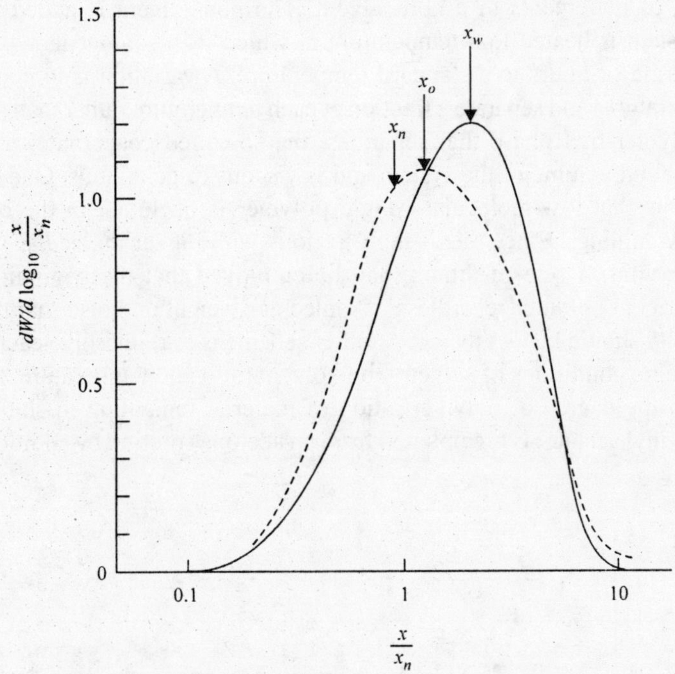

Fig. 12.12. Differential distribution curves for models of Fig. 12.11.

The differential weight distribution w_x then is

$$w_x = \frac{dW(x)}{dx} = 0.01 \quad \text{... (12.53)}$$

The degree of polymerisation is obtained by integration and normalisation.

$$\bar{x}_n = \frac{\int_1^{\infty} w_x dx}{\int_1^{\infty} w_x / x dx} = \frac{\int_{10}^{110} 0.01 dx}{\int_{10}^{110} 0.01 / x dx} = 41.7 \quad \text{... (12.54)}$$

$$\bar{x}_w = \frac{\int_1^{\infty} x w_x dx}{\int_1^{\infty} w_x dx} = \frac{\int_{10}^{110} 0.01 x dx}{\int_{10}^{110} 0.01 dx} = 60.0 \quad \text{... (12.55)}$$

The continuous models are particularly useful when analysing experimental data on distributions where a cumulative weight is obtained first and must be converted mathematically to a differential distribution.

MEASUREMENT OF DISTRIBUTION

Both x_n and x_w can be measured independently in order to get some idea of polydispersity. The ultracentrifuge, when applicable, gives a distribution directly. However, until the advent of gel permeation chromatography, fractionation of polymer from dilute solution was used. Usually there is a decrease in solubility with increasing molecular weight. Polymer can be put in dilute solution and progressively precipitated by addition of increments of a non-solvent. The non-solvent is added to give a visible precipitate. Then the system is heated to a temperature at which it is homogeneous, after which it is slowly cooled over a period of hours to a standard temperature. A variation is to cool the solution to successively lower temperatures and separate a fraction at each temperature rather than at each increment of non-solvent. The polymer-rich phase that separates, the so-called coacervate, usually is a small volume compared to the total volume of the system and is viscous or gelatinous. One drawback of this process of coacervation is that low-molecular-weight polymer is occluded in the coaceravate even under conditions of slow cooling. Since this is more serious with higher molecular weights and with higher starting concentrations, a rule of thumb may be employed that the beginning concentration should be less than 1 per cent polymer regardless of molecular weight and also that the concentration times the intrinsic viscosity should be less than 1. Another scheme is to re-dissolve each fraction and re-precipitate it under the some conditions of solvent-non-solvent ratio and temperature. Fig. 12.13 shows the relationship between the degree of polymerisation of material remaining in solution and solvent composition when poly(vinyl acetate) is precipitated from an acetone solution by addition of a methanol-water mixture.

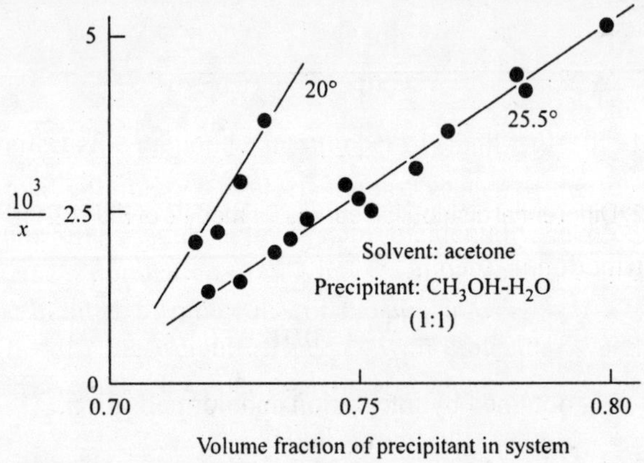

Fig. 12.13. Decrease in solubility of poly(vinyl acetate) with added non-solvent, increasing degree of polymerisation x, and decreasing temperature.

The procedure is tedious, because hours or days are required for clean separation of each fraction. Even then, each fraction is heterogeneous with x_w/x_n = 1.05 to 1.2 at best. Rather than precipitate the polymer from a dilute solution, it is possible to dissolve a solid polymer fractionally. Diffusion through the solid is slow, so it is essential that the material be spread out in a very thin film. A typical technique for fractionating polyethylene uses a 2-inch diameter column about 2 ft long packed with ordinary sand (40 to 200 mesh). A hot solution of polyethylene in xylene is introduced and cooled so that the entire charge of polymer, a few grams, is precipitated on the surface of the sand, giving a thin film. Then a solvent or a solvent mixture is added to elutriate the low-molecular-weight portion of the polymer. When no more polymer is elutriated at that solvent composition or temperature, the solvent-to-non-solvent ratio is increased or the temperature is raised so that a higher molecular weight fraction can be extracted. Columns of this type can be automated with a few solenoid-operated valves and a fraction collector in order that fractionation of a sample into 10 to 20 fractions can be carried out overnight without immediate supervision. In this column method one does not have to wait for gravitational settling out of coacervate, and it is the speed with which fractionation can proceed that is its advantage.

Both in precipitation and dissolving, a single equilibrium stage of separation is achieved. Multistaged separations are not usually convenient. In either case, the raw data consists of a series of *i* fractions each of weight w_i and an average molecular size $\bar{x}_i$. A cumulative curve can then be plotted in which the ordinate is $\Sigma\, w_{i-1} + w_i/2$ and the abscissa is $\bar{x}_i$. A smooth line through the points gives a curve that is a good approximation to $W(x)$, the cumulative molecular weight distribution function. Differentiation of this curve then gives $w_x = dW(x)/dx$, the differential molecular weight distribution function. The reason that the raw values of w_i and x_i cannot be plotted as a histogram is that the intervals of neither variable are uniform. An example (Table 12.3), (Fig. 12.14) shows the long high-molecular-weight tail typical of most polymers. The uncertainty in the slope at the high end also is apparent. In Fig. 12.15 the advantage of a logarithmic plot in showing greater symmetry is seen.

Table 12.3. Analysis of typical fractionation data.

Raw experimental data			*Calculated results*		
Fraction no. *i*	Fraction of total weight recovered w_i	Degree of polymerisation $x_i (\times 10^{-3})$	$\Sigma\, w_i$	$\Sigma\, w_{i-1} + w_i/2$ W_i	$(dW/dx)_i$ $(\times 10^3)$
1	0.050	1.1	0.050	0.025	0.040
2	0.100	2.5	0.150	0.100	0.060
3	0.120	4.1	0.270	0.210	0.063
4	0.083	5.8	0.353	0.313	0.052
5	0.075	7.4	0.428	0.390	0.043
6	0.110	9.4	0.538	0.483	0.039
7	0.120	12.5	0.658	0.598	0.030
8	0.130	17.5	0.788	0.723	0.0165
9	0.130	29.0	0.918	0.853	0.0072
10	0.082	64.0	1.000	0.959	0.0013

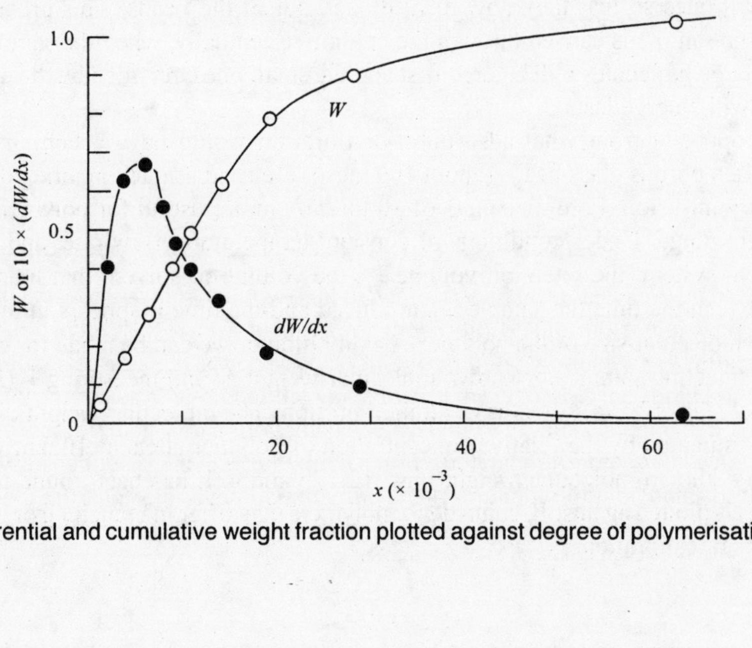

Fig. 12.14. Differential and cumulative weight fraction plotted against degree of polymerisation for distribution of Table 12.3.

A packed column technique is now popular under the names 'gel filtration' and 'gel permeation chromatography' (GPC). The first name was applied by *Flodin* to the separation obtained when water-soluble polymers are eluted from a column packed with swollen, cross-linked, hydrophilic dextran (Sephadex) beads. The second name was attached by *Moore* to the separation of organic soluble polymers in columns packed with swollen, cross-linked, lyophilic polystyrene beads. Both names are misleading, since the gel is not necessary, porous glass beads being equally applicable; filtration is not part of the process, small molecules being retarded and large ones passed rather than the other way around; and chromatography classically implies formation of colours, which seldom occurs in either technique. The terms liquid exclusion chromatography has been recommended.

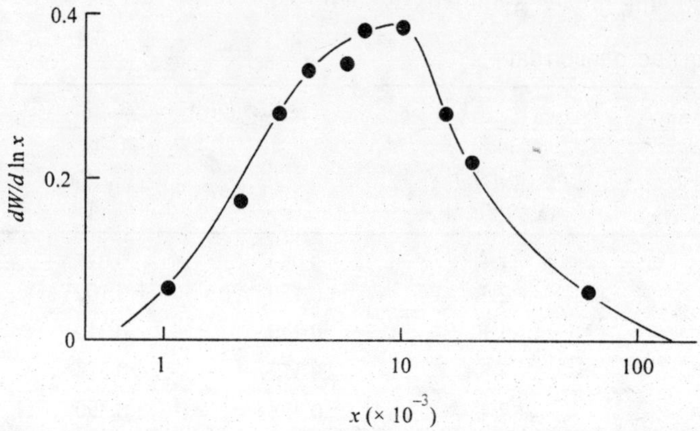

Fig. 12.15. Data of Table 12.3 plotted as differential distribution on a logarithmic basis.

The principle of the method is illustrated in Fig. 12.16, where a mixture of small and large molecules has been deposited at one end of a column packed with porous beads. Initially there is a concentration gradient causing diffusion of polymer into the bead. However, the large molecules cannot penetrate the beads. A continuous flow of solvent sweeps the large molecules along and reverses the concentration gradient for the small ones so that they now diffuse back out of the beads. This process is repeated again and again as the sample is carried through the column. Eventually, when the sample is elutriated from the other end, large molecules will emerge first and the small ones, retarded by the diffusion in and out of the packing, will emerge last.

This is just the opposite from what adsorption or filtration would have accomplished as in gas chromatography. When porous glass beads (about 100 mesh) are used as packing, the pore sizes can be measured by mercury intrusion. Pore diameters of 10 to 250 nm are useful for polymers in the 10^3 to 10^7 molecular weight range. Under conditions of constant temperature, flow rate, and concentration, for the same physical system, the retention volume V_r, the volume of solvent that must be elutriated from the system between the time the sample is introduced and the time it appears in the effluent, is a function only of the molecular size of the polymer. A calibration curve can be made for each individual polymer species by injecting samples of known molecular weight M and measuring $V_r(M)$ for each. A semi-logarithmic plot (Fig. 12.17a) often has a straight portion. The molecular weight distribution of a new sample can be estimated by the relative concentration of polymer at each effluent volume, which now corresponds to a known molecular weight Figs. 12.17b and c. It has been found that calibration curves in which $M[\eta]$ is plotted against V_r coincide for polymers that differ in branching or in composition, thus giving a 'universal' calibration.

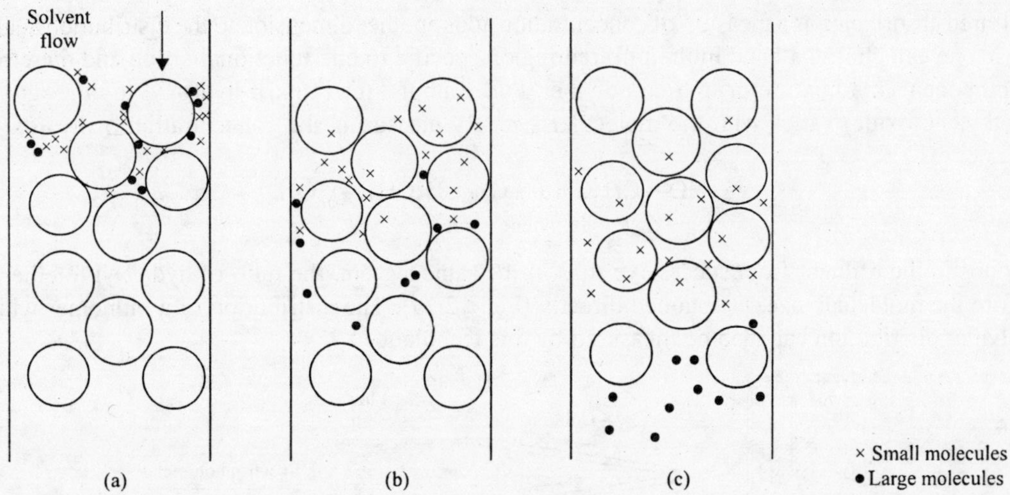

Fig. 12.16. Process of gel permeation chromatography: (a) sample injection; (b) elution; and (c) continued elution.

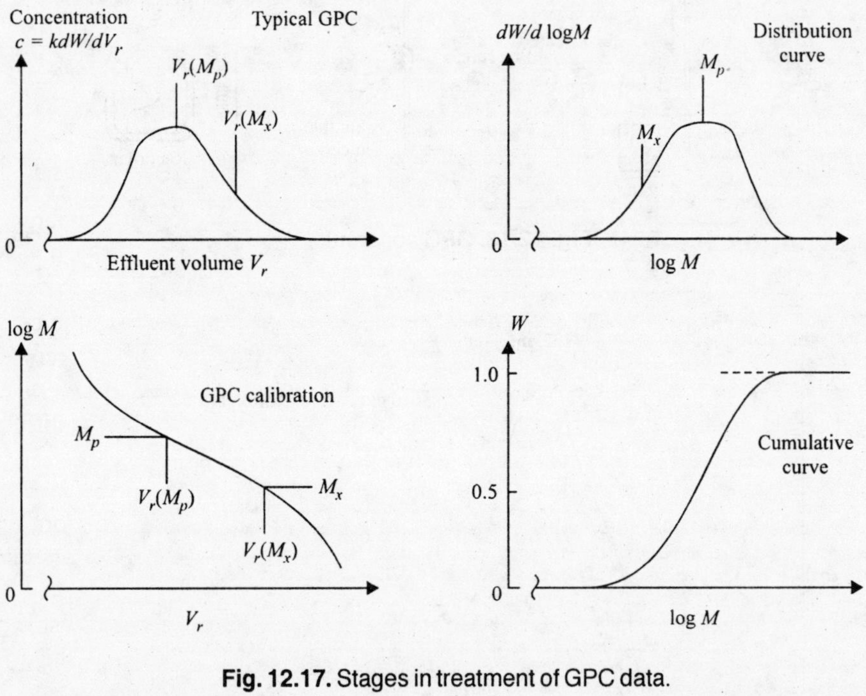

Fig. 12.17. Stages in treatment of GPC data.

A typical system is shown in Fig. 12.18. The packed column may be anywhere from 1 to 20 ft. long and packed with swollen beads, which leave about 30 to 50 per cent void space. Solvent flows continuously by gravity or constant-volume pumping through the column and then to an analyser. Samples can be collected and concentration analysed by various means, or a continuous-flow cell may be used to measure some property that can be correlated with concentration, such as conductivity, radioactivity, light absorbance, or refractive index.

Infrared absorbance as a measure of concentration adds another dimension to the distribution analysis, because one can choose a band in the infrared region specific to one functional group and therefore to one component in a mixture or in a co-polymer. For example, the ratio of hydroxyl to ether group in poly(ethylene oxide) varies with the molecular size. By measuring the concentration of hydroxy and ether

$$HO—CH_2CH_2(—O—CH_2CH_2—)_n\,OH$$

Poly(ethylene oxide)

groups in the effluent from successive runs in the same system, the ratio of hydroxyl to ether, and therefore the molecular size, is obtained directly (Fig. 12.19). The distribution of a monomer within a co-polymer distribution can also be measured by this technique.

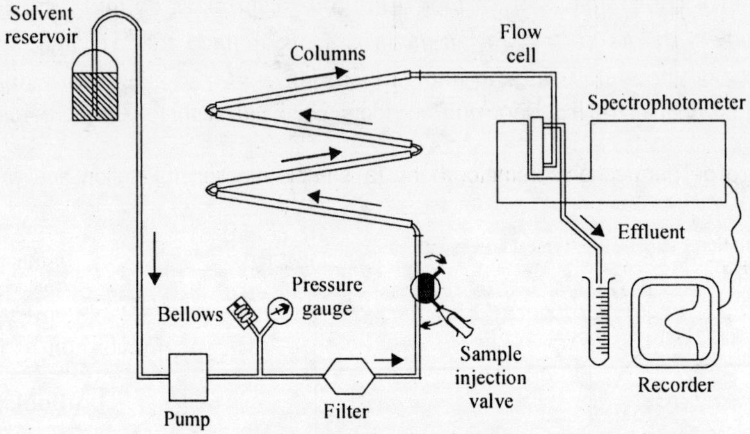

Fig. 12.18. GPC apparatus.

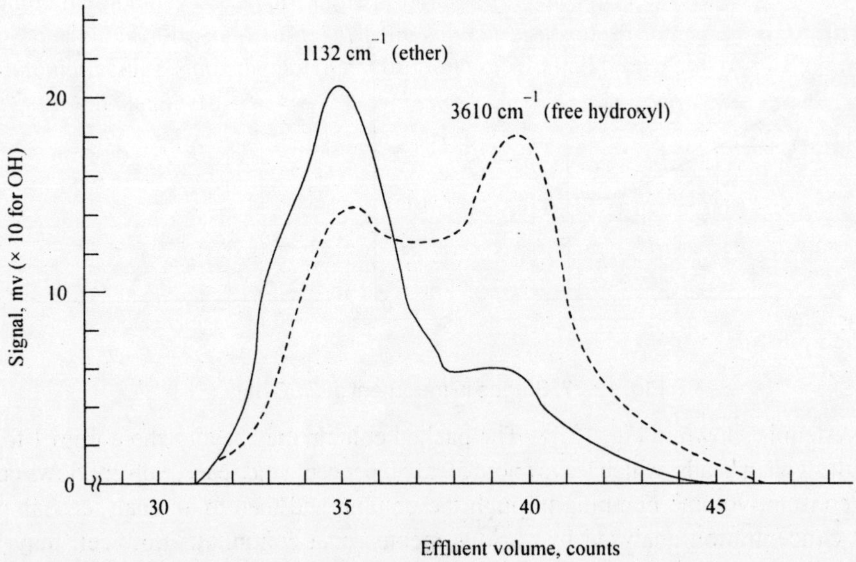

Fig. 12.19. Mixture of poly(ethylene oxide)s with molecular weights of 300 and 4000 analysed at ether and end group wave numbers (maximum signal for OH, 1.8 mV).

In a conventional apparatus for liquid exclusion chromatography, a single high-molecular-weight sample usually can be analysed in about two hours. It was found that shorter columns, finer packing, and much higher pressures lead to faster analyses, so that the time could approach 15 or 20 min per sample. In HPLC (high-performance liquid chromatography), pressures on the order of 100 atm often are used, compared to the 1 to 10 atm used in conventional GPC or gel filtration. Although similar to GPC in principle, HPLC requires precision, constant-flow-rate pumps, and special sample injection ports because of the small volumes and high pressures involved. In any chromatographic method the eluting liquid can be programmed to change in temperature or in composition over the total analysis time. Such methods are not as commonly used in measuring molecular weight distribution as they are in analysing mixtures of drugs, natural products, or other complex materials.

MEASUREMENT OF MOLECULAR WEIGHT

We accept now-a-days as the valid indication of molecular size the molecular weights of polymers measured by the same techniques that have been used for small molecules. Before the 1920s, many chemists were reluctant to believe that such large, covalently bonded structures were possible. The molecular weights were regarded as 'apparent' values brought about by a physical association of smaller molecules in a 'colloidal' state. Perhaps we should not dismiss all the reservations that those scientists had when we look at experimental results. No measurement is foolproof, and it is a brave person indeed who lists a molecular weight of a polymer in the hundred-thousand range to four or five figures. Even the so-called absolute methods we are about to discuss depend on extrapolations that are not completely objective, since some mathematical model or form invariably must be postulated.

Methods for determining molecular weight can be relative or absolute. Many properties of polymers that depend on molecular weight, such as solubility, elasticity, adsorption on solids, and tear strength, can be correlated with an average molecular weight. Once correlated, the property can be used as a measure of molecular weight. In practice, viscosity of melts and dilute solutions is the most often used relative method. Absolute methods can be categorised by the kind of average they measure.

NUMBER-AVERAGE MOLECULAR WEIGHT METHODS

The measurement of the concentration of end groups when one knows the exact number per molecule is a method of counting the number of molecules. In the hydroxy-acid ester example, a simple titration by standard sodium hydroxide would give the total number of free carboxyl groups. Since there is but one such group for each polymer molecule, we have a molecule-counting method. End group counting can be more sophisticated, as in the use of infrared or isotopic analysis. In any of these we derive a number-average molecular weight $\overline{M}_n$ from Eq. 12.35 together with Eq. 12.36.

$$\overline{M}_n = \bar{x}_n M_0 \qquad \text{... (12.56)}$$

where M_0 is the molecular weight of a repeating unit in the polymer.

The magnitude of the depression of the melting point or the elevation of the boiling point of a solvent by a non-ionised solute depends on the number of solute molecules present but not on their chemical nature. These colligative properties are used to estimate the molecular weight of a solute. The melting point depression is widely used in organic chemistry for the characterisation of small molecules. The lowering of the melting temperature, $T_f - T$, is related to the normal melting temperature of the solvent T_f, its heat of fusion ΔH_f, and molecular weight M_1, and also to the molecular weight of the solute M_2 and the relative weights of solvent and solute w_1 and w_2, respectively.

$$T_f - T = \frac{RT_f^2}{\Delta H_f} \frac{M_1}{M_2} \frac{w_2}{w_1} \quad \text{... (12.57)}$$

where R is the gas constant. For a solution of 1 g of polyethylene with M_2 = 500 in 100 g of camphor (T_f= 178.4°C, ΔH_f= 2.58 kcal/mol, M_1 = 152), $T_f - T$ is 0.48°C. As the molecular weight of the solute increases, measurement of the temperature lowering demands more sensitive equipment. In order to measure the number of molecules by colligative methods, it is necessary to use dilute solutions. In practice, we often extrapolate results to infinite dilution by some suitable linearisation techniques. Osmotic pressure is a useful measuring method. We place on opposite sides of a semi-permeable membrane (typically cellophane) a polymer solution and pure solvent (Fig. 12.20). The membrane is selected to pass solvent but not polymer, the diffusion rate of polymer above some low molecular weight being negligibly small. To establish equilibrium or by measuring and compensating for pressures automatically, osmotic pressure π can be measured at several concentrations. According to thermodynamic theory:

$$\left(\frac{\pi}{c}\right)_{c=0} = \frac{RT}{M} \quad \text{... (12.58)}$$

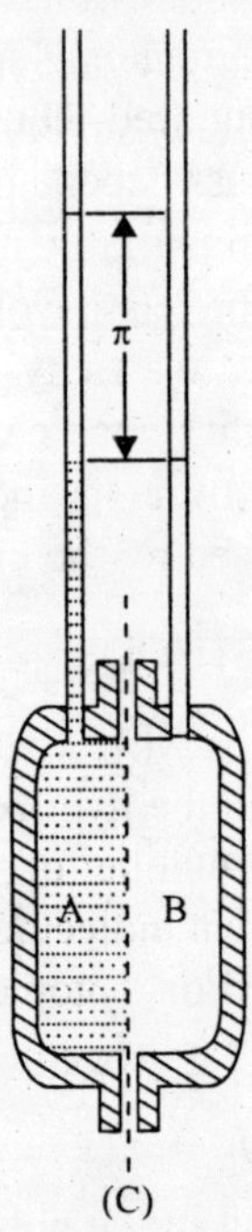

Fig. 12.20. Osmotic pressure cell containing polymer solution A and solvent B separated by semi-permeable membrane C. The osmotic pressure is π.

where

π = osmotic pressure (g/cm²) = $h\rho$

h = difference in liquid levels at equilibrium, cm

ρ = density of solvent, g/cm³

c = concentration, g/cm^3
T = absolute temperature, K
M = molecular weight, g/mol
R = gas constant, 8.48×10^4 g.cm/mol·K

The difference between osmotic pressure and vapour pressure lowering can be illustrated by considering a situation where we take pure solvent at total pressure $P(0)$, increase the vapour pressure by increasing the total pressure to $P(0) + \pi$, then decrease the vapour pressure by adding a solute with a mole fraction N_2 (Table 12.4). Assuming ideal solutions and perfect gases, the criterion for equilibrium of solvent between the first and last conditions reduces to having the vapour pressure of solvent p in pure solvent at $P(0)$ equal to that in solution at $P(0) + \pi$. The change in vapour pressure of solvent with total pressure $P(T)$ is given by the *Poynting* equation.

$$\frac{dp}{dP(T)} = \frac{V_1}{V_g} \quad \text{... (12.59)}$$

where V_1 is the molar volume of liquid solvent and V_g is the molar volume of gaseous solvent. Assuming a perfect gas, we can replace V_g by RT/p and integrate:

$$RT\int_{p(0)}^{p(\pi)} d(\ln p) = V_1\int_{P(0)}^{P(0)+\pi} dP(T) \quad \text{... (12.60)}$$

$$RT \ln \frac{p(\pi)}{p(0)} = V_1\pi \quad \text{... (12.61)}$$

where the vapour pressure is increased as indicated by the limits when the total pressure is increased by the osmotic pressure π. The subsequent decrease in vapour pressure is accomplished by adding solute according to *Raoult's* law, $p(N_2, \pi) = N_1 p(0, \pi)$. Our final condition is that $P(N_2, \pi)$ be equal to $p(0)$. Also, when N_1 is nearly 1, $-\ln N_1$ can be approximated by $1-N_1$, which is N_2, the mole fraction of solute. If we replace N_2 by cV_1/M_2 with appropriate attention to units, we produce Eq. 12.61.

$$RT \ln \frac{p(\pi)}{p(0)} = RT \ln \frac{p(0,\pi)}{p(N_2,\pi)} = -RT \ln N_1 = RTN_2 = \frac{RTcV_1}{M_2} = V_1\pi \quad \text{... (12.62)}$$

Thus it becomes clear that the osmotic pressure is large compared to the vapour pressure lowering because it corresponds to the change in total pressure that causes the lowering. Rearranging Eq. 12.59 and putting in finite changes in vapour pressure for a change in total pressure equal to the osmotic pressure, it is seen that the ratio is that of the volumes of equal weights of vapour and liquid, usually a factor of several hundredfold.

$$\frac{\pi}{\Delta p} = \frac{V_g}{V_1} \quad \text{... (12.63)}$$

Equation 12.61 holds only at infinite dilution (or in a theta solvent). At finite concentrations the osmotic pressure is best represented by an attenuated power series

$$\frac{\pi}{c}\frac{M}{RT} = 1 + A_2 Mc + A_3 M^2 c^2 \quad \text{... (12.64)}$$

where A_2 and A_3 are the second and third virial coefficients. Often, A_2 can be taken as equal to $(A_2/2)^2$, so that Eq. 12.64 can be re-written:

$$\left(\frac{\pi}{c}\right)^{1/2} = \left(\frac{RT}{M}\right)^{1/2}\left(1+\frac{A_2 Mc}{2}\right) \quad \text{... (12.65)}$$

Table 12.4. Characterisation of osmotic pressure*.

Condition	*A*	*B*	*C*
Mole fraction of solvent	1.0	1.0	N_1
Mole fraction of solute (polymer)	0	0	N_2
Vapour pressure of solvent	$p(0, 0)$	$p(0, \pi)$	$p(N_2, \pi)$
Total pressure on system	$P(0)$	$P(0) + \pi$	$P(0) + \pi$

* Criterion for equilibrium between *A* and *C*: $p(0, 0) = p(N_2, \pi)$ when partial molar free energy of solvent is same in *A* and *B*. *Raoult's* law: $p(N_2, p) = N_1 p(0, \pi)$.

Plots of $(\pi/c)^{1/2}$ versus concentration often are linear (Fig. 12.21) and allow extrapolation to infinite dilution. The second virial coefficient is obtained from such plots by dividing the slope by the intercept and by $M/2$.

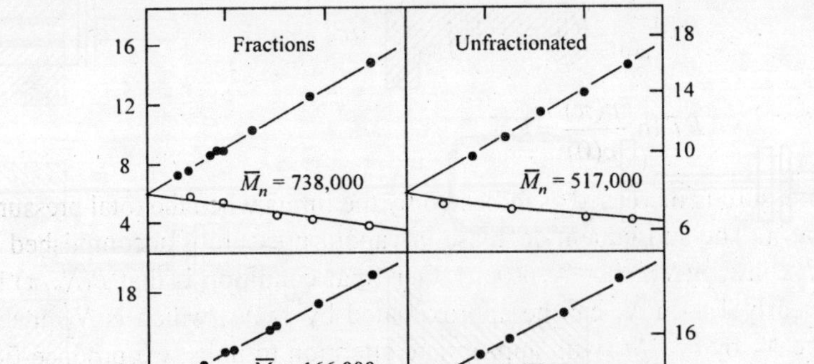

Fig. 12.21. Osmotic pressures plotted according to Eq. 12.68 for poly(methyl methacrylate) in toluene [•(t)], acetone (•), or acetonitrile (o).

Osmotic pressure probably is the most popular of the colligative methods. On the other hand, membrane preparation is tricky, equilibrium may be slow to be reached, and the effect observed, π, decreases as M increases. A commercially available automatic osmometer operates on the null-point principle; that is, solvent pressure is adjusted by a servomechanism until a sensor can detect no tendency for solvent to flow through the membrane in either direction (Fig. 12.22). Equilibrium can be reached in 5 or 10 min, so that the convenience of measurement approaches that of viscometry. With care, molecular weights of 10,000 to 500,000 can be measured with about 1 per cent accuracy. Below 10,000 many polymers can penetrate the common membrane materials. When a low-molecular-weight polymer is placed in an osmometer, the rate of diffusion of the polymer through the membrane is discernible, since it results in a drift in the pressure necessary to maintain a null flow of solvent. Because the rate of diffusion decreases with increasing molecular size, the drift in osmotic pressure with time can be calibrated to yield the molecular weight of a specific polymer despite membrane penetration. This is not an absolute method, of course.

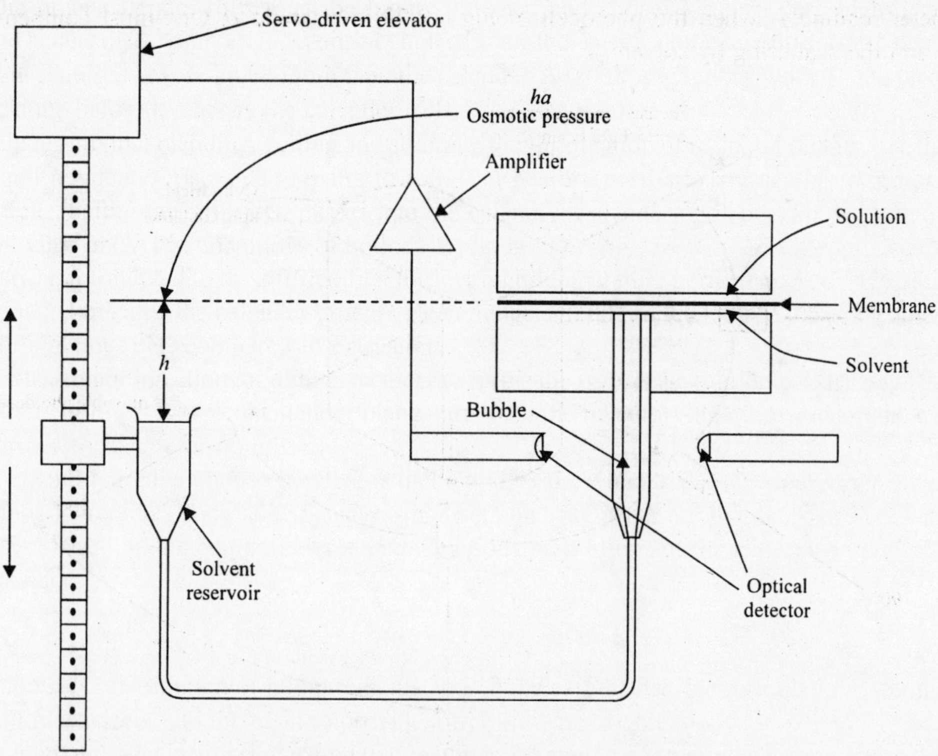

Fig. 12.22. Automatic membrane osmometer.

For molecular weights less than about 20,000, another technique has been automated. In the so-called vapour pressure osmometer there is no membrane. A drop of solution and a drop of pure solvent are placed on adjacent thermistors. The difference in solvent activity brings about a distillation of solvent from the solvent bead to the solution. The temperature change that results from the differential evaporation and condensation can be calibrated in terms of the number-average molecular weight of the solute. The method is rapid, although multiple concentrations and extrapolation to infinite dilution are still required.

WEIGHT-AVERAGE MOLECULAR WEIGHT METHODS

The scattering of light by small particles is a familiar phenomenon as, for example, the appearance of dust particles in a beam of sunlight. Similarly, if polymer molecules are dissolved in a solvent, the light scattered by the polymer far exceeds that scattered by the solvent and is an absolute measure of molecular weight (Fig. 12.23). The *Debye* relationship is

$$\frac{H'c}{\kappa} = \frac{1}{MP(\theta)} + 2A_2c \qquad \text{... (12.66)}$$

where H' is a lumped constant including geometric factors and also the change in refractive index with polymer concentration for the particular system being investigated. This latter relationship usually is established in a separate experiment using a differential refractometer. The intensity of light is measured at angle θ and concentration c. The second virial coefficient A_2 and a complex function of molecular shape $P(\theta)$ usually are derived from the data. The light intensity factor κ is derived from the raw galvanometer reading I_g when the photocell of Fig. 12.23 is at angle θ. One must compensate for geometry and the scattering by solvent, I_{gs}.

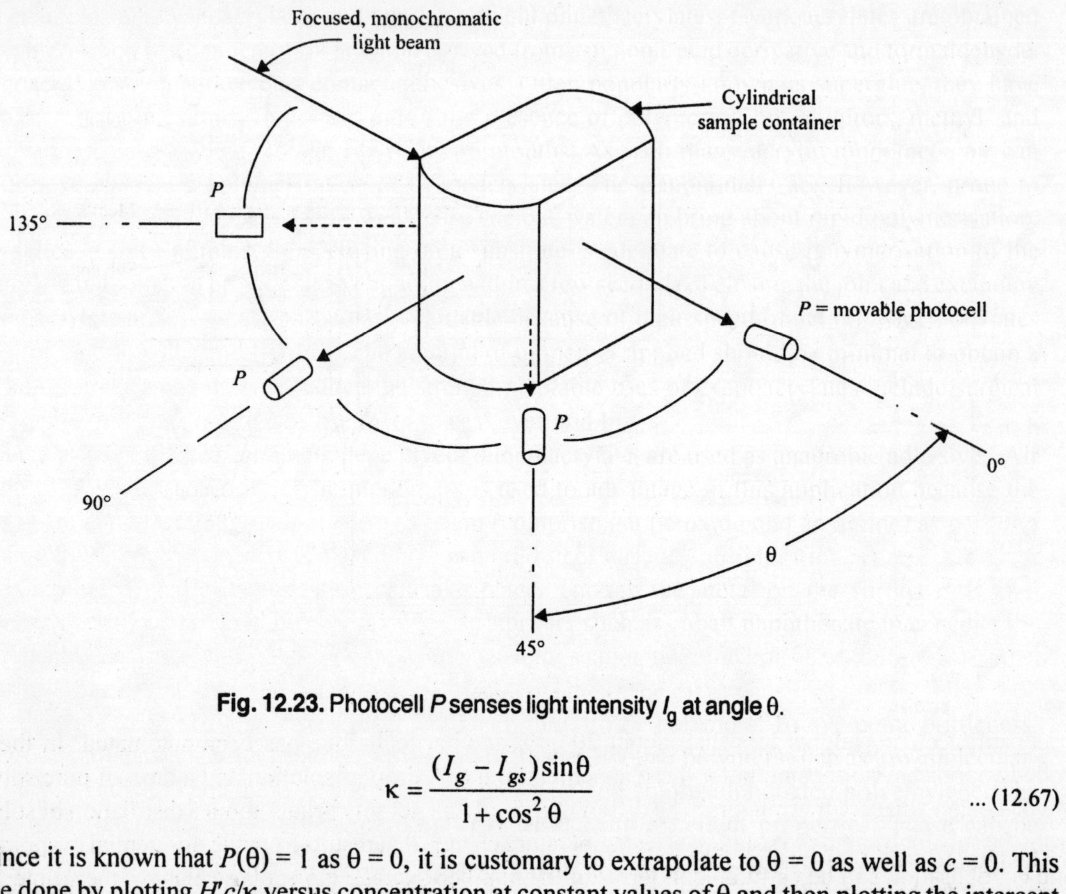

Fig. 12.23. Photocell *P* senses light intensity I_g at angle θ.

$$\kappa = \frac{(I_g - I_{gs})\sin\theta}{1+\cos^2\theta} \qquad \text{... (12.67)}$$

Since it is known that $P(\theta) = 1$ as $\theta = 0$, it is customary to extrapolate to $\theta = 0$ as well as $c = 0$. This can be done by plotting $H'c/\kappa$ versus concentration at constant values of θ and then plotting the intercept $1/MP(\theta)$ versus $\sin^2(\theta/2)$ to give the intercept $1/M$. Both can be done simultaneously in a *Zimm* plot

(Fig. 12.24), where a network results. As in osmotic pressure, measurements in a poor solvent give a low value for A_2. *Margerison* and *East* have worked out a complete example for polystyrene in benzene.

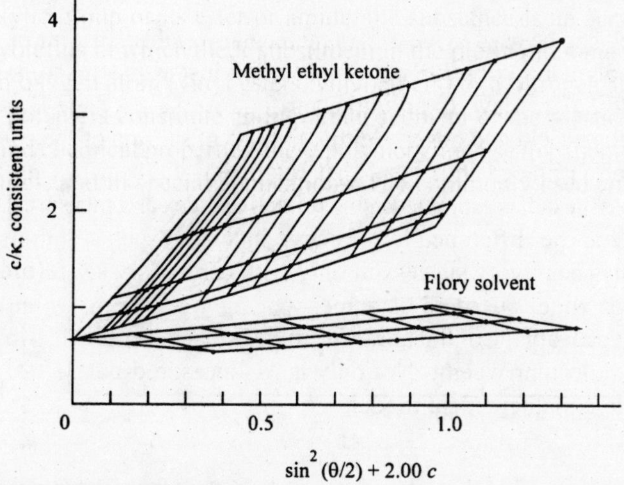

Fig. 12.24. *Zimm* plots for scattering of light by poly(vinyl acetate) with $M_w = 3.0 \times 10^6$ in a good solvent, methyl ethyl ketone, and a poor one, methyl isopropyl ketone: *n*-heptane, 73.2 : 26.8. The ordinate has been adjusted to the same intercept with each solvent.

The extrapolation to $\theta = 0$ can be eliminated if scattering can be measured at forward angles of less than 2° from the incident beam. This becomes possible when the light source is a small, single transverse mode laser. If the sample cell volume is made sufficiently small, laser beam light scattering can be used as the detector in column chromatography (GPC or HPLC).

With the proper auxiliary apparatus and computational devices attached, the column effluent can be analysed in terms of molecular weight directly without recourse to a separate calibration with molecular weight standards.

To see what kind of average molecular weight we measure by light scattering, we can examine the contributions made by individual species to the factor κ. For a monodisperse polymer *i* at $\theta = 0$:

$$\frac{H' c_i}{\kappa_i} = \frac{1}{M_i} \qquad \text{... (12.68)}$$

For a polydisperse polymer

$$\kappa = \Sigma\, \kappa_i = \Sigma\, H' c_i M_i \qquad \text{... (12.69)}$$

and

$$c = \Sigma\, c_i \qquad \text{... (12.70)}$$

The average molecular weight $\overline{M}$ is obtained from the overall scattering and concentration:

$$\overline{M} = \frac{\kappa}{H' c} = \frac{\Sigma H' c_i M_i}{H' c} = \frac{\Sigma c_i M_i}{c} = \Sigma w_i M_i \qquad \text{... (12.71)}$$

Thus it is seen that since c_i/c is the weight fraction of component *i*, w_i, the average molecular weight is M_w, the weight-average molecular weight.

A major practical problem in light scattering is the presence of dust particles or gelled polymer particles. A small concentration of these obscures the scattering by ordinary polymer particles and results in $\overline{M}_w = \infty$. Removal of dust by ultracentrifugation may be necessary, making this a rather expensive tool.

The ultracentrifuge itself can be used to measure a molecular weight distribution. Under exceptionally high gravitational fields, polymers will settle out of a solution just as macroscopic beads fall through a viscous liquid according to *Stokes'* law. In an ultracentrifuge we balance the Brownian motion of the molecules against a centrifugal field and observe either the rate of settling or the equilibrium in the field. A cell holding the solution is placed in a head that rotates on air bearings up to 70,000 rpm. The concentration gradient across the cell is estimated using refractive index as a measure. Because extrapolation to $c = 0$ is still necessary and the difference in refractive index between solution and solvent must be large, the ultracentrifuge has been very successful only with certain classes of polymers, for example, proteins. It has not been as widely used as osmometry or light scattering as an absolute measure of molecular weight. A definite advantage of the ultracentrifuge is that the entire distribution is obtained, not just one lumped average molecular weight. Not only is M_w measured but M_n, M_z, etc., are measured directly without recourse to a mathematical model.

CHAPTER 13

Structure of Polymeric Materials

INTRODUCTION

The difference in structure between thermosetting and thermoplastic materials is so great, that it is not usually possible to apply the same considerations to both. To eliminate the need to continually state which type of material is being considered, it will be assumed that the material is thermoplastic, unless otherwise stated.

A solid differs from a liquid and a gas is that it has a definite volume and shape. The distribution of the molecules within a liquid is completely random, and the molecules move relative to one another, the thermal energy being sufficient to overcome the intermolecular forces. For materials consisting of small molecules, the change from liquid to solid is a change from disorder to order. In the solid the intermolecular forces are large enough to hold the molecules in an ordered array, despite the thermal energy which causes them to vibrate about their equilibrium position in that array. The molecular arrangement is regular throughout. Such solids are crystalline, and melt at a fixed temperature corresponding to a thermal energy just sufficient to overcome the intermolecular forces.

Materials consisting of molecules which are very large or have extremely complex structure show no sharp melting point of this sort. The material, which is said to be solid because of its rigidity, may be completely amorphous, having no regular molecular arrangement. Such a material can be considered equally well as a very viscous liquid, whose viscosity decreases with increase in temperature.

The molecules in polymeric materials are large, complex and of varying length. Single crystals of some polymeric materials have been produced under special conditions, but normally such materials do not exist in a completely crystalline state. Many are completely amorphous. Examination of some materials shows ordered as well as amorphous regions. These ordered regions are known as crystallites, and consist of a regular pattern of polymer chains. A study of crystallites shows that their dimensions are less than the length of the polymer molecules when extended. The chain sections in a crystallite are in extended form, so only part of a molecule can contribute to a given crystallite. The same molecule can be part of several crystallites passing through amorphous regions between them. The transition from ordered to random is not abrupt, the crystallites being surrounded by semi-ordered transition regions. The relative percentages of order and disorder vary widely with the nature of the material. Considerable variations are possible in samples of the same material, depending on the history of the sample. Some materials consists mainly of crystallites surrounded by small disordered or semi-ordered regions, while others are mainly amorphous with a few, widely scattered, ordered regions. In some cases, the order of the system is further increased by groups of crystallites forming ordered patterns. These are known as spherulites.

The term crystalline, when applied to polymeric materials, merely indicates that some degree of crystallinity is possible. The extent of this depends on the regularity of the polymer. The more regular the molecular structure, the easier it is for the chains to pack in an ordered manner. Thus, polymers like polythene $-(CH_2-CH_2)_n-$ and polytetrafluoroethylene $-(CF_2-CF_2)_n-$ are highly crystalline.

Polymers of the type $-(CR_2-CH_2)_n-$ can be arranged in a head to tail form

$$-CH_2-CR_2-CH_2-CR_2-CH_2-CR_2-CH_2-CR_2-CH_2-$$

or a head to head form

$$-CH_2-CR_2-CR_2-CH_2-CH_2-CR_2-CR_2-CH_2-CH_2-$$

or they can have a random head to tail, head to head form. Materials containing regular molecules of the head to tail, or the head to head type, will usually be crystalline, while those containing molecules with the random arrangement will be amorphous. Most polymerisation methods give a head to tail arrangement. Thus, commercially made polyvinylidene chloride $-(CCl_2-CH_2)_n-$ is crystalline.

Polymers of the type $-(CHR-CH_2)_n-$ are normally in the head to tail form. Despite this, they will only crystallise if they are isotactic or syndiotactic. Most polymerisation processes give atactic polymers and polyvinyl chloride $-(CHCl-CH_2)_n-$ and polystyrene $-(CHC_6H_5-CH_2)_n-$ are usually amorphous. Commercially made polypropylene $-(CHCH_3-CH_2)_n-$, on the other hand, is crystalline, as a stereo specific catalyst is used in the polymerisation, giving an isotactic polymer.

Condensation polymers are crystalline, if the repeating units have a regularity in configuration. The size of the repeating unit may alter the ease with which the crystallisation occurs, but it does not affect its extent. Many of the common condensation polymers, such as nylon, polyethylene terephthalate and polycarbonate, are crystalline.

The regularity of the molecules determines the extent to which the chains can pack in an ordered pattern. Once in this pattern, they must be held there by the intermolecular forces, which oppose the thermal agitation of the molecules. The regularity of the crystallite leads to higher intermolecular forces that are present in amorphous regions. The intermolecular forces are greater if polar bonds are present in the polymer, particularly if these can lead to hydrogen bonding as in the polyamides.

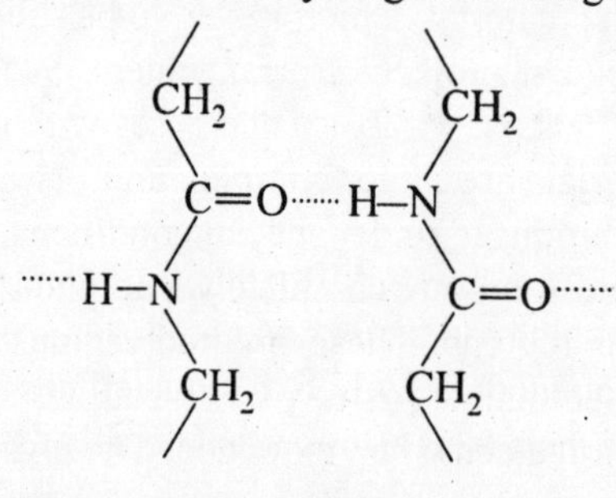

The strength of the intermolecular forces determines the temperatures at which crystallites will form.

POLYMER STRUCTURE

In addition to the size of molecules and the distribution of sizes in a polymer, the shapes or structures of individual polymer molecules also play a major role in determining the properties of a plastic. It has been found that polymers form by aligning themselves into long chains of molecules without any side protrusions or branches; or lateral connections, between molecules. Some polymers do largely this and nothing more; however, it is also possible for polymers to form more complex structures. Thus, polymer molecules may form in the shape of branched molecules, in the form of giant three-dimensional networks, in the form of linear molecules with regular lateral connections to form 'ladder-type' polymers, in the form of two dimensional networks or platelets, and so forth, depending on how many connections or bonds can exist between the mono-disperse monomeric molecules that were used to form the polymer and between sites on the forming or already formed polymer molecules (Fig. 13.1).

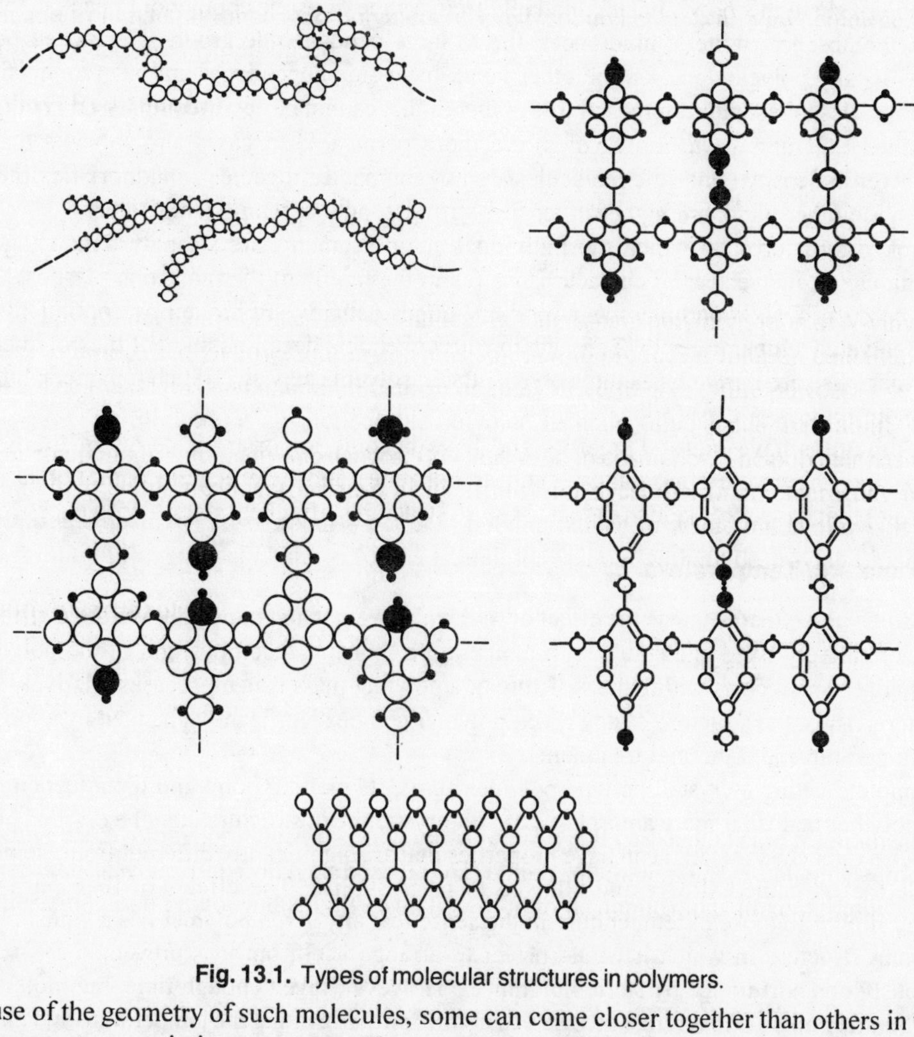

Fig. 13.1. Types of molecular structures in polymers.

Because of the geometry of such molecules, some can come closer together than others in which the structure prevents more intimate contact. The structural obstruction to close approach is called steric hindrance. Thus, polymers that can be packed closely or exhibit little steric hindrance ordinarily can more easily form crystalline structures in which the molecules are aligned in some regular pattern; others, such as polymers that are cross-linked prior to crystallisation, are prohibited from aligning themselves in crystals because of the hindrance created by the multiple interconnections and hence tend to be amorphous, or non-crystalline. Amorphous polymers do not have melting points, but rather softening ranges; are normally transparent; and undergo only small volume changes when solidifying from the melt, or when the solid softens and becomes fluid. Crystalline polymers, on the other hand, have considerable order to the molecules in the solid state, indicating that many of the other atoms are regularly spaced; have a true melting point with a latent heat of fusion associated with the melting and freezing process; and have a relatively large volume change during the transition from melt to solid.

Thus, many different structures are possible with plastics—and each will affect the basic properties of the polymer. For example, linear polymers, such as high-density polyethylene, are made of molecules that resemble spaghetti in a bowl and are relatively free to slide over one another or to pack more closely together (in the absence of steric hindrances due to large pendant side groups). Branched polymers, such as low-density polyethylene, on the other hand, have side appendages and interconnections that cause the molecules to resemble clumps of tree branches that cannot be easily compressed or compacted. Thus, branched polymers (with more voids) are more permeable to gases and solvents than linear polymers, lower in density (since the molecules are not compacted together), and more flexible. Linear polymers, on the other hand, are higher in tensile, stiffness, and softening temperatures.

Cross-linked structures, in which the individual chain segments are strongly bound together by chemical unions, also have special characteristics (as in the family of thermosetting plastics). They do not exhibit crepe or relaxation unless such primary bonds actually are broken by continually applied stress or by elevated temperatures high enough to cause chemical decomposition of the polymer. Cross-linked polymers also are fairly resistant to solvent attack; solvents may swell such polymers, but seldom cause complete rupture or dissolution.

Ladder structures have unusual stability and have become important in terms of the new heat-resistant plastics. Aromatic compounds (such as benzene) and heterocyclic compounds (such as benzimidazole) that have semi-ladder, ladder, or spiro structures offer heat stabilities in excess of 900°F.

Effect of Time and Temperature

It should also be noted that whereas the effect of molecular weight and molecular weight distribution on properties is relatively fixed and stable with temperature (barring decomposition of the polymer), the arrangement of the molecules within the structure of a polymer mass is in most cases relatively sensitive to temperature. Thus, the structure of any given polymer can be significantly changed by exposing it to different temperatures and thermal treatments.

For example, heating a crystalline-type polymer above its melting point and then quenching it can produce a polymer that is far more amorphous or non-crystalline in structure than the original polymeric sample. Such a quenched material can have properties that are significantly different from the properties of a sample that is cooled slowly and allowed to re-crystallise. The effects of time on a polymer structure are similar to those of temperature in the sense that any given polymer has a "most preferred" or equilibrium structure in which it would prefer to arrange itself, but it is prevented from doing so instantaneously on short notice by steric hindrances. However, given enough time, the molecules in a polymer ultimately migrate to arrange themselves in this form. Elevating the temperature and making the molecules more mobile or spreading them apart allows them to accomplish this in a shorter time and vice versa.

Thus, over an extended period of time, the properties of a polymer can become significantly different from those measured earlier if the structure of the polymer was in an unstable form when the properties were initially evaluated.

Chemical composition

In addition to all the variations in the make-up of the polymer discussed above and having to do largely with extra or intermolecular phenomena, polymer properties also are heavily dependent on such factors as the intramolecular chemical composition of the polymer. For example, as shown in Fig. 13.2, polyethylene consists (except for catalyst or extraneous impurities) completely of atoms of carbon and hydrogen. This internal make-up of the polymer molecule affects in turn all the previously discussed variables and hence contributes its own basic characteristics to the overall properties of the polymer. Going further, if one takes every fourth hydrogen atom occurring in polyethylene and replaces it with a methyl (CH_3) group in regular intervals along the length of the chainlike molecule as shown in Fig. 13.2, the polyethylene is transformed into polypropylene.

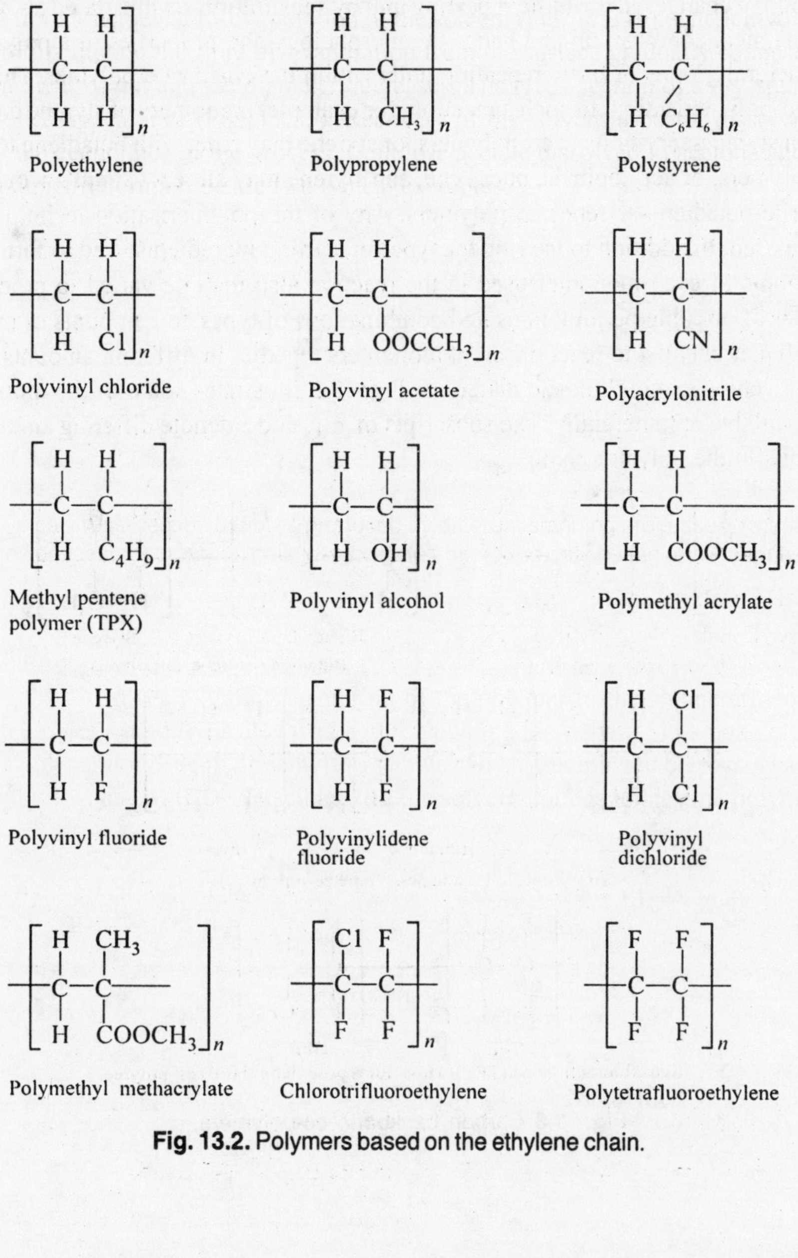

Fig. 13.2. Polymers based on the ethylene chain.

In such a case the degree of polymerisation and the molecular weight distribution remain roughly the same; but the spacing of the molecules in the polymer matrix or the morphology of the polymer changes, and thus its macroscopic physical properties also change. Similarly, if every fourth hydrogen atom of the polyethylene were substituted with a benzene structure (C_6H_6), or with a chlorine atom (Cl),as shown in Fig. 13.2, the ethylene would become either polystyrene or polyvinyl chloride, respectively.

The simple substitution of a single hydrogen atom with another atom or chemical group can cause a drastic change in the properties of a polymer. Whereas polyethylene is translucent, flexible, and crystalline, polystyrene is transparent, brittle, and amorphous.

Co-polymers

In addition to making changes in the basic repeating unit by substitution, as illustrated above it is possible to change the chemical composition and hence the morphology and properties of a polymer by mixing the types of structural groups or basic repeating units within the chain of a polymer. This is done by a process called co-polymerisation. In such a reaction, for example, monomers of styrene and acrylonitrile can react to form styrene-acrylonitrile co-polymers; or styrene may react with butadiene to form styrene-butadiene co-polymers; or acrylonitrile, butadiene, and styrene may all react simultaneously to form the ABS (acrylonitrile-butadiene-styrene) co-polymers. Any of the polymerisation techniques previously discussed can be used. In addition to varying the types of starting ingredients used to form co-polymers, the relative amounts of each monomer used in the reaction also may be varied to produce a literally unlimited number of possible permutations and combinations of types and amounts of monomers. The ability of the polymer chemist to react different monomers together in different amounts has given rise to the concept of 'tailor-making' plastic materials. Fig. 13.3 illustrates some of the more common co-polymers now available commercially. The subscripts of x, y, and z denote differing amounts of each of the structural units in the polymer chain.

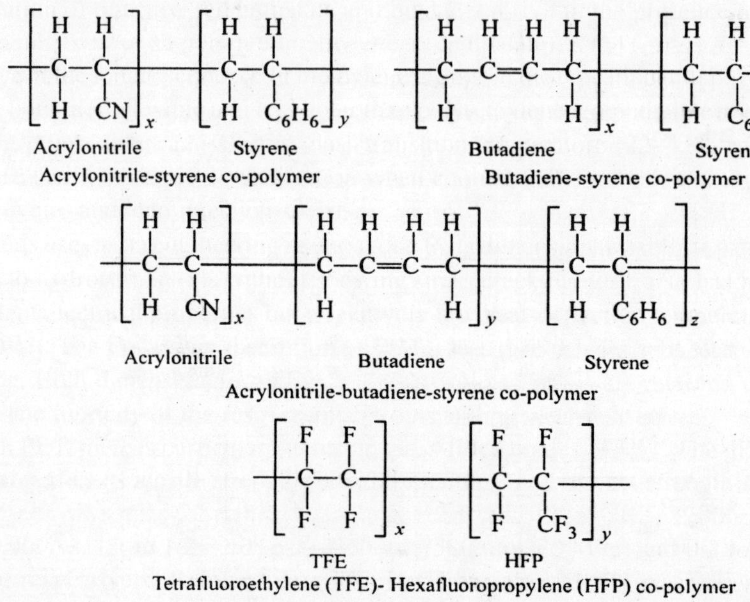

Fig. 13.3 Carbon 'backbone' co-polymers.

It should be noted that the structural formulae in Fig. 13.3 are somewhat fictional, in the sense that there is an implication that a long chain of one group is connected to a long chain of another group, and successively to a long chain of still a third type of polymer. Actually, the distribution of each of the species within the chain will depend both on the amount present and how the polymerisation was carried out.

For example, an acrylonitrilestyrene co-polymer might be made by first reacting a good portion of the acrylonitrile to form polyacrylonitrile and then adding the styrene and remaining acrylonitrile to complete the polymerisation of the desired end-product. Or, in another case, both ingredients might be added to a reactor simultaneously and allowed to react concurrently.

Other polymers

From a study of Fig. 13.3 and Fig. 13.4, it will be noted that all of the polymers listed there are formed by combinations made possibly by reaction of the double bonds that exist between adjacent carbon atoms in the monomers used as starting materials. However, as mentioned earlier, this is not the only way in which polymers are formed. When polymers are created by rearrangements or condensation reactions, other types of polymeric chains can be formed.

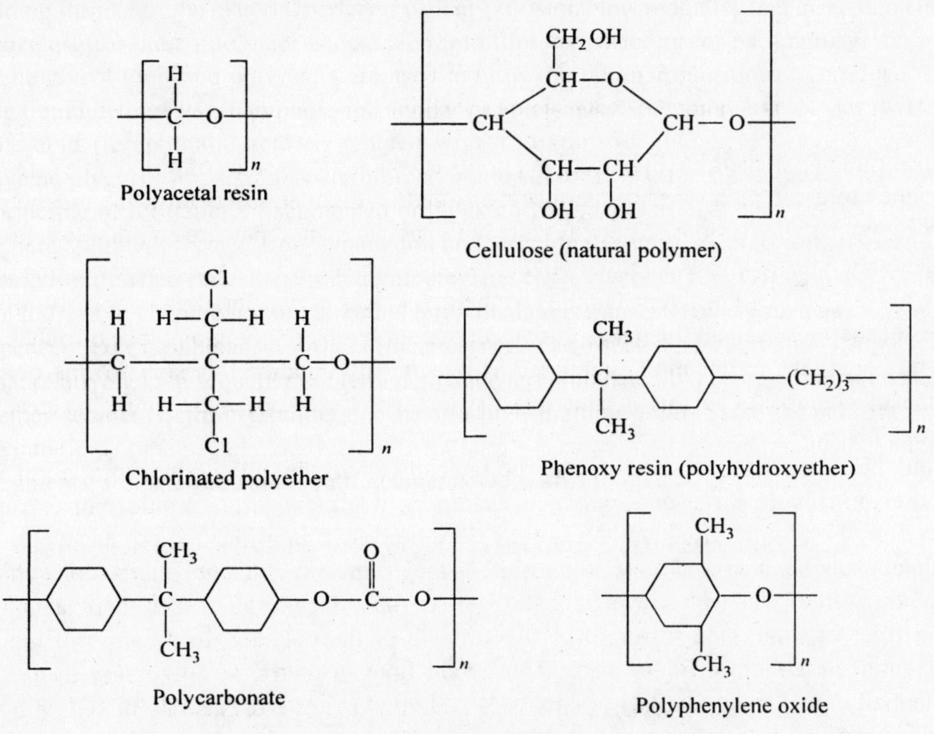

Fig. 13.4. Polymers with oxygen in the chain.

Fig. 13.4 lists the structural formulae of several commercially important polymers in which oxygen (O) is an integral part of the bonds holding the polymer together. Similarly, Fig. 13.5 lists several types of polyamides and a polyurethane in which nitrogen (N) atoms form the bonds between portions of the polymer molecule. In some cases the polymers are formed by molecular rearrangements (in the case of polyurethanes and epoxies) and in other cases by condensation reactions (some nylons).

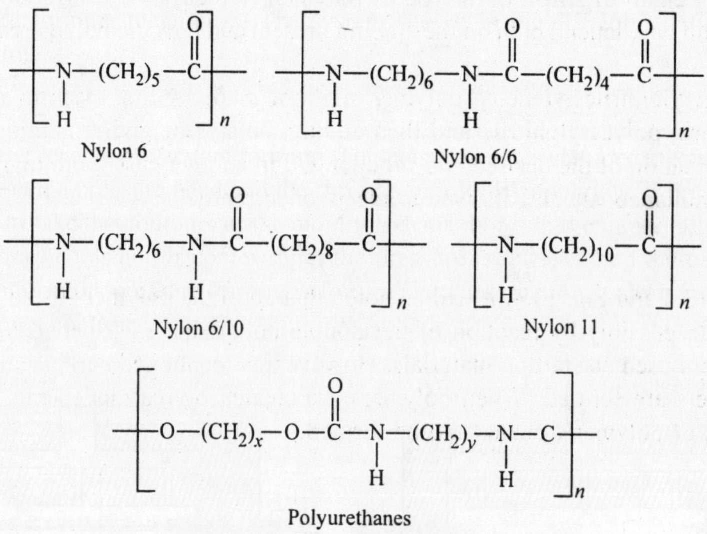

Fig. 13.5. Polymers with nitrogen in the chain.

Other condensation polymers are illustrated in Fig. 13.6, which shows some of the complex structures proposed for the more common types of thermosets. Note that the thermosets always have more than two linkages connecting the various structural units in the polymer, and they are commonly referred to as cross-linked materials involving a network like molecular structure.

There are at least 30 to 40 different families of thermoplastics now offered and about 10 different basic families of thermosets. And this takes into account only the more common types of polymers. It should also be noted that co-polymers, mixtures, and chemically modified versions of the polymers listed make the total number of polymeric materials that are theoretically possible entirely too great to list them in any one place.

In addition, the properties of a polymer may be radically altered by mixing it with non-polymeric materials and chemicals. In fact, some polymers would be worthless to the moulder or extruder if additives were not used to modify them before they were processed into finished products.

For example, phenolic, urea, and melamine resins, as they come out of the polymerisation kettle, are largely brittle, frangible solids with low impact strength. To make them usable, such materials must be mixed with a filler of some kind to reinforce the strength of the polymer. In the case of the three thermosets mentioned, cellulose fillers made from wood flour or cotton are necessary to make the resins commercially mouldable. Similarly, cellulosics and vinyl resins must also be mixed with semi-solvents, called plasticisers, to soften them to some degree (so they may be moulded or extruded) or to modify their properties (to make them soft and pliable, rather than hard and stiff).

In actual practice, the number of different types of chemical additives and fillers that are mixed with the fundamental polymer is so large that the chemistry of plastic compounds becomes extremely complex. It is also possible to vary existing polymers by any number of chemical, mechanical, or irradiation techniques to change the structure of the polymer. During processing, for example, it is possible to align or orient the molecules in a polymer to significantly increase the strength of the polymer in the direction of the orientation.

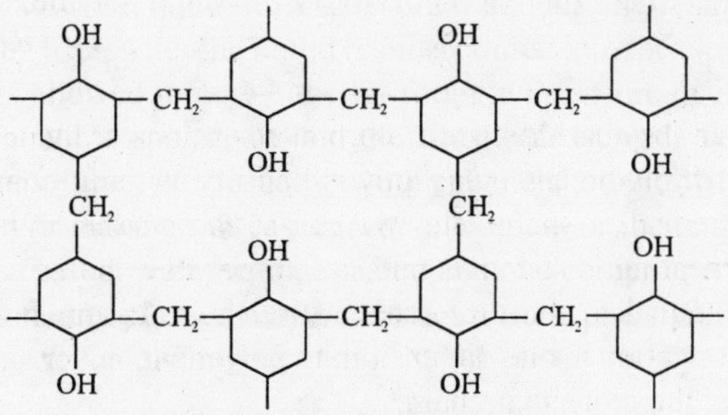

Phenol-formaldehyde resin

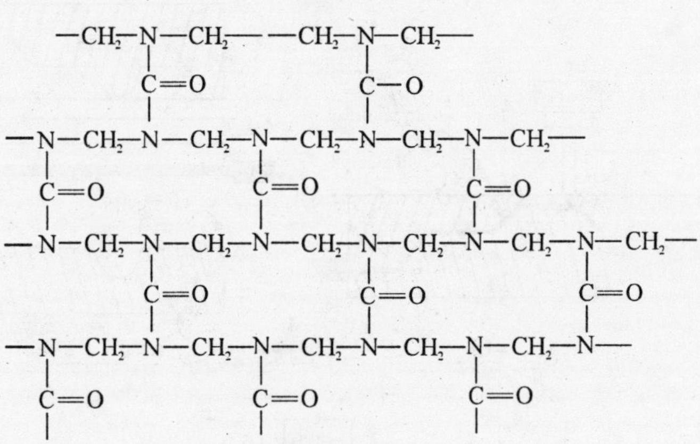

Urea-formaldehyde

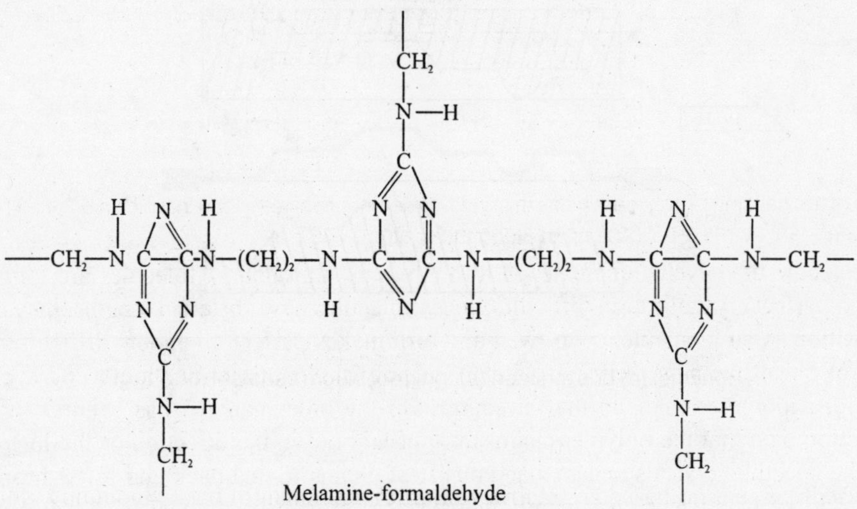

Melamine-formaldehyde

Fig. 13.6. Thermoset structures.

Chemical or irradiation methods also can be used to cross-link a plastic (such as polyethylene) to improve its toughness and chemical resistance. Nucleation techniques and chemical modification (as in the case of chlorinated polyethylene) are further adaptable to changing polymer structures.

POLYMER CRYSTALLINITY

Metals invariably crystallise when they solidify under normal conditions. Polymers behave differently. Some solidify without crystallising, whereas many others solidify without complete crystallisation. Atoms in metals are almost spherical in shape, and we can view a metal crystal as a collection of individual balls arranged in a repeating long-range pattern. No polymer molecule has equiaxed shape with dimensions approximately equal in the three space directions. Linear polymer molecules are characterised by their extreme elongation in one dimension. In polyethylene, for example, the molecules have strong covalent bonds along the molecular chain.

The bonds between chains are much weaker but still present, so the chains tend to line up as shown in Fig. 13.7.

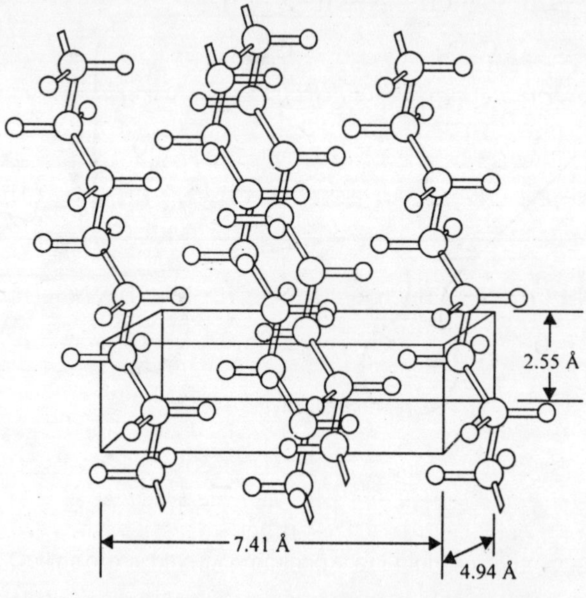

Fig. 13.7. Model of the packing of polymer chains in crystal structure of polyethylene in which a = 7.41 Å, b = 4.94 Å, and successive pendant atoms are 2.55 Å apart along the chains axis.

A single molecule of polyethylene extends through many unit cells. Molecules are long, and that makes obtaining perfect crystallisation difficult. Since molecules have different lengths, they do not end at the same position. Also the tendency to twist and turn makes perfect alignment difficult to achieve. The thickness of sheet-like single-crystal platelets of polyethylene is only about 100 Å, clearly not enough room for a polymer chain in that direction. On the other hand, it has been shown through electron diffraction data that the polymer chains are not laid out in the direction of the longer axes of these plates. The sheetlike crystals or lamellae consist of chains folded back and forth throughout the entire thickness of the platelets. Consequently, the surface of the crystal consists of hair-pin turns, as illustrated in Fig. 13.8.

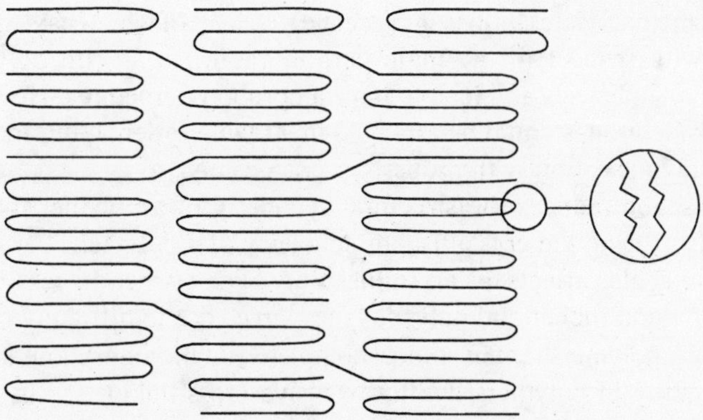

Fig. 13.8. Folded-chain lamellae of polyethylene.

The folded-chain picture is well suited for highly crystalline polymers. Most polymers are, however, semi-crystalline, having the characteristics of both crystalline solids and highly viscous liquids. This structure is revealed by X-ray and electron diffraction patterns of polymers, which often show the sharp features characteristic of crystalline solids as well as the diffuse features typical of liquids.

For semi-crystalline polymers it is often more advantageous to adopt the fringed-micelle picture or a model that uses the features of both the fringed-micelle and folded-chain concepts. Thus, as in the fringed-micelle concept, polymers may be pictured as two-phase systems composed of crystallites embedded in an amorphous matrix, whereas the structure of the crystallites may be that of the folded-chain lamella. The crystallites, promote rigidity, hardness, and heat resistance, and the amorphous regions give rise to flexibility. A certain proportion of the amorphous region is indeed necessary to provide needed flexibility for absorbing impact shocks; a completely crystalline polymer is too brittle.

Determinants of Polymer Crystallinity

The extent to which polymer molecules will crystallise depends on their structures and on the magnitudes of the secondary bond forces among the polymer chains: the greater the structural regularity and symmetry of the polymer molecule and the stronger the secondary forces, the greater the tendency toward crystallisation. We give a few examples:

1. Linear polyethylene has essentially the best structure for chain packing. Its molecular structure is very simple and perfectly regular, and the small methylene groups fit easily into a crystal lattice. Linear polyethylene (high density) therefore crystallises easily and to a high degree (over 90 per cent) even though its secondary forces are small. Branching impairs the regularity of the structure and makes chain packing difficult. Branched polyethylene (low density) is thus only partially (50–60 per cent) crystalline. Most of the differences in properties between low-density and high-density polyethylenes can be attributed to the higher crystallinity of the latter. Thus, linear polyethylenes have higher density than the branched material (density range of 0.95–0.97 g/cm^3 vs. 0.91–0.94 g/cm^3), higher melting point (typically 135° vs. 115°C), greater stiffness (modulus of 100,000 psi vs. 20,000 psi), greater tensile strength, greater hardness, and less permeability to gases and vapours.
2. Substituents hanging off polymer chains lead to difficulties in packing and generally decrease the tendency toward crystallisation. Moreover, crystallisation does not take place easily when polymer molecules have a low degree of symmetry. Thus, polymers such as polystyrene,

poly(methyl methacrylate), poly(vinyl acetate), etc., all of which have bulky side groups oriented at random with respect to the main carbon chain (in atactic polymers), show very poor crystallisation tendencies and tend to have amorphous structures. However, crystallinity would result if the side groups could be arranged in a regular orientation. Indeed, this can be done by controlled polymerisation with properly chosen catalysts.

3. Co-polymerisation reduces the structural symmetry of a polymer. Thus it is a very effective method of decreasing the crystallisation tendency of a polymer.
4. Chain flexibility also affects the crystallisability of a polymer. Excessive flexibility in a polymer chain, as in natural rubber and polysiloxanes, gives rise to difficulty in chain packing, with the result that such polymers remain almost completely in the amorphous state. In the other extreme, excessive rigidity in polymers due to extensive cross-linking, as in thermosetting resins like phenol-formaldehyde and urea-formaldehyde, also results in an inability to crystallise.
5. The presence of polar groups—such as amide, carboxyl, hydroxyl, chlorine, fluorine, and nitrile —along the polymer chains greatly increases the intermolecular or secondary attraction forces, which is favourable for crystallisation. However, high secondary forces alone may not give rise to high crystallinity unless the chain segments are aligned. Mechanical stretching of the polymer makes this alignment easier. For example, nylon-6,6 (a polyamide) has less than the expected degree of crystallinity of the unstretched condition and is used as a plastics. Highly crystalline strong fibres are produced by stretching (cold-drawing) the polyamide polymer 400 to 500 per cent. Mechanical stretching also makes it possible to develop a degree of order and crystallinity in several other thermoplastic resins that do not ordinarily crystallise. An unusual example of alignment and crystallisation on stretching is rubber.
6. The degree of crystallinity of polymeric materials is reduced by adding plasticisers. Crystallisation in many synthetic resins is not always desirable because it makes shaping more difficult and reduces transparency by closely packing the polymer molecules. Therefore, plasticisers which cause partial neutralisation of the intermolecular forces of attraction by coming between polymer molecules, thus enhancing flexibility and plasticity, are often added to the polymeric mass before shaping. The oldest example is celluloid, made by plasticising nitrocellulose (ordinarily a crystalline material) with camphor. Cellophane (regenerated cellulose film produced by a viscose process) is plasticised with glycerine to prevent crystallisation and loss of transparency. Polyvinyl chloride (PVC) is made flexible by adding plasticisers, such as dioctyl phthalate, for use as wire coating, upholstery, film and tubing. The unplasticised rigid PVC is used for the production of pipe, sheet, and moulded parts. The disadvantage of plasticisers is that they reduce the tensile strength and chemical resistance of the material.

AMORPHOUS STATE

The amorphous state is the characteristic of all polymers at temperatures above their melting points (except under special circumstances where liquid crystals may form). If a molten polymer retains its amorphous nature on cooling to the solid state, the process is called vitrification. In the vitrified amorphous state, the polymer resembles a glass. It is characteristic of those polymers in the solid state that, for reasons of structure, exhibit no tendency towards crystallisation. The amorphous solid state is characterised by glass transition (T_g), which is described in a latter section. We consider below only the behaviour of polymer melt.

When an amorphous polymer achieves a certain degree of rotational freedom, it can be deformed. If there is sufficient freedom, the molecules begin to move past one another and the polymer flows. The science of deformation and flow is called rheology. It is of fundamental importance in industrial applications since it is usually in the molten state that polymers are moulded into useful objects.

To cause a polymer to deform or flow requires the application of a force. If a force is applied and then withdrawn quickly, the polymer molecules tend to revert to their previous undeformed configuration, a process called relaxation. In other words, the polymer melt exhibits a certain elastic quality. This elasticity comes about because the molecules were disturbed from what was a thermodynamically favourable arrangement. If, however, the force is applied gradually and consistently, the molecules begin to flow irreversibly. (Silly putty, a siloxane polymer, is ideal for demonstrating this effect. If dropped, it bounces; but it can be shaped by the slow application of pressure). Because of entanglement of the polymer chains and frictional effects, the flowing liquid will be very viscous. This combination of properties, namely elasticity and viscous flow, is why polymers are referred to as viscoelastic materials.

A consequence of viscoelasticity of polymer melt is die swell, which refers to the fact that the thickness of the melt emerging through a narrow orifice or die is greater than the width of the die opening. This is explained as follows: as the molecules flow rapidly through the die opening, they are compressed, and when they emerge, the resultant reduction in pressure causes the molecules to "rebound" to a degree. This dimensional increase of the extrudate must be taken into account by engineers who design polymer processing machinery.

Chain entanglement that contributes to the high viscosity of polymer melt is clearly going to change as the molecular weight increases. Molecular weight is thus a critical variable in polymer rheology. Studies have shown that with flexible chain polymers, there exists a critical molecular weight ($\overline{M}_c$)for entanglement to begin. For most common polymers, $\overline{M}_c$ falls in the range 4000 to 15,000. Although $\overline{M}_c$ varies from one polymer to another, it has been shown that elimination of mass effects arising from substituents on the chains and calculation of chain length leads to a value that is remarkably constant from one polymer to another and corresponds to a $\overline{DP}$ of about 600. In other words, a critical chain length, rather than a critical molecular weight *per se*, is necessary for entanglement.

STRUCTURAL SHAPE OF POLYMER MOLECULES

Polymers can be classified, based on the structural shape of polymer molecules, as linear, branched, or cross-linked. Schematic representation are given in Fig. 13.9. Linear polymers have repeating units linked together in a continuous length (Fig. 13.9a). When branches protrude from the main polymer chain at irregular intervals, the polymer is termed a branched polymer. Branches may be long or short, forming a comblike structure (Fig. 13.9b), or divergent (Fig. 13.9c), forming a dendritelike structure. [Regularly repeating side groups which are a part of the monomer structure are not considered as branches. Thus polypropylene is a linear polymer, as are polystyrene and poly(methyl methacrylate)].

Both linear and branched polymers are thermoplastic; that is, they can be softened and hardened reversibly changing the temperature. Fabricating processes like injection moulding, extrustion moulding, casting, and blowing takes advantage of this feature to shape thermoplastic resins. The rigidity of thermoplastic resins at low temperatures is attributed to the existence of secondary bond forces between the polymer chains. These bonds are destroyed at higher temperatures, thereby causing fluidity of the resin.

Polymers used as textile fibres are linear. However, they must satisfy two additional requirements: (i) high molecular weight; and (ii) a permanent orientation of the molecules parallel to the fibre axis. The molecules must have a high degree of order and/or strong secondary forces to permit orientation and crystallisation. The chain orientation necessary to develop sufficient strength by crystallisation is achieved by a process known as cold drawing, in which the initially formed filaments (unoriented or only slightly oriented) are drawn at a temperature above the glass transition temperature (discussed later), which is the temperature at which sufficient energy is available to the molecular segments to cause them to begin to rotate.

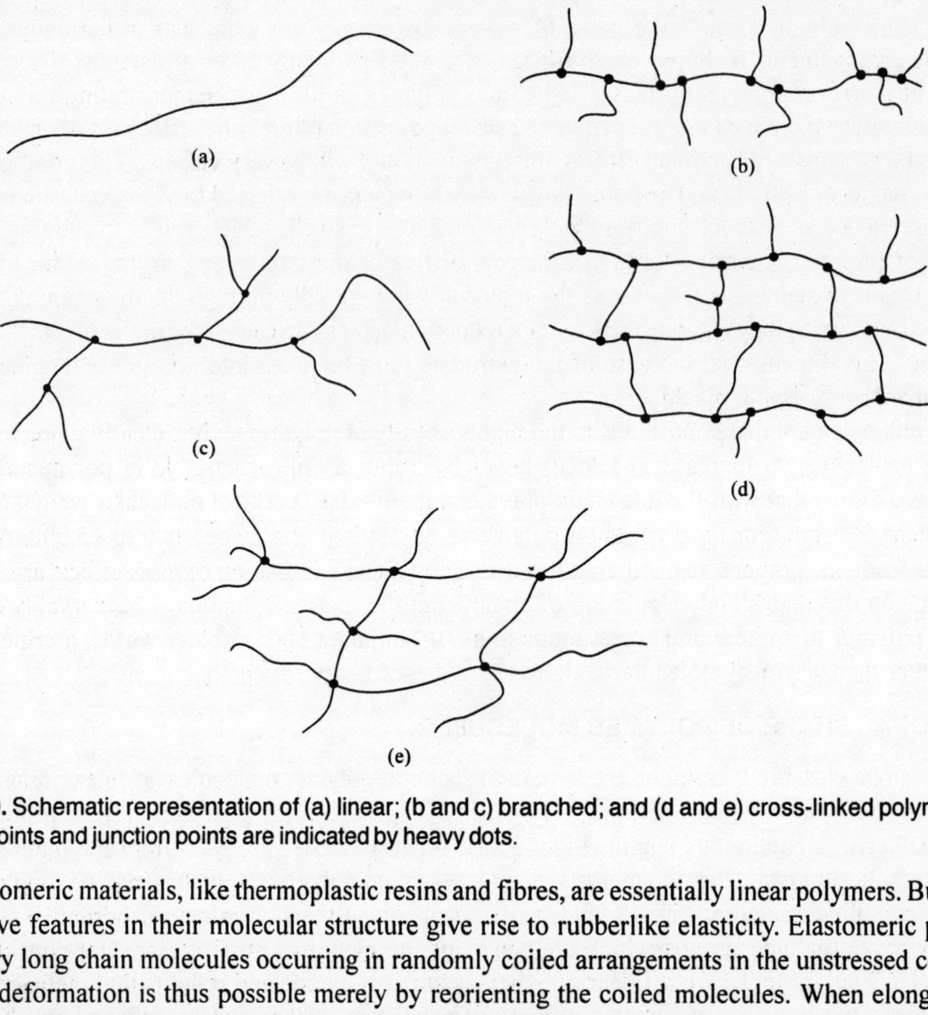

Fig. 13.9. Schematic representation of (a) linear; (b and c) branched; and (d and e) cross-linked polymers. The branch points and junction points are indicated by heavy dots.

Elastomeric materials, like thermoplastic resins and fibres, are essentially linear polymers. But certain distinctive features in their molecular structure give rise to rubberlike elasticity. Elastomeric polymers have very long chain molecules occurring in randomly coiled arrangements in the unstressed condition. A large deformation is thus possible merely by reorienting the coiled molecules. When elongated, the molecular coils partially open up and become aligned more or less parallel to the direction of elongation. The aligned configuration represents a less probable state or a state of lower entropy than a random arrangement. The aligned polymer chains therefore have a tendency to return to their original randomly coiled state. The large deformability of elastomeric materials is due to the presence of a certain internal mobility that allows rearranging the chain orientation, the absence of which in linear chain plastic materials (at normal temperatures) constitutes the essential difference between the two groups.

Although the aforesaid requirements are necessary conditions for ensuring a large extent of deformability, the remarkable characteristic of the rubbery state—namely, nearly complete recovery—cannot be obtained without a permanent network structure, since permanent deformation rather than elastic recovery will occur. A small amount of cross-linkage is necessary to provide this essential network structure. Natural rubber (polyisoprene), for example, simply flows like an extremely viscous liquid at room temperature if it is not cross-linked. Corss-linking is achieved by vulcanisation, the process of introducing cross-links between long-chain molecules. In commercial vulcanisation, sulphur cross-links are introduced into a rubber by heating raw rubber with sulphur (1–2 per cent by weight) and accelerating agents. Sulphur reacts with the double-bonded carbon atoms to produce a network structure, as shown schematically in Fig. 13.10. The amount of cross-linkage must be as small as possible to retain the structure; excessive cross-linkages will make the internal structure too stiff to permit even the required rearrangement of chain orientation during both deformation and recovery—in other words, it will destroy the rubbery state. An example of this is best furnished by ebonite, which is a rigid plastic made by vulcanising natural rubber with large quantities of sulphur.

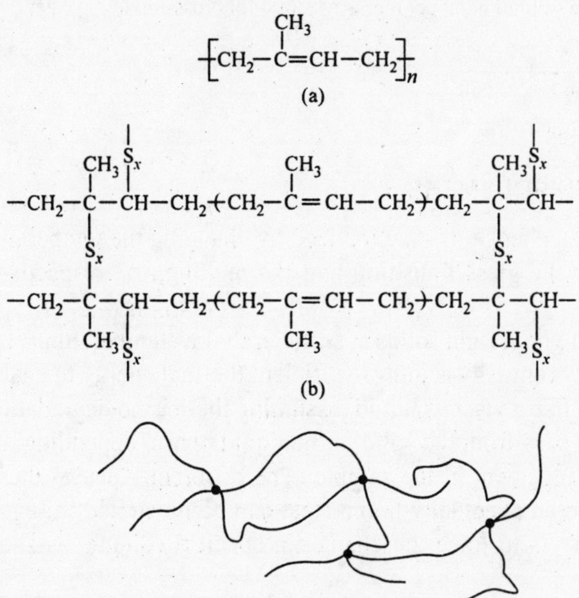

Fig. 13.10. Vulcanisation of natural rubber with sulphur. (a) linear polyisoprene (natural rubber); (b) vulcanised structure (idealised) showing cross-links of sulphide groups. The number, x, of sulphur atoms in a cross-link is 1 or 2 in efficient vulcanisation systems but may be as high as 8 under conditions where cyclic and other structures are also formed in the reaction; (c) the effect of cross-linking is to introduce points of linkage or anchor points between chain molecules, restricting their slippage. The cross-links in elastomers are typically a few hundred carbon atoms apart.

Example 13.1. (a) How much sulphur is required to fully cross-link natural rubber? (b) What is the sulphur content of vulcanised natural rubber that is 50% cross-linked ? (Assume that each cross-link contains one sulphur atom).

Solution. Mer weight of isoprene (Fig. 13.10a):

$C_5H_8 = (5)(12) + (8)(1) = 68$ g/mer

Assuming $x = 1$ in Fig. 13.10b, one sulphur atom, on the average, is required for cross-linking per mer of isoprene. Therefore,

(a) Amount of sulphur $= \frac{32}{68} x 100$

$= 47$ g/100 g of raw rubber

$$\text{(b) Sulphur content} = \frac{100(0.5)(32)}{(0.5)(32) + 68}$$
$$= 19\%$$

Example 13.2. A rubber contains 60% butadiene, 30% isoprene, 5% sulphur, and 5% carbon black. What fraction of possible cross-links are joined by vulcanisation? (Assume that all the sulphur is used in cross-linking).

Solution. 1 mer weight of butadiene (C_4H_6)

$= (4)(12) + (6)(1) = 54$ g/mer

1 mer weight of isoprene (C_5H_8)

$= (5)(12) + (8)(1) = 68$ g/mer

1 atomic weight of sulphur = 32

Assuming that, on the average, one sulphur atom per mer is required for cross-linking, we get

$$\text{Fraction of cross-links} = \frac{5/32}{60/54 + 30/68}$$
$$= 0.101 \text{ or } 10.1\%$$

THERMAL TRANSITIONS IN POLYMERS

The term 'transition' refers to a change of state induced by changing the temperatures or pressure. Two major thermal transitions are the glass transition and the melting, the respective temperatures being called T_g and T_m.

T_g and T_m: All polymers are hard rigid solids at sufficiently low temperatures, but as the temperature rises a thermoplastic polymer eventually acquires sufficient thermal energy to enable its chains to move freely enough for it to behave like a viscous liquid (assuming there is no degradation). There are several ways in which a polymer can pass from the solid to the liquid phase, depending on the structure of the polymer chains and their arrangement in the sample. The different types of thermal response in this transition from a rigid solid to an eventually liquid state can be illustrated in several ways. One of the simplest and most satisfactory is to trace the change in specific volume, as shown schematically in Fig. 13.11.

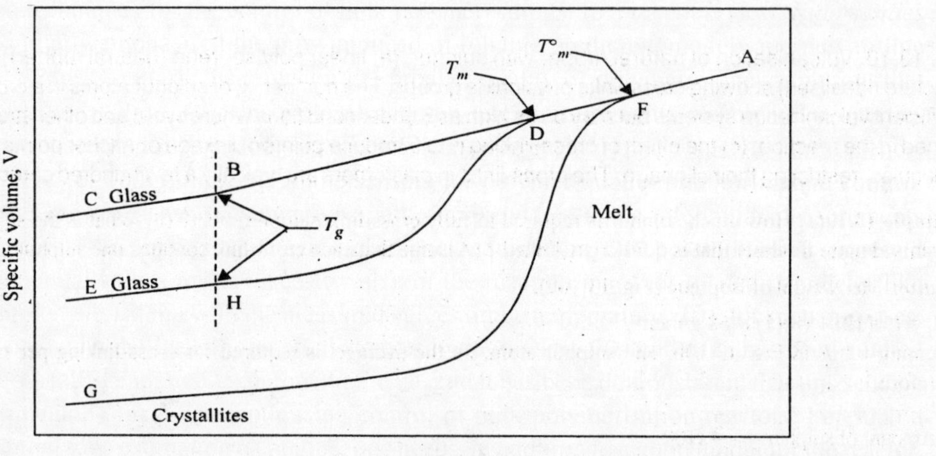

Fig. 13.11. Schematic representation of the change of specific volume of a polymer with temperature for a completely amorphous sample (A–B–C), a semi-crystalline sample (A–D–E), and a perfectly crystalline material (A–F–G).

In first-order transitions, such as melting, there is a discontinuity in the volume-temperature plot (Fig. 13.11) or enthalpy-temperature plot at the transition temperature. In second-order transitions, only a change in slope occurs and thus there is a marked change in the first derivative or temperature coefficients, as illustrated in Fig. 13.12. The glass transition is not a first-order transition, as no discontinuities are observed at T_g when the specific volume or entropy of the polymer is measured as a function of temperature. However, the first derivative of the property-temperature curve, i.e., the temperature coefficient of the property, exhibits a marked change in the vicinity of T_g; for this reason it is sometimes called a second-order transition.

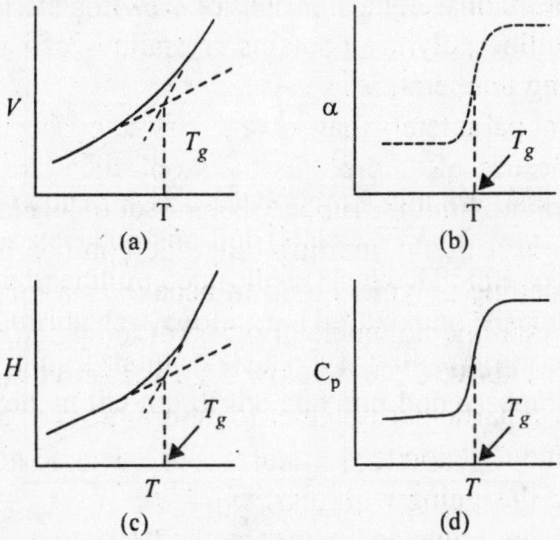

Fig. 13.12. Idealised variations in volume (V) and enthalpy (H). Also shown are α, the volume coefficient of expansion, and C_p, the heat capacity, which are, respectively, the first derivatives of V and H with respect to temperature (T).

The volumetric coefficient of expansion, α, is defined as

$$\alpha = \frac{1}{V}\left(\frac{\partial V}{\partial T}\right)_P \qquad \text{... (13.1)}$$

where V is the volume of the material, and α has the units °K^{-1}. While this quantity increases rather sharply at T_g, the increase actually occurs over a range (10–30°C). Similar changes occur in the heat capacity at constant pressure (C_p), which is the first derivative of enthalpy H with respect to temperature.

A thermoplastics polymer may be completely amorphous in the solid state, which means that the polymer molecular chains in the specimen are arranged in a totally random fashion. Referring to Fig. 13.11, the volume change in amorphous polymers follows the curve ABC. In the region C–B, the polymer is a glassy solid and has the characteristics of glasses, including hardness, stiffness and brittleness. In the glassy region, the available thermal energy (RT energy units/mol) is insufficient to allow rotation about single bonds in the polymer backbone overcoming intramolecular energy barriers, and movements of large (some 10–50 consecutive chain atoms) segments of the polymer chain cannot take place. But as the sample is heated, it passes through a temperature T_g, called the glass transition temperature, about which it softens and becomes rubber-like. This is an important temperature and marks the onset of extensive molecular motion which is reflected in marked changes in properties, such as specific volume, refractive index, stiffness, and hardness. Above T_g, the material may be more easily deformed or become ductile. A continuing increase in temperature along B–A leads to a change of the rubbery polymers to a viscous liquid without any sharp transition.

In a perfectly crystalline polymer, all the chains would be contained in regions of three-dimensional order, called crystallites, and no glass transition would be observed, because of the absence of disordered chains in the sample. A perfectly crystalline polymer, on heating, would follow curve G–F–A, melting at $T°_m$ to become a viscous liquid.

Perfectly crystalline polymers are, however, not encountered in practice and real polymers may instead contain varying proportions of ordered and disordered regions in the sample. These semi-crystalline polymers usually exhibit both T_g and T_m (not $T°_m$) corresponding to the disordered and ordered regions, respectively, and follow curves similar to E–H–D–A. As $T°_m$ is the melting temperature of a perfectly crystalline polymer of high molecular weight, T_m is lower than $T°_m$ and more often represents a melting range, because the semi-crystalline polymer contains crystallites of various sizes with many defects which act to depress the melting temperature.

Both T_g and T_m are important parameters that serve to characterise a given polymer. While T_g sets an upper temperature limit for the use of amorphous thermoplastics like poly(methyl methacrylate) or polystyrene and a lower temperature limit for rubbery behaviour of an elastomer like SBR rubber or 1,4-*cis*-polybutadiene, T_m or the onset of the melting range determines the upper service temperature. Between T_m and T_g, semi-crystalline polymers tend to behave as a tough and leathery material. As a general rule, however, semi-crystalline polymers are used at temperatures between T_g and a practical softening temperature which lies above T_g and below T_m. (The onset of softening is usually measured as the temperature required for a particular polymer to deform a given amount under a specified load. These values are known as heat deflection temperatures. Such data do not have any direct relation with T_m, but they are widely used in designing with plastics).

The T_g and T_m values for some polymers are shown in Table 13.1. In general, both T_g and T_m are affected in the same manner by considerations of polymer structure. Thus, both T_g and T_m increase with higher molecular symmetry, structural rigidity, and secondary forces of polymer chains.

The T_g and T_m values of a polymer determine the temperature range in which it can be employed. Amorphous elastomeric polymers, for example, must be used at temperatures (region B–D in Fig. 13.11) well above T_g to permit the high, local, segmental mobility required in such materials. Thus styrene-butadiene (25/75) co-polymer (T_g = –57°C), polyisoprene (T_g = –73°C), and polyisobutylene (T_g = –73°C) can be used as rubbers at ambient temperatures. Amorphous structural polymers, such as polystyrene and poly(methyl methacrylate), depend on their glasslike rigidity below T_g for their utility; they should therefore have high T_g values so that under ambient conditions they are well below T_g.

Table 13.1. Glass transition temperatures (T_g) and crystalline melting temperatures (T_m) of polymers.

Polymer	T_g *(°C)*	T_m *(°C)*
Polyethylene (high density)	– 115	137
Polyoxymethylene	– 85	181
Polyisoprene (natural rubber)	– 73	28
Polyisobutylene	– 73	44
Polypropylene	– 20	176
Poly(vinylidene chloride)	– 19	190
Poly(chlorotrifluoroethylene) (kel-F)	45	220
Poly(hexamethylene adipamide) (nylon-6,6)	53	265

(Contd ...)

Polymer	T_g (°C)	T_m (°C)
Poly(ethylene terephthalate) (Terylene, Dacron)	69	265
Poly(vinyl chloride)	81	212
Polystyrene	100	240
Poly(methyl methacrylate) (Perspex, Lucite)	105	200
Cellulose triacetate	105	306
Polytetrafluoroethylene (Teflon)	127	327

Tough, leatherlike polymers are limited for use in the immediate vicinity of their T_g. Such behaviour is observed in vinyl chloride-based plastics, which are used as substitutes for leather in automobile seat covers, travel luggage, and ladies' handbags. Highly crystalline fibre-forming polymers must be used at temperatures substantially below T_m (about 100°C), since changes in crystal structure can occur as T_m is approached. The T_m of a fibre must therefore be above 200°C to remain unaffected at use temperatures encountered in cleaning and ironing. (T_m should not, however, be excessively high—not more than 300°C; otherwise spinning of the fibre by melt-spining processes may not be possible). The T_g of a fibre, on the other hand, should have an intermediate value, because too high a value of T_g would interfere with the stretching operation as well as with ironing, and too low a value of T_g would not permit crease retention in fabrics. Nylon and terylene, as may be seen from Table 13.1, therefore have optimal values of T_m and T_g. Semi-crystalline polymers with about 50 per cent crystallinity are used at temperatures between T_g and T_m, since in this range the material exhibits moderate rigidity and a high degree of toughness, somewhat analogous to reinforced rubber. Branched polyethylene (low density), with $T_g = -120$°C and $T_m = 115$°C, used at ambient temperatures is a typical example.

Regions of Viscoelastic Behaviour

As we have seen above, the transition that separates the glassy state from the viscous state is known as the glass-rubber transition. This transition attains the properties of a second-order transition at very slow rates of heating or cooling. In order to clearly locate the region of this transition and to provide a broader picture of the temperature dependence of polymer properties, the principal regions of viscoelastic behaviour of polymers will be briefly discussed.

Broadly, there are five regions of viscoelastic behaviour for linear amorphous polymers as shown in Fig. 13.13. In region I (a to b), the polymer is glassy and frequently brittle. Typical examples are polystyrene and poly(methyl methacrylate) at room temperature. Young's modulus in this region just below the glass transition temperature is approximately 3×10^{10} dyne/cm^2 (3×10^9 Pa) and it is nearly the same for a wide range of polymers.

Range II (b to c) is the glass transition region. Typically, the modulus drops a factor of about a thousand in a 20–30°C range. The behaviour of polymers in this region is best described as leathery. For static or quasistatic measurements, such as illustrated in Fig. 13.13, the glass transition temperature, T_g, is often taken at the maximum rate of turndown of the modulus at the elbow, i.e., where d^2E/dT^2 is at a maximum. A few T_g values are shown in Table 13.1.

Region III (c to d) in Fig. 13.13 is the rubbery plateau region. After a sharp drop that the modulus exhibits in the glass transition region, it again becomes almost constant in the rubbery plateau region, with typical values of 2×10^7 dyne/cm^2 (2×10^6 Pa). In this region, polymers exhibit high rubber elasticity, so much so that an elastomer in this region can be stretched perhaps several hundred per cent and substantially snap back to its original length on being released.

In region III (c to d) of Fig. 13.13, two cases need be distinguished. First if the polymer is linear, the solid line is followed on which the modulus drops off slowly with increasing temperature. The width of the plateau is governed primarily by the molecular weight of the polymer; the higher the molecular weight, the longer the plateau. Second, if the polymer is cross-linked, the dashed line in Fig. 13.13 is followed, and improved rubber elasticity is observed, with the crepe portion suppressed. A common example of a cross-linked polymer above its glass transition temperature is the ordinary rubber band. For cross-linked polymers region III remains in effect at higher temperatures up to the decomposition temperature, and regions IV and V do not occur.

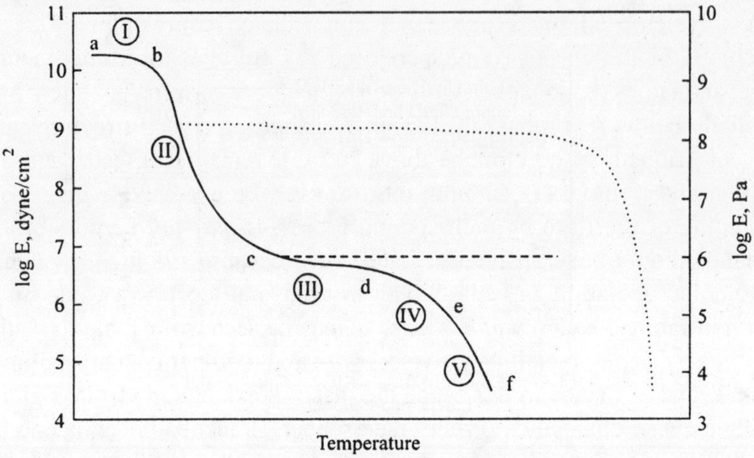

Fig. 13.13. Five regions of viscoelastic behaviour for a linear, amorphous polymer: I (a to b); II (b to c); III (c to d); IV (d to e); and V (e to f). Also illustrated are effects of crystallinity (dotted line) and cross-linking (dashed line).

The discussion above has been limited to amorphous polymers. However, if the polymer is semi-crystalline, the dotted line in Fig. 13.13 is followed. Since the crystalline regions in the polymer matrix tend to behave as a filler phase and also as a type of physical cross-link between the chains, the height of the plateau (i.e., the modulus) will be governed by the degree of crystallinity.

In the case of linear amorphous polymers, raising the temperature past the rubbery plateau region brings them to the rubbery flow region—region IV. In this region, the polymer exhibits both rubber elasticity and flow properties depending on the time scale of the experiment. For experiments performed in a short time, the physical entanglements of polymer chains are not able to relax and the material still behaves rubbery. For longer duration experiments, the increased molecular motion imparted by the increased temperature cause chains to move, resulting in visible flow. An example of a material in the rubbery flow region at ambient temperature is Silly Putty which bounces like a ball when thrown (quick experiment) but deforms like a taffy when pulled (a much slower experiment).

Region V (e to f) in Fig. 13.13 is the liquid flow region which is reached at still higher temperatures where the increased kinetic energy of the chains permits them wriggle out through entanglements rapidly and move as individual molecules, often producing highly viscous flow. This is the melting temperature and it is always above the glass transition temperature.

Example 13.3. A new polymer was reported to soften at 60°C but it is not known for sure whether the softening was a glass transition or a melting point. Describe a simple experiment to distinguish between the two posibilities. It is known, however, that the new polymer is essentially a linear polymer.

Solution. If 60°C is a glass transition, then heating the polymer slowly past 60°C would take it to the rubbery plateau region (region III in Fig. 13.13), where the modulus, E, and hence hardness, would remain fairly constant with increase of temperature. For a melting transition, however, the modulus would drop rapidly and the polymer would become increasingly softer in a similar experiment.

Pressing one's thumb in an object is a simple way to gauze the object's hardness. Its scientific analog is the measurement of hardness by indentation. In practice, the point of a weighted needle is allowed to rest on the polymer surface as the temperature is raised. The movement of the needle as it penetrates the surface can be monitored by means of an amplification gauge. Though less accurate than other more sophisticated methods, it is useful for the preliminary engineering-oriented examination of systems.

Factors Affecting T_g

As T_g marks the onset of molecular motion, a number of factors which affect rotation about links (necessary for movement of polymer chains) will also influence the T_g of a polymer. These include (i) chain flexibility; (ii) molecular structure (steric effects); (iii) molecular weight; and (iv) branching and cross-linking.

Chain flexibility

The flexibility of the chain is undoubtedly the most important factor influencing the T_g of a polymer. The chain flexibility depends more on the rotation or torsion of skeletal bonds than on changes in bond angles. When a randomly coiled chain is pulled out into an elongated conformation, the skeletal bonds 'unwind' rather than undergo angular distortion (Fig. 13.14). Thus, flexibility on a macroscopic scale depends on torsional mobility at the molecular level. If a highly flexible chain is present, T_g will generally be low and if the chain is rigid, the T_g value will be high.

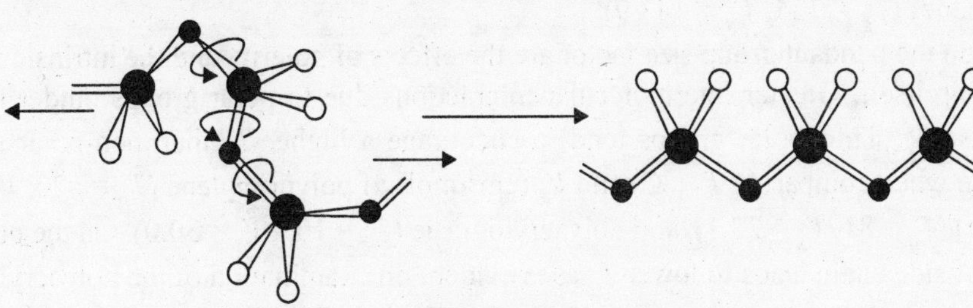

Fig. 13.14. Elasticity of a polymer such as silicone rubber depends on the ease with which a random coil chain can be stretched out. The stretching is a consequence of the unwinding of bonds rather than a marked widening of bond angles.

For symmetrical polymers, the chemical nature of the backbone chain is the important factor determining the chain flexibility and hence T_g. Chains made up of bond sequences that are able to rotate easily are flexible, and hence polymers containing —(—CH_2—CH_2—)—, —(—CH_2—O—CH_2—)—, or —(—Si—O—Si—)— links will have correspondingly low values of T_g. For example, poly (dimethyl siloxane) has one of the lowest T_g values known (–123°C) presumably because the Si—O bonds have considerable torsional mobility. The value of T_g is raised markedly by the insertion of groups that stiffen the chain by impeding rotation, so that more thermal energy is required to set the chain in motion. Particularly effective in this respect is the *p*-phenylene ring.

Thus, a chain consisting entirely of *p*-phenylene rings, namely, poly*p*-phenylene) (II) has a highly intractable, rigid structure with no softening point.

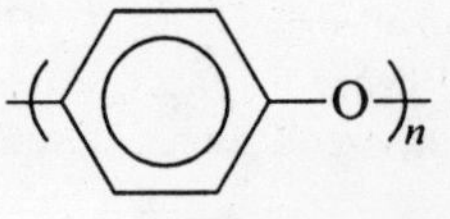

(II)

This sturucture can be modified by introducing flexible groups in the chain to produce tractable polymers with high values of T_g.

Some examples are poly(phenylene oxide) (III), $T_g = 83°C$ and poly(*p*-xylylene) (IV), $T_g \simeq 280°C$ as compared to polyethylene, $T_g = -93°C$, and poly(ethylene oxide), $T_g = -67°C$.

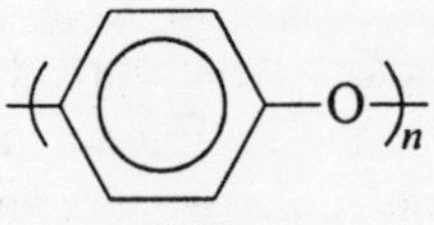

(III)

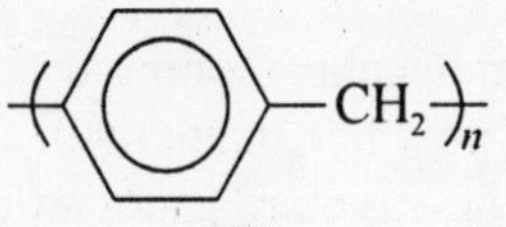

(IV)

Steric effects

When the polymer chains are unsymmetrical, with repeat units of the type —(—CH_2—CHX—)—, an additional restriction is imposed by steric effects depending on the size of the pendant group X. Bulky pendant groups hinder the rotation about the backbone and cause T_g to increase. There is some evidence of a correlation between T_g and the size of the pendant group measured by its molar volume V_x. For example, T_g (°C) increases with the increasing V_x (cm^3/mol) in the progressive series, polyethylene ($T_g = -93$, $V_x = 3.7$), polypropylene ($T_g = -20$, $V_x = 25.9$), polystyrene ($T_g = 100$, $V_x = 92.3$), and poly(vinylnaphthalene) ($T_g = 135$, $V_x = 143.9$).

Superimposed on the pendant group size factor are the effects of polarity and the intrinsic flexibility of the pendant group itself. Greater intermolecular interactions due to polar groups hinder molecular motion and increase T_g. Thus polar groups tend to encourage a higher T_g than non-polar groups of similar size, as seen when comparing T_g (°C) and V_x (cm^3/mol) of polypropylene ($T_g = -20$, $V_x = 25.9$), poly(vinyl chloride) ($T_g = 81$, $V_x = 22.1$), and polyacrylonitrile ($T_g = 105$, $V_x = 30.0$). On the other hand, greater flexibility of side chain leads to lower T_g as is evident on examination of the polyacrylate series from methyl through butyl: poly(ethylacrylate) ($T_g = -24$, $V_x = 82.3$), poly(propyl acrylate) ($T_g = -48$, $V_x = 104.5$), and poly(butyl acrylate) ($T_g = -55$, $V_x = 126.7$), where T_g is in °C and V_x is in cm^3/mol.

A further increase in steric hindrance is caused by the presence of an α-methyl group, which restricts rotation even further, thus causing T_g to increase. Typical examples are the pair polystyrene ($T_g = 100°C$)–poly(α-methylstyrene) ($T_g = 172°C$) and the pair poly(methyl acrylate) ($T_g = 6°C$)–poly (methyl methacrylate) ($T_g = 105°C$).

Configurational effects

It should be noted that the steric effects of the pendant groups considered above are simply additional contributions to the main chain effects. Similarly *cis-trans* isomerism in polydienes and tacticity variations in certain α-methyl substituted polymers alter chain flexibility and hence affect T_g. Well-known examples of *cis-trans* variations are polybutadiene *cis* ($T_g = -108°C$) and *trans* ($T_g = -18°C$) or polyisoprene *cis* ($T_g = -73°C$) and *trans* ($T_g = -53°C$). An example of tacticity variation is poly(methyl methacrylate) for which the isotactic, atactic, and syndiotactic stereostructures are associated with T_g values of 45, 105, and 115°C, respectively.

Effect of cross-linking

When cross-links are introduced into a polymer, the molecular motion in the sample is restricted and T_g rises. This transition is ill-defined for a high cross-like density, but at lower values T_g is found to increase linearly with the number of cross-links.

Factors Affecting T_m

The application of macroscopic thermodynamics leads to some useful generalisations of factors affecting T_m. At T_m, the free energy change is zero, i.e.,

$$\Delta G_m = \Delta H_m - T_m \Delta S_m = 0$$

where $T_m = \Delta H_m / \Delta S_m$. This expression for T_m predicts that a high melting point can be the result of a high value of the enthalpy change ΔH_m and/or a small value of the entropy change ΔS_m in melting. The former corresponds to stronger binding of adjacent but unbounded units in the polymer lattice and thus to higher degree of crystallinity. The factors that affect crystallinity and T_m can be classified as symmetry, intermolecular bounding, tacticity, branching, and molecular weight. These are discussed below. If ΔS_m is small, melting does not result in a large gain of conformational entropy and, to some degree, the structure of the solid must persist in the melt. For example, molecules of isotactic polypropylene crystallise in the form of helices and these are thought to occur in the melt as well, thus making ΔS_m small and T_m high.

Symmetry

The formation of stable crystalline regions (crystallites) in a polymer requires (i) that an efficient close-packed arrangement of the polymer chains can be achieved in three dimensions; and (ii) that a favourable change in internal energy is obtained during this process. This imposes restrictions on the type of chain that can form crystallites easily and one would expect linear chains with high degree of symmetry such as polyethylene, polytetrafluoroethylene, various polyesters, and polyamides to crystallise most readily.

Linear symmetrical molecules such as polyethylene, polytetrafluoroethylene, and other linear molecules with more complex backbones containing —(—O—)—, —(—COO—)—, and —(—CONH—)— groups, such as polyethers, polyesters, and polyamides, all possess a suitable symmetry for crystallite formation and usually assume extended zigzag conformations when aligned in the lattice. On the other hand, chains containing irregular units, which impair the symmetry, reduce the ability of the polymer to crystallise. Thus *cis* double bonds (V), *o*-phenylene groups (VI), and *m*-phenylene groups (VII) all encourage bending and twisting in the chains and make close packing very difficult.

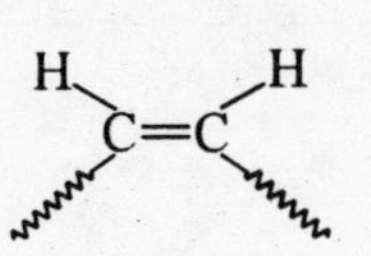

(V)

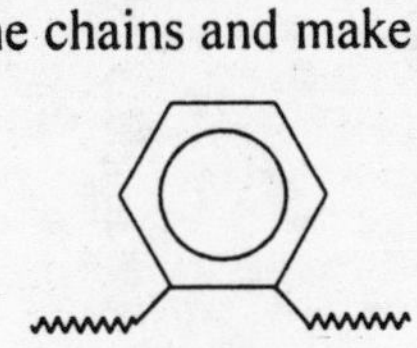

(VI)

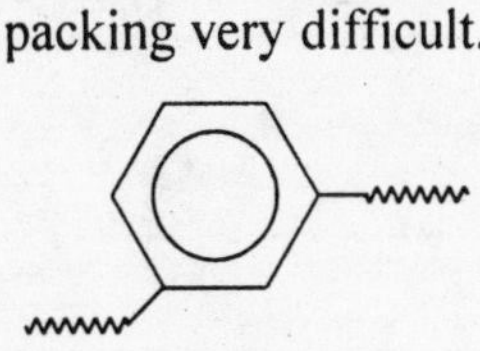

(VII)

If, however, the phenylene rings are *para*-oriented, the chains retain their axial symmetry and can crystallise more readily. Similarly, double bonds in *trans* configuration maintain the chain symmetry thus allowing for crystallite formation.

This is highlighted by a comparison of the amorphous elastomeric *cis*-polyisoprene (T_m = 28°C) with highly crystalline *trans*-polyisoprene (T_m = 74°C), which is a non-elastomeric rigid polymer, or *cis*-1,4-polybutadiene (T_m = −11°C) with *trans*-1,4-polybutadiene (T_m = 148°C).

Another aspect of high chain symmetry is the possibility of molecular motion within the crystal lattice contributing to higher T_m. For example, polyethylene and polytetrafluoroethylene are both sufficiently symmetrical to be considered as smooth, stiff cylindrical rods. In the crystal, these rods tend to roll over each other and change position when thermally agitated. This motion within the crystal lattice, called pre-melting, effectively stabilises the lattice. Consequently, more thermal energy is required to break down the crystallite into a disordered melt, and T_m is raised. Irregularly shaped polymers with bends and bumps in the chain cannot move in this way without disrupting the crystal lattice, and so have lower T_m values.

Intermolecular bonding

Any interaction between polymer chains in the crystal lattice serves to hold the structure more firmly and raise the melting temperature. In polyethylene, the close packing in crystallites achieved due to high chain symmetry and lack of substituents on the chains allows the van der Waals' forces to act cooperatively and provide additional stability to the lattice. In polymers containing polar substituents, for example, Cl, CN, or OH, the chains can be aligned and held rigidly by the strong dipole-dipole interactions between the substituents. This effect is more obvious in the symmetrical polyamides. These polymers can form intermolecular hydrogen bonds which greatly enhance crystallite stability and raise the T_m significantly. This is illustrated in Fig. 13.15 for nylon-6,6, where the extended zigzag structure greatly facilitates the formation of regular intermolecular hydrogen bonding. The increased stability is reflected in the value of T_m, which for nylon-6,6 is 267°C, compared with 137°C for high-density polyethylene.

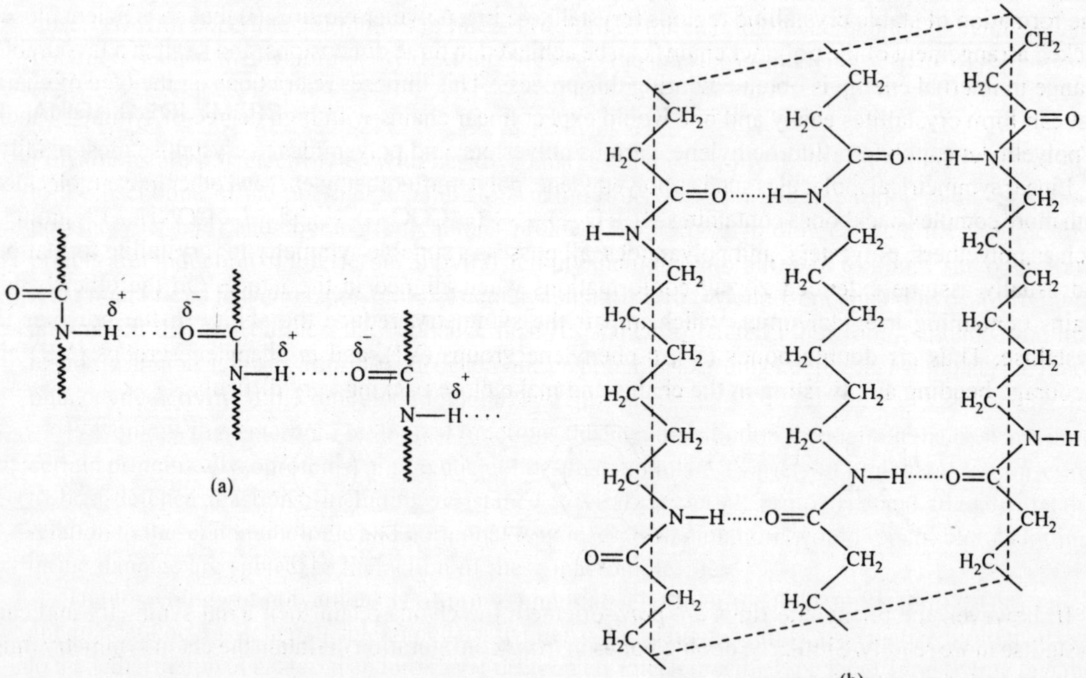

Fig. 13.15. (a) Hydrogen bonds between neighbouring chains of polyamide. (b) Arrangement of chains in hydrogen-bonded sheets in the crystal structure of nylon-6,6.

Tacticity

Stiffening of the polymer chain due to the presence of large pendant groups tends to raise T_m, but this will also increase the difficulty of close packing to form a crystalline array and so tend to lower T_m. However, this latter problem can be overcome if the pendant groups are arranged in a regular fashion along the chain. Thus isotactic polymers tend to form helices to accommodate the substituents in the most stable steric positions, and being of regular forms these helices are capable of regular alignment promoting crystallite formation. Thus isotactic polystyrene is semi-crystalline (T_m = 240°C), though atactic polystyrene is amorphous. Syndiotactic polymers are also sufficiently regular to crystallise, not necessarily as a helix, but rather in glide planes.

Branching, chain flexibility, and molecular weight

Chain flexibility has a direct bearing on the melting point. Insertion of groups that stiffen the chain increases T_m, while introducing flexible groups into the chain lowers the value of T_m (see 'Factors Affecting T_g'). Branching in the side group tends to stiffen the chain and raise T_m, as shown in the series poly(but-1-ene) (T_m = 126°C), poly(3-methyl but-1-ene) (T_m = 145°C), poly(3,3'- dimethyl but-1-ene) (T_m > 320°C). However, if the side group is flexible and non-polar, T_m is lowered. Also, if the chain is substantially branched to reduce the packing efficiency, the crystalline content is lowered, as is the melting point. A good example is low-density polyethylene where extensive branching lowers the density and T_m of the polymer. Molecular weight in relatively lower order can have significant effect on T_m. Since chain ends are relatively free to move and the number of chain ends increases with the decrease in molecular weight, T_m is lowered because less energy is then required to stimulate chain motion and melting. For example, polypropylene with molecular weight 2000 has T_m = 114°C, whereas a sample with molecular weight 30,000 has T_m = 170°C.

Relation between T_m and T_g

While T_m is a first-order transition, T_g is a second-order transition and this precludes the possibility of a simple relation between them. There is , however, a crude relation between T_m, and T_g. Thus the ratio T_g/T_m ranges from 0.5 to 0.75 when the temperatures are expressed in degrees Kelvin. The relation is represented in Fig. 13.16 where a broad band covers most of the results for linear homopolymers and the ratio (T_g/T_m) lies between 0.5 and 0.75 for about 80% of these. The ratio is closer to 0.5 for symmetrical polymers such as polyethylene and polybutadiene, but closer to 0.75 for unsymmetrical polymers, such as polystyrene and polyisoprene. The difference in these values may be related to the fact that in unsymmetrical chains with repeat units of the type —(CH_2—CHX—)— an additional restriction to rotation is imposed by steric effects causing T_g to increase, and conversely, an increase in symmetry lowers T_g.

GLASS TRANSITION TEMPERATURE AND MOLECULAR WEIGHT

The glass transition temperature of a polymer is influenced by its molecular weight, at least up to around a value of 20,000; beyond this, the effect of the molecular weight is not pronounced and this can be seen from Fig. 13.17. Many attempts have been made to compute mathematical relationships, among which the following two are frequently used:

$$T_g = T_g^{\infty} - \frac{K}{M_n} \qquad \text{... (13.2)}$$

$$\frac{1}{T_g} = \frac{1}{T_g^{\infty}} + \frac{A}{M_n} \qquad \text{... (13.3)}$$

where T_g^{∞} is the glass transition temperature at infinite molecular weight and K and A are arbitrary constants.

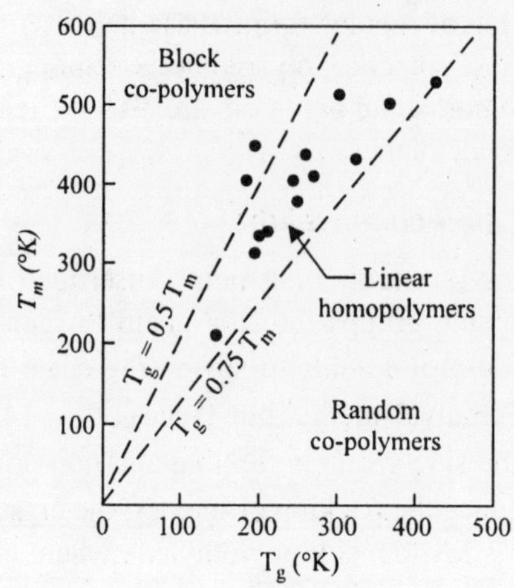

Fig. 13.16. Plot of T_m against T_g for linear homopolymers with T_g/T_m in the range 0.5 and 0.75 broadly demarcating $T_m - T_g$ domains covered by homopolymers and co-polymers.

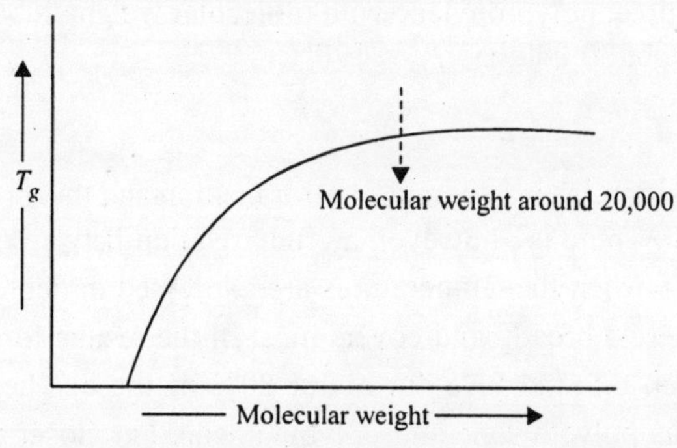

Fig. 13.17. Plot showing effect of molecular weight on glass transition temperature of polymers.

Now, let us see how the molecular weight can affect the T_g values. We know that polymer chains made of hundreds of monomeric units comprise several segments, each made of a few tens of monomeric units. We also know from earlier discussions that the mobility of chain segments influences the glass transition temperature. The segments that are away from the chain ends, although part of the main chain, can be thought of as connected at the two ends to the main chain (Fig. 13.18) and, hence, their mobility is rather restricted. The two end segments containing the chain ends, on the other hand, are connected to the main chain at one of their ends only and, hence, have more freedom for motion. For a given weight of the polymer, a low molecular weight sample will have more chain end segments than a high molecular weight sample. The larger the number of chain end segments, the larger will be the effective segmental motion. Thus, the T_g value will be lower for low molecular weight polymers.

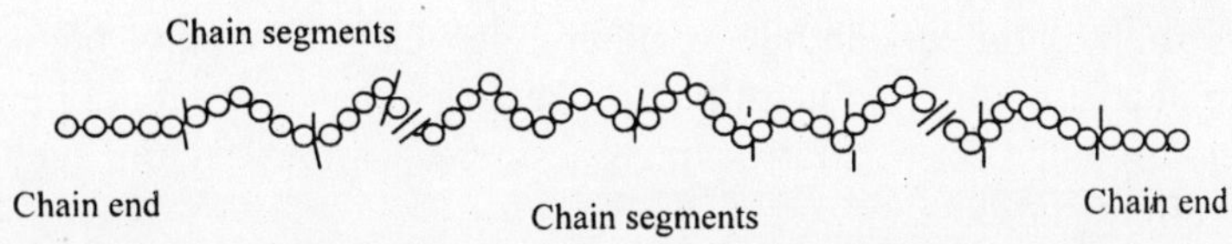

Fig. 13.18. Schematic representation of a polymer chain showing chain ends and chain segments.

GLASS TRANSITION TEMPERATURE AND PLASTICISERS

Plasticisers are low molecular weight non-volatile substances (mostly liquids), which when added to a polymers improve its flexibility, processibility and, hence, utility. The plasticiser substantially reduces the brittleness of many amorphous polymers because its addition even in small quantities markedly reduces the T_g of the polymer. This effect is due to a reduction in cohesive forces of attraction between polymer chains. Plasticiser molecules (being relatively small in size compared to polymer molecules) penetrate into the polymer matrix and establish polar attractive forces between it and the chain segments. These attractive forces reduce the cohesive forces between the polymer chains and increase the segmental mobility, thereby reducing the T_g value.

The reduction in T_g with the addition of different plasticisers to a polystyrene sample can be seen in Fig. 13.19. It would appear that although all plasticisers reduce T_g considerably, the extent of reduction (i.e., the 'plasticiser efficiency') is different, and depends on factors such as solubility parameter, polarity and density.

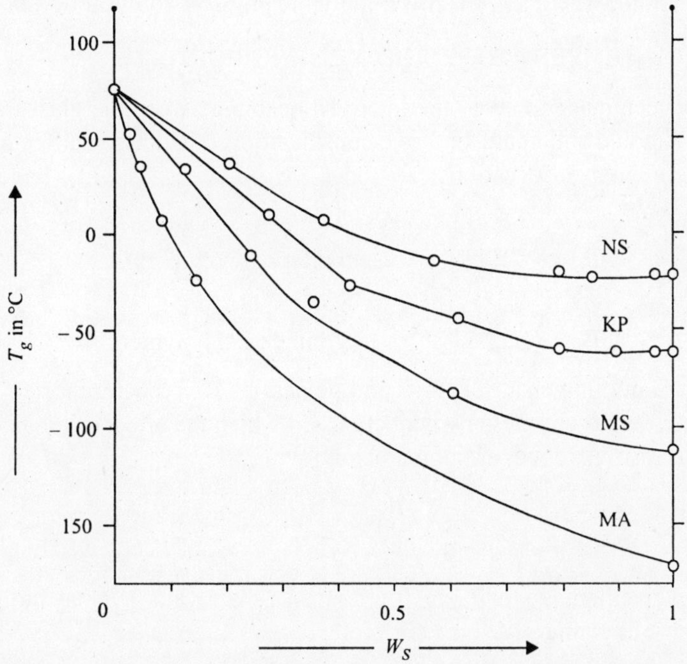

Fig. 13.19. Plot showing lowering of glass transition temperature of a polystyrene sample by different weight fractions (W_s) of various plasticisers. (NS= naphthyl salicylate, KP = tricresyl phosphate, MS = methyl salicylate and MA = methyl acetate).

Many plasticisers such as dibutyl phthalate, dioctyl sebacate, diisooctyl phthalate and tricresyl phosphate are used to decrease the T_g value of PVC. Selecting a plasticiser for commercial application is obviously a matter of compromise, after having considered various factors such as compatibility, efficiency, migration effect and cost.

GLASS TRANSITION TEMPERATURE OF CO-POLYMERS

We will take the case of a random co-polymer. The net effect of the interaction between two monomeric components of co-polymer on the T_g value is as though the co-polymer is a mixture of two homopolymers. The T_g value of random co-polymers, therefore, depends on the T_g value of the individual homopolymers and on the ratio of the two monomeric components. This leads to the formulation of an empirical relation:

$$\frac{1}{T_{g(AB)}} = \frac{W_A}{T_{g(A)}} + \frac{W_B}{T_{g(B)}} \qquad \text{... (13.4)}$$

where $T_{g(A)}$, $T_{g(B)}$ and $T_{g(AB)}$ are the glass transition temperatures of homopolymers A and B and the co-polymer AB, respectively, and W_A and W_B are the weight fractions of the respective monomer components in the co-polymer. It can be seen from this equation that the glass transition temperature of a co-polymer will be in between those of the respective homopolymers. The glass transition temperature of many homopolymers could thus be lowered by co-polymerisation with small quantities of another suitable monomer. This process is generally referred to as 'internal plasticisation', which is in contrast to 'external plasticisation', where in T_g is lowered by the physical addition of a plasticiser.

GLASS TRANSITION TEMPERATURE AND MELTING POINT

Many attempts have been made to study the interrelationship between the glass transition temperature and melting point. We could anticipate, on basic considerations, that factors affecting T_g will also affect T_m. Based on experimental observations, T_g and T_m (in degrees Kelvin), have been shown to be interconnected as follows:

$$T_g = \frac{1}{2} T_m \qquad \text{(for symmetrical polymers)} \qquad \text{... (13.5)}$$

$$T_g = \frac{2}{3} T_m \qquad \text{(for unsymmetrical polymers)} \qquad \text{... (13.6)}$$

However, these are not universal relations and deviations have been reported in the case of many polymers. A combined version of these two equations, in which the effect of molecular symmetry is not taken into account, is also in vogue, as shown below:

$$\frac{1}{2} < \frac{T_g}{T_m} < \frac{2}{3} \qquad \text{... (13.7)}$$

Since this relation gives a range for T_g/T_m (rather than a sharp value), it may more realistically reflect the thermal behaviour of polymers.

IMPORTANCE OF GLASS TRANSITION TEMPERATURE

The glass transition temperature is an important parameter of a polymeric material. It is used as a measure for evaluating the flexibility of a polymer molecule and the type of response the polymeric material would exhibit to mechanical stress. The T_g value of a polymer, therefore, decides whether a polymer at the 'use temperature' will behave like rubber or plastic. At present, it would suffice to know that polymers above their T_g will be soft and flexible and exhibit a delayed elastic response (viscoelasticity), while those below their T_g will be hard and brittle and will possess dimensional stability.

The T_g value along with the T_m value gives an indication of the temperature region at which a polymeric material transforms from a rigid solid to a soft viscous state. This helps in choosing the right processing temperature, i.e., the temperature region in which the material can be converted into finished products through different processing techniques such as moulding, calendering and extrusion.

HEAT DISTORTION TEMPERATURE

Closely associated with T_g is the heat distortion temperature. While T_g gives a temperature range over which a hard polymeric material tends to become soften under static conditions, the heat distortion temperature indicates the temperature range over which the polymer begins to soften and get deformed under the influence of a load. In the polymer industry, the heat distortion temperature is, in fact, more often used (than T_g) as an empirical method of comparing the transition behaviour of different polymers from their hard solid state to the soft viscous state. The heat distortion temperature can be expressed as the narrow temperature range over which a standard test specimen at a standard rate of heating gets distorted to a fixed extent under a given load. It will be close to T_g for amorphous polymers and close to T_m for highly crystalline polymers.

ROTATIONAL TRANSITIONS

Many of the bonds in a polymer chain are single bonds, which are capable of rotation. Such rotation will be opposed by forces due to the interaction between adjacent atoms and groups in the chain and neighbouring chains. If these are greater than the rotational energy, no rotation will occur. Consider the chain shown in Fig. 13.20(a). This represents the last seventeen units of a polymer molecule. If the only rotation were one of 180° about bond 1, all seventeen chain units would have to change their position relative to adjacent molecules. Not only would the intermolecular forces be much too large to allow this, but there would not be room in the mass of the material for it to occur. If, however, two 180° rotations, one about bond 1 and the other about bond 2, were to occur simultaneously, only four units of the chain would have to be displaced. This would not require so much work against inter-molecular forces, and there may well be sufficient room for it to occur. The new orientation of the molecule would be as shown in Fig.13.20(b). A change of this nature would alter the position of the molecule relative to its neighbours.

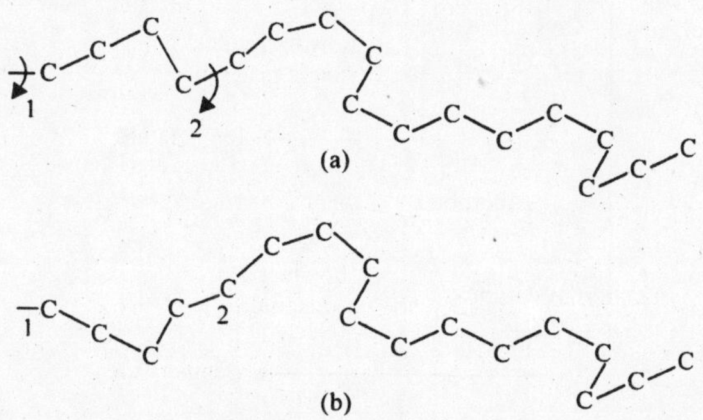

Fig. 13.20. Effect of two 180° rotations on the spatial arrangement of a carbon chain.

Provided there is sufficient rotational energy available, this type of rotational transition will be occurring in all the molecules throughout a material. The chains are in a state of continual motion relative to one another. The analogy to a compact mass of intertwined, wriggling, worms can be drawn. For a given temperature, the activity of the chains will depend on the size and nature of the inter-molecular forces opposing rotation, and also on the compactness of the chain units. Molecules, like polythene can rotate easily due to their compactness. Molecules with large bulky chain units cannot easily rotate, considerable room being required for even small rotational transitions.

CRYSTALLINE MELTING POINT

Most polymeric materials break down chemically before they reach a temperature at which they become free-flowing liquids. Nevertheless, most of them can be heated to a stage where they are viscous liquids, rather than solids. In this state, the molecules move easily by rotational transitions, and have a continually changing, random arrangement. If a material in this state is cooled, movement of the molecules relative to one another become increasingly difficult. The liquid becomes more and more viscous, but the arrangement of the molecules is still random. If the material is not capable of crystallite formation, this process will continue, and the viscosity will become so great that the material becomes an amorphous solid. If the material can become crystalline, a much more drastic change occurs. When a low molecular weight material is cooled from the liquid state, a temperature is reached at which there is a sudden change in many physical properties, such as specific volume and heat capacity. The structural change is from complete disorder to complete order, as the material changes from a liquid to a completely crystalline solid. By very rapid cooling, it is sometimes possible to supercool the liquid, but this state is extremely unstable, and complete crystallisation occurs, following the slightest impetus.

A polymeric material, cooling from the liquid state, will exhibit similar changes, if it is capable of crystallising. The changes in properties are not so marked, and usually occur over a range of temperature, possibly as much as 10°C. The temperature at which these changes occur is known as the crystalline melting point T_m. The change in properties is due to the growth of crystallites, and the system changes, from one in which the molecules are all moving relative to one another, to one in which segments of the molecules are bound in ordered arrays. The use of the term melting point is misleading, because the onset of crystallisation may not occur until the viscosity of the material is so high that it is effectively a solid. Typical plots of specific volume against temperature for the three types of material are shown in Fig. 13.21.

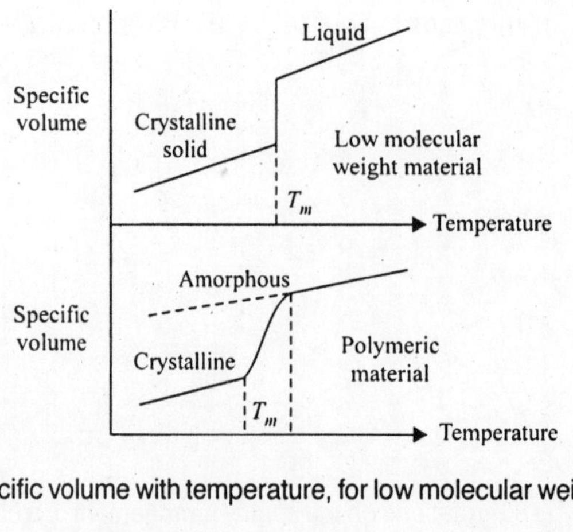

Fig. 13.21. Variation of specific volume with temperature, for low molecular weight and linear polymeric materials.

The following relationship between crystalline melting point and molecular weight for a polymer has been suggested.

$$\frac{1}{T_m} = a + \frac{b}{M} \quad \text{... (13.8)}$$

When the molecular weight is high, the crystalline melting point is virtually independent of the molecular weight. This is true for most polymeric materials as normally used. The crystalline melting point will depend on the inter-molecular forces holding the segments in the crystallite. The presence of polar bonds, and particularly hydrogen bonds, leads to high values. Chain flexibility, i.e., the ease with which chain segments can rotate, also affects the value of the crystalline melting point. The more flexible the chain the lower the value.

This chain flexibility has an important effect on the rate at which crystallisation occurs. In a low molecular weight liquid at its melting point, the molecules can move easily relative to one another and crystallisation is rapid. Super cooling can occur, but only under extreme conditions. In a polymeric material at its crystalline melting point, movement of the molecules relative to one another is difficult, especially if the chain is made inflexible by the presence of long repeating units of low symmetry. Development of crystallites is not rapid, and continues as the temperature falls below the crystalline melting point. The crystalline structure of a material can vary from one sample to another, depending on the rate of cooling.

If the crystallisation occurs in the absence of external forces, the orientation of the crystallites will be random as shown in Fig. 13.22. If the material is cooled past its crystalline melting point under the influence of an external stress, the direction of the crystallites will be orientated, as shown in Fig. 13.22. If a material has been super cooled, the application of a stress can increase the rate of crystallisation. The resultant crystallites will be orientated. If a crystalline material below its crystalline melting point is rapidly stretched, the crystallites become orientated. When this occur, the cross-section is reduced in the line of the stress. If the material is initially in fibre form, 'necking' occurs at a point, as shown in Fig. 13.22, and a thinner fibre can be drawn off from this point. The properties of a polymeric material vary with the degree of orientation of the crystallites. The mechanical strength is increased in the direction of crystallite alignment.

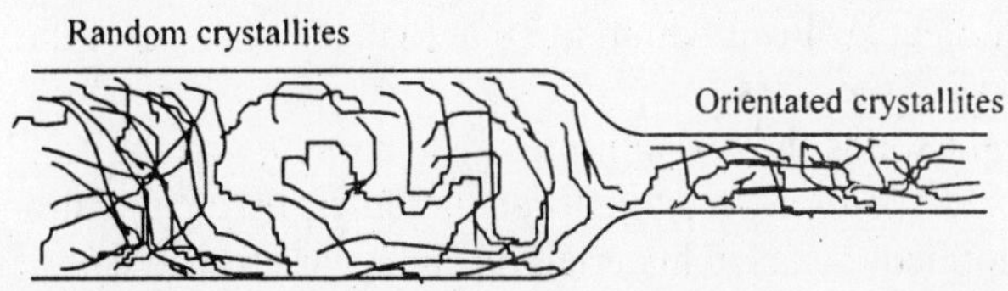

Fig. 13.22. Orientation of crystallites resulting from cold drawing.

RUBBER GLASS TRANSITIONS

Cooling a polymeric material from the liquid will produce a solid which is either completely amorphous or partially crystalline. In the former, the molecules are moving relative to one another throughout the material, while in the latter they are fixed in the crystallites, but moving in the amorphous regions. In either case, cooling the solid slows down and limits these movements. At some stage in the cooling, the thermal energy will become so small that all rotation about the bonds will cease. The chains in the amorphous regions then become set in their random patterns. The temperature at which this occurs is known as the glass transition point T_g. This is a second-order transition. It is not accompanied by a change in specific volume and other properties, merely by an alternation in their rate of change with temperature. The size of the inter-molecular forces and the chain flexibility affect the glass transition, just as they affect the crystalline melting point. In general materials with a high T_m will have a high T_g and vice versa.

Some values of crystalline melting point and glass transition point are given below.

Polymer	T_m °C	T_g °C
Branched polythene	105 to 120	Less than –50
Linear polythene	125 to 135	Less than –50
Polyethylene terephthalate	265	65 to 70
Polymethyl methacrylate	–	60 to 110
Nylon 6,6	255 to 265	40 to 50
Isotactic polypropylene	176	0
Polystyrene	–	100
Polytetrafluoroethylene	327	–
Polyvinyl chloride	–	70 to 80
Rubbers	–	–40 to –100

CROSS-LINKED MATERIALS

If the degree of cross-linking is high, rotational transitions of segments will be very restricted and can be neglected. The structure is rigid and has no order. As the degree of cross-linking is reduced, the extent of the motion of the segments of the structure increases, and lightly cross-linked materials show definite transition points, and are similar in behaviour to thermoplastic materials.

MECHANISM OF DISTORTION

When a stress is applied to a material a distortion occurs, and this can be measured, if the stress is sufficiently large. The distortion of a piece of material is the sum of the much smaller distortions which occur on an atomic and molecular level within the material. The nature of these distortions determines the reaction of the material to the stress.

In low molecular-weight solids the crystallisation is complete and the solids can be considered as a mass of atoms held in fixed positions, relative to one another, by chemical bonds and inter-molecular forces. When a stress is applied to such a material, the atoms are forced away from these positions, and the distances between them change until the new forces set up, balance the stress. If the distortion is not too great the atoms will return to their original positions when the stress is removed. Large stresses produce a permanent distortion. Provided the stress does not reach these values, the overall change in the material can be considered as instantaneous and elastic.

When a stress is applied to a plastic, this type of distortion occurs wherever two atoms are held in fixed positions, relative to one another, by either chemical bonds or inter-molecular forces. In highly cross-linked materials and in thermoplastics below their glass transition points, all the atoms can be considered as being fixed in this way, and this is the only type of distortion possible. It also occurs in thermoplastics above their glass transition points, as the atoms in the molecules are held by chemical bonds, and if there are crystallites, atoms are held by inter-molecular forces as well. In the amorphous regions the molecules are not held and move relative to one another. This molecular movement gives rise to a second type of distortion under stress which is not instantaneous, and depends upon the viscosity of the material. This is viscoelastic distortion.

In the linear polymer there are a very large number of arrangements of the chain. These vary from the fully extended to the tightly curled. As rotational transitions are occurring all the time, the shape of the chain will be changing continuously. For every molecule which tends to straighten as a result of these changes, there will be one which tends to curl up. Thus, the average extension of the chains remains constant. If a stress is applied to the material, the changes in molecular shape cease to be random, and all tend towards, chain extension in the direction of the stress. Thus, a tensile stress produces an extension of the material. This is shown in Fig. 13.23. When the stress is removed the changes in shape continue, but most of them cause the chains to curl up, until the original average extension results. The material returns to its original size. Thus, the distortion of the material due to chain straightening is elastic. It is viscoelastic because the changes of shape of the chain are time dependent. The extent of this time dependence is determined by the size of the inter-molecular forces opposing the rotational transitions.

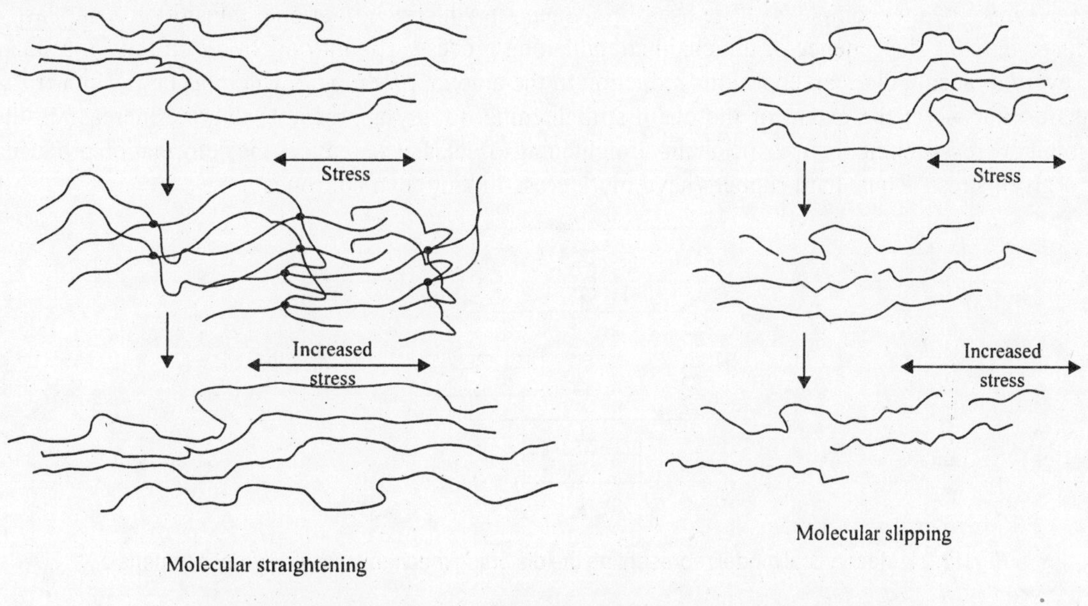

Fig. 13.23. Effect of stress on molecular arrangement in a linear polymer.

The application of a stress affects the movements in an individual chain. The overall movement of the molecules relative to one another is also affected. The result is molecular slipping. This is shown in Fig. 13.23. This slipping of the molecules is not reversible, so the resultant distortion is not elastic. Molecular slipping is also time dependent, being governed by the rotational transitions. The extent of the rotational transitions needed to produce appreciable movement of the molecules relative to one another, must be much greater than those needed for the segment shifts which lead to chain straightening. The straightening process is, therefore, much quicker than the slipping process.

A completely amorphous thermoplastic, above its glass transition point, is rubbery. It can be represented by a simple mechanical model as shown in Fig. 13.24. The spring S_1 represents the instantaneous elastic deformation of the bonds. The spring S_2 represents the elastic deformation due to chain straightening, the dash pot DP_1 representing the time dependence of that deformation. The dash pot DP_2 represents the non-elastic, time-dependent deformation due to molecular slipping. If a stress is suddenly applied to a material of this type there will be a large elastic deformation due, in the main, to chain straightening and, to a much lesser extent, to bond distortion. This deformation is not instantaneous, but is virtually complete before any molecular slipping occurs. If the stress is quickly removed, the material will recover. If the rotational transitions are rapid, the material responds to short duration stresses like a rubber, snapping back to its original state when the stress is removed. If, however, the original deformation is maintained, the stress gradually falls to zero. Molecular slipping occurs to such an extent that the chains can return to their original average extension, even though their positions, relative to one another, have been altered. In terms of the mechanical model, this corresponds to an initial deformation of both springs, followed by their return to their original extension, while the deformation is taken up by movement at the dash pot DP_2. The stress decay takes longer as the plastic is cooled towards its glass transition point. It also takes longer if the frictional entanglement of the molecules is increased, by increasing the molecular weight.

The requirements of an ideal rubber are an instantaneous elastic extension, and no stress decay. These cannot be met completely. The use of polymers, in which the rotational transitions occur easily, reduces the time dependence of the chain straightening process. The introduction of a few cross-links for every molecule gives an enormous reduction in the amount of slipping which occurs. A few cross-links do not affect the extent of the chain straightening, to any appreciable degree. Increase in the amount of cross-linking reduces the chain-straightening effect, and so reduces the deformation produced by a given stress. Thus, hard rubbers have more cross-linking than soft ones.

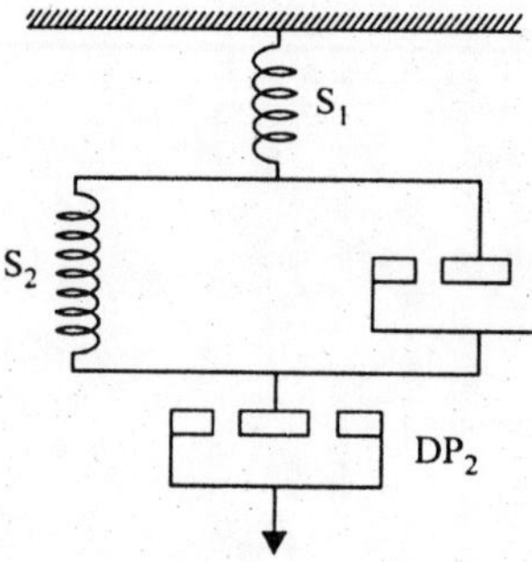

Fig. 13.24. Mechanical model representing the distortion mechanisms in polymeric materials.

DESIGNING A POLYMER STRUCTURE FOR IMPROVED PROPERTIES

The three principles applied to give strength and resistance to polymers are (i) crystallisation; (ii) cross-linking; and (iii) increasing inherent stiffness of polymer molecules. Combinations of any two or all of the three strengthening principles have proved effective in achieving various properties with polymers (Fig. 13.25). For polymers composed of inherently flexible chains, crystallisation and cross-linking are the only available means to enhance polymer properties. The factors affecting the crystallinity of polymers have been discussed previously, and the method of introducing cross-links between polymer molecules are discussed later.

The third, and relatively new, strengthening principle is to increase chain stiffness. One possible way of stiffening a polymeric chain is to hang bulky side groups on the chain to restrict chain bending. For example, in polystyrene, benzene rings are attached to the carbon backbone of the chain, this causes stiffening of the molecule sufficient to make polystyrene a hard and rigid plastic with a softening temperature higher than that of polyethylene, even though the polymer is neither cross-linked nor crystalline. The method is advantageous because the absence of crystallinity makes the material completely transparent, and the absence of cross-linking makes it readily mouldable. A similar example is poly(methyl methacrylate).

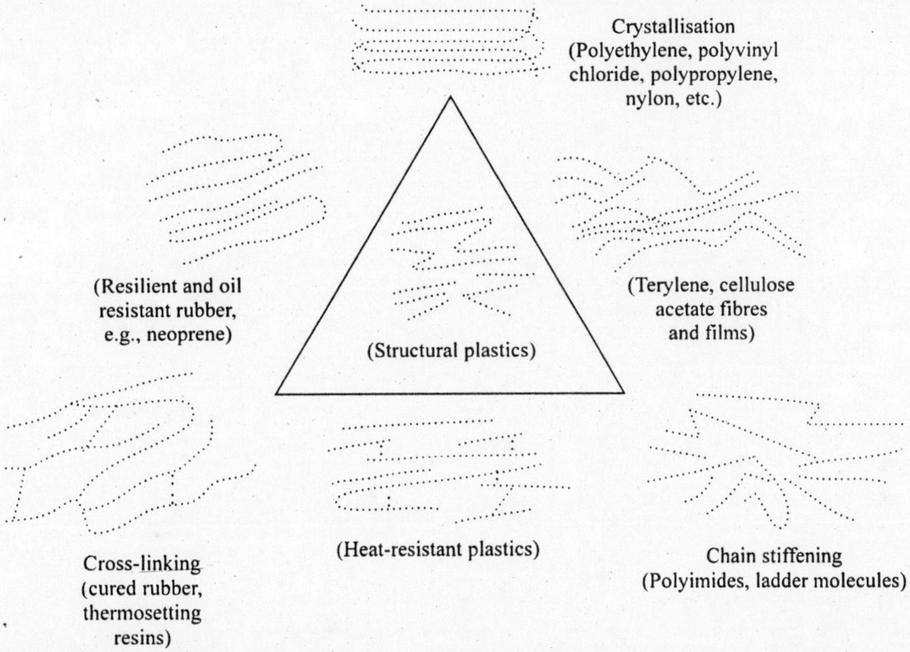

Fig. 13.25. Three basic principles—crystallisation, cross-linking, and chain stiffening—for making polymers strong and temperature resistant are represented at the three corners of the triangle. The sides and the centre of the triangle indicate various combinations of the principles.

However, the disadvantage of attaching bulky side groups is that the material dissolves in solvent fairly easily and undergoes swelling, since the bulky side groups allow ready penetration by solvents and swelling agents. This problem can be eliminated by stiffening the backbone of the chain itself. One way to do this is to introduce rigid ring structures in the polymer chain. A classic example of such a polymer is cellulose, which is the structural framework of wood and the most abundant organic material in the world. Its chain molecule consisting of a string of ring-shaped condensed glucose molecules has an intrinsically stiff backbone. Cellulose therefore has a high tensile strength. In poly(ethylene terephthalate) fibre the chains are only moderately stiff, but the combination of chain stiffness and crystallisation suffices to give the fibre high strength and a high melting point (265°C). The new plastic polycarbonate containing benzene rings in the backbone of the polymer chain is so tough that it can withstand the blows of a hammer. Ladder polymers based on aromatic chains consist of double-stranded chains made up of benzene-type rings. These hard polymers are completely unmeltable and insoluble.

The combination of all three principles has led to the development of new and interesting products with enhanced properties. Composites based on epoxy and urethane-type polymers may be cited as an example. Thus, the stiff polymeric chains of epoxy and urethane types are cross-linked by curing reactions, and fillers are added to produce the equivalent of crystallisation.

SECTION IV

Industrial Polymers

CHAPTER 14

Industrial Polymers: An Overview

INTRODUCTION

The amazing scope of wartime applications accelerated the development and growth of polymers to meet the diverse needs of special materials in different fields of activity.

The development of new polymeric materials proceeded at an even faster pace after the war. Epoxies were developed in 1947, and acrylonitrile butadiene-styrene (ABS) terpolymer in 1948. The polyurethanes, introduced in Germany in 1937, saw rapid development in the United States as the technology became available after the war. The discovery of Ziegler–Natta catalysts in the 1950s brought about the development of linear polyethylene and stereoregular polypropylene. These years also saw the emergence of acetal, polyethylene terephthalate, polycarbonate, and a host of new co-polymers. The next two decades saw the commercial development of a number of highly temperature-resistant materials, which included poly (phenylene oxide) (PPO), polysulphones, polyimides, polyamide-imides, and polybenzimidazoles.

More recently, we have seen other new materials arrive, including several exotic materials which are mostly development products today and are very expensive. They will undoubtedly find their first uses, if any, in the aerospace industry, which can afford to pay for them. The growth of all plastics, in general, and engineering plastics, in particular, has more than kept pace with the general progress in technology. On a volume basis, US plastics production in 1991 have surpasses US steel production, heralding the arrival of the plastics age in the United States. Plastics are still evolving, changing, and improving as the industry moves into the 21st century.

PLASTICS, RUBBER AND FIBRES

Numerous plastics and fibres are produced from synthetic polymers: containers from polypropylene, coating materials from PVC, packaging film from polyethylene, experimental apparatus from Teflon, organic glasses from poly (methyl methacrylate), stockings from nylon fibre—there are simply too many to mention them all. The reason why plastics materials are popular is that they may offer such advantages as transparency, self-lubrication, light weight, flexibility, economy in fabricating and decorating. Properties of plastics can be modified through the use of fillers, reinforcing agents, and chemical additives. Plastics have thus found many engineering applications, such as mechanical units under stress, low-friction components, heat and chemical-resistant units, electrical parts, high-light-transmission applications, housing, building construction functions, and many others. Although it is true that in these applications plastics have been used in a proper manner according to our needs, other ways of utilising both natural and synthetic polymers may still remain. To investigate these further possibilities, active research has been initiated in a field called speciality polymers. This field relates to synthesis of new polymers with high additional value and specific functions.

Many of the synthetic plastic material have found established uses in a number of important areas of engineering involving mechanical, electrical, telecommunication, aerospace, chemical, biochemical, and biomedical applications. There is, however, no single satisfactory definition of engineering plastics. According to one definition, 'engineering plastics are those which posses physical properties enabling them to perform for prolonged use in structural applications, over a wide temperature range, under mechanical stress and in difficult chemical and physical environments'. In the most general sense, however, all polymers are engineering materials, in that they offer specific properties which we judge quantitatively in the design of end-use applications.

Cellulose have extremely large molecules, the molecules are made up of recurring units. For instance, it may be recalled that the formula for cellulose may be written $(C_6H_{10}O_5)_n$. There is a unit, $C_6H_{10}O_5$, that is repeated a large number of times in the molecule. Cellulose is an example of a naturally occurring high polymer. It will be recalled that cellulose is in some contrast to a substance such as glucose that has a rather modest molecular weight. Similarly cellulose nitrate and cellulose acetate may be regarded as plastics, or high polymers. These are examples of synthetic polymers, at least to the extent that the original molecule has been considerably modified.

The formation of a polymer may be regarded in the following way. Suppose we have a molecule X, and that this molecule is able to unite with itself to form X_2. (Such a case actually occurs in the equilibrium 2NO I N_2O_4). But suppose that the molecule X is able to continue uniting with itself to form very long chains which might be indicated thus.

$$X{-}X{-}X{-}X{-}X{-}X{-}X{-}X$$

The resulting large molecule is called a polymer of X. The word 'polymer' is derived from the Greek meaning 'many times'. In speaking of polymers it is the custom to refer to the single unit (X in the above case) as a monomer. If the molecular weight of the polymer is very large, the substance is said to be a high polymer.

Plastics are high polymer, usually synthetic, combined with other ingredients, such as curatives, fillers, reinforcing agents, colourants, plasticisers, etc.; the mixture can be formed or moulded under heat and pressure in its raw state, and machined to high dimensional accuracy, trimmed and finished in its hardened state. The thermoplastic type can be resoftened to its original condition by heat; the thermosetting type cannot.

Plastics in general (including all forms) are sensitive to high temperatures, among the more resistant being fluorocarbon resins, nylon, phenolics, though even these soften or melt above 260°C. Other types are combustible when exposed to flame for a short time (polyethylene, acrylic polymers, polystyrene), and still others burn with evolution of toxic fumes (polyurethane).

Engineering plastics are those to which standard metal engineering equations can be applied; they are capable of sustaining high loads and stresses, and are machinable and dimensionally stable. They are used in construction, as machine parts, automobile components, etc. Among the more important are nylon, acetals, polycarbonates, ABS resins, PPO/Styrene, and polybutylene terephthalate.

Fibres, films, and bristles are examples of extruded forms. Plastics may be shaped by either compression moulding (direct pressure on solid material in a hydraulic press) or injection moulding (ejection of a measured amount of material into a mould in liquid form). The latter is most generally used, and articles of considerable size can be produced. Because of their dielectric and non-toxic properties, plastics are essential components of electrical and electronic equipment (especially for use within the human body).

Plastics can be made into flexible and rigid foams by use of a blowing agent; they are light and strong, and the rigid type is machinable. These forms are collectively called cellular plastics. Plastics can also be reinforced, usually with glass or metallic fibres for added strength. They are laminated to paper, cloth, wood, etc. for many uses in the packaging, electrical and furniture industries; they also can be metal-plated. Plastic pipe is widely used for underground transportation of gases and liquids over long distances as well as intraplant.

Several natural materials (waxes, clays, and asphalts) have rheological properties similar to synthetic products, but as they are not polymeric, they are not considered true plastics. Certain proteins (casein, zein) are natural high polymers from which plastics are made (buttons, and other small items), but they are of decreasing importance.

Plastics have permeated industrial technology. Not only have they replaced and improved upon many materials formerly used, but also have made possible industrial and medical applications that would have been impracticable with older technologies.

Synthetic resins are man-made higher polymer resulting from a chemical reaction between two (or more) substances, usually with heat or a catalyst. This definition includes synthetic rubbers and silicones (elastomers), but excludes modified, watersoluble polymers (often called resins). Distinction should be made between a synthetic resin and a plastic. The former is the polymer itself, whereas the latter is the polymer plus such additives as fillers, colourant, plasticisers, etc.

The first truly synthetic resin was developed by Backeland (phenol-formaldehyde). This was soon followed by a petroleum-derived product called coumarine-indene, which did indeed have the properties of a resin. The first synthetic elastomer was polychloroprene, and later called neoprene. Since then many new types of synthetic polymers have been synthesised, perhaps the most sophisticated of which are nylon and its congeners (polymamides, and the inorganic silicone group). Other important types are alkyds, acrylics, aminoplasts, polyvinyl halides, polyesters, epoxides and polyolefins.

In addition to their many applications in plastics, textiles, and paints, special types of synthetic resins; are useful as ion-exchange media.

Rubber is the name given to elastic, high-molecular compounds capable of changing their shape (deforming) when acted upon by an external force and of rapidly resuming their original shape when the force is removed. The elastic properties of rubber are due to the fact that the linear macro-molecules of the material which normally are of a bent, curled or curved shape, easily stretch out of contract under the effect of an applied force; when the force is removed, the molecules return to their initial shape and length.

If sulphur is added to natural rubber which is then heated, the sulphur will form links between the linear macromolecules of the rubber, i.e., form bridges between them; by this larger macro-molecules of a web structure will be produced. This process is called vulcanisation of the rubber, and the product of reaction between the rubber and sulphur is vulcanised rubber. Vulcanisation improves the mechanical properties of the rubber and increases the range of temperatures within which the elasticity of the material is retained; it also increases the resistance of the rubber to organic solvents, etc.

Natural and synthetic rubber is the basic material in the fabrication of rubber-textile and rubber-metal articles which are so extensively used in various branches of the economy. These are motor car and aircraft tyres, drive belts and conveyer belts, flexible hoses and sleeves, machine parts and various gaskets, electric insulating materials, rubberised fabrics and rubber footwear, articles of personal hygiene and many other rubber products. The consumption of rubber has increased year by year. This explains the attention devoted to developing the rubber industry in India and in many other countries. Rubbers are classified as natural rubber and synthetic rubber. Until the nineteen thirties only natural rubber, obtained from the latex of rubber plants which grow in countries with a damp tropical climate (mainly in Malaysia, Indonesia, etc.), was used.

EQUIPMENTS USED FOR POLYMER PROCESSING

The equipment used for polymer processing tends to be more complex, mainly because molten plastic is a very viscous liquid whereas molten metal is far more free flowing. In simple compression-moulding processes this is of little consequence as the raw plastic is in the form of granules which are introduced directly into the mould cavity. In injection or transfer moulding, however, the granules are plasticised (changed to a molten state) before moulding takes place and viscosity then becomes an important consideration.

All polymer processing techniques require the input of heat to soften the polymer and enable the forming process to produce the required shape. The temperature and period of heating are critical, and are the main source of processing defects. If the polymer is exposed to excessive or prolonged heating, thermal degradation will take place. This will show itself in a reduction of viscosity, oxidation or hydrolysis and, in extreme cases, depolymerisation.

Polymers can be broadly categorised as thermoplastic, thermosetting and elastomers, the names being synonymous with their main characteristic. Thermoplastic polymer softens under the application of heat and so it is heated to make it mouldable, then cooled to hold the shape. The process is reversible and there is virtually no limit to the number of times that this softening/hardening cycle can be carried out. Consequently clean scrap can easily be recycled.

Thermosetting polymer hardens under the application of heat and so it is heated to hold its shape. The polymer undergoes an irreversible chemical change during the heating cycle and cannot therefore be reused. Elastomers exhibit a highly elastic, rubberlike behaviour. They are really a subgroup of either thermoplastic or thermosetting polymers, but behave differently in that they take up their original shape upon removal of a deforming force.

Polymerisation is the reaction in which the raw materials (the monomers) are converted into the products (the polymers). As the reaction takes place, the material is constrained by a mould or die so that the desired shape or form is achieved. The material used in all the moulding processes is supplied in the form of liquids, sheets, granules or powder. This makes the handling of them very convenient and problems do not arise until the material is being manipulated in the mould.

Both viscosity and elasticity can be reduced by increasing the working temperature, but there is an upper limit at which the polymer will begin to degenerate in some way. In addition different types of polymers have different working temperature ranges. PVC for example has a very narrow range in comparison to polythene, whose viscosity can be varied by a factor of ten. Viscosity can also be reduced by the addition of plasticisers. These are liquids which enter the structure of the polymer and enable the molecular chains to slide more easily over one another. The visco-elastic characteristic can also be modified by compounding two or more different polymers.

Apart from altering the visco-elastic nature of the polymer, a number of additives can be introduced either at the mixing stage or during processing, to achieve other desired effects.

For the purpose of this discussion, we will classify polymers into three broad groups: addition polymers, condensation polymers, and special polymers. By convention, polymers whose main chains consist entirely of C–C bonds are addition polymers, whereas those in which hetero atoms (e.g.,O, N, S,Si) are present in the polymer backbone are considered to be condensation polymers. Grouped as special polymers are those products which have special properties, such as temperature and fire resistance, photosensitivity, electrical conductivity, and piezoelectric properties, or which posses specific reactivates to serve as functional polymers.

Further classification of polymers in the groups of additional polymers and condensation polymer has been on monomer composition, because this provides an orderly approach, whereas classification based on polymer uses, such as plastics, elastomers, fibres, coatings, etc. would result in too much overlap. For example, polyamides are used not only as synthetic fibres but also as thermoplastic moulding compound and polypropylene, which is used as thermoplastic moulding compound has also found uses as a fibre-forming material.

All vinyl polymers are addition polymers. To differentiate them, the homopolymers have been classified by the substituents attached to one carbon atoms of the double bone. If the substituent is hydrogen, alkyl, aryl, or halogen, the homopolymers are listed under polyolefines. Olefin homopolymers with other substituents are described under polyvinyl compounds, except where the substituent is a nitrile, a carboxylic acid, or a carboxylic acid ester or amide. The monomers in the latter cases being derivatives of acrylic acid, the derived polymers are listed under acrylics. Under olefin co-polymer are listed products which are produced by co-polymerisation of two or more monomers. Condensation polymers are classified as polyesters, polyamides, polyurethanes, and ether polymers, based on the internal functional group being ester (—COO—), amide (—CONH—), urethane (—OCONH—), or ether (—O—). Polymers with special properties have been classified into three groups: heat-resistant polymers, silicones and other inorganic polymers, and functional polymers. Discussions in all cases are centered on important properties and main applications of polymers.

CHAPTER 15

Addition Polymers

INTRODUCTION

Addition polymers are produced in largest tonnages among industrial polymers. The most important monomers are ethylene, propylene, and butadiene. They are based on low-cost petrochemicals or natural gas and are produced by cracking or refining of crude oil. Polyethylene, polypropylene, poly(vinyl chloride), and polystyrene are the four major addition polymers and are by far the least-expensive industrial polymers on the market. In addition to these four products, a wide variety of other addition polymers are commercially available.

For addition polymers four types of polymerisation processes are known: free-radical-initiated chain polymerisation, anionic polymerisation, cationic polymerisation, and co-ordination polymerisation (with Ziegler-Natta catalysts).

By far the most extensively used process is the free-radical-initiated chain polymerisation. However, the more recent development of stereoregular polymers using certain organometallic co-ordination compounds called Ziegler-Natta catalysts, which has added a new dimension to polymerisation processes, is expected to play a more important role in coming years. The production of linear low-density polyethylene (LLDPE) is a good example. Ionic polymerisation is used to a lesser extent. Thus, anionic polymerisation is used mainly in the co-polymerisation of olefins, such as the production of styrene-butadiene elastomers, and cationic polymerisation is used exclusively in the production of butyl rubber.

Different processes are used in industry for the manufacture of polymers by free-radical chain polymerisation. Among them homogeneous bulk polymerisation is economically the most attractive and yields products of higher purity and clarity. But it has problems associated with the heat of polymerisation, increases in viscosity, and removal of unreacted monomer. This method is nevertheless used for the manufacture of PVC, polystyrene, and poly(methyl methacrylate). More common processes are homogenous solution polymerisation and heterogeneous suspension polymerisation.

Solution polymerisation is used for the manufacture of polyethylene, polypropylene, and polystyrene, but by far the most widely used process for polystyrene and PVC is suspension polymerisation. In the latter process (also known as bead, pearl, or granular polymerisation because of the form in which the final products may be obtained), the monomer is dispersed as droplets (0.01–0.05 cm in diameter) in water by mechanical agitation. Various types of stabilisers, which include water-soluble organic polymers, electrolytes, and water-insoluble inorganic compounds, are added to prevent agglomeration of the monomer droplets. Each monomer droplet in the suspension constitutes a small bulk polymerisation system and is transformed finally into a solid bead. Heat of polymerisation is quickly dissipated by continuously stirring the suspension medium, which makes temperature control relatively easy.

POLYOLEFINS

Polyethylene

$-[CH_2-CH_2]_n-$

Monomer	*Polymerisation*	*Major uses*
Ethylene	LDPE: Free-radial-initiated chain polymerisation HDPE: Ziegler-Natta or metal oxide catalysed chain polymerisation	LDPE: Film and sheet (55%), housewares and toys (16%), wire and cable coating (5%) HDPE: Bottles (40%), housewares, containers, toys (35%), pipe and fittings (10%), film and sheet (5%)

Polyethylene is the most widely used thermoplastic material and is composed of ethylene. The two main types are low-density polyethylene (LDPE) and high-density polyethylene (HDPE).

Manufacturing processes

LDPE is manufactured by polymerisation of ethylene under high pressures (15,000–50,000 psi, i.e., 103–345 MPa) and elevated temperatures (200–350°C) in the presence of oxygen (0.03–0.1 per cent) as free-radical initiator. Ethylene is a supercritical fluid with density 0.4–0.5 g/cm^3 under these conditions. Polyethylene remains dissolved in ethylene at high pressures and temperatures but separates in the lower ranges. Branch polyethylene is produced due to chain transfer to polymer. The type and extent of branching depends on the local reaction temperature and concentrations of monomer and polymer. The molecular weight distributions and the frequencies of long and short branches on polymer chains depend strongly on reactor geometry and operation. The branched products (LDPE) are less crystalline and rigid than higher density species (HDPE) made by low pressure co-ordination polymerisation.

Linear polyethylenes are produced in solution, slurry, and increasingly, gas-phase low-pressure processes. The Phillips process developed during the mid 1950s used supported chromium trioxide catalysts in a continuous slurry process (or particle-form process) carried out in loop reactors. Earlier, Standard Oil of Indiana patented a process using a supported molybdenum oxide catalysts. The polyethylenes made by both these processes are high-density polyethylene (HDPE) with densities of 0.950–0.965 g/cm^3 and they are linear with very few side-chain branches and have a high degree of crystallinity. During the late 1970s, Union Carbide developed a low-pressure polymerisation process (Unipol process) capable of producing polyethylene in the gas phase that required no solvents. The process employed a chromium based catalyst. In this process (Fig. 15.1) ethylene gas and solid catalysts are fed continuously to a fluidised bed reactor. The fluidised material is polyethylene powder which is produced as a result of polymerisation of the ethylene on the catalyst. The ethylene, which is recycled, supplies monomer for the reaction, fluidises the solid, and serves as a heat-removal medium. The reaction is exothermic and is normally run at temperatures 25–50°C below the softening temperatures of the polyethylene powder in the bed. This operation requires very good heat transfer to avoid hot spots and means that the gas distribution and fluidisation must be uniform.

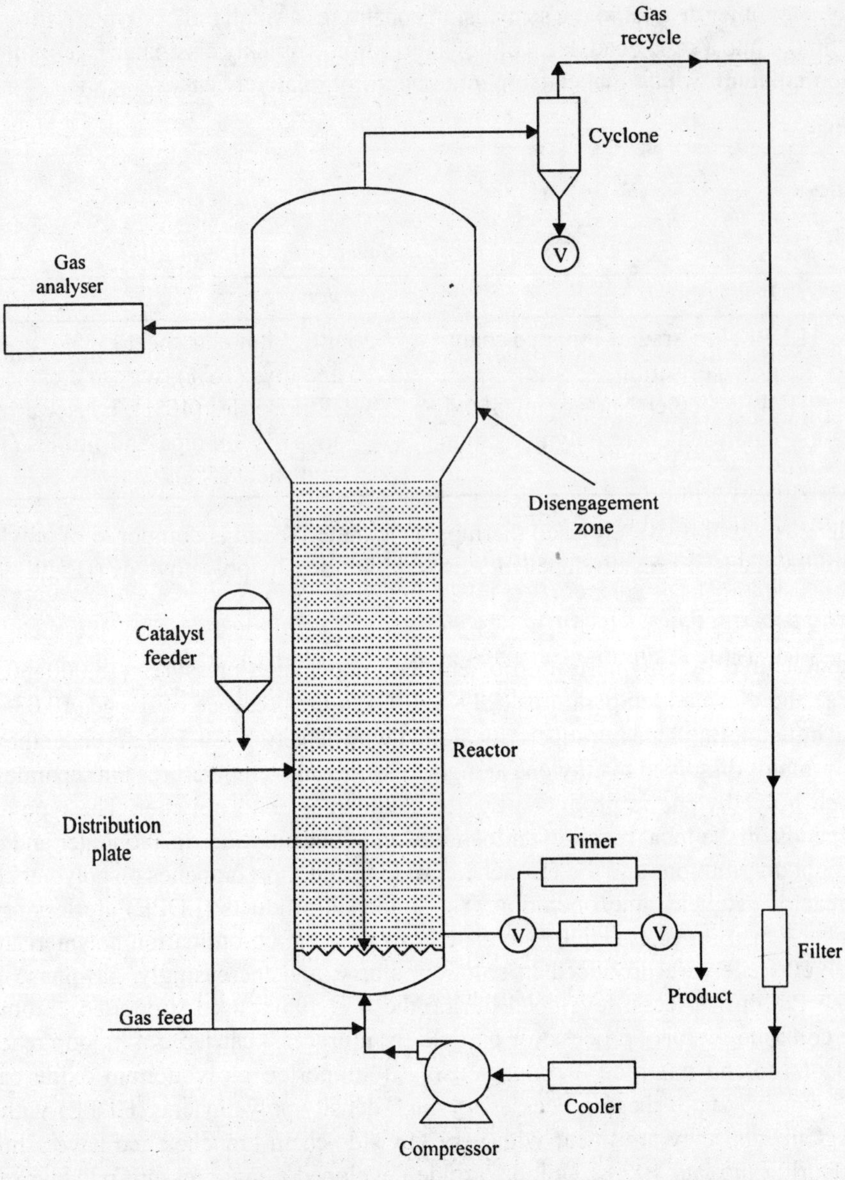

Fig. 15.1. Union carbide gas phase process for the production of LDPE.

The keys to the process are active catalysts. These are special organo-chromium compounds on particular supports. The catalysts yield up to about 10^6 kg of polymer per kilogram of metallic chromium. Branching is controlled by the use of co-monomers like propylene or 1-butene, and hydrogen is used as a chain transfer agent. The catalyst is so efficient that its concentration in the final product is negligible. The absence of a solvent and a catalyst removal step makes the process less expensive. The products marketed as linear low-density polyethylenes (LLDPE) can be considered as linear polyethylenes having a significant number of branches (pendant alkyl groups). The linearity imparts strength, the branches impart toughness.

During the late 1970s, the Dow Chemical Co. also began producing polyethylene using a proprietary solution process based on Ziegler-Natta type catalysts. Resins are made at low pressures and with lower densities in a system derived essentially from high-density resin technology (Ziegler-Natta). Higher boiling co-monomers (1-hexene and 1-octene) are used to produce LLDPE having superior mechanical properties and film-drawing tendencies. A difference of the *Dow* solution products from gas-phase products is that they are produced in standard pellet form with any needed additives incorporated into the pellet.

Polyethylene is partially amorphous and partially crystalline. Linearity of polymer chains affords more efficient packing of molecules and hence a higher degree of crystallinity. On the other hand, side-chain branching reduces the degree of crystallinity. Increasing crystallinity increases density, stiffness, hardness, tensile strength, heat and chemical resistance, crepe resistance, barrier properties, and opacity, but it reduces stress-crack resistance, permeability, and impact strength. Table 15.1 shows a comparison of three types of polyethylene.

Table 15.1. Types of polyethylene.

Material	*Chain structure*	*Density (g/cm^3)*	*Crystallinity*	*Process*
LDPE	Branched	0.912–0.94	50%	High pressure
LLDPE	Linear/less branched	0.92–0.94	50%	Low pressure
HDPE	Linear	0.958	90%	Low pressure

Polyethylene has excellent chemical resistance and is not attacked by acids, bases, or salts. (It is, however, attached by strong oxidising agents). The other characteristics of polyethylene which have led to its widespread use are low cost, easy processablility, excellent electrical insulation properties, toughness and flexibility even at low temperatures, freedom from odour and toxicity, reasonable clarity of thin films, and sufficiently low permeability to water vapour for many packaging, building, and agricultural applications.

Major uses for LDPE are in packaging and sheeting, whereas HDPE is used mainly in blow-moulded products (milk bottles, household and cosmetic bottles, fuel tanks), and in pipe, wire, and cable applications. Ultrahigh-molecular-weight polyethylene materials, which are the toughest of plastics, are doing an unusual job in the textile machinery field.

Chlorinated polyethylene

Low chlorination of polyethylene, causing random substitution, reduces chain order and thereby also the crystallinity. The low chlorine products (22–26 per cent chlorine) of polyethylene are softer, more rubber-like, and more compatible and soluble than the original polyethylene. However, much of the market of such materials has been taken up by chlorosulphonated polyethylene, produced by chlorination of polyethylene in the presence of sulphur dioxide, which introduces chlorosulphonyl groups in the chain. Chlorosulphonated LDPE containing about 27 per cent chlorine and 1.5 per cent sulphur has the highest elongation.

Chlorosulphonated polyethylene rubbers, designated as CSM rubbers, have very good heat, ozone, and weathering resistance together with a good resistance to oils and a wide spectrum of chemicals. The bulk of the output is used for fabric coating, film sheeting, and pit liner systems in the construction industry, and as sheathing for nuclear power cables, for offshore oil rig cables, and in diesel electric locomotives.

Cross-linked polyethylene

Cross-linking polyethylene enhances its heat resistance (in terms of resistance to melt flow) since the network persists even about the crystalline melting point of the uncross-linked material. Cross-linked polyethylene thus finds application in the cable industry as a dielectric and as a sheathing material. Three main approaches used for cross-linking polyethylene are (i) radiation cross-linking; (ii) peroxide cross-linking; and (iii) vinyl silane cross-linking.

Radiation cross-linking is most suitable for thin sections. The technique, however, requires expensive equipment and protective measures against radiation. Equipment requirements for peroxide curing are simpler, but the method requires close control. The peroxide molecules break up at elevated temperatures, producing free radicals which then abstract hydrogen from the polymer chain to produce a polymer free radical. Two such radicals can combine and thus cross-link the two chains. It is important, however, that the peroxide be sufficiently stable thermally so that premature cross-linking does not take place during compounding and shaping operations. Dicumyl peroxide is often used for LDPE but more stable peroxides are necessary for HDPE. For cross-linking polyethylene in cable coverings, high curing temperature, using high-pressure steam in a long curing tube set into the extrusion line, are normally employed.

Co-polymers of ethylene with a small amount of vinyl acetate are often preferred for peroxide cross-linking because the latter promotes the cross-linking process. Large amounts of carbon black may be incorporated into polyethylene that is to be cross-linked. The carbon black is believed to take part in the cross-linking process, and the mechanical properties of the resulting product are superior to those of the unfilled material.

In the vinyl silane cross-linking process (*Sioplas* process) developed by Dow, an easily hydrolysable trialkoxy vinyl silane, $CH_2=CHSi(OR)_3$, is grafted into the polyethylene chain, the site activation having been achieved with the aid of a small amount of peroxide. The material is then extruded onto the wire. When exposed to hot water or low-pressure steam, the alkoxy groups hydrolyse and then condense to form a siloxane cross-link:

$$(-\overset{|}{\underset{|}{Si}}-O-\overset{|}{\underset{|}{Si}}-)$$

The cross-linking reaction is facilitated by the use of a cross-linking catalyst, which is typically an organo-tin compound. There are several variations of the silane cross-linking process. In one process, compounding, grafting, and extrusion onto wire are carried out in the same extruder.

Cross-linked LDPE foam has been produced by using either chemical cross-linking or radiation cross-linking. These materials have been used in the automotive industry for carpeting, boot mats, sound deadening, and pipe insulation, and as flotation media for oil-carrying and dredging hose.

Linear low-density polyethylene (LLDPE)

Chemically, LLDPE can be described as linear polyethylene co-polymers with alpha-olefin co-monomers in the ethylene chain. They are produced primarily at low pressures and temperatures by the co-polymerisation of ethylene with various alpha-olefins such as butene, hexene, octene, etc., in the presence of suitable catalysts. Either gas-phase fluidised-bed reactors or liquid-phase solution-process reactors are used. (In contrast, LDPE is produced at very high pressures and temperatures either in autoclaves or tubular reactors.) Polymer properties such as molecular weight, molecular-weight distribution, crystallinity, and density are controlled through catalyst selection and control of reactor conditions. Among the

LLDPE processes, the gas-phase process has shown the greatest flexibility to produce resins over the full commercial range.

The molecular structure of LLDPE differs significantly from that of LDPE— LDPE has a higly branched structure, but LLDPE has the linear molecular structure of HDPE, though it has less crystallinity and density than the latter (Table 15.1).

The stress-crack resistance of LLDPE is considerably higher than that of LDPE with the same melt index and density. Similar comparisons can be made with regard to puncture resistance, tensile strength tensile elongation, and low-and high-temperature toughness. Thus LLDPE allows the processor to make a stronger product at the same gauge or an equivalent product at a reduced gauge.

LLDPE is now replacing conventional LDPE in many architectures because of the combination of favourable production economics and product performance characteristics. For many architectures (blow moulding, injection moulding, rotational moulding, etc.) existing equipment for processing LDPE can be used to process LLDPE. LLDPE film can be treated, printed, and sealed by using the same equipment used for LDPE. Heat-sealing may, however, require slightly higher temperatures.

LLDPE films provide superior puncture resistance, high tensile strength, high impact strength, and outstanding low-temperature properties. The resins can be drawn down to thicknesses below 0.5 mil without bubble breaks. Slot-cast films combine high clarity and gloss with toughness. LLDPE films are being increasingly used in food packaging for such markets as ice bags and retail merchandise bags, and as industrial liners and garment bags.

Good flex properties and environmental stress-crack resistance combined with good low-temperature impact strength and low warpage make LLDPE suitable for injection-moulding parts for housewares, closures, and lids. Extruded pipe and tubing made from LLDPE exhibit good stress-crack resistance and good bursting strength. Blow-moulded LLDPE parts such as toys, bottles, and drum liners provide high strength, flex life, and stress-crack resistance. Light-weight parts and faster blow-moulding cycle times can be achieved. The combination of good high and low-temperature properties, thoughness, environmental stress-crack resistance, and good dielectric properties suit LLDPE for wire and cable insulation and jacketing applications.

A new class of linear polyethylene co-polymers with densities ranging between 0.890 and 0.915 g/cm^3, known as very low-density polyethylene (VLDPE), was introduced commercially in late 1984 by Union Carbide. These resins are produced by co-polymerisation of ethylene and alpha-olefins in the presence of a catalyst.

VLDPE provides flexibility previously available only in lower-strength materials, such as ethylene-vinyl acetate (EVA) co-polymer and plasticised PVC, together with the toughness and broader operating temperature range of LLDPE. Its unique combination of properties makes VLDPE suited for a wide range of applications.

Generally, it is expected that VLDPE will be widely used as an impact modifier. Tests suggest that it is suited as blending resin for polypropylene and in HDPE films for improved tear strength.

On its own, VLDPE should find use in applications requiring impact strength, puncture resistance, and dart drop resistance combined with flexibility. The drawdown characteristics of VLDPE allow for very thin films to be formed without pinholing. Soft flexible films for disposable gloves, furniture films, and high-performance stretch and shrink film are potential markets.

High-molecular-weight high-density polyethylene

High-molecular-weight high-density polyethylene (HMW–HDPE) is defined as a linear homopolymer or co-polymer with a weight-average molecular weight ($\overline{M}_w$) in the range of approximately 200,000 to

500,000. HMW-HDPE resins are manufactured using predominantly two basic catalyst systems: Ziegler-type catalysts and chromium oxide-based catalysts. These catalysts produce linear polymers which can be either homopolymers when higher-density products are required or co-polymers with lower density. Typical co-monomers used in the latter type of products are butene, hexene, and octenes.

HMW-HDPE resins have high viscosity because of their high molecular weight. This presents problems in processing and, consequently, these resins are normally produced with broad molecular-weight distribution (MWD).

The combination of high molecular weight and high density imparts the HMW-HDPE good stiffness characteristics together with above-average abrasion resistance and chemical resistance. Because of the relatively high melting temperature, it is imperative that HMW-HDPE resins be specially stabilised with antioxidant and processing stabilisers. HMW-HDPE products are normally manufactured by the extrusion process; injection moulding is seldom used.

The principal applications of HMW-HDPE are in film, pressure pipe, large blow-moulded articles, and extruded sheet. HMW-HDPE film now finds application in T-shirt grocery sacks (with 0.6 to 0.9 mil thick sacks capable of carrying 30 lb of produce), trash bags, industrial liners, and speciality roll stock. Sheets 20 to 100 mils thick and 18 to 20 ft wide are available that can be welded *in situ* for pond and tank liners. HMW-HDPE piping is used extensively in gas distribution, water collection and supply, irrigation pipe, industrial effluent discharge, and cable conduit. The availability of pipe materials with significantly higher hydrostatic design stress (800 psi compared with 630 psi of the original HDPE resins) has given added impetus to their use. Large-diameter HMW-HDPE piping has found increasing use in sewer re-lining. Large-blow-moulded articles, such as 55-gal shipping containers, are produced. Equipment is now available to blow mould very large containers, such as 200 and 500-gal capacity industrial trash receptacles and 250-gal vessels to transport hazardous chemicals.

Ultrahigh-molecular-weight polyethylene

Ultrahigh-molecular-weight polyethylene (UHMWPE) is defined by ASTM as 'polyethylene with molecular weight over three million (weight average)'. The resin is made by a special Ziegler-type polymerisation. Being chemically similar to HDPE, UHMWPE shows the typical polyethylene characteristics of chemical inertness, lubricity, and electrical resistance, while its very long substantially linear chains provide greater impact strength, abrasion resistance, toughness, and freedom from stress cracking. However, this very high molecular weight also makes it difficult to process the polymer by standard moulding and extrusion techniques. Compression moulding of sheets and ram extrusion of profiles are the normal manufacturing techniques.

Compression moulding is by far the most commonly employed process. Forms produced by compression moulding or speciality extrusion can be made into final form by machining, sintering, or forging. Standard wood-working techniques are employed for machining; sharp tools, low pressures, and good cooling are used. Forging can be accomplished by pressing a preform and billet and the forging to the final shape. Parts are also formed from compression-moulded or skived sheets by heating them above 300°F (~150°C) and stamping them in typical metal-stamping equipment. Such UHMWPE items have better abrasion resistance than do steel or polyurethanes. They also have high impact strength even at very low temperatures, high resistance to cyclic fatigue and stress cracking, low coefficient friction, good corrosion and chemical resistance, good resistance to nuclear radiation, and resistance to boiling water. Fillers such as graphite, talc, glass beads or fibres, mica, and powdered metals can be incorporated to improve stiffness or to reduce deformation and deflection under load. Resistance to abrasion and deformation can be increased by peroxide cross-linking, described earlier.

UHMWPE was first used in the textile machinery field picker blocks and throw sticks, for example. Wear strips, timing wheels, and gears made of the UHMW polymer are used in material handling, assembly, and packaging lines. Chemical resistance and lubricity of the polymer are important in its applications in chemical, food, beverage, mining, mineral processing, and paper industries. All sorts of self-unloading containers use UHMWPE liners to reduce wear, prevent sticking and speed up the unloading cycles. The polymer provides slippery surfaces that facilitate unloading even when the product is wet or frozen.

The polymer finds applications in transportation, recreation, lumbering, and general manufacturing. Metal equipment parts in some cases are coated or replaced with UHMWPE parts to reduce wear and prevent corrosion. Sewage plants have used this polymer to replace cast-iron wear shoes and rails, bearings, and sprockets. There is even an effort to use UHMW polymer chain to replace metal chain, which is corroded by such environments

Porous UHMW polymer is made by sintering to produce articles of varied porosity. It has found growing use for controlled-porosity battery separators. Patents have been issued recently on the production of ultrahigh strength, very lightweight fibres from UHMW polymer by gel spinning.

Polypropylene

$$\left[CH_2-\underset{\displaystyle CH_3}{\underset{|}{CH}} \right]_n$$

Monomer	*Polymerisation*	*Major uses*
Propylene	*Ziegler-Natta* catalysed chain polymerisation	Fibre products (30%), housewares and toys (15%), automotive parts (15%), appliance parts (5%)

Since its conception in the late 1950s, the propylene polymerisation is being revolutionised in terms of both manufacturing hardware and, more importantly, catalyst technology. New catalyst technologies, in conjunction with state-of-the-art polymerisation, have established polypropylene as the kingpin in the field of polyolefins. A brief description of manufacturing processes and catalyst technologies are presented below.

Manufacturing processes

The workhorse process for commercial production of PP has been slurry polymerisation in liquid hydrocarbon diluent, for example, hexane or heptane. These are carried out either in stirred batch or continuous reactors.

High purity (>99.5 per cent) propylene is fed to the reactor containing diluent, as a suspension of solid (Ziegler-Natta) catalyst particles is metered in. The reaction is carried out at 50–80°C and at 5 to 20 atmospheric pressure. The crystalline polymer produced is insoluble and forms a finely divided granular solid enveloping the solid catalyst particles. Monomer addition is continued until the slurry reaches 20–40 per cent solids. Residence time varies from minutes to several hours, depending on the catalyst concentration and activity, as well as the specific reaction conditions. Molecular weight is controlled preferentially by the addition of hydrogen as the chain transfer agent.

The reactor slurry of PP is discharged to a stripping unit where the unreacted monomer flashes out for recycling. The catalyst is then deactivated and solubilised by the addition of alcohol. The bulk of the diluent, solubilised catalysts, and atactic polypropylene in solution are removed at this point by centrifuging. The crystalline polymer is purified by steam distillation and/or by water washing with surface active agents, followed by filtration and centrifuging and then drying. The dried polymer can be stored, transported, or pre-mixed with stabilisers to be used with or without pelletisation.

The efficiency of the Ziegler-Natta catalysts is of the order to 1,500 grams polymer formed per gram of transition metal. Residual catalyst has adverse effects on the corrosiveness, colour, and light stability of the polymer, and extraction processes must be used to remove it from the product. However, by utilising state-of-the-art, high-mileage (supported) catalyst systems, polymer yields are obtained which are several orders of magnitude higher than those obtained with first generation, Ziegler catalysts. Typically the high-mileage catalyst produce about 300,000 grams of polymer per gram of transition metal. Since there are only about 3 ppm of residual metal in the polymer, catalyst removal is unnecessary. The expensive catalyst removal (deashing) steps required for the products made by earlier Ziegler catalyst systems are thus eliminated.

Propylene is readily polymerised in bulk; that is, in the liquid monomer itself. *Arco, El Paso, Phillips,* and *Shell* are practitioners of bulk processing in stirred or loop reactor systems. In either case, liquid propylene (and ethylene, if random co-polymer is desired) is continuously metered to the polymerisation reactor along with a high-activity/high-stereospecificity catalyst system. Polymerisation temperatures are normally in the range of 45–80°C with pressures sufficient to maintain propylene in the liquid phase (250–500 psi, that is, 1.7–3.5 MPa). Hydrogen is used for molecular weight control. The polymer slurry (approximately 30–50 per cent solids in liquid propylene) is continuously discharged from loop reactors through a series of sequence valves into a zone maintained essentially at atmospheric pressure and containing terminating agents. Technologies also exist for production of propylene/ethylene block co-polymers via bulk polymerisation employing stirred or loop reactors.

Modern vapour-phase polymerisation is represented in one form by the stirred gas-phase process originally developed by BASF. BASF process reactors contain a spiral or double-helical agitator to stir the polymer bed. Cooling of the bed is maintained by continuous injection of fresh, high-purity propylene in a liquid or partly liquefied state into the reaction zone. The unreaced propylene is removed from the top of the reactor during polymerisation, condensed, and re-injected with fresh propylene. Evaporation of the unreacted propylene absorbs the heat of polymerisation and also brings about intense mixing of the solid polymer particles with the gas phase. Energy costs of the process are economically attractive. The diluent-free BASF process provides sufficiently high yield of polymer per unit of catalyst so that deashing is not required. Although products made in this way contain relatively high levels of titanium and aluminium residues, a unique finishing step during extrusion pelletising reduces active chlorides to an innocuous level.

With dramatic improvements in Ziegler-Natta catalyst technology, the *Spheripol* process, first developed by Montedison and Mitsui Petrochemical with simplified bulk (liquid propylene) process technology operating with loop reactors, is capable of directly producing a relatively large round bead with suitable density to eliminate the need for pelletising for many applications. Subsequently, Montedison and Hercules, assumed responsibility for all polypropylene operations and technology of the parent Companies.

The *Himont Spheripol* loop reactor process is initiated by injecting specially prepared supported catalyst and co-catalyst into liquid propylene circulated in a relatively simple high L/D ratio loop reactor, followed by monomer removal (Fig. 15.2). The homopolymer so produced can be circulated through ethylene and ethylene/propylene gas phase reactors for insertion of co-polymer fractions before final monomer stripping.

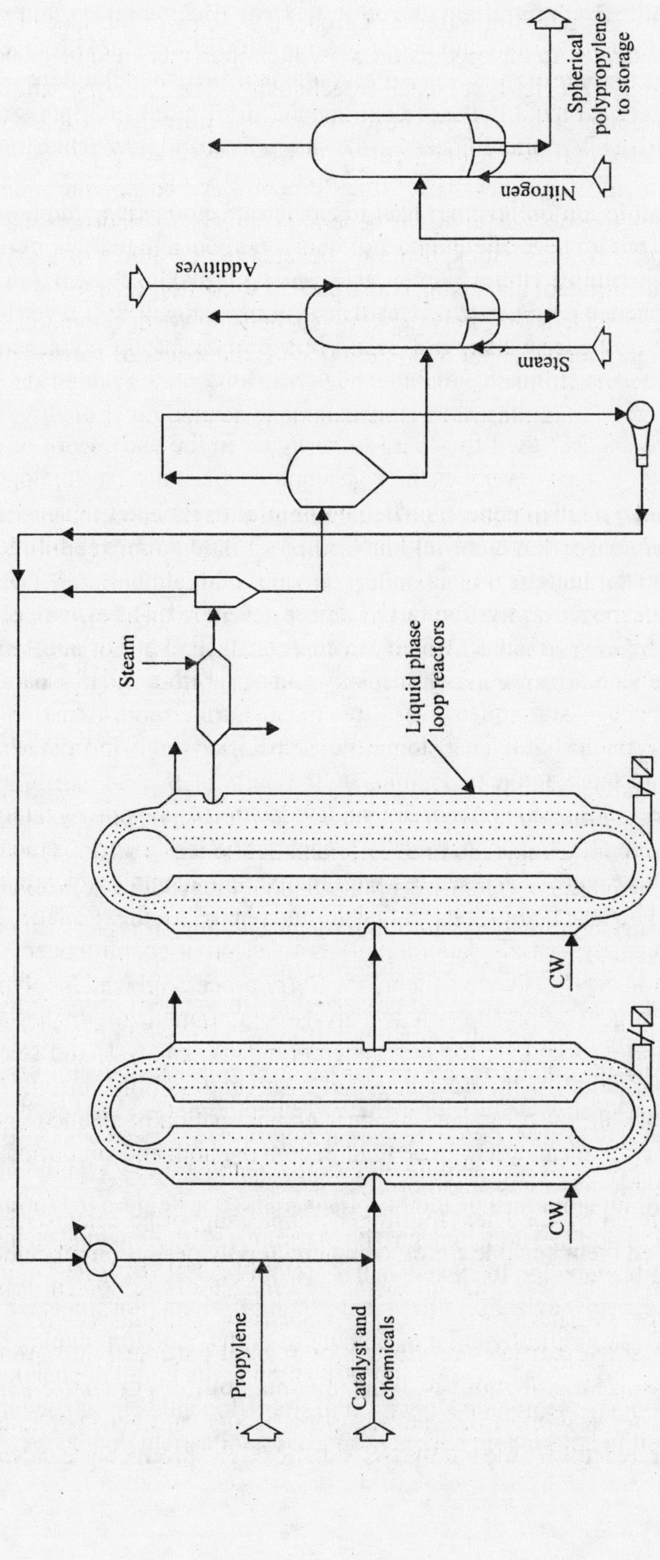

Fig. 15.2. A simplified flow diagram of the *Himont Spheripol* loop reactor process.

The *Unipol* low-pressure gas-phase fluidised-bed process, which was introduced by Union Carbide for LLDPE, has also been adapted to the production of PP homopolymers and block co-polymers using *Shell* Chemicals' high-activity (Ziegler-type) catalyst technology. The *Spheripol* and *Unipol* processes are capable of producing polymer in 'crumb' bead or granular forms, with the potential for direct marketing without any pelletising finishing operation.

Regardless of the polymerisation process used, the PP homo- and co-polymer must be stabilised to some degree to prevent oxidative degradation. The general practice is to incorporate a small quantity of stabiliser in the polymer prior to the first exposure to elevated temperatures of a drying operation for long-term storage. Inert-gas (nitrogen) blanketing is also used in some storage/transfer systems. Additional stabilisers, up to 1 per cent, are added to the polymer during pelletising. Most commercial PP compositions contain mixtures of hindered phenols and a hydroperoxide decomposers or various phosphites.

Catalyst technology

The origin of sophisticated catalyst used today are to be found in the early work of *Karl Ziegler* and *Guilio Natta*. In the last four decades, several distinct 'generations' of catalyst technologies have emerged. The earliest commercial catalysts (first generation) were essentially titanium trichloride, simply prepared by reducing $TiCl_4$ with alkylaluminiums to yield brown (β) $TiCl_3$, which was subsequently heated to convert it to the stereospecific purple (γ) form.

In the 1970s, improved or second-generation catalysts were developed. The essence of the improvement was that catalyst poisons $AlCl_3$ or $AlEtCl_2$, which are co-crystallised with or absorbed onto the $TiCl_3$ catalyst, were removed by using dialkyl ethers (especially di-*n*-butyl ether and di-isoamyl ether) either in the $TiCl_4$ reduction step or as a subsequent wash treatment. Furthermore excess $TiCl_4$ was used to 'catalyse' the phase transformation of β-$TiCl_3$ to the stereospecific γ- or δ-form at lower temperatures, leading to smaller and hence more active catalyst crystallites.

The 1980s heralded the widespread commercial implementation of supported catalysts. These third-generation catalysts comprise of $TiCl_4$ on a specially prepared $MgCl_2$ support. Commercially available $MgCl_2$ is converted to 'active $MgCl_2$' by treating with 'activating agents', which are electron donors (Lewis bases) such as ethyl benzoate, diisobutylphthalate, and phenyl triethoxy silane. These are so used in conjunction with the co-catalyst (trialkylaluminium) as a 'selectivity control agent'.

While the first-generation catalysts were suitable for slurry process, in which polymerisation occurs in a paraffinic solvent, third-generation supported catalysts with dramatically higher activity (typically 1500 kg PP/g Ti compared to 15 kg/g Ti for first generation $TiCl_3$ catalyst) and stereospecificity not only allowed the full exploitation of the advantages of a solventless polymerisation, but also made substantial simplification of slurry process possible through elimination of atactic removal and catalyst de-ashing. These third-generation catalysts can be used in both the bulk and simplified slurry processes. They are, however, unsuitable for a gas-phase process.

The major advance offered by the later (fourth generation) supported catalysts is their controlled morphology, which rendered them suitable for all commercial polymerisation processes. These are the only catalysts suitable for the full exploitation of the advantages of the solvent-less polymerisation including the gas-phase process. These catalysts come in a variety of regular shapes, such as spherical, cubical, or cylindrical, as single particles or clusters of several particles, and are characterised by sufficiently narrow particle size distributions with the minimum of fines or coarse particles.

During polymerisation, replication of the catalyst shape occurs, and hence spherical PP beads are obtained directly from the reactor, if the catalyst is spherical. By producing catalysts with a dense, spherical, and uniform particle shape it is thus possible to generate PP particles that, for many applications, can be shipped and used without a granulation or extrusion step to generate the 'nibs' desired by most customers. This concept is already part of the *Himont Spheripol* process mentioned earlier.

While attractive in terms of economy, since the energy consuming extrustion step is obviated in the above process, there are various disadvantages such as the difficulty of homogeneously administrating additives/stabilisers, the loss of the options for molecular weight modifications via extruder operation, and the fact that the process yields particles of smaller size than what is preferred by the industry simply because such large-size particles cannot be kept in suspension during polymerisation.

Montecatini has developed spherical morphology $MgCl_2$ supported catalysts for ethylene and propylene polymerisation. The name of this process is 'supherilene'. The major advantage in such system is the preparation of marketable powders directly from polymerisation plants and the catalyst is suitable for slurry, bulk, and gas-phase processes.

In the so-called reactor granule technology developed by *Montecatini*, the growing particle itself serves as the reactor within which the polymerisation takes place. Thus alloys and blends previously possible by mixing and melt extrusion of polyolefins can be made with this technology directly in the reactor from the individual monomers, dramatically decreasing the production cost by eliminating the need for compounding equipment. A variety of speciality polyolefins and polyolefin alloys can now be made directly in the reactor taking advantage of the new technology. Examples are the *catalloy* materials from Himont, which are polyolefin alloys made by synthesis and not by the conventional route of compounding. Hivalloy is a polypropylene/polystyrene alloy made by synthesis and combines the properties of both crystalline and amorphous engineering polymers. Such materials could challenge the established positions of several thermoplastic elastomers.

A desired approach reducing the costs (energy consumption) of an extruder step would be to execute the polymerisation in an extruder. A major advantage of this concept is that the enormous heat of polymerisation could itself be used to melt the polymer, resulting in considerable energy saving. In terms of catalyst technology, this involves the development of a catalyst that is able to polymerise propylene to high conversions with good selectivity and activity even at temperatures in excess of the melting point of polypropylene (that is, temperatures of around 180°C in the final stages of the process).

Despite the domination of supported catalysts in polyolefin production, industrial research work has shifted toward new generation, single-site, homogeneous, Ziegler-Natta catalyst systems, the main compound of which is the Group IVB transition metallocenes (titanocene, zirconocenes, and hafnocenes). The molecular structure of the two famous *Brintzinger* catalysts, ethylenebis (indenyl) zirconium dichloride $Et(ind)_2ZrCl_2$ and ethylenebis (tetrahydroindenyl) zirconium dichloride $Et(H_2Ind)_2ZrCl_2$ are shown in Fig. 15.3. Methylalumoxane (MAO) is the most important co-catalyst that activates the group IVB metallocenes is homogeneous Ziegler-Natta polymerisation.

Before the discovery of the MAO co-catalyst, the homogeneous Ziegler-Natta catalyst Cp_2TiCl_2 (Cp = cyclopentadiene) was activated with alkyl aluminium chloride, which led to poor catalyst activity. The use of MAO co-catalyst raised the catalyst activity by several orders of magnitude. Possible structures of MAO are shown in Fig. 15.4. Using the chiral zirconocene catalyst in combination with the MAO co-catalyst, polypropylene can be obtained in high purity and high yield even at relatively high temperatures (room temperature and above).

The zirconium catalysts are three orders of magnitude more active than their titanium counterparts. Unlike the heterogeneous Ziegler-Natta catalysts in which usually several types of active catalyst sites are present, metallocene catalyst are single-site in nature, that is, they have a single active catalyst site on the catalyst structure and thus they make one type of polymer, ensuring a high degree of purity.

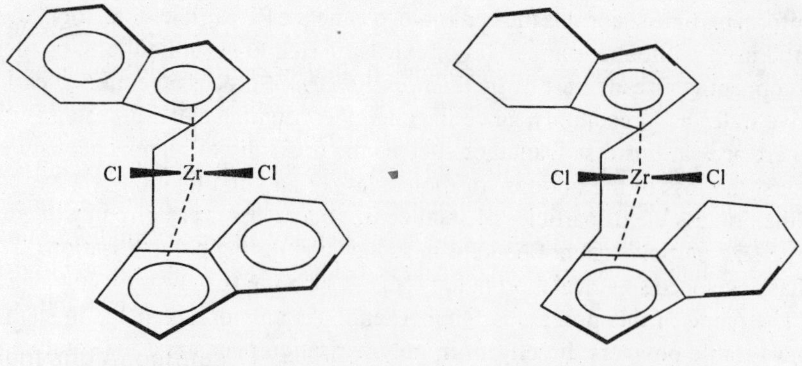

Et (Ind)$_2$ ZrCl$_2$ Et (H$_4$ Ind)$_2$ ZrCl$_2$

Fig. 15.3. Structures of the Brintzinger catalysts.

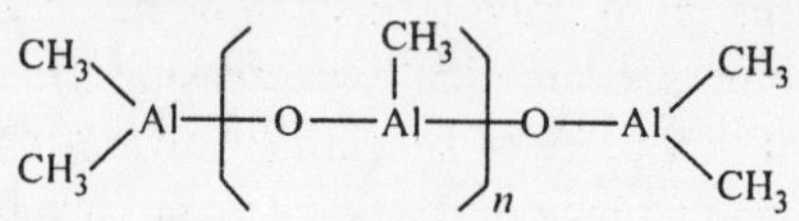

Linear

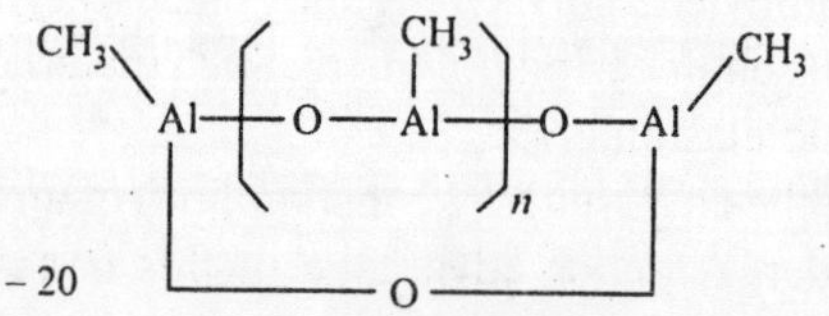

$n = 4 - 20$

Cyclic

Fig. 15.4. Possible structures of methylalumoxane.

The very high selectivity of the zirconocene single-site catalysts and their high activity, which approaches that of the $MgCl_2$-supported catalysts described earlier, makes them a serious contender for future processes. It can be expected that the relative ease of 'tailoring' of homogeneous catalysts compared to complicated heterogeneous systems will enable these catalysts to be further improved and exploited in terms of activity and selectivity (isotacticity, molecular weight, and distribution).

Researches in the field of metallocene catalysts which are at the heart of 'single-site' catalyst (SSC) technology, are experiencing enormous progress. The exploding development in this field is reflected by the increasing amount of patents issued in the last few years. Instead of four decades of development from first generation to fourth generation superactive Ziegler-Natta systems, approximately a decade led to the creation of superactive ultraselective catalysts for the production of every possible grade of polyethylene and polypropylene with tailored molecular-weight distributions and a very high degree of co-monomer composition homogeneity.

LLDPE product manufactured from metallocene catalysts has dramatically superior properties as compared to conventional LLDPE. These include high clarity, high tear strength, low heat-seal temperature, and very low amounts of extractables. The technology of metallocene catalysis is expensive, still young and yet to mature. It is anticipated that higher yield would offset the higher cost of metallocene catalysts. The real benefits to the processes are yet to come and are expected in the near future.

It is worthwhile to mention that homogeneous Ziegler-Natta catalysts other than metallocene-based catalysts have also been developing rapidly in recent years. The catalyst precursors of these systems are non-metallocene organometallic compounds, such as monocyclopentadienyl derivatives. A representative of monocyclopentadienyl catalysts is the constrained geomery (CG) catalyst.

This new type of homogeneous catalyst was developed by Dow Plastics. The catalyst system is based on group IVB transition metals such as Ti, covalently bonded to a cyclopentadienyl group bridged with a heteroatom such as nitrogen. The components are linked in such a way that a constrained cyclic structure is formed with Ti at the centre. The bond angle between the monocyclopentadienyl group, Ti centre, and heteroatom is less than 115°. The catalyst is activated by strong Lewis acid systems to a highly efficient cationic form.

The CG catalysts produce highly processable polyolefins with a unique combination of narrow MWD and long chain branches. Ethylene-octene co-polymers produced with CG catalysts have useful properties across a range of densities and melting. These novel co-polymer families are called polyolefin plastomers (POP) and polyolefin elastomers (POE). POPs prosses plastic and elastic properties while POEs containing greater than 20 wt per cent octene co-monomer units have higher elasticity.

The CG catalyst technology is not limited to the typical selection of C_2–C_8 α-olefins, but can include higher α-olefins. The open structure of the CG catalyst significantly increases the flexibility to insert higher α-olefin co-monomers into the polymer structure. This technology also allows addition of vinyl-ended polymer chains to produce long chain branching.

Properties

'Polypropylene' is not one or even 100 products. Rather it is a multidimensional range of products with properties and characteristics interdependent on the type of polymer (homopolymer, random, or block co-polymer), molecular weight and molecular weight distribution, morphology and crystalline structure, additives, fillers and reinforcing fillers, and fabrication techniques.

Commercial homopolymers are usually about 90 to 95 per cent isotactic, the other structures being atactic and syndiotactic (a rough measure of isotacticity is provided by the 'isotactic index'—the per centage of polymer insoluble in heptane): the greater the degree of isotacticity the greater the crystallinity and hence the greater the softening point, stiffness, tensile strength, modulus, and hardness.

Although very similar to HDPE, PP has a lower density (0.90 g/cm^3) and a higher softening point, which enables it to withstand boiling water and many steam sterilising operations. It has a higher brittle point and appears to be free from environmental stress-cracking problems, except with concentrated sulphuric acid, chromic acid, and aqua regia. However, because of the presence of tertiary carbon atoms occurring alternately on the chain backbone, PP is more susceptible to UV radiation and oxidation at elevated temperatures. Whereas PE cross-links on oxidation, PP undergoes degradation to form lower-molecular-weight products. Substantial improvement can be made by the inclusion of antioxidants, and such additives are used in all commercial PP compounds. The electrical properties of PP are very similar to those of HDPE.

Because of its reasonable cost and good combination of the foregoing properties, PP has found many applications, ranging from fibres and filaments to films and extrusion coatings. A significant portion of the PP produced is used in mouldings, which include luggage, stacking chairs, hospital sterilisable equipment, toilet cisterns, washing machine parts, and various auto parts, such as accelerator pedals, battery cases dome lights, kick panels, and door frames.

Although commercial PP is a highly crystalline polymer, PP mouldings are less opaque when unpigmented than are corresponding HDPE mouldings, because the differences between amorphous and crystal densities are less with PP (0.85 and 0.94 g/cm^3, respectively) than with polyethylene (0.84 and 1.01 g/cm^3, respectively).

A particularly useful property of PP is the excellent resistance of thin sections to continued flexing. This has led to the production of one-piece mouldings for boxes, cases, and accelerator pedals in which the hinge is an integral part of the moulding.

Monoaxially oriented polypropylene film tapes have been widely used for carpet backing and for woven sacks (replacing those made from jute). Combining strength and lightness, oriented PP straps have gained rapid and widespread acceptance for packaging.

Non-oriented PP film, which is glass clear, is used mainly for textile packaging. However, biaxially oriented PP film is more important because of its greater clarity, impact strength, and barrier properties. Coated grades of this material are used for packaging potato crisps, for wrapping bread and biscuits, and for capacitor dielectrics. In these applications PP has largely replaced regenerated cellulose. (The high degree of clarity of biaxially oriented PP is caused by layering of the crystalline structures. Layering reduces the variations in refractive index across the thickness of the film, which thus reduces the amount of light scattering.)

Polypropylene, produced by an oriented extrusion process, has been uniquely successful as a fibre. Its excellent wear, inertness to water, and micro-organisms, and its comparatively low cost have made it extensively used in functional applications, such as carpet backing, upholstery fabrics, and interior trim for automobiles.

Random ethylene-propylene co-polymers, another important variety of polypropylene, are noted for high clarity, a lower and broader melting range than homopolymer grades, reduced flexural modulus, and higher melt strengths. They are produced by the random addition of ethylene to a polypropylene chain as it grows. The melt-flow rate of random co-polymers ranges from 1 g/10 min for a blow-moulding grade to 35 g/10 min for an injection-moulding grade. The density is about 0.90 g/cm^3 and the notched Izod impact strength of the materials ranges from under 1 to more than 5 ft-lb/inch.

The blow-moulded bottles capitalise on the good clarity provided by the random co-polymer. The high gloss and very broad heat sealing range of this resins is useful in such cast-film applications as trading cards and documents protectors. Polypropylene co-polymers with a melt-flow rate of 35 g/10 min or above find applications in thin wall parts, usually for injection moulded food packaging such as delicatessen containers or yogurt cups. Such containers have walls with a length-to-thickness ratio as high as 400:1; yet they retain the properties of top-load strength, impact resistance, and recyclability that are typical of polypropylene.

The full potential of the many varieties of polypropylene has hardly been tapped. Extrusion coaters can make use of high-melt-strength polypropylene resins with melt-flow rates as high as 40 g/10 min. These materials resist pinholes and curling, while providing very low neck-in. Producers of large blow-moulded or thermoformed parts can utilise grades with improved melt strength. In applications such as automotive parts, household appliances, industrial food trays, and water-storage tanks, the resins provide resistance to chemicals, thermal shock, and heat distortion up to 100°C at 66 psi. The resins thus seem poised for further penetration of markets with wider range of applications.

Block co-polymers, preferably with ethylene, are classed as having medium, high, or extra-high impact resistance with particular respect to sub-zero temperatures. Block co-polymers consist of a crystalline PP matrix containing segments of EPR-type elastomer and/or crystalline PE for energy impact absorption in the rubber phase. The level of the ethylene co-monomer as well as the size of these segments has an important bearing on the physical properties of the final block co-polymer.

Use pattern

Few materials are as compatible with as many processing techniques or are used in as many commercial applications as polypropylene. It is found in everything from flexible and rigid packaging to fibres and large moulded parts for automotive and consumer products. Largely conforming to this diversity of applications is the fact that the material can be processed by most methods, including extrusion, extrusion coating, blown and cast film, blow moulding, injection moulding, and thermoforming.

Polypropylene fibres and filaments form the largest market area, which is comprised of several segments with carpetting applications being the largest; these include primary and secondary woven and non-woven uses, carpet backing face yarns, indoor/outdoor constructions, automotive interior mats and trunk linings, and synthetic turf. The good wickability of PP is utilised in such non-woven applications as disposable diaper. Other applications include clothing inner liners, drapes and gowns, sleeping bags, wall coverings wiping cloths, and tea bags. Furniture and automotive upholstery fabrics and produced from both continuous monofilament and staple fibres.

The second largest PP market is film, both oriented (OPP) and cast. Large users of OPP films are packaging for snack foods, bakery products, dry foods, candy, gum, cheese, tobacco products, and electrical capacitors. Cast (unoriented) film is used for packaging textile soft goods, cheese, snack foods, and bakery products.

Largest users of injection moulding PP are in transportation, particularly automotive and truck battery cases. PP co-polymers have secured about 90 per cent of this market as a result of a drive by automotive manufacturers to reduce weight and cost. In addition to being lightweight, PP also provides outstanding resistance to crepe and fatigue, high temperature rigidity, impact strength, and resistance to corrosion.

The next largest moulded product market for PP is packaging, especially closures and containers. Child-resistance, tamperproof, linerless features are important design factors as also inherent chemical resistance, stress-crack resistance, and high productivity at low cost. Housewares utilise random co-polymers for refrigerator and shelf-storage containers and lids. Medium impact co-polymers are used for hot/cold thermos containers, lunch boxes, coolers, and picnic ware.

Medical applications of PP such as disposable syringes, hospital trays, and labware are contingent on sterilisability, either autoclaving or radiation. Disposable syringes that are sterilised by radiation require special formulations to prevent discoloration (yellowing) or brittleness as a consequence of degradation and cross-linking.

PP finds highly successful uses in both major and small appliances. Washing machines, dish washers, tub liners, agitators, bleach and detergent dispensing units, valve and control assemblies, drain tubes, pump housings, door liners, coffee makers, hair dryers, vacuum cleaners, can openers, knife sharpeners, room humidifiers and dehumidifiers, floor and ceiling fans, and window air-conditioner units are some examples.

Polyallomer

$$\left(CH_2-\underset{\displaystyle CH_3}{\underset{|}{CH}} \right)_m - \left(CH_2-CH_2 \right)_n$$

A proprietary polymerisation process, developed in the mid 1960s by staff researchers of Eastman Chemical Products, produces co-polymers of 1-olefins that give a degree of crystallinity normally obtained only with homopolymers. The term 'polyallomer' was coined to identify the polymers manufactured by this process and to distinguish them from conventional co-polymers. The polyallomer materials available today are based on block co-polymers of propylene and ethylene.

Polyallomers combine the most desirable properties of both crystalline polypropylene and high-density polyethylene (HDPE) and can offer impact strengths three or four times that of polypropylene. Resistance to heat distortion is better than that of HDPE but not quite as good as that of polypropylene. Polyallomer has better abrasion resistance than polypropylene and comparable hinge-forming characteristics.

Polyallomer lightweight cases can be moulded entirely in one piece. Back, front, hinges, handles, and snap clasps can be moulded in at the same time in a wide range of colours. Polyallomer is thus used in such injection moulded items as fishing tackle boxes, typewriter cases, gas-mask cases, and bowling-ball bags.

Shoe toes moulded of polyallomer resist cracking and denting under repeated hammer blows. They withstand temperatures from –40 to 150°C and can withstand up to 300 pounds of force (1335 N). Polyallomers can be processed easily on conventional moulding and extruding equipment. Polypropylene colour concentrates can be used to colour polyallomer, since these two polymers are compatible.

Poly(Vinyl Chloride)

$$\left[CH_2-\underset{\displaystyle Cl}{\underset{|}{CH}} \right]_n$$

Monomer	*Polymerisation*	*Major uses*
Vinyl chloride	Free-radical-initiated chain polymerisation	Pipe and fittings (35%), film and sheet (15%), flooring materials (10%), wire and cable insulation (5%), automotive parts (5%), adhesives and coatings (5%)

Poly(vinyl chloride)(PVC) is produced by polymerisation of vinyl chloride by free-radical mechanisms, mainly in suspension and emulsion, but bulk and solution processes are also employed to some extent. (The control of vinyl chloride monomer escaping into the atmosphere in the PVC production plant has become important because cases of angiosarcoma, a rare type of liver cancer, were found among workers exposed to the monomer. This lead to setting of stringent standards by governments and modification of manufacturing processes by the producers to comply with the standards).

At processing temperatures used in practice (150–200°C), sufficient degradation may take place to render the product useless. Evidence points to the fact that dehydrochlorination occurs at an early stage in the degradation process and produces polyene structures:

$$\sim\sim CH_2-\underset{\displaystyle Cl}{\underset{|}{CH}}-CH_2-\underset{\displaystyle Cl}{\underset{|}{CH}}\sim\sim \xrightarrow{-HCl} \sim\sim CH=CH-CH=CH\sim\sim$$

It is believed that the liberated hydrogen chloride can accelerate further decomposition and that oxygen also has an effect on the reaction. However, incorporation of certain materials known as stabilisers retards or moderates the degradation reaction so that useful processed materials can be obtained. Many stabilisers are also useful in improving the resistance of PVC to weathering, particularly against degradation by UV radiation.

Characterisation of commercial resins

Commercial PVC polymers are largely amorphous, slightly branched molecules with the monomer units arranged in a head-to-tail sequence. Individual grades of PVC do 'however' differ in average molecular weight, molecular weight distribution, and particle shape, size, and size distribution.

The molecular weights of the bulk commercial polymers are in the range $\overline{M}_w$ = 100,000–200,000 and $\overline{M}_n$ = 45,000 – 64,000, although values may be as low as 40,000 and as high as 480,000 for $\overline{M}_w$. In practice, however, it has been common to characterise the molecular weight of the PVC polymer by its *Fikentscher K*-value rather than to quote an actual figure for molecular weight. The *K*-value is obtained from the following relationship:

$$\log_{10} \eta_{rel} = \left[\left(\frac{75K^2 \times 10^{-6}}{1 + 1.5Kc \times 10^{-3}}\right) + (K \times 10^{-3})\right] c$$

where η_{rel} = relative viscosity = η/η_o, η = viscosity of the polymer solution, η_o = viscosity of the pure solvent, *K* = *K*-value, and *c* = concentration in g/100 mL.

Since the *K*-value is rather dependent on the method of measurement and the solvent used, the ISO viscosity number is often used to characterise the molecular weight of a PVC polymer. Table 15.2 compares typical correlations between number and weight average molecular weights with ISO numbers and *K*-values. Most general purpose polymer for use in plasticised PVC compound have ISO numbers of about 125. Because of processing problems the polymer used for unplasticised PVC compounds have lower molecules weights, typical ISO numbers being 105 for pipe, 85–95 for rigid sheet, and as low as 70 for injection moulding compounds.

Table 15.2. Molecular weight characterisation of PVC.

Average mol. wt.		*ISO/R174-1961(E):*	*K-value*		
Weight	*Number*	*viscosity number*	*0.5% in ethylene dichloride*	*1% in cyclohexanone*	*0.4% in nitrobenzene*
54000	26000	57	45	47	52
70000	36000	70	50	54	57.5
100000	45500	87	55	60	62.5
140000	55000	105	60	65	68
200000	64000	125	65	70	70.5
260000	73000	145	70	74	78
340000	82000	165	75	–	–

With commercial polymers that major differences are in the characteristics of the particle, i.e., its shape, size, size distribution, and porosity. Such differences considerably affect the processing behaviour of a polymer.

Considerable effect has been expanded to develop suitable processes to control porosity, surface area, and diffusivity of PVC particles and this has led to great improvements over the years in the processability of PVC.

If PVC polymer particles are mixed, at room temperature, with plasticisers, the immediate product may take one of two forms. If the plasticiser quantity is insufficient to fill all the gaps between the particles, a 'mush' will be produced. If all the voids are filled then the particles will become suspended in the excess plasticiser and a paste will be formed.

It is found that the viscosity of a PVC paste made from a fixed polymer- plasticiser ratio depends to a great extent on the particle size and size distribution. To obtain a low viscosity paste, the less the amount of plasticiser required to fill the voids between particles the better. Any additional plasticiser present is then available to act as a lubricant for the particles, facilitating their general mobility in suspension. Thus in general PVC pastes in which the polymer has a wide particle-size distribution (but within limits set by problems of significant plasticiser absorption even at room temperature by very small particles and settling caused by large particles) so that particles pack efficiently and leave less voids (Fig. 15.5a) will have lower viscosity than those of constant particle size (Fig. 15.5b).

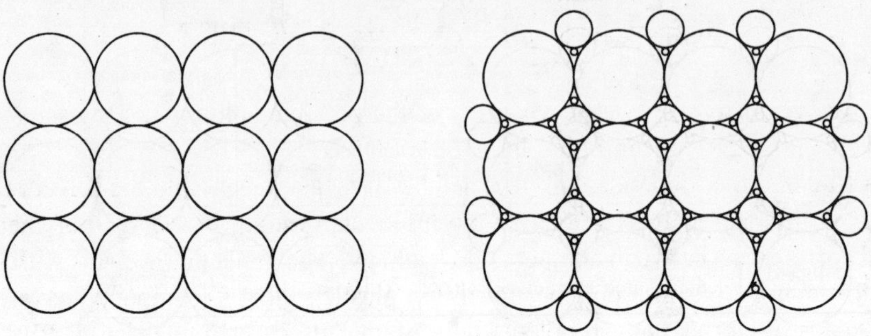

(a) (b)

Fig. 15.5. (a) PVC paste polymer particles with homogeneous particles size—less efficient packing; and (b) PVC paste polymer particles with distribution of size—efficient packing.

The use of 'filler' polymers in increasing quantities in PVC paste technology can be considered as an extension of this principle. These filler polymers are made by suspension (granular, dispersion) polymerisation and by themselves the particles are too large to make stable pastes. However, in the presence of much smaller paste polymer particles they remain in stable suspension. As shown in Fig. 15.6, the replacement in space of a mixture of paste-polymer particles and plasticiser by a large granular polymer particle releases plasticiser which then acts as a lubricant, i.e., a viscosity depressant.

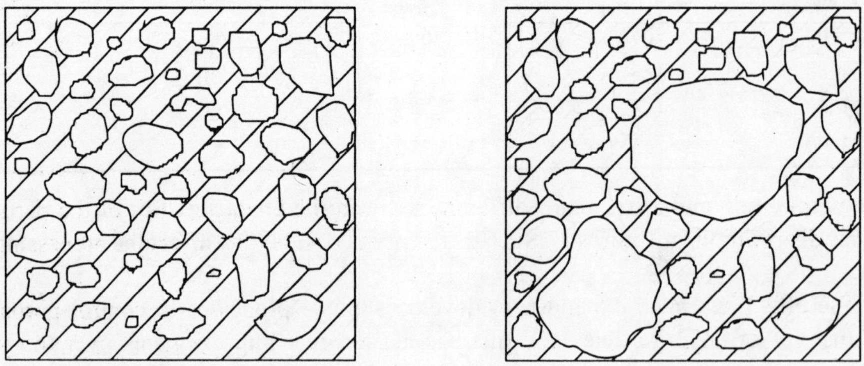

(a) (b)

Fig. 15.6. (a) PVC paste polymer suspended in plasticiser; and (b) PVC paste containing filler polymer. Less plasticiser is required to fill voids in unit volume.

PVC pastes exhibit complex rheological behaviour with the viscosities showing dependence on the shear rate and on the time of shear. A paste viscosity may increase with shear rate (dilatancy) or decrease (shear thinning or pseudoplasticity). Some pastes may show dilatant tendencies over one range of shear rates but be shear thinning over another range. The viscosities may also decrease with time of stirring (thixotropy) or increase with it (rheopexy).

It has been observed that spherical particles with distribution of size giving a high degree of packing are closest to Newtonian liquids in their behaviour. Spherical particles of homogeneous size, however, give shear thinning pastes. This may be due to the fact that these particles tend to aggregate at rest while shearing causes disaggregation and hence easier movement of particles. Very coarse and lumpy uneven granules do not slide past each other in pastes and tend to become more entangled as shear rate increases. Such pastes commonly show dilatant behaviour.

Compounding ingredients

PVC is a colourless rigid material with limited heat stability and with a tendency to adhere to metallic surfaces when heated. For these and other reasons it is necessary to compound the polymer with other ingredients to make useful products. It is possible in this way to make a wide range of products including rigid piping and soft elastic cellular materials.

A PVC compound may contain, besides the polymer, the following ingredients: stabilisers, plasticisers, extenders, lubricants, fillers, pigments, and polymeric processing aids. Other ingredients also used occasionally include impact modifiers, fire retardants, optical bleaches, and blowing agents.

Stabilisers

The most important class of stabilisers are the lead compounds which form lead chloride on reaction with the hydrogen chloride evolved during decomposition. Basic lead carbonate (white lead), which has a low weight cost, is more commonly used. A disadvantage of lead carbonate is that it may decompose with the evolution of carbon dioxide at higher processing temperatures and lead to porous product. For this reason, tribasic lead sulphate, which gives PVC products with better electrical insulation properties than lead carbonate, is often used despite its somewhat higher weight cost. Other lead stabilisers are of much more specific applications. For example, dibasic lead phthalate, which is an excellent heat stabiliser, is used in heat-resistant insulation compounds (e.g., in 105°C wire), in high-fidelity gramophone records, in PVC coatings for steel, and in expanded PVC formulation.

The use of lead compounds as stabilisers has been subjected to regulation because of its toxicity. Generally, lead stabilisers are not allowed in food-packaging PVC materials, but in most countries they are allowed in PVC pipes for conveying drinking water, with reduction in the level of use of such stabilisers.

Today the compounds of cadmium, barium, calcium, and zinc have gained prominence as PVC stabilisers. A modern stabilising system may contain a large number of components. A typical cadmium-barium 'packaged' stabiliser may have the following compositions: cadmium-barium phenate 2 to 3 parts, epoxidised oils 3 to 5 parts, trisonyl phenyl phosphite 1 part, stearic acid 0.5 to 1 part, and zinc octoate 0.5 part by weight. For flooring compositions, calcium-barium, magnesium-barium, and copper-barium compounds are sometimes used in conjuction with pentaerythritol (which has the function of reducing colour by chelating iron present in asbestos).

Another group of stabilisers are the organo-tin compounds. Development of materials with low toxicity, excellent stabilising performance, and improving relative price situation has led to considerable growth in the organo-tin market during the last decade. Though the level of toxicity of butyltins is not sufficiently low for application in contact with foodstuffs, many of the octyltins, such as dioctyltin dilaurate and dioctyltin octylthioglycolate, meet stringent requirements for use in contact with foodstuffs. Further additions to the class of organo-tins include the estertins characterised by low toxicity, odour, and volatility, and the methyltins having higher efficiency per unit weight compared with the more common organo-tins.

Organo-tin compounds that are salts of alkyltin oxides with carboxylic acids (e.g., dioctyltin dilaurate) are usually called organo-tin carboxylates. Organo-tin compounds with at least one tin-sulphur bond (e.g., dioctyltin octylthioglycolates) are generally called organo-tin mercaptides. The latter are considered to be the most efficient and most universal heat stabilisers.

One of the most important properties—not only of the sulphur containing tin stabilisers but also of the whole group of organo-tin stabilisers—is absolute crystal clarity, which can be achieved by means of a proper formulation. Clarity is required for bottles, containers, all kinds of packaging films, corrugated sheets, light panels, and also for hose, profiles, swinging doors, and transparent top coats of floor or wall coverings made from plasticised PVC.

Plasticisers

In addition to resin and stabilisers, a PVC compound may contain ingredients such as plasticisers, extenders, lubricants, fillers, pigments, polymeric processing aids, and impact modifiers.

Plasticisers are essentially non-volatile solvents for PVC. At the processing temperature of about 150°C, molecular mixing occurs in a short period of time to give products of greater flexibility. Phthalates prepared from alcohols with about eight carbon atoms are by far the most important class and constitute more than 70 per cent of plasticisers used. For economic reasons, diisooctyl phthalate (DIOP), di-2-ethylhexyl phthalate (DEHP or DOP), and the phthalate ester of the C_7–C_9 oxoalcohol, often known as dialphanyl phthalate (DAP) are used. DIOP has somewhat less odour, whereas DAP has the greatest heat stability. Dibutyl phthalate and diisobutyl phthalate are also efficient plasticisers and continue to be used in PVC (except in thin sheets) despite their high volatility and water extractability.

Phosphate plasticisers such as tritolyl phosphate (TTP) and trixylyl phosphate (TXP) are generally used where good flame resistance is required, such as in insulation and mine belting. These materials, however, are toxic and give products with poor low-temperature resistance, i.e., with a high cold flex temperature (typically, –5°C).

For applications where it is important to have a compound with good low-temperature resistance, aliphatic ester plasticisers are of great value. Dibutyl sebacate (DBS), dioctyl sebacate (DOS), and, more commonly, cheaper esters of similar effect derived from mixed acids produced by the petrochemical industry are used. These plasticisers give PVC products with a cold flex temperature of –42°C.

Esters based on allyl alcohol, such as diallyl phthalate and various polyunsaturated acrylates, have proved useful in improving adhesion of PVC to metal.

They may be considered as polymerisable plasticisers. In PVC pastes they can be made to cross-link by the action of peroxides or perbenzoates when the paste is spread on to metal, giving a 'cured coating' with a high degree of adhesion.

The high adhesion of these rather complex compounds has led to their development as metal-to-metal adhesives used, for example, in car manufacture. Metal coatings may also be provided from plasticised powders containing polymerisable plasticisers by means of fluidised bed or electrostatic spraying techniques.

Extenders

In the formulation of PVC compounds it is not uncommon to replace some of the plasticiser with an extender, a material that is not in itself a plasticiser because of its very low compatibility but that can be used in conjunction with a true plasticiser. Commercial extenders are cheaper than plasticisers and can often be used to replace up to one-third of the plasticiser without seriously affecting the properties of the compound. Three commonly employed types of extenders are chlorinated paraffin waxes, chlorinated liquid paraffinic fractions, and oil extracts.

Lubricants

In plasticised PVC it is common practice to incorporate a lubricant whose main function is to prevent sticking of the compound to processing equipment. The material used should have limited compatibility such that it will sweat out during processing to form a film between the bulk of the compound and the metal surfaces of the processing equipment. The additives used for such a purpose are known as external lubricants. In the United States normal lead stearate is commonly used. This material melts during processing and lubricates like wax. Also used is dibasic lead stearate, which does not melt but lubricates like graphite and improves flow properties. In Britain, stearic acid is mostly used with transparent products, calcium stearate with non-transparent products.

An unplasticised PVC formulation usually contains at least one other lubricant, which is mainly intended to improve the flow of the melt, i.e., to reduce the apparent melt viscosity. Such materials are known as internal lubricants. Unlike external lubricants they are reasonably compatible with the polymer and are more like plasticisers in their behaviour at processing temperatures, whereas at room temperature this effect is negligible. Among materials usually classified as internal lubricants are montan wax derivatives, glyceryl monosterate, and long-chain esters such as cetyl palmitate.

Fillers

Fillers are commonly employed in opaque PVC compounds to reduce cost and to improve electrical insulation properties, to improve heat deformation resistance of cables, to increase the hardness of a flooring compound, and to reduce tackiness of highly plasticised compounds. Various calcium carbonates (such as whiting, ground limestone, precipitated calcium carbonate) are used for general-purpose work, china clay is commonly employed for electrical insulation, and asbestos for flooring applications. Also employed occasionally are the silicas and silicates, talc, light magnesium carbonate, and barytes (barium sulphate).

Pigments

Many pigments are now available commercially for use with PVC. Pigment selection should be based on the pigment's ability to withstand process conditions, its effect on stabiliser and lubricant, and its effect on end-use properties, such as electrical insulation.

Impact modifiers and processing aids

Unplasticised PVC presents some processing difficulties due to its high melt viscosity; in addition, the finished product is too brittle for some applications. To overcome these problems and to produce toughening, certain polymeric additives are usually added to the PVC. These materials, known as impact modifiers, are generally semi-compatible and often somewhat rubbery in nature. Among the most important impact modifiers in use today are butadiene-acrylonitrile co-polymer (nitrile rubber), acrylonitrile-butadiene-styrene (ABS) graft terpolymers, methacrylate-butadiene-styrene (MBS) terpolymers, chlorinated polyethylene, and some polyacrylates.

ABS materials are widely used as impact modifiers, but they cause opacity and have only moderate ageing characteristics. Many grades also shown severe stress whitening, a phenomenon advantageously employed in labeling tapes, such as Dymotape. MBS modifiers have been used such as where tough PVC materials of high clarity are desired (e.g., bottles and film). Chlorinated polyethylene has been widely used as an impact modifier where good ageing properties are required.

A number of polymeric additives are also added to PVC as processing aids. They are more compatible with PVC and are included mainly to ensure more uniform flow and thus improve the surface finish. In chemical constitution they are similar to impact modifiers and include ABS, MBS, acrylate-methacrylate co-polymers, and chlorinated polyethylene.

Properties and applications

PVC may reasonably be considered as the most versatile of plastics; its usage ranges from building construction to toys and footwear. PVC compounds are made in a wide range of formulations, which makes it difficult to make generalisations about their properties. Mechanical properties are considerably affected by the type and amount of plasticiser. Table 15.3 illustrates differences in some properties of three distinct types of compound. To a lesser extent, fillers also affect the physical properties.

Unplasticised PVC (UPVC) is a rigid material, whereas the plasticised material is tough, flexible, and even rubbery at high plasticiser loadings. Relatively high plasticiser loadings are necessary to achieve any significant improvement in impact strength. Thus incorporation of less than 20 per cent plasticiser does not give compounds with impact strength higher than that of unplasticised grades. Lightly plasticised grades are therefore used when the ease of processing is more important than achieving good impact strength.

Table 15.3. Properties of three types of PVC compounds.

Property	*Unplasticised PVC*	*PVC + DIOP (50 parts per 100 resin)*	*Vinyl chloride-vinyl acetate co-polymer (sheet)*
Specific gravity	1.4	1.31	1.35
Tensile strength			
lbf/in^2	8500	2700	7000
MPa	58	19	48
Elongation at break (%)	5	300	5
Vicat softening (°C)	80	Flexible at room temperature	70

PVC is resistant to most aqueous solutions, including those of alkalies and dilute mineral acids. The polymer also has a good resistance to hydrocarbons. The only effective solvents appear to be those which are capable of some form of interaction with the polymer. These include cyclohexanone and tetrahydrofuran.

At ordinary temperatures, PVC compounds are reasonably good electrical insulators over a wide range of frequencies, but above the glass transition temperature their value as an insulator is limited to low-frequency applications. The volume resistivity decreases as the amount of plasticiser increases.

PVC has the advantage over other thermoplastic polyolefins of built-in fire retardancy because of its 57 per cent chlorine content.

Co-polymers of vinyl chloride with vinyl acetate have lower softening points, easier processing, and better vacuum-forming characteristics than the homopolymer. They are soluble in ketones, esters, and certain chlorinated hydrocarbons, and have generally inferior long-term heat stability.

About 90 per cent of the PVC produced is used in the form of homopolymers, the other 10 per cent as co-polymers and terpolymers. The largest application of homopolymer PVC compounds, particularly unplasticised grades, is for rigid pipes and fittings, most commonly as suspension homopolymers of high bulk density compounded as powder blends. In addition to the aforesaid properties, UPVC has an excellent resistance to weathering. Moreover, when the cost of installation is taken into account, the material frequently turns out to be cheaper. UPVC is therefore becoming used increasingly in place of traditional materials. Important uses include translucent roof sheathing with good flame-retarding properties, window frames, and piping that neither corrodes nor rots.

As a pipe material PVC is widely used in soil pipes and for drainage and above-ground applications. Piping with diameters of up to 60 cm is not uncommon.

UPVC is now being increasingly used as a wood replacement due to its more favourable economics, taking into account both initial cost and installation. Specific applications include bench-type seating at sports stadia, window fittings, wall-cladding, and fencing. UPVC bottles have better clarity, oil resistance, and barrier properties than those made from polyethylene. Compared with glass they are also lighter, less brittle, and possess greater design flexibility. These products have thus made extensive penetration into the packaging market for fruit juices and beverages, as well as bathroom toiletry. Sacks made entirely of PVC enable fertilisers and other products to be stored outdoors.

The largest applications of plasticised PVC are wire and cable insulation and as film and sheet. PVC is of great value as an insulator for direct-current and low-frequency alternating-current carriers. It has almost completely replaced rubber in wire insulation. PVC is widely used in cable sheathing where polyethylene is employed as the insulator.

Other major outlets of plasticised PVC include floor coverings, leathercloth, tubes and profiles, injection mouldings, laminates, and paste processes.

When a thin layer of plasticised PVC is laminated to a metal sheet, the bond may be strong enough that the laminate can be punched, cut, or shaped without parting the two layers. A pattern may be printed or embossed on the plastic before such fabrication. Typewriter cases and appliance cabinets have been produced with such materials.

PVC leathercloth has been widely used for many years in upholstery and trim in car applications, house furnishings, and personal apparel. The large-scale replacement of leather by PVC initiated in the 1950s and 1960s was primarily due to the greater abrasion resistance, flex resistance, and washability of PVC. Ladies handbags are frequently made from PVC leathercloth. House furnishing applications include kitchen upholstery, printed sheets, and bathroom curtains. Washable wallpapers are obtained by treating paper with PVC compounds.

Special grades of PVC are used in metal-finishing applications, for example, in stacking chairs. Calendered plasticised PVC sheet is used in making plastic rainwear and baby pants by the high-frequency welding technique. The application of PVC in mine belting is still important in terms of the actual tonnage of material consumption. All-PVC shoes are useful as beachwear and standard footwear. PVC has also proved to be an excellent abrasion-resistant material for shoe soles. PVC adhesives, generally containing a polymerisable plasticiser, are useful in many industries.

The two main applications of vinyl chloride-vinyl acetate co-polymers are phonograph records and vinyl floor tiles. The co-polymers contain an average of about 13 per cent of vinyl acetate. They may be processed at lower temperatures than those used for the homopolymer. Phonograph records contain only a stabiliser, lubricant, pigment, and possibly, an antistatic agent; there are no fillers. Preformed resin biscuits are normally moulded in compression presses at about 130 to 140°C. The press is a flash mould that resembles a waffle iron. The faces of the mould may be nickel negatives of an original disc recording that have been made by electrodeposition.

Floor tiles contain about 30 to 40 parts plasticiser per 100 parts co-polymer and about 400 parts filler (usually a mixture of asbestos and chalk). Processing involves mixing in an internal mixer at about 130°C, followed by calendering at 110 to 120°C.

Pastes

A PVC paste is obtained when the voids between the polymer particles in a powder are completely filled with plasticiser so that the particles are suspended in it. To ensure a stable paste, there is an upper limit and a lower limit to the order of particle size. PVC paste polymers have an average particle size of about 0.2 to 1.5 μm. The distribution of particle sizes also has significant influences on the flow and fluxing characteristics of the paste.

The main types of PVC pastes are plastisols, organosols, plastisols incorporating filler polymers (including the rigisols), plastigels, hot-melt compounds, and compounds for producing cellular products. Typical formulations of the first three types are shown in Table 15.4.

Table 15.4. Typical formulations[a] of three types of PVC pastes.

Ingredient	*Plastisol*	*Organosol*	*Plastigel*
PVC paste polymer	100	100	100
Plasticiser (e.g., DOP)	80	30	80
Filler (e.g., china clay)	10	10	10
Stabiliser (e.g., white lead)	4	4	4
Naphtha	—	50	—
Aluminium stearate	—	—	4

[a] Parts by weight

Plastisols are of considerable importance commercially. They are converted into tough, rubbery products by heating at about 160°C (gelation). Organosols are characterised by the presence of a volatile organic diluent whose sole function is to reduce the paste viscosity. The diluent is removed after application and before gelling the paste.

Another method is reducing paste viscosity is to use a filler polymer to replace a part of the PVC paste polymer. The filler polymer particles are too large to make stable pastes by themselves, but in the presence of paste-polymer particles they remain in stable suspension. Being very much larger than paste-polymer particles and having a low plasticiser absorption, they take up large volumes in the paste and make more plasticiser available for particle lubrication, thus reducing paste viscosity. The use of filler polymers has increased considerably in recent years. Pastes prepared using filler polymers and only small quantities of plasticiser (approximately 20 parts per 100 parts of polymer) are termed rigisols.

The incorporation of such materials as aluminium stearate, fumed silicas, or certain bentonites gives a paste that shows pronounced Bingham Body behaviour (i.e., it only flows on application of shearing stress above a certain value). Such putty-like materials (called plastigels), which are usually thixotropic may be hand-shaped and subsequently gelled.

Plastigels are often compared with hot-melt PVC compounds. These later materials are prepared by fluxing polymer with large quantities of plasticisers and extenders. They melt at elevated temperatures and become very fluid, so they may be poured. These compounds are extensively used for casting and prototype work.

Sigma-blade trough mixers are most commonly used for mixing PVC pastes. It is common practice to mix the dry ingredients initially with part of the plasticiser so that the shearing stresses are high enough to break down the aggregates. The remainder of the plasticiser is then added to dilute the product. The mix is preferably deaerated to remove air bubbles before final processing.

A large proportion of PVC paste is used in the manufacture of leather cloth by a spreading technique. A layer of paste is smeared on the cloth by drawing the latter between a roller or endless belt and a doctor blade against which there is a rolling bank of paste. The paste is gelled by passing through a heated tunnel or under infrared heaters. Embossing operations may be carried out by using patterned rollers when the gelled paste is still hot. The leathercloth is then cooled and wound up. Where it is desired that the paste should enter the interstices of the cloth, a shear-thinning (pseudo-plastic) paste is employed. Conversely, where strike-through should be minimised, a dilatant paste (viscosity increases with shear rate) is employed.

Numerous methods exist for producing cellular products from PVC pastes. Closed-cell products can be made if a blowing agent such as azodiisobutyronitrile is incorporated into the paste. The paste is then heated in a mould to cause the blowing agent to decompose and the compound to gel. Since the mould is full, expansion does not take place at this stage. The unexpanded block is removed after thoroughly cooling the mould and is heated in an oven at about 100°C to produce uniform expansion. One method of producing a flexible, substantially open-cell product is to blend the paste with carbon dioxide (either as dry ice or under pressure). The mixture is heated to volatilise the carbon dioxide to produce a foam, which is then gelled at a higher temperature.

Chlorinated PVC

Closely related to PVC, but with distinct properties of its own, is chlorinated poly(vinyl chloride) (CPVC), a polymer produced by post-chlorination of PVC. The effect of adding more chlorine to the PVC molecule is to raise the T_g of the base resin to 115–135°C (239–275°F) range and the heat deflection temperature under load to around 115°C (239°F). CPVC also has higher tensile strength, higher modulus, and greater resistance to combustion and smoke generation.

The compounding process for CPVC is similar to that used in PVC compounding, but is more complex. Processing of CPVC is done by the traditional thermoplastic operations of extrusion, calendering, and injection moulding. However, because of the high temperature of CPVC polymer melt (205–230°C), extrusion of the resin requires chrome-plated or stainless steel dies. Injection moulding of CPVC requires low-compression screws with good exit depth and moulds should be stainless steel or chrome or nickel-plated. Traditional applications of CPVC compounds are hot and cold water distribution piping and fillings, as well as industrial pipe, fittings, and valves that can handle industrial liquids and chemicals. The increasing popularity of CPVC in its application in hot and cold water pipes in residential units stems from its continuous-use rating of 80°C (176°F) and 100 psi.

The high-heat capability, low combustion ratings, and resistance to grease and oil-induced cracking have made CPVC a strong contender for applications in automotive interiors. With its combination of excellent properties of PVC and the added ability to perform at elevated temperatures, CPVC is beginning to penetrate markets formerly dominated by metals or the more expensive engineering polymers.

Poly(Vinylidene Chloride)

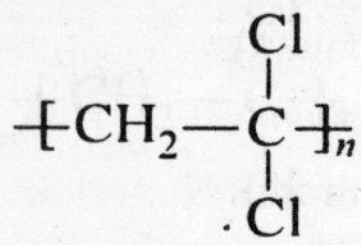

Monomer	*Polymerisation*	*Major uses*
Vinylidene chloride	Free-radical-initiated chain polymerisation	Film and sheeting for food packaging

The polymer may be prepared readily by free-radical mechanisms in bulk, emulsion, and suspension; the latter technique is usually preferred on an industrial scale. Co-polymers of vinylidene chloride with vinyl chloride, acrylates, and acrylonitrile are also produced.

Since the poly(vinylidene chloride) molecule has an extremely regular structure (and the question of tacticity does not arise), the polymer is capable of crystallisation. Because of the resultant close packing and presence of heavy chloride atoms the polymer has a high specific gravity (1.875) and a low permeability to vapours and gases. The chlorine present gives a self-extinguishing polymer.

Vinylidene chloride-vinyl chloride co-polymers are also self-extinguishing and possess very good resistance to a wide range of chemicals, including acids and alkalies. Because of a high degree of crystallisation, even in the co-polymers, high strengths are attained even though the products have low molecular weights (~20,000–50,000). Both poly(vinylidene chloride) and co-polymers containing vinylidene chloride are used to produce flexible films and coatings. Flexible films are used extensively for food packaging because of their superior barrier resistance to water and oxygen. The coating resins are used for cellophane, polyethylene, paper, fabric, and container liner applications. Dow's trade name for a co-polymer of vinylidene chloride (87 per cent) and vinyl chloride (13 per cent) is Saran. Biaxially stretched Saran film is a useful, though expensive, packaging material possessing exceptional clarity, brilliance, toughness, and impermeability to water and gases. Vinylidene chloride-vinyl chloride co-polymers are used in the manufacture of filaments. The filaments have high toughness, flexibility, durability, and chemical resistance. They find use in car upholstery, deck-chair fabrics, decorative radio grilles, doll hair, filter presses, and other applications. A flame-resisting fibre said to be a 50:50 vinylidene chloride-acrylonitrile co-polymer is marketed by Courtaulds with the name Teklan.

Polytetrafluoroethylene and Other Fluoropolymers

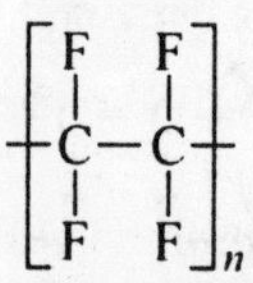

Monomers	*Polymerisation*	*Major uses*
Tetrafluoroethylene, hexafluoropropylene, chlorotrifluoroethylene, vinyl fluoride, vinylidene fluoride, perfluoroalkyl vinyl ether	Free-radical-initiated chain polymerisation	Coatings for chemical process equipment, cable insulation, electrical components, non-sticking surfaces for cookware

Polytetrafluoroethylene (PTFE) probably accounts for at least 85 per cent of the fluorinated polymers and, in spite of its high cost, has a great diversity of applications. It is produced by the free-radical chain polymerisation of tetrafluoroethylene. With a linear molecular structure of repeating $–CF_2–CF_2–$units, PTFE is a highly crystalline polymer with a melting point of 327°C. Density is 2.13 to 2.19 g/cm^3. Commercially PTFE is made by two major processes–one leading to the so-called granular polymer, the second to a dispersion of polymers of much finer particle size and lower molecular weight.

Since the carbon-fluorine bond is very stable and the only other bond present in PTFE is the stable C–C bond, the polymer has a high stability, even when heated above its melting point. Its upper-use temperature is given as 260°C. It is reported to give ductile rather than brittle failures at temperatures just above absolute zero, signifying a useful temperature range of more than 500°C. In many instances PTFE has been used satisfactorily as a totally enclosed gasket for considerable periods of time at temperatures well above the recommended upper-use temperature.

Because of its high crystallinity (>90 per cent) and incapability of specific interaction, PTFE has exceptional chemical resistance and is insoluble in all organic solvents. (PTFE dissolves in certain fluorinated liquids such as perfluorinated kerosenes at temperature approaching the melting point of the polymer). At room temperature it is attacked only by alkali metals and, in some cases, by fluorine. Treatment with a solution of sodium metal in liquid ammonia sufficiently alters the surface of a PTFE sample to enable it to be cemented to other materials by using epoxide resin adhesives. PTFE has good weathering resistance. The polymer is not wetted by water and has negligible water absorption (0.005 per cent). The permeability of gases is low—the rate of transmission of water vapour is approximately half that of poly(ethylene terephthalate) and low-density polyethylene.

PTFE is, however, degraded by high-energy radiation. For example, the tensile strength of a given sample may be halved by exposure to a dosage of 70 Mrad. PTFE is a tough, flexible material of moderate tensile strength (2500–3800 psi, i.e., 17–21 MPa) at 23°C. Temperature has a considerable effect on its properties. It remains ductile in compression at temperatures as low as 4 K (–269°C). The crepe resistance is low in comparison to other engineering plastics. Thus, even at 20°C unfilled PTFE has a measurable crepe with compression loads as low as 300 psi (2.1 MPa).

The coefficient of friction is unusually low and is lower than almost any other material. The values reported are usually in the range 0.02 to 0.10 for polymer to polymer and 0.09 to 0.12 for polymer to metal. The polymer has a high oxygen index and will not support combustion.

PTFE has outstanding insulation properties over a wide range of temperatures and frequencies. The volume resistivity exceeds 10^{20} ohm-m. The power factor (< 0.003 at 60 Hz and < 0.0003 at 10^6 Hz) is negligible in the temperature range – 60 to + 250°C. PTFE's low dielectric constant (2.1) is unaffected by frequency. The dielectric strength of the polymer is 16 to 20 kV/mm (short time on 2-mm thick sheet).

Processing

PTFE is commonly available in three forms: (i) granular polymers with average particle size of 300 to 600 μm; (ii) dispersion polymers (obtained by coagulation of a dispersion) consisting of agglomerates with an average diameter of 450 μm, which are made up of primary particles 0.1 μm in diameter; and (iii) dispersion (lattices) containing about 60 per cent polymer in the form of particles with an average diameter of about 0.16 μm.

ASTM Standard Specification D1457 covers the forms sold as powders and defines seven types of PTFE moulding and extrusion materials ranging from general-purpose granular resin (Type I) to presintered resin (Type VII). ASTM D4441 describes eight types of aqueous dispersion that differ, primarily, in the solids content and the amount of surfactant.

PTFE cannot be processed by the usual thermoplastics processing technique because of its exceptionally high melt viscosity (~10^{10}–10^{11} poises at 350°C). Granular polymers are processed by press and sinter methods used in powder metallurgy. In principle, these methods involve preforming the sieved powder by compressing in a mould at 1.0 to 3.5 tonf/in^2 (16 to 54 MPa) usually at room temperature or at 100°C, followed by sintering at a temperature above the melting point (typically at about 370°C), and then cooling. Free-sintering of the preform in an oven at about 380°C, is also satisfactory. The sintering period depends on the thickness of the sample. For example, a 0.5-inch (1.25-cm)-thick sample will need sintering for 3.5 hour. Granular polymers may also be extruded, though at very low rate (2.5–16 cm/min), by screw and ram extruders. The extrudates are reasonably free of voids.

Billets in sizes varying from less than a kilogram up 700 kg (among the largest moulding made of any plastic material) are made by preform sintering. The powder is placed in a mould at or slightly above room temperature and compressed at pressures from 2000–5000 psi (14–34 MPa). After being removed from the mould, the preform is sintered by heating it unconfined in an oven at temperatures in the range of 360–380°C for times ranging from a few hours to several days. The time-temperature schedule depends on the size and shape of the billet. The billets are often in the form of cylinder, which are then mounted on a mandrel. Sheeting is prepared by skiving, much like plywood is cut from large logs. The sheeting is cut in the thickness range of 0.025 mm (0.001 inch) to 2.5 mm (0.1 inch).

With sheet moulding, the procedures are very similar to those used as described above for billet moulding except for the shape and size of the moulding. This process is used for sheeting above 2.5 mm (0.1 inch) up to large blocks. The latter are used for a block method skiving operation similar to that used traditionally for cellulose nitrate.

With ram extrusion, the PTFE powder is fed into a cavity at one end of a heated tube. A reciprocating ram compacts the powder and forces it into the tube. While it is being transported down the length of the tube, the PTFE is melted and coalesced. Continuous lengths of sintered rod or shapes comes out of the other end of the extruder.

PTFE mouldings and extrudates may be machined without difficulty. Continuous film may be obtained by peeling a pressure-sintered ring and welding it to a similar film by heat sealing under pressure at about 350°C.

A PTFE dispersion polymer leads to products with improved tensile strength and flex life. Preforms are made by mixing the polymer with 15 to 25 per cent of a lubricant and extruded. This step is followed by lubricant removal and sintering. In a typical process a mixture of PTFE dispersion polymer (83 parts) and petroleum ether (17 parts) with a 100 to 120°C boiling range is compacted into a preform billet which is then extruded by a vertical ram extruder. The extrudate is heated in an oven at about 105°C to remove the lubricant and is then sintered at about 380°C. Because of the need to remove the lubricant, only thin sections can be produced by this process. Thin-walled tubes with excellent fatigue resistance can be produced, or wire can be coated with very thin coatings of PTFE.

Tapes also may be made by a similar process. However, in this case the lubricant used is a non-volatile oil. The preform is extruded in the shape of a rod, which is then passed between a pair of calendered rolls at about 60 to 80°C. The unsintered tape finds an important application in pipe-thread sealing. If sintered tape is required, the calendered product is first degreased by passing through boiling trichloroethylene and then sintered by passing through a salt bath. The tape made in this way is superior to that obtained by machining granular polymer mouldings.

A convenient way to apply PTFE dispersions to surfaces is by use of aqueous dispersion coatings. For this is dispersion as provided by the manufacturer may be used. There are also many coating formulations that can be purchased for use in such applications as cookware and bakeware. Coating is done as dip coating and, with thickened dispersions, as roller coating.

The dispersion should have a concentration of 45–50 per cent PTFE and 6–9 per cent of a wetting agent, based on the amount of PTFE. Usually a non-ionic agent is used. The amount of PTFE deposited on each coat must be restricted so as to prevent 'mud cracking' when the coating is dried. For drying the coating, infrared lamps or forced-convention ovens at 85–95°C are usually used. Multiple coats are used to obtain thicker films.

Drying is followed by baking and sintering to remove the wetting agent and then coalesce the PTFE. Temperatures of 260–315°C are used for the baking and 360–400°C for the sintering. The time required for each of these steps varies from several seconds to a few minutes, depending on the shape of the surface and thickness of the coating. To obtain good homogeneity, multiple dips are used, with baking and sintering after each.

Casting is used to make this films for use in such applications as heart-lung machines, special electronic equipment, and various speciality applications. Often it is done by allowing the dispersion to flow onto a support surface, usually a polished stainless-steel belt, that carries the film through the successive steps of coating, drying, baking, sintering, re-coating, and finally, stripping from the belt.

Glass-coated PTFE laminates may be produced by piling up layers of glass cloth impregnated with PTFE dispersions and pressing at about 330°C. Asbestos-PTFE laminates may be produced in a similar way. The dispersions can also be used for producing filled PTFE moulding material. The process typically involves stirring fillers into the dispersion, coagulating with acetone, drying at 280° to 290°C, and disintegrating the resulting cake of material.

Applications

The exceptional properties of PTFE makes it highly useful. It is selected for a wide range of applications that affect every person. The applications fall in the five areas that require one or more of its chemical, mechanical, electrical, thermal and surface properties. In essentially every instance, a successful application employs at least two of the outstanding properties of this polymer. However, because of its high volume cost, PTFE is not generally used to produce large objects.

In many cases it is possible to coat a metal object with a layer of PTFE to meet the particular requirement. Nonstick home cookware is perhaps the best-known example. PTFE is used for lining chutes and coating other metal objects where low coefficients of friction, chemical inertness, non-adhesive characteristics are required. The same properties make possible coverings on rollers in food processing equipment, xerographic copiers and saw blades, coatings on snow shovels, and many other similar applications. The Alaskan oil pipeline, for example, rests on PTFE-coated steel plates. Most new bridges and tunnels use similar supports.

Because of its exceptional chemical resistance over a wide temperature range, PTFE is used in a variety of seals, gaskets, packings, valve and pump parts, and laboratory equipment.

Its excellent electrical insulation properties and heat resistance lead to its use in high-temperature wire and cable insulation, moulded electrical components, insulated transformers, hermetic seals for condensers, laminates for printed circuitry, and many other electrical applications.

Reinforced PTFE applications include bushings and seals in compressor hydraulic applications, automotive applications, and pipe liners. A variety of mouldings are used in aircraft and missiles and also in other applications where use at elevated temperatures is required.

An important application for aqueous dispersions of PTFE is the architectural fabric market. This product consists of fibreglass fabric coated with special forms of the aqueous dispersion. The resulting material is used as roofs in a wide variety of buildings, especially where a large area must be covered with minimum support.

Co-polymers of tetrafluoroethylene were developed in attempts to provide materials with the general properties of PTFE and the melt processability of the more conventional thermoplastics. Two such co-polymers are tetrafluoroethylene-hexafluoropropylene (TFE-HFP) co-polymers (Teflon FEP resins by Du Pont; FEP stands for fluorinated ethylene propylene) with a melting point of 290°C and tetrafluoroethylene-ethylene (ETFE) co-polymers (Tefzel by Du Pont) with a melting point of 270°C. These products are melt processable. A number of other fluorine containing melt processable polymers have been introduced.

Polychlorotrifluoroethylene (PCTFE) was the first fluorinated polymer to be produced on an experimental scale and was used in the United States early in World War II. It was also used in the handling of corrosive materials, such as uranium hexafluoride, during the development of the atomic bomb. PCTFE is a crystalline polymer with a melting point of 218°C and density of 2.13 g/cm^3. The polymer is inert to most reactive chemicals at room temperature. However, above 100°C a few solvents dissolve the polymer and a few, especially chlorinated types, swell it. The polymer is melt processable, but processing is difficult because of its high melt viscosity and its tendency to degrade, resulting in deterioration of its properties.

PCTFE is used in chemical processing equipment and cryogenic and electrical applications. Major applications include wafer boats, gaskets, O-rings, seals, and electrical components. PCTFE has outstanding barrier properties to gases, the PCTFE film has the lowest water-vapour transmission of any transparent plastic film. It is used in pharmaceutical packaging and other applications for its vapour barrier properties, including electroluminescent lamps.

Other melt-processable thermoplastics include ethylene-chlorotrifluoroethylene (ECTFE) co-polymer (melting point 240°C), polyvinylidene fluoride (PVDF) (melting point 170°C) and polyvinyl fluoride (PVF), which is commercially available only as film.

PVF is tough and flexible, has good abrasion and staining resistance, and has outstanding weathering resistance. It maintains useful properties over a temperature range of –70 to 110°C. PVF can be laminated to polywood, hardboard, vinyl, reinforced polyesters, metal foils, and galvanised steel. These laminates are used in aircraft interior panels, lighting panels, wall coverings, and a variety of building applications. PVF is also used as glazing in solar energy collectors. PVF film is marketed by Du Pont as Tedlar.

PVDF is correctly named poly(1,1-difluoroethylene) and represented by $(—CF_2CH_2—)_n$. It is a hard, tough thermoplastic fluoropolymer. PVDF is prepared by free-radical initiated polymerisation, either in suspension or (usually) in emulsion systems. The basic raw material for PVDF is vinylidene fluoride ($CH_2 = CF_2$), a preferred synthesis of which is dehydrochlorination of chlorodifluoroethane.

PVDF has the lowest melting point of any of the commerical fluoropolymers. As a result, its upper use temperature is limited to about 150°C, compared to values of 200°C for FEP and 260°C for PTFE. At temperatures in its useful range, however, PVDF maintains its stiffness and toughness very well.

PVDF exhibits the excellent resistance to harsh environments characteristics of fluoropolymers. It is widely used in the chemical processing industry, in piping systems, valves, tanks (both moulded and lined), and other areas where its combination of excellent mechanical properties and superb resistance to most chemicals make its an ideal material for fluid handling equipment. Increasingly important is use of PVDF as the base resin for long-life, exterior coatings on aluminium, steel, masonry, wood, and plastics.

The high dielectric loss and high dielectric constant of PVDF (8 to 9) both restrict its use in some electrical applications and provide superior performance in others. PVDF has very unusual piezoelectric and pyroelectric properties which are opening up many new applications with a very high value in use.

PVDF's cost, about the lowest of the melt-processible fluoropolymers, is an important advantage. Essentially all the common procedures available for thermoplastic polymers can be used with PVDF.

Perfluoroalkoxy (PFA) resins are the newest commercially available class of melt-processable fluoroplastics. Its general chemical structure is

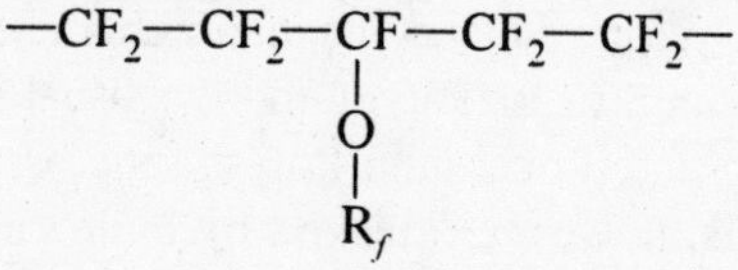

where $R_f = —C_nF_{2n+1}$.

PFA resin has somewhat better mechanical properties than FEP above 150°C and can be used up to 260°C. In chemical resistance it is about equal to PTFE. PFE resin is sold by Du Pont under the Teflon trademark.

Polyisobutylene

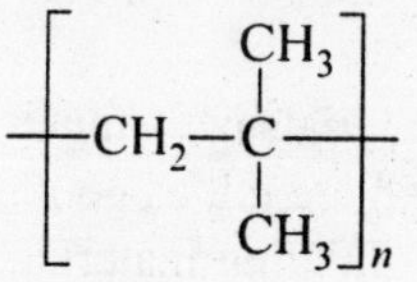

Monomer	*Polymerisation*	*Major uses*
Isobutylene	Cationic-initiated chain polymerisation	Lubricating oils, sealants

High-molecular-weight polyisobutylene (PIB) is produced by cationic chain polymerisation in methyl chloride solution at –70°C using aluminium chloride as the catalyst. Such polymers are currently available from Esso(Vistanex) and BASF(Oppanol).

PIB finds a variety of uses. It is used as a motor oil additive to improve viscosity characteristics, as a blending agent for polyethylene to improve its impact strength and environmental stress-cracking resistance, as a base for chewing gum, as a tackifier for greases; it is also used in caulking compounds and tank linings.

Because of its high cold flow, it has little use as a rubber in itself, but co-polymers containing about 2 per cent isoprene to introduce unsaturation for cross-linking are widely used in butyl rubber.

Polystyrene

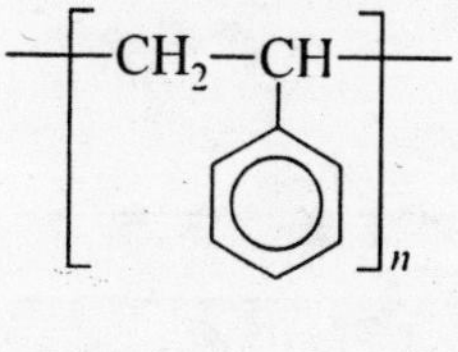

Monomer	*Polymerisation*	*Major uses*
Styrene	Free-radical-initiated chain polymerisation	Packaging and containers (35%), housewares, toys and recreational equipment (25%), appliance parts (10%), disposable food containers (10%)

Polystyrene is made by bulk or suspension polymerisation of styrene. Polystyrene is very low cost and is extensively used where price alone dictates. Its major characteristics include rigidity, transparency, high refractive index, no taste, odour, or toxicity, good electrical insulation characteristics, low water absorption, and ease of colouring and processing. Polystyrene has excellent organic acid, alkali, salts, and lower alcohol resistance. It is, however, attacked by hydrocarbons, esters, ketones, and essential oils. A more serious limitation of polystyrene in many applications is its brittleness. This limitation led to the development of rubber modified polystyrenes (containing usually 5–15 per cent rubber), the so-called high impact polystyrenes (HIPS). The most commonly used are styrene-butadiene rubber and *cis*-1,4-polybutadiene. The method of mixing the polystyrene and rubber has a profound effect on the properties of the product. Thus, much better results are obtained if the material is prepared by polymerisation of styrene in the presence of the rubber rather than by simply blending the two polymers. The product of the former method contains not only polystyrene and straight rubber but also a graft co-polymer in which polystyrene side chains are attached to the rubber.

Compared to straight or general-purpose polystyrene, high-impact polystyrene materials have much greater toughness and impact strength, but clarity, softening point, and tensile strength are not as good. Expanded or foamed polystyrene, which has become very important as a thermal insulating material, has a low density, has a low weight cost, is less brittle, and can be made fire retarding.

End-uses for all types of polystyrene are packaging, toys, and recreational products, housewares, bottles, lenses, novelties, electronic appliances, capacitor dielectrics, low-cost insulators, musical instrument reeds, light-duty industrial components, furniture, refrigeration, and building and construction uses (insulation). Packaging is by far the largest outlet: bottle caps, small jars and other injection-moulded containers, blow-moulded containers, toughened polystyrene liners (vacuum formed) for boxed goods, and oriented polystyrene film for foodstuffs are some of its uses. The second important outlet is refrigeration equipment, including door liners and inner liners (made from toughened polystyrene sheet), mouldings for refrigerator furnishings, such as flip lids and trays, and expanded polystyrene for thermal insulation. Expanded polystyrene products have widely increased the market for polystyrene resin. With as light a weight as 2 lb/ft^3 (0.032 g/cm^3), the thermal conductivity of expanded polystyrene is very low, and its cushioning value is high. It is an ideal insulation and packaging material. Common applications include ice buckets, water coolers, wall panels, and general thermal insulation applications. Packaging uses of expanded polystyrene range from thermoformed egg boxes and individually designed shipping packages for delicate equipment, such as cameras and electronic equipment, to individual beads (which may be about 1 cm in diameter and up to 5 cm long) for use as a loose fill material in packages.

Polybutadiene(Butadiene Rubber)

$$\left[CH_2—CH=CH—CH_2 \right]_n$$

Monomer	*Polymerisation*	*Major uses*
Butadiene	Ziegler-Natta-catalysed chain polymerisation	Tyres and tyre products (90%)

Polybutadiene is made by solution polymerisation of butadiene using Ziegler-Natta catalysts. Slight changes in catalyst composition produce drastic changes in the stereoregularity of the polymer. For example, polymers containing 97 to 98 per cent of *trans*-1,4 structure can be produced by using ET_3Al/VCl_3 catalyst, those with 93 to 94 per cent *cis*-1,4 structure by using ($Et_2AlCl/CoCl_2$, and those with 90 per cent 1,2-polybutadiene by using $Et_3Al/Ti(OBu)_4$. The stereochemical composition of polybutadiene is important if the product is to be used as a base polymer for further grafting. For example, a polybutadiene with 60 per cent *trans*-1,4, 20 per cent *cis*-1,4, and 20 per cent 1,2 configuration is used in the manufacture of ABS resin.

Polybutadiene rubbers generally have a higher resilience than natural rubbers at room temperature, which is important in rubber applications. On the other hand, these rubbers have poor tear resistance, poor tack, and poor tensile strength. For this reason polybutadiene rubbers are usually used in conjunction with other materials for optimum combination of properties. For example, they are blended with natural rubber in the manufacture of truck tyres and with styrene-butadiene rubber (SBR) in the manufacture of automobile tyres. Polybutadiene is also produced in low volume as speciality products. These include low-molecular-weight, liquid 1,2-polybutadienes (60–80 per cent, 1,2 content) used as potting compounds for transformers and submersible electric motors and pumps, liquid *trans*-1,4-polybutadienes used in protective coatings inside metal cans, and hydroxy-terminated polybutadiene liquid resins for use as a binder and in polyurethane and epoxy resin formulations.

Polyisoprene

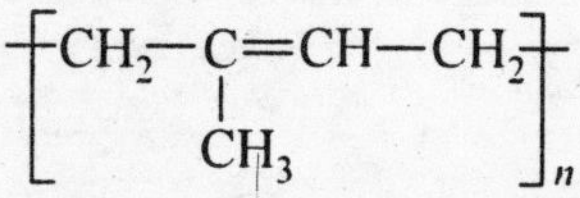

Monomer	*Polymerisation*	*Major uses*
Isoprene	Ziegler-Natta-catalysed chain polymerisation	Car tyres (55%), mechanical goods, sporting goods, footwear, sealants, and caulking compounds

Polyisoprene is produced by solution polymerisation using Ziegler-Natta catalysts. The *cis*-1,4-polyisoprene is a synthetic equivalent of natural rubber. However, the synthetic polyisoprenes have *cis* contents of only about 92 to 96 per cent; consequently, these rubbers differ from natural rubber in several ways. The raw synthetic polyisoprene is softer than raw natural rubber (due to a reduced tendency for a stress-induced crystallisation because of the lower *cis* content) and is therefore more difficult to mill. On the other hand, the unvulcanised synthetic material flows more readily; this feature makes it easier to injection mould. The synthetic product is somewhat more expensive than natural rubber. Polyisoprene rubbers are used in the construction of automobile tyre carcasses and inner liners and truck and bus tyre treads. Other important applications are in mechanical goods, sporting goods, footwear, sealants, and caulking compounds.

Polychloroprene

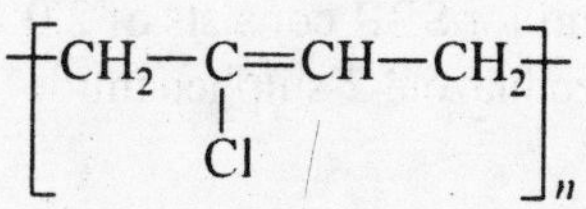

Monomer	*Polymerisation*	*Major uses*
Chloroprene (2-chlorobuta-1,3-diene)	Free-radical-initiated chain polymerisation (mostly emulsion polymerisation)	Conveyor belts, hose, seals and gaskets, wire and cable sheathing

A comparison of polychloroprene and natural rubber or polyisoprene molecular structures shows close similarities. However, while the methyl groups activates the double bond in the polyisoprene molecule, the chlorine atom has the opposite effect in polychloroprene. Thus polychloroprene is less prone to oxygen and ozone attack than natural rubber is. At the same time accelerated sulphur vulcanisation is also not a feasible proposition, and alternative vulcanisation or curing systems are necessary.

Vulcanisation of polychloroprene rubbers is achieved with a combination of zinc and magnesium oxide and added accelerators and antioxidants. The vulcanisates are broadly similar to those of natural rubber in physical strength and elasticity. However, the polychloroprene vulcanisates show much better heat resistance and have a high order of oil and solvent resistance (though less resistant than those of nitrile rubber). Aliphatic solvents have little effect, although aromatic and chlorinated solvents cause some swelling. Because of chlorine content chloroprene rubber is generally self extinguishing.

Because of their greater overall durability, chloroprene rubbers are used chiefly where a combination of deteriorating effects exists. Products commonly made of chloroprene rubber include conveyor belts, V-belts, diaphragms, hoses, seals, gaskets, and weather strips. Some important construction uses are highway joint seals, pipe gaskets, and bridge mounts and expansion joints. Latexes are used in gloves, balloons, foams, adhesives, and corrosion-resistant coatings.

OLEFIN CO-POLYMERS

Styrene-Butadiene Rubber

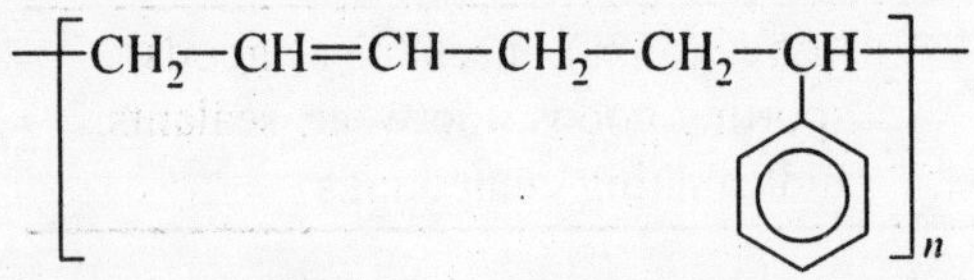

Monomers	*Polymerisation*	*Major uses*
Styrene, butadiene	Free-radical-initiated chain polymerisation (mostly emulsion polymerisation)	Tyres and tread (65%), mechanical goods (15%), latex (10%), automotive mechanical goods (5%)

In tonnage terms SBR is the world's most important rubber. Its market dominance is primarily due to three factors: low cost, good abrasion resistance, and a higher level of product uniformity that can be achieved with natural rubber.

Reinforcement of SBR with carbon black leads to vulcanisates which resemble those of natural rubber, and the two products are interchangeable in most applications. As with natural rubber, accelerated sulphur systems consisting of sulphur and an activator comprising a metal oxide (usually zinc oxide) and a fatty acid (commonly stearic acid) are used. A conventional curing system for SBR consists of 2.0 parts sulphur, 5.0 parts zinc oxide, 2.0 parts stearic acid, and 1.0 part *N-t*-butylbenzothiazole-2-sulphenamide (TBBS) per 100 parts polymers.

The most important applications of SBR is in car tyres and tyre products, but there is also widespread use of the rubber in mechanical and industrial goods. SBR latexes, which are emulsions of styrene-butadiene co-polymers (containing about 23–25 per cent styrene), are used for the manufacture of foam rubber backing for carpets and for adhesive and moulded foam applications.

Nitrile Rubber

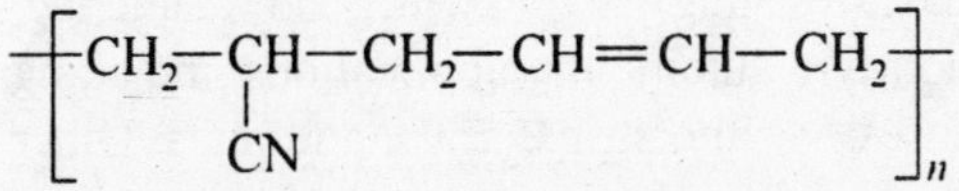

Monomers	*Polymerisation*	*Major uses*
Acrylonitrile, butadiene	Free-radical-initiated chain polymerisation (mostly emulsion polymerisation)	Gasoline hose, seals, gaskets, printing tools, adhesive, footwear

Nitrile rubber (acrylonitrile-butadiene co-polymer) is a unique elastomer. The acrylonitrile content of the commercial elastomers ranges from 25 to 50 per cent with 34 per cent being a typical value. This non-hydrocarbon monomer imparts to the co-polymer very good hydrocarbon oil and gasoline resistance. The oil resistance increases with increasing amounts of acrylonitrile in the co-polymer. Nitrile rubber is also noted for its high strength and excellent resistance to abrasion, water, alcohols, and heat. Its drawbacks are poor dielectric properties and poor resistance to ozone.

Because of the diene component, nitrile rubbers can be vulcanised with sulphur. A conventional curing system consists of 2.5 parts sulphur, 5.0 parts zinc oxide, 2.0 parts stearic acid, and 0.6 parts *N-t*-butylbenzothiazole-2-sulphenamide (TBBS) per 100 parts polymer.

Nitrile rubbers (vulcanised) are used almost invariably because of their resistance to hydrocarbon oil and gasoline. They are, however, swollen by aromatic hydrocarbons and polar solvents such as chlorinated hydrocarbons, esters, and ketones.

Ethylene-Propylene-Elastomer

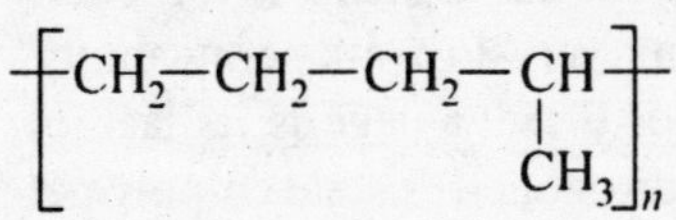

Monomers	*Polymerisation*	*Major uses*
Ethylene, propylene	Ziegler-Natta-catalysed chain polymerisation	Automotive parts, radiator and heater hoses, seals

Two types of ethylene-propylene elastomers are currently being produced; ethylene-propylene binary co-polymers (EPM rubbers) and ethylene-propylene-diene ternary co-polymers (EPDM rubbers). Because of their saturated structure, EPM rubbers cannot be vulcanised by using accelerated sulphur systems, and the less convenient vulcanisation with free-radical generators (peroxide) is required. In contrast, EPDM rubbers are produced by polymerising ethylene and propylene with a small amount (3–8 per cent) of a diene monomer, which provides a cross-link site for accelerated vulcanisation with sulphur. A typical vulcanisation system for EPDM rubber consists of 1.5 parts sulphur, 5.0 parts zinc oxide, 1.0 parts stearic acid, 1.5 parts 2-mercaptobenzothiazole (MBT), and 0.5 part tetramethylthiuram disulphide (TMTD) per 100 parts polymer.

The EPDM rubbers, though hydrocarbon, differ significantly from the diene hydrocabon rubbers considered earlier in that the level of unsaturation in the former is much lower, giving rubbers much better heat, oxygen, and ozone resistance. Dienes commonly used in EPDM rubbers include dicyclopentadiene, ethylidene norbornene, and hexa-1,4-diene.

The use of EPDM rubbers for the manufacture of automobile and truck tyres has not been successful, mainly because of poor tyre cord adhesion and poor compatibility with most other rubbers. However, EPDM rubbers have become widely accepted as a moderately heat-resisting material with good weathering, oxygen, and ozone resistance. They find extensive use in non-tyre automobile applications, including body and chassis parts, car bumpers, radiator and heater hoses, weatherstrips, seals, and mats. Other applications include wire and cable insulation, appliance parts, hoses, gaskets and seals, and coated fabrics. These rubbers are now also being blended on a large scale with polyolefin plastics, particularly polypropylene, to produce an array of materials, ranging from tough plastics at one end to the thermoplastic polyolefin rubbers at the other.

Butyl Rubber

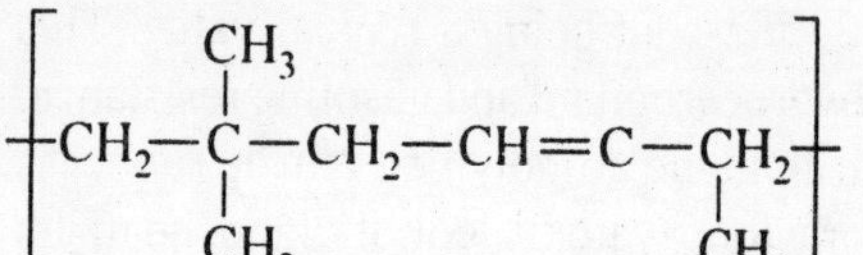

Monomers	*Polymerisation*	*Major uses*
Isobutylene, isoprene	Cationic chain polymerisation of isobutylene with 0.5 to 2.5 mol% of isoprene	Tyre inner tubes and inner liners of tubeless tyres (70%), inflatable sporting goods

Although polyisobutylene described earlier is a non-rubbery polymer exhibiting high cold flow, the co-polymer containing about 2 per cent isoprene can be vulcanised with a powerful accelerated sulphur system to give rubbery polymers. Being almost saturated, they are broadly similar to the EPDM rubbers in many properties. The most outstanding property of butyl rubber is its very low air permeability, which has led to its extensive use in tyre inner tubes and liners. A major disadvantage is its lack of compatibility with SBR, polybutadiene, and natural rubber. More recently an ozone-resistant co-polymer of isobutylene and cyclopentadiene has been marketed.

Thermoplastic Elastomers

Monomers	*Polymerisation*	*Major uses*
Butadiene, isoprene, styrene	Anionic block polymerisation	Footwear, automotive parts, hot-melt adhesives

Conventional rubbers are vulcanised, that is, cross-linked by primary valence bonding. For this reason vulcanised rubbers cannot dissolve or melt unless the network structure is irreversibly destroyed. These products cannot therefore be re-processed like thermoplastics. Hence, if a polymer could be developed which showed rubbery properties at normal service temperatures but could be re-processed like thermoplastics, it would be of great interest. During the past two decades a few groups of materials have been developed that could be considered as being in this category. Designated as thermoplastic elastomers, they include (i) styrene-diene-styrene triblock co-polymers; (ii) thermoplastic polyester elastomers and thermoplastic polyurethane elastomers; and (iii) thermoplastic polyolefin rubbers (polyolefin blends).

Styrene-diene-styrene triblock elastomers

The styrene-diene-styrene triblocks consist of a block of diene units joined at each end to a block of styrene units and are made by sequential anionic polymerisation of styrene and a diene. In this way two important triblock co-polymers have been produced—the styrene-butadiene-styrene (SBS) and styrene-isoprene-styrene (SIS) materials. The commercial thermoplastic rubbers, are produced by joining styrene-butadiene or styrene-isoprene diblocks at the active ends by using a difunctional coupling agent. Similarly, co-polymer molecules in the shape of a T, X, or star have been produced by using coupling agents of higher functionality.

The outstanding behaviour of these rubbers arises from the natural tendency of two polymer species to separate. However, this separation is restrained in these polymers since the blocks are covalently linked to each other. In a typical commercial SBS triblock co-polymer with about 30 per cent styrene content, the styrene blocks congregate into rigid, glassy domains which act effectively to link the butadiene segments into a network (Fig. 15.7) analogous to that of cross-linked rubber.

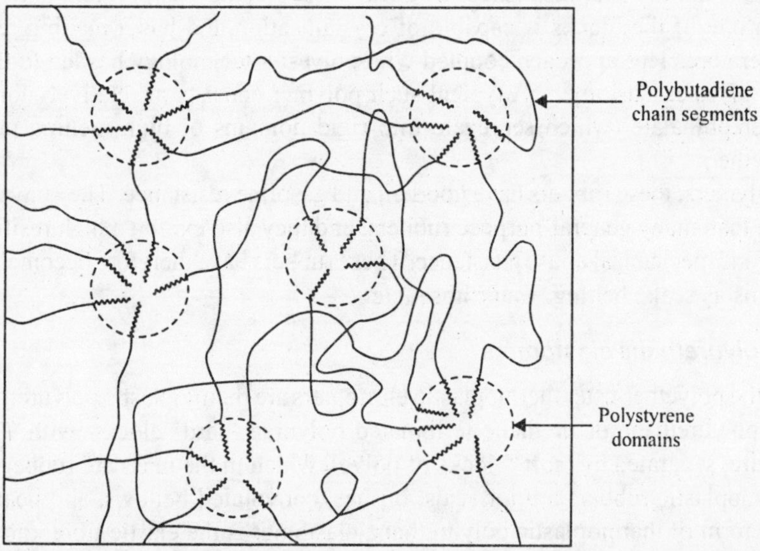

Fig. 15.7. Schematic representation of polystyrene domain structure in styrene-butadiene-styrene triblock co-polymers.

As a SBS elastomer is heated above the glass transition temperature (T_g) of the polystyrene, the glassy domains disappear and the polymer begins to flow like a thermoplastic. However, when the molten material is cooled (below T_g), the domains re-harden and the material once again exhibits properties similar to those of a cross-linked rubber.

Below T_g of polystyrene the glassy domains also fulfill another useful role by acting like a reinforcing particulate filler. It is also an apparent consequence of this role that SBS polymers behave like carbon-black-reinforced elastomers with respect to tensile strength.

The styrene-diene-styrene triblock co-polymers are not used extensively in traditional rubber applications because they show a high level of crepe. The block co-polymers can, however, be blended with many conventional thermoplastics, such as polystyrene, polyethylene, and polypropylene, to obtain improved properties. A major area of use is in footwear, where blends of SBS and polystyrene have been used with remarkable success for crepe soles.

Other important uses are adhesives and coatings. A wide variety of resins, plasticisers, fillers, and other ingredients commonly used in adhesives and coatings can be used with styrene-diene-styrene triblock co-polymers. With these ingredients properties such as tack, stiffness, softening temperatures, and cohesive strength can be varied over a wide range. With aliphatic resin additives the block co-polymers are used for permanently tacky pressure-sensitive adhesives, and in conjunction with aromatic resins they are used for contact adhesives. The co-polymers can be compounded into these adhesives by solution or hot-melt techniques.

The block co-polymers are also used in a wide variety of sealants, including construction, industrial, and consumer-grade products. They are unique in that they can be formulated to produce a clear, water white product. Other applications include bookbinding and product assembly and chemical milling coatings.

Thermoplastic polyester elastomers

Because of the relatively low T_g of the short polystyrene blocks, the styrene-diene-styrene triblock elastomers have very limited heat resistance. One way to overcome this problem is to use a block co-polymer in which one of the blocks is capable of crystallisation and has a melting temperature well above room temperature. This approach coupled with polyester technology has led to the development of thermoplastic polyester elastomers. A typical such polymer consists of relatively long sequences of tetra-methylene terephthalate (which segregate into rigid domains of high melting point) and softer segments of polyether.

Being polar polymers, these rubbers have good oil and gasoline resistance. They have a wider service temperature range than many general-purpose rubbers, and they also exhibit a high resilience, good flex fatigue resistance, and mechanical abuse resistance. These rubbers have therefore become widely accepted in such applications as seals, belting, water hose, etc.

Thermoplastic polyurethane elastomers

Closely related to the polyether-ester thermoplastic elastomers are thermoplastic polyurethane elastomers, which consist of polyurethane or urethane terminated polyurea 'hard' blocks, with T_g above normal ambient temperature, separated by 'soft' blocks of polyol, which in the mass are rubbery in nature. The main uses of thermoplastic rubbers are for seals, bushes, convoluted bellows, and bearings.

One particular form of thermoplastic polyurethane elastomer is the elastic fibre known as Spandex. Spandex fibres have higher modulus, tensile strength, and resistance to oxidation, and are able to produce finer deniers than natural rubber. They have enabled lighter-weight garments to be produced. Staple fibre blends of Spandex fibre with non-elastic fibres have also been introduced.

Thermoplastic polyolefin elastomers

Blends of EPDM rubbers with polypropylene in suitable ratios have been marketed as thermoplastic polyolefin rubbers. Their recoverable high elasticity is believed to be due to short propylene blocks in the EPDM rubber co-crystallising with segments of the polypropylene molecules so that these crystalline domains act like cross-linking agents. Having good weathering properties, negligible toxicity hazards, and easy processability, these rubbers have received rapid acceptance for use in a large variety of non-tyre automotive applications, such as bumper covers, headlight frames, radiator grilles, door gaskets, and other auto parts. They have also found use in cable insulation.

Ionic elastomers

Ionic elastomers have been obtained using sulphonated EPDM. In one case an EPDM terpolymer consisting of 55 per cent ethylene units, 40 per cent propylene units, and 5 per cent ethylidene norbornene units is sulphonated to introduce about 1 mol per cent sulphonate groups (appended to some of the unsaturated) groups of the EPDM). The sulphonic acid group is then neutralised with zinc acetate to form the zinc salt. The ionised sulphonic groups create ionic cross-links in the intermolecular structure (Fig. 15.8), giving properties normally associated with a cross-linked elastomer.

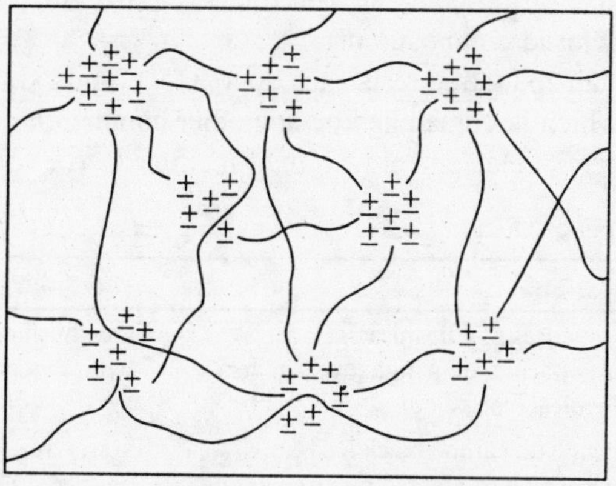

Fig. 15.8. Schematic representation of domain structure in ionic elastomers.

However, being a thermoplastic material, it can be processed in conventional moulding machines. This rubber, however, has a very high melt viscosity, which must be reduced by using a polar flow promoter, such as zinc stearate, at levels of 9.5 to 19 per cent.

Fluoroelastomers

Monomers	*Polymerisation*	*Major uses*
Vinylidene fluoride ($CH_2 = CF_2$) chlorotrifluoroethylene ($CF_2 = CFCl$), tetrafluoroethylene ($CF_2 = CF_2$), hexafluoropropylene ($CF_3CF = CF_2$), perfluoromethyl vinyl ether ($CF_2 = CFOCF_3$)	Free-radical-initiated chain polymerisation	Aerospace industry (20%), industrial equipment, wire and cable jacketing, and other insulation applications

The fluoroelastomers are a general family of fluorinated olefin co-polymers. To be rubbery, the co-polymer must have a flexible backbone and be sufficiently irregular in structure to be non-crystalline. A number of important fluororubbers are based on vinylidene fluoride (CH_2=CF_2).

The most important of these are the co-polymers of vinylidene fluoride and hexafluoropropylene (VF_2-HFP). The terpolymer of these two monomers together with tetrafluoroethylene (VF_2-HFP-TFE) is also of importance.

This terpolymer is the best among oil-resistant rubbers in its resistance to heat ageing, although its actual strengths are lower than for some other rubbers. The co-polymers of vinylidene fluoride and chlorotrifluoroethylene (VF_2-CTFE) are notable for their superior resistance to oxidising acids such as fuming nitric acid.

Fluoroelastomers with no C—H groups will be expected to exhibit a higher thermal stability. *Du Pont* thus developed a terpolymer of tetrafluoroethylene, perfluoro (methyl vinyl ether) and, in small amounts, a cure site monomer of undisclosed composition.

This product has excellent air-oxidation resistance up to 315°C and exhibits extremely low swelling in a wide range of solvents, which is unmatched by any other commercial fluoroelastomer. Table 15.5 lists the presently used commercial elastomers with their main properties and applications.

Table 15.5. Commercial elastomer products.

Type	*Properties*	*Major uses*
Natural rubber	Excellent properties of vulcanisates under conditions not demanding high levels of heat, oil, and chemical resistance	Tyres, bushings, couplings, seals, footwear and belting; second place in global tonnage
Styrene-butadiene (SBR)	Reinforcement with carbon black leads to vulcanisates which resemble those of natural rubber: more effectively stabilised by antioxidants than natural rubber	Tyres, tyre products, footwear, wire and cable covering, adhesives; highest global tonnage
Polybutadiene	Higher resilience than similar natural rubber compounds, good low-temperature behaviour and adhesion to metals, but poor tear resistance, poor tack, and poor tensile strength	Blends with natural rubber and SBR; manufacture of high impact polystyrene
Polyisoprene	Similar to natural rubber, but excellent flow characteristics during moulding	Tyres, belting, footwear, flooring
Butyl	Outstanding air-retention property, but low resiliency and poor resistance to oils and fuels	Tyre inner tubes, inner liners, seals, coated fabrics
Ethylene-propylene	Outstanding resistance to oxygen and ozone, poor fatigue resistance, poor tyre-cord adhesion	Non-tyre automotive parts, radiator and heater hoses, wire and cable insulation
Nitrile	Excellent resistance to oils and solvents, poor low-temperature flexibility and poor resistance to weathering	Hoses, seals, gaskets, footwear
Chloroprene	High order of oil and solvent resistance (but less than nitrile rubber), good resistance to most chemicals, oxygen and ozone, good heat resistance, high strength but difficult to process	Mechanical automotive goods, conveyor belts, diaphragms, hose seals and gaskets

(Contd ...)

Type	Properties	Major uses
Silicone	Outstanding electrical and high-temperature properties, retention of elasticity at low temperature, poor tear and abrasion resistance, relatively high price	Gaskets and sealing rings for jet engines, ducting, sealing strips, vibration dampers and insulation equipment in aircraft, cable insulation in naval craft, potting and encapsulation
Polyurethane	High tensile strength, tear, abrasion and oil resistance, relatively high price	Oil seals, shoe soles and heels, forklift truck tyres, diaphragms, fibre coatings resistant to dry cleaning, variety of mechanical goods
Chlorosulphonated polyethylene	Very good heat, ozone, and weathering resistance, good resistance to oil and a wide range of chemicals, high elasticity, good abrasion resistance	Wire and cable coating, chemical plant hose, fabric coating, film sheeting, footwear, pond liners
Polysulphide	Excellent oil, solvent, and water resistance, high impermeability to gases, low strength, unpleasant odour (particularly during processing)	Adhesive, sealants, binders, hose
Epichlorohydrin	Low air permeability, low resilience, excellent ozone resistance, good heat resistance, flame resistance, and weathering resistance	Seals, gaskets, wire and cable coating
Fluoroelastomers	Outstanding heat resistance, superior oil, chemical, and solvent resistance, highest-priced elastomer	Aerospace applications, high quality seals, and gaskets

Styrene-Acrylonitrile Co-polymer

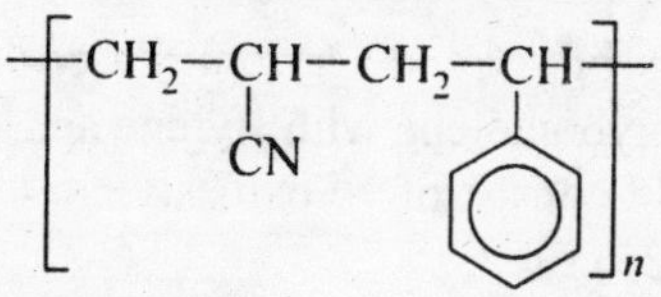

Monomers	Polymerisation	Major uses
Acrylonitrile, styrene	Free-radical-initiated chain polymerisation	Components of domestic appliances, electrical equipment and car equipment, picnic ware, housewares

Because of the polar nature of the acrylonitrile molecule, styrene-acrylonitrile (SAN) co-polymers have better resistance to hydrocarbons, oils, and greases than polystyrene. These co-polymers have a higher softening point, a much better resistance to stress cracking and crazing, and a higher impact strength than the homopolymer polystyrene, yet they retain the transparency of the latter. The toughness and chemical resistance of the co-polymer increases with the acrylonitrile content but so do the difficulty in moulding and the yellowness of the resin. Commercially available SAN co-polymers have 20 to 30 per cent acrylonitrile content. They are produced by emulsion, suspension, or continuous polymerisation.

Due to their rigidity, transparency, and thermal stability, SAN resins have found applications for dials, knobs, and covers for domestic appliances, electrical equipment, car equipment, dishwasher-safe housewares, such as refrigerator meat and vegetable drawers, blender bowls, vacuum cleaner parts, humidifier parts, plus other industrial and domestic applications with requirements more stringent than can be met by polystyrene.

SAN resins are also reinforced with glass to make dashboard components and battery cases. Over 35 per cent of the total SAN production is used in the manufacture of ABS blends.

Acrylonitrile-Butadiene-Styrene Terpolymer

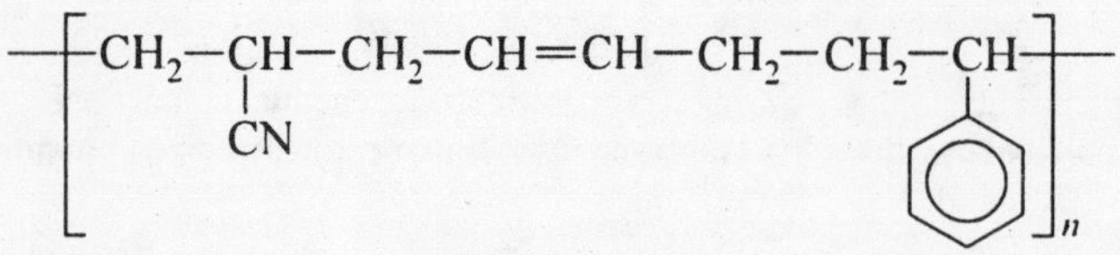

Monomers	*Polymerisation*	*Major uses*
Acrylonitrile, butadiene styrene	Free-radical-initiated chain polymerisation	Pipe and fittings (30%), automotive and appliance (15%), telephones and business machine housings

A range of materials popularly referred to as ABS polymers are formed basically from three different monomers: acrylonitrile, butadiene, and styrene. Acrylonitrile contributes chemical resistance, heat resistance, and high strength; butadiene contributes toughness, impact strength, and low-temperature property retention; styrene contributes rigidity, surface appearance (gloss), and processability. Not only may the ratios of the monomers be varied, but the way in which they can be assembled into the final polymer can also be the subject of considerable variations: The range of possible ABS-type polymers is therefore very large.

The most important ways of producing ABS polymers are (i) blends of styrene-acrylonitrile co-polymers with butadiene-acrylonitrile rubber; and (ii) interpolymers of polybutadiene with styrene and acrylonitrile, which is now the most important type. A typical blend would consist of 70 parts styrene-acrylonitrile (70:30) co-polymer and 40 parts butadiene-acrylonitrile (65:35) rubber.

Interpolymers are produced by co-polymerising styrene and acrylonitrile in the presence of polybutadiene rubber (latex) by using batch or continuous emulsion polymerisation. The resultant materials are a mixture of polybutadiene, SAN co-polymer, and polybutadiene grafted with styrene and acrylonitrile. The mixture is made up of three phases: a continuous matrix of SAN, a dispersed phase of polybutadiene, and a boundary layer of SAN graft.

ABS polymers are processable by all techniques commonly used with thermoplastics. They are slightly hygroscopic and should be dried 2 to 4 hr at 180 to 200°F (82 to 93°C) just prior to processing. A dehumidifying circulating air-hopper dryer is recommended. ABS can be hot stamped, painted, printed, vacuum metallised, electroplates, and embossed. Common fabrication techniques are applicable, including sawing, drilling, punching, riveting, bonding, and incorporating metal inserts and threaded and non-threaded fasteners. The machining characteristics of ABS are similar to those of non-ferrous metals.

ABS materials are superior to the ordinary styrene products and are commonly described as tough, hard, and rigid. This combination is unusual for thermoplastics. Moreover, the moulded specimens generally have a very good surface finish, and this property is particularly marked with the interpolymer type ABS polymers. Light weight and the ability to economically achieve a one-step finished appearance part have contributed to large-volume applications of ABS.

Adequate chemical resistance is present in the ABS materials for ordinary applications. They are affected little by water, alkalies, weak acids, and inorganic salts. Alcohol and hydrocarbon may affect the surfaces. ABS has poor resistance to outdoor UV light; significant changes in appearance and mechanical properties will result after exposure. Protective coatings can be applied to improve resistance to UV light. ABS materials are employed in thousand of applications, such as house-hold appliances, business machine and camera housings, telephone hand-sets, electrical hand tools (such as drill housings), handles, knobs, cams, bearings, wheels, gears, pump impellers, automotive trim and hardware, bathtubs, refrigerator liners, pipe and fittings, shower heads, and sporting goods. Business machines, consumer electronics, and telecommunications applications represent the fastest-growing areas for ABS. Painted, electroplated, and vacuum-metallised parts are used throughout the automotive, business machine, and electronics markets.

Multilayered laminates with an ABS outer layer can be produced by co-extrusion. In this process two or three different polymers may be combined into a multilayered film or sheet. Adhesion is enhanced by cooling the extruded laminate directly from the melt rather than in a separate operation after the components of the sheet have been formed and cooled separately. In one process flows from individual extruders are combined in a flow block and then conveyed to a single manifold die. All the polymer streams should have approximately the same viscosity so that laminar flow can be maintained.

Multilayered films and sheets have the advantage that a chemically resistant sheet can be combined with a good barrier to oxygen and water diffusion, or a decorative glossy sheet can be placed over a tough, strong material. One commercial example is an ABS–high-impact-polystyrene sheet, which can be thermoformed to make the inside door and food compartment of a refrigerator. Another example is a four-layered sheet comprising ABS, polyethylene, polystyrene, and rubber-modified polystyrene for butter and margarine packages. However, not all combinations adhere equally well, so there are limits to the design of such structures.

In recent years there has been an increased demand for a variety of special ABS grades. These include high-temperature-resistant grades (for automotive instrument panels, power tool housings), fire-retardant grades (for appliance housings, business machines, television cabinets), electroplating grades (for automotive grilles and exterior decorative trim), high-gloss, low-gloss, and matte-finish grades (for moulding and extrusion applications), clear ABS grades (using methyl methacrylate as the fourth monomer), and structural foam grades (for moulded parts with high strength-to-weight ratio). The structural foam grades are available for general-purpose and flame-retardant applications. The cellular structure can be produced by injecting nitrogen gas into the melt just prior to entering the mould or by using chemical blowing agents in the resin.

Ethylene-Methacrylic Acid Co-polymers (Ionomers)

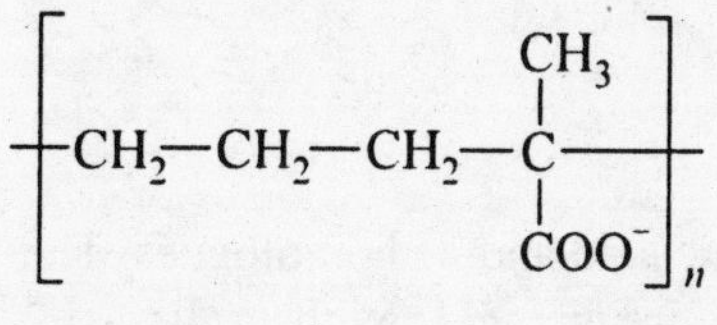

Monomers	*Polymerisation*	*Major uses*
Ethylene, methacrylic acid	Free-radical-initiated chain polymerisation	Packaging film, golf ball covers, automotive parts, footwear

Ionomers

Ionomer is a generic name for polymers containing interchain ionic bonding. They have the characteristics of both thermoplastics and thermosetting materials and are derived by co-polymerising ethylene with a small amount (1–10 per cent in the basic patent) of an unsaturated acid, such as methacrylic acid, using the high-pressure process. The carboxyl groups in the co-polymer are then neutralised by monovalent and divalent cations, resulting in some form of ionic cross-links (Fig. 15.8) which are stable at normal ambient temperatures but which reversibly break down on heating. These materials thus posses the advantage of cross-linking, such as enhanced toughness and stiffness, at ambient temperatures, but they behave as linear polymers at elevated temperatures, so they may be processed and even re-processed without undue difficulty.

Co-polymerisation used in making ionomers has had the effect of depressing crystallinity, although not completely eliminating it, so the materials are also transparent. Ionomers also have excellent oil and grease resistance, excellent resistance to stress cracking, and a higher water vapour permeability than does polyethylene.

The principal uses of ionomers are for film lamination and co-extrusion for composite food packaging. The ionomer resin provides an outer layer with good sealability and significantly greater puncture resistance than an LDPE film. Sporting goods utilise the high-impact toughness of ionomers. Most major golf ball manufacturers use covers of durable ionomer. Such covers are virtually cut proof in normal use and retain a greater resiliency over a wider temperature range; they are superior to synthetic *trans*-polyisoprene in these respects.

Automotive uses (bumper pads and bumper guards) are based on impact toughness and paintability. In footwear applications, resilience and flex toughness of ionomers are advantages in box toes, counters, and shoe soles. Ski boot and ice skate manufacturers produce light weight outer shells of ionomers. Sheet and foamed sheet products include carpet mats, furniture tops, ski lift seat pads, boat bumpers, and wrestling mats.

Ionomers should be differentiated from polyelectrolytes and ion-exchange resins, which also contain ionic groups. Polyelectrolytes show ionic dissociation in water and are used, among other things, as thickening agents. Common examples are sodium polyacrylate, ammonium polymethacrylate (both anionic polyacrylate) and poly-(*N*-butyl-4-vinyl-pyridinium bromide), a cationic polyelectrolyte. Ion-exchange resins used in water softening, in chromatography, and for various industrial purposes, are cross-linked polymers containing ionic groups.

Polyelectrolytes and ion-exchange resins are, in general, intractable materials and not processable on conventional plastics machinery. In ionomers, however, the amount of ionic bonding is limited to yield useful and tractable plastics. Using this principle, manufacturers can produce rubbers which undergo ionic cross-linking to give the effect of vulcanisation as they cool on emergence from an extruder or in the mould of an injection-moulding machine.

ACRYLICS

Acrylic polymers may be considered structurally as derivatives of acrylic acid and its homologues. The family of acrylics includes a range of commercial polymers based on acrylic acid, methacrylic acid, esters of acrylic acid and of methacrylic acid, acrylonitrile, acrylamide, and co-polymers of these compounds. By far the best known applications of acrylics are acrylic fibres and acrylonitrile co-polymer such as NBR, SAN, and ABS.

Polyacrylonitrile

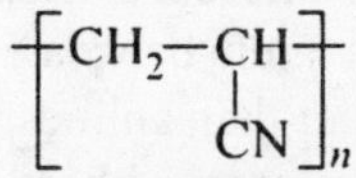

Monomer	*Polymerisation*	*Major uses*
Acrylonitrile	Free-radical-initiated chain polymerisation	Fibres in apparel (70%) and house furnishings (30%)

Polyacrylonitrile and closely related co-polymers have found wide use as fibres. In polyacrylonitrile appreciable electrostatic forces occur between the dipoles of adjacent nitrile groups on the same polymer molecule. This restricts the bond rotation and leads to a stiff, rodlike structure of the polymer chain. As a result, polyacrylonitrile has a very high crystalline melting point (317°C) and is soluble in only a few solvents, such as dimethylformamide and dimethylacetamide, and in concentrated aqueous solutions of inorganic salts, such as calcium thiocyanate, sodium perchlorate, and zinc chloride. Polyacrylonitrile cannot be melt processed because its decomposition temperature is close to the melting point. Fibres are therefore spun from solution by either wet or dry spinning.

Fibres prepared from straight polyacrylonitrile are difficult to dye. To improve dyeability, manufacturers invariably add to commercial fibres minor amounts of one or two co-monomers, such as methyl acrylate, methyl methacrylate, vinyl acetate, and 2-vinyl-pyridine. Small amounts of ionic monomers (sodium styrene sulphonate) are often included for better dyeability. Modacrylic fibres are composed of 35 to 85 per cent acrylonitrile and contain co-monomers, such as vinyl chloride, to improve fire retardancy.

Acrylic fibres are more durable than cotton, and they are the best alternative to wool, for sweaters. A major portion of the acrylic fibres produced are used in apparel (primarily hosiery). Other uses include pile fabrics (for simulated fur), craft yarns, blankets, draperies, carpets, and rugs.

Polyacrylates

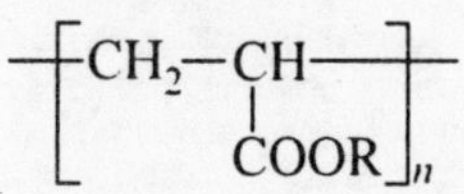

Monomer	*Polymerisation*	*Major uses*
Acrylic acid esters	Free-radical-initiated chain polymerisation	Fibre modification, coatings, adhesives, paints

The properties of acrylic ester polymers depend largely on the type of alcohol from which the acrylic acid ester is prepared. Solubility in oils and hydrocarbons increases as the length of the side chain increases. The lowest member of the series, poly (methyl acrylate), has poor low-temperature properties and is water sensitive. It is therefore restricted to such applications as textile sizes and leather finishes. Poly(ethyl acrylate) is used in fibre modifications and in coatings; and poly(butyl acrylate) and poly(2-ethylhexyl acrylate) are used in the formulation of paints and adhesives.

The original acrylate rubbers first introduced in 1948 by *B. F. Goodrich* and marketed as Hycar 4021 were a co-polymer of ethyl acrylate with about 5 per cent of 2-chloroethyl vinyl ether ($CH_2=CH-O-CH_2-CH_2Cl$) acting as a cure site monomer. Such polymers are vulcanised through the chlorine atoms by amines (such as triethylenetetramine), which are, however, not easy to handle. Therefore, 2-chloroethyl vinyl ether has not been replaced with other cure site monomers, such as vinyl and allyl chloroacetates; the increased reactivity of the chlorine in these monomers permits vulcanisation with ammonium benzoate (which decomposes on heating to produce ammonia, the actual cross-linking agent) rather than amines.

Acrylate rubbers have good oil resistance. In heat resistance they are superior to most rubbers, exceptions being the fluororubbers, the silicones, and the fluorosilicones. It is these properties which account for the major use of acrylate rubbers, i.e., in oil seals for automobiles. They are, however, inferior in low-temperature properties.

A few acrylate rubbers are based on butyl acrylate. These materials are generally co-polymers of butyl acrylate and acrylonitrile (~10 per cent) and may be vulcanised with amines. They have improved low-temperature flexibility compared to ethyl acrylate co-polymers but swell more in aromatic oils.

Ethylene co-polymers with acrylates represent a significant segment of the ethylene co-polymer market, as many LDPE producers use co-polymerisation as a strategy to obtain products more resistant to displacement by HDPE and LLDPE. Ethylene co-polymers with methyl methacrylate and ethyl, butyl, and methyl acrylates are similar to ethylene-vinyl acetate (EVA) co-polymers in properties (discussed later) but have improved thermal stability during extrusion and increased low-temperature flexibility. The commercial products generally contain 15–30 per cent of the acrylate or methacrylate co-monomer. Applications include medical packaging, disposable gloves, hoses, tubing, gaskets, cable insulation, and squeeze toys. Use of ethylene-ethyl acrylate co-polymers for making vacuum cleaner hoses demonstrates the increased flexibility and long flex life that is possible with such materials.

Co-polymers in which the acrylate monomer is the major component are useful as ethylene-acrylate elastomers (trade name: Vamac). These are terpolymers containing a small amount of an alkenoic acid to introduce sites (C=C) for subsequent cross-linking via reaction with primary diamines. These elastomers have excellent oil resistance and stability over a wide temperature range (–50 to 200°C), being superior to chloroprene and nitrile rubbers. Although not superior to silicone and fluoroelastomers, they are less costly; uses include automotive (hydraulic systems seals, hoses) and wire and cable insulation.

Polymethacrylates

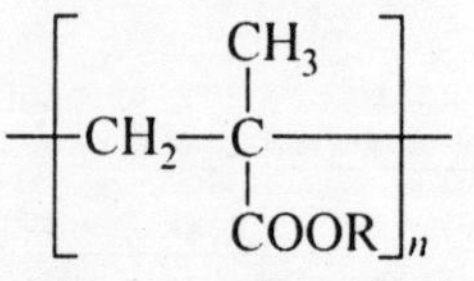

A large number of alkyl methacrylates, which may be considered as esters of poly(methacrylic acid), have been prepared. By far the most important of these polymers is poly(methyl methacrylate), which is an established major plastics material. As with other linear polymers, the mechanical and thermal properties of polymethacrylates are largely determined by the intermolecular attraction, spatial symmetry, and chain stiffness.

As the size of the ester alkyl group increases in a series of poly(*n*-alkyl methacrylates), the polymer molecules become spaced further apart and the intermolecular attraction is reduced. Thus, as the length of the side chain increases, the softening point decreases, and the polymers become rubbery at progressively lower temperatures (Fig. 15.9). However, when the number of carbon atoms in the side chain exceeds 12, the polymers become less rubbery, and the softening point, brittle point, and other properties related to the glass transition temperature rise with an increase in chain length.

As with the polyolefins, this effect is due to side-chain crystallisation. Glass transition temperatures of polymethacrylates is given in Table 15.6).

Table 15.6. Glass transition temperatures of polymethacrylates.

Ester group	T_g *(°C)*
Methyl	105
Ethyl	65
n-Butyl	20
n-Decyl	– 70
n-Hexadecyl	– 9

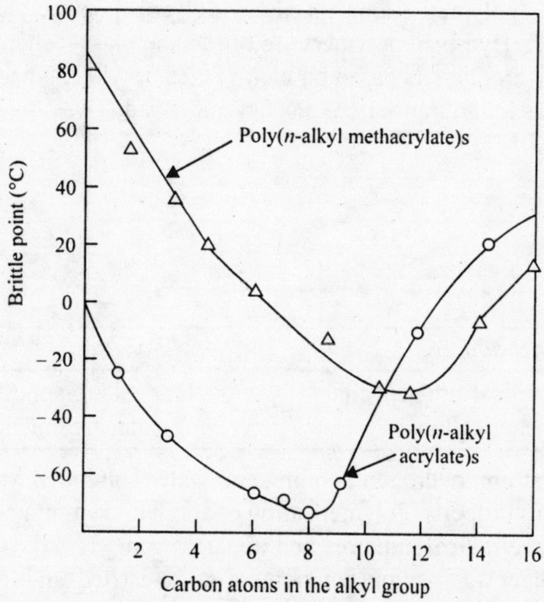

Fig. 15.9. Brittle points of poly (*n*-alkyl acrylate)s and poly (*n*-alkyl methacrylate)s.

Poly(alkyl methacrylate)s in which the alkyl group is branched have higher softening points and are harder than their unbranched isomers. This effect is not simply due to the better packing possible with the branched isomers. The lumpy branched structures impede rotation about the carbon-carbon bond on the main chain, thus contributing to stiffness of the molecule and consequently a higher transition temperature. Similarly, since the α-methyl group in polymethacrylates reduces chain flexibility, the lower polymethacrylates have higher softening points than the corresponding polyacrylates do.

Poly(methyl methacrylate) (PMMA) is by far the predominant polymethacrylate used in rigid applications because it has crystal clear transparency, excellent weatherability (better than most other plastics), and a useful combination of stiffness, density, and moderate toughness. The glass transition temperature of the polymer is 105°C (221°F), and the heat deflection temperatures range from 75 to 100°C (167 to 212°F). The mechanical properties of PMMA can be further improved by orientation of heat-cast sheets.

PMMA is widely used for signs, glazing, lighting, fixtures, sanitary wares, solar panels, and automotive tail and stoplight lenses. The low index of refraction (1.49) and high degree of uniformity make PMMA an excellent lens material for optical applications.

Methyl methacrylate has been co-polymerised with a wide variety of other monomers, such as acrylates, acrylonitrile, styrene, and butadiene. Co-polymerisation with styrene gives a material with improved melt-flow characteristics. Co-polymerisation with either butadiene or acrylonitrile, or blending PMMA with SBR, improves impact resistance. Butadiene-methyl methacrylate co-polymer has been used in paper and board finishes.

Higher *n*-alkyl methacrylate polymers have commercial applications. Poly(*n*-butyl-), poly(*n*-octyl) and poly(*n*-nonyl methacrylate)s are used as leather finishes; poly (lauryl methacrylate) is used to depress the pour point and improve the viscosity-temperature property of lubricating oils.

Mention may also be made here of the 2-hydroxyethyl ester of methacrylic acid, which is the monomer used for soft contact lenses. Co-polymerisation with ethylene glycol dimethacrylate produces a hydrophilic network polymer (a hydrogel). Hydrogel polymers are brittle and glassy when dry but become soft and plastic on swelling in water. Terpolymers based on methyl methacrylate, butadiene, and styrene (MBS) are being increasingly used as tough transparent plastics and as additives for PVC.

Polyacrylamide

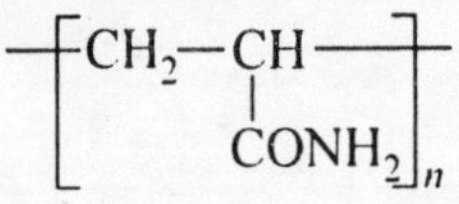

Monomer	*Polymerisation*	*Major uses*
Acrylamide	Free-radical-initiated chain polymerisation	Flocculant, adhesives, paper treatment, water treatment, coatings

Polyacrylamide exhibits strong hydrogen bonding and water solubility. Most of the interest in this polymer is associated with this property. Polymerisation of acrylamide monomer is usually conducted in an aqueous solution, using free-radical initiators and transfer agents.

Co-polymerisation with other water-soluble monomers is also carried out in a similar manner. Cationic polyacrylamides are obtained by co-polymerising with ionic monomers such as dimethylaminoethyl methacrylate, dialkyldimethylammonium chloride, and vinylbenzyltrimethylammonium chloride. These impart a positive charge to the molecule. Anionic character can be imparted by co-polymerising with monomers such as acrylic acid, methacrylic acid, 2-acrylamido-2-methyl-propanesulphonic acid, and sodium styrene sulphonate. Partial hydrolysis of polyacrylamide, which converts some of the amide groups to carboxylate ion, also results in anionic polyacrylamides.

Polyacrylamides have several properties which lead to a multitude of uses in diverse industries. Table 15.7 lists the main functions of the polymers and their uses in various industries.

Polyacrylamides are used as primary flocculants or coagulant aids in water clarification and mining application. They are effective for clarification of raw river water. The capacity of water clarifiers can be increased when the polymer is used as a secondary coagulant in conjunction with lime and ferric chloride. Polyacrylamides, and especially cationic polyacrylamide, are used for conditioning municipal and industries sludges for dewatering by porous and empty sand beds, vacuum filters, centrifuges, and other mechanical devices.

Table 15.7. Applications of polyacrylamide.

Function	*Application*	*Industry*
Flocculation	Water clarification	General
	Waste removal	Sewage
	Solids recovery	Mining
	Retention aid	Paper
	Drainage aid	Paper
Adhesion	Dry strength	Paper
	Wallboard cementing	Construction
Rheology control	Waterflooding	Petroleum
	Viscous drag reduction	Petroleum
		Fire fighting
		Irrigation pumping

Certain anionic polyacrylamides are approved by the U.S. Environmental Protection Agency for clarification of potable water. Polymer treatment also allows filters to operate at higher hydraulic rates. The function of clarification is not explained by a simple mechanism. The long-chain linear polymer apparently functions to encompass a number of individual fine particles of the dispersed material in water, attaching itself to the particles at various sites by chemical bonds, electrostatic attraction, or other attractive forces. Relatively stable aggregates are thus produced, which may be removed by filtration, settling, or other convenient means.

Polyacrylamides are useful in the paper industry as processing aids, in compounding and formulating, and as filler-retention aids. Polyacrylamides and co-polymers of acrylamides and acrylic acid are used to increase the dry strength of paper.

Polyacrylamides are used as flooding aids in secondary oil recovery from the producing oil well. Water, being of low viscosity, tends to finger ahead of the more viscous oil. However, addition of as little as 0.05 per cent polyacrylamide to the waterflood reduces oil bypass and give significantly higher oil to water ratios at the producing wellhead. Greatly increased yields of oil result from adding polymer to waterflooding. Solutions containing polyacrylamide are very slippery and can be used for water-based lubrication. Small amounts of polymer, when added to an aqueous solution, can significantly reduce the friction in pipes, thereby increasing the throughput or reducing the power consumption.

Other applications include additives in coatings and adhesives and binders for pigments.

Poly(Acrylic Acid) and Poly(Methacrylic Acid)

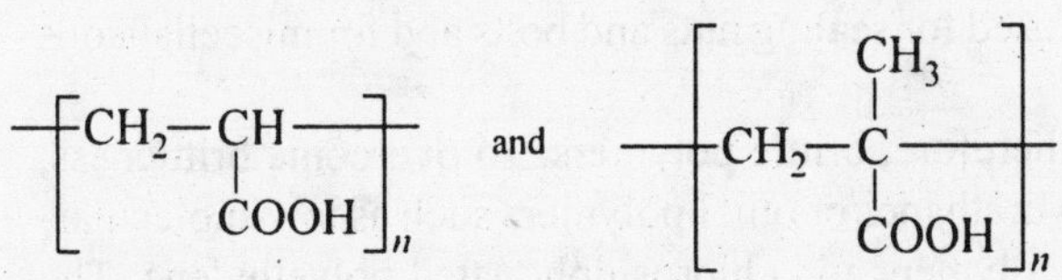

Monomers	*Polymerisation*	*Major uses*
Acrylic acid and methacrylic acid	Free-radical-initiated chain polymerisation	Sodium and ammonium-salts as polyelectrolytes, thickening agents

Poly(acrylic acid) and poly(methacrylic acid) may be prepared by direct polymerisation of the appropriate monomer, namely, acrylic acid or methacrylic acid, by conventional free-radical techniques, with potassium persulphate used as the initiator and water as the solvent (in which the polymers are soluble); or if a solid polymer is required, a solvent such as benzene, in which the polymer is insoluble, can be used, with benzoyl peroxide as a suitable initiator.

The multitude of applications of poly(acrylic acid) and poly(methacrylic acid) is reminiscent of the fable of the man who blew on his hands to warm them and blow on his porridge to cool it. They are used as adhesives and release agents, as flocculants and dispersants, as thickeners and fluidisers, as reaction inhibitors and promoters, as permanent coatings and removable coatings, etc. Such uses are the direct result of their varied physical properties and their reactivity. Many of the applications thus depend on the ability of these polymers to form complexes and to bond to substrates. Monovalent metal and ammonium salts of these polymers are generally soluble in water. These materials behave as anionic polyelectrolytes and are used for a variety of purposes, such as thickening agents, particularly for rubber latex.

Acrylic Adhesives

Acrylic adhesives are essentially acrylic monomers which achieve excellent bonding upon polymerisation. Typical examples are cyanoacrylates and ethylene glycol dimethacrylates. Cyanoacrylates are obtained by depolymerisation of a condensation polymer derived from a malonic acid derivative and formaldehyde.

Cyanoacrylates are marketed as contact adhesives. Often popularly known as superglue, they have found numerous applications. In dry air and in the presence of polymerisation inhibitors, methyl- and ethyl-2-cyanoacrylates have a storage life of many months. As with many acrylic monomers, air can inhibit or severely retard polymerisation of cyanoacrylates. These monomers are, however, prone to anionic polymerisation, and even a very weak base such as water can bring about rapid polymerisation.

In practice, a trace of moisture occurring on a substrate is adequate to cause polymerisation of the cyanoacrylate monomer to provide strong bonding within a few seconds of closing the joint and excluding air. Cyanoacrylate adhesives are particularly valuable because of their speed of action, which obviates the need for complex jibs and fixtures. The amount of monomer applied should be minimal to obtain a strong joint. Larger amounts only reduce the strength. Notable uses of cyanoacrylates include surgical glue and dental sealants; morticians use them to seal eyes and lips.

Dimethacrylates, such as tetramethylene glycol dimethacrylate, are used as anaerobic adhesives. Air inhibition of polymerisation of acrylic monomers is used to advantage in this application because the monomers are supplied along with a curing system (comprising a peroxide and an amine) as part of a one-part pack. When this adhesive is placed between mild steel surfaces, air inhibition is prevented since the air is excluded, and polymerisation can take place. Though the metal on the surface acts as a polymerisation promoter, it may be necessary to use a primer such as cobalt naphthenate to expedite the polymerisation. The anaerobic adhesives are widely used for sealing nuts and bolts and for miscellaneous engineering purposes.

Dimethacrylates form highly cross-linked and, therefore, brittle polymers. To overcome brittleness, manufacturers often blend dimethacrylates with polyurethanes or other polymers such as low-molecular-weight vinyl-terminated butadiene-acrylonitrile co-polymers and chlorosulphonated polyethylene. The modified dimethacrylate systems provide tough adhesives with excellent properties. These can be formulated as two-component adhesives, the catalyst component being added just prior to use or applied separately to the surface to be bonded. One-component systems also have been formulated which can be conveniently cured by ultraviolet radiation.

VINYL POLYMERS

If the R substituent in an olefin monomer ($CH_2 = CHR$) is either hydrogen, alkyl, aryl, or halogen, the corresponding polymer in the present discussion is grouped under polyolefins. If the R is a cyanide group or a carboxylic group or its ester or amide, the substance is an acrylic polymer. Vinyl polymers include those polyolefins in which the R substituent in the olefin monomer is bonded to the unsaturated carbon through an oxygen atom (vinyl esters, vinyl ethers) or a nitrogen atom (vinyl pyrrolidone, vinyl carbazole). Vinyl polymers constitute an important segment of the plastics industry. Depending on the specific physical and chemical properties, these polymers find use in patients and adhesives, in treatments for paper and textiles, and in special applications. The commonly used commercial vinyl polymers are described next.

Poly(Vinyl Acetate)

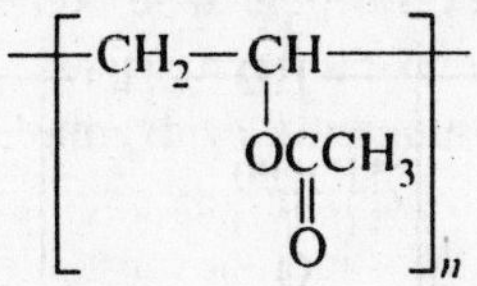

Monomers	*Polymerisation*	*Major uses*
Vinyl acetate	Free-radical-initiated chain polymerisation (mainly emulsion polymerisation)	Emulsion paints, adhesives, sizing

Since poly(vinyl acetate) is usually used in an emulsion form, it is manufactured primarily by free-radical-initiated emulsion polymerisation. The polymer is too soft and shows excessive cold flow, which precludes its use in moulded plastics. The reason is that the glass transition temperature of 28°C is either slightly above or (at various times) below the ambient temperatures.

Vinyl acetate polymers are extensively used in emulsion paints, as adhesives for textiles, paper, and wood, as a sizing material, and as a 'permanent starch'. A number of commercial grades are available which differ in molecular weight and in the nature of co-monomers (e.g., acrylate, maleate, fumarate) which are often used. Two vinyl acetate co-polymers of particular interest to the plastics industry are ethylene-vinyl acetate and vinyl chloride-vinyl acetate co-polymers.

Ethylene vinyl acetate (EVA) co-polymers represent the largest-volume segment of ethylene co-polymer market and are the products of low-density polyethylene (LDPE) technology. Commercial preparation of EVA co-polymer is based on the same process as LDPE with the addition of controlled co-monomer stream into the reactor. EVA co-polymers are thermoplastic materials consisting of an ethylene chain incorporating 5 to 20 mol per cent vinyl acetate (VA), in general. The VA produces a co-polymer with lower crystallinity than conventional ethylene homopolymer.

These lower crystallinity resins have lower melting points and heat seal temperatures, along with reduced stiffness, tensile strength, and hardness. EVAs have greater clarity, low-temperature flexibility, stress-crack resistance, and impact strength than LDPE. EVA resins are more permeable to oxygen, water vapour, and carbon dioxide. Chemical resistance is similar to that of LDPE, with somewhat better resistance to oil and grease for EVA resins of higher VA content. The VA groups contribute to improved adhesion in extrusions or hot-melt adhesive formulations.

The outdoor stability of EVA resins is superior to that of LDPE by virtue of their greater flexibility. Addition of UV stabilisers can extend the outdoor life of clear compounds to three to five years, depending on the degree of exposure. Outdoor life expectancy is also enhanced by the addition of carbon black.

In addition to speciality applications involving film and adhesives production, EVAs are used in a variety of moulding, compounding, and extrusion applications. Some typical end uses include flexible hose and tubing, footwear components, toys and athletic goods, wire and cable compounding, extruded gaskets, moulded automotive parts (such as energy-absorbing bumper components), cap and closure seals, and colour concentrates. In footwear applications, EVA resins are used in a canvas box toes and flocked or fabric-laminated contours. Foamed and cross-linked EVA is used in athletic or leisure shoe midsoles and in sandals.

With increasing VA content, EVA resin properties range from those of LDPE to those of highly plasticised PVC. High VA resins are soft and flexible, with excellent toughness and good stress-crack resistance. With EVA resin these properties are permanent and do not dissipate with time because of the loss of the liquid plasticiser. EVAs also have exceptional low temperature toughness and flexibility.

The EVA co-polymers are slightly less flexible than normal rubber compounds but have the advantage of simpler processing since no vulcanisation is necessary. The materials have thus been largely used in injection moulding in place of plasticised PVC or vulcanised rubber. Typical applications include turntable mats, based pads for small items of office equipment, buttons, car door protection strips, and for other parts where a soft product of good appearance is required.

A substantial use of EVA co-polymers is as wax additives and additives for hot-melt coatings and adhesives. Cellular cross-linked EVA co-polymers are used in shoe parts.

EVA co-polymers with only a small vinyl acetate content (~3 mol per cent) are best considered as a modification of low-density polyethylene. These co-polymers have less crystallinity and greater flexibility, softness, and, in case of film, surface gloss.

EVA can be processed by all standard plastics processing techniques, including injection and blow moulding, thermoforming, and extrusion into sheet and shapes. They accommodate high loadings of fillers, pigments, and carbon blacks. They are compatible with other thermoplastics, and thus are frequently used for impact modification and improvement of stress crack resistance. This combination of properties makes EVA highly adaptable vehicles for colour concentrates. EVA resins can be formulated with blowing agents and cross-linking to produce low density foams via compression moulding.

Hydrolysis of EVA co-polymers yields ethylene-vinyl alcohol co-polymers (EVOH). EVOH has exceptional gas barrier properties as well as oil and organic solvent resistance. The poor moisture resistance of EVOH is overcome by coating, co-extrusion, and lamination with other substrates. Applications include containers for food (ketchup, jelly, mayonnaise) as well as chemicals and solvents.

Poly(Vinyl Alcohol)

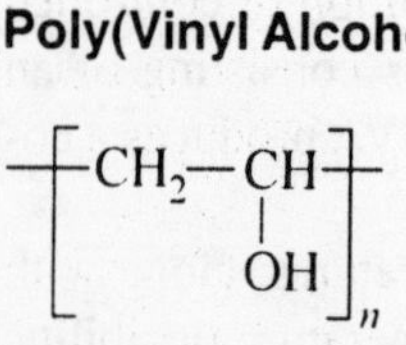

Manufacture	*Major uses*
Alcoholysis of poly(vinyl acetate)	Paper sizing, textile sizing, cosmetics

Poly(vinyl alcohol) (PVA) is produced by alcoholysis of poly(vinyl acetate), because vinyl alcohol monomer does not exist in the free state. (The term 'hydrolysis' is often used incorrectly to describe this process). Either acid or base catalysts may be employed for alcoholysis. Alkaline catalysts such as sodium hydroxide or sodium methoxide give more rapid alcoholysis. The degree of alcoholysis, and hence the residual acetate content, is controlled by varying the catalyst concentration.

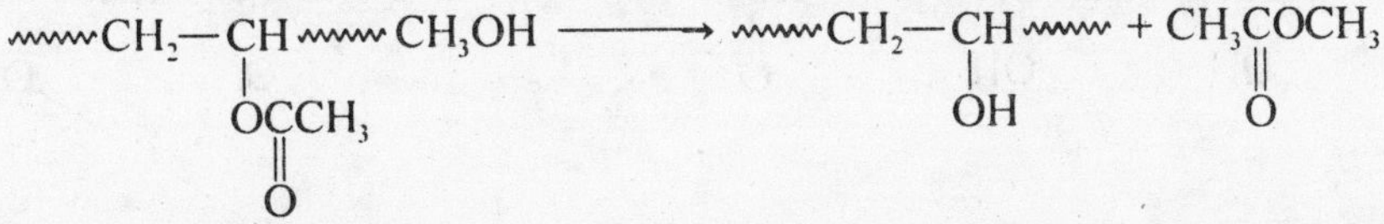

The presence of hydroxyl groups attached to the main chain renders the polymer hydrophilic. PVA therefore dissolves in water to a greater or lesser extent according to the degree of hydrolysis. Polymers with a degree of hydrolysis in the range of 87 to 89 per cent readily dissolve in cold water. Solubility decreases with an increase in the degree of hydrolysis, and fully hydrolysed polymers are water soluble only at higher temperatures (>85°C). This apparently anomalous behaviour is due to the higher degree of crystallinity and the greater extent of hydrogen bonding in the completely hydrolysed polymers.

Commercial PVA is available in a number of grades which differ in molecular weight and degree of hydrolysis. The polymer finds a variety of uses. It functions as a non-ionic surface active agent and is used in suspension polymerisation as a protective colloid. It also serves as a binder and thickener and is widely used in adhesives, paper coatings, paper sizing, textile sizing, ceramics, and cosmetics.

Completely hydrolysed grades of PVA find use in quick-setting, water resistant adhesives. Combinations of fully hydrolysed PVA and starch are used as a quick-setting adhesive for paper converting. Borated PVA, commonly called 'tackified', are combined with clay and used in adhesive applications requiring a high degree of wet tack. They are used extensively to glue two or more plies of paper together to form a variety of shapes such as tubes, cans, and cores.

Since PVA film has little tendency to adhere to other plastics, it can be used to prevent sticking to mould. Films cast from aqueous solution of PVA are used as release agents in the manufacture of reinforced plastics.

Partially hydrolysed grades have been developed for making tubular blown film (similar to that with polyethylene) for packages for bleaches, bath salts, insecticides, and disinfectants. Use of water-soluble PVA film for packaging pre-weighed quantities of such materials permits their addition to aqueous systems without breaking the package or removing the contents, thereby saving time and reducing material losses. Film made from PVA may be used for hospital laundry bags that are added directly to the washing machine.

A process has been developed in Japan for producing fibres from poly(vinyl alcohol). The polymer is wet spun from a warm aqueous solution into a concentrated aqueous solution of sodium sulphate containing sulphuric acid and formaldehyde, which insolubilises the alcohol by formation of formal groups. These fibres are generally known as vinal or vinylon fibres.

Poly(Vinyl Acetals)

Poly(vinyl acetals) are produced by treating poly(vinyl alcohol) with aldehydes. (They may also be made directly from poly(vinyl acetate) without separating the alcohol). Since the reaction with aldehyde involves a pair of neighbouring hydroxyl groups on the polymer chain and the reaction occurs at random, some hydroxyl groups become isolated and remain unreacted. A poly(vinyl acetal) molecule will thus contain acetal groups and residual hydroxyl groups. In addition, there will be residual acetate groups due to incomplete hydrolysis of poly(vinyl acetate) to the poly(vinyl alcohol) used in the acetalisation reaction. The relative proportions of these three types of groups may have a significant effect on specific properties of the polymer.

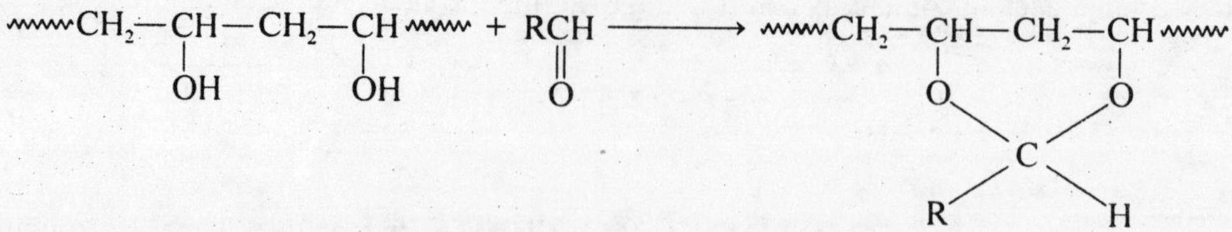

When the aldehyde in this reaction is formaldehyde, the product is poly(vinyl formal). This polymer is, however, made directly from poly(vinyl acetate) and formaldehyde without separating the alcohol. The product with low hydroxyl (5–6 per cent) and acetate (9.5–13 per cent) content (the balance being *formal*) is used in wire enamel and in structural adhesives (e.g., Redux). In both applications the polymer is used in conjunction with phenolic resin and is heat cured.

When the aldehyde in the acetalisation reaction is butyraldehyde, i.e., $R = CH_3CH_2CH_2—$, the product is poly(vinyl butyral). Sulphuric acid is the catalyst in this reaction. Poly(vinyl butyral) is characterised by high adhesion to glass, toughness, light stability, clarity, and moisture insensitivity. It is therefore extensively used as an adhesive interlayer between glass plates in the manufacture of laminated safety glass and bullet-proof composition.

Poly(Vinyl Cinnamate)

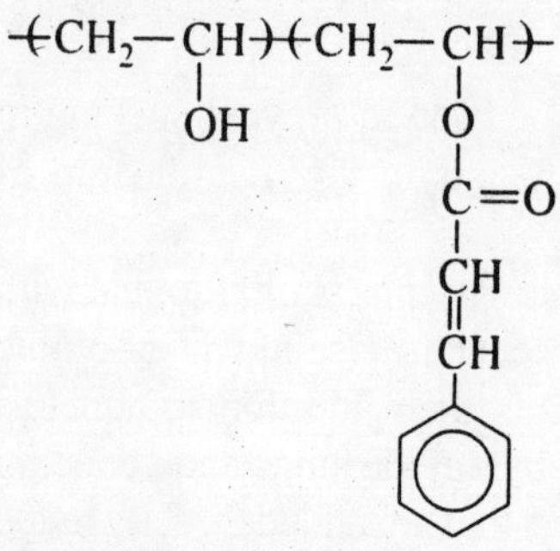

Poly(vinyl cinnamate) is conveniently made by the *Schotten-Baumann* reaction using poly(vinyl alcohol) in sodium or potassium hydroxide solution and cinnamoyl chloride in methyl ethyl ketone. The product is, in effect, a co-polymer of vinyl alcohol and vinyl cinnamate, as shown. The polymer has the ability to cross-link on exposure to light, which has led to its important applications in photography, lithography, and related fields as a photoresist.

Poly(Vinyl Ethers)

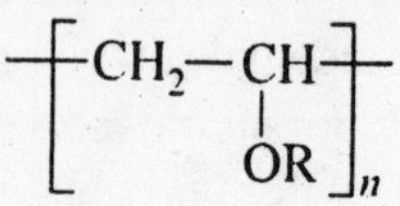

Commercial uses have developed for several poly (vinyl ethers) in which R is methyl, ethyl, and isobutyl. The vinyl alkyl ether monomers are produced from acetylene and the corresponding alcohols, and the polymerisation is usually conducted by cationic initiation using Friedel-Craft-type catalysts.

Poly(vinyl methyl ether) is a water-soluble viscous liquid which has found application in the adhesive and rubber industries. One particular applications has been as a heat sensitiser in the manufacture of rubber latex dipped goods.

Ethyl and butyl derivatives have found use as adhesives. Pressure-sensitive adhesive tapes made from poly(vinyl ethyl ether) incorporating antioxidants are said to have twice the shelf life of similar tapes made from natural rubber. Co-polymers of vinyl isobutyl ether with vinyl chloride, vinyl acetate, and ethyl acrylate are also produced.

Poly(Vinyl Pyrrolidone)

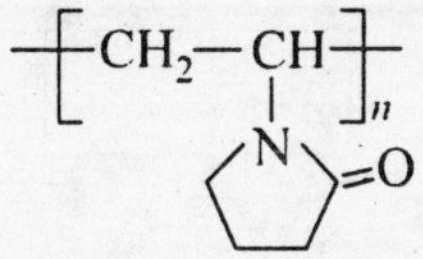

Poly(vinyl pyrrolidone) is produced by free-radical-initiated chain polymerisation of *N*-vinyl pyrrolidone. Polymerisation is usually carried out in aqueous solution to produce a solution containing 30 per cent polymer. The material is marketed in this form or spray dried to give a fine powder.

Poly(vinyl pyrrolidone) is a water-soluble polymer. Its main value is due to its ability to form loose addition compounds with many substances. It is thus used in cosmetics. The polymer has found several applications in textile treatment because of its affinity for dyestuffs. In an emergency it is used as a blood plasma substitute. Also, about 7 per cent polymer added to whole blood allows it to be frozen, stored at liquid nitrogen temperatures for years, and thawed out without destroying blood cells.

Poly(Vinyl Carbazole)

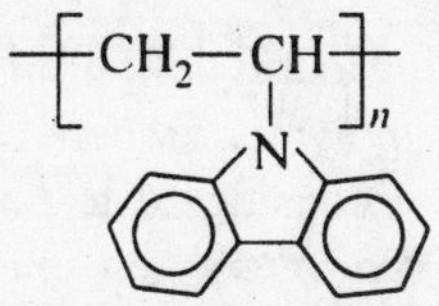

Poly (vinyl carbazole) is produced by polymerisation of vinyl carbazole using free-radical initiation or Ziegler-Natta catalysis. Poly(vinyl carboazole) has a high softening point, excellent electrical insulating properties, and good photoconductivity, which has led to its applications in xerography.

CHAPTER 16

Condensation Polymers

INTRODUCTION

According to the original classification of *Carothers*, condensation polymers are formed from bi- or polyfunctional monomers by reactions which involve elimination of some smaller molecule. A condensation polymer, according to this definition, is one in which the repeating unit lacks certain atoms which were present in the monomer(s) from which the polymer was formed.

With the development of polymer science and synthesis of newer polymers, this definition of condensation polymer was found to be inadequate. For example, in polyurethanes, which are classified as condensation polymers, the repeat unit has the same net composition as the two monomers—that is, a diol and a diisocyanate, which react without the elimination of any small molecule. Similarly the polymers produced by the ring-opening polymerisation of cyclic monomers, such as cyclic ethers and amides, are generally classified as condensation polymers based on the presence of functional groups, such as the ether and amide linkages, in the polymer chains, even though the polymerisation occurs without elimination of any small molecule.

To overcome such problems an alternative definition has been introduced. According to this definition, polymers whose main chains consist entirely of C—C bonds are classified as addition polymers, whereas those in which heteratoms (O, N, S, Si) are present in the polymer backbone are considered to be condensation polymers. A polymer which satisfies both the original definition (of *Carothers*) and the alternative definition or either of them, is classified as a condensation polymer. Phenol-formaldehyde condensation polymers, for example, satisfy the first definition but not the second.

POLYESTERS

Polyesters were historically the first synthetic condensation polymers. Commercial polyesters were manufactured by polycondensation reactions, the methods commonly used being melt polymerisation of diacid and diol, ester interchange of diester and diol, and interfacial polymerisation (*Schotten-Baumann* reaction) of diacid chloride and diol. In a polycondensation reaction a by-product is generated which has to be removed as the reaction progresses.

Thermoplastic saturated polyesters are widely used in synthetic fibres and also in films and moulding applications. The production of polyester fibres accounts for nearly 30 per cent of the total amount of synthetic fibres. Unsaturated polyesters are mainly used in glass-fibre reinforced plastic products.

Poly(Ethylene Terephthalate)

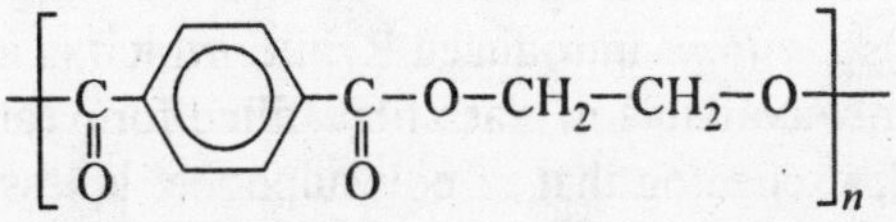

Monomers	*Polymerisation*	*Major uses*
Dimethyl terephthalate or terephthalic acid, ethylene glycol	Bulk polycondensation	Apparel (61%), home furnishings (18%), tyre cord (10%)

Ester interchange (also known as ester exchange or alcoholysis) was once the preferred method for making polyethylene terephthalate (PET) because dimethyl terephthalate can be readily purified to the high quality necessary for the production of the polymer. The process is carried out in two steps. Dimethyl terephthalate is heated at 150°C with an excess of ethylene glycol (mole ratio 1:2.1–2.2) in the presence of catalysts such as antimony trioxide and cobaltous acetate to give a mixture consisting of dihydroxyethyl terephthalate and small amounts of higher oligomers. Further heating to 270°C under vacuum in the presence of a catalyst produces the final polymer. In this second stage an ester interchange reaction occurs in which the dihydroxyethyl terephthalate serves as both ester and alcohol, and successive interchanges result in the formation of a polyester.

In recent years methods have been developed to produce terephthalic acid with satisfactory purity, and direct polycondensation reaction with ethylene glycol is now the preferred route to this polymer.

PET is widely used in synthetic fibres designed to simulate wool, cotton, or rayon, depending on the processing conditions. They have good wash-and-wear properties and resistance to wrinkling. In the production of fibre the molten polymer is extruded through spinnerets and rapidly cooled in air. The filaments thus formed are, however, largely amorphous and weak. They are therefore drawn at a temperature (80°C) above T_g and finally heated at 190°C under tension, whereby maximum molecular orientation, crystallinity, and dimensional stability are achieved. The melting point of highly crystalline PET is 271°C. Crystalline PET has good resistance to water and dilute mineral acids but is degraded by concentrated nitric and sulphuric acids. It is soluble at normal temperatures only in proton donors which are capable of interaction with the ester group, such as chlorinated and fluorinated acetic acids, phenols, and anhydrous hydrofluoric acid. PET is also used in film form and as a moulding material. The manufacture of PET film closely resembles the manufacture of fibre. The film is produced by quenching extruded sheet to the amorphous state and then re-heating and stretching the sheet approximately threefold in the axial and transverse directions at 80 to 100°C. To stabilise the biaxially oriented film, it is annealed under restraint at 180 to 210°C. This operation increases the crystallinity of PET film and reduces its tendency to shrink on heating. The strength of PET in its oriented form is outstanding.

The principal uses of biaxially oriented PET film are in capacitors, in slot liners for motors, and for magnetic tape. Although a polar polymer, its electrical insulation properties at room temperature, are good (even at high frequencies) because at room temperature, which is well below T_g (69°C), dipole orientation is severely restricted. The high strength and dimensional stability of the polyester film have also led to its use for X-ray and photographic film and to a number of graphic art and drafting applications. The film is also used in food packaging, including boil-in-bag food pouches. Metallised polyester films have many uses as a decorative material.

Because of its rather high glass transition temperature, only a limited amount of crystallisation can occur during cooling after injection moulding of PET. The idea of moulding PET was thus for many years not a technical proposition. Toward the end of the 1970s *Du Pont* introduced Rynite, which is a PET nucleated with an ionomer, containing a plasticiser and only available in glass-fibre-filled form (at 30, 45, and 55 per cent fill levels). The material is very rigid, exceeding that of polysulphone, is less water sensitive than an unfilled polymer, and has a high heat-deflection temperature (227°C at 264 psi).

Later the benefits of biaxial stretching PET were extended from film to bottle manufacture. Producing carbonated beverages PET bottles by blow moulding has gained prominence (particularly in the United States) because PET has low permeability to carbon dioxide. The process has been extended, particularly in Europe, to produce bottles for other purposes, such as fruit juice concentrates and saucés, wide-necked jars for coffee, and other materials. Because of its excellent thermal stability, PET is also used material for microwave and conventional ovens.

Poly(Butylene Terephthalate)

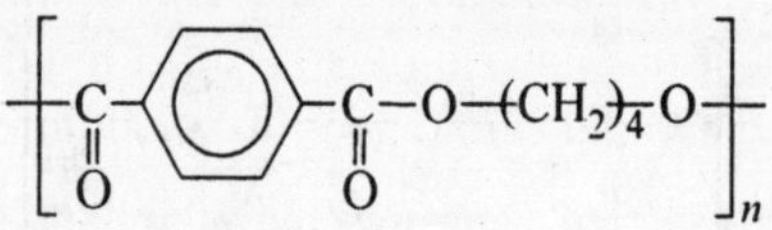

Monomers	*Polymerisation*	*Major uses*
Dimethyl terephthalate or terephthalic acid, butanediol	Bulk polycondensation	Machine parts, electrical applications, small appliances

Poly(butylene terephthalate), often abbreviated to PBT or PBTP, is manufactured by condensation polymerisation of dimethyl terephthalate and butane-1,4-diol in the presence of tetrabutyl titanate. The polymer is also known as poly(tetramethylene terephthalate), PTMT in short.

Because of the longer sequence of methylene groups in the repeating unit poly(butylene terephthalate) chains are both more flexible and less polar than poly(ethylene terephthalate). This leads to lower values for melting point (about 224°C) and glass transition temperature (22–43°C). The low glass transition temperature facilitates rapid crystallisation when cooling in the mould, and this allows short injection-moulding cycles and high injection speeds.

PBT finds uses as an engineering material due to its dimensional stability, particularly in water, and its resistance to hydrocarbon oils without showing stress cracking. PBT also has high mechanical strength and excellent electrical properties but a relatively low heat-deflection temperature 130°F (54°C) at 264 psi (1.8 MPa). The low water absorption of PBT—less than 0.1 per cent after 24-hours immersion—is outstanding. Both dimensional stability and electrical properties are retained under conditions of high humidity. The lubricity of the resin results in outstanding wear resistance.

As with PET, there is particular interest in glass-filled grades of PBT. The glass has a profound effect on such properties as tensile strength, flexural modulus, and impact strength, as can be seen from the values of these properties for unfilled and 30 per cent glass-filled PBT: 8200 vs. 17,000 psi (56 vs. 117 MPa) 340,000 vs. 1.1 to 1.2 × 10^6 psi (2350 vs. 7580 to 8270 MPa) and 0.8 to 1.0 vs. 1.3 to 1.6 ft-lbs/in^2 (Izod), respectively. Reinforcing with glass fibre also results in an increase in heat-deflection temperature to over 400°F (204°C) at 264 psi (1.8 MPa).

Typical applications of PBT include pump housings, impellers, bearing bushings, gear wheels, automotive exterior and under-the-hood parts, and electrical parts such as connectors and fuse cases.

Poly(Dihydroxymethylcyclohexyl Terephthalate)

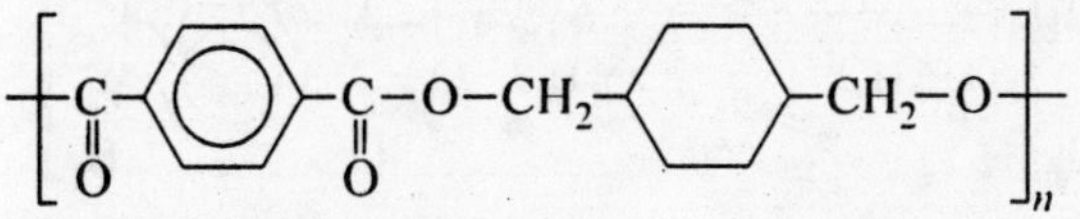

Eastman Kodak introduced a more hydrophobic polyester fibre under the trade name Kodel. The raw material for this polyester is dimethyl terephthalate. Reduction leads to 1,4-cyclohexylene glycol, which is used with dimethyl terephthalate in the polycondensation (ester exchange) reaction.

Eastman Kodak also introduced later a co-polyester based on 1,4-cyclohexylene glycol and a mixture of terephthalic and isophthalic acids. Being irregular in structure, the polymer is amorphous and gives products of brilliant clarity. In spite of the presence of the heterocyclic ring, the deflection temperature under load is as low as that of the poly(butylene terephthalate)s, and the polymer can be thermoformed at draw ratios as high as 4:1 without blustering or embrittlement. Because of its good melt strength and low moulding shrinkage, the material performs well in extrusion blow moulding and in injection moulding. The primary use for the co-polymer is extrusion into film and sheeting for packaging.

Ethylene glycol-modified polyesters are used in blow-moulding applications to produce bottles for packaging liquid detergents, shampoos, and similar products. One such product is Kodar PETG 6703 in which one acid (terephthalic acid) is reacted with a mixture of glycols (ethylene glycol and 1,4-cyclohexylene glycol). A related glass-reinforced grade (Ektar PCTG) has also been offered.

The principle of formation of segmented or block co-polymers has also been applied to polyesters, with the 'hard' segment formed from butanediol and terephthalic acid, and the 'soft' segment provider by a hydroxyl-terminated polyether [polytetramethylene ether glycol (PTMEG)] with molecular weight 600 to 3000. In a typical preparation, dimethyl terephthalate is transesterified with a mixture of PTMEG and a 50 per cent excess of butane-1,4-diol in the presence of an ester exchange catalyst. The stoichiometry is such that relatively long sequences of tetramethylene terephthalate (TMT) are produced which, unlike the polyether segments, are crystalline and have a high melting point. Since the sequences of TMT segregate into rigid domains, they are referred to as 'hard' segments, and the softer polyether terephthalate (PE/T) segments are referred to as 'soft' segments (Fig. 16.1).

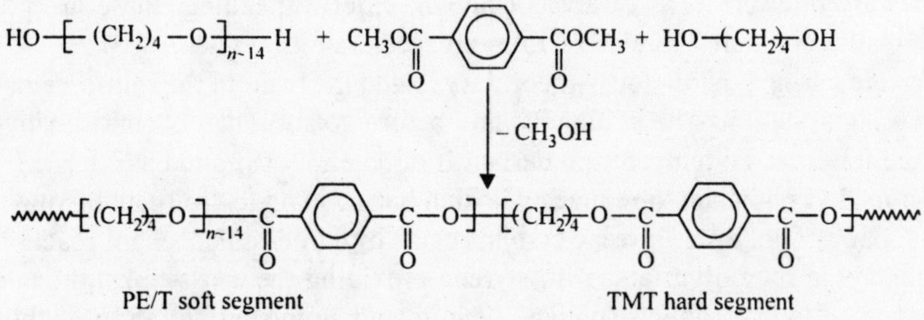

Fig. 16.1. Formation of polyester-type segmented or block co-polymer.

These elastomers are available in a range of stiffnesses. The harder grades have up to 84 per cent TMT segments and a melting point of 214°C, and the softest grades contain as little as 33 per cent TMT units and have a melting point of 163°C.

Processing of these thermoplastic rubbers is quite straightforward. The high crystallisation rates of the hard segments facilitate injection moulding, while the low viscosity at low shear rates facilitates a low shear process, such as rotational moulding.

These materials are superior to conventional rubbers in a number of properties. Consequently, in spite of their relatively high price they have become widely accepted as engineering rubbers in many applications.

Unsaturated Polyesters

Monomers	*Polymerisation*	*Major uses*
Phthalic anhydride, maleic anhydride, fumaric acid, isophthalic acid, ethylene glycol, propylene glycol, diethylene glycol, styrene	Bulk polycondensation followed by free-radical-initiated chain polymerisation	Construction, automotive applications, marine applications

Unsaturated polyester laminating resins are viscous materials of a low degree of polymerisation (i.e., oligomers) with molecular weights of 1500 to 3000. They are produced by condensing a glycol with both an unsaturated dicarboxylic acid (maleic acid) and a saturated carboxylic acid (phthalic or isophthalic acid). The viscous polyesters are dissolved in styrene monomer (30–50 per cent concentration) to reduce the viscosity. Addition of glass fibres and curing with peroxide initiators produces a cross-linked polymer (solid) consisting of the original polyester oligomers, which are not interconnected with polystyrene chains. The unsaturated acid residues in the initial polyester oligomer provide a site for cross-linking in this curing step, while the saturated acid reduces the brittleness of the final cross-linked product by reducing the frequency of cross-links. In practice, the peroxide curing system is blended into the resin before applying the resin to the reinforcement, which is usually glass fibre (E type), as preform, cloth, mat, or rovings, but sisal or more conventional fabrics may also be used. The curing system may be so varied (in both composition and quality) that curing times may range from a few minutes to several hours, and the cure may be arranged to proceed either at ambient or elevated temperatures. The two most important peroxy materials used for room temperature curing of polyester resins are methyl ethyl ketone peroxide (MEKP) and cyclohexanone peroxide. These are used in conjunction with a cobalt compound such as a naphthenate, octoate, or other organic-solvent-soluble soap. The peroxides are referred to as catalysts (though, strictly speaking, these are polymerisation initiators) and the cobalt compound is referred to as an accelerator.

In room-temperature curing it is obviously necessary to add the resin to the reinforcement as soon as possible after the curing system has been blended and before gelation can occur. Benzoyl peroxide is most commonly used for elevated-temperature curing. It is generally supplied as a paste (~50 per cent) in a liquid such as dimethyl phthalate to reduce explosion hazards and to facilitate mixing.

Since the cross-linking of polyester-styrene system occurs by a free-radical chain-reaction mechanism across the double bonds in the polyesters, with styrene providing the cross-links, the curing reaction does not give rise to volatile by-products (unlike phenolic and amino resins) and it is thus possible to cure without applying pressure. This fact as well as that room temperature cures are also possible makes unsaturated polyesters most useful in the manufacture of large structures such as boats and car bodies. Unsaturated polyesters find applications mainly in two ways: polyester-glass-fibre laminates and polyester moulding compositions.

Polyester-glass-fibre laminates (GRP, FRP)

The major process today is the hand layup technique in which the resin is brushed or rolled into the glass mat (or cloth) by hand. Since unsaturated polyesters are susceptible to polymerisation inhibition by air, surfaces of the hand layup laminates may remain undercured, soft, and, in some cases, tacky if freely exposed to air during the curing. A common way of avoiding this difficulty is to blend a small amount of paraffin wax (or other incompatible material) in with the resin. This blooms out on the surface and forms a protective layer over the resin during cure.

For mass production purposes, matched metal moulding techniques involving higher temperatures and pressures are employed. A number of intermediate techniques also exist involving vacuum bag, pressure bag, pultrusion, and filament winding.

Glass fibres are the preferred form of reinforcement for polyester resins. Glass fibres are available in a number of forms, such as glass cloth, chopped strands, mats, or rovings.

Being relatively cheaper, polyesters are preferred to epoxide and furan resins for general-purpose laminates. Polyesters thus account for no less than 95 per cent of the low-pressure laminates produced. The largest single outlet is in sheeting for roofing and building insulation. For the greatest transparency of the laminate the refractive indices of glass-cured resins and binder should be identical.

The second major outlet is in land transport. Polyester-glass laminates are used in the binding of sports car bodies, translucent roofing panel in lorries, and in public transport vehicles. In such applications the ability to construct large polyester-glass mouldings without complicated equipment is used to advantage. Polyester resins in conjunction with glass cloth or mat are widely used in the manufacture of boat hulls up to 153 ft (~46 m) in length. Such hulls are competitive in price and are easier to maintain and to repair. The high strength-to-weight ratio of polyester-glass laminates—particularly when glass is used—microwave transparency, and corrosion resistance of the laminates have led to their use in air transport applications as in aircraft radomes, ducting, spinners, and other parts. Land, sea, and air transport applications account for nearly half the polyester resin produced. Other applications include such diverse items as chemical storage vessels, chemical plant components, swimming pools, stacking chairs, trays, and sports equipment.

Polyester moulding compositions

Four types of polyester moulding compounds may be recognised: (i) dough-moulding compound (DMC); (ii) sheet-moulding compound (SMC); (iii) alkyd-moulding compositions, sometimes referred to as polyester alkyds; and (iv) diallyl phthalate (DAP) and diallyl isophthalate (DAIP) compounds.

Dough-moulding compounds of puttylike consistency are prepared by blending resins, catalyst, powdered mineral filler, reinforcing fibre (chopped strand), pigment, and lubricant in a dough mixer, usually of the Z-blade type.

The tendency of thick sections of DMC structural parts to crack has been overcome by using low-profile polyester resins (or low-shrink resins). These are prepared by making a blend of a thermoplastic (e.g., acrylic) polymer-styrene system with a polyester-styrene system. Mouldings of this blend cured at elevated temperatures exhibit negligible shrinkage and minimal warpage and have very smooth surfaces, to which paint may be applied with very little pre-treatment. A wide spectrum of properties may be obtained by varying the ratios of thermoplastics, polyester, and styrene in the blend. Among the thermoplastics, for such blending are poly(methyl methacrylate), polystyrene, PVC, and polyethylene. High-gloss DMCs using low-shrink resins have found uses in kitchen appliances such as toaster end plates, steam iron bases, and casings for electric heaters.

Since the manufacture of DMC involves intensive shear that causes extensive damage to fibres, DMC mouldings have less strength than GRP laminates. This problem is largely avoided with the sheet-moulding compounds (SMC). In the SMC process, unsaturated polyester resin curing systems, filler thickening agents, and lubricant are blended together and coated onto two polyethylene films. Chopped-glass rovings are supplied between the resin layers, which are then sandwiched together and compacted as shown in Fig. 16.2. Thickening occurs by the reaction of free carboxyl end groups with magnesium oxide. This converts the soft, sticky mass to a handleable sheet, which takes usually a day or two. A typical formulation consists of 30 per cent chopped-glass fibre, 30 per cent ground limestone, and resin. For moulding, the sheet may be easily cut to the appropriate weight and shape and placed between the halves of the heated mould. The main applications of SMCs are in car parts, baths, and doors.

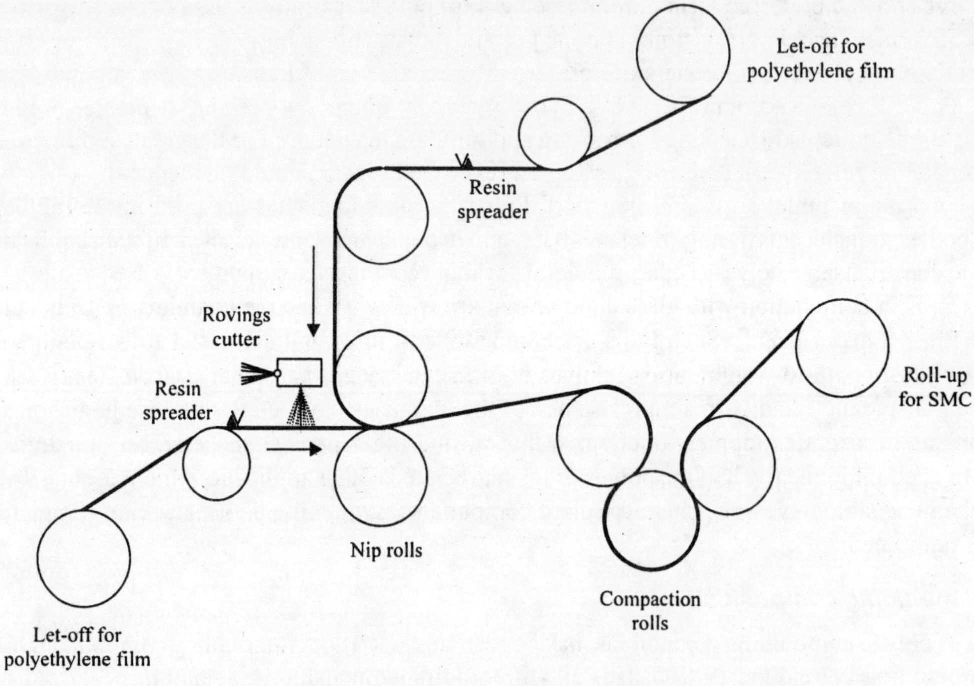

Fig. 16.2. Outline of machine for preparing sheet-moulding compounds.

The polyester alkyd moulding compositions are also based on polyester resin similar to those used for laminating. (The term 'alkyd' is derived from *al*cohol and aci*d*).

They are prepared by blending the resin with cellulose pulp, mineral filler, pigments, lubricants, and peroxide curing systems on hot mills to the desired flow properties. The mix is then removed, cooled, crushed, and ground.

Diallyl phthalate (DAP) is a diester of phthalic acid and allyl alcohol and contains two double bonds.

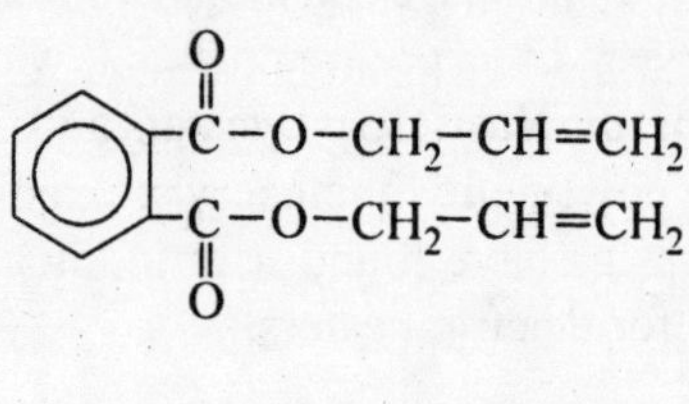

On heating with a peroxide, DAP therefore polymerises and eventually cross-links, forming an insoluble network polymer. However, it is possible to heat the DAP monomer under carefully controlled conditions, to give a soluble and stable partial polymer in the form of a white powder. The powder may then be blended with peroxide catalysts, fillers, and other ingredients to form a moulding powder in the same manner as polyester alkyds. Similar products can be obtained from diallyl isophthalate (DAIP).

Both DAP and DAIP alkyd mouldings are superior to the phenolics in their tracking resistance and in their availability in a wide range of colours; however, they tend to show a higher shrinkage on cure. The DAIP materials are more expensive than DAP but have better heat resistance. They are supposed to be able to withstand temperature as high as 220°C for long periods. The polyester alkyd resins are cheaper than the DAP alkyd resins but are mechanically weaker and do not maintain their electrical properties as well under severe humid conditions. Some pertinent properties of the polyester moulding composition are compared in Table 16.1 along with those of a GP phenolic composition. The alkyd moulding compositions are used almost entirely in electrical applications where the cheaper phenolic and amino resins are not suitable.

Table 16.1. Properties of thermosetting polyester mouldings.

Property	*Phenolic (GP)*	*DMC (GP)*	*Polyester alkyd*	*DAP alkyd*	*DAIP alkyd*
Moulding temperature (°C)	150–170	140–160	140–165	150–165	150–165
Cure time (cup flow test) (sec)	60–70	25–40	20–30	60–90	60–90
Shrinkage (cm/cm)	0.007	0.004	0.009	0.009	0.006
Specific gravity	1.3	2.0	1.7	1.6	1.8
Impact strength (ft-lb)	0.12–0.2	2.0–4.0	0.13–0.18	0.12–0.18	0.09–0.13
Volume resistivity (ohm-m)	10^{12}–10^{14}	10^{16}	10^{16}	10^{16}	10^{16}
Dielectric constant (10^6 Hz)	4.5–5.5	5.6–6.0	4.5–5.0	3.5–5.0	4.0–6.0
Dielectric strength (90°C) (kV/cm)	39–97	78–117	94–135	117–156	117–156
Power factor (10^6 Hz)	0.03–0.05	0.01–0.03	0.02–0.04	0.02–0.04	0.04–0.06
Water absorption (mg)	45–65	15–30	40–70	5–15	10–20

Aromatic Polyesters

Monomers	*Polymerisation*	*Major uses*
p-Hydroxy benzoic acid, bisphenol A, diphenyl isophthalate	Bulk polycondensation	High-temperature engineering thermoplastics, plasma coatings, abradable seals

Recently several polyesters with high aromatic content have been introduced mainly because of their outstanding heat resistance. In the 1960s the Carborundum Company introduced the homopolymer of *p*-hydroxybenzoic acid under the trade name Ekonol. It is used in plasma coating. This wholly aromatic homopolyester is produced in practice by the self-ester exchange of the phenyl ester of *p*-hydroxybenzoic acid.

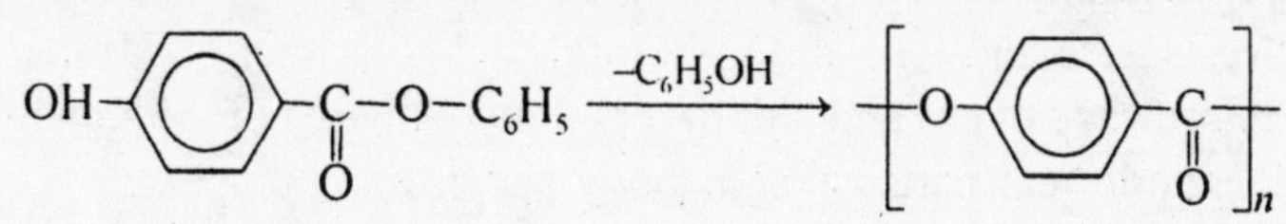

The homopolyester (mol. wt. 8,000–12,000) is insoluble in dilute acids and bases and all solvents up to their boiling points. It melts at about 500°C and is difficult to fabricate. It can be shaped only by hammering (like a metal), by impact moulding and by pressure sintering (420°C at 35 MPa). The difficulty in fabrication has severely limited the wider application of these polymers.

The homopolyester is available as finely divided powder in several grades, based on particle size. The average particle size ranges from 35 to 85 μm. The material can be blended with various powdered metals, such as bronze, aluminium, and nickel-chrome, and is used in flame-spray compounds. Plasma-sprayed coatings are thermally stable, self-lubricating, and wear and corrosion resistant. Applications include abradable seals for jet aircraft engine parts.

The polymer can also be blended up to 25 per cent with PTFE. Such blends have good temperature and wear resistance and are self-lubricating. Applications include seals, bearings, and rotors.

Co-polymeric aromatic polyesters, though possessing a somewhat lower level of heat resistance are easier to fabricate than are the wholly aromatic polymers; they also possess many properties that make them of interest as high-temperature engineering materials. These materials, called polyacrylates, are co-polyesters from bisphenols and dicarboxylic acids. One such material is a co-polyester of terephthalic acid, isophthalic acid, and bisphenol A in the ratio of 1:1:2.

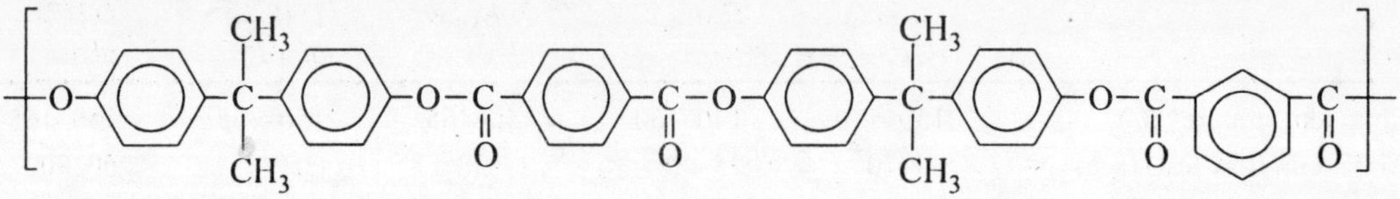

The use of two isomeric acids leads to an irregular chain which inhibits crystallisation. This allows the polymer to be processed at much lower temperatures than would be possible with a crystalline homopolymer. Nevertheless the high aromatic content of these polyesters ensures a high T_g (~90°C). The polymer is self-extinguishing with a limiting oxygen index of 34 and a self-ignition temperature of 545°C. The heat-deflection temperature under load (1.8 MPa) is about 175°C.

Among other distinctive properties of the polyacrylate are its good optical properties (luminous light transmission 84 to 88 per cent with 1 to 2 per cent haze, refractive index 1.61), high impact strength between that of polycarbonate and polysulphone, exceptionally high level of recovery after deformation (important in applications such as clips and snap fasteners), good toughness at both elevated and low temperatures with very little notch sensitivity, and high abrasion resistance which is superior to that of polycarbonates. Polyacrylate's weatherability and flammability (high oxygen index, low flame spread) are inherent and are achieved without additives. The weatherability properties therefore do not deteriorate significantly with time. (Tests show that over 5000 hr of accelerated weathering results in virtually no change in performance with respect to luminous light transmittance, haze, gloss, yellowness, and impact). Having no flame-retardant additives, the combustion products of polyacrylate are only carbon dioxide, carbon monoxide, and water, with no formation of toxic gas. Several companies have marketed polyacrylates under the trade names. These are non-crystallising co-polymers of mixed phthalic acids with a bisphenol and have repeat units of the type shown above. They are melt processable with T_g and heat distortion temperatures in the range of 150–200°C and have similar mechanical properties to polycarbonate and polyethersulphones. The polymers are useful for electrical and mechanical components that require good heat resistance and for lighting fixtures and consumer goods that operate at elevated temperatures, such as microwave ovens and hair dryers. Potential uses of the somewhat cheaper Arylon type polyesters include exterior car parts, such as body panels and bumpers.

A newer approach to obtaining polymers with good melt processability coupled with high softening point has led to another type of aromatic co-polyester, the so-called liquid crystalline polymers (LCPs). The retention of liquid crystalline order in the melt gives lower melt viscosities than would otherwise be achieved. Heat distortion temperatures are also in the high range of 180–240°C. LCPs may thus herald a new era of readily moulded engineering and electrical parts for high temperature use. Typical properties of aromatic polyesters cited above are shown in Table 16.2.

Table 16.2. Properties of unfilled aromatic polyesters.

Property	*Ekonol*	*Arylef or Ardel*	*Arylon*	*Vectra (range for various grades)*
Tensile strength				
10^3 lbf/in.^{2}	11	10	10	20–35
MPa	74 (flexural)	70 (yield)	70	140–240
Elongation at break (%)	–	50	25	1.6–7
Tensile modulus				
10^5 lbf/in.^{2}	–	3.0	2.9	14–58
GPa	–	2.1	2.0	10–40
Flexural modulus				
10^5 lbf/in.^{2}	10	2.9	3.0	14–51
GPa	7.1	2.0	2.1	10–35
Heat distortion temperature (°C)	>550	175	155	180–240
Impact strength, notched Izod				
ft-lb/in.	–	2.8–4.7	5.4	1–10
J/m	–	150–250	288	53–530
Limiting oxygen index (%)	–	34	26	35–50

Wholly Aromatic Co-polyester

A high-performance, wholly aromatic co-polyester suitable for injection moulding was commercially introduced in late 1984 by Dartco Manufacturing under the trade name Xydar. Xydar injection-moulding resins are based on terephthalic acid, *p, p'*-dihydroxybiphenyl, and *p*-hydroxybenzoic acid.

Polymers of this class contain long, relatively rigid chains which are thought to undergo parallel ordering in the melt, resulting in low melt viscosity and good injection-moulding characteristics, although at relatively high melt temperatures—750 to 806°F (400 to 430°C). The melt solidifies to form tightly packed fibrous chains in the moulded parts, which give rise to exceptional physical properties. The tensile modulus of the moulded unfilled resin is 2.4×10^6 psi (16,500 MPa) at room temperature and 1.2×10^6 psi (8300 MPa) at 575°F (300°C). Tensile strength is about 20,000 psi (138 MPa), compressive strength is 6000 psi (41 MPa), and elongation is approximately 5 per cent. Mechanical properties are claimed to improve at sub-zero temperatures. The wholly aromatic co-polyester is reported to have outstanding thermal oxidative stability, with a decomposition temperature in air of 1040°F (560°C) and 1053°F (567°C) in a nitrogen atmosphere. The resin is inherently flame retardant and does not sustain combustion. Its oxygen index is 42, and smoke generation is extremely low.

The resin is extremely inert, resists attack by virtually all chemicals, including acids, solvents, boiling water, and hydrocarbons. It is attacked by concentrated, boiling caustic but is unaffected by 30 days of immersion in 10 per cent sodium hydroxide solution at 127°F (53°C). It withstands a high level of UV radiation and is transparent to microwaves.

The wholly aromatic co-polyester for injection moulding is available in filled and unfilled grades. It can be moulded into thin-wall components at high speeds. The high melt flow also enables it to be moulded into heavy-wall parts. No mould release is required because of the inherent lubricity and non-stick properties. No post-curing is necessary because the material is completely thermoplastic in nature. The material is expected to have many applications because of its mouldability and its resistance to high temperatures, fire, and chemicals.

Polycarbonates

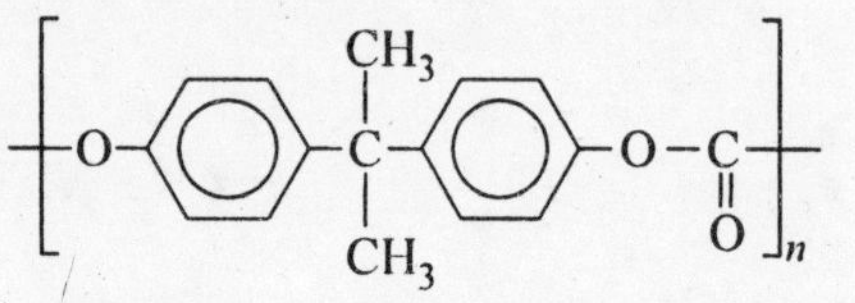

Monomers	Polymerisation	Major uses
Bisphenol A, phosgene	Interfacial polycondensation, solution polycondensation, transesterification	Glazing (37%), electrical and electronics (15%), appliances (15%), compact discs

The major processes for polycarbonate manufacture include:

(i) Transesterification of bisphenol A with diphenyl carbonate:

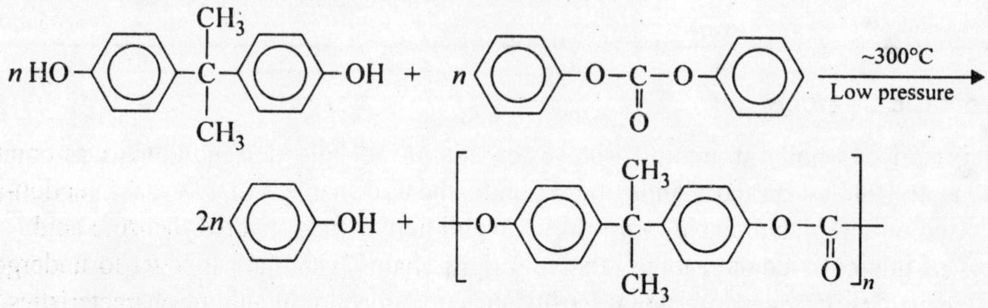

(ii) Solution phosgenation in the presence of an acid acceptor such as pyridine:

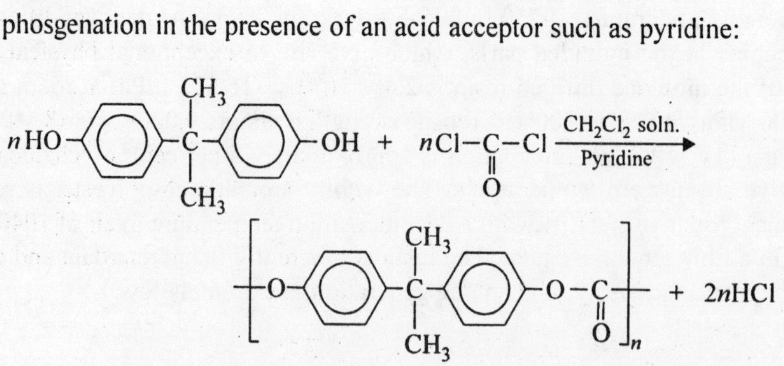

and (iii) interfacial phosgenation in which the basic reaction is the same as in solution phosgenation, but it occurs at the interface of an aqueous phase and an organic phase. Here the acid acceptor is aqueous sodium hydroxide, which dissolves the bisphenol A and a monohydric phenol used for molecular-weight control (without which very high-molecular-weight polymers of little commercial value will be obtained), and the organic phase is a solvent for phosgene and the polymer formed. A mixture of methylene chloride and chlorobenzene is a suitable solvent. The interfacial polycondensation method is the most important process af present for the production of polycarbonate. Interestingly, polycarbonate represents the first commercial application of interfacial polycondensation. Fire-retardant grades of polycarbonates are produced by using tetrabromo-bisphenol A as co-monomer.

Polycarbonate resin is easily processed by all thermoplastic-moulding methods. Although it is most often injection moulded or extruded into flat sheets, other options include blow moulding, profile extrusion, and structural foam moulding. Polycarbonate sheet can be readily thermoformed. The resin should be dried to less than 0.02 per cent moisture before processing to prevent hydrolytic degradation at the high temperatures necessary for processing.

The chemical resistance of polyester materials is generally limited due to the comparative ease of hydrolysis of the ester groups, but the bisphenol A polycarbonates are somewhat more resistant. This resistance may be attributed to the shielding of the carbonate group by the hydrophobic benzene rings on either side. The resin thus shows resistance to dilute mineral acids; however, it has poor resistance to alkali and to aromatic and chlorinated hydrocarbons.

Polycarbonates have an unusual combination of high impact strength (12 to 16 ft-lbf per inch notch for 1/2-in. × 1/8-in. bar), heat-distortion temperature (132°C), transparency, very good electrical insulation characteristics, virtually self-extinguishing nature, and physiological inertness. As an illustration of the toughness of polycarbonate resins, it is claimed that an 1/8-inch-thick moulded disc will stop a 0.22 caliber bullet, causing denting but not cracking. In crepe resistance, polycarbonates are markedly superior to acetal and polyamide thermoplastics.

Because of a small dipole polarisation effect, the dielectric constant of polycarbonates (e.g., 3.0 at 10^3 Hz) is somewhat higher than that for PTFE and the polyolefins (2.1 to 2.5 at 10^3 Hz). The dielectric constant is also almost unaffected by frequency changes up to 10^6 Hz and temperature changes over the normal range of operations. (Note that for satisfactory performance electrical insulating materials should have a low dielectric constant for low dissipation factor but high dielectric strength. For dielectrics used in capacitors, however, a high dielectric constant is desirable). At low frequencies (60 Hz) and in the ordinary temperature range (20–100°C), the power factor of polycarbonates (~0.0009) is remarkably low for a polar polymer. It increases, however, at the higher frequencies, reaching a value of 0.010 at 10^6 Hz. The polycarbonates have a high volume resistivity (2.1×10^{20} ohm-cm at 23°C) and a high dielectric strength (400 kV/in., 1/8-in. sample). Because of the low water absorption, these properties are affected little by humidity. Polycarbonates, however, do have a poor resistance to tracking.

Although the electrical properties of polycarbonates are not as impressive as those observed with polyethylene, they are adequate for many purposes. These properties, coupled with the high impact strength, heat and flame resistance, transparency, and toughness have led to the extensive use of these resins in electronics and electrical engineering, which remains the largest single field of their application. Polycarbonate is the only material that can provide such a combination of properties, at least at a reasonable cost.

Polycarbonate covers for time switches, batteries, and relays utilise the good electrical insulation characteristics in conjunction with transparency, toughness, and flame resistance of the polymer. Its combination of properties also accounts for its wide use in making coil formers. Many other electrical and electronic applications include mouldings for computers, calculating machines, and magnetic disc pack housing, contact strips, switch plates, and starter enclosures for fluorescent lamps. Polycarbonate films of high molecular weight are used in the manufacture of capacitors.

Traditional applications of polycarbonate in the medical market, such as filter housings, tubing connectors, and surgical staplers, have relied on the material's unique combination of strength, purity, transparency, and ability to stand all sterilisation methods (steam, ethylene oxide gas, and gamma radiation). Recently developed polycarbonate-based blends and co-polymers have further extended the material's usefulness to medical applications.

Recent years have seen a continuing growth of the market for polycarbonate glazing and light transmission units. Applications here include lenses and protective domes as well as glazing. The toughness and transparency of polycarbonates have led to many successful glazing applications of the polymer, such as bus shelters, telephone kiosks, gymnasium windows, lamp housings for street lighting, traffic lights, and automobiles, strip-lighting covers at ground level, safety goggles, riot-squad helmets, armour, and machine guards. The limited scratch and weathering resistance of the polycarbonates is a serious drawback in these applications, and much effort is being directed at overcoming these problems. One approach is to coat the polycarbonate sheet with a glasslike composition by using a suitable priming material to ensure good adhesion between coating and the base plastic.

Polycarbonates modified with ABS (acrylonitrile and styrene grafted onto polybutadiene) and MBS (methyl methacrylate and styrene grafted onto polybutadiene) resins have been available for many years. Usually used to the extent of 2 to 9 per cent, the styrene-based terpolymers are claimed to reduce the notch sensitivity of the polycarbonate and to improve its resistance to environmental stress cracking while retaining for some grades the high impact strength of the unmodified polycarbonate. These materials find use in the electrical industry, in the automotive industry (instrument panels and glove compartment flaps), and for household appliances (coffee machine housings, hair drier housings, and steam handles). Elastomer modified polycarbonates have been used for automobile front ends and bumpers.

Polycarbonate is presently used for making compact audio discs, which are based on digital recording and playback technology and can store millions of bits of information in the form of minute pits in an area only a few inches in diameter. It is the presence or absence of the pits which is ready by the laser. Each 'track' comprising a spiral of these pits is laid in polycarbonate which is backed with reflective aluminium and coated with a protective acrylic layer. The processability of the polycarbonate, or any other material used as the substrate, is crucial in the manufacture of all optical discs.

POLYAMIDES

The early development of polyamides started nylon-6,6 was first synthesised—a polyamide of hexamethylene diamine and adipic acid—after extensive and classical researches into condensation polymerisation.

Later polyamides was developed by ring-opening polymerisation of cyclic lactams, and nylon-6 derived from caprolactam. Today nylon-6,6 and nylon-6 account for nearly all of the polyamides produced for fibre applications.

Nylon-6,6 and nylon-6 are also used for plastics applications. Besides these two polyamides very many other aliphatic polyamides have been prepared in the laboratory, and a few of them (nylon-11, nylon-12, and nylon-6,10 in particular) have become of specialised interest as plastics materials. However, only about 10 per cent of the nylons produced are used for plastics production. Virtually all of the rest goes for the production of fibres where the market is shared, roughly equally, between nylon-6 and nylon-6,6. (Nylon is the trade name of the polyamides from unsubstituted, non-branched aliphatic monomers. A polyamide made from either an amino acid or a lactam is called nylon-x, where x is the number of carbon atoms in the repeating unit. A nylon made from a diamine and a dibasic acid is designated by two numbers, in which the first represents the number of carbons in the diamine chain and the second the number of carbons in the dibasic acid).

For a variety of technical reasons the development of aromatic polyamides was much slower in comparison. The aromatic polyamides have expanded the maximum temperature well above 200°C. High-tenacity, high-modulus polyamide fibres (aramid fibres) have provided new levels of properties ideally suited for tyre reinforcement.

More recently there has been considerable interest in some new aromatic glassy polymers, in thermoplastic polyamide elastomers, and in a variety of other novel materials.

Aliphatic Polyamides

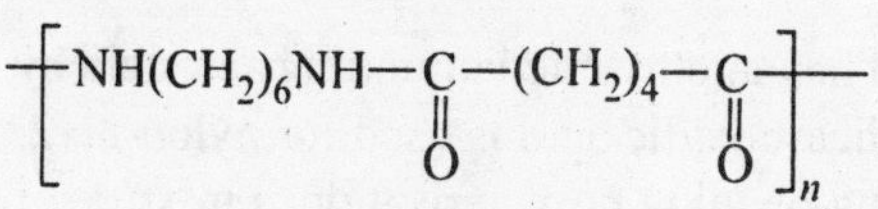

Nylon-6,6

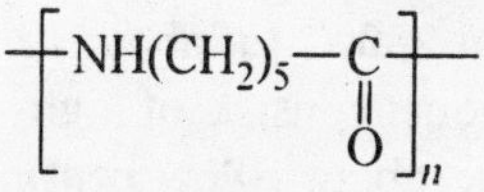

Nylon-6

Monomers	*Polymerisation*	*Major uses*
Adipic acid, hexamethylenediamine, caprolactam	Bulk polycondensation	Home furnishings, apparel, tyre cord

Aliphatic polyamides are produced commercially by condensation of diamines with dibasic acids, by self-condensation of an amino acid, or by ring-opening polymerisation of a lactam. To obtain polymers of high molecular weight, it is important that there should be stoichiometric equivalence of amine and acid groups of the monomers.

For amino acids and lactams the stoichiometric balance is ensured by the use of pure monomers; for diamines and dibasic acids this is readily obtained by the preliminary formation of a 1:1 ammonium salt, often referred to as a nylon salt. Small quantities of monofunctional compounds are often used to control the molecular weight.

The nylon-6,6 salt (melting point 190–191°C) is prepared by reacting hexamethylenediamine and adipic acid in boiling methanol, the comparatively insoluble salt precipitating out. A 60 per cent aqueous slurry of the salt together with a trace of acetic acid to limit the molecular weight to the desired level (9000–15000) is heated under a nitrogen blanket at about 220°C in a closed autoclave under a pressure of about 20 atmospheres (atm).

The polymerisation proceeds to approximately 80 to 90 per cent without removal of by-product water. The autoclave temperature is then raised to 270 to 300°C, and the steam is continuously driven off to drive the polymerisation to completion.

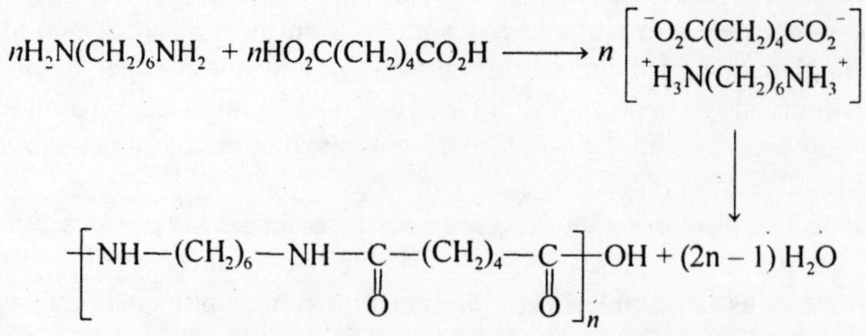

The later stages of the polymerisation reaction constitute a melt polycondensation, since the reaction temperature is above the melting point of the polyamide. The molten polymer is extruded by nitrogen pressure on to a water-cooled casting wheel to form a ribbon which is subsequently disintegrated. In a continuous process for the production of nylon-6,6 similar reaction conditions are used, but the reaction mixture moves slowly through various zones of a reactor.

Nylon-6,10 is prepared from the salt (melting point 170°C) of hexamethylenediamine sebacic acid by a similar technique. Nylon-6,9 uses azelaic acid. Decane-1,10-dicarboxylic acid is used for nylon-6,12.

In a typical batch process for the production of nylon-6 by ring-opening polymerisation, a mixture of caprolactam, water (5–10 per cent by weight), which acts as a catalyst, and a molecular-weight regulator [e.g., acetic acid (~0.1 per cent)] is heated in a reactor under a nitrogen blanket at 250°C for about 12 hours, a pressure of about 15 atm being maintained by venting off steam. The product consists of high-molecular-weight polymer (about 90 per cent) and low-molecular-weight material (about 10 per cent), which is mainly monomer. To obtain the best physical properties, the low-molecular-weight materials may be removed by leaching and/or by vacuum distillation. In the continuous process, similar reaction conditions are used. In one process a mixture of molten caprolactam, water, and acetic acid is fed continuously to a reactor operating at about 260°C. Residence time is 18 to 20 hours.

A newer technique for the preparation of nylon-6 is the polymerisation casting of caprolactam *in situ* in the mould. In this process rapid formation of polymer is achieved by anionic polymerisation, initiated by strong bases such as metal amides, metal hydrides, and alkali metals. However, the anionic polymerisation of lactams by strong bases alone is relatively slow because it is associated with an induction period due to a slow step in the initiation sequence leading to an *N*-acyl lactam which participates in the propagation reaction. The induction period may be eliminated by adding along with the strong base a preformed *N*-acyl lactam or related compound at the start of the reaction.

A typical system for the polymerisation casting of caprolactam thus uses as a catalyst 0.1 to 1 mol per cent *N*-acetyl caprolactam and 0.15 to 0.50 mol per cent of the sodium salt of caprolactam. The reaction temperature is initially about 150°C, but during polymerisation it rises to about 200°C. The technique is especially applicable to the production of large, complex shapes that could not be made by the more conventional plastics processing techniques. Important advantages of the process are the low heats and low pressures involved. Although the polymerisation process is exothermic, the relatively low heat of polymerisation of caprolactam, coupled with its low melting point, makes the process easy to control and simplifies the heat transfer problem generally associated with the production of massive parts. Mouldings of cast nylon-6 up to one tonne are claimed to have been produced by these techniques.

Nylon parts made by polymerisation casting of caprolactam exhibit higher molecular weights and a highly crystalline structure and are, therefore, slightly harder and stiffer than conventionally moulded nylon-6.

Applications for cast nylon-6 include huge gears (e.g., a 150-kg nylon gear for driving a large steel drum drier) and bearings, gasoline and fuel tanks, buckets, building shutters, and various components for paper production machinery and mining and construction equipment. Later development has centred on adding reinforcing materials to the monomer before polymerisation to produce parts with higher heat distortion temperature, impact strength and tensile strength.

Nylon-12 is produced by the ring-opening polymerisation of laurolactam (dodecyl lactam) such as by heating the lactam at about 300°C in the presence of aqueous phosphoric acid. Unlike the polymerisation of caprolactam, the polymerisation of dodecyl lactam does not involve an equilibrium reaction. Hence, an almost quantitative yield of nylon-12 polymer is obtained by the reaction, and the removal of low-molecular-weight material is unnecessary. Nylon-11 is produced by the condensation polymerisation of ω-aminoundecanoic acid at 200–220°C with continuous removal of water. The latter stages of the reaction are conducted under reduced pressure to drive the polymerisation to completion.

Nylon co-polymers can be obtained by heating a blend of two or more different nylons above the melting point so that amide interchange occurs. Initially, block co-polymers are formed, but prolonged reaction leads to random co-polymers. For example, a blend of nylon-6,6 and nylon-6,10 heated for 2 hours gives a random co-polymer (nylon-6,6—nylon-6,10) which is identical with a co-polymer prepared directly from the mixed monomers. Other co-polymers of this type are available commercially.

Properties

Aliphatic polyamides are linear polymers containing polar —CONH— groups spaced at regular intervals by aliphatic chain segments. The principal structural difference between the various types of nylon is in the length of aliphatic chain segments separating the adjacent amide groups. The polar amide groups give rise to high interchain attraction in the crystalline zones, and the aliphatic segments impart a measure of chain flexibility in the amorphous zones. This combination of properties yields polymers which are tough above their glass transition temperatures. The high intermolecular attraction also accounts for high melting points of nylons, which are usually more than 200°C. The melting point, however, decreases (which facilitates processing) as the length of the aliphatic segment in the chain increases, as indicated in Table 16.3.

Table 16.3. Melt temperatures of aliphatic polyamides.

Polyamide	T_m (°C)	*Polyamide*	T_m (°C)
Nylon-6.6	265	Nylon-6	230
Nylon-6.8	240	Nylon-7	223
Nylon-6.10	225	Nylon-11	188
Nylon-6.12	212	Nylon-12	180

Because of the high cohesive energy density and their crystalline state, the nylons are resistant to most solvents. They have exceptionally good resistance to hydrocarbons and are affected little by esters, alkyl halides, and glycols. There are only a few solvents for the nylons, of which the most common are formic acid, glacial acetic acid, phenols, and cresols. Alcohols generally have some swelling action and may dissolve some co-polymers (e.g., nylon-6,6, nylon-6,10, nylon-6). Nylons have very good resistance to alkalies at room temperature. Mineral acids attack nylons, but the rate of attack depends on the nature and concentration of acids and the type of nylon. Nitric acid is generally active at all concentrations.

Because of the presence of amide groups, the nylons absorb water. Fig. 16.3 shows how the equilibrium water absorption of different nylons varies with humidity at room temperature, and Fig. 16.4 shows how the rate of moisture absorption of nylon-6,6 is affected by the environmental conditions. Since dimensional changes may occur as a result of water absorption, this effect should be considered when dimensional accuracy is required in a specific application.

Manufacturers commonly supply data on the dimensional changes of their products with ambient humidity.

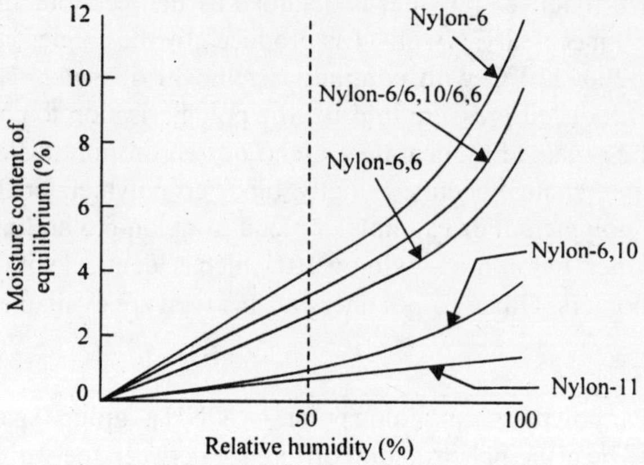

Fig. 16.3. Effect of relative humidity on the equilibrium moisture absorption of the nylons.

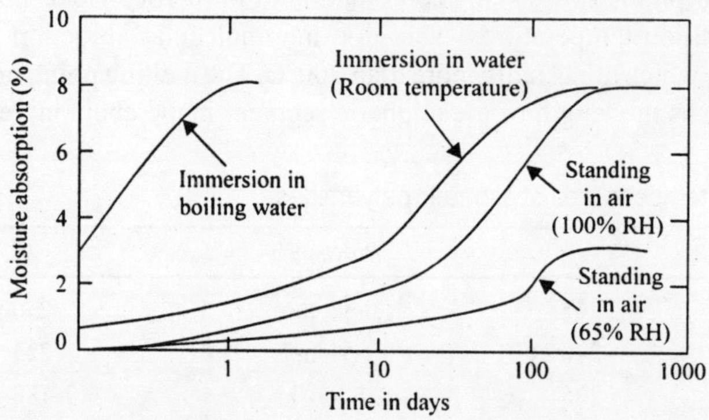

Fig. 16.4. Effect of environmental conditions on rate of moisture absorption of nylon-6,6 (1/8-inch thick specimens).

The various types of nylon have generally similar physical properties, being characterised by high toughness, impact strength, and flexibility (Table 16.4). Mechanical properties of nylons are affected significantly by the amount of crystallisation in the test piece, ambient temperature (Fig. 16.5), and humidity (Fig. 16.6), and it is necessary to control these factors carefully in the determination of comparative properties.

Moisture has a profound plasticising influence on the modulus. For example, the Young's modulus values for nylon-6,6 and nylon-6 decreases by about 40 per cent with the absorption of 2 per cent moisture.

Table 16.4. Comparative properties[a] of typical commercial grades of nylon.

Property	*6,6*	*6*	*6,10*	*11*	*12*	*6,6/6,10/6 (40:30:30)*
Specific gravity	1.14	1.13	1.09	1.04	1.02	1.09
Tensile stress at yield						
10^3 lbf/in.^{2}	11.5	11.0	8.5	5.5	6.6	–
MPa	80	76	55	38	45	–
Elongation at break (%)	80–100	100–200	100–150	300	200	300
Tensile modulus						
10^5 lbf/in.^{2}	4.3	4	3	2	2	2
10^2 MPa	30	28	21	14	14	14
Impact strength						
ft-lbf/1/2-in. notch	1.0–1.5	1.5–3.0	1.6–2.0	1.8	1.9	–
Rockwell hardness	R118	R112	R111	R108	R107	R83
Heat distortion temperature						
(264 lbf/in.^{2}) (°C)	75	60	55	55	51	30
Coefficient of linear expansion						
10^{-5} cm/cm-°C	10	9.5	15	15	12	–
Volume resistivity,						
ohm-m (dry)	> 10^{17}	> 10^{17}	> 10^{17}	–	–	–
ohm-m (50% RH)	10^{15}	–	10^{16}	–	–	10^{15}
Dielectric constant						
(10^3 Hz dry)	3.6–6.0	3.6–6.0	3.6–6.0	–	–	–
Power factor (10^3 Hz dry)	0.04	0.02–0.06	0.02	–	–	–
Dielectric strength						
(kV/cm) (25°C, 50% RH)	>100	>100	>100	–	–	–

[a] ASTM tests for mechanical and thermal properties.

Nylons have extremely good abrasion resistance. This property can be further enhanced by addition of external lubricants and by providing a highly crystalline hard surface to the bearings. The surface crystallinity can be developed by the use of hot injection moulds and by annealing in a non-oxidising fluid at an elevated temperature (e.g., 150–200°C for nylon-6,6).

The coefficient of friction of nylon-6,6 is lower than mild steel but is higher than the acetal resins. The frictional heat buildup, which determines the upper working limits for bearing applications, is related to the coefficient of friction under working conditions.

The upper working limits measured by the maximum *LS* value (the product of load *L* in psi on the projected bearing area and the peripheral speed *S* in ft/min) are 500 to 1000 for continuous operation of unlubricated nylon-6,6.

For intermittent operation initially oiled nylon bearings can be used at *LS* values of 8000. Higher *LS* values can be employed with continuously lubricated bearings.

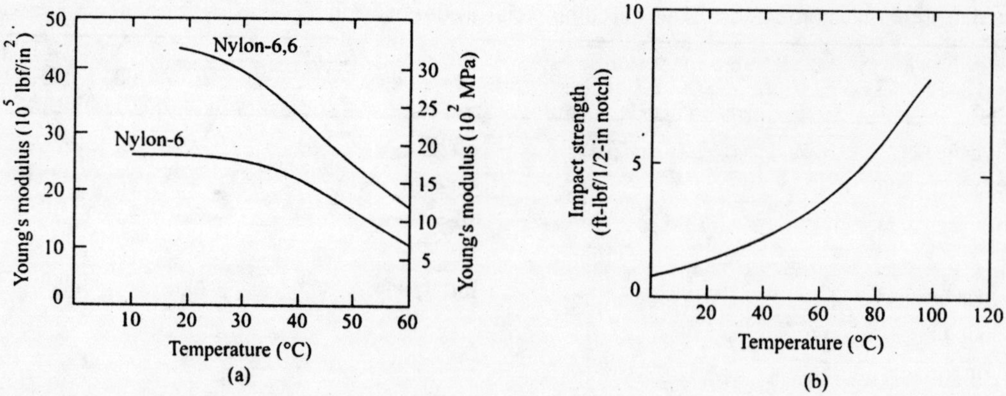

Fig. 16.5. Effect of temperature on (a) Young's modulus of nylon-6,6 and nylon-6; and (b) impact strength of nylon-6,6.

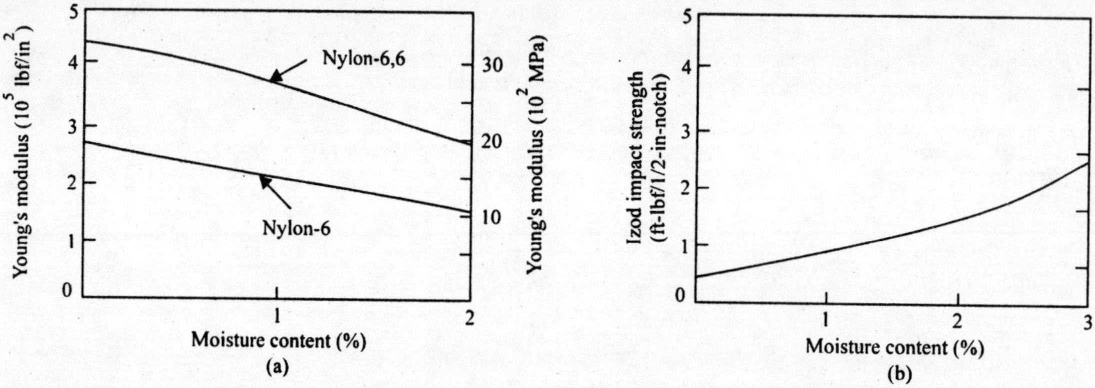

Fig. 16.6. Effect of moisture content on (a) Young's modulus of nylon-6,6 and nylon-6; and (b) impact strength of nylon-6,6.

The electrical insulation properties of the nylons are reasonably good at room temperature, under conditions of low humidity, and at low frequencies. Because of the presence of polar amide groups, they are not good insulators for high-frequency work, and since they absorb water, the electrical insulation properties deteriorate as the humidity increases (Fig. 16.7).

The properties of nylons are considerably affected by the amount of crystallisation and by the size of morphological structures, such as spherulites, which in turn, are generally influenced by the processing conditions. Thus, a moulding of nylon-6, slowly cooled and subsequently annealed, may be 50 to 60 per cent crystalline, whereas a rapidly cooled thin-walled moulding may be only 10 per cent crystalline.

Slowly cooled melts may form bigger spherulites, but rapidly cooled polymers may form only fine aggregates. Consequently, in an injection moulding the morphological form of rapidly cooled surface layers may be quite different from that of the more slowly cooled centres. The use of nucleating agents (e.g., about 0.1 per cent of a fine silica) can give smaller spherulites and thus a more uniform structure in an injection moulding. Such a product may have greater tensile strength, hardness, and abrasion resistance at the cost of some reduction in impact strength and elongation at break: the higher the degree of crystallinity the less the water absorption, and hence the less will be the effect of humidity on the properties of the polymer.

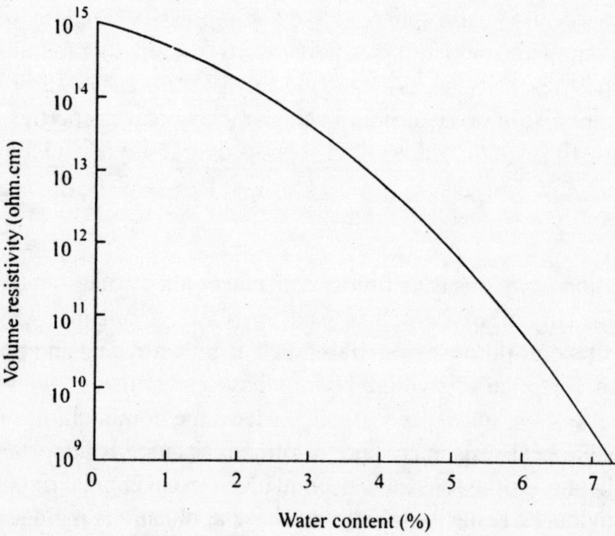

Fig. 16.7. Effect of moisture content on the volume resistivity of nylon-6,6.

Nylon moulding materials are available in a number of grades which may differ in molecular weight and/or in the nature of additives which may be present. The various types of additives used in nylon can be grouped as heat stabilisers, light stabilisers, lubricants, plasticisers, pigments, nucleating agents, flame retarders, and reinforcing fillers.

Heat stabilisers include copper salts, phosphoric acid esters, mercaptobenzothiazole, mercapto-benzimidazole, and phenyl-β-naphthyl-amine. Among light stabilisers are carbon black and various phenolic materials. Self-lubricating grades of nylon which are of value in some gear and bearing applications incorporate lubricants such as molybdenum disulphide (0.2 per cent) and graphite (1 per cent).

Plasticisers may be added to nylon to lower the melting point and to improve toughness and flexibility particularly at low temperatures. A plasticiser used commercially is a blend of *o*- and *p*-toluene ethyl sulphonamide.

Substances used as nucleating agents include silica and phosphorus compounds. Nucleating agents are used to control the size of morphological structures of the moulding.

There have been substantial efforts to improve the flame resistance of nylons. Various halogen compounds (synergised by zinc oxide or zinc borate) and phosphorus compounds have been used. They are, however, dark in colour. Glass-reinforced nylons have become available in recent years. Two main types of glass fillers used as glass fibres and glass beads. From 20 to 40 per cent glass is used. Compared to unfilled nylons, glass-fibre reinforcement leads to a substantial increase in tensile strength (160 vs. 80 MPa), flexural modulus (8000 vs. 3000 MPa), hardness, crepe resistance (at least three times as great), and heat-distortion temperature under load (245 vs. 75°C under 264 psi), and to a significant reduction in coefficient of expansion (2.8×10^{-5} vs. 9.9×10^{-5} cm/cm-°C).

The glass-fibre-filled types can be obtained in two ways. One route involves passing continuous lengths of glass fibre (as rovings) through a polymer melt or solution to produce glass-reinforced nylon strand that is chopped into pellets. Another route involves blending a mixture of resin and glass fibres about 1/4 in. (0.6 cm) long in an extruder. Usually E-grade glass with a diameter of about 0.001 cm treated with a coupling agent, such as a silane, to improve the resin-glass bond is used.

Nylons filled with 4.0 per cent of glass spheres have a compressive strength about eightfold higher than unfilled grades, besides showing good improvement in tensile strength, modulus, and heat-distortion temperature. Having low melt viscosity, glass-bead-filled nylons are easier to process than the glass-fibre-filled varieties. They are also more isotropic in their mechanical properties and show minimum warpage. Glass fillers, both fibres and beads, tend to improve self-extinguishing characteristics of nylons.

Applications

The most important application of nylons is as fibres, which account for nearly 90 per cent of the world production of all nylons. Virtually all of the rest is used for plastic applications. Because of their high cost, they have not become general-purpose materials, such as polyethylene and polystyrene, which are available at about one-third the price of nylons. Nylons have nevertheless found steadily increasing application as plastics materials for speciality purposes where the combination of toughness, rigidity, abrasion resistance, reasonable heat resistance, and gasoline resistance is important.

The largest plastics applications of nylons have been in mechanical engineering—nylon-6, nylon-6,6, nylon-6,10, nylon-11, and nylon-12 being mainly used. These applications include gears, cams, bushes, bearings, and valve seats. Zippers made of nylon last longer than traditional ones of fabric or metal. Nylon moving parts have the advantage that they may often be operated without lubrication, and they may often be moulded in one piece, they are smooth running. Among the aforesaid nylons, nylon-11 and nylon-12 have the lowest water absorption and are easy to process, but there is some loss in mechanical properties. For the best mechanical properties, the nylon-6,6 would be considered, but this material is also the most difficult to process and has high water absorption. Nylon-6 is easier to process but has even higher water absorption as already shown in Fig. 16.3.

Other applications include sterilisable nylon mouldings in medicine and pharmacy, nylon hair combs, and nylon film for packaging foodstuffs and pharmaceutical products. The value of nylon in these later applications is due to its low odour transmission and the boil-in-the-bag feature. Nylons have reasonable heat resistance. Spatula blades and spoons of nylon-6,6 withstand highest cooking temperatures.

Besides film, other extruded applications of nylons are as monofilaments, which have found applications in surgical sutures, brush tufting, wigs, sports equipment, braiding, outdoor upholstery, and angling.

Production of mouldings by polymerisation casting of caprolactam and the ability to produce large objects in this way have widened the use of nylon plastics in engineering and other applications. The process gives comparatively stress-free mouldings having a reasonably consistent morphological structure with a 45 to 50 per cent crystallinity, which is higher than melt-processed materials, and thus leads to higher tensile strength, modulus, hardness, and resistance to crepe. Products made by polymerisation casting include main drive gears for use in the textile and papermaking industries, conveyor buckets used in the mining industry, liners for coal-washing equipment, and propellers for small marine craft.

Glass-filled nylons form the most important group of glass-filled varieties of thermoplastics. Glass-reinforced nylon plastics have high rigidity, excellent crepe resistance, low coefficient of friction, high heat-deflection temperature, good low-frequency electrical insulation properties, and they are non-magnetic in nature. Therefore they have replaced metals in many applications.

Nylons reinforced with glass fibres are thus widely used in domestic appliances, in housings and casing, in car components, including radiator parts, and in the telecommunication field for relay coil formers and tag blocks. Glass-bead-filled nylons have found use in bobbins. Carbon-fibre reinforcement has been used with nylon-6 and nylon-6–nylon-12 mixtures. These materials have found use in the aerospace field and in tennis rackets. A more recent development is the appearance of supertough nylon plastics, which are blends of nylon-6,6 with other resins, such as an ionomer resin used in the initial grades or a modified ethylene-propylene-diene terpolymer rubber (EPDM rubber) used in more recent materials.

The production of nylon has been improved by blending with PPO. Nylon has also been blended with PC, HDPE, PP, SAN and ABS, PBT and polyarylates. These blends have lower water absorption than nylon-6,6. Nylon-SAN, which has a high impact strength of 16 ft-lb/in. of notch (854 J/m), has received a UL (*Underwriters*' Laboratory) rating of 104°C. The nylon-arylate blend is transparent and has a heat-distortion temperature under load (264 lbf/in^2.) value of 154°C.

Interest has been aroused by the appearance of novel elastomeric polyamides. The products introduced may be considered as the polyether-amide analogue of the polyether-ester thermoplastic elastomers. The polyether-amide is a block co-polymer prepared by the condensation of polytetramethylene ether glycol (i.e., polytetrahydrofuran) with laurin lactam and decane-1,10-dicarboxylic acid. The elastomeric polyamide is reported to have a specific gravity of 1.02, a yield stress of 24 MPa, a modulus of elasticity of 300 MPa, and an elongation at break of 360 per cent.

Aromatic Polyamides

In recent years there has been considerable interest in aromatic polyamide fibres, better known as aramid fibres. These fibres have been defined as 'a long chain synthetic polyamide in which at least 85 per cent of the amide linkages are attached directly to two aromatic rings'. It is poly(*m*-phenyleneiso-phthalamide), prepared from *m*-phenylenediamine and iso-phthaloyl chloride by interfacial polycondensation.

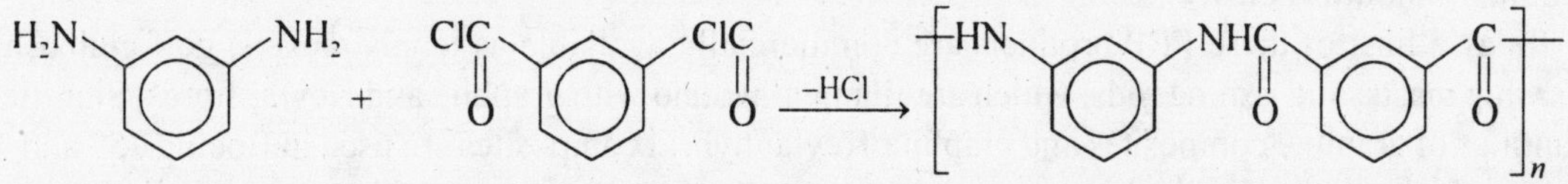

The fibre may be spun from a solution of the polymer in dimethylformamide containing lithium chloride. In 1973, *Du Pont* commenced production of another aromatic polyamide fibre, a poly(*p*-phenylene terephthalamide) marketed as Kevlar. It is produced by the reaction of *p*-phenylenediamine with terephthaloyl chloride in a mixture of hexamethylphosphoramide and *N*-methyl pyrrolidone (2:1) at –10°C.

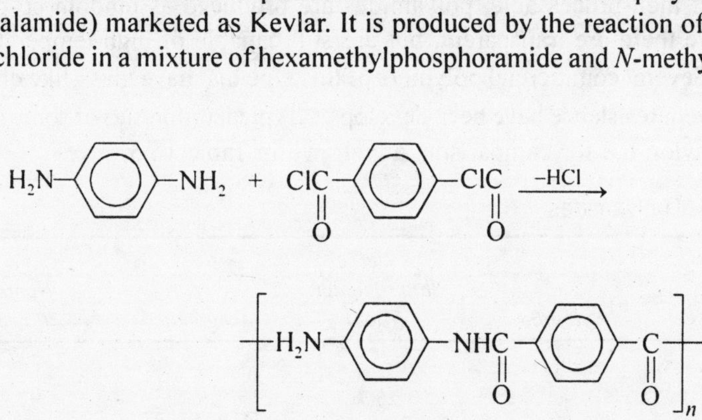

These fibres are as strong as steel but have one-fifth the weight. These are thus ideally suited as tyre cord materials and for ballistic vests. The fibres have a high T_g (> 300°C) and can be heated without decomposition to temperatures exceeding 500°C. The dimensional stability of these fibres is outstanding: It shows essentially to crepe or shrinkage as high as 200°C. In view of the high melting temperatures of the aromatic polyamides and their poor solubility in conventional solvents, special techniques are required to produce the fibres. For example, these are wet spun from a solution in concentrated sulphuric acid.

Polymers have also been prepared from cyclic amines such as piperazine and bis(*p*-aminocyclohexyl) methane. The latter amine is condensed with decanedioic acid to produce the silklike fibre Qiana.

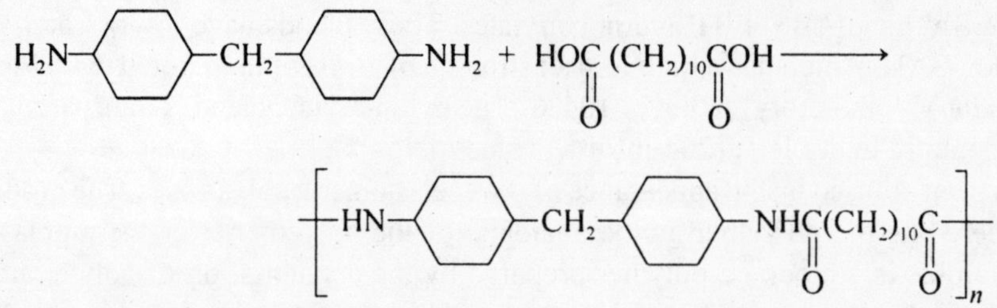

Qiana fibres have a high glass transition temperature (135°C, as compared to 90°C for nylon-6,6), which assures that the polymer will remain in the glassy state during fabric laundering and resist wrinkles and creases.

Synthetic fibres range in properties from low-modulus, high-elongation fibres to high-modulus high-tenacity fibres such as Kevlar. Another high-performance fibre in commercial application is graphite. The use of these new fibres has resulted in the development of superior composite materials, generally referred to as fibre-reinforced plastics or FRPs, which have shown promise as metal-replacement materials by virtue of their low density, high specific strength (strength/density), and high specific modulus (modulus/density).

Today a host of these FRP products are commercially available as tennis rackets, golf clubs shaft, skis, ship masts, and fishing rods, which are filament wound with graphite and Kevlar fibres. Significant quantities of graphite composites and graphite/Kevlar hybrid composites are used in Boeing 757 and 767 planes, which make possible dramatic weight saving. Boron, alumina, and silicon carbide fibres are also high-performance fibres but they are too expensive for large-scale commercial applications.

Partially aromatic, melt processable, polyamides are produced as random co-polymers, which do not crystallise and are therefore transparent, but are still capable of high-temperature use because of their high T_g values. Several commercial polymers of this type that have glass-like clarity, high softening point, and oil and solvent resistance have been developed. Typical properties of some of these polyamides, along with those of nylon-6,6 for comparison, are shown in Table 16.5.

Table 16.5. Properties of polyamides.

Property	*Nylon-6,6*	*Ixef (high-impact grade)*	*Trogamid T*	*Nomex (fibre)*	*Kevlar*
Tensile strength					
10^3 lbf/in.^{2}	9.4	25.7	8.7	97.2	435
MPa	65	177	60	670	3000
Tensile modulus					
10^5 lbf/in.^{2}	4.6	19.4	4.4	25.5	194
GPa	3.2	13.4	3.0	17.6	134

(Contd ...)

Property	*Nylon-6,6*	*Ixef (high-impact grade)*	*Trogamid T*	*Nomex (fibre)*	*Kevlar*
Elongation at break (%)	100	2.7	132	22	2.6
Flexural modulus					
10^5 lbf/in.^{2}	4.8	14.8	–	–	–
GPa	3.3	10.2	–	–	–
Impact strength, notched Izod					
ft-lb/in.	1.3	3.0	–	–	–
J/m	69	159	–	–	–

Co-polymers containing amide and imide units, the polyamideimides, are described in the following section on polyimides.

Polyimides

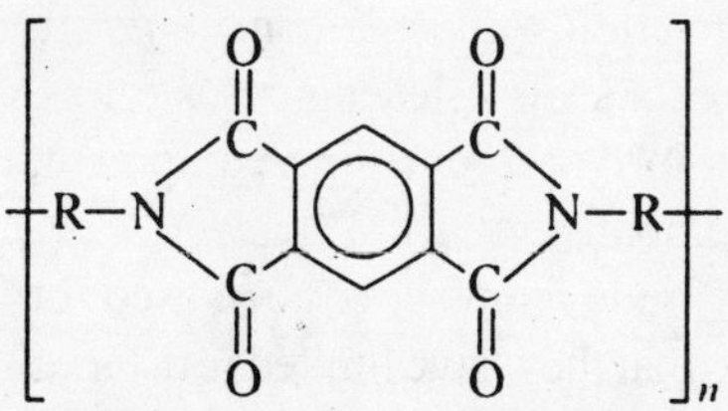

The polyimides have the characteristic functional group

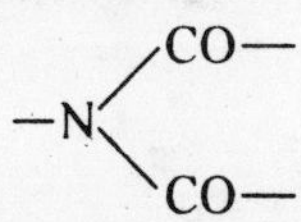

and are thus closely related to amides. The branched nature of the imide functional group enables production of polymers having predominantly ring structures in the backbone and hence high softening points. Many of the structures exhibit such a high level of thermal stability that they have become important for application at much higher service temperatures than had been hitherto achieved with polymers.

The use of tetracarboxylic acid anhydride instead of the dicarboxylic acids used in the manufacture of polyamides yields polyimides. A number of diamines have been investigated, and it has been found that certain aromatic amines, which include *m*-phenylenediamine, benzidine, and di-(4-aminophenyl)ether, give polymers with a high degree of oxidative and thermal stability.

The aromatic amine di(4-aminophenyl)ether is employed in the manufacture of polyimide film. Other commercial materials of this type included a coating resin and a machinable block form. In spite of their high price these materials have found established uses because of their exceptional heat resistance and good retention of properties at high temperatures.

Since the polyimides are insoluble and infusible, they are manufactured in two stages. The first stage involves an amidation reaction carried out in a polar solvent (such as dimethylformamide and dimethylacetamide) to produce an intermediate poly(amic acid) which is still soluble and fusible. The poly(amic acid) is shaped into the desired physical form for the final product (e.g., film, fibre, coating, laminate) and then the second stage of the reaction is carried out. In the second stage the poly(amic acid) is cyclised in the solid state to the polyimide by heating at moderately high temperatures above 150°C. A different approach, avoiding the intermediate poly(amic acid) step, has been pioneered by Upjohn. The Upjohn process involves the self-condensation of the isocyanate of trimellitic acid, and the reaction product is carbon dioxide.

Polypyromellitimides have many outstanding properties: flame resistance, excellent electrical properties, outstanding abrasion resistance, exceptional heat resistance, and excellent resistance to oxidative degradation, most chemicals (except strong bases), and high-energy radiation. After 1000 hours of exposure to air at 300°C the polymers retained 90 per cent of their tensile strength, and after 1500 hours exposure to a radiation of about 10 rad at 175°C, they retained form stability, although they became brittle.

The first commercial applications of polypyromellitimides were as wire enamels, as insulating varnishes, as coating for glass cloth. A fabricated solid grade was marketed as Vespel (*Du Pont*). Laminates were produced by impregnation of glass and carbon fibre, with the polyimide precursor followed by pressing and curing at about 200°C and further curing at temperatures of up to 350°C. Such laminates could be used continuously at temperatures up to 250°C and intermittently to 400°C. The laminates have thus found important application in the aircraft industry, particularly in connection with supersonic aircraft.

At the present time the applications of the polyimides include compressor seals in jet engines, sleeves, bearings, pressure discs, sliding and guide rolls, and friction elements in data processing equipment, value shafts in shutoff valves, and parts in soldering and welding equipment.

Polyimides have also found a number of specialist applications. Polyimide foams have been used for sound deadening of jet engines. A particular drawback of the polyimides is that they have limited resistance to hydrolysis and may crack in water or steam at temperatures above 100°C. Consequently, polyimides have now encountered competition from polyetheretherketones (PEEK), which are not only superior in this regard but are also easier to mould.

Modified polyimides

The application potential of polyimides is quite limited because, being infusible, they cannot be moulded by conventional thermoplastics techniques. In trying to overcome this limitation, scientists, in the early 1970s, developed commercially modified polyimides, which are more tractable materials than polyimides but still possessing significant heat resistance. The important groups of such modified polyimides are the polyamide-imides, the polybismaleiimides, the polyester-imides and the polyether-imides.

Other possible routes to this type of product involve the reaction of trimellitic anhydride with diisocyanates or diurethanes. Closely related is the Upjohn process for polyimide by self-condensation of the isocyanate of trimellitic acid, although the product in this case is a true polyimide rather than a polyamide-imide. Polyamide-imides may also be produced by reacting together pyromellitic dianhydride, a diamine, and a diacid chloride. Alternatively, it may be produced in a two-stage process in which a diacid chloride is reacted with an excess of diamine to produce a low-molecular-weight polyamide with amine end groups which may then be chain extended by reaction with pyromellitic dianhydride to produce imide linkages.

The Torlon (materials produced by Amoco Chemicals has high strength, stiffness, and crepe resistance, shows good performance at moderately high temperatures, and has excellent resistance to radiation. The polymers are unaffected by all types of hydrocarbons (including chlorinated and fluorinated products), aldehydes, ketones, ethers, esters, and dilute acids, but resistance to alkalies is poor.

Uses of polyamide-imides include pumps, valves, refrigeration plant accessories, and electronic components. The polymers have low coefficient of friction, e.g., 0.2 (to steel), which is further reduced to as little as 0.02 to 0.08 by blending with graphite and Teflon. In solution form in *N*-methyl-2-pyrrolidone, Torlon has been used as a wire enamel, as a decorative finish for kitchen equipment, and as an adhesive and laminating resin in spacecraft.

The polyimides and polyamide-imides are produced by condensation reactions which give off volatile low-molecular-weight by-products. The polybismaleiimides may be produced by rearrangement polymerisation with no formation of by-products. The starting materials in this case are the bismaleiimides, which are synthesised by the reaction of maleic anhydride with diamines.

The bismaleiimides can be reacted with a variety of bifunctional compounds to form polymers by rearrangement reactions. These include amines, mercaptans, and aldoximes. If the reaction is carried out with a deficiency of the bifunctional compound, the polymer will have terminal double bonds to serve as a cure site for the formation of a cross-linked polymer via a double bond polymerisation mechanism during moulding. The cross-linking in this case occurs without the formation of any volatile by-products.

Polybismaleiimides are used for making laminates with glass- and carbon-fibre fabrics, for making printed circuit boards, and for filament winding. Filled grades of polybismaleiimides are available with a variety of fillers such as asbestos, glass fibre, carbon fibre, graphite, Teflon, and molybdenum sulphide. They find application in aircraft, spacecraft, and rocket and weapons technology. Specific uses include fabrication of rings, gear wheels, friction bearings, cam discs, and brake equipment.

The polyester-imides constitute a class of modified polyimide. Polyether-imides form yet another class of modified polyimide. These are high-performance amorphous thermoplastics based on regular repeating ether and imide linkages. The aromatic imide units provide stiffness, while the ether linkages allow for good melt-flow characteristics and processability. One of several synthetic routes to polyetherimides of a general structure involves a cyclisation reaction to form the imide rings and a displacement reaction to prepare the ether linkages and form the polymer.

The first step of this synthesis is a bis-imide monomer formed by the reaction of nitrophthalic anhydride and a diamine. The second step of polyetherimide synthesis involves the formation of a bisphenol dianion by treatment of a diphenol with two equivalents of base, followed by removal of water. The polymerisation step involves displacement of the nitro groups of the bis-imide by the bisphenol dianion to form the ether linkages of the polymer. A large number of polyetherimides can be prepared by this synthetic route.

Polyetherimides are suitable for applications that require high temperature stability, high mechanical strength, inherent flame resistance with extremely low smoke evolution, outstanding electrical properties over a wide frequency and temperature range, chemical resistance to aliphatic hydrocarbons, acids and dilute bases, UV stability, and ready processability on conventional equipment.

FORMALDEHYDE RESINS

The phenol-formaldehyde and urea-formaldehyde resins are the most widely used thermoset polymers. The phenolic resins were the first truly synthetic polymers to be produced commercially. Both phenolic and urea resins are used in the highly cross-linked final form (C-stage), which is obtained by a stepwise polymerisation process. Lower-molecular-weight pre-polymers are used as precursors (A-stage resins), and the final form and shape are generated under heat and pressure. In this process water is generated in the form of steam because of the high processing temperatures. Fillers are usually added to reduce resin content and to improve physical properties. The preferred form of processing is compression moulding. The phenolics and urea resins are high-volume thermosets which owe their existence to the relatively low cost of the starting materials and their superior thermal and chemical resistance. Today these resins are widely used in moulding applications, in surface coatings and adhesives, as laminating resins, casting resins, binders and impregnants, and in numerous other applications. However, as with all products based on formaldehyde, there is concern about the toxicity of these resins during processing and about the residual traces of formaldehyde in the finished product.

Phenol-Formaldehyde Resins

Monomers	*Polymerisation*	*Major uses*
Phenol, formaldehyde	Base- or acid-catalysed stepwise polycondensation	Plywood adhesives (34%), glass-fibre insulation (19%), moulding compound (8%)

The reaction of phenol with formaldehyde involves a condensation reaction which, under appropriate conditions, leads to a cross-linked or network polymer structure. Since the cross-linked polymer is insoluble and infusible, it is necessary for commercial applications to produce first a tractable and fusible low-molecular-weight pre-polymer which may, when desired, be transformed into the cross-linked polymer. The initial phenol-formaldehyde products (pre-polymers) may be of two types: resols and novolacs.

Resols

Resols are produced by reacting a phenol with a molar excess of formaldehyde (commonly about 1:1.5 to 2) by using a basic catalyst (ammonia or sodium hydroxide). This procedure corresponds to *Baekeland's* original technique. Typically, reaction is carried out batchwise in a resin kettle equipped with stirrer and jacketed for heating and cooling. The resin kettle is also fitted with a condenser such that either reflux or distillation may take place as required. A mixture of phenol, formalin, and ammonia (1 to 3 per cent on the weight of phenol) is heated under reflux at about 100°C for 0.25 to 1 hour, and then the water formed is removed by distillation, usually under reduced pressure to prevent heat hardening of the resin.

Two classes of resins are generally distinguished. Resols prepared with ammonia as catalysts are spirit-soluble resins having good electrical insulation properties. Water-soluble resols are prepared with caustic soda as catalyst. In aqueous solutions (with a solids content of about 70 per cent) these are used mainly for mechanical grade paper and cloth laminates and in decorative laminates.

The reaction of phenol and formaldehyde in alkaline conditions results in the formation of *o*- and *p*-methylol phenols. These are more reactive towards formaldehyde than the original phenol and undergo rapid substitution with the formation of di- and trimethylol derivatives. The methylol phenols obtained are relatively stable in an alkaline medium but can undergo self-condensation to form dinuclear and polynuclear phenols (of low molecular weight) in which the phenolic nuclei are bridged by methylene groups. Thus in the base-catalysed condensation of phenol and formaldehyde, there is a tendency for polynuclear phenols, as well as mono-, di-, and trimethylol phenols to be formed. Liquid resols have an average of less than two phenolic nuclei per molecule, and a solid resol may have only three or four. Because of the presence of methylol groups, the resol has some degree of water tolerance. However, for the same reason, the shelf life of resols is limited.

Resols are generally neutralised or made slightly acidic before cure (cross-linking) is carried out. Network polymers are then obtained simply by heating, which results in cross-linking via the uncondensed methylol groups or by more complex mechanisms. Above 160°C it is believed that quinone methide groups are formed by condensation of the ether linkages with the phenolic hydroxyl groups. These quinone methide structures can be cross-linked by cycloaddition and can undergo other chemical reactions. It is likely that this formation of quinone methide and other related structures is responsible for the dark colour of phenolic compression mouldings made at higher temperatures. Note that cast phenol-formaldehyde resins, which are cured at much lower temperatures, are water white in colour. If they are heated to about 180°C, they darken considerably.

Novolac

The resols we have described are sometimes referred to as one-stage resins, since cross-linked products may be made from the initial reaction mixture only by adjusting the pH. The resol process is also known as the one-stage process. On the other hand, the novolacs are sometimes referred to as two-stage resins because, in this case, it is necessary to add, as we will show, some agent to enable formation of cross-linked products.

Novolac resins are normally prepared by the reaction of a molar excess of phenol with formaldehyde (commonly about 1.25:1) under acidic conditions. The reaction is commonly carried out batchwise in a resin kettle of the type used for resol manufacture. Typically, a mixture of phenol, formalin, and acid is heated under reflux at about 100°C. The acid is usually either hydrochloric acid (0.1 to 0.3 per cent on the weight of phenol) or oxalic acid (0.5 to 2 per cent). Under acidic conditions the formation of methylol phenols is rather slow, and the condensation reaction thus takes approximately 2 to 4 hour. When the resin reaches the requisite degree of condensation, it become hydrophobic, and the mixture appears turbid. Water is then distilled off until a cooled sample of the residual resin shows a melting point of 65 to 75°C. The resin is then discharged and cooled to give a hard, brittle solid (novolac).

Unlike resols, the distillation of water for novolac is normally carried out without using a vacuum. Therefore the temperature of the resin increases as the water is removed and the reaction proceeds, the temperature reaching as high as 160°C at the end. At these temperatures the resin is less viscous and more easily stirred.

The mechanism of phenol-formaldehyde reaction under acidic conditions is different from that under basic conditions described previously. In the presence of acid the products *o*- and *p*-methylol phenols, which are formed initially, react rapidly with free phenol to form dihydroxy diphenyl methanes. The latter undergo slow reaction with formaldehyde and phenolic species, forming polynuclear phenols by further methylolation and methylol link formation. Reactions of this type continue until all the formaldehyde has been used up. The final product thus consists of a complex mixture of polynuclear phenols linked by *o*- and *p*-methylene groups. The average molecular weight of the final product (novolac) is governed by the initial molar ratio of phenol and formaldehyde. A typical value of average molecular weight is 600, which corresponds to about six phenolic nuclei per chain. The number of nuclei in individual chains is usually 2 to 13.

A significant feature of novolacs is that they represent completed reactions and as such have no ability to continue increasing in average molecular weight. Thus there is no danger of gelation (cross-linking) during novolac production. Resols, however, contain reactive methylol groups and so are capable of cross-linking on heating.

To convert novolacs into network polymers, the addition of a cross-linking agent (hardener) is necessary. Hexamethylenetetramine (also known as hexa or hexamine) is invariably used as the hardener. The mechanism of the curing process is complex.

Because of the exothermic reaction on curing and the accompanying shrinkage, it is necessary to incorporate inert fillers to reduce resin content. Fillers also serve to reduce cost and may give additional benefits, such as improving the shock resistance. Commonly used fillers are wood flour, cotton flock, textile shreds, mica, and asbestos. Wood flour, a fine sawdust preferably from soft woods, is the most commonly used filler. Good adhesion occurs between the resin and the wood flour, and some chemical bonding may also occur. Wood flour reduces exotherm and shrinkage, improves the impact strength of the mouldings, and is cheap. For better impact strength cotton fabric or chopped fabric may be incorporated. Asbestos may be used for improved heat and chemical resistance, and iron-free mica powder may be used for superior electrical insulation resistance characteristics.

Other ingredients which may be incorporated into a phenolic moulding powder include accelerators (e.g., lime or magnesium oxide) to promote the curing reaction, lubricants (e.g., stearic acid and metal stearates) to prevent sticking to moulds, plasticisers (e.g., naphthalene, furfural, and dibutyl phthalate) to improve flow properties during cure, and pigments or dyes (e.g., nigrosine) to colour the product. Some typical formulations of phenolic moulding powders are given in Table 16.6.

Table 16.6. Typical formulations[a] by phenolic moulding grades.

Ingredient	*General-purpose grade*	*Medium shock-resisting grade*	*High shock-resisting grade*	*Electrical grade*
Novolac resin	100	100	100	100
Hexa	12.5	12.5	17	14
Magnesium oxide	3	2	2	2
Magnesium stearate	2	2	3.3	2
Nigrosine dye	4	3	3	3
Wood flour	100	–	–	–
Cotton flock	–	110	–	–
Textile shreds	–	–	150	–
Asbestos	–	–	–	40
Mica	–	–	–	120

[a] Parts by weight.

Since the phenolic resins cure with evolution of volatiles, compression moulding is performed using moulding pressures of 1 to 2 tonne/in.^{2} (15 to 30 MPa) at 155 to 170°C. Phenolic moulding compositions may be pre-heated by high frequency or other methods. Pre-heating reduces cure time, shrinkage, and required moulding pressures. It also enhances the ease of flow, with consequent reduction of mould wear and danger of damage to inserts. Moulding shrinkage of general-purpose grades is about 0.005 to 0.08 in./in. Highly loaded mineral-filled grades exhibit lower shrinkage.

Phenol-formaldehyde moulding compositions are traditionally processed on compression- and transfer-moulding machines with a very small amount being extruded. However, the injection-moulding process as modified for thermosetting plastics is being increasingly used and it is today the most common method used to process phenolic compounds.

The shorter cycle times and low waste factor available with screw injection moulding, which contribute to the lowest unit costs for extended runs, have induced phenolic moulding compounders to develop products for this moulding process.

Properties and applications

Since the polymer in phenolic mouldings is highly cross-linked and interlocked, the mouldings are hard, infusible, and insoluble. The chemical resistance of the mouldings depends on the type of resin and filler used. General-purpose PF grades are readily attacked by aqueous sodium hydroxide, but cresol- and xylenol-based resins are more resistant. Phenolic mouldings are resistant to acids except formic acid, 50 per cent sulphuric acid, and oxidising acids. The resins are ordinarily stable up to 200°C.

The mechanical properties of phenolic mouldings are strongly dependent on the type of filler used (Table 16.7). Being polar, the electrical insulation properties of phenolics as compared to aminoplasts and alkyds is their poor tracking resistance under high humidity, but this problem is not serious, as will be evident from the wide use of phenolics for electrical insulation applications.

Table 16.7. Properties of phenolic mouldings[a].

Property	*General-purpose grade*	*Medium shock-resisting grade*	*High shock-resisting grade*	*Electrical grade*
Specific gravity	1.35	1.37	1.40	1.85
Shrinkage (cm/cm)	0.006	0.005	0.002	0.002
Tensile strength				
lbf/in.^{2}	8000	7000	6500	8500
MPa	55	48	45	58
Impact strength				
ft-lbf	0.16	0.29	0.8–1.4	0.14
J	0.22	0.39	1.08–1.9	0.18
Dielectric constant				
at 800 Hz	6.0–10.0	5.5–5.7	6.0–10.0	4.0–6.0
at 10^6 Hz	4.5–5.5	–	–	4.3–5.4
Dielectric strength (20°C)				
V/mil	150–300	200–275	150–250	275–350
kV/cm	58–116	78–106	58–97	106–135
Power factor				
at 800 Hz	0.1–0.4	0.1–0.35	0.1–0.5	0.03–0.05
at 10^6 Hz	0.03–0.05	–	–	0.01–0.02
Volume resistivity (ohm-m)	10^{12}–10^{14}	10^{12}–10^{14}	10^{11}–10^{13}	10^{13}–10^{16}
Water absorption (24 hr, 23°C) mg	45–65	30–50	50–100	2–6

[a] Testing according to BS 2782.

Perhaps the most well-known applications of PF moulding compositions are in domestic plugs and switches. However, in these applications PF has now been largely replaced by urea-formaldehyde plastics because of their better antitracking property and wider range of colour possibility. (Because of the dark colour of the phenolic resins moulded above 160°C, the range of pigments available is limited to relatively darker colours—blacks, browns, deep blues, greens, reds, and oranges). Nevertheless, phenolics continue to be used as insulators in many applications because their properties have proved quite adequate. There are also many applications of phenolics where high electrical insulation properties are not as important, and their heat resistance, adequate shock resistance, and low cost are important features: for example, knobs, handles, telephones, and instrument cases. In some of these applications phenolics have been replaced by urea-formaldehyde, melamine-formaldehyde, alkyd, or newer thermoplastics because of the need for brighter colours and tougher products.

In general, phenolics have better heat and moisture resistance than urea-formaldehyde mouldings. Heat-resistance phenolics are used in handles and knobs of cookware, welding tongs, electric iron parts, and in the automobile industry for fuse box covers, distributor heads, and other applications where good electrical insulation together with good heat resistance is required. Bottle caps and closures continue to be made in large quantities from phenolics. The development of machines for injection moulding of thermosetting plastics and availability of fast-curing grades of phenolics have stimulated the use of PF for many small applications in spite of competition from other plastics.

Among the large range of laminated plastics available today, the phenolics were the first to achieve commercial significance, and they are still of considerable importance. In these applications one-stage resins (resols) are used, since they have sufficient methylol groups to enable curing without the need of a curing agent. Caustic soda is commonly used as the catalyst for the manufacture of resols for mechanical and decorative laminates. However, it is not used in electrical laminates because it adversely affects the electrical insulation properties. For electrical-grade resols ammonia is the usual catalyst, and the resins are usually dissolved in industrial methylated spirits. The use of cresylic acid (*m*-cresol content 50–55 per cent) in place of phenol yields laminating resins of better electrical properties.

In the manufacture of laminates for electrical insulation, paper (which is the best dielectric) is normally used as the base reinforcement. Phenolic paper laminates are extensively used for high-voltage insulation applications.

Besides their good insulation properties, phenolic laminates also possess good strength, high rigidity, and machinability. Sheet, tubular, and moulded laminates are employed. Phenolic laminates with cotton fabric reinforcement are used to manufacture gear wheels which are quiet running but must be used at lower working stresses than steel. Phenolic-cotton or phenolic-asbestos laminates have been used as bearings for steel rolling mills to sustain bearing loads as high as 3000 psi (21 MPa). Because of the advent of cheaper thermoplastics, cast phenolic resins (resols) are no longer an important class of plastics materials.

Urea-Formaldehyde Resins

Monomers	*Polymerisation*	*Major uses*
Urea, formaldehyde	Stepwise polycondensation	Particle-board binder resin (60%), paper and textile treatment (10%), moulding compounds (9%), coatings (7%)

Aminoresins or aminoplastics covers a range of resinous polymers produced by reaction of amines or amides with aldehydes. Two such polymers of current commercial importance in the field of plastics are the urea-formaldehyde and melamine-formaldehyde resins. Formaldehyde reacts with the amino groups to form aminomethylol derivatives which undergo further condensation to form resinous products. In contrast to phenolic resins, products derived from urea and melamine are colourless.

Urea and formaldehyde resins are usually prepared by a two-stage reaction. In the first stage, urea and formaldehyde (mole ratio in the range 1:1.3 to 1:1.5) are reacted under mildly alkaline (pH 8) conditions, leading to the production of monomethylol urea [Fig. 16.8(II)] and dimethylol urea

[Fig. 16.8 (II)]. If the product of the first stage, which in practice usually also contains unreacted urea and formaldehyde, is subjected to acid conditions at elevated temperatures (stage 2), the solution increases in viscosity and sets to an insoluble and irreversible gel. The gel eventually converts with evolution of water and formaldehyde to a hard, colourless, transparent, insoluble, and infusible mass having a network molecular structure.

The precise mechanisms involved during the second stage are not fully understood. It does appear that in the initial period of the second stage methylol ureas condense with each other by reaction of a —CH_2OH group of one molecule with an —NH_2 of another molecule, leading to linear polymers of the form shown in Fig. 16.8(III). These polymers are relatively less soluble in aqueous media and tend to form amorphous white precipitates on cooling to room temperature. More soluble resins are formed on continuation of heating. This probably involves the formation of pendant methylol groups [Fig. 16.8(IV)] by reactions of the —NH— groups with free formaldehyde. These methylol groups and the methylol groups on the chain ends of the initial reaction product can then react with each other to produce ether linkages, or with amine groups to give methylene linkages [Fig. 16.8(III)]. The ether linkages may also break down on heating to methylene linkages with the evolution of formaldehyde [Fig. 16.8(V)].

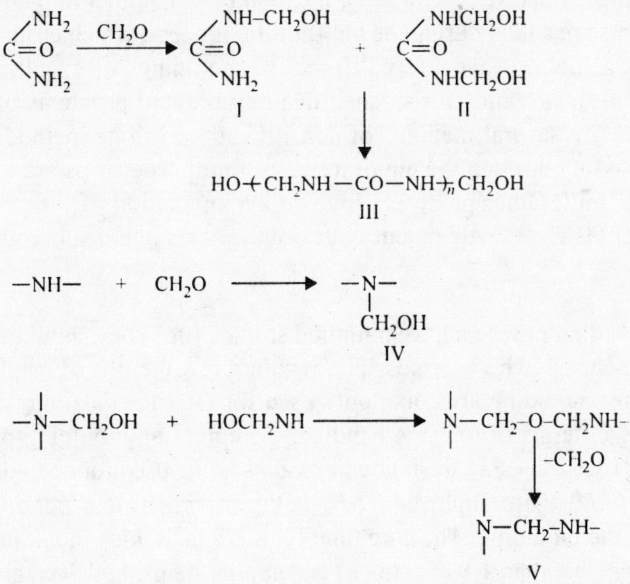

Fig. 16.8. Reactions in the formation of urea-formaldehyde resins.

Moulding powder

The urea-formaldehyde (UF) moulding powder will contain a number of ingredients. Most commonly these include resin, fiiller, pigment, accelerator, stabiliser, lubricant, and plasticiser.

Bleached wood pulp is employed as a filler for the widest range of bright colours and in slightly translucent mouldings. Wood flour, which is much cheaper, may also be used.

A wide variety of pigments is now used in UF moulding compositions. Their principal requirements are that they should be stable to processing conditions and be unaffected by service conditions of the moulding. To obtain a sufficiently high rate of cure at moulding temperatures, it is usual to add about 0.2 to 2.0 per cent of an accelerator (hardener)—a latent acid catalyst which decomposes at moulding temperatures to yield an acidic body that will accelerate the rate of cure. Many such materials have been described, the most prominent of them being ammonium sulphamate, ammonium phenoxyacetate, trimethyl phosphate, and ethylene sulphite. A stabiliser such as hexamine is often incorporated into the moulding powder to improve its shelf life.

Metal stearates, such as zinc, magnesium, or aluminium stearates are commonly used as lubricants at about 1 per cent concentration. Plasticisers (e.g., monocresyl glycidyl ether) are used in special grades of moulding powders. They enable more highly condensed resins to be used in the moulding powder and thus reduce curing shrinkage while maintaining good flow properties.

In a typical manufacturing process the freshly prepared UF first-stage reaction production is mixed with the filler (usually with a filler-resin dry weight ratio of 1:2) and other ingredients except pigment in a trough mixer at about 60°C for about 2 hour. Thorough impregnation of the filler with the resin solution and further condensation of the resin takes place during this process. Next, the wet mix is in a turbine or rotary drier for about 2 hour at 100°C or about 1 hour in a countercurrent of air at 120 to 130°C. The drying process reduces the water content from about 40 per cent to about 6 per cent and also causes further condensation of the resin.

After it is removed from the drier, the product is ground in a hammer mill and then in a ball mill for 6 to 9 hour. The pigments are added during the ball-milling process, which ensures a good dispersion of the pigment and gives a fine powder that will produce mouldings of excellent finish. The powder, however, has a high bulk factor and needs densitification to avoid problems of air and gas trappings during moulding. There are several methods of densitification. In one method, the heated powder is formed into strips by passing through the nip of a two-roll mill. The strips are then powdered into tiny flat flakes in a hammer mill. Other processes involve agglomeration of the powder by heating in an internal mixer at about 100°C or by treatment with water or steam and subsequent drying.

Processing

Urea-formaldehyde moulding powders have a limited storage life. They should therefore be stored in a cool place and should be used, wherever possible, within a few months of manufacture. Conventional compression and transfer moulding are commonly used for UF materials, the former being by far the most important process in terms of tonnage handled. Compression moulding pressures usually range from 1 to 4 tonnes/in.^{2} (15 to 60 MPa), the higher pressures being used for deep-draw articles. Moulding temperatures from 125 to 160°C are employed. The cure time necessary depends on the mould temperature and on the thickness of the moulding. The cure time for a 1/8 inch thick moulding is typically about 55 sec at 145°C. Bottle caps (less than 1/8 inch thick) and similar items, however, are moulded industrially with much shorter cure times (~10 to 20 sec) at the higher end of the moulding temperature range. For transfer moulding of UF moulding powders, pressures of 4 to 10 tonne/inch^2 (60 to 150 MPa), calculated on the area of the transfer pot, are generally recommended.

Special injection grades of UF moulding powder have been developed for injection-moulding applications which call for moulding materials with good flow characteristics between 70 and 100°C and that are unaffected by long residence time in the barrel but capable of almost instant cure in the mould cavity at a higher temperature. Although the transition from compression moulding to injection moulding has been extensive for phenolics, the same cannot be said for UF materials, because they are more difficult to mould, possibly because the UF are more brittle than a phenolic resin and so are less able to withstand the stress peaks caused by filler orientation during moulding. A combination of compression and injection processes has therefore been developed in which a screw pre-plasticising unit delivers pre-heated and softened material directly to a compression-mould cavity.

Properties and applications

The wide colour range possible with UF moulding powders has been an important reason for the widespread use of the material. These mouldings have a number of other desirable features: low cost, good electrical insulation properties, and resistance to continuous heat up to a temperature of 70°C. Some typical values of physical properties of UF moulding compositions are given in Table 16.8. They do not impart taste and odour to foodstuffs and beverages with which they come in contact and are resistant to detergents and dry-cleaning solvents.

Table 16.8. Properties of urea-formaldehyde and melamine-formaldehyde mouldings[a].

Property	*Urea-formaldehyde (α-cellulose filled)*	*Melamine-formaldehyde (cellulose filled)*
Specific gravity	1.5–1.6	1.5–1.55
Tensile strength		
10^3 lbf/in.^{2}	7.5–11.5	8–12
MPa	52–80	55–83
Impact strength (ft-lbf)	0.20–0.35	0.15–0.24
Dielectric strength (90°C) V/0.001 in.	120–200	160–240
Volume resistivity (ohm-m)	10^{13}–10^{15}	10^9–10^{10}
Water absorption (mg)		
24 hr at 20°C	50–130	10–50
30 min at 100°C	180–460	40–110

[a] Testing according to BS 2782.

The foregoing properties account for major uses of UF in two applications, namely, bottle caps and electrical fittings. It is also used for coloured toilet seats, vacuum flasks, cups and jugs, hair drier housings, toys, knobs, meat trays, switches, lamp shades, and ceiling light bowls. In the latter applications it is important to ensure adequate ventilation to prevent overheating and consequent cracking of the moulded articles.

However, only about 3 per cent of UF resins are used for moulding powders. The bulk (about 85 per cent) of the resins are used as adhesives in the particle-board, plywood, and furniture industries. Resins for these applications are commonly available with UF molar ratios ranging from 1:1.4 to 1:2.2.

To prepare a suitable resin for adhesive applications, urea is dissolved in formalin (initially neutralised to pH 7.5) to give the desired UF molar ratio. After boiling under reflux for about 15 minutes to give demethylol urea and other low-molecular products, the resin is acidified, conveniently with formic acid, to pH 4, and reacted for a further period of 5 to 20 minutes. The resulting water-soluble resin with approximately 50 per cent solids content is stabilised by neutralising to a pH 7.5 with alkali. For use as an aqueous solution, as is normally the case, the resin is then partially dehydrated by vacuum distillation to give a 70 per cent solids content.

Phosphoric acid, or more commonly ammonium chloride, is used as a hardener for UF resin adhesives. Ammonium chloride reacts with formaldehyde to produce hexamine and hydrochloric acid, and the latter catalyses the curing of the resin. In the manufacture of plywood a resin (with UF molar ratio typically 1:1.8) mixed with hardener is applied to wood veneers, which are then plied together and pressed at 95 to 110°C under a pressure of 200 to 800 psi (1.38 to 5.52 MPa). The UF resin-bonded plywood is suitable for indoor applications phenol-formaldehyde, resorcinol-formaldehyde, or melamine-formaldehyde resins are more suitable.

Large quantities of UF resin are used in general wood assembly work. For joining pieces of wood the resin-hardener solution is usually applied to the surfaces to be joined and then clamped under pressure while hardening occurs. Alternatively, the resin may be applied to one surface and the hardener to the other, allowing them to come into contact *in situ*. This method serves to eliminate pot-life problems of the resin-hardener mixture. Gap-filling resins are produced by incorporating into UF resins plasticisers, such as furfuryl alcohol, and fillers to minimise shrinkage and consequent cracking and crazing.

In the manufacture of wood chipboard, which now represents one of the largest applications of UF resins, wood chips are mixed with about 10 per cent of a resin-hardener solution and pressed in a multidaylight press at 150°C for about 8 minutes. Since some formaldehyde is released during the opening of the press, it is necessary to use a resin with a low formaldehyde content. Because it has no grain, a wood chipboard is nearly isotropic in its behaviour and so does not warp or crack. However, the water resistance of chipboard is poor.

Melamine-Formaldehyde Resins

Monomers	*Polymerisation*	*Major uses*
Melamine (trimerisation of cyanamide), formaldehyde	Stepwise polycondensation	Dinnerware, table tops, coatings

Reaction of melamine (2,4,6-triamino-1,3,5-triazine) with neutralised formalin at about 80 to 100°C leads to the production of a mixture of water-soluble methylol melamines. The methylol content of the mixture depends on the initial ratio of formaldehyde to melamine and on the reaction conditions. Methylol melamines possesing up to six methylol groups per molecule are formed.

On further heating, the methylol melamines undergo condensation reactions, and a point is reached where hydrophobic resin separates out. The rate of resinification depends on pH. The rate is minimum at about pH 10.0 to 10.5 and increases considerably both at lower and higher pH. The mechanism of resinification and cross-linking is similar to that observed for urea-formaldehyde (Fig. 16.8) and involves methylol-amine and methylol-methylol condensations.

$$\sim\sim NH{\bullet}CH_2OH + H_2N\sim\sim \longrightarrow \sim\sim NH{\bullet}CH_2{\bullet}NH\sim\sim + H_2O$$

$$\sim\sim NH{\bullet}CH_2OH + HO{\bullet}CH_2NH\sim\sim \longrightarrow \sim\sim NH{\bullet}CH_2{\bullet}O{\bullet}CH_2{\bullet}NH\sim\sim + H_2O$$

$$\sim\sim NH{\bullet}CH_2{\bullet}O{\bullet}CH_2{\bullet}NH\sim\sim \longrightarrow \sim\sim NH{\bullet}CH_2{\bullet}NH\sim\sim + CH_2O$$

In industrial practice, resinification is carried out to a point close to the hydrophobe point. This liquid resin is either applied to the substrate or dried and converted into moulding powder before proceeding with the final cure. In a typical process a jacketed resin kettle fitted with stirrer and reflux condenser is charged with 240 parts of 40% w/v formalin (pH adjusted to 8.0 to 8.5 using a sodium carbonate solution) and 126 parts of melamine (to give a melamine-formaldehyde ratio of 1:3), and the temperature is raised to 85°C. The melamine forms methylol derivatives and goes into solution. This water-soluble A-stage resin may be used for treatment of paper, leather, and fabrics to impart crease resistance, stiffness, shrinkage control, water repellency, and fire retardance. It may be spray dried to give a more stable, water-soluble product.

For laminating and other purposes the initial product is subjected to further condensation reactions at about 85°C with continuous stirring for more than 30 minutes. The hydrophilicity of the resin, as shown by its water tolerance, decreases with increasing condensation. The reaction is usually continued until a stage is reached when addition of 3 cm^3 of water will cause 1 cm^3 of resin to become turbid. The condensation reactions may be carried out at higher temperatures and lower pH values to achieve this stage more rapidly.

In aqueous solutions the hydrophobic resins have a shelf life of just a few days. The resin may be diluted with methylated spirit to about 50 per cent solids content and pH adjusted to 9.0 to 9.5 to achieve greater stability. The addition of about 0.1 per cent borax (calculated on the weight of the solids content) as an aqueous solution is useful in obtaining this pH and maintaining it for several months. The stabilised resin is stored preferably at 20 to 35°C, because too low a storage temperature will cause precipitation and too high at temperature will cause gelation.

Melamine-formaldehyde moulding powders are generally prepared by methods similar to those used for UF moulding powders. In a typical process an aqueous syrup of MF resin with melamine-formaldehyde ratio of 1:2 is compounded with fillers, pigments, lubricants, stabilisers, and accelerators in a dough-type mixer. The product is then dried and ball-milled by processes similar to those used for UF moulding powders. Alpha-cellulose is used as a filler for the more common decorative moulding powders. Industrial-grade MF materials use fillers such as asbestos, silica, and glass fibre. These fillers are incorporated by dry blending methods. The use of glass fibre gives mouldings of higher mechanical strength, improved dimensional stability, and higher heat resistance than other fillers. The mineral-filled MF mouldings have superior electrical insulation and heat resistance and may be used when phenolics and UF compositions are unsuitable.

MF mouldings are superior to UF products in lower water absorption (Table 16.8), greater resistance to staining by aqueous solutions such as fruit juices and beverages, better retention of electrical properties in damp conditions, better heat resistance, and greater hardness. Compared with the phenolic resins, MF resins have better colour range, track resistance, and scratch resistance. MF resins, are however, more expensive than general-purpose UF and PF resins.

MF compositions are easily moulded in conventional compression- and transfer-moulding equipment. Moulding temperatures from 145 to 165°C and moulding pressures 2 to 4 tonne f/in.^{2} (30 to 60 MPa) are usually employed. In transfer/moulding pressures of 5 to 10 tonne f/in.^{2} (75 to 150 MPa) are used. The cure time for an 1/8-inch-thick moulding is typically 2½ min at 150°C.

Largely because of their wide colour range, surface hardness, and stain resistance, MF resins are used as moulding compositions for a variety of mechanical parts or household goods and as laminating resins for tops for counters, cabinets, and tables. The mineral-filled moulding powders are used in electrical application and knobs and handles for kitchen utensils. An interesting application of MF resins in compression moulding involves decorative foils made by impregnating a printed or decorated grade of paper with resin and then drying. The foil may be applied to a compression moulding shortly before the cure is complete, and the resin in the foil may be cured in that position to produce a bonding.

In a typical process of laminating paper layers to make materials useful in electrical applications as well as decorative laminates (best known as formica), kraft paper, about the weight used in shopping bags, is run through a solution of melamine-formaldehyde pre-polymer. Drying out water or driving off the solvent leaves an impregnated sheet that can be handled easily, since the brittle polymer does not leave the surface sticky. As many as a dozen or more layers are piled up. For decorative purposes a printed rag or decorated cloth paper is put on top and covered with a translucent paper layer. The entire assembly is heated between smooth plates in a high-pressure press to carry out the thermosetting (curing) reaction that binds the sheets together into a strong, solvent-resistant, heat-resistant, and scratch-resistant surfacing material. The laminate, which is only about 1.5 mm thick, can be glued to a plywood base for use in tabletops, countertops, and the like.

POLYURETHANES

A urethane linkage (—NHCOO—) is formed by the reaction of an isocyanate (—NCO) and an alcohol: RNCO + R′OH→NHCOOR′. By the same reaction, polyhydroxy materials will react with polyisocyanates to yield polyurethanes. Subsequent work resulted in the production of other useful products by using polymeric hydroxyl-containing compounds such as polyesters to give rubbers, foams, adhesives, and coatings. Today polyurethanes are by far the most versatile group of polymers, because products ranging from soft thermoplastic elastomers to hard thermoset rigid foams are readily produced from liquid monomers.

The basic building blocks for polyurethanes are polyisocyanates and macroglycols, also called polyols. The commonly used polyisocyanates are toluene-diisocyanate (TDI), diphenylmethane diisocyanate or methylenediphenyl isocyanate (MDI), and polymeric methylenediphenyl isocyanate (PMDI) mixtures manufactured by phosgenating aromatic polyamines derived from the acid-catalysed condensation of aniline and formaldehyde. MDI and PMDI are produced by the same reaction, and separation of MDI is achieved by distillation. Polyether diols are low-molecular-weight polymers prepared by ring-opening polymerisation of olefin oxides, and commonly used polyester polyols are polyadipates. A polyol produced by ring-opening polymerisation of caprolactone, initiated with low-molecular-weight glycols, is also used.

Isocyanates are highly reactive materials and enter into a number of reactions with groups or molecules containing active hydrogen, such as water, amine, amide, and also urethane. Isocyanates are also toxic and care should be exercised in their use. Their main effect is on the respiratory system.

The initial research on polyurethanes was directed towards the preparation of fibre-forming polymers. Many diisocyanates and glycols were used, and the properties of the polymers were compared. From the consideration of properties and the commercial availability of reactants with the desired purity, 6,4-polyurethane produced by the reaction of hexamethylene diisocyanate and butane-1,4-diol was chosen for commercial production in Germany during World War II. The linear polyurethanes used to make fibres (known as Perlon U) can also be used as thermoplastics (marketed as Durethan U by Farbenfabriken Bayer) and may be processed by injection-moulding and extrusion techniques.

The greatest difference in properties between 6,4-polyurethane and nylon-6,6, both for fibres and moulding compositions, is in water absorption: the 6,4-polyurethane moulding absorbs only about one-sixth the moisture of nylon-6,6 under comparable conditions. Thus it has better dimensional stability and a good retention of electrical insulation properties in conditions of high humidity. However, the thermoplastic polyurethanes are costlier than nylon-6,6, and nylon-6. For engineering applications requiring low water absorption, nylon-11, acetal resin, and, in certain instances, polycarbonates are cheaper and usually as good. Polyurethane crystalline fibres and the corresponding thermoplastics are thus no longer of importance. However, elastic polyurethane fibres, commonly known as spandex fibres, are significant. They are considered in a later section.

Major polyurethane products today include cellular materials such as water-blown flexible foams or fluorocarbon-blown rigid foams, elastomers, coatings, and elastic fibres, which are described subsequently. Closely related to polyurethanes is an isocyanate-based product called isocyanurate foam.

Polyurethane Rubbers and Spandex Fibres

By careful formulations it is possible to produce polyurethane rubbers with a number of desirable properties. The rubbers can be thermoplastic (linear) or thermoset (slightly cross-linked) products.

Cross-linked polyurethane rubbers (CLPR)

The starting point in the preparation of this type of rubber is a polyester prepared by reacting a glycol such as ethylene or propylene glycol with adipic acid. The glycol is in excess so that the polyester formed has hydroxyl end groups. This polyester macroglycol is then reacted with an excess of a diisocyanate such as 1,5-naphthalene diisocyanate or MDI. The molar excess of diisocyanate is about 30 per cent, so the number of polyesters joined together is only about 2 to 3, and the resulting pre-polymer has isocyanate end groups. The pre-polymer can be chain extended with water, glycols, or amines which link up pre-polymer chains by reacting with terminal isocyanate groups. The water reaction liberates carbon dioxide, so it must be avoided in the production of elastomers, but it is important in the manufacture of foams. The urea and urethane linkages formed in the chain extension reactions also provide sites for branching and cross-linking, since these groups can react with free isocyanate or terminal isocyanate groups to form biuret and allophanate linkages, respectively. Biuret links, however, predominate since the urea group reacts faster than the urethane groups. The degree of cross-linking can to some extent be controlled by adjusting the amount of excess isocyanate, whereas more highly cross-linked structures may be produced by the use of a triol in the initial polyester.

CLPR suffer from the disadvantage that the pre-polymers are unstable and must be used within a day or two of their production. Moreover, these rubbers cannot be processed with conventional rubber machiners, so the products are usually made by a casting process. Attempts were then made to develop other polyurethane rubbers which could be processed by conventional techniques.

One approach was to react the diisocyanate with a slight excess of polyester so that the pre-polymer produced has terminal hydroxyl groups. The pre-polymers are rubberlike gums and can be compounded with other ingredients on two-roll mills. Final curing can be done by the addition of a diisocyanate or, preferably, a latent diisocyanate, i.e., a substance which produces an active diisocyanate under the conditions of moulding. Polyurethane rubbers, in general, CLPR in particular, possess certain outstanding properties. They usually have higher tensile strengths than other rubbers and possess excellent tear and abrasion resistance. The urethane rubbers show excellent resistance to ozone and oxygen (in contrast to diene rubbers) and to aliphatic hydrocarbons. However, they swell in aromatic hydrocarbons and undergo hydrolytic decomposition with acids, alkalies, and prolonged action of water and steam.

Though urethane rubbers are more costly than most of other rubbers, they are utilised in applications requiring superior toughness and resistance to tear, abrasion, ozone, fungus, aliphatic hydrocarbons, and dilute acids and bases. In addition, they excel in low-temperature impact and flexibility. Urethane rubbers have found increasing use for forklift tyres, shoes soles and heels, oil seals, diaphragms, chute linings, and a variety of mechanical applications in which high elasticity is not an important prerequisite.

Emphasis on reaction injection-moulding (RIM) technology in the automotive industry to produce automotive exterior parts has created a large potential for thermoset polyurethane elastomers. Reaction injection moulding, originally known as reaction casting, is a rapid, one-step process to produce thermoset polyurethane products from liquid monomers. In this process liquid monomers are mixed under high pressure prior to injection into the mould. The polymerisation occurs in the mould. Commercial RIM polyurethane products are produced from MDI, macroglycols, and glycol or diamine extenders. The products have the rigidity of plastics and the resiliency of rubber.

Rubber advances include short, glass-fibre reinforced, high-modulus (flexural modulus greater than 300,000 psi, i.e., 2070 MPa) polyurethane elastomers produced by the reinforced RIM process. These reinforced high-modulus polyurethane elastomers are considered for automotive door panels, trunk lids, and fender applications.

Though originally developed for the automotive industry for the production of car bumpers, the RIM process has found its greatest success in the shoe industry, where semi-flexible polyurethane foams have proved to be good soling materials.

Thermoplastic polyurethane rubbers

The reactions of polyols, diisocyanates, and glycols, as described, do tend to produce block co-polymers in which hard blocks with glass transition temperatures well above normal ambient temperature are separated by soft rubbery blocks. These polymers thus resembles the SBS triblock elastomers and, more closely, the polyether-ester thermoplastic elastomers of the Hytrel-type described earlier.

In a typical process of manufacturing thermoplastic polyurethane elastomers, a pre-polymer is first produced by reacting a polyol, such as linear polyester with terminal hydroxyl groups, or a hydroxyl-terminated polyether, of molecular weights in the range of 800 to 2500, with an excess of diisocyanate (usually of the MDI type) to give a mixture of isocyanate-terminated polyol pre-polymer and free (unreacted) diisocyanate. This mixture is then reacted with a chain extender such as 1,4-butanediol to give a polymer with long polyurethane segments whose block length depends on the extent of excess isocyanate and the corresponding stoichiometric glycol.

The polymers, thus, produced are mainly thermoplastic in nature. Though it is possible that an excess of isothiocyanate may react with urethane groups in the chain to produce allophanate cross-links, these cross-links do not destroy the thermoplastic nature of the polymer because of their thermal ability, breaking down on heating and reforming on cooling. However, where amines have been used as chain extenders, urea groups are produced which, on reaction with excess isocyanate may give the more stable biuret cross-links.

Many of the commercial materials designated as thermoplastic polyurethanes are in reality slightly cross-linked. This cross-linking may be increased permanently by a post-curing reaction after shaping. The polyurethane product may, however, be regarded as truly thermoplastic. The thermoplastic rubbers have properties similar to those of CLPR-type cast polyurethane rubbers, but they have higher values for compression set.

Thermoplastic polyurethane elastomers can be moulded and extruded to produce flexible articles. Applications include wire insulation, hose, tracks for all-terrain vehicles, solid tyres, roller skate wheels, seals, bushings, convoluted bellows, bearings, and small gears for high-load applications. In the automobile industry thermoplastic polyurethanes are used primarily for exterior parts. Their ability to be painted with flexible polyurethane-based paints without pre-treatment is valuable.

Spandex fibres

One particular form of thermoplastic polyurethane elastomers is the elastic fibre known as Spandex. It is a relatively high-priced elastomeric fibre made on the principle of segmented co-polymers. Here again the soft block is a polyglycol, and the hard block is formed from MDI (4,4′-diisocyanatodiphenyl methane) and hydrazine. This fibre-forming polymer contains urethane and semi-carbazide linkages in the chain. The product is soluble in amide solvent, and the fibre is produced by dry spinning from a solution.

Possessing higher modulus, tensile strength, resistance to oxidation, and ability to be produced at finer deniers, spandex fibres have made severe inroads into the natural rubber latex thread market. Major end uses are in apparel. Staple fibre blends with non-elastic fibres have also been introduced.

Flexible Polyurethane Foam

Although polyurethane rubbers are speciality products, polyurethane foams are well known and widely used materials. About half of the weight of plastics in modern cars is accounted for by such forms.

Flexible urethane foam is made in low densities of 1 to 1.2 lb/ft^3 (0.016 to 0.019 g/cm^3), intermediate densities of 1.2 to 3 lb/ft^3 (0.019 to 0.048 g/cm^3), high resilience (HR) foams of 1.8 to 3 lb/ft^3 (0.029 to 0.048 g/cm^3), and semi-flexible foams of 6 to 12 lb/ft^3 (0.096 to 0.192 g/cm^3). Filled foams of densities as high as 45 lb/ft^3 (0.72 g/cm^3) have also been made.

Polyols usually are polyether based, since these are cheaper, give greater ease of foam processing and provide more hydrolysis resistance than polyesters. Polyether polyols can be diols, such as polypropylene glycols with molecular weight of about 2000, or triosl with molecular weight of 3000 to 6000. The most common type of the latter is triol adduct of ethylene oxide and propylene oxide with glycerol with a molecular weight of 3000. For HR foams polyether triosl with molecular weight of 4500 to 6000, made by reacting ethylene oxide with polypropylene oxide-based triols, are used.

The reaction of isocyanate and water that evolves carbon dioxide is utilised for foaming in the production of flexible foams. The density of the product, which depends on the amount of gas evolved, can be reduced by increasing the isocyanate content of the reaction mixture and by correspondingly increasing the amount of water to react with the excess isocyanate. For greater softe ness and lower density some fluorocarbons may be added in addition to water.

Applications

The largest-volume use of flexible foam is furniture and bedding, which account for nearly 47 per cent of the flexible polyurethane foam market. Almost all furniture cushioning is polyurethane. Automobile seating is either made from flexible slabstock or poured directly into frames, so-called deep seating, using HR chemical formulations. Semi-flexible polyurethane foams find use in crash pads, arm and head rests, and door panels.

Much flexible foam is used in carpet underlay. About 43 per cent is virgin foam, and 57 per cent is scrap re-bonded with urethane adhesives under heat and pressure.

Thermal interlining can be made by flame bonding thin-sliced, low-density (1 to 1.5 lb/ft^3) polyester-based foam to fabric. The flame melts about one-third of the foam thickness, and the molten surface adheres to the fabric. Flexible foams also find use in packaging applications. Die-cut flexible foam is used to package costly goods such as delicate instruments, optical products, and pharmaceuticals. Semi-flexible foam lining is used for cart interiors to protect auto and machine parts during transportation.

Rigid and Semi-rigid Polyurethane Foams

Most rigid foam is made from polymeric isocyanate (PMDI) and difunctional polyether polyols. PMDI of functionality 2.7 (average number of isocyanate groups per molecule) is used in insulation foam manufacturing. Functionalities greater than two contribute rigidity through cross-linking. Higher-functionality, low-molecular-weight polyols are sometimes added because they contribute rigidity by cross-linking and short chain length. Such polyols are made by reaction of propylene oxide with sucrose, pentaerythritol, or sorbitol.

Rigid insulation foams are usually hydrocarbon blown to produce a closed-cell foam with excellent insulation properties. High-density foams are water blown where structural and screw-holding strength is needed, and halocarbons are used as blowing agents when high-quality, decorative surfaces are required. A compromise between these two aims can be achieved by the addition of water to halocarbon-blown formulations.

Major uses of these foams are in building and construction (56 per cent), transportation (12 per cent), furniture, and packaging. Low-density, rigid foam is the most efficient thermal insulation commercially available and is extensively used in building construction. In transportation, urethane insulation is used in rail cars, containers, truck trailer bodies, and in ships for transporting liquefied natural gas. This requires thinner insulation and so yields more cargo space. Rigid foam is also used to give flotation in barge compartments.

High-density rigid foam, 5 to 15 lb/ft^3 (0.08 to 0.24 g/cm^3), is used for furniture items such as TV and stereo cabinets, chair shells, frames, arms, and legs, cabinet drawer fronts and doors, and mirror and picture frames. RIM-moulded, integral-skin, high-density foams with core densities of 10 to 20 lb/ft^3 (0.16 to 0.32 g/cm^3) and skin densities of 55 to 65 lb/ft^3 (0.88 to 1.04 g/cm^3) are used in electronic, instrument, and computer cabinets.

Industrial uses of rigid foams include commercial refrigeration facilities as well as tanks and pipelines for cryogenic transport and storage.

Polyisocyanurates

Though closed-cell rigid polyurethane foams are excellent thermal insulators, they suffer from the drawback of unsatisfactory fire resistance even in the presence of phosphorus- and halogen-based fire retardants. In this context, polyisocyanurates, which are also based on isocyanates, have shown considerable promise.

Isocyanurate has greater flame resistance than urethane. Although rigid polyurethane is specified for use temperatures up to 200°F (93°C), rigid polyisocyanurate foams, often called trimer foams, withstand use temperatures to 300°F (149°C). Physical properties and insulation efficiency are similar for both types.

The underlying reaction for polyisocyanurate formation is the trimerisation of an isocyanate under the influence of specific catalysts. The most commonly used isocyanate is a polymeric isocyanate (PMDI) prepared by reacting phosgene with formaldehyde-aniline condensates. PMDIs are less reactive than monomeric diisocyanate but are also less volatile. Some of the catalysts used for the polytrimerisation reactions are alkali-metal phenolates, alcoholates and carboxylates, and the compounds containing *o*-(dimethylaminomethyl)-phenol groups.

To produce foams, fluorocarbons such as trichlorofluoromethanes are used as the sole blowing agents. Polyisocyanurate foams may be prepared by using standard polyurethane foaming equipment and a two-component system, with isocyanate and fluorocarbon forming one component and the activator or activator mixture forming the second component. Because of the high cross-link density of polyisocyanurates, the resultant foam tends to be brittle. Consequently, there has been a move toward making polyisocyanurate-polyurethane combinations.

For example, the isocyanate trimerisation reaction has been carried out with isocyanate end-capped TDI-based pre-polymers to make isocyanurate-containing polyurethane foams. Isocyanate trimerisation in the presence of polyols of molecular weights less than 300 has also been employed to produce foams by both one-shot and pre-polymer methods.

Polyurethane Coatings

Polyurethane systems are also formulated for surface-coating applications. A wide range of such products has become available in recent years. These include simple solutions of finished polymer (linear polyurethanes), one-component systems containing blocked isocyanates, two-component systems based on polyester and isocyanate or polyether and isocyanate, and a variety of pre-polymer and adduct systems.

Coatings based on TDI and MDI gradually discolour upon exposure to light and oxygen. In contrast, aliphatic diisocyanates such as methylene dicyclohexyl isocyanate, hexamethylene diisocyanate derivatives, and isophorone diisocyanate all produce yellowing resistant, clear, or colour-stable pigmented coatings. Of the polyols used, half are polyester type and half are polyether type. The coatings can vary considerably in hardness and flexibility, depending on formulation.

Polyurethane coatings are used wherever applications require toughness, abrasion resistance, skin flexibility, fast curing, good adhesion, and chemical resistance. Uses include metal finishes in chemical plants, wood finishes for boats and sports equipment, finishes for rubber goods, and rain-erosion resistant coatings for aircraft. The one-component coating systems require blocking of the isocyanate groups to prevent polymerisation in the container.

Generation of the blocking agent upon heating to cause polymerisation is a disadvantage of blocked one-component systems. This problem can be overcome by using a masked aliphatic diisocyanate system. The cyclic bisurea derivative used in such a system is stable in the polyol or in water emulsion formulated with the polyol at ordinary temperatures. Upon heating, ring opening occurs, generating the diisocyanate, which reacts instantaneously with the macroglycol to form a polyurethane coating.

Polyurethane adhesives involving both polyols and isocyanates are used. These materials have found major uses in the boot and shoe industry.

ETHER POLYMERS

Ether polymers or polyethers are defined as polymers which contain recurring ether groupings in their backbone structure:

$$-\overset{|}{\underset{|}{C}}-O-\overset{|}{\underset{|}{C}}-$$

Polyethers are obtained from three different classes of monomers, namely, carbonyl compounds, cyclic ethers, and phenols. They are manufactured by a variety of polymerisation processes, such as ionic polymerisation (polyacetal), ring-opening polymerisation (polyethylene oxide, polypropylene oxide, and epoxy resins), oxidative coupling (polyphenylene oxide), and polycondensation (polysulphone).

Polyacetal and polyphenylene oxide are widely used as engineering thermoplastics, and epoxy resins are used in adhesive and coating applications. The main uses of poly(ethylene oxide) and poly(propylene oxide) are as macroglycols in the production of polyurethanes. Polysulphone is one of the high-temperature-resistant engineering plastics.

Polyacetal

$$-\!\!\left[-OCH_2-\right]_n$$

Monomers	*Polymerisation*	*Major uses*
Formaldehyde, trioxane	Cationic or anionic chain polymerisation	Appliances, plumbing and hardware, transportation

Polyoxymethylene (polyacetal) is the polymer of formaldehyde and is obtained by polymerisation of aqueous formaldehyde or ring-opening polymerisation of trioxane (cyclic trimer of formaldehyde, melting, point 60 to 62°C), the latter being the preferred method. The polymerisation of trioxane is conducted in bulk with cationic initiators. In contrast, highly purified formaldehyde is polymerised in solution using either cationic or anionic initiators.

Polyacetal strongly resembles polyethylene in structure, both polymers being linear with a flexible chain backbone. Since the structures of both the polymers are regular and the question of tacticity does not arise, both polymers are capable of a high degree of crystallisation. However, the acetal polymer molecules have a shorter backbone (—C—O—) bond and so pack more closely together than polyethylene molecules. The acetal polymer is thus harder and has a higher melting point (175°C for the homopolymer).

Being crystalline and incapable of specific interaction with liquids, acetal homopolymer resins have outstanding resistance to organic solvents. No effective solvent has yet been found for temperatures below 70°C (126°F). Above this temperature, solution occurs in a few solvents such as chlorophenols. Swelling occurs with solvents of similar solubility parameter to that of the polymer [$\delta = 11.1\ (\text{cal/cm}^3)^{1/2} = 22.6\ \text{MPa}^{1/2}$]. The resistance of polyacetal to inorganic reagents is not outstanding, however. Strong acids, strong alkalies, and oxidising agents cause a deterioration in mechanical properties.

The ceiling temperature for the acetal polymer is 127°C. Above this temperature the thermodynamics indicate that depolymerisation will take place. Thus it is absolutely vital to stabilise the polyacetal resin sufficiently for melt processing at temperature above 200°C. Stabilisation is accomplished by capping the thermolabile hydroxyl end groups of the macromolecule by etherification or esterification, or by co-polymerising with small concentrations of ethylene oxide. These expedients retard the initiation or propagation steps of chain reactions that could cause the polymer to 'unzip' to monomer (formaldehyde). End-group capping is more conveniently achieved by esterification using acetic anhydride.

$$\text{HOCH}_2\!+\!\text{OCH}_2\!+_n\!\text{O—CH}_2\text{OH} \xrightarrow{\text{Ac}_2\text{O}}$$

$$\underset{\underset{\text{O}}{\|}}{\text{CH}_3\text{C}}\text{—O—CH}_2\!+\!\text{OCH}_2\!+_n\!\text{O—CH}_2\text{—O—}\underset{\underset{\text{O}}{\|}}{\text{CCH}_3}$$

If formaldehyde is co-polymerised with a second monomer, which is a cyclic ether such as ethylene oxide and 1,3-dioxolane, end-group capping is not necessary. The co-polymerisation results in occasional incorporation of molecules containing two successive methylene groups, whereby the tendency of the molecules to 'unzip' is markedly reduced.

Degradation of polyacetals may also occur by oxidative attack at random along the chain leading to chain scission and subsequent depolymerisation ('unzipping').

The principal features of acetal polymers which render them useful as engineering thermoplastics are high stiffness, mechanical strength over a wide temperature range, high fatigue endurance, resistance to crepe, and good appearance. Although similar to nylons in many respects, acetal polymers are superior to them in fatigue resistance, crepe resistance, stiffness, and water resistance (24-hour water absorption at saturation 0.22 per cent for acetal co-polymer vs. 8.9 per cent for nylon-6,6). The nylons (except under dry conditions) are superior to acetal polymers in impact toughness. Various tests indicate that the acetal polymers are superior to most other plastics and die cast aluminium.

The electrical properties of the acetal polymers may be described as good but not outstanding. They would thus be considered in applications where impact toughness and rigidity are required in addition to good electrical insulation characteristics.

The end-group capped acetal homopolymer and the trioxane-based co-polymers are generally similar in properties. The co-polymer has greater thermal stability, easier mouldability, better hydrolytic stability at elevated temperatures, and much better alkali resistance than the homopolymer. The homopolymer, on the other hand, has slightly better mechanical properties, e.g., higher flexural modulus, higher tensile strength, and greater surface hardness.

Acetal polymers and co-polymers are engineering materials and are competitive with a number of plastics materials, nylon in particular, and with metals. Acetal resins are being used to replace metals because of such desirable properties as low weight (sp. gr. 1.41 to 1.42), corrosion resistance, resistance to fatigue, low coefficient of friction, and ease of fabrication. The resins may be processed without difficulty on conventional injection-moulding, blow-moulding, and extrusion equipment. The acetal resins are used widely in the moulding of telephone components, radios, small appliances, links in conveyor belts, moulded sprockets and chains, pump housings, pump impellers, carburettor bodies, blower wheels, cams, fan blades, check valves, and plumbing components such as valve stems and shower heads.

Because of their light weight, low coefficient of friction, absence of slipstick behaviour, and ability to be moulded into intricate shapes in one piece, acetal resins find use as bearings. The lowest coefficient of friction and wear of acetal resins are obtained against steel. With other metals, aluminium in particular, high friction and greater wear are known to occur. The use of acetal to acetal in bearings is not desirable because of the tendency to build up heat except with very light loads. Where a non-metallic material has to be used, acetal and nylon in conjunction are found to give better results compared to either on its own.

Though counted as one of the engineering plastics, acetal resins with their comparatively high cost cannot, however, be considered as general-purpose thermoplastics in line with polyethylene, polypropylene, PVC, and polystyrene.

Poly(Ethylene Oxide)

$+\!\!\left[OCH_2CH_2 \right]\!\!+$

Monomer	*Polymerisation*	*Major uses*
Ethylene oxide	Ring-opening polymerisation	Molecular weights 200 to 600—surfactants, humectants, lubricants
		Molecular weights > 600—pharmaceutical and cosmetic bases, lubricants, mould release agents
		Molecular weights 10^5 to 5×10^6—water-soluble packaging films and capsules

Poly(ethylene oxide) of low molecular weight, i.e., below about 3000, are generally prepared by passing ethylene oxide into ethylene glycol at 120 to 150°C (248 to 302°F) and about 3 atm pressure (304 kPa) by using an alkaline catalyst such as sodium hydroxide. Polymerisation takes place by an anionic mechanism. The polymers produced by these methods are thus terminated mainly by hydroxyl groups and are often referred to as polyethylene glycols (PEGs). Depending on the chain length, PEGs range in physical form at room temperature from water white viscous liquids (mol. wt. 200 to 700), through waxy semi-solids (mol. wt. 1000 to 2000), to hard, waxlike solids (mol. wt. 3000 to 20,000 and above). All are completely soluble in water, bland, non-irritating, and very low in toxicity; they possess good stability, good lubricity, and wide compatibility.

Since PEGs form a homologous series of polymers, many of their properties vary in a continuous manner with molecular weight. The freezing temperature, which is less than –10°C for PEG of molecular weight 300, rises first rapidly with molecular weight through the low-molecular-weight grades, then increases more gradually through the solids while approaching 66°C, the true crystalline melting point for very high-molecular-weight poly(ethylene oxide) resins. Other examples of such continuous variations are the increase in viscosity, flash points, and fire points with an increase in molecular weight and a slower increase in specific gravity. In a reverse relationship, hygroscopicity decreases as molecular weight increases, as does solubility in water.

Commercial poly(ethylene oxide) resins supplied in the molecular weight range 1×10^5 to 5×10^6 are dry, free-flowing, white powders soluble in an unusually broad range of solvents. The resins are soluble in water at temperatures up to 98°C and also in a number of organic solvents, which include chlorinated hydrocarbons such as carbon tetrachloride and methylene chloride, aromatic hydrocarbons such as benzene and toluene, ketones such as acetone and methyl ethyl ketone, and alcohols such as methanol and isopropanol.

Polyethylene glycol (PEG)

The unusual combination of properties of PEGs has enabled them to find a very wide range of commercial uses as cosmetic creams, lotions, and dressings; textile sizes; paper-coating lubricants; pharmaceutical salves, ointments, and suppositories, softeners and modifiers; metal-working lubricants; detergent modifiers; and wood impregnants. In addition, the chemical derivatives of PEGs, such as the mono- and diesters of fatty acids, are widely used as emulsifiers and lubricants.

The PEGs themselves show little surface activity, but when converted to mono- and diesters by reaction with fatty acids, they form a series of widely useful non-ionic surfactants. The required balance of hydrophilic-hydrophobic character can be achieved by suitable combination of the molecular weight of the PEG and the nature of the fatty acid. For large-volume items a second production route of direct addition of ethylene oxide to the fatty acids is often preferred. End uses for the fatty esters are largely as textile lubricants and softeners and as emulsifiers in food products, cosmetics, and pharmaceuticals.

The PEGs have found a variety of uses in pharmaceutical products. Their water solubility, blandness, good solvent action for many medicaments, pleasant and non-greasy feel on the skin, and tolerance of body fluids are the reasons why they are frequently the products of choice. Blends of liquid and solid grades are often selected because of their desirable petrolactumlike consistency. An especially important example of pharmaceutical application of PEGs is as bases for suppositories, where the various molecular grades can be blended to provide any desired melting point, degree of stability, and rate of release of medication. The fatty acid esters of the PEGs are often used in pharmaceuticals as emulsifiers and suspending agents because of their non-ionic, blandness, and desirable surface activity. The solid PEGs also find use as lubricants and binders in the manufacture of medicinal tablets.

The PEGs, providing they contain not over 0.2 per cent ethylene and diethylene glycols, are permitted as good additives. The PEG fatty acid esters are especially useful emulsifying agents in food products.

For many of the same reasons which account for their use in pharmaceuticals, the PEGs and their fatty acid esters find many applications in cosmetics and toiletries. Their moisturising, softening, and skin-smoothing characteristics are especially useful. Typical examples of applications are shaving creams, vanishing creams, toothpastes, powders, shampoos, hair rinses, suntan lotions, pomades and dressings, deodorants, stick perfumes, rouge, mascara, and so on. The non-ionic surface-active PEG fatty acid esters find use in a variety of detergents and cleaning compositions.

Liquid PEGs, solid PEGs, or their solutions are often used in a variety of ink preparations, such as thixotropic inks for ballpoint pens, water-based stencil inks, steam-set printing inks, and stamp-pad inks.

There are numerous other industrial uses for the PEGs in which they serve primarily as processing aids and do not remain as integral components of the products. The PEGs add green strength and good formability to various ceramic compositions to be stamped, extruded, or moulded. They can be burned out cleanly during subsequent firing operations. In electroplating baths small amounts of PEGs are effective lubricants in paper-coating compositions and promote better gloss and smoothness in calendering operations. PEGs, and more particularly their fatty acid esters, are quite widely used as emulsifiers, lubricants, and softeners in textile processing. PEGs and their esters find use as components of metal corrosion inhibitors in oil wells where corrosive brines are present. Fatty acid esters of PEGs are useful demulsifiers for crude oil-water separation.

The water solubility, non-volatility, blandness, and good lubricating abilities are some of the reasons for the use of PEGs in metal-working operations. Metal-working lubricants for all but the most severe forming operations are made with the PEGs or with their esters or other derivatives.

Poly(ethylene oxide)

The reported and established applications for high-molecular-weight poly(ethylene oxide) resins have been numerous and diversified. Table 16.9 summarises in alphabetical order the main applications of these resins in various industries. In addition to their water solubility and blandness, the main functions and effects of poly(ethylene oxide) resins which lead to these diverse applications are lubrication, flocculation, thickening, adhesion, hydrodynamic drag reduction, and formation of association complexes. The resins are relatively non-toxic and have a very low level of biodegradability (low BOD). Extensive testing has indicated that poly(ethylene oxide) resins with molecular weights from 1×10^5 to 1×10^7 have a very low level of oral toxicity and are not readily absorbed from the gastrointestinal tract. The resins are relatively non-irritating to the skin and have a low sensitising potential.

Table 16.9. Applications of poly(ethylene oxide) resins.

Industry	*Applications*
Agriculture	Water-soluble seed tapes
	Water-soluble packages for agricultural chemicals
	Hydrogels as soil amendments to increase water retention
	Soil stabilisation using association complexes with poly(acrylic acid)
	Drift control agent for sprays
Ceramics and glass	Binders for ceramics
	Size for staple glass-fibre yarns
Chemical	Dispersant and stabiliser in aqueous suspension polymerisation
Electrical	Water-soluble, fugitive binder for microporous battery and fuel-cell electrodes
Metals and mining	Flocculant for removal of silicas and clays in hydrometallurgical processes
	Flocculant for clarification of effluent streams from coal-washing plants
Paper	Filler retention and drainage aid in the manufacture of paper
	Flocculant for clarification of effluent water

(Contd ...)

Industry	*Applications*
Personal-care products	Lubricant and toothpaste
	Thickener in preparation of shaving stick
	Ophthalmic solution for wetting, cleaning, cushioning, and lubricating contact lenses
	Adhesion and cushioning ingredient in denture fixatives
	Hydrogels as adsorptive pads for catamenial devices and disposable diapers
Petroleum	Thickener for bentonite drilling muds
	Thickener for secondary oil-recovery fluids in waterflooding process
Pharmaceutical	Water-soluble coating for tablets
	Suspending agent to inhibit settling of ceramic lotion
Printing	Microencapsulation of inks
Soap and detergent	Emollient and thickener for detergent bars and liquids
Textile	Additive to improve dyeability and anti-static properties of polyolefin, polyester, and polyamide fibres

Poly(ethylene oxide) resins can be formed into various shapes by using conventional thermoplastic processing techniques. Commercially, thermoplastic processing of these resins has been, however, limited almost exclusively to the manufacture of film and sheeting. Generally, the medium-molecular-weight resins (4×10^5 to 6×10^5) possess melt rheology best suited to thermoplastic processing. The films are produced by calendering or blown-film extrusion techniques. Usually produced in thicknesses from 1 to 3 mils, the films have very good mechanical properties combined with complete water solubility.

Poly(ethylene oxide) films have been used to produce seed tapes, which consist of seeds sandwiched between two narrow strips of film scaled at the edges. When the seed tape is planted, water from the soil dissolves the water-soluble film within a day or two, releasing the seed for germination. Because the seeds are properly spaced along the tape, the process virtually eliminates the need for thinning of crops.

Films for poly(ethylene oxide) are also used to manufacture water-soluble packages for pre-weighed quantities of fertilisers, pesticides, insecticides, detergents, dyestuffs, and the like. The packages dissolve quickly in water, releasing the contents. They eliminate the need for weighing and offer protection to the user from toxic or hazardous substances.

Poly(ethylene oxide) forms a water-insoluble association complex with poly(acrylic acid). This is the basis of microencapsulation of non-aqueous printing inks. Dry, free-flowing powders obtained by this process can be used to produce 'carbonless' carbon papers. When pressure is applied to the paper coated with the microencapsulated ink, the capsule wall ruptures and the ink is released.

Various water-immiscible oils and solids can be microencapsulated in the poly(ethylene oxide)-polycarboxylic acid association complex by taking advantage of the pH dependence of the complex formation. The material to be encapsulated is emulsified with a suitable non-ionic surfactant in an aqueous solution of poly(ethylene oxide). The emulsion is acidified to form the association complex at the interface, producing a coating around the water-immiscible phase and precipitation of the microcapsules, which are then dried to produce free-flowing powders.

The formulation of a water-insoluble association complex of poly(ethylene oxide) and poly(acrylic acid) is also the basis for a soil stabilisation process to prevent erosion of soil on hillsides and river banks.

A variety of different types of adhesives can be produced by forming association complexes of poly(ethylene oxide) with tannin or phenolic resins. Examples include wood glue, water-soluble quickset adhesive, and pressure-sensitive adhesives.

High-molecular-weight poly(ethylene oxide) resins are effective flocculants for many types of clays, coal suspensions, and colloidal silica, and so find application as process aids in mining and hydrometallurgy.

The turbulent flow of water through pipes and hoses or over surfaces causes the effect known as hydrodynamic drag. High-molecular-weight poly(ethylene oxide) resins are most effective in reducing the hydrodynamic drag and thus find use in fire fighting, where small concentrations of these resins (50 to 100 ppm) reduce the pressure loss in fire hoses and make it possible to deliver as much as 60 per cent more water through a standard 2.5 inch (6.35-cm)-diameter fire hose.

The ability of poly(ethylene oxide) to reduce hydrodynamic drag has also led to its use in fluid-jet systems used for cutting soft goods, such as textiles, rubber, foam, cardboard, etc. In these systems specially designed nozzles produce a very-small-diameter water jet at a pressure of 30,000 to 60,000 psi (200 to 400 MPa). Although a plain water disperses significantly as it leaves the nozzle, with poly(ethylene oxide) addition the stream becomes more cohesive and maintains its very small diameter up to 4 inch (10 cm) from the nozzle.

Chemical or irradiation cross-linking of poly(ethylene oxide) resins yield hydrogels, which are not water soluble but water absorptive, capable of absorbing 25 to 100 times their own weight of water. These hydrogels are reportedly useful in the manufacture of absorptive pads for catamenial devices and disposable diapers. The water absorbed by the hydrogels is also readily desorbed by drying the hydrogel. This characteristics is the basis for the use of these hydrogels as so-called soil amendments. When mixed with ordinary soil in a concentration of about 0.001 to 5.0 wt per cent of the soil, these hydrogels will reduce the rate of moisture loss due to evaporation but will still release water to the plants and thus eliminate the need for frequent watering of the soil.

Poly(Propylene Oxide)

$$\left[-OCH_2-\underset{\substack{|\\CH_3}}{CH}- \right]_n$$

Monomer	*Polymerisation*	*Major uses*
Propylene oxide	Base-catalysed ring-opening polymerisation	Polyols for polyurethane foams, surfactants, lubricants, cosmetic bases

Propylene oxide is polymerised by methods similar to those described in the preceding section for poly(ethylene oxide). Like the latter, low- and high-molecular-weight polymers are of commercial interest.

Poly(propylene oxide)s of low molecular weight (i.e., from 500 to 3500), often referred to as polypropylene glycols (PPGs), and important commercial materials mainly because of their extensive use in the production of polyurethane foams. PPGs are less hydrophilic and lower in cost and may be prepared by polymerising propylene oxide in the presence of propylene glycol as an initiator and sodium hydroxide as a catalyst at about 160°C. The polymers have the general structure

$$HO-\underset{\substack{|\\CH_3}}{CH}-CH_2-O\left[-CH_2-\underset{\substack{|\\CH_3}}{CH}-O- \right]_n CH_2-\underset{\substack{|\\CH_3}}{CH}-OH$$

The end hydroxy groups of the polymer are secondary groups and are ordinarily rather unreactive in the urethane reaction. Initially, this limitation was overcome by the preparation of isocyanate-terminated pre-polymer and by the use of block co-polymers with ethylene oxide. The latter products are known as tipped polyols and are terminated with primary hydroxy groups of enhanced activity.

$$HO \left[CH_2-CH_2-O \right]_x \left[CH_2-\underset{}{\overset{CH_3}{\overset{|}{CH}}}-O \right]_y \left[CH_2-CH_2-O \right]_z CH_2-CH_2-OH$$

(Note that straight PEG is not satisfactory for polyurethane foam production due to its water sensitivity and tendency to crystallise).

The latter advent of more powerful catalysts, however, made it possible for straight PPG to be used in the preparation of flexible polyurethane foams without recoruse to the foregoing procedures. Also, today, the bulk of the polyethers used are triols rather than diols, since these lead to slightly cross-linked flexible foams with improved load-bearing characteristics. The polyether triols are produced by polymerising propylene oxide by using 1,1,1-trimethylolpropane, 1,2,6-hexane triol, or glycerol as the initiator. The use of, for example, trimethylolpropane leads to the following polyether triol.

For flexible polyurethane foams, polyether triols of molecular weights from 3000 to 3500 are normally used because they give the best balance of properties. For the production of rigid foams, polyether triols of lower molecular weight (about 500) are used to increase the degree of cross-linking. Alternatively, polyether polyols of higher functionality, such as produced by polymerising propylene oxide with pentaerythritol or sorbitol, may be used.

Co-polymerisation of ethylene oxide and propylene oxide yields quite valuable functional fluids of various sorts. The random co-polymers of ethylene and propylene oxides of relatively low molecular weights are water soluble when the proportion of ethylene oxide is at least 40 to 50 per cent by weight. The block co-polymers consist of sequence of 'blocks' of all-oxypropylene or all-oxyethylene groups.

Properties vary considerably, depending on the lengths and arrangements of these blocks. The block co-polymers comprise unique and valuable surface-active agents. They can act as breakers for water-in-oil emulsions, as defoamers, and as wetting and dispersing agents.

Epoxy Resins

Monomers	*Polymerisation*	*Major uses*
Bisphenol A, epichlorohydrin	Condensation and ring-opening polymerisation	Surface coating (44%), laminates and composites (18%), moulding (9%), flooring (6%), adhesives (5%)

Epoxide or epoxy resins contain the epoxide group, also called the epoxy, oxirane, or ethoxyline group, which is a three-membered oxide ring.

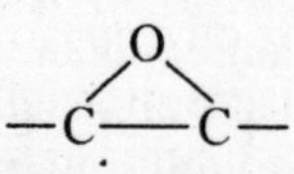

(The simplest compound in which the epoxy group is found is ethylene oxide). In the uncured stage epoxies are polymers with a low degree of polymerisation. They are most often used as thermosetting resins which cross-link to form a three-dimensional non-melting matrix. A curing agent (hardener) is generally used to achieve the cross-linking. In room-temperature curing the hardener is generally an amine such as diethylene triamine or triethylenetetramine. For elevated temperature curing a number of different curing agents could be utilised, including aromatic amines and acid anhydrides.

Epoxy resins first developed commercially and still completely dominating the worldwide markets are those based on 2,2-*bis*-(4′-hydroxyphenol) propane, more commonly known as bisphenol A, and 1-chloro-2,3-epoxy-propane, also known as epichlorohydrin. It can be seen from the general formula of these resins that the molecular species concerned is a linear polyether with terminal glycidyl ether group

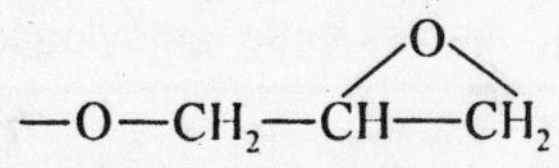

and secondary hydroxyl groups occurring at regular intervals along the length of the macromolecule. The number of repeating units (n) depends essentially on the molar ratio of bisphenol A and epichlorohydrin. When $n = 0$, the product is diglycidyl ether and the molecular weight is 340. When $n = 10$, the molecular weight is about 3000. Commercial liquid epoxy resins based on bisphenol A and epichlorohydrin have average molecular weights from 340 to 400, and it is therefore obvious that these materials are composed largely of diglycidyl ether. The liquid resins is thus often referred to as DGEBA (diglycidyl ether of bisphenol A).

DGEBA may be reacted with additional quantities of bisphenol A in an advancement reaction. This advancement produces higher-molecular-weight solid resins possessing a higher melting point (> 90°C). Advancement generally increases flexibility, improves salt fog corrosion resistance, and increases hydroxyl content which can be utilised later for cross-linking. Possessing generally low functionality (number of epoxy groups), their major use is in coatings. They provide outstanding adhesion and good salt fog corrosion resistance.

Commercial solid epoxy resins seldom have average molecular weights exceeding 4000, which corresponds to an average value of n of about 13. Resins with molecular weights above 4000 are of limited use since their high viscosity and low solubility make subsequent processing difficult.

Resin preparation

The molecular weights of epoxy resins depend on the molar ratio of epichlorohydrin and bisphenol A used in their preparation (Table 16.10). In a typical process for the production of liquid epoxy resins, epichlorohydrin and bisphenol A in the molar ratio of 10:1 are added to a stainless steel kettle fitted with a powerful anchor stirrer. The water content of the mixture is reduced to below 2 per cent by heating the mixture until the epichlorohydrin-water distils off. After condensation, the epichlorohydrin layer is returned to the kettle, the water being discarded.

When the necessary water content of the reaction mixture is reached, the reaction is started by the slow addition of sodium hydroxide (2 mole per mole of bisphenol A) in the form of 40 per cent aqueous solution, the temperature being maintained at about 100°C. The water content of the reaction mixture should be maintained between 0.31 and 2 per cent by weight throughout the reaction if high yields (90 to 95 per cent) are to be obtained. (No reaction occurs under anhydrous conditions, and undesired by-products are formed if the water content is greater than 2 per cent). Besides distilling off the water as an azeotrope with the epichlorohydrin, the water content can also be partly controlled by the rate of addition of the caustic soda solution.

When all the alkali has been added, which may take 2 to 3 hours, the excess epichlorohydrin is recovered by distillation at reduced pressure. A solvent such as toluene or methyl isobutyl ketone is then added to the cooled reaction product to dissolve the resin and leave the salt formed in the reaction.

Table 16.10. Effect of reactant ratio on molecular weight of epoxy resins.

Molar ratio epichlorohydrin/ bisphenol A	*Molecular weight*	*Epoxide equivalent*[a]	*Softening point (°C)*
10:1	370	192	9
2:1	451	314	43
1.4:1	791	592	84
1.33:1	802	730	90
1.25:1	1133	862	100
1.2:1	1420	1176	112

[a] This is the weight of resin (in grams) containing one epoxide equivalent. For a pure diglycidyl ether (mol. wt. 340) with two epoxy groups per molecule, epoxide equivalent = 340/2 = 170.

The resin solution is washed with hot water, and after filtration and further washing, the solvent is removed by distillation and the resin is dried by heating under vacuum.

Though the pure diglycidyl ether of bisphenol A is a solid (m.p. 43°C), the commercial grades of the resin which contain a proportion of high-molecular-weight materials are supercooled liquids with viscosities of about 100 to 140 poise at room temperature.

The high-molecular-weight epoxy resins usually manufactured have values of n in the general formula ranging from 2 to 12. These resins are synthesised by allowing epichlorohydrin and bisphenol A to interact in the presence of excess sodium hydroxide.

In the *taffy process* usually employed, a mixture of bisphenol A and epichlorohydrin (the molar ratio of the reactants used depends on the resin molecular weight required, (Table 16.10) is heated to 100°C and aqueous sodium hydroxide (NaOH-epichlorohydrin molar ratio 1.3:1) is added slowly with vigorous stirring, the reaction being completed in 1 to 2 hour.

A white puttylike taffy (which is an emulsion of about 30 per cent water in resin and also contains salt and sodium hydroxide) rises to the top of the reaction mixture. The lower layer of brine is removed; the resinous layer is coagulated and washed with hot water until free from alkali and salt. The resin is then dried by heating and stirring at 150°C under reduced pressure, poured into cooling pans, and subsequently crushed and bagged.

To avoid the difficulty of washing highly viscous materials, higher-molecular-weight epoxy resins may also be prepared by a two-stage process (advancement reaction, described earlier). This process consists of fusing a resin of lower molecular weight with more bisphenol A at about 190°C. The residual base content of the resin is usually sufficient to catalyse the second-stage reaction between the phenolic hydroxyl group and the epoxy group.

Solid resins have been prepared with narrow molecular-weight distributions. These resins melt sharply to give low-viscosity liquids, which enables incorporation of larger amounts of fillers with a consequent reduction in cost and coefficient of expansion. These resins are therefore useful in casting operations.

Curing

To convert the epoxy resins into cross-linked structures, it is necessary to add a curing agent. Most of the curing agents in common use can be classified into three groups: tertiary amines, polyfunctional amines, and acid anhydrides.

Examples of tertiary amines used as curing agents for epoxy resins include: (i) triethylamine (TEA); (ii) benzyldimethylamine (BDA); (iii) dimethylaminomethylphenol (DMAMP); (iv) tri(dimethylaminomethyl) phenol (TDMAMP); and (v) tri-2-ethylhexoate salt of TDMAMP.

A number of aromatic amines also function as epoxy hardeners. Since they introduce the rigid benzene ring structure into the cross-linked network, the resulting products have significantly higher heat-distortion temperatures than are obtainable with the aliphatic amines. For example, metaphenylenediamine (MPDA) gives cure resins with a heat distortion temperature of 150°C and very good chemical resistance. The hardener finds use in the manufacture of chemical resistance laminates.

Cyclic acid anhydrides are widely employed as curing agents for epoxy resins. Compared with amines they are less skin sensitive and generally give lower exotherms in curing reaction. Some acid curing agents provide cured resins with very high heat-distortion temperatures and with generally good mechanical, electrical, and chemical properties.

The acid-cured resins, however, show less alkali resistance than amine-cured resins because of the susceptibility of ester groups to hydrolysis. Anhydride curing agents find use in most of the important applications of epoxy resins, particularly in casting and laminates.

In practice, acid anhydrides are preferred to acids since the latter are generally less soluble in the resin and also release more water on cure, leading to foaming of the product. Care must be taken, however, during storage since the anhydrides in general are somewhat hygroscopic.

The mechanism of hardening by anhydride is complex. In general, however, two types of reactions occur: (i) opening of the anhydride ring with the formation of carboxy groups; and (ii) opening of the epoxy ring.

Phthalic anhydride is the cheapest anhydride curing agent, but it has the disadvantage of being rather difficult to mix with the resin. Liquid anhydrides (e.g., dodecenylsuccinic anhydride and nadic methyl anhydride), low-melting anhydrides (e.g., hexahydrophthalic anhydride), and eutectic mixtures are more easily incorporated into the resin.

Since maleic anhydride produces brittle products, it is seldom used by itself and is used as a secondary hardener in admixture with other anhydrides. Dodecenylsuccinic anhydride imparts flexibility into the casting, whereas chlorendic anhydride confers flame resistance. Anhydride-cured resins generally have better thermal stability. Pyromellitic dianhydride with higher functionality produces tightly cross-linked products of high heat-distortion temperatures. Heat-distortion temperatures as high as 290°C have been quoted.

In addition to the amine and anhydride hardeners, many other curing agents have been made available. Among them are the so-called fatty polyamides. These polymers are of low molecular weight (2000 to 5000) and are prepared by treating dimer acid (which is a complex mixture consisting of 60 to 75 per cent dimerised fatty acids together with lesser amounts of trimerised acids and higher polymers) with stoichiometric excess of ethylendiamine or diethylenetriamine so that the resultant amides have free amine groups.

Fatty polyamides are used to cure epoxy resins where a more flexible product is required, particularly in adhesive and coating applications. An advantage of the system is that roughly similar quantities of hardener and resin are required and, because it is not critical, metering can be done visually without the need of measuring aids.

They thus form the basis of some domestic adhesive systems. Also used with epoxy resins for adhesives is dicyanodiamide, $H_2N{\cdot}C(=NH)NH{\cdot}CN$. It is insoluble in common resins at room temperature but is soluble at elevated temperatures and thus forms the basis of a one-pack system.

Other epoxies

Resins containing the glycidyl ether group

$$(\text{—O—CH}_2\text{—}\overset{\text{O}}{\widehat{\text{CH—CH}_2}})$$

result from the reaction of epichlorohydrin and hydroxy compounds. Although bisphenol A is the most commonly used hydroxyl compound, a few glycidyl ether resins based on other hydroxy compounds are also commercially available, including novolac epoxies, polyglycol epoxies, and halogenated epoxies.

A typical commercial novolac epoxy resin produced by the epoxidation of phenolic hydroxyl groups of novolac by treatment with epichlorohydrin has an average molecular weight of 650 and contains about 3.6 epoxy groups per molecule. Because of higher functionality, the novolac epoxy resins give, on curing, more tightly cross-linked products than the bisphenol A-based resins. This results in greater thermal stability, higher heat-deflection temperatures, and improved chemical resistance. Their main applications have been in high-temperature adhesives, heat-resistant structural laminates, 'electrical' laminates resistant to solder baths, and chemical-resistant filament-wound pipes. The use of novolac epoxies has been limited, however, by their high viscosity and consequent handling difficulties.

Polymeric glycols such as polypropylene glycol may be epoxidised through the terminal hydroxy groups to give diglycidyl ethers.

Epoxies containing halogens have flame-retardant properties and may be prepared from halogenated hydroxy compounds. Halogenated epoxies are available based on tetrabromobisphenol A and tetrachlorobisphenol A.

The brominated resins are more effective than the chlorinated resins and have become more predominant commercially. The ability of the resins to retard or extinguish burning is due to the evolution of hydrogen halide at elevated temperatures. Brominated epoxy resins are generally blended with other epoxy resins to impart flame retardance in such applications as laminates and adhesives.

Non-glycidyl ether epoxy resins are usually prepared by treating unsaturated compounds with peracetic acid.

Generally, acid anhydrides are the preferred curing agents since amines are less effective. A hydroxy compound, such as ethylene glycol, is often added as initiator.

Because of their more compact structure, cycloaliphatic resins produce greater density of cross-links in the cured products than bisphenol A-based glycidyl resins. This generally leads to higher heat-distortion temperatures and to increased brittleness. The products also are clearly superior in arc resistance and are track resistant. Thus although bisphenol A-based epoxies decompose in the presence of a high-temperature arc to produce carbon which leads to tracking and insulation failure, cycloaliphatic epoxies oxidise to volatile products which do not cause tracking. This has led to such applications as heavy-duty electrical castings and laminates, tension insulators, rocket motor cases, and transformer encapsulation.

Acyclic aliphatic resins differ from cyclic aliphatic resins in that the basic structure of the molecules in the former is a long chain, whereas the latter, as shown, contains ring structures. Two types of acyclic aliphatic epoxies are commercially available, namely, epoxidised diene polymers and epoxidised oils. Typical of the epoxidised diene polymers are the products produced by treatment of polybutadiene with peracetic acid. Epoxidised diene polymers are not very reactive toward amines but may be cross-linked with acid hardeners. Cured resins have substantially higher heat-distortion temperatures (typically, 250°C) than do the conventional amine-cured diglycidyl ether resins.

Epoxidised oils are obtained by treatment of drying and semi-drying oils (unsaturated), such as linseed and soybean oils, with peracetic acid. Epoxidised oils find use primarily as plasticisers—stabilisers for PVC.

Applications

Epoxy resins have found a wide range of applications and a steady rate of growth over the years mainly because of their versatility. Properties of the cured products can be tailored by proper selection of resin, modifier, cross-linking agent, and the curing schedule. The main attributes of properly cured epoxy systems are outstanding adhesion to a wide variety of substrates, including metals and concrete; ability to cure over a wide temperature range; very low shrinkage on cure; excellent resistance to chemicals and corrosion; excellent electrical insulation properties; and high tensile, compressive, and flexural strengths. In general, the toughness, adhesion, chemical resistance, and corrosion resistance of epoxies suit them for protective coating applications. It is not surprising that about 50 per cent of epoxy resins are used in protective coating applications.

Two types of epoxy coatings are formulated: those cured at ambient temperature and those that are heat cured. The first type uses amine hardening systems, fatty acid polyamides, and polymer-captans as curing agents. Very high cure rates may be achieved by using mercaptans. Heat-cured types use acid anhydrides and polycarboxylic acids as well as formaldehyde resins as curing agents.

Typical coating applications for phenol-formaldehyde resin-modified epoxies include food and beverage can coatings, drum and tank liners, internal coatings for pipes, wire coating, and impregnation varnishes. Urea-formaldehyde resin-modified epoxies offer better colour range and are used as appliance primers, can linings, and coatings for hospital and laboratory furniture.

Environmental concerns have prompted major developmental trends in epoxy coating systems. Thus epoxy coating systems have been developed to meet the high-temperature and hot-acid environments to which SO_2 scrubbers and related equipment are subjected. Ambient curable epoxy systems have been developed to provide resistance to concentrated inorganic acids.

In the development of maintenance and marine coatings, the emphasis has been on the development of low-solvent, or solvent-free, coatings to satisfy EPA volatile organic content standards. This liquid epoxy systems based on polyamidamines have been developed. Epoxies have been formulated as powdered coatings, thus completely eliminating solvents.

Pipe coatings still represent a major market for epoxies. High-molecular-weight powdered formulations are used in this application.

The important maintenance coating area, particularly for pipe and tank coatings, is served by epoxy systems cured with polyamine or polyamidamines. Two-component, air-dried, solventless systems used in maintenance coatings provide tough, durable, non-porous surfaces with good resistance to water, acids, alkalies, organic solvents, and corrosion.

Emulsifiable epoxies of varying molecular weights, water-dilutable modified epoxies, and dispersions of standard resins represent promising developments for coating applications. Two-component waterborne systems are also finding use as architectural coatings.

For electrical and electronic applications epoxy formulations are available with low or high viscosity, unfilled or filled, slow or fast curing at low or high temperatures. Potting, encapsulation, and casting of transistors, integrated circuits, switches, coils, and insulators are a few electrical applications of epoxies. With their adhesion to glass, electrical properties, and flexural strength, epoxies provide high-quality printed circuit boards. Epoxies have been successfully used in Europe for outdoor insulators, switchgear, and transformers for many years. In these heavy electrical applications, the advantage of cycloaliphatic epoxies over porcelain has been demonstrated.

In a relatively new development, epoxy photopolymers have been used as solder masks and photoresists in printed circuit board fabrications.

Adhesion properties of epoxies, complete reactivity with no volatiles during cure, and minimal shrinkage make the materials outstanding for adhesives, particularly in structural applications. The most commonly known adhesive applications involve the two-component liquids or pastes, which cure at room or elevated temperatures. A novel, latent curing system, which gives more than one year pot life at room temperature, has increased the use of epoxies for speciality adhesives and sealants, and for vinyl plastisols. The one-pack system provides fast cure when heated—for example 5 minute at 100°C.

Epoxies are used in fibre-reinforced composites, providing high strength-to-weight ratios and good thermal and electrical properties. Filament-wound epoxy composites are used for rocket motor casings, pressure vessels, and tanks. Glass-fibre reinforced epoxy pipes are used in the oil, gas, mining, and chemical industries.

Sand-filled epoxies are used in industrial flooring. Resistance to a wide variety of chemicals and solvents and adhesion to concrete are key properties responsible for this use. Epoxies are also used in patching concrete highways. Decorative flooring and exposed aggregate systems make use of epoxies because of their low curing shrinkage, and the good bonding of glass, marble, and quartz chip by the epoxy matrix.

Poly(Phenylene Oxide)

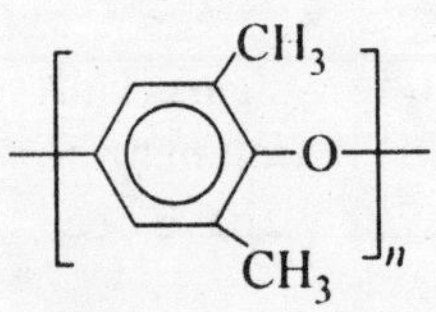

Monomer	*Polymerisation*	*Major uses*
2,6-dimethylphenol	Condensation polymerisation by oxidative coupling	Automotive, appliances, business machine cases, electrical components

Poly(2,6-dimethyl-1,4-phenylene oxide), commonly called poly(phenylene oxide) or PPO, was introduced commercially in 1964. PPO is manufactured by oxidation of 2,6-dimethyl phenol in solution using cuprous chloride and pyridine as catalyst. The monomer is obtained by the alkylation of phenol with methanol. End-group stabilisation with acetic anhydride improves the oxidation resistance of PPO.

PPO is counted as one of the 'engineering plastics'. The rigid structure of the polymer molecules leads to a material with a high T_g of 208°C. It is characterised by high tensile strength, stiffness, impact strength, and crepe resistance, and low coefficient of thermal expansion. These properties are maintained over a broad temperature range (–45 to 120°C). One particular feature of PPO is its exceptional dimensional stability among the so-called engineering plastics. The polymer is self-extinguishing.

PPO has excellent resistance to most aqueous reagents and is unaffected by acids, alkalies, and detergents. The polymer has outstanding hydrolytic stability and has one of the lowest water absorption rates among the engineering thermoplastics. PPO is soluble in aromatic hydrocarbons and chlorinated solvents. Several aliphatic hydrocarbons cause environmental stress cracking.

PPO has low moulding shrinkage. The polymer is used for the injection moulding of such items as pump components, domestic appliance and business machines, and electrical parts such as connectors and terminal blocks.

The high price of PPO has greatly restricted its application and has led to the introduction of the related and cheaper thermoplastic materials. If PPO (T_g = 208°C) is blended with the polystyrene (T_g ~ 90°C) in equal quantities, a transparent polymer is obtained with a single T_g of about 150°C, which apparently indicates a molecular level of mixing. Noryl thermopolastics may be considered as being derived from such polystyrene-PPO blends. Since the electrical properties of the two polymers are very similar, the blends also have similar electrical characteristics. In addition to Noryl blends produced by General Electric, grafts of styrene onto PPO are also now available (Xyron by Asahi-Dow).

The styrenic component in polystyrene-PPO blends may not necessarily be straight polystyrene (PS) but instead high-impact polystyrene (HIPS) or some other related material. The most widely used blend is the blend of PPO and HIPS.

Like polystyrene, these blends have the following useful characteristics: (i) good dimensional stability and low moulding shrinkage, thus allowing close dimensional tolerance in the production of mouldings; (ii) low water absorption; (iii) excellent resistance to hydrolysis; and (iv) very good dielectric properties over a wide range of temperature. In addition, unlike polystyrene, the blends have heat-distortion temperatures above the boiling point of water and in some grades this is as high as 160°C.

The range of Noryl blends available comprises a broad spectrum of materials superior in many respects, particularly heat deformation resistance, to the general purpose thermoplastics but at a lower price than the more heat resistant materials such as polycarbonates, polyphenylene sulphides, and polysulphones. The materials that come close to them in properties are the ABS/polycarbonate blends. Noryl is also characterised by high dielectric strength (192 V/mil) and low dissipation factors (4.7×10^3 at 100 Hz and 3.9×10^3 at 10^6 Hz).

In common with other engineering thermoplastics, there are four main groups of modified PPOs available. They are: (i) non-self-extinguishing grades with a heat-distortion temperature in the range 110–160°C and with a notched Izod impact strength of 200–500 J/m; (ii) self-extinguishing grades with slightly lower heat-distortion temperatures and impact strengths; (iii) non-self-extinguishing glass-reinforced grades (10 per cent, 20 per cent, 30 per cent glass fibre) with heat-distortion temperatures in the range of 120–140°C; and self-extinguishing glass-reinforced grades. Among the special grades that should be mentioned are those containing blowing agents for use in the manufacture of structural foams. Noryls may be extruded, injection moulded and blow-moulded without undue difficulty. Processing conditions depend on the grade used but in injection moulding a typical melt temperature would be in the range 250–300°C.

The introduction of self-extinguishing, glass-reinforced, and structural foam grades has led to steady increase in the use of these materials in five main application areas. These are: (i) the automotive industry; (ii) the electrical industry; (iii) radio and television; (iv) business machines and computer housings; and (v) pumps and other plumbing applications. Use in the automotive industries largely arises from the availability of high impact grades with heat-distortion temperatures above those of the general purpose thermoplastics. Specific uses include instrument panels, steering column cladding, central consoles, loudspeaker housings, ventilator grilles and nozzles, and parcel shelves. In cooling systems, glass-reinforced grades have been used for radiator and expansion tanks. Several components of car heating systems are also produced from modified PPOs. The materials have been increasingly used for car exterior trim such as air inlet and outlet grills and outer mirror housings.

In the electrical industry, well-known applications include fuse boxes, switch cabinets, housing for small motors, transformers, and protective circuits.

Uses in radio and television arise largely from the ability to produce components with a high level of dimensional accuracy coupled with good dielectric properties, high heat-distortion temperatures, and the availability of self-extinguishing grades. Specific uses include coil formers, picture tube deflection yokes, and insert card mountings.

In the manufacture of business machine and computer housings, structural foam grades have found use and mouldings weighing as much as 50 kg have been reported.

Glass-reinforced grades have widely replaced metals in pumps and other functional parts in washing equipment and central heating systems.

Another PPO, called PPE, is produced by the oxidative coupling of a mixture of 2,6-dimethylphenol and 2,3,6-trimethylphenol. This stiff polymer, like Noryl, is available from General Electric. It is usually modified by blending with PS or HIPS.

Blends of PPO and polyamide (PA, nylon) are incompatible but good properties can be obtained through the use of compatibilising agents. The PPO is dispersed in a continuous nylon matrix in these blends. Because of the incompatibility of the two phases, the modulus decreases very little at the T_g of the PA (71°C) and is maintained up to the T_g of the PPO phase (208°C).

The PPO-PA blends show a mould shrinkage of 0.001 in./in. The moulding pressure is 15×10^3 lbf/in.^{2} (103 MPa) and the processing temperature 260°C. The heat-deflection temperature of the moulded specimen under flexural load of 264 lbf/in.^{2} is typically 190°C and the maximum resistance to continuous heat is 175°C. The coefficient of linear expansion is 10^{-5} cm/cm.°C. The mechanical properties of the PPO-PA blends are shown in Table 16.11.

Table 16.11. Properties of typical PPO-PA blends.

Property	*Value*
Tensile strength	
10^3 lbf/in.^{2}	13
MPa	90
Tensile modulus	
10^5 lbf/in.^{2}	2.0
GPa	1.4
Elongation at break (%)	10
Flexural strength	
10^3 lbf/in.^{2}	19
MPa	131
Flexural modulus	
10^5 lbf/in.^{2}	2.25
GPa	1.6
Impact strength, notched Izod	
ft-lb/in.	1.5
J/m	80

CELLULOSIC POLYMERS

Cellulose is a carbohydrate with molecular formula $(C_6H_{10}O_5)_n$, where n is a few thousand. Complete hydrolysis of cellulose by boiling with concentrated hydrochloric acid yields D-glucose, $C_6H_{12}C_6$, in 95 to 96 per cent yield. Cellulose can thus be considered chemically as a polyanhydroglucose. Careful chemical analysis of stepwise degradation products of cellulose reveals that the anhydroglucose units making up a cellulose molecule are joined together by beta-glucosidic (1,4) linkages. Thus since the linkage between the anhydroglucose units occurs at the reducing carbon atom, cellulose is non-reducing, unlike glucose which is a strongly reducing sugar.

The regularity of the cellulose chain and extensive hydrogen bonding between hydroxyl groups of adjacent chains cause cellulose to be a tightly packed crystalline material which is insoluble and infusible. As a result, cellulose cannot be processed in the melt or in solution. However, cellulose derivatives in which there is less hydrogen bonding are processable. The most common means of preparing processable cellulose derivatives are esterification and etherification of the hydroxyl groups. Here, the more important commercial cellulosic polymers are described.

Regenerated Cellulose

As mentioned, many derivatives of cellulose are soluble, though cellulose itself is insoluble. A solution of a cellulose derivative can be processed (usually by extrusion) to produce the desired shape (commonly fibre or film) and then treated to remove the modifying groups to reform or regenerate unmodified cellulose. Such material is known as regenerated cellulose.

Cellulose can be rendered soluble by xanthate formation by treatment with sodium hydroxide and carbon disulphide and regenerated by acidification of the xanthate solution. This process is known as the viscose process. The reactions can be indicated schematically as

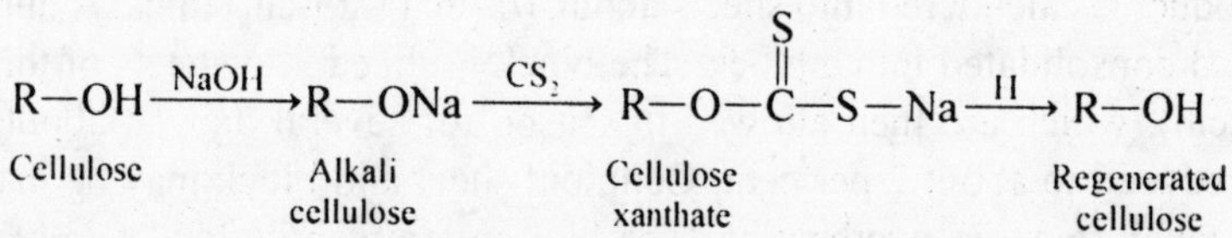

The viscose process is used for the production of textile fibres, known as viscose rayon, and transparent packaging film, known as cellophane (the name is coined from cellulose and diaphane, which is French for 'transparent').

A suitably aged solution of cellulose xanthate, known as viscose, is fed through spinnerets with many small holes (in the production of fibre), through a slot die (in the production of film), or through a ring die (in the production of continuous tube used as sausage casing) into a bath containing 10 to 15 per cent sulphuric acid and 10 to 20 per cent sodium sulphate at 35 to 40°C, which coagulates and completely hydrolyses the viscose. Cellulose is thus regenerated in the desired shape and form.

It is possible to carry out a drawing operation on the fibre as it passes through the coagulating bath. The stretching (50 to 150 per cent) produces crystalline orientation in the fibre. The product, known as high-tenacity rayon, has high strength and low elongation and is used in such application as tyre cord and conveyor belting.

For the production of cellophane, the regenerated cellulose film is washed, bleached, plasticised with ethylene glycol or glycerol, and then dried; sometimes a coating of pyroxylin (cellulose nitrate solution) containing dibutyl phthalate as plasticiser is applied to give heat sealability and lower moisture permeability.

Cellophane has been extensively and successfully used as a wrapping material, particularly in the food and tobacco industries. However, the advent of polypropylene in the early 1960s has produced a serious competitor to this material.

Cellulose Nitrate

Cellulose nitrate or nitrocellulose (as it is often erroneously called) is the doyen of cellulose ester polymers. It is prepared by direct nitration with nitric and sulphuric acid mixtures at about 30 to 40°C for 20 to 60 minutes. Complete substitution at all three hydroxyl groups on the repeating anhydroglucose unit will give cellulose trinitrate containing 14.14 per cent nitrogen:

$$[C_6H_7O_2(OH)_3]_n \xrightarrow{HNO_3/H_2SO_4} [C_6H_7O_2(ONO_2)_3]_n + 3nH_2O$$

This material is explosive and is not made commercially, but esters with lower degrees of nitration are of importance. The degree of nitration may be regulated by the choice of reaction conditions. Industrial nitrocelluloses have a degree of substitution somewhere between 1.9 and 2.7 and are generally characterised for their various uses by their nitrogen content, usually about 11 per cent for plastics, 12 per cent for lacquer and cement base, and 13 per cent for explosives.

The largest use of cellulose nitrate is as a base for lacquers and cements. Butyl acetate is used as a solvent. Plasticisers such as dibutyl phthalate and tritolyl phosphate are necessary to give films of acceptable flexibility and adhesion.

For use as plastic in bulk form, cellulose nitrate is plasticised with camphor. The product is known as celluloid. In a typical process alcohol-wet cellulose nitrate is kneaded at about 40°C with camphor (about 30 per cent) to form a viscous plastic mass. Pigments or dyes may be added at this stage. The dough is then heated at about 80°C on milling rolls until the alcohol content is reduced to about 15 per cent. The milled product is calendered into sheets about 1/2-in. (1.25-cm) thick. A number of sheets are laid up in a press and consolidated into a block. The block is sliced into sheets of thickness 0.005 to 1 inch (0.012 to 2.5 cm), which are then allowed to season for several days at about 50°C so that the volatile content is reduced to about 2 per cent. Celluloid sheet and block may be machined with little difficulty if care is taken to avoid overheating. The high inflammability and relatively poor chemical resistance of celluloid severely restrict its use in industrial applications. Consequently the material is used because of its desirable characteristics, which include rigidity, dimensional stability, low water absorption, reasonable toughness, after-shrinkage around inserts, and ability of forming highly attractive coloured sheeting. Today the principal outlets of celluloid are knife handles, table-tennis balls, and spectacle frames.

Cellulose Acetate

The acetylation of cellulose is usually carried out with acetic anhydride in the presence of sulphuric acid as catalyst. It is not practicable to stop acetylation short of the essentially completely esterified triacetate. Products of lower acetyl content are thus produced by partial hydrolysis of the triacetate to remove some of the acetyl groups.

$$\underset{\text{Cellulose}}{[C_6H_7O_2(OH)_3]_n} \xrightarrow{HOAc,\ Ac_2O} \underset{\text{Triacetate (44.8\% acetyl)}}{[C_6H_7O_2(OAc)_3]_n} \xrightarrow{H_2O} \underset{\text{Diacetate (34.9\% acetyl)}}{[C_6H_7O_2(OAc)_2OH]_n}$$

Cellulose triacetate is often known as primary cellulose acetate, and partially hydrolysed material is called secondary cellulose acetate. Many physical and chemical properties of cellulose acetylation products are strongly dependent on the degree of esterification, which is measured by the acetyl content (i.e., the weight of acetyl radical (CH_3CO—) in the material) or acetic acid yield (i.e., the weight of acetic acid produced by complete hydrolysis of the ester). The commercial products can be broadly distinguished as cellulose acetate (37–40 per cent acetyl), high-acetyl cellulose acetate (40–42 per cent acetyl), and cellulose triacetate (43.7–44.8 per cent acetyl).

Cellulose acetate containing 37 to 40 per cent acetyl is usually preferred for use in general-purpose injection-moulding compounds. Cellulose acetate, however, decomposes below is softening point, and it is necessary to add a plasticiser (e.g., dimethyl phthalate or triphenyl phosphate), usually 25 to 35 per cent, to obtain mouldable material. The use of cellulose acetate for moulding and extrusion is becoming small owing largely to the competition of polystyrene and other polyolefins. At the present time the major outlets of cellulose acetate are in the fancy goods trade as toothbrushes, combs, hair slides, etc.

Cellulose acetate with a slightly higher degree of esterification (38.7–40.1 per cent acetyl) is usually preferred for the preparation of fibres, films, and lacquers because of the greater resistance. A significant application of cellulose acetate film has been found in sea-water desalination by reverse osmosis.

High-acetyl cellulose acetate (40–42 per cent acetyl) has found occasional use in injection-moulding compounds where greater dimensional stability is required. However, processing is more difficult.

Cellulose triacetate (43.7–44.8 per cent acetyl) finds little use in moulding compositions because its very high softening temperature is not greatly reduced by plasticisers. It is therefore processed in solution. A mixture of methylene chloride and methanol is the commonly used solvent. The sheeting and fibres are made from cellulose triacetate by casting or by extruding a viscous solution and evaporating the solvent. The sheeting and film are grainless, have good gauge uniformity, and good optical clarity. The products have good dimensional stability and are highly resistant to water, grease, oils, and most common solvents such as alcohol and acetone. They also have good heat resistance and high dielectric constant. Sheeting and film of cellulose triacetate are used in the production of visual aids, graphic arts, greeting cards, photographic albums, and protective folders. Cellulose triacetate is extensively used for photographic, X-ray, and cinematographic films. In these applications cellulose triacetate has displaced celluloid mainly because the triacetate does not have the great inflammability of celluloid.

Other Cellulose Esters

Homologues of acetic acid have been employed to make other cellulose esters. Of these, cellulose propionate, cellulose acetate-propionate, and cellulose acetate-butyrate are produced on a commercial scale. They are produced in a manner similar to that described previously for cellulose acetate. The propionate and butyrate esters are made by substituting propionic acid and propionic anhydride or butyric acid and butyric anhydride for some of the acetic acid and acetic anhydride.

Cellulose acetate-butyrate (CAB) has several advantages in properties over cellulose acetate: lower moisture absorption, greater solubility and compatibility with plasticiser, higher impact strength, and excellent dimensional stability. CAB used in plastics has about 13 per cent acetyl and 37 per cent butyryl content. It is an excellent injection-moulding material (Tenite Butyrate by Kodak, Cellidor B by *Bayer*). Principal end products of CAB have been for tabulator keys, automobile parts, toys, pen and pencil barrels, steering wheels, and tool handles. In the United States CAB has been used for telephone housings, and extruded CAB piping has been used for conveying water, oil, and natural gas, CAB sheet is readily vacuum formed and is especially useful for laminating with thin-gauge aluminium foil. It also surveys particularly well for vacuum metallising.

Cellulose propionate is very similar in both cost and properties to CAB. It has been used for similar purposes as CAB. Cellulose acetate-propionate is similar to cellulose propionate. It finds wide use in blister packages and formed containers, safety goggles, motor covers, metallised flash cubes, brush handles, steering wheels, face shields, displays, and lighting fixtures.

Cellulose Ethers

Of cellulose ethers only ethyl cellulose has found application as a moulding material. Methyl cellulose, hydroxyethyl cellulose, and sodium carboxymethyl cellulose are useful water-soluble polymers. The first step in the manufacture of each of these materials is the preparation of alkali cellulose (soda cellulose) by treating cellulose with concentrated sodium hydroxide. Ethyl cellulose is made by reacting alkali cellulose with ethyl chloride.

Ethyl cellulose is produced in pellet form for moulding and extrusion and in sheet form for fabrication. It has good processability, is tough, and is moderately flexible; its outstanding feature is its toughness at low temperatures. The principal uses of ethyl cellulose mouldings are thus in those applications where good impact strength at low temperatures are required, such as refrigerator bases and flip lids and ice-crusher parts. Ethyl cellulose is often employed in the form of hot melt for strippable coatings used for protection of metal parts against corrosion and marring during shipment and storage. A recent development is the use of ethyl cellulose gel lacquers for permanent coatings.

Methyl cellulose is prepared by reacting alkali cellulose with methyl chloride at 50 to 100°C (ethyl cellulose). With a degree of substitution of 1.6 to 1.8, the resultant ether is soluble in cold water but not in hot water. It is used as a thi ckening agent and emulsifier in cosmetics, pharmaceuticals, ceramics, as a paper size, and in leather-tanning operations. Hydroxyethyl cellulose, produced by reacting alkali cellulose with ethylene oxide, is employed for similar purposes. Reaction of alkali cellulose with sodium salt of chloroacetic acid yields sodium carboxymethyl cellulose (SCMC).

Polysulphides

Monomers	*Polymerisation*	*Major uses*
Aliphatic dihalide, sodium sulphide	Polycondensation	Sealing, caulks, gaskets

Polysulphide elastomers are produced by the reaction of an aliphatic dihalide, usually bis (2-chloroethyl) formal, with sodium polysulphide under alkaline conditions:

$$\mathrm{ClCH_2CH_2OCH_2OCH_2CH_2Cl + Na_2S_x \longrightarrow}$$
$$\mathrm{\text{+}CH_2CH_2OCH_2OCH_2CH_2S_x\text{+} + NaCl}$$

The reaction is carried out with the dihalide suspended in an aqueous magnesium hydroxide phase. The value of x is slightly above 2. A typical polymerisation occurs readily yielding a polymer with a very high molecular weight.

High molecular weight, however, is not desirable until its end-use application. The molecular weight is therefore lowered, and the polysulphide rank (value of x) is simultaneously brought close to 2, by reductive treatment with NaSH and Na_2SO_3 followed by acidification. The result is a liquid, branched polysulphide with terminal thiol end group and a molecular weight in the range of 1000–8000. Curing to a solid elastomer is accomplished by oxidation of thiol to disulphide links by oxidants such as lead dioxide and *p*-quinone dioxime:

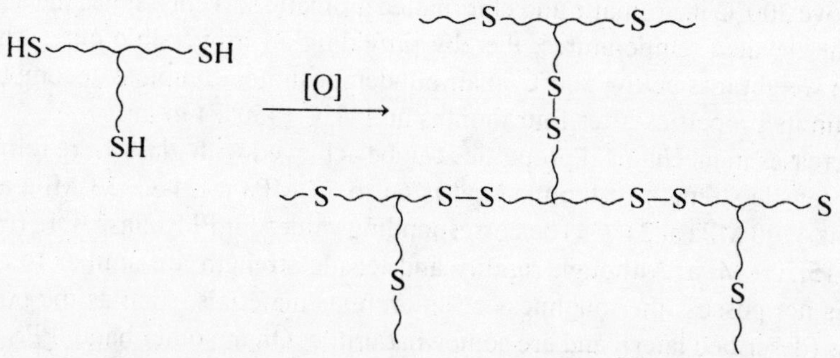

These materials are widely used as sealants, binders for solid propellants, caulking materials, and cements for insulating glass and fuel tanks.

Polysulphides, often referred to as thiokols, are produced at low volumes as speciality materials geared toward a narrow market. The advantages and disadvantages of polysulphides both reside in the disulphide linkage. Thus they possess low-temperature flexibility and very good resistance to ozone, oil, solvent (hydrocarbons as well as polar solvents such as alcohols, ketones, esters, etc.), and weathering. However, polysulphides have poor thermal stability and crepe resistance, have low resilience, and are malodorous.

Thiokols are amorphous polymers which do not crystallise when stretched and hence reinforcing fillers, such as carbon black, must be added to obtain relatively high tensile strengths. Thiokol may be vulcanised in the presence of zinc oxide and thiuram accelerators, such as tetramethyl-thiuram disulphide (Tuads). The accelerators modify the sulphur links and serve as chemical plasticisers.

A typical thiokol with 60 parts carbon black per 100 of polymer has a tensile strength of 1200 lbf/in^2, an elongation of 300 per cent, a specific gravity of 1.25, and a Shore A hardness of 68. Thiokol has excellent resistance to ozone (O_3) and ultraviolet radiation. It has a low permeability to solvents, such as gasoline; esters, such as ethyl acetate; and ketones, such as acetone.

The principal application of solid Thiokol elastomers is as gaskets, O-rings, gasoline, and fuel hose lines, gas meter diaphragms, and as rollers, which are used for lacquering cans.

Poly(Phenylene Sulphide)

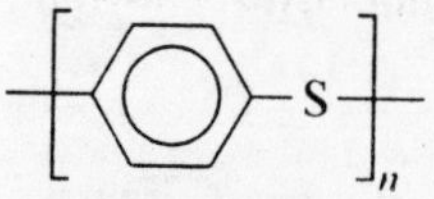

Monomers	*Polymerisation*	*Major uses*
p-Dichlorobenzene, sodium sulphide	Polycondensation	Electrical components, mechanical parts

Poly(phenylene sulphide) (PPS) is the thio analogue of poly(phenylene oxide) (PPO). The first commercial grades were introduced by Phillips Petroleum in 1968 under the trade name Ryton. Other manufacturers also have introduced PPS (e.g., Tedur by *Bayer*). The commercial process involves the reaction of *p*-dichlorobenzene with sodium sulphide in a polar solvent.

PPS is an engineering plastic. The thermoplastic grades of PPS are outstanding in heat resistance, flame resistance, chemical resistance, and electrical insulation characteristics. The linear polymers are highly crystalline with melting point in the range of 285 to 295°C and T_g of 193 to 204°C. The material is soluble only above 200°C in aromatic and chlorinated aromatic solvents. It has the ability to cross-link by air-oxidation at elevated temperatures, thereby providing an irreversible cure. Thermogravimetric analysis shows no weight loss below 500°C in air but demonstrates complete decomposition by 700°C. It is found to retain its properties after four months at 233°C (450°F) in air.

Significant increases in mechanical properties can be achieved with glass-fibre reinforcement. In the unfilled form the tensile strength of the material is 64 to 77 MPa at 21°C, 33 MPa at 204°C, and the flexural modulus is 4200 MPa at 21°C. The corresponding values for PPS-glass fibre (60:40) composites are 150, 33 and 15,500 MPa. Although rigidity and tensile strength are similar to other engineering plastics, PPS does not possess the toughness of amorphous materials, such as the polycarbonates and the polysulphones (described later), and are somewhat brittle. On the other hand, PPS does show a high level of resistance to environmental stress cracking.

Being one of the most expensive commercial mouldable thermoplastics, the use of PPS is heavily dependent on its particular combination of properties. Good electrical insulation characteristics, including good arcing and arc-tracking resistance has led to PPS replacing some of the older thermosets in electrical parts. These include connectors, terminal blocks, relay components, brush holders, and switch components. PPS is used in chemical process plants for gear pumps. More recently, it has found application in the automotive sector, in such specific uses as carburettor parts, ignition plates, flow control valves for heating systems, and exhaust-gas return valves to control pollution. The material also finds uses in sterilisable medical, dental, and general laboratory equipment, cooking appliances, and hair dryer components.

Injection-moulded products include high-temperature lamp holders and reflectors, pump parts, valves, and, especially when filled for example with PTFE or graphite, bearings. Processing temperatures are 300–350°C with mould temperatures of up to 200°C. PPS is also used for encapsulation of electronic components and as a high temperature surface coating material.

PPS is resistant to neutron and gamma radiation. Therefore, in nuclear installations, its flexural strength and modulus are essentially unchanged when it is exposed to gamma radiation of 5×10^9 rad and neutron radiation of 1×10^9 rad.

Polysulphone

Polysulphones are a family of engineering thermoplastics with excellent high-temperature properties. The simplest aromatic polysulphone, poly(*p*-phenylene sulphone) does not show thermoplastic behaviour, melting with decomposition above 500°C.

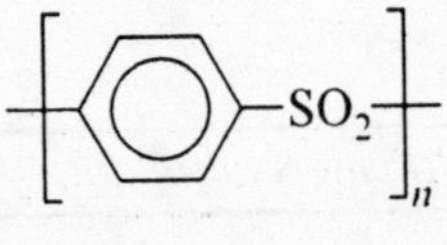

Hence, to obtain a material capable of injection moulding in conventional machines, the polymer chain is made more flexible by incorporating ether links into the backbone. The polymers have different degrees of spacing between the *p*-phenylene groups and thus have a spectrum of glass transition temperatures which determine the heat-distortion temperature (or deflection temperature under load), since the materials are all amorphous.

In principle, there are two main routes to the preparation of polysulphones: (i) polysulphonylation; and (ii) polyetherification.

Polysulphonylation reactions are of the following general types:

$$H\text{—}Ar\text{—}H + ClSO_2\text{—}Ar'\text{—}SO_2Cl \xrightarrow[\text{catalyst}]{\text{Friedel-crafts}}$$

$$\text{—}(\!-Ar\text{—}SO_2\text{—}Ar'\text{—}SO_2-\!)_n\text{—} + HCl$$

$$H\text{—}Ar\text{—}SO_2Cl \longrightarrow \text{—}(\!-Ar\text{—}SO_2-\!)_n\text{—} + HCl$$

The Ar and/or Ar′ group(s) will contain an ether oxygen, and if Ar = Ar′, then basically identical products may be obtained by the two routes.

In the polyetherification route the condensation reaction proceeds by reactions of types

$$HO\text{—}Ar\text{—}OH + Cl\text{—}Ar'\text{—}Cl \xrightarrow{\text{NaOH}} \text{—}(\!-O\text{—}Ar\text{—}O\text{—}Ar'-\!)_n\text{—} + NaCl$$

$$HO\text{—}Ar\text{—}Cl \xrightarrow{\text{NaOH}} \text{—}(\!-O\text{—}Ar-\!)_n\text{—} + NaCl$$

The Ar and or Ar′ group(s) will contain sulphone groups, and if Ar = Ar′, then identical products may be obtained by the two routes.

Polyetherification processes form the basis of current commercial polysulphone production methods. For example, the Udel-type polymer (Union Carbide) is prepared by reacting 4,4′-dichlorodiphenylsulphone with an alkali salt of bisphenol A. The polycondensation is conducted in highly polar solvents, such as dimethylsulphoxide or sulpholane.

Properties

In spite of their linear and regular structure the commercial polysulphones are amorphous. This property might be attributed to the high degree of chain stiffness of polymer molecules which make crystallisation difficult. Because of their high-in-chain aromaticity and consequent high chain stiffness, the polymers have high values of T_g, which means that the processing temperatures must be above 300°C.

Commercial polymers generally resist aqeuous acids and alkalies but are attacked by concentrated sulphuric acid. Being highly polar, the polymer is not dissolved by aliphatic hydrocarbons but dissolves in dimethyl formamide and dimethyl acetamide.

In addition to the high heat-deformation resistance, the polymers also exhibit a high degree of chemical stability. This has been ascribed to an enhanced bond strength arising from the high degree of resonance in the structure. The polymers are thus capable of absorbing a high degree of thermal and ionising radiation without cross-linking.

The principal features of commercial polysulphones are their rigidity, transparency, self-extinguishing characteristics, exceptional resistance to crepe, especially at somewhat elevated temperatures, and good high-temperature resistance.

The use temperatures of the major engineering thermoplastics are compared in Fig. 16.9. Polysulphones are among the higher-priced engineering thermoplastics and so are only considered when polycarbonates or other cheaper polymers are unsuitable. In brief, polysulphones are more heat resistant and have greater resistance to crepe, whereas polysulphonates have a somewhat higher Izod and tensile impact strength besides being less expensive.

In many fields of use polysulphones have replaced or are replacing metals, ceramics, and thermosetting plastics, rather than other thermoplastics. Since commercial polysulphones can be injection moulded into complex shapes, they avoid costly machining and finishing operations. Polysulphones can also be extruded into film and foil. The latter is of interest for flexible printed circuitry because of its high-temperature performance.

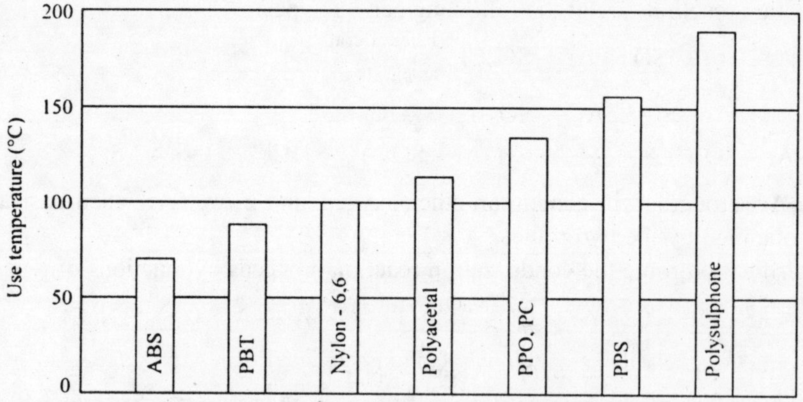

Fig. 16.9. Use temperatures of major engineering thermoplastics.

Polysulphones have found widespread use where good dimensional stability at elevated temperatures is required and fabrication is done by injection moulding. Some products made from polysulphones are electrical components, connectors, coil bobbins, relays, and appliances operating at high temperatures (e.g., hair driers, fan heaters, microwave ovens, lamp housings and bases). The materials are transparent (though often slightly yellow), have low flammability (limiting oxygen index typically 38), and burn with little smoke production.

Polyether Ether Ketone

The chemistry and technology of aromatic polyether ketones may be considered as an extension to those of the polysulphones. The two polymer classes have strong structural similarities, and they are strong parallels in preparative methods. Preparations reported for aromatic polyether ketones are analogous to the polysulphonylation and polyetherification reactions with the polysulphones. Several aromatic polyether ketones have been prepared. The polyether ether ketone (PEEK) is made by the reaction of the potassium salt of hydroquinone with difluorobenzophenone in a high boiling solvent, diphenylsulphone, at temperatures close to the melting point of the polymer.

PEEK is semi-crystalline with a melting temperature (T_m) of 335°C and a glass transition temperature (T_g) of 145°C. The degree of crystallinity can vary from 40 per cent (slow cooling) to essentially amorphous (quenching), but is usually about 35 per cent. PEEK is a high-temperature-resistant thermoplastic suitable for wire coating, injection moulding, film, and advanced structural composite fabrication. The wholly aromatic structure of PEEK contributes to its high-temperature performance. The polymer exhibits very low water absorption, very good resistance to water at 125°C (under which conditions other heat-resisting materials, such as aromatic polyamides, are liable to fail), and is resistant to attack over a wide pH range, from 60 per cent sulphuric acid to 40 per cent sodium hydroxide at elevated temperatures, although attack can occur with some concentrated acids. PEEK has outstanding resistance to both abrasion and dynamic fatigue. Its tensile strength decreases less than 10 per cent after 10^7 cycles at 23°C. It has low flammability with a limiting oxygen index of 35 per cent and generates an exceptionally low level of smoke. Other specific features are excellent resistant to gamma radiation and good resistance to environmental stress cracking. PEEK has superior heat resistance to poly(phenylene sulphide) and is also markedly tougher (and markedly more expensive). PEEK is melt processable and may be injection moulded and extruded on conventional equipment.

Typical applications of PEEK include coating and insulation for high-performance wiring, particularly for the aerospace and computer industries, military equipment, nuclear plant applications, oil wells, and compressor parts.

Since it is a crystalline polymer, the strength and thermal resistance of PEEK are increased dramatically by incorporation of reinforcing agents. Composite prepegs with carbon fibres have been developed for structural aircraft components. Typical properties of PEEK are shown in Table 16.12.

Table 16.12. Properties of polyether ether ketone.

Property	*Unfilled*	*30% glass fibre filled*
Tensile strength		
10^3 lbf/in.^{2}	13.3	22.8
MPa	92	157
Elongation at break (%)	4.9 (yield)	2.2
Flexural modulus		
10^5 lbf/in.^{2}	5.4	14.9
GPa	3.7	10.3
Impact strength, notched Izod		
ft-lb/in.	1.6	1.8
J/m	83	96
Heat distortion temperature (°C)	140	315
Limiting oxygen index (%)	35	–

Polybenzimidazole

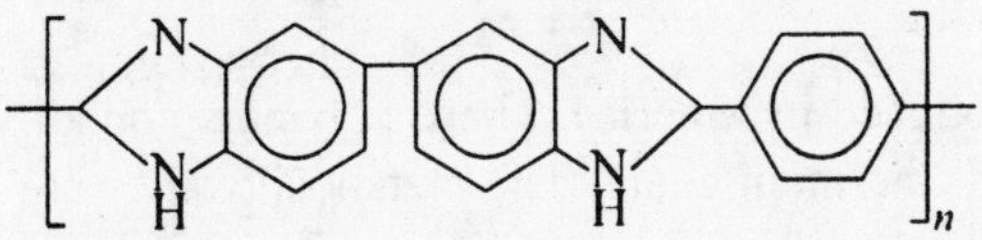

Monomers	*Polymerisation*	*Major uses*
Tetraaminobiphenyl, terephthalic acid	Polycondensation	Fibre

Polybenzimidazole (PBI) is the most well-known commercial example of aromatic heterocycles used as high-temperature polymers. The synthesis of PBI is carried out as follows. The tetraaminobiphenyl required for the synthesis of PBI is obtained from 3,3′-dichloro-4,4′-diaminodiphenyl (a dye intermediate) and ammonia. Many other tetraamines and dicarboxylic acids have been condensed to PBI polymeric systems.

The high thermal stability of PBI (use temperature about 400°C compared to about 300°C for polyimides) combined with good stability makes it an outstanding candidate for high-temperature application despite its relatively high cost. Fibres have been wet spun from dimethylacetamide solution, and a deep gold woven cloth has been made from this fibre. The cloth is said to be more comfortable than cotton (due to high moisture retention) and has greater flame resistance than Nomex (oxygen index of 29 per cent for PBI compared to 17 per cent for Nomex). Other applications of PBI are in drogue parachutes and lines for military aircraft as well as ablative heat shields. The PBI fibres have also shown promise as reverse osmosis membranes and in graphitisation to high-strength, high-modulus fibres for use in composites. The development of ultra-fine PBI fibres for use in battery separator and fuel cell applications has been undertaken.

A new technique of simple precipitation has been used to process PBI polymers into films. High-strength molecular-composite films have been produced with tensile strength in the region of 20,000 psi (137 MPa). The PBI polymer has also been fabricated as a foam. The material provides a low-weight, high-strength, thermally stable, machinable insulation, much needed in the aerospace industry. PBIs exhibited good adhesion as films when cast from solution onto glass plates. This property leads to their use in glass composites, laminates, and filament-wound structures.

SILICONES AND OTHER INORGANIC POLYMERS

The well-known thermal stability of minerals and glasses, many of which are themselves polymeric, has led to intensive research into synthetic inorganic and semi-inorganic polymers. Numerous such polymers have been synthesised, but only a few have found industrial acceptance, due to the difficulties encountered in processing them. These polymers can be classified into the following main groups: (i) polymers containing main-chain silicon atoms, (ii) polymetallosiloxanes; (iii) polymetalloxanes; (iv) phosphorus-containing polymers; (v) boron-containing polymers; and (vi) sulphur-containing polymers.

Among the polymers containing main-chain silicon atoms most important commercially are the silicones, which are described in a later section. Several other polymers which contain silicon atoms in the main chain have been studied in recent years. These include silicon sulphides and polymers containing silicon and nitrogen in the main chain. Unfortunately, neither of these materials show hydrolytic stability. Of greater interest and potential value are the polymetallosiloxanes. Where the metal is more electropositive than silicon, the metal-oxygen bond will be more polar than the silicon-oxygen bond, and the material will be expected to show greater thermal stability. Thermally stable polyorganosiloxymetalloxanes have been prepared which are soluble in organic solvents and may therefore be cast on films and lacquers, or they may be used in laminating resins.

A different class of polymetallosiloxanes have been produced in America. These polymers contain both the metal atom and the silicon atom incorporated into the main chain. The metallosiloxanes are liable to hydrolysis, but the rate of hydrolysis depends on the metal incorporated in the polymer chain. For example, the relative rates of hydrolysis of tin, aluminium, and titanium derivatives of low molecular weight are 2220:27:1.

Polymetalloxanes are akin to the silicones but contain, for example, tin, germanium, and titanium instead of silicon. Of the polyorganostannoxanes, dibutylin oxide finds use as a stabiliser for PVC and as a silicone cross-linking agent. Titanium does not form a stable bond with carbon. However, organotitanium polymers involving a Ti—O bond, such as polyalkoxytitanoxanes and polymeric titanate esters, have been prepared. Butyl titanate polymers are used in surface coatings.

Among phosphorus-based polymers, several phosphonitrilic polymers have proved to be of some interest. Though many polymers containing boron in the main chain have been prepared, most of them have either low molecular weight or interactable cross-linked structures. Polyboron nitride is a heteropolymer of boron and nitrogen. It is obtained readily from borax and ammonium chloride.

$$Na_2B_4O_7 + 2NH_4Cl \xrightarrow{1000°C} 2 \text{-}[B{=}N]\text{-} + B_2O_3 + 2NaCl + 4H_2O$$

Polyboron nitride melts at 3000°C under pressure. Two structural modifications are known: as hexagonal layer structure similar to graphite, and a tetrahedral structure isomorphic and isoelectronic with diamond. The former is used in insulators, and the latter is used as an abrasive.

Sulphur-containing polymers generally lack hydrolysis stability or, alternatively, have a tendency to revert to simple forms. Although some inorganic polymers have found limited use as laminating resins, surface-coating resins, and wire enamels, the fruits of research to date on inorganic polymers have been rather disappointing. Polymers of high thermal stability have often lacked hydrolytic stability and in other cases have been brittle or interactable.

Silicones

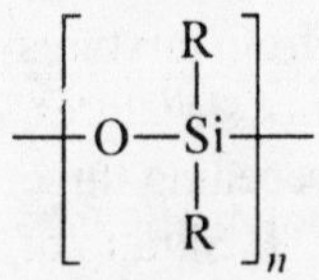

Monomer	*Polymerisation*	*Major uses*
Chlorosilanes	Polycondensation	Elastomer, sealants, and fluids

Silicones are by far the most important inorganic polymers and are based on silicon, an element abundantly available on our planet. The silicone polymers are available in a number of forms, such as fluids, greases, rubbers, and resins. Because of their general thermal stability, water repellency, antiadhesive characteristics, and constancy of properties over a wide temperature range, silicones have found many and diverse applications. [The structure used as the basis of the nomenclature of the silicon compounds is silane SiH_4, corresponding to methane CH_4. Alkyl-, aryl-, alkoxy-, and halogen-substituted silanes are referred to by prefixing 'silane' by the specific group present. For example, $(CH_3)_2SiH_2$ is dimethyl silane, and CH_3SiCl_3 is trichloromethylsilane. Polymers in which the main chain consists of repeating —Si—O— units together with predominantly organic side groups are referred to as polyorganosiloxanes or, more loosely, as silicones].

The commercial production of a broad variety of products from a few basic monomers followed the development of an economically attractive direct process for chlorosilanes, involves reaction of alkyl or aryl halides with elementary silicon in the presence of a catalyst, e.g., copper for methyl- and silver for phenylchlorosilanes. The basic chemistry can be described as

$$SiO_2 + C \longrightarrow Si + 2CO$$

$$Si + RX \longrightarrow R_nSiX_{4-n} \qquad (n = 0-4)$$

In the alkylation of silicon with methyl chloride, mono-, di-, and trimethyl chlorosilanes are formed. The reaction products must then be fractionated. Because the dimethyl derivative is bifunctional, it produces linear methylsilicone polymers on hydrolysis.

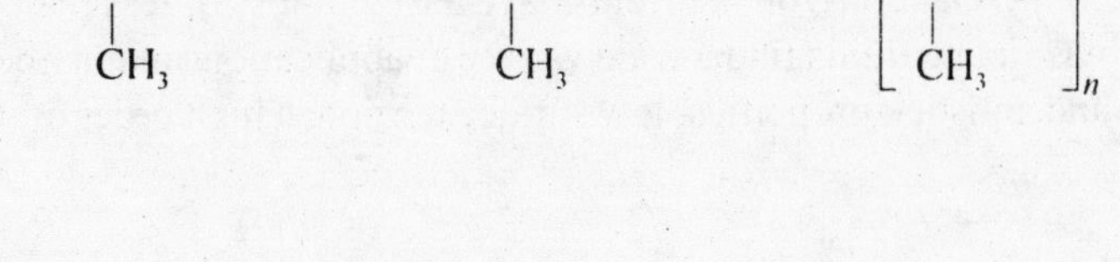

Since monomethyl trichlorosilane has a functionality of 3, the hydrolysis leads to the formation of a highly cross-linked gel.

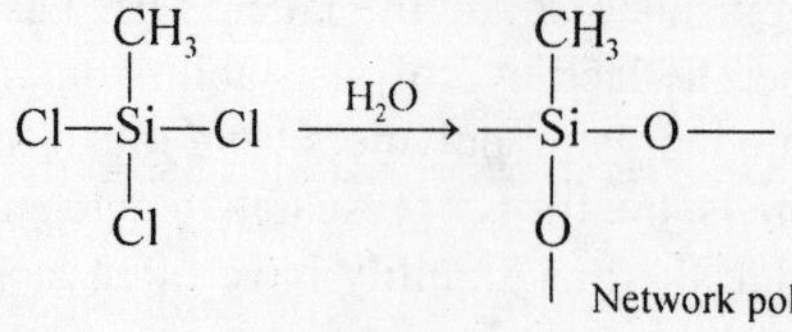

Network polymer

Since the trimethyl monochlorosilane is monofunctional, it forms only a disiloxane.

$$(CH_3)_3SiCl \xrightarrow{H_2O} (CH_3)_3Si{-}O{-}Si(CH_3)_3$$

Products of different molecular-weight ranges and degrees of cross-linking are obtained from mixtures of these chlorosilanes in different ratios. In characterising commercial branched and network structures, the CH_3/Si ratio (or, generally, R/Si ratio) is thus a useful parameter. For example, the preceding three idealised products have CH_3/Si ratios of 2:1, 1:1, and 3:1, respectively. A product with a CH_3/Si ratio of 1.5:1 will thus be expected to have a moderate degree of cross-linking.

Many different silicon products are available today. The major applications are listed in Table 16.13.

Table 16.13. Major applications of silicones.

Mould-release agents	Greases and waxes
Water repellants	Cosmetics
Antifoaming agents	Insulation
Glass-sizing agents	Dielectric encapsulation
Heat-exchange fluids	Caulking agents (RTV)
Hydraulic fluids	Gaskets and seals
Surfactants	Laminates
Coupling agents	Biomedical devices

Silicone fluids

The silicone fluids form a range of colourless liquids with viscosities from 1 to 1,000,000 centistokes (cs). The conversion of chlorosilane intermediates into polymer is accomplished by hydrolysis with water, which is followed by spontaneous condensation. In practice, the process involves three important stages: (i) hydrolysis, condensation, and neutralisation (of the HCl evolved on hydrolysis); (ii) catalytic equilibrium; and (iii) devolatilisation.

The product after the first stage consists of an approximately equal mixture of cyclic compounds, mainly the tetramer, and linear polymer. To achieve a more linear polymer and also to stabilise the viscosity, it is common practice to equilibrate the products of hydrolysis by heating with a catalyst such as dilute sulphuric acid. For fluids of viscosities below 1000 cs, this equilibrium reaction is carried out for hours at 100 to 150°C. After addition of water, the oil is separated from the aqueous acid layer and neutralised. To produce non-volatile silicone fluids, volatile low-molecular products are removed by using a vacuum still. Commercial non-volatile fluids have a weight loss of less than 0.5 per cent after 24 hours at 150°C. Dimethylsilicone fluids find a wide variety of applications mainly because of their water repellency, lubrication and antistick properties, low surface tension, a high order of thermal stability, and a fair constancy of physical properties over a wide range of temperature (–70 to 200°C). As a class the silicone fluids have no colour or odour, have low volatility, and are non-toxic. The fluids have reasoanable chemical resistance but are attacked by concentrated mineral acids and alkalies. They are soluble in aliphatic, aromatic, and chlorinated hydrocarbons.

A well-known application of the dimethylsilicone fluids is as a polish additive. The value of the silicone fluid in this application is due to its ability to lubricate, without softening, the microcrystalline wax plates.

Dilute solutions or emulsions containing 0.5 to 1 per cent of a silicone fluid are extensively used as a release agent for rubber moulding. However, their use has been restricted with thermoplastics because of the tendency of the fluids to cause stress cracking in polymers.

Silicone fluids are used in shock absorbers, hydraulic fluids, dashpots, and in other damping systems in high-temperature operations.

The silicones have established their value as water-repellent finishes for a range of natural and synthetic textiles. Techniques have been developed which result in the pick up of 1 to 3 per cent of silicone on the cloth. Leather also may be made water repellent by treatment with solutions or emulsions of silicone fluids. These solutions are also used for paper treatment.

Silicone fluids and greases are useful as lubricants for high-temperature operations for applications depending on rolling friction. Greases may be made by blending silicone with an inert filler such as fine silicas, carbon black, or a metallic soap. The silicone/silica greases are used as electrical greases for such applications as aircraft and car ignition systems. Silicone greases have also found uses in the laboratory for lubricating stopcocks and for high-vacuum work.

Silicone fluids are used extensively as antifoams, although concentration needed is normally only a few parts per million. The fluids have also found a number of uses in medicine. Barrier creams based on silicone fluids are particularly useful against cutting oils used in metal machinery processes.

High-molecular-weight dimethylsilicone fluids are used as stationary phase for columns in vapour-phase chromatographic apparatus.

Surfactants based on block co-polymers of dimethylsilicone with poly(ethylene oxide) are unique in regulating the cell size in polyurethane foams. One route to such polymers used the reaction between a polysiloxane and an allyl ether of poly(ethylene oxide).

Silicone resins

Silicone resins are manufactured batchwise by hydrolysis of a blend of chlorosilanes. For the final product to be cross-linked, a certain amount of trichlorosilane must be incorporated into the blend. (In commercial practice, R/Si ratios are typically in the range of 1.2:1 to 1.6:1.) The cross-linking of the resin is, of course, not carried out until it is *in situ* in the finished product. The cross-linking takes place by heating the resin at elevated temperatures with a catalyst, (e.g., triethanolamine and metal octoates).

The resins have good heat resistance but are mechanically much weaker than cross-linked organic plastics. The resins are highly water repellent and are good electrical insulators particularly at elevated temperatures and under damp conditions. The properties are reasonably constant over a fair range of temperature and frequency.

Methyl phenyl silicone resins are used in the manufacture of heat-resistant glass-cloth laminates particularly for electrical applications. These are generally superior to PF and MF glass-cloth laminates. The dielectric strength of silicone-bonded glass-cloth laminates is 100 to 120 kV/cm compared to 60 to 80 kV/cm for both PF and MF laminates. The insulation resistance (dry) of the former (500,000 Ω) is significantly greater than those for the PF and MF laminates (10,000 to 20,000 Ω, respectively). The corresponding values after water immersion are 10,000, 10, and 10 Ω.

Silicone laminates are used principally in electrical applications such as printed circuit boards, transformers, and slot wedges in electric motors. Compression-moulding powders based on silicone resins are available and have been used in the moulding of switch parts, brush ring holders, and other electrical applications that need to withstand high temperatures.

Silicone rubbers

Silicone elastomers are either room-temperature vulcanisation (RTV) or heat-cured silicone rubbers, depending on whether cross-linking is accomplished at ambient or elevated temperature. (The term vulcanisation is a synonym for cross-linking. While curing is also a synonym for cross-linking, it often refers to a combination of additional polymerisation plus cross-linking). RTV and heat-cured silicone rubbers typically involve polysiloxanes with degrees of polymerisations of 200–1500 and 2500–11,000, respectively. While the lower-molecular-weight polysiloxanes can be synthesised by the hydrolytic step polymerisation process, the higher-molecular-weight polymers are synthesised by ring-opening polymerisation using ionic initiators.

Before fabrication it is necessary to compound the gum with fillers, a curing agent, and other special additives on a two-roll mill or in an internal mixer. Unfilled polymers have negligible strength, whereas reinforced silicone rubbers may have strengths up to 2000 psi (14 MPa).

Silica fillers are generally used with silicone rubbers. These materials with particle sizes in the range 0.003 to 0.03 μm are prepared by combustion of silicone tetrachloride (fume silicas), by precipitation or as an aerogel.

Heat-curing of silicone rubbers usually involve free-radical initiators such as benzoyl peroxide, 2,4-dichlorobenzoyl peroxide, and *t*-butyl perbenzoate, used in quantities of 0.5 to 3 per cent. These materials are stable in the compounds at room temperature for several months but will start to cure at about 70°C. The curing (cross-linking) is believed to take place by the sequence of reactions.

The process involves the formation of polymer radicals via hydrogen abstraction by the peroxy radicals formed from the thermal decomposition of the peroxide and subsequent cross-linking by coupling of the polymer radicals.

The cross-linking efficiency of the peroxide process can be increased by incorporating small amounts of a co-monomer containing vinyl groups into the polymer, e.g., by co-polymerisation with small amounts of vinyl-methyl silanol.

Dimethyl silicone rubbers show a high compression set. (For example, normal cured compounds have a compression set of 20 to 50 per cent after 24 hour at 150°C.) Substantially reduced compression set values may be obtained by using a polymer containing small amounts of methylvinylsiloxane. Rubbers containing vinyl groups can be cross-linked by weaker peroxide catalysts. Where there is a high vinyl content (4 to 5 per cent molar), it is also possible to vulcanise with sulphur.

Room-temperature vulcanising silicone rubbers (RTV rubbers) are low-molecular-weight liquid silicones with reactive end groups and loaded with reinforcing fillers. Several types are available on the market. 'One-component' RTV rubbers consist of an air-tight package containing silanol-terminated polysiloxane, cross-linking agent (methylacetoxysilane), and catalyst (e.g., dibutyltin laurate). Moisture from the atmosphere converts the cross-linking agent to the corresponding silanol (acetic acid is a by-product), $CH_3Si(OH)_3$, which brings about further polymerisation combined with cross-linking of the polysiloxane,

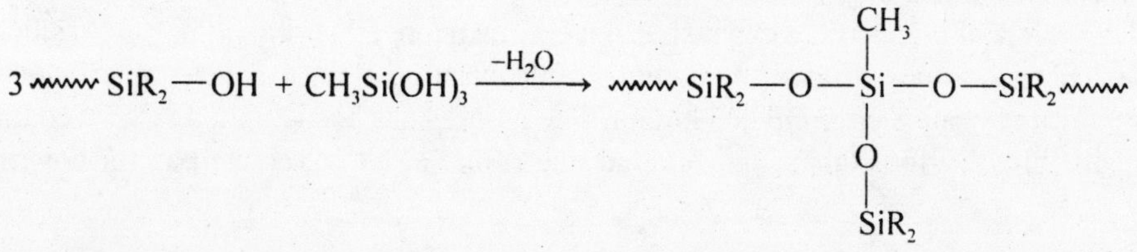

Hydride functional siloxanes can also be cross-link silanol-terminated polysiloxanes. The reaction is catalysed by tin salts and involves elimination of H_2 between Si–H and Si–O–H groups.

RTV rubbers have proved of considerable value as they provide a method for producing rubbery products with the simplest equipment. These rubbers find use in the building industry for caulking and in the electrical industry for encapsulation.

Non-tacky self-adhesive rubbers (fusible rubbers) are obtained if small amounts of boron (~1 boron atom per 300 silicon atoms) are incorporated into the polymer chain. They may be obtained by condensing dialkylpolysiloxanes end-blocked with silanol groups with boric acid or by reacting ethoxyl end-blocked polymers with boron triacetate.

Bouncing putty is somewhat similar in that the Si—O—B bond occurs occasionally along the chain. It is based on a polydimethylsiloxane polymer modified with boric acid, additives, fillers, and plasticisers to give a material that shows a high elastic rebound when small pieces are dropped on a hard surface but flows like a viscous fluid on storage or slow application of pressure.

The applications of the rubbers stem from their important properties, which include thermal stability, good electrical insulation properties, non-stick properties, physiological inertness, and retention of elasticity at low temperatures. The temperature range of general-purpose material is approximately –50 to +250°C, and the range may be extended with special rubbers. Silicone rubbers are, however, used only as special-purpose materials because of their high cost and inferior mechanical properties at room temperature as compared to conventional rubbers (e.g., natural rubbers and SBR).

Modern passenger and military aircraft each use about 500 kg of silicone rubber. This is to be found in gaskets and sealing rings for jet engines, vibration dampers, ducting, sealing strips, and electrical insulators. Silicone cable insulation is also used extensively in naval craft since the insulation is not destroyed in the event of fire but forms an insulating layer of silica.

The rubbers find use in diverse other applications which include electric iron gaskets, domestic refrigerators, antibiotic container closures, and for non-adhesive rubber-covered rollers for handling such materials as confectionary and adhesive tape.

Due to their relatively inertness, new applications have emerged in the biomedical field. A silicone rubber ball is used in combination with a fluorocarbon seal to replace defective human heart valves. Silicone rubber has had many applications in reconstructive surgery on or near the surface of the body. Prosthetic devices are very successfully used in all parts of the body.

The cold-curing silicone rubbers are of value in potting and encapsulation.

In recent years there has been considerable interest in a new form of silicone materials, namely, the liquid silicone rubbers. These may be considered as a development from the RTV silicone rubbers but have a better pot life and improved physical properties, including heat stability (in the cured state) similar to that of conventional silicone elastomers. Liquid silicone rubbers range from a flow consistency to a paste consistency and are usually supplied as a two-pack system, which requires simple blending before use. The materials cure rapidly above 110°C. In injection moulding of small parts at high temperatures

(200 to 250°C), cure times may be as small as a few seconds. One example of application is in baby bottle nipples, which, although more expensive, have a much longer working life. Liquid silicone rubbers have also been used in some extruded applications. Vulcanisation of the extruded material may be carried out by using infrared heaters or circulated hot air. The process has been applied to wire coating, ignition cables, optical fibres, various tapes, and braided glass-fibre-sleeving, as well as for covering delicate products.

Polyphosphazenes

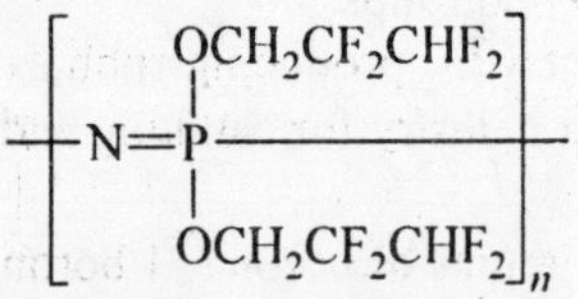

Monomers	*Polymerisation*	*Major uses*
Phosphorus pentachloride, ammonium chloride, fluorinated alcohols	Polycondensation followed by nucleophilic replacement of chlorogroups	Aerospace, military, oil exploration applications

Polyphosphazenes containing nitrogen and phosphorus have been synthesised by replacing the chlorine atoms on the backbone chain of polymeric phosphonitrilic chloride (dichlorophosphazene) by alkoxy or fluoroalkoxy groups. These derivative polymers do not exhibit the hydrolytic instability of the patient polymer.

With mixtures of alkoxy substituents with longer alkyl chains, crystallisation can be avoided to produce an amorphous rubber. The product is referred to as phosphonitrilic fluoroelastomer, a 'semi-organic' rubber. The rubber can be cross-linked with free-radical initiators or by radiation. A commercial rubber is based on alkoxides of trifluoroethyl alcohol or heptafluoroisobutyl alcohol.

The polyphosphazene rubbers have excellent resistance to oils and chemicals (except alcohols and ketones), good dynamic properties, good abrasion resistance, and a broad range of use temperatures (–65°C to +117°C). Disadvantages are water resistance that is only fairly good.

Polythiazyl

Monomer	*Polymerisation*	*Major uses*
S_4N_4	Ring-opening polymerisation	Semi-conductive polymers

The solid four-member ring reaction product S_2N_2 obtained by pyrolysis of gaseous S_4N_4 under vacuum is polymerised at room temperature by ring-opening polymerisation to give the linear chain polythiazyl

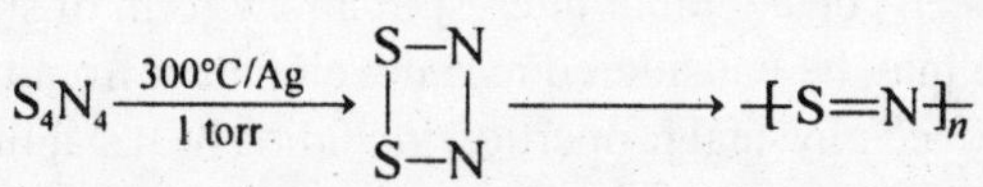

The product is a brasslike solid material that behaves like a metal or alloy but is lighter and more flexible. Polythiazyl has electrical conductivity (3700 reciprocal ohm-cm, or siemens/cm) at room temperature and super-conductivity at 0.3 K. Some doped polymers are photoconductive.

POLYBLENDS

The concept of physically blending two or more existing polymers or co-polymers to obtain new products or for problem solving is now attracting widespread interest and commercial utilisation. By definition, any physical mixture of two or more different polymers or co-polymers that are not linked by covalent bonds is a polymer blend or polyblend.

Blends have been classified as miscible one-phase systems and partially miscible or immiscible two-phase systems. A negative change in free energy (ΔG) is required for miscibility. The change in entropy (ΔS) is essentially negligible and hence the change in enthalpy (ΔH) must be negative or zero for the formation of miscible blends as shown by the Gibbs free energy equation:

$$\Delta G = \Delta H - T\Delta S$$

where T is the absolute temperature. The requirement for a negative ΔH value can be met when there is a physical attraction, such as hydrogen bonding, between the component polymers.

A polymer blend (PB) that is homogeneous down to the molecular level and is associated with the negative value of the free energy of mixing, $\Delta G \simeq \Delta H \leq 0$, is termed a miscible polymer blend. An immiscible polymer blend, on the other hand, is any PB for which $\Delta G \simeq \Delta H > 0$. To make an analogy with metals polymer blends are sometimes referred to as polymer alloys. Thus the term 'alloys' has been used to describe miscible or immiscible mixtures of polymers that are usually blended as melts. Another definition often used for a polymer alloy is that it is an immiscible PB having a modified interface and/or morphology. The general relationship between blends and alloys is shown in Fig. 16.10. The term compatibilisation in this figure refers to a process of modification of interfacial properties (discussed later) of an immiscible PB, leading to the creation of a polymer alloy.

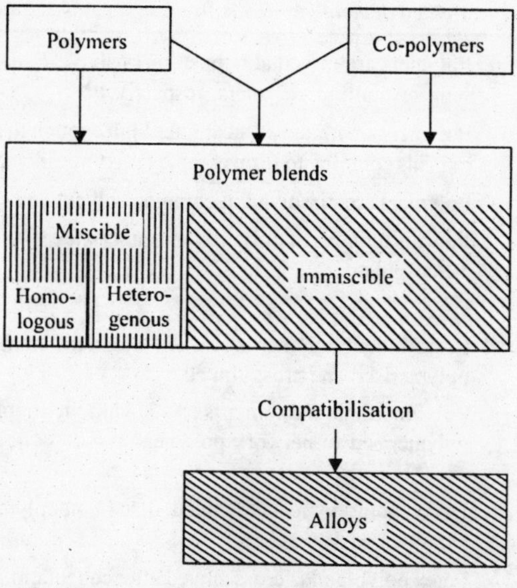

Fig. 16.10. Interrelations in polymer blend nomenclature.

The concept of blending polymers is not new; the rubber industry has used it for decades. In recent years, however, there has been a resurgence of interest arising primarily from the demand for engineering plastics and speciality polymers. There are sound economic reasons for this interest. Blending technology offers many advantages. Gaps in thermoplastics technology can be filled more easily by compounding polymers that are already available than by development of new products. Development of a new polymer to meet a specific need is a costly enterprise. If the desired properties can be realised simply by mixing two or more existing polymers, there is an obvious pecuniary advantage. Thus the time normally required to develop a new blend ready for the market is 3–5 years, compared to 8–10 years required for new plastics materials. This reduces the cost of development of a new product. Cost is also reduced if an expensive polymer can be diluted by a less expensive one.

Blending provides a convenient way of combining the mechanical, physical, or thermal properties of more than one material. One example is the blend of ABS and PVC. The ABS contributes high heat-distortion temperature, toughness, and easy mouldability, while the PVC imparts weatherability and flame retardance, as well as reducing the cost of the blend. Applications of the blend include automotive interior trim, luggage shells, and canoes.

Important disadvantages of a basic polymer (for example, difficulties of fabrication) may be overcome by blending with one or more other polymers. In some cases, synergistic improvement in properties, exceeding the value for either polymer alone, is achieved by blending, though such cases are relatively rare. A good example is the improved Izod notched impact strength shown when ABS and polycarbonate (PC) are blended. At sub-zero temperatures, the blend has better notched impact strength than either of its component polymers.

A number of technologies have been devised to prepare polyblends; these are summarised in Table 16.14. For economic reasons, however, mechanical blending predominates.

Table 16.14. Types and methods of producing of polyblends.

Type	*Method of blending*
Mechanical blends	Polymers are mixed at temperatures above T_g or T_m for amorphous and semi-crystalline polymers, respectively.
Mechanochemical blends	Polymers are mixed at shear rates high enough to cause degradation. Resultant free radicals combine to form complex mixtures including block and graft components.
Solution-cast blends	Polymers are dissolved in common solvent and solvent is removed.
Latex blends	Fine dispersions of polymers in water (latexes) are mixed, and the mixed polymers are coagulated.
Chemical blends	
Interpenetrating polymer networks (IPN)	Cross-linked polymer is swollen with a different monomer; the monomer is then polymerised and cross-linked.
Semi-interpenetrating polymer networks (semi-IPN, also called pseudo IPN)	Polyfunctional monomer is mixed with thermoplastic polymer, then monomer is polymerised to network polymer.
Simultaneous interpenetrating polymer networks (SIN)	Different monomers are mixed, then homopolymerised and cross-linked simultaneously.
Interpenetrating elastomeric networks (IEN)	Latex polyblend is cross-linked after coagulation.

It so happens that most polymers are not miscible; rather they separate into discrete phases on being mixed. Differences between miscible and immiscible polyblends are manifested in appearance (miscible blends are usually clear, immiscible blends are opaque) and in such properties as glass transition temperature (miscible blends exhibit a single T_g intermediate between those of the individual components, whereas immiscible blends exhibit separate T_gs characteristic of each component).

Though miscibility is by no means a prerequisite to commercial utility, homogeneous polymer blends are more convenient from the standpoint of being able to predict properties or processing characteristics. For example, if additives are used, there are no problems of migration from one phase to another. Physical or mechanical properties usually reflect, to a degree, the weighted average of the properties of each component. For a binary homogeneous blend, this can be expressed quantitatively for a particular property (P) by means of a semi-empirical relationship :

$$P = \phi_1 P_1 + \phi_2 P_2 + \phi_1 \phi_2 I \quad \text{... (16.3)}$$

where ϕ is the volume fraction in the mix and I is an interaction term that may be negative, zero, or positive. For $I = 0$, the properties are strictly additive. If I is positive, the property in question is better than the weighted average and the blend is said to be synergistic for that property. However, if I is negative, the property is worse than the weighted average (imcompatible, non-synergistic). This is illustrated schematically in Fig. 16.11 by a plot of property versus composition.

In the case of complete miscibility, we would expect properties to follow a simple monotomic function, more or less proportional to the contents of the two polymer components in the blend (Fig. 16.11a). This is particularly useful for processors because it permits them to inventory a few commodity polymers and simply blend them in different proportions to meet the specific requirements of each product they manufacture. Noryl, for example, is composed of polystyrene, an inexpensive polymer, and poly(phenylene oxide) or PPO, a relatively expensive polyether. For the most, the properties of Noryl are additive. For example, Noryl has poorer thermal stability than the polyether alone, but is easier to process. Its single glass transition temperature increases with increasing polyether content. In terms of tensile strength, however, the polyblend is synergistic.

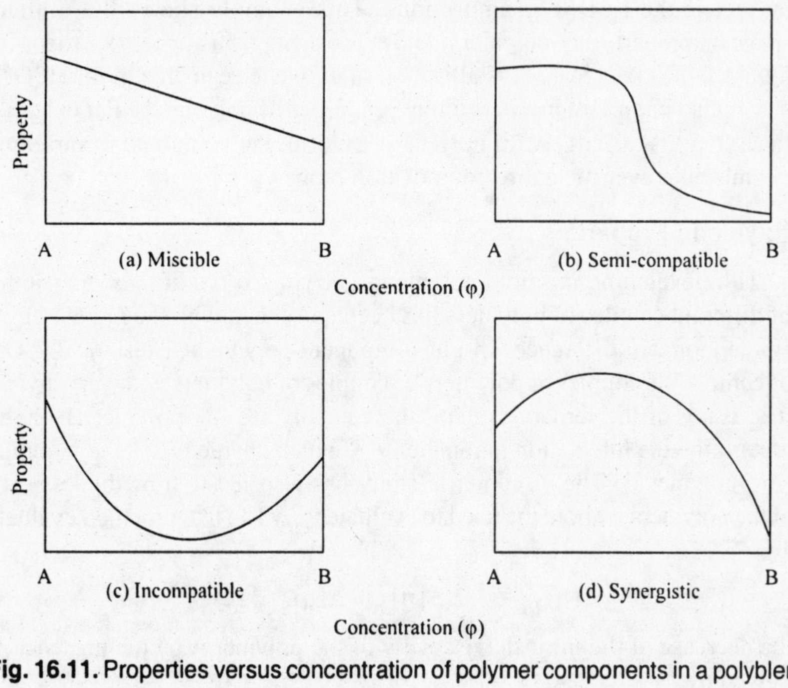

Fig. 16.11. Properties versus concentration of polymer components in a polyblend.

For two polymers to be completely miscible, the optimum requirements are similar polarity, low molecule weight, and hydrogen bonding or other strong intermolecular attraction. Most polymer pairs do not meet these requirements for complete theoretical miscibility. The free energy of mixing is positive, and they tend to separate into two phases.

If they are slightly immiscible, each phase will be a solid solution of minor polymer in major polymer, and the phases will separate into sub-microscopic domains with the polymer present in major amount forming the continuous matrix phase and contributing most toward its properties. Plots of properties versus composition will be S-shaped showing an intermediate transition region where there is a phase inversion from one continuous phase to the other (Fig. 16.11b). Most commercial polyblends are of this type, with the major polymer forming the continuous phase and retaining most of its useful properties, while the minor polymer forms small discrete domains, which contribute synergistically to certain specific properties.

When the polymer components in a blend are less miscible, phase separation will form larger domains with weaker interfacial bonding between them. The interfaces will therefore fail under stress and properties of polyblends are thus likely to be poorer than for either of the polymers in the blend. U-shaped property curves (Fig. 16.11c) thus provide a strong indication of immiscibility. In most cases they also signify practical incompatibility, and hence lack of practical utility.

A fourth type of curve for properties versus polyblend composition representing synergistic behaviour (Fig. 16.11d) has been obtained in a few cases of polymer blending. This is synergistic improvement of properties, beyond what would be expected from simple monotonic proportionality, and sometimes far exceeding the value for either polymer alone. Synergism may result from a very favourable dipole-dipole attraction between the polymer components.

A blend of low-density polyethylene (LDPE) with the terpolymer ethylene-propylene-diene monomer rubber (EPDM) exhibits a synergistic effect on tensile strength if EPDM is partially crystalline, but a non-synergistic effect if the EPDM is amorphous. This example shows the dramatic effect that morphology can have on properties of polymer blends. The synergism apparently arises from a tendency for crystallites in the LDPE to nucleate crystallisation of ethylene segments in the EPDM.

Stereochemistry is also important in determining properties of polyblends. For example, syndiotactic poly(methyl methacrylate) is miscible with poly(vinyl chloride) at certain concentrations, whereas the isotactic form is immiscible over the entire composition range.

Prediction of Polyblend Properties

A major problem with developing miscible polyblend is trying to predict polymer immiscibility. The incompatibility of different couples of polymers has been correlated with the mutual effects on intrinsic viscosities and dipole moment differences of the component polymers. These results can give a guide for the research of compatible couples of polymers or couples of polymers with very low incompatibility.

The viscometric study of the ternary system, that is, polymer A—polymer B—solvent, has been used in order to determine the interaction parameter χ_{AB} which characterises the incompatibility of the couple polymer A—polymer B. The viscometric study has also led to a method of estimation of the incompatibility of the polymers without the need to evaluate χ_{AB}. The latter method evaluates the quantity $\Delta[\eta]_{AB}$ defined by

$$\Delta(\eta)_{AB} = 1/2(\Delta[\eta]_A + \Delta[\eta]_B)\ \% \qquad \text{... (16.4)}$$

where $\Delta[\eta]_A$ is the decrease of the intrinsic viscositiy of the polymer A by the presence of the polymer B in the solvent, given by

$$\Delta[\eta]_A = ([\eta]_A - [\eta]_{AB})/[\eta]_A \qquad \text{... (16.5)}$$

The decrease is due to the incompatibility existing between polymer A and polymer B. The quantity $\Delta[\eta]_B = ([\eta]_B - [\eta]_{BA})/[\eta]_B$ similarly gives the decrease of the intrinsic viscosity of polymer B due to the presence of polymer A in the solvent. The constant quantity of the polymer A or B in the mixture, solvent plus polymer, is always the same (typically 0.125×10^{-2} g/cm^3). Then the quantity $\Delta[\eta]_{AB}$ given by Eq. (16.4) expresses a measure of the incompatibility existing between polymer A and polymer B. A high value of $\Delta[\eta]_{AB}$ for a polymer couple indicates a high value of incompatibility and vice versa.

The incompatibility of polymer couples has been correlated with the dipole moment μ of the two polymers. As it is known the value of μ of each polymer is given per monomer repeating unit and is given by each reaction $\mu = (\overline{\mu^2} / \overline{DP}_n)^{1/2}$, where $\overline{DP}_n$ is the number average degree of polymerisation and $\overline{\mu^2}$ is the mean square dipole moment of the long-chain molecule.

The values of $\Delta[\eta]_{AB}$ and $\Delta\mu$ for several couples of vinyl polymers drawn from polystyrene, poly(methyl methacrylate), poly(vinyl chloride) and poly(vinyl acetate) are given in Table 16.15. It is seen that when the dipole moment of polymer A is very close to the dipole moment of polymer B, the incompatibility of the mixture, polymer A —polymer B presents a high incompatibility when the dipole moment of polymer A is too different from the dipole moment of polymer B.

Table 16.15. Values of $\Delta(\eta)_{AB}$ and $\Delta\mu$ for six polymer couples.

Couples of polymers	$\Delta(\eta)_{AB}$	$\Delta\mu$
PS–PMMA	7.00	0.99
PS–PVC	10.55	1.19
PS–PVA	11.20	1.34
PMMA–PVA	3.6	0.35
PMMA–PVC	10.55	1.19
PMMA–PVA	3.30	0.15

In the case of two polymers presenting a high value of the difference $|\mu_A - \mu_B|$, the dissimilarity is high and the monomers of each polymer give interactions of the type $\delta^- - \delta^+$ with other monomers that belong to the same polymer and this leads to the phase separation. In the contrary, in the case of two polymers presenting a very low value of the difference $|\mu_A - \mu_B|$, the dissimilarity is low and the monomers of each polymer give interactions with other monomers that belong to the same or to the other polymer indifferently; hence the polymers are compatible.

Attempts to predict miscibility using simple solubility parameters have been largely unsuccessful because strong dipolar interactions are not taken into account. The importance of these interactions has been demonstrated, for example, with miscible mixtures of poly(acrylic esters) and poly(vinyl fluoride), where the compatibility has been attributed to dipolar interactions of the type.

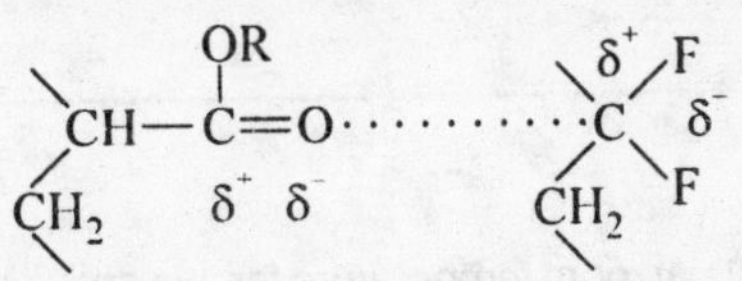

Predictions of properties for immiscible polyblends is much more complicated. This is partly due to the effects of varying morphologies that might arise as a result of processing variables. Frequently, one polymer will form a continuous phase with the second being dispersed as a non-continuous phase in the form of spheres, lamellae, fibrils, and so on. It is, however, the polymer in the continuous phase that largely determines the properties of the polyblend. For example, a 50:50 blend of polystyrene (a hard, glassy polymer at ordinary temperature) and polybutadiene (an elastomer) will be hard if polystyrene is the continuous phase, but soft if polystyrene is the dispersed phase. In some cases, however, an immiscible polyblend may have both components dispersed as continuous phases. Evidently, a proper control of phase morphology is of utmost importance with immiscible blends. The size of the dispersed phase should be optimised considering the final performance of the blend.

The major problem with immiscible blends is the poor physical attraction at phase boundaries that can lead to phase separation under stress resulting in poor mechanical properties. A number of ingenious approaches have been adopted to overcome this problem by improving compatibility between immiscible phases. One is through formation of interpenetrating polymer networks (IPN) as described in Table 16.14.

In IPNs the polymers are physically 'locked' together by the interdispersed three-dimensional network, a phenomena referred to as topological bonding. Such mixtures, however, still undergo phase separation into microdomains that vary in size according to the degree of immiscibility. Controlling the kinetics of phase separation during the formation of IPN [or the semi-interpenetrating polymer networks (SINs)] provides the method of generation of desired properties.

Another approach is to incorporate compatibilisers or interfacial agents into the blend to improve adhesion between phases. The concept of compatibilisation of polymers is described in a later section.

Discovery of practical synergistic polyblends is difficult and is not predictable with the present state of understanding. In some cases, the blend properties are truly exceptional beyond expectation. One of these rare cases is presented in Fig. 16.12, in which the room temperature notched Izod impact strength is plotted as a function of composition for PBT/PC/co-polymer (SAN-EPDM) system.

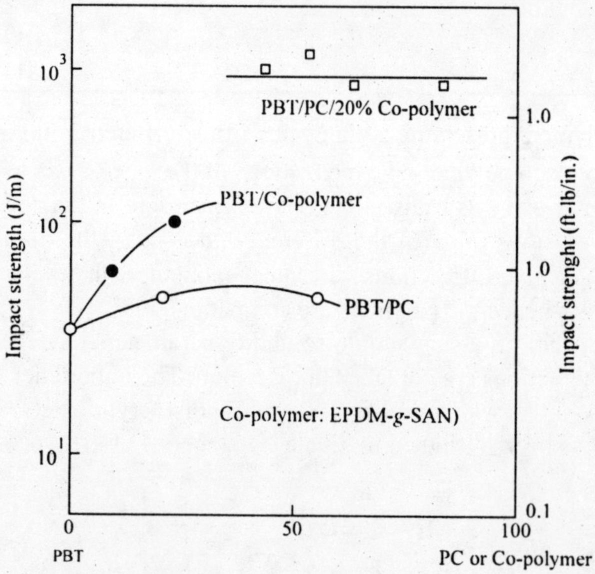

Fig. 16.12. Notched Izod impact strength at room temperature for two and three component blends of poly(butylene terephthalate) (PBT) with polycarbonate (PC) and styrene-acrylonitrile (SAN) grafted on ethylene-propylene-ethylidenenorbornene (EPDM) co-polymer.

Addition to PBT of either PC or a co-polymer of styrene-acrylonitrile grafted on ethylene-polypropylene-diene (EPDM) resulted only in a small improvement of impact strength. However, blending PBT with a polyblend of 80 per cent PC and 20 per cent co-polymer improved the impact strength by 15 times. Since each constituent of the polyblend showed only a small effect, the large increase in impact strength would be attributed to three pairwise interaction coefficients: PC-PBT, PBT-SAN, and SAN-PC. A 'window of miscibility' may be responsible for the effect.

Selection of Blend Components

The properties of engineering polymer blends and alloys that are sought to be improved by blending are impact strength, processability, tensile strength, rigidity/modulus, heat-deflection temperature, flame retardancy, thermal stability, dimensional stability, and chemical/solvent resistance. Among them, toughening and processability are of major concern.

The second group of importance (27 per cent) includes strength, modulus, and heat-deflection temperature. The third group (14 per cent) concerns flame retardancy, solvent resistance, as well as thermal and dimensional stability. Other properties such as elongation, and gloss are seldom mentioned (8 per cent).

The patent literature also provides information on means of achieving these goals. As Table 16.16 indicates, the blending effect in most cases is non-specific, e.g., addition of an elastomer to any engineering resin will improve its impact strength (Fig. 16.13), or addition of a high modulus material to the lower modulus matrix will increase rigidity. Optimisation of performance would involve a multivariable analysis where the molecular parameters of the ingredients, the interphase, and the morphology interact. This obviously cannot be gleaned from the data in Table 16.16. The latter should only be considered a general guide to blending. The fact that two polymers with desired properties are immiscible should not be a deterrent since the modern compatibilisation and reactive processing methods (discussed later) can overcome such problems.

Table 16.16. How to modify properties by blending.

Property	*Matrix resin*	*Modifying polymer*
Impact strength	PVC, PP, PE, PC, PA, PPE, TPE	ABS, ASA, SBS, EPR, EPDM, PBR, SAN, SMA, MBA, Polyolefin, HIPS
HDT, stiffness	PC, PA	TPEs, PEI, PPE
	ABS, SAN	PC, PSO
Flame retardancy	ABS, Acrylics	PVC, CPE
	PA, PC	Aromatic-PA, PSO, co-polysiloxanes or phosphazanes
Chemical/solvent resistance	PC, PA, PPO	TPEs
Barrier properties	Polyolefins	PA, EVOH, $PVCl_2$
Processability	PPO	Styrenics
	PET, PA, PC	PE, PBR, MBS, EVOH
	PVC	CPE, Acrylics
	PSO	PA
	PO	PTFE, SI

There is no doubt that the main reason for polymer blending is economy. If a material can be generated at a lower cost with properties meeting specifications, the manufacturer must use it to remain competitive. In general, the following economy related reasons can be cited: (i) extending engineering resin performance by diluting it with a low-cost polymer (Noryl being an outstanding example); (ii) developing materials with a full set of desired properties; (iii) forming a high-performance blend from synergistically interacting polymers; (iv) adjusting the composition of the blend to customer specifications; and (v) recycling industrial and municipal plastics scrap.

The main and most difficult task in polyblend production is the development of materials with a full set of desired properties. This is usually achieved by selecting blend components in such a way that the principal advantages of one polymer will compensate for deficiencies of the second and vice versa. For example, the disadvantages of PPO with regard to processability and impact strength (Table 16.17) are compensated by advantageous properties of either PA or HIPS. Due to PPO/PS miscibility, the original Noryls were formulated as PPO/HIPS polyblend.

The reactive methods of compatibilisation developed subsequently allowed the second generation Noryl GTX (a blend of PPO with PA) to be developed. The compositions claimed usually cover 30–70 per cent of each of the main ingredients, PPO and PA, with additionally up to four parts of such modifier as polycarboxylic acid, trimellitic anhydride acid chloride, quinone, oxidised polyolefin wax, and so on. In most cases, PA forms a matrix with spherical inclusions of PPO acting as compatibilised low-density filler.

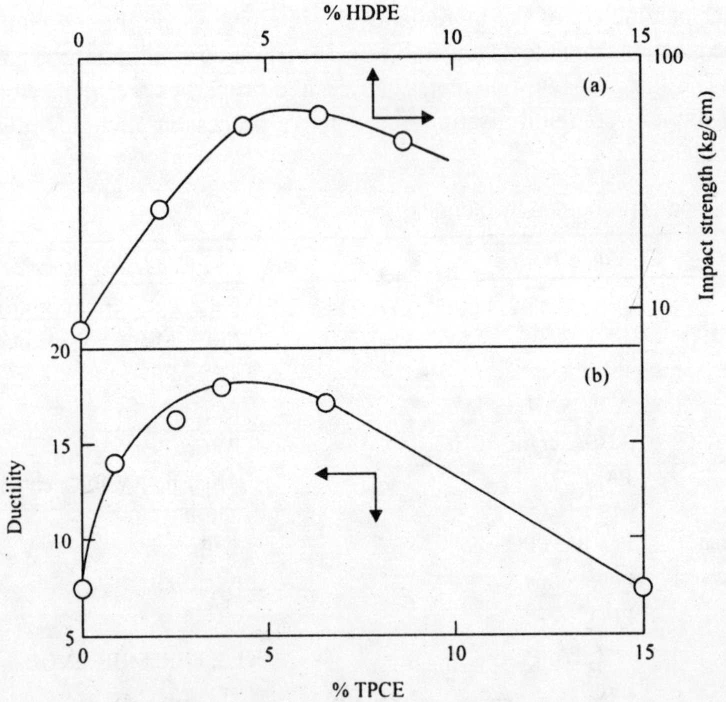

Fig. 16.13. Concentration dependence of (a) notched Izod impact strength at room temperature of polycarbonate/ high-density polyethyleneblend and (b) ductility of acrylonitrile-butadiene-styrene (ABS) co-polymer blend with thermoplastic co-polyether ester elastomer (TPCE).

Table 16.17. Advantages and disadvantages of some engineering polymers and modifiers.

Polymer	*Advantages in*	*Disadvantages in*
Polyamide (PA)	Processability, impact strength, crystallinity	Water absorption, HDT
Polycarbonate (PC)	Low-temperature toughness, HDT	Stress-crack sensitivity, solvent and chemical resistance
Polyoxymethylene (POM)	Tensile strength, modulus	Stress-crack sensitivity, impact strength
Polyphenylene oxide (PPO)	HDT, rigidity, flame retardancy	Processability, impact strength
Thermoplastic polyesters (PET, PBT)	Chemical and solvent resistance	Shrinkage, low-temperature toughness, processability
High-impact polystyrene (HIPS)	Processability, impact strength	HDT
Acrylonitrile-butadiene-styrene co-polymer (ABS or ASA)	Impact strength, processability, weatherability	HDT

Similarly the disadvantages of PC are the stress cracking and chemical sensitivity. Stress cracking can be treated as a part of impact properties and a simple solution may thus be addition of ABS or ASA. On the other hand, to improve the solvent resistance—a property that is particularly important in automobile applications—a semi-crystalline polymer may be added. From Table 16.17, it is apparent that TPEs (e.g., PBT, PET) can provide that property, but they also lack warp resistance and impact strength. Hence an ideal blend for automobile application based on PC and TPEs should be impact modified with, for example, an acrylic latex co-polymer. A schematic of preparation of this type of toughened blends Xenoy is shown in Fig. 16.14.

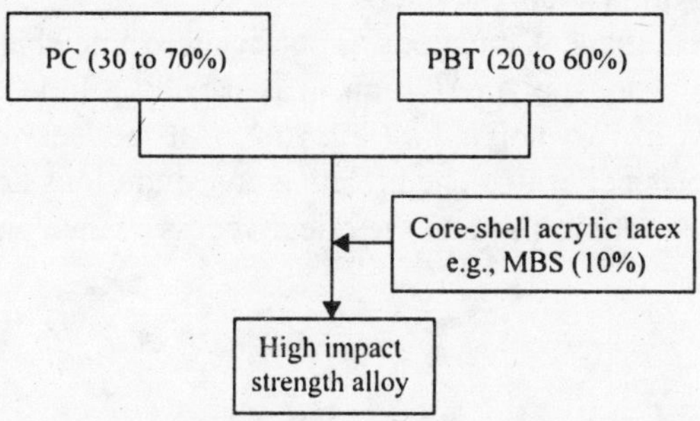

Fig. 16.14. Schematic preparation of polycarbonate/poly(butylene terephthalate) blend.

As exemplified by PC/TPE/latex system, modern blends are increasingly required to play multiple roles so as to provide a balance of such diverse properties as mechanical behaviour, chemical/solvent resistance, dimensional stability, paintability, weatherability, and, of course, economy. Such a complex balance is usually achieved by means of multicomponent blending often with unavoidable compromises.

Compatibilisation of Polymers

Blends made from incompatible polymers are usually weak. These poor blends are a result of high interfacial tension and poor adhesion between the two phases. For many years, blending of polymers was unsuccessful due to the fact that many polymers were incompatible. Lately new polymer blends have been successfully made by the use of compatibilising agents. This has yielded polymers of unique properties not attainable from either of the polymer components of the blends. Sometimes these blends are preferred over completely miscible blends since they may combine important characteristics of both polymers. Their performance greatly depends on the size and morphology of the dispersed phase.

Polymers can be compatibilised in a number of ways. Polymers that act as mutual solvents can be used as compatibilising agent, e.g., polycaprolactone in a blend of polycarbonate and poly(styrene-co-acrylonitrile). Polymers can also be compatibilised by the use of functionalised polymers. Thus polymers can be post-reacted with a reactive monomer to form functional groups that react with a second polymer to form grafts in a blend, e.g., EPDM-*g*-maleic anhydride/nylon-6,6 blends are produced by grafting maleic anhydride onto EPDM, followed by blending with nylon-6,6. The grafted maleic anhydride reacts with the terminal —NH_2 groups of the nylon to form the following co-polymer:

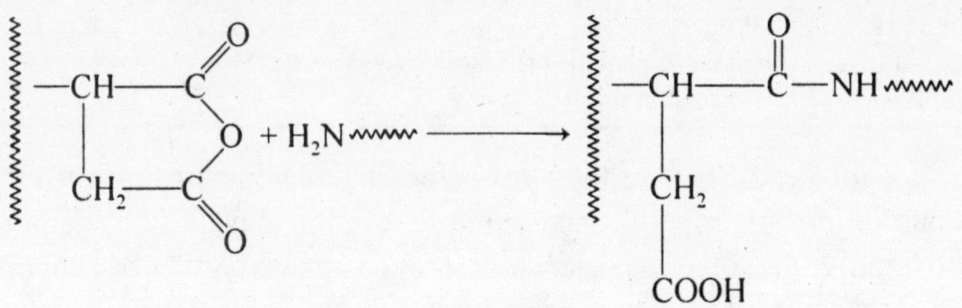

Compatibilisation can also be realised by *in situ* formation of block or graft co-polymers. For example, blends of EPDM and PMMA are obtained by the reactive extrusion of the homopolymers and a peroxide in a twin screw extruder. Block or graft co-polymers can also be added separately to produce blends with controlled particle sizes. This type of compatibilising agent is called an interfacial agent since it acts at the boundary between two immiscible polymers.

An interfacial compatibilising agent functions as surfactant, somewhat the same as low molecular weight surfactant used in emulsion polymerisation. It is usually a block or graft co-polymer and is located at the interface between two polymers (Fig. 16.15). It reduces interfacial tension, alleviates gross segregation, and promotes adhesion. The effective concentrations are from 0.1 per cent to 5 per cent depending on the desired particle size and the effectiveness of the compatibilising agent.

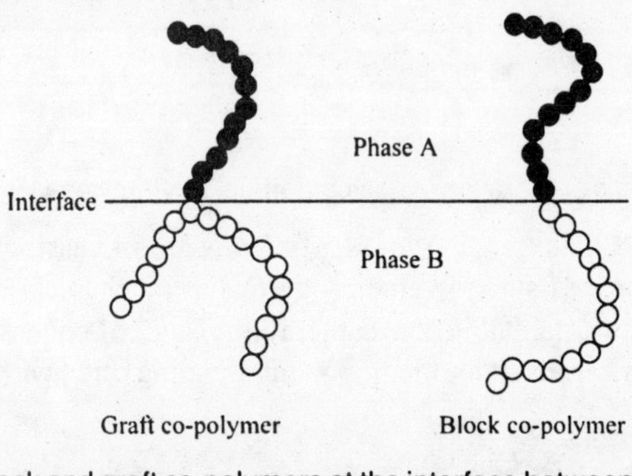

Fig. 16.15. Ideal location of block and graft co-polymers at the interface between polymers phases A and B in a polyblend.

The block or graft co-polymer used as a compatibilising agent must have the ability to separate into their respective phases and not be miscible as a whole molecule in one phase. The best choice of block or graft co-polymers would be to choose polymers identical to the blended polymers, e.g., poly(A-*b*-B) or poly(A-*g*-B) co-polymers for blending poly(A) and poly(B). Another alternative is to use poly(C-*b*-D) or poly(C-*g*-D) where C and D are compatible with A and B, respectively or where C and D adhere to A and B, respectively, but are not miscible with them. Examples of such compatibilised systems that have been studied include EPDM/PMMA blends compatibilised with EPDM-*g*-MMA, polypropylene/polyethylene with EPM or EPDM, polystyrene/nylon-6 with polystyrene/nylon-6 block co-polymer, and poly(styrene-co-acrylonitrile)/poly(styrene-co-butadiene) with butadiene rubber/PMMA block co-polymer. Reactive compatibilising agents of the type A-C can also compatibilise an A/B blend as along as C can chemically react with B. Recently studied systems include polyethylene (PE)/nylon-6 blends compatibilised with carboxyl functional PE, polypropylene (PP)/poly(ethylene-terephthalate), with PP-*g*-acrylic acid, nylon-6,6/EPDM with poly(styrene-co-maleic anhydride), and nylon-6/PP with PP-*g*-maleic anhydride. For the last cited system, for example, maleic anhydride was grafted onto PP and the later was then blended with nylon-6 in order to compatibilise an immiscible blend of nylon-6/PP by A-C method through anhydride-amine reaction.

In several commercial polymer blends, 'modifiers' that also act as compatibilising agents are used. The modifier is usually an interactive co-polymer containing a rubbery component. Acrylic-based co-polymers, chlorinated polyolefins, ethylene-propylene-diene, poly(ethylene-*co*-vinyl acetate), etc., are frequently used. These play a dual role, compatibilising and toughening the blend. For this reason, they are used at much higher loading than pure compatibilisers. Thus, while 0.1 to 5 wt per cent of the latter may be sufficient, 10 to 40 wt per cent of a modifier may be needed (Fig. 16.14).

Co-reaction of blends to improve the performance has for decades been a practice in the rubber industry. In high-shear mixers some of the chains in rubbers are broken and are reformed by the free-radical mechanism. A similar phenomenon occurs during intensive mixing of polyolefins (Table 16.14). To enhance this process, sometimes a free-radical source, e.g., peroxides, can be added.

Industrial Polyblends

The automotive sector is an important area for polymer blends. Alloys of polycarbonate (PC) and poly(butylene terephthalate) (PBT), which combine the high impact strength of PC with the resistance of gasoline and oil and processing ease of PBT, are being increasingly used in car bumpers. General Electric's PC/PBT alloy is used for bumpers in several cars made in the United States and Europe. Bayer's PC/PBT blends, called Markrolon PR, are used for several automobile impact-resistant parts. Markrolon EC900 is also used for crash helmets.

An important requirement for plastics materials in the automotive industry is paintability alongside metal components in high temperature areas. Noryl GTX series of General Electric has been developed to meet this requirement. The blend consisting of polyphenylene oxide (PPO) in a nylon matrix combines the heat and dimensional stability and the low water absorption of PPO with the flow and semi-crystalline properties of nylons.

Its impact strength, however, is not as high as PC/PBT blends. Several grades of Noryl GTX are available, some of these being glass-filled. The unfilled grades are used for automobile exterior components such as fenders, spoilers and wheeltrims, whereas the glass-reinforced grades are used in under-the-bonnet applications, such as cooling fans, radiator end caps, and impeller housings, where in addition to high heat resistance extra stiffness is required. Bayer's Bayblend PC/ABS alloys are used in dashboard housings, interior mirror surrounds, and steering column covers. The high stability of the alloy allows pigmentation in pastel colours.

Some alloys of PPO have been developed specifically for office equipment and housings. Noryl AS alloys, which are blends of PPO/PS and special fillers, are expected to replace metals in such applications as structural chassis for typewriters and printers.

Monsanto Chemical Co. produce injection moulded Nylon/ABS alloys (Triax) that are characterised by good toughness, high chemical resistance, and easy mouldability. Applications of these alloys include industrial power tools, lawn and garden equipment housings and handles, gears, pump impellers, and car fascia.

Bayer and BASF have developed a PC/ASA blend where ASA is the rubber variant of ABS in which the polybutadiene phase is completely replaced by a polyacrylate rubber. Compared to PC/ABS alloys, the PC/ASA blend has greater resistance to weathering, which permits outdoor applications and an extra heat stability that reduces the risk of yellowing in processing. BASF's alloy, Terblend S, is recommended for outdoor applications as it has six times the UV resistance of PC/ABS and an even higher resistance when pigmented.

A number of polyphenylene ether (PPE) blends have been introduced by Borg-Warner, Hulls, and BASF. These blends have the advantage that they can be made flame retardant without halogens. Borg-Warner offers many grades of PPE/high impact polystyrene (HIPS) blend under the trade name Prevex. The materials are characterised by high impact resistance and high heat distortion temperatures, Applications are for office equipment, business machines, and telecommunication equipment. Luranyl PPE/HIPS blends of BASF include general purpose types, reinforced grades, impact modified types, and flame-proof blends. BASF also has a PPE/nylon blend tradenamed Ultranyl, which is aimed at a similar market.

Blends have also been used to a limited extent in packaging. Du Pont's Solar is a nylon/polyethylene blend that can be blow-moulded into bottles and can contain volatile hydrocarbons found in household cleaners and agricultural chemicals. Further developments in such barrier blends are expected.

INTERPENETRATING POLYMER NETWORKS

An interpenetrating polymer network (IPN) has been defined as 'an intimate combination of two polymers both in network form, at least one of which is synthesised or cross-linked in the presence of the other'. There are no induced covalent bonds between the two polymers, i.e., monomer A reacts only with other molecules of monomer A. A schematic representation of an ideal IPN (full IPN) in which both the polymers are cross-linked is shown in Fig. 16.16a. If one of the two polymers is in network for (cross-linked) and the other a linear polymer, i.e., not cross-linked, the product is called a semi-IPN (Fig. 16.16b). An IPN is distinguished from simple polymer blends and block or graft co-polymers in two ways: it swells but does not dissolve in solvents, and crepe and flow are suppressed.

IPNs represent a third mechanism, in addition to mechanical blending and co-polymerisation, by which different polymers can be intimately combined. IPNs can be synthesised either by sequential polymerisation or simultaneous polymerisation of two monomers. In a sequential synthesis, monomer I is polymerised using an initiator and a cross-linking agent to form network I. Network I is then swollen in monomer II containing cross-linking agent and initiator to form network II by polymerisation *in situ*. Such polymers are called sequential interpenetrating networks (SIPN). On the other hand, polymer networks synthesised by mixing monomers or linear polymers (pre-polymers) of the monomers together with their respective cross-linking agents and catalysts in melt, solution, or dispersion, followed, usually immediately, by simultaneous polymerisation by non-interfering modes, are called simultaneous interpenetrating networks (SIN). In the latter process, the individual monomers are polymerised by chain or stepwise polymerisation, while reaction between the polymers is usually prevented due to different modes of polymerisation.

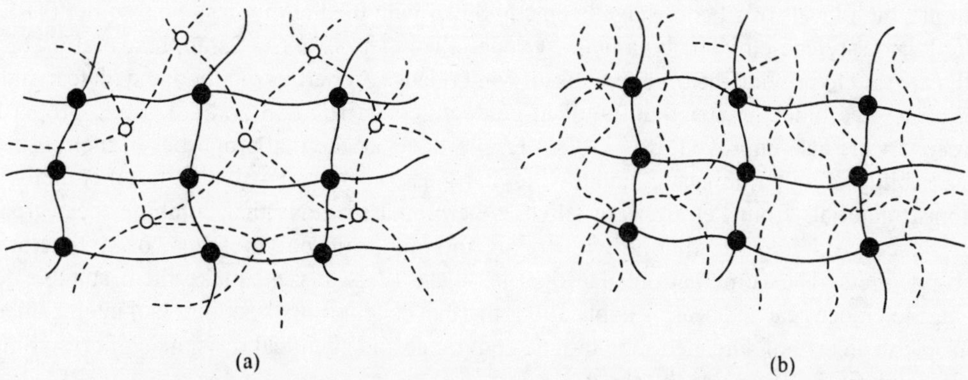

Fig. 16.16. Schematic representation of (a) a full IPN and (b) a semi-IPN.

The advantage of SINs is the ease of processing; for example, polyurethane-epoxy or unsaturated polyester SINs are used in reaction injection moulding (RIM) applications. An A-component consists of the polyurethane polyol, while liquid MDI blended with liquid epoxy or unsaturated polyester forms the B-component. The two components may be mixed together for about 30 seconds and elastomeric sheets cast in heated aluminium moulds.

The modulus of such SIN-RIM part is enhanced by modifying the polyurethane with glassy epoxy or unsaturated polyester, unlike conventional reinforced RIM (RRIM), which utilises glass fibres to increase the modulus (glass has the disadvantage of increasing the viscosity and wearing down machine pumps by abrasion).

Interpenetrating polymerisation is the only way of combining cross-linked polymers. This technique can be used to combine two or more polymers into a mixture in which phase separation is not as extensive as it would be otherwise. Normal blending or mixing, it may be noted, result in a multiphase morphology unless the polymers are completely compatible. However, complete compatibility, which is almost impossible, is not necessary for complete phase mixing by interpenetrating polymerisation, since permanent entanglements produced by interpenetration prevent phase separation. With moderate compatibility, intermediate and complex phase behaviour results.

Thus IPNs with dispersed phase domains have been reported ranging from a few micrometers (the largest) to a few hundred nanometers (intermediate) to those without a resolvable domain structure (complete mixing). With highly incompatible polymers, on the other hand, the thermodynamics of phase separation is so powerful that it occurs before it can be prevented by cross-linking.

Combining polymeric networks in different compositions often results in IPNs with controlled different morphologies showing synergistic behaviour. For example, a glassy polymer (T_g above room temperature) combined with an elastomeric polymer (T_g below room temperature) gives a reinforced rubber, if the elastomeric phase is predominant and continuous, or a high impact plastic, if the glassy phase is continuous. More complete phase mixing enhances mechanical properties due to increased physical cross-link density. IPNs have thus been synthesised in various network compositions with optimum bulk properties such as tensile strength, impact strength, and thermal resistance.

Industrial Interpenetrating Polymer Networks (IPNs)

Commercial IPNs have been developed to combine useful properties of two or more polymer systems. For example, high levels of silicone have been combined with the thermoplastic elastomer (TPE) based on *Shell's* Kraton styrene-ethylene/butadiene-styrene TPE and *Monsanto's* Santoprene olefin TPE. These IPN TPEs are said to provide the biocompatibility and release properties of silicone with tear and tensile strength up to five times greater than medical-grade silicone. Thermal and electronic properties and elastic recovery are also improved. IPN TPEs offer physical and thermal properties of thermoset rubber, the processability of a thermoplastic, and a wider hardness range than available to other TPEs.

Rimplast materials from Petrarch are silicone/thermoplastic IPNs that combine the warpage and wear resistance properties of silicone with nylon, thermoplastic polyurethane, or styrene/butadiene block co-polymers. The combination of properties these IPNs possess make them suitable for high quality, high tolerance gear and bearing applications that can benefit from the addition of internal lubrication and isotropic shrinkage of silicone to the fatigue endurance and chemical resistance of crystalline resin, while processability still remains the same.

Interpenetrating polymer/metal networks are now on the market. One of these, produced by Princeton Polymer Laboratories, imparts conductive properties to the IPN. Low melting alloys such as zinc and tin are melt-blended with a resin, forming a fibrous network within the resin matrix. The process of metal addition eliminates the problem of breaking up of fibre or flake additives and has very little effect on processability.

SECTION V

Polymer Processing and Fabrication of Polymers

Section V

Polymer Processing and Fabrication of Polymers

CHAPTER 17

Plastic Processing and Fabrication Processes

INTRODUCTION

The various processes are used to fabricate all types and shapes of plastic products, ranging from household convenience packages to electronic devices and many others—including the strongest products in the world, used in space vehicles, aircraft, building structures, and so on. Proper process selection depends upon the nature and requirements of the plastic, the properties desired in the final product, the cost of the process, its speed, and product volume. (Note that a plastic also may be called a polymer or a resin).

Some materials can be used with many kinds of processes; others require a specific or specialised machine. Numerous fabrications process variables play an important role and can markedly influence a product's aesthetics, performance, and cost.

This chapter contains information applicable to all processes characterised by certain common variables or behaviours, such as plastic melt flow, heat controls, and so forth. It is essential to recognise that for any change in a processing operation, there can be advantages and/or disadvantages. The old rule still holds for every action there is a reaction. A gain in one area must not be allowed to cause a loss in another; changes must be made that will not be damaging in any respects.

All processes fit into an overall scheme that requires interaction and proper control of different operations. This FALLO (Follow ALL Opportunities) approach can be used in any process by including those 'blocks' that pertain to the fabricated product's requirements.

The FALLO concept has been used by many manufacturers to produce acceptable products at the lowest cost. Computer programmes featuring this type of layout are available. The FALLO approach makes one aware that many steps are involved in processing, and all must be coordinated.

The specific process (injection, etc.) is an important part of the overall scheme and should not be problematic. The process depends on several interrelated factors, such as: (i) designing a part to meet performance and manufacturing requirements at the lowest cost; (ii) specifying the plastic; (iii) specifying the manufacturing process, which basically requires (a) designing a tool 'around' the part, (b) putting the 'proper performance' fabricating process around the tool, (c) setting up necessary auxiliary equipment to interface with the main processing machine, and (d) setting up 'completely integrated' controls to meet the goal of zero defects; and 'properly' purchasing equipment and materials, and warehousing the materials.

TYPES OF PROCESSES

The family of polymers is extraordinarily large and varied. There are, however, some fairly broad and basic approaches that can be followed when designing or fabricating a product out of polymers or, more commonly, polymers compounded with other ingredients. The type of fabrication process to be adopted depends on the properties and characteristics of the polymer and on the shape and form of the final product.

In the broad classification of plastics there are two generally accepted categories: thermoplastic resins and thermosetting resins.

Thermoplastic resins consist of long polymer molecules, each of which may or may not have side chains or groups. The side chains or groups, if present, are not linked to other polymer molecules (i.e., are not cross-linked). The thermoplastic resins, usually obtained as a granular polymer, can therefore be repeatedly melted or solidified by heating or cooling. Heat softens or melts the material so that it can be formed; subsequent cooling then hardens or solidifies the material in the shape. No chemical change usually takes place during this shaping process.

In thermosetting resins the reactive groups of the molecules form cross-links between the molecules during the fabrication process. The cross-linked or 'cured' materials cannot be softened by heating. Thermoset materials are usually supplied as a partially polymerised moulding compound or as a liquid monomer-polymer mixture. In this uncured condition they can be shaped with or without pressure and polymerised to the cured state with chemicals or heat.

With the progress of technology the demarcation between thermoplastic and thermoset processing has become less distinct. For thermosets processes have been developed which make use of the economic processing characteristics of thermoplastics. For example, cross-linked polyethylene wire coating is made by extruding the thermoplastic polyethylene, which is then cross-linked (either chemically or by irradiation) to form what is actually a thermoset material that cannot be melted again by heating. More recently, modified machinery and moulding compositions have become available to provide the economics of thermoplastic processing to thermosetting materials. Injection moulding of phenolics and other thermosetting materials are such examples. Nevertheless, it is still a widespread practice in industry to distinguish between thermoplastic and thermosetting resins. Examples of common thermoplastics are ABS, PVC, SAN, acetals, acrylics, cellulosics, polyethylenes, polypropylenes, polystyrenes, polycarbonates, polyesters, nylons, and fluoropolymers. Some of the more common thermosets are phenolics, ureas, melamines, epoxies, alkyds, polyesters, silicones, and urethanes.

Compressing and transfer moulding are the most common methods of processing thermosetting plastics. For thermoplastics, the more important processing techniques are extrusion, injection, blow moulding, and calendering; other processes are thermoforming, slush moulding, and spinning.

TOOLING FOR PLASTICS PROCESSING

Tooling for plastics processing defines the shape of the part. It falls into two major categories, moulds and dies. A mould is used to form a complete three-dimensional plastic part. The plastics processes that use mould are compression moulding, injection moulding, blow moulding, thermoforming, and reaction injection moulding. A die, on the other hand, is used to form two of the three dimensions of a plastic part. The third dimension, usually thickness or length, is controlled by other process variables. The plastics processes that use dies are extrusion, pultrusion and thermoforming. Many plastics processes do not differentiate between the terms mould and die. Moulds, however, are the most predominant form of plastics tooling.

Types of Mould

The basic types of mould, regardless of whether they are compression, injection, transfer, or even blow moulds, are usually classified by the type and number of cavities they have. For example, Fig. 17.1 illustrates three mould types: (i) single-cavity; (ii) dedicated multiple-cavity; and (iii) family multiple-cavity.

Single-cavity mould (Fig. 17.1a) represents one of the simplest mould concepts. This design lends itself to low-volume product and to large plastic part designs. The multiple-cavity moulds may be of two types. A dedicated multiple-cavity mould (Fig. 17.1b) has cavities that produce the same part. This type of mould is very popular because it is easy to balance the plastic flow and establish a controlled process. In a family multiple-cavity mould (Fig. 17.1c), each cavity may produce a different part. Historically, family mould designs were avoided because of difficulty in filling uniformly; however, recent advances in mould making and gating technology make family moulds appealing. This is the case especially when a processor has a multiple-part assembly and would like to keep inventories balanced.

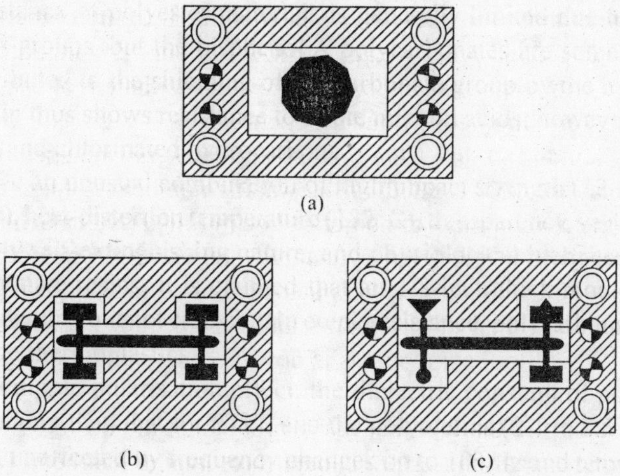

Fig. 17.1. Three basic types of moulds: (a) single-cavity; (b) dedicated multiple-cavity; and (c) family multiple-cavity.

Types of Dies

Within the plastics industry, the term die is most often applied to the processes of extrusion. Extrusion dies may be categorised by the type of product being produced (e.g., film, sheet, profile, or co-extrusion), but they all have some common features, as described below.

1. Steel: The extrusion process being continuous, both erosion and corrosion are significant factors. Hence the dies must be made of a high-quality tool steel, hardened so that the areas that contact the plastic material do not erode. Additionally, many dies have a dense, hard chrome plating in the area where plastic melt contacts the die.
2. Heaters: Extrusion dies are to be heated in order to maintain a melt flow condition for the plastic material. Most of the heaters are cartridge-type elements that slip fit into the die at particular locations. In addition to the heaters, the dies have to accommodate temperature sensors, such as thermocouples.

3. Melt pressure: Many sophisticated dies are equipped with sensors that monitor melt pressure. This allows the processor to better monitor and control the process.
4. Parting line: Large extrusion dies must be able to separate at the melt flow line for easier fabrication and maintenance. Smaller extrusion dies may not have a parting area, because they can be constructed in one piece.
5. Die swell compensation: The polymer melt swells when it exits the die, as explained previously. This is die swell is a function of the type of plastic material, the melt temperature, the melt pressure, and the die configuration. The die must be compensated for die swell so that the extruded part has the corrected shape and dimensions.

Moulds and dies for different fabrication processes will be described later in more detail when the processes are discussed.

Tool Design

The design of the tooling to produce a specific plastics part must be considered during the design of the part itself. The tool designer must consider several factors that may affect the fabricated part, such as the plastics material, shrinkage, and process equipment. Additionally, competitive pressures within the plastics industry require the tool designer to consider how to facilitate tool changeovers, optimise tool maintenance, and simplify (or eliminate) secondary operations.

Historically, plastics moulds and dies were built by tool-makers who spent their lives learning and perfecting their craft. Today the void created by the waning numbers of these classically trained tool-makers is being filled by the development of numerically controlled (NC) machinery centres, computer-based numerically controlled (CNC) machinery centres, and computer-aided design (CAD) systems. Moulds and dies can now be machined on computer-controlled mills, lathes, and electric discharge machines that require understanding of computers and design, rather than years of experience and machining skills. The quality of tool components is now more a function of the equipment than of the tool-makers's skill.

The high costs of moulds and the fact that many production moulds are built under extreme time constraints leave no room for trial and error. Though prototyping has been widely used to evaluate smaller part designs when circumstances and time allow, prototyping is not always feasible for larger part designs. There are, however, several alternatives to prototyping, e.g., computer-aided design (CAD), finite-element analysis (FEA), and rapid prototyping.

While CAD allows a tool designer to work, with a three-dimensional computer model of the tool being designed and to analyse the design, FEA allows the tool to be evaluated (on a computer) for production worthiness. The mould is then fabricated from the computer model, a process called computeraided manufacturing (CAM). Rapid prototyping is a relatively new method of producing a plastics part by using a three-dimensional computer drawing. A sophisticated prototyping apparatus interprets the drawing and guides an articulating lesser beam across a specific medium such as a photopolymer plastic or laminated paper, the result being a physical representation of the computer-based drawing.

Prototyped parts can be produced in less than 24 hours, and part designs can be scaled to fit the size of the prototyping equipment. Another trend is the introduction of moulds that accept interchangeable modules.

Modules take less time to manufacture, and in turn, cut down on the delivery time and costs. In addition, it usually takes less time to change the modules than the entire mould frame.

Compression moulding

Compression moulding is normally used for thermosetting plastics, although certain thermoplastic articles are made by this process. A measured amount of polymer is charged to the mould, which is then closed and heated. The polymer can be in the form of a loose material charge (i.e. powder of granules) or be a preformed shape, as shown in Fig 17.2. The preform is a low density moulding that has been lightly compacted in a cold mould.

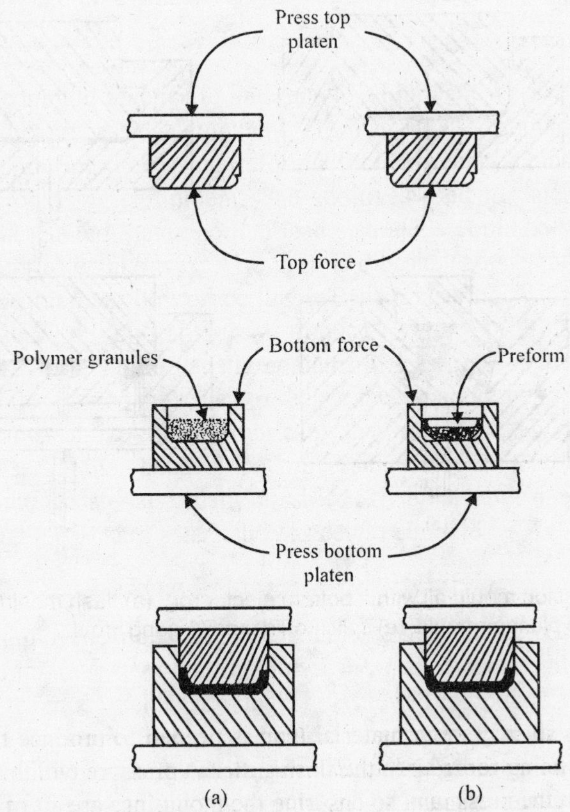

Fig. 17.2. Compression moulding using (a) a loose material charge; and (b) a low density preform.

The pressure exerted by the plunger is usually pre-determined to give the required density of the finished article. The heat and pressure cause the polymer to plasticise and flow throughout the cavity. The pressure required may be governed initially by certain additives in the polymer. Heavy, coarse fillings will cause poor flow characteristics which may be impossible to counteract by temperature alone. The shape of the component may also affect the polymer flow. This is particularly true of intricate shapes and deep sections. For this reason, compression moulding is better suited to relatively shallow components having no intricate shapes or inserts. Preforming helps to overcome the flow problem as the uniform, low density preform will not need to flow as much as loose material charged directly into the heated mould. The design of compression moulds varies quite markedly as shown in Fig 17.3. All these would be used in upstroking or downstroking presses incorporating heated platens. They each have various advantageous and limitations as outlined below.

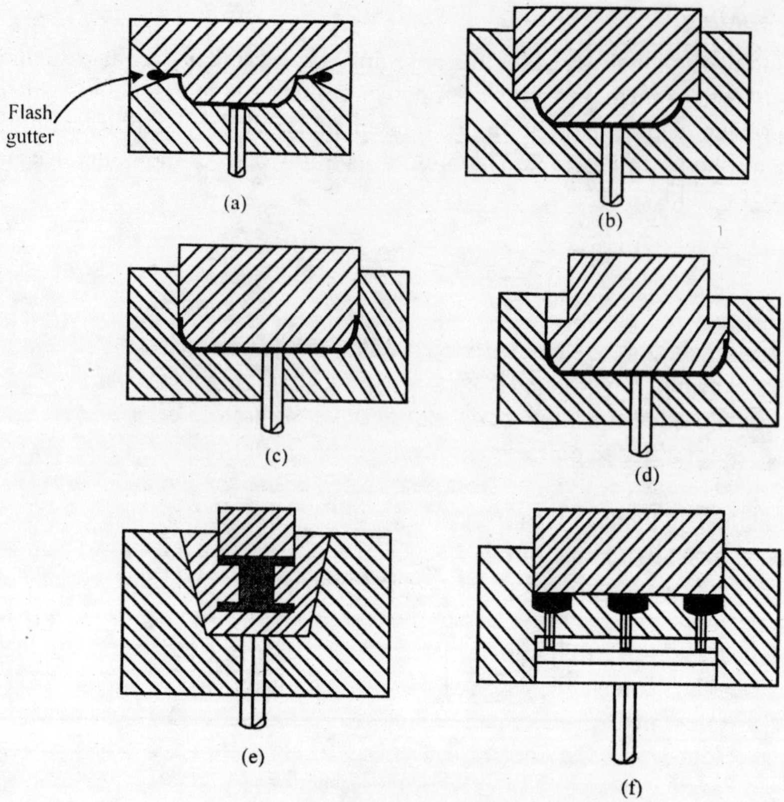

Fig. 17.3. Types of compression mould all with a bottom ejector rod: (a) flash mould; (b) semi-positive mould; (c) positive mould; (d) landed plunger mould; (e) split mould; and (f) gang mould.

Flash mould

The cavity is charged with slightly more material than is needed to produce the finished component, which results in the excess being forced into the flash gutter. A pressure will thus be set up in the cavity balancing the pressure from the press ram, so ensuring the mouldings are all of uniform density. This is the simplest type of mould and therefore the cheapest to manufacture.

Semi-positive mould

Also known as positive flash moulds, because excess material will escape vertically until the plunger engages with the cavity walls. Consequently the mould can be overcharged to ensure the mouldings are all of uniform density. This type of mould is particularly suitable for high bulk factor materials or deep mouldings, and the vertical flash is usually easier to trim then horizontally produced flash.

Positive mould

This will overcome the problem of a moulding whose density varies, which can happen when the leak-off in a semi-positive mould varies around the cavity. However, any variation in the amount of material charged into the cavity will result in density variations between individual moulding. The charge in usually preformed to help overcome the problem.

Landed plunger mould

These can be used for high bulk materials because the reduced land width in contact with the inside face will permit much deeper cavities than other mould designs. This also enables mouldings to be produced of a much higher density than most other moulds. The thin line of vertical flash that inevitable ensures is less costly to remove than a horizontal line from the flash mould.

Split mould

The inner ring forming the actual mould cavity is usually made up of three segments. The inner ring is ejected along with the moulding and the ring segments then stripped away. This allows shapes to be moulded which would otherwise be impossible to strip from the mould. This design of mould is a much cheaper way to produce such components than alternative methods (e.g. machining).

Gang mould

This is an economical method of producing a number of small components in a single shot. The cavities can be machined in a single block as shown in Fig. 17.3, or as separate mating blocks enabling cheaper replacements to be made if damage occurs. The mouldings are connected by a thin line of flash if the measured charge is correct, but if this cannot be controlled accurately, the land between adjacent cavities is kept to an absolute minimum.

Transfer moulding

This overcomes the main disadvantage of compression moulding by having a separate heated chamber outside the mould cavity. The polymer is first charged to the heated transfer pot and when sufficiently molten, forced by the plunger into the mould below. After cooling the moulding has to be broken from the sprue, the feed pipe from the transfer gate (Fig 17.4). Unfortunately this creates an increase in the downtime of the process, as the material left in the transfer gate (the cull) has to be removed after each moulding cycle and discarded. This adds to the curing time of between one and five minutes for which the polymer has to remain in the mould.

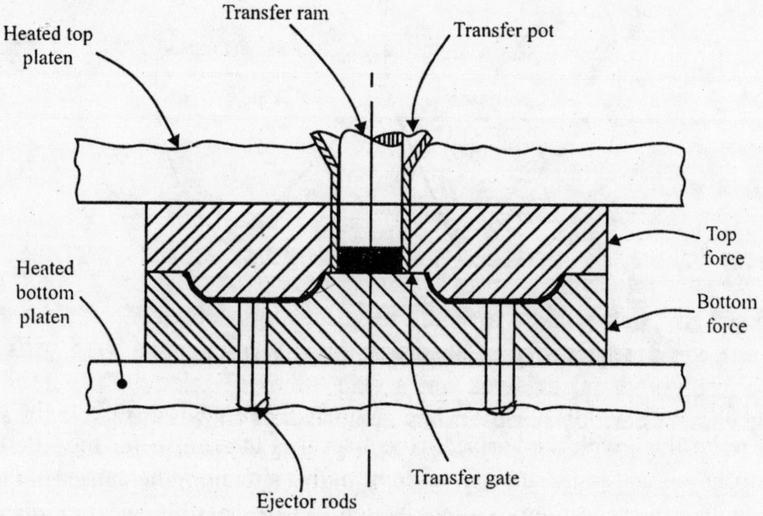

Fig. 17.4. Transfer mould.

The main criterion in the design of the mould is the positioning of the transfer gate. For articles of consistent shape the gate should be positioned so that the polymer is forced to travel the shortest possible distance in filling the mould cavity. However, due to the polymer being plasticised before it enters the mould cavity, thick sections do not suffer from density variations. As this leads to transfer moulds being used for articles having both thick and thin sections, it is better to place the transfer gate above a thick rather than thin section. Just like the multiple or gang compression mould, a multi-cavity transfer mould can be used. This has the cavities arranged radially around the central transfer plot.

The closure of the transfer mould can be done by operating the transfer pot with a separate press cylinder to the ram, or allowing the ram pressure to hold the mould cavity closed. In the latter case the area of the transfer chamber should be at least 25 per cent greater than the area of the contact face of the mould, otherwise there is a danger of the internal pressure forcing open the cavity.

Inserts in plastics

A fairly common feature of polymer components is some form of insert which has been incorporated during the forming process. Most inserts are metallic and are usually of the form of an external pin or internal bush. The most common type of insert is the threaded insert for although threads can be easily formed in the polymer, the material is often too weak to accommodate the forces exerted on the thread form. Three different designs for a threaded insert are shown in Fig 17.5. Each illustrates a different method of preventing the insert from being pulled out or turning. Feathered edges alone are to be avoided as the thickness of polymer preventing the insert from being pulled out can easily be sheared. Similarly square or hexagonal inserts can cause the stress, which is developed as a result of the tendency to rotation, to be concentrated at the corners. This can cause the polymer to crack or split.

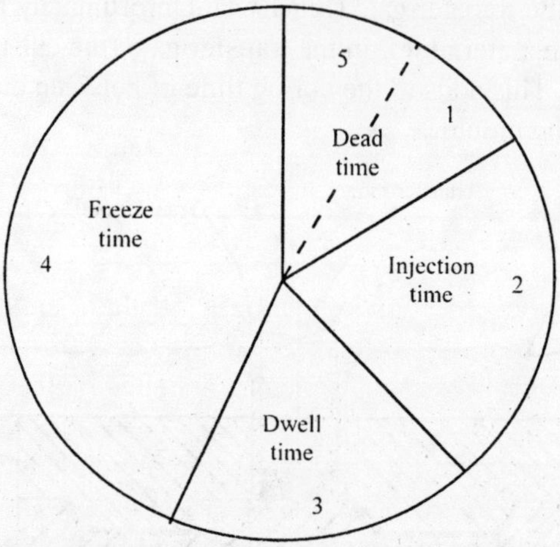

Fig. 17.5. Diagram of the injection moulding cycle.

Consideration has to be given to the possibility of the polymer encroaching on the face of the insert if this is required to be flush with the surface. One answer is to arrange for the insert to be the same diameter as the carrier pin and to be screwed on to it. In this situation, the carrier pin would need flats on the shank to enable it to be unscrewed after ejection. Where possible, ejection pins are arranged to push on the insert carrier, or are designed as a combined insert carrier/ejector pin as shown in Fig 17.6.

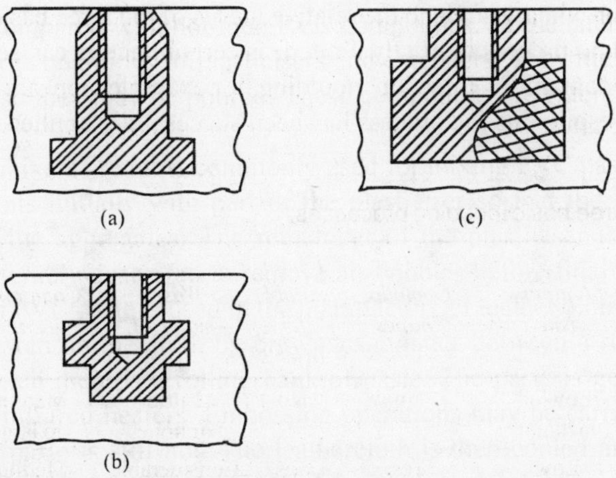

Fig. 17.6. Types of threaded insert: (a) feathered edge; (b) knurled periphery; and (c) squared ends.

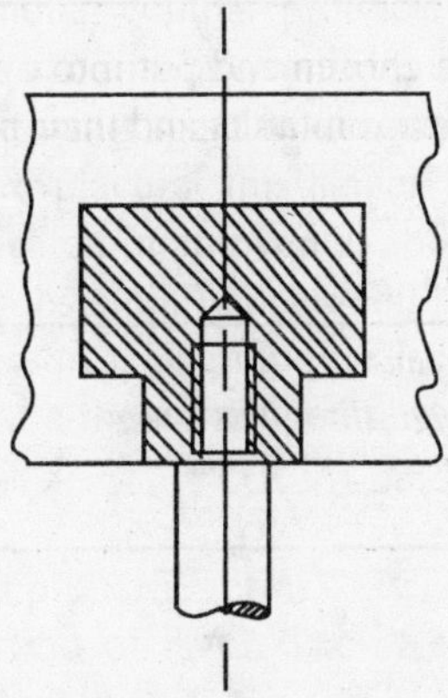

Fig. 17.7. Insert located on a carrier or ejector.

Selection of production processes

The correct choice of production process for a particular component often requires a deep understanding of polymer processing and a wide range of experience. A number of factors and variables have to be considered, such as:

1. The type of polymer—thermosetting or thermoplastic.
2. The form—liquid, granule, powder or sheet.
3. The type of additives—solid, liquid, gaseous or metallic.
4. The size of the article.
5. The complexity of its shape.
6. The quantity required.
7. Whether inserts are required.
8. The design of the tooling.
9. The tool cost.

Table 17.1 gives an approximate guide to the relative merits of the three basic processes. It should be appreciated, however, that no process is strictly limited, as certain factors can sometimes be modified to give greater freedom and adaptability. Injection moulding, for example, appears more suited to very high output of small articles. Despite this the process has been successfully applied to the production of car bodies and boat hulls.

Table 17.1. A guide to the three basic mouding processes.

Process	*Polymer type*	*Output rate*	*Complex shapes*	*Factor suitable for inserts*	*Waste material*	*Component size*	*Machine capital cost*	*Mould cost*
Compression moulding	Mainly thermosetting	Low or medium	No	No	Little or none	Medium to large	Medium	Low
Transfer moulding	Mainly thermosetting	Low	Yes	Yes	Considerable	Medium to large	Medium	Medium
Injection moulding	Thermo Plastic	High	Yes	Yes	None	Small to medium	High	High

After the production process has been chosen and put into operation, there are numerous faults that can occur in the component. The most common faults and their probable causes (★) are shown in Table 17.2.

Table 17.2. Common injection-moulding faults.

Fault	*Probable cause*					
	Operating temperature	*Injection pressure*	*Injection time*	*Cooling time*	*Mould locking force*	*Gate design*
Brittleness	★	★				
Bubbles		★	★			★
Crazing	★	★				
Distortion	★	★	★	★		
Flashing	★	★			★	
Gloss	★					
Hollows	★	★		★		★
Sticking	★	★				

Other Processing Techniques

Compression, transfer and injection moulding are the three most widely used methods of polymer processing. There are a number of other basic techniques used by industry for producing the vast array of consumer products, not the least of which is packaging, which accounts for so much of the polymer that is manufactured.

Calendering

Three or four heated rolls in a single vertical stand are used for producing continous thermoplastic sheet. The sheet can be used as the raw material for further processing or as a finished product.

Extrusion

This uses a machine similar to the screw injection machine with an extrusion die in place of the split mould. As well as solid sections, tubes of various diameters can be extruded, including tubular film for making bags and sacks.

INJECTION MOULDING OF THERMOPLASTICS

Injection moulding is the most important moulding method for thermoplastics. It is based on the ability of thermoplastic materials to be softened by heat and to harden when cooled. The process thus consists essentially of softening the material in a heated cylinder and injecting it under pressure into the mould cavity, where it hardens by cooling. Each step is carried out in a separate zone of the same apparatus in a cyclic operation.

A diagram of a typical injection-moulding machine is shown in Fig. 17.8. Granular material (the plastic resin) falls from the hopper into the barrel when the plunger is withdrawn. The plunger then pushes the materials into the heating zone, where it is heated and softened (plasticised or plasticated). Rapid heating takes place due to spreading of the polymer into a thin film around a torpedo. The already molten polymer displaced by this new material is pushed forward through the nozzle, which is in intimate contact with the mould. The molten polymer flows through the sprue opening in the die, down the runner, past the gate, and into the mould cavity. The mould is held tightly closed by the clamping action of the press platen. The molten polymer is thus forced into all parts of the mould cavities, giving a perfect reproduction of the mould. The material in the mould must be cooled under pressure below T_m or T_g before the mould is opened and the moulded part is ejected. The plunger is then withdrawn, a fresh charge of material drops down, the mould is closed under a locking force, and the entire cycle is repeated. Mould pressure of 8000 to 40,000 psi (562 to 2812 kg/cm^2) and cycle times as low as 15 sec are achieved on some machines.

Note that the feed mechanism of the injection moulding machine is activated by the plunger stroke. The function of the torpedo in the heating zone is to spread the polymer melt into thin film in close contact with the heated cylinder walls. The fins, which keep the torpedo centred, also conduct heat from the cylinder walls to the torpedo, although in some machines the torpedo is heated separately.

Injection-moulding machines are rated by their capacity to mould polystyrene in a single shot. Thus a 2 oz machine can melt and push 2 oz of general-purpose polystyrene into a mould in one shot. This capacity is determined by a number of factors such as plunger diameter, plunger travel, and heating capacity. The main components of an injection-moulding machine are (i) the injection unit which melts the moulding material and forces it into the mould; (ii) the clamping unit which opens the mould and closes it under pressure; (iii) the mould used; and (iv) the machine controls.

INJECTION MOULDING PROCESS

Fundamentally the injection moulding of thermoplastics, compared to the compression, moulding of thermosets, is a simple process. Thermoplastic pellets are heated until fluid, then forced into a mould where the plastic re-solidifies to reproduce the shape of the mould. In practice, however, injection moulding is not a simple and easy process, and many details must be successfully worked out to circumvent difficulties. Time, pressure, and temperature are the basic variables that must be controlled, given the poor thermal conductivity of the thermoplastics. Heating a thermoplastic until fluid is not necessarily easy; it is possible to degrade the surface by heat before the interior of the plastic mass is warm. That the plastic mass will necessarily fill the mould is also an innocent assumption; it may have voids, blisters, sinks, and shorts, and some cavities in a multicavity mould may not receive any plastic at all. Finally, of course, the part must be ejected, and thermoplastics, like thermosets, are adhesive and, given any scope for doing so, will stick.

The plunger type of injection moulding machine is shown in Fig. 17.8. The cycle begins with a feed mechanism. This meters out a constant amount of material for each moulding cycle.

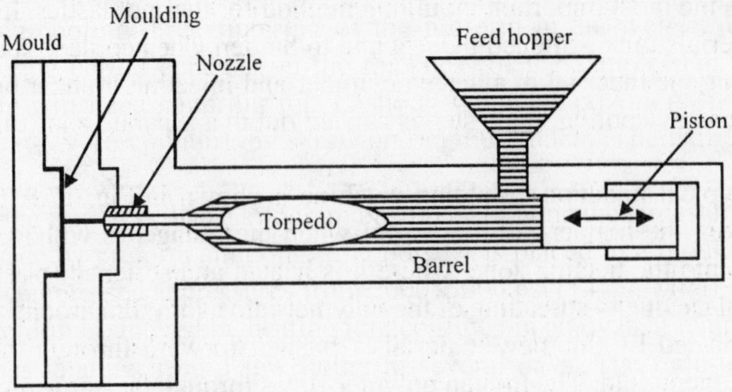

Fig. 17.8. Injection moulding machine, plunger type.

The hydraulic plunger pushes the plastic through the plasticating chamber, which is a heating cylinder, and then through the nozzle into the mould. The torpedo or spreader (Fig. 17.9), provides an increased surface-to-mass ratio by reducing the cross-section of the passages, thus improving the rate of heat transfer into the plastic.

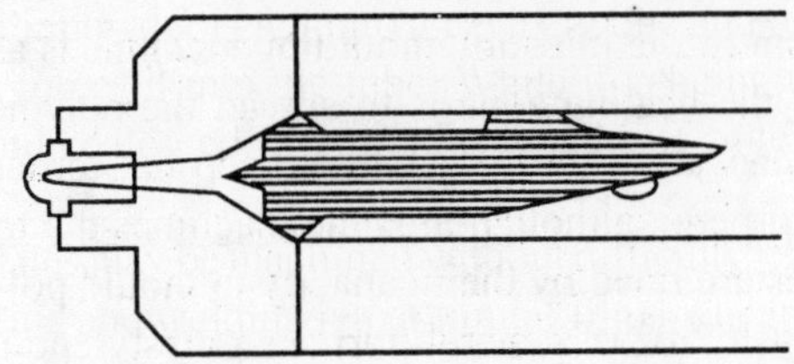

Fig. 17.9. The spreader (torpedo).

The fluid plastic is forced into the mould though a nozzle connecting the cylinder and the mould. The area of contact between mould and cylinder is small to minimise the flow of heat from the cylinder to the mould. Besides a cavity or cavities, the mould must contain passages to connect the cavities to the nozzle. These connecting passages which supply the cavities are the sprue, runners, and gates. The sprue is the passage between the nozzle and the runners. It is usually tapered. Each cavity is connected to the sprue by a runner. The gates are very short, restricted passages at the entrance to the cavity.

The plastic that freezes in the sprue and runners is removed from the part and re-ground to be fed again into the moulding machine. Time, temperature, and pressure controls indicate whether performance requirements of a moulded part are met. Time factors include rate of injection, duration or ram pressure, time of cooling, time of plastication, and screw RPM. Pressure factors are injection high and low pressure, back pressure on the extruder screw, and pressure loss before the plastic enters the cavity, which can be caused by a variety of restrictions in the mould. Temperature factors are mould (cavity and core), barrel, and nozzle temperatures, as well as the melt temperature due to back pressure, screw speed, frictional heat, and so on. The principal variables that must be controlled are nine in number. There may be additional controls also, depending on the uses to which the machine is put:

1. The amount of plastic introduced into the cylinder.
2. The pressure applied to the plunger.
3. Plunger speed.
4. Temperature of the heating cylinder. (Nozzle temperature also may be controlled).
5. Temperature of the mould.
6. Plunger-forward time.
7. Mould closed time.
8. Mould clamping force.
9. Mould open time.

The temperature of the plastic is a critical variable, since viscosity is controlled by temperature. Mould temperature is controlled by a circulating fluid and a means of heating and cooling the fluid. The quality of the moulded product is highly dependent on the ability to adjust these control settings.

Many of the above variables influence the others; for example, if the temperature of the heating cylinder is increased, then the gate freezes more slowly and a longer plunger-forward time may be needed if sinks and voids are to be prevented.

Immediately after the mould is filled with plastic, the mould pressure is greatly increased, and packing of the mould occurs. The compressibility of the plastic under high pressure allows some flow into the mould after the filling stage. Also, the plastic in the cold mould will at once begin to cool and therefore to occupy less volume and allow more plastic to enter from the gate. The usual practice is to time the ram to return after the gate to the cavity is sealed, in order to prevent any discharge from the mould.

The plastic granules are metered into the process either by volume or by weight. The usual arrangement is a volumetric feed, in which the plastic falls from the hopper into a box, which may be adjusted to the desired volume. The measured volume of material then drops into the heating cylinder in front of the ram. Volumetric feeding is sufficiently precise for most injection-moulding operations. More complex parts, or high-speed operation may require more precise method of feeding by weight. Slight difference in granule geometry, granule friction, or even static electricity may cause variations in volume.

The plasticating chamber is a steel cylinder heated by electric resistance elements in the outer jacket. The discharge end of the cylinder is fitted with a nozzle that mates with the sprue bushing of the mould. Nozzles may be changed to suit the requirements of the mould. To promote more uniform and more rapid heating, a torpedo or spreader is placed in the plasticating chamber. Heat is transmitted from the cylinder wall to the torpedo by fins.

Some machines are equipped with a pre-plasticator, which is an auxiliary heating cylinder in which the plastic is melted. The plunger of the pre-plasticator forces the plastic into the shooting cylinder, which shoots it into the mould (Fig. 17.10). The pre-plasticator makes it possible to use lower injection moulding pressures, a higher rate of injection, lower cylinder temperature, and a larger shot capacity with increased heating capacity. Some pre-plasticators use an extruder screw instead of a ram plunger.

The newer designs of injection moulding machines use an extruder screw for melting the plastic. The screw is also used as a plunger to feed the mould (Fig. 17.11). The extruder screw has a number of important advantages. Mechanical energy from the screw is converted into heat, which is distributed uniformly throughout the plastic as the screw works and mixes it. Heating time and thermal degradation is reduced, and mixing of pigment in the plastic is more uniform.

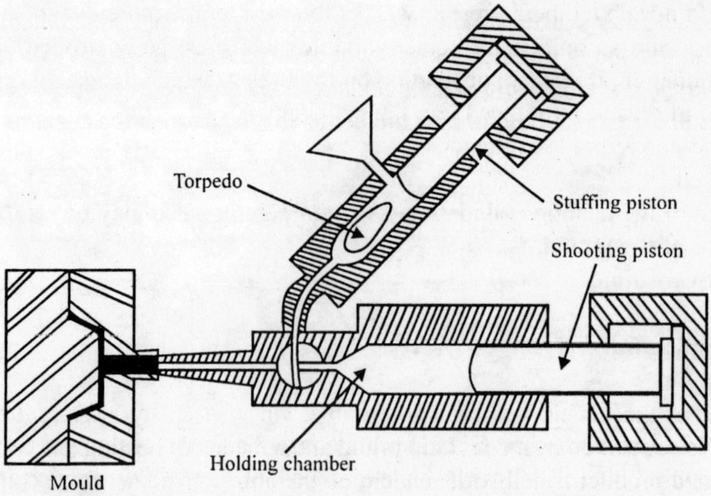

Fig. 17.10. Piston type pre-plasticating machine.

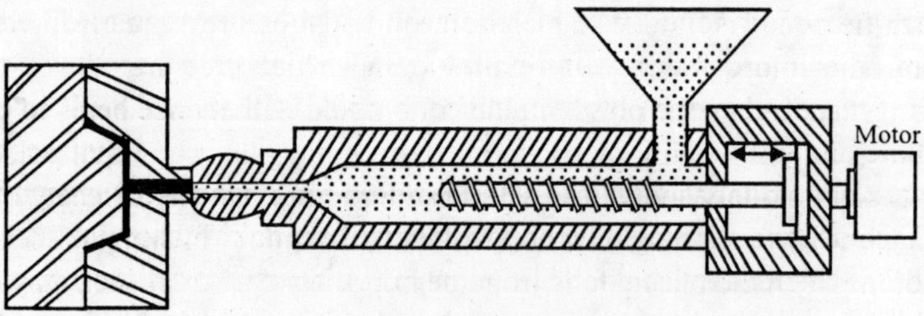

Fig. 17.11. Reciprocating screw injection moulding machine.

Clamping Units

The clamping unit keeps the mould closed while plasticised material is injected into it and opens the mould when the moulded article is ejected. The pressure produced by the injection plunger in the cylinder is transmitted through the column of plasticised material and then through the nozzle into the mould. The unlocking force, that is, the force which tends to open the mould, is given by the product of the injection pressure and the projected area of the moulding. Obviously, the clamping force must be greater than the unlocking force to hold the mould halves closed during injection.

Several techniques can be used for the clamping unit: (i) hydraulic clamps, in which the hydraulic cylinder operates on the movable parts of the mould to open and close it; (ii) toggle or mechanical clamps, in which the hydraulic cylinder operates through a toggle linkage to open and close the mould; and (iii) various types of hydraulic mechanical clamps that combine features of (i) and (ii).

Clamps are usually built as horizontal units, with injection taking place through the centre of the stationary platen, although vertical clamp presses are also available for special jobs.

Moulds

The mould is probably the most important element of a moulding machine. Although the primary purpose of the mould is to determine the shape of the moulded part, it performs several other jobs. It conducts the hot plasticised material from the heating cylinder to the cavity, vents the entrapped air or gas, cools the part until it is rigid, and ejects the part without leaving marks or causing damage. The mould design, construction, the craftsmanship largely determine the quality of the part and its manufacturing cost. The injection moulding is normally described by a variety of criteria, including (i) number of cavities in the mould; (ii) material of construction, e.g., steel, stainless steel, hardened steel, beryllium copper, chrome-plated aluminium, and epoxy steel; (iii) parting line, e.g., regular, irregular, two-plate mould, and three-plate mould; (iv) method of manufacture, e.g., machining, hobbing, casting, pressure casting, electroplating, and spark erosion; (v) runner system, e.g., hot runner and insulated runner; (vi) gating type, e.g., edge, restricted (pinpoint), submarine, sprue, ring, diaphragm, tab, flash, fan, and multiple; and (vii) method of ejection, e.g., knockout pins, stripper ring, stripper plate, unscrewing cam, removable insert, hydraulic core pull, and pneumatic core pull.

Mould designs

Moulds used for injection moulding of thermoplastic resins are usually flash moulds, because in injection moulding, as in transfer moulding, no extra loading space is needed. However, there are many variations of this basic type of mould design. The design most commonly used for all types of materials is the two plate design (Fig. 17.12). The cavities are set in one plate, the plungers in the second plate. The sprue bushing is incorporated in that plate mounted to the stationary half of the mould. With this arrangement it is possible to use a direct centre gate that leads either into a single-cavity mould or into a runner system for a multi-cavity mould. The plungers are ejector assembly and, in most cases, the runner system belongs to the moving half of the mould. This is the basic design of an injection mould, though many variations have been developed to meet specific requirements.

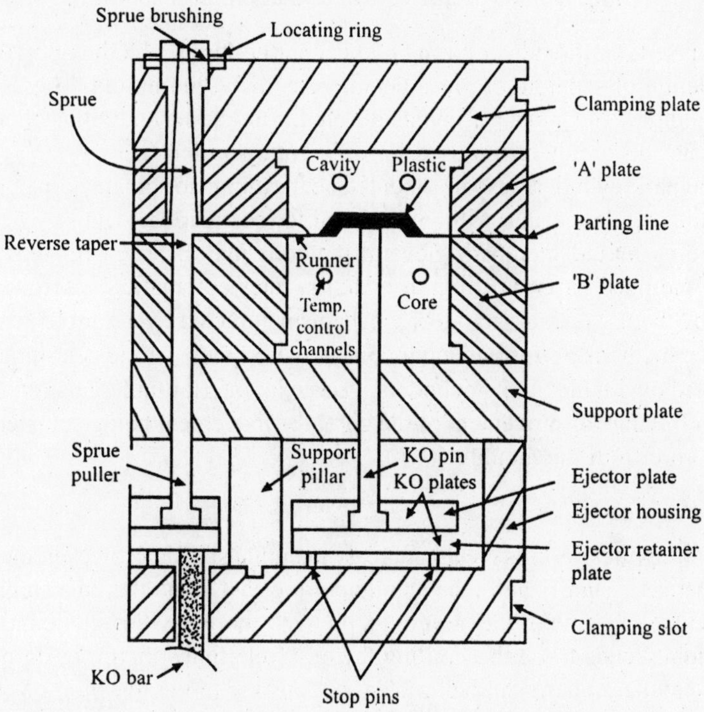

Fig. 17.12. Schematic drawing of a 'two' plate mould.

A three-plate mould design (Fig. 17.13) features a third, movable, plate which contains the cavities, thereby permitting centre or offset gating into each cavity for multicavity operation. When the mould is opened, it provides two openings, one for ejection of the moulded part and the other for removal of the runner and sprue.

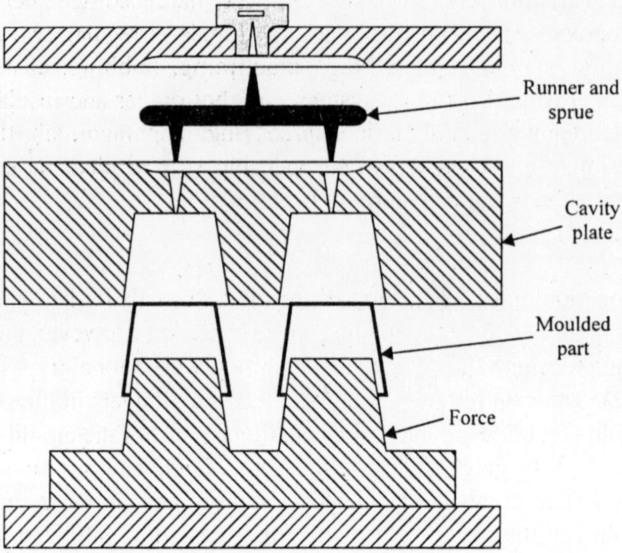

Fig. 17.13. A diagram of a three-plate injection mould.

Mouldings with inserts or threads or coring that cannot be formed by the normal functioning of the press require installation of separate or loose details or cores in the mould. These loose members are ejected with the moulding. They must be separated from the moulding and re-installed in the mould after every cycle. Duplicate sets are therefore used for efficient operation.

Hydraulic or pneumatic cylinders may be mounted on the mould to actuate horizontal coring members. It is possible to mould angular coring, without the need for costly loose details, by adding angular core pins engaged in sliding mould members. Several methods may be used for unscrewing internal or external threads on moulded parts: For high production rates automatic usnscrewing may be done at relatively low cost by the use of rack-and-gear mechanism attracted by a double-action hydraulic long-stroke cylinder. Other methods of unscrewing involve the use of an electric gear-motor drive or friction-mould wipers actuated by double-acting cylinders. Parts with interior undercuts can be made in a mould which has provision for angular movement of the core, the movement being actuated by the ejector bar that frees the metal core from the moulding.

Number of mould cavities

Use of multiple mould cavities permits greater increase in output speeds. However, the greater complexity of the mould also increases significantly the manufacturing cost. Note that in a single-cavity mould the limiting factor is the cooling time of the moulding, but with more cavities in the mould the plasticising capacity of the machine tends to be the limiting factor. Cycle times therefore do not increase prorate with the number of cavities. There can be no clear-cut answer to the question of optimum number of mould cavities, since it depends on factors such as the complexity of the moulding the size and type of the machine, cycle time, and the number of mouldings required. If a fairly accurate estimate can be made of the costs and cycle time for moulds with each possible number of cavities and a cost of running the machine (with material) is assumed, a break-even quantity of the number of mouldings per hour can be calculated and compared with the total production required.

Runners

The channels through which the plasticised material enters the gate areas of the mould cavities are called runners. Normally, runners are either full round or trapezoidal in cross-section. Round cross-section offers the least resistance to the flow of material but requires a duplicate machining operation in the mould, since both plates must be cut at the parting line. In three-plate mould designs, however, trapezoidal runners are preferred, since sliding movements are required across the parting-line runner face.

One can see from Fig. 17.13 that a three-plate mould operation necessitates removal of the runner and sprue system, which must be re-ground, and the material re-used. It is possible, however, to eliminate the runner system completely by keeping the material in a fluid state. This mould is called a hot-runner mould (Fig. 17.14). The material is kept fluid by the hot-runner manifold, which is heated with electrical cartridges. The advantage of a hot-runner mould is that in a long-running job it is the most economical way of moulding—there is no re-grinding, with its attendant cost of handling and loss of material, and the mould runs automatically, eliminating variations caused by operators. A hot-runner mould also presents certain difficulties: It takes considerably longer to become operational, and in multicavity moulds balancing the gate and the flow and preventing drooling are difficult. These difficulties are partially overcome in an insulated-runner mould, which is a cross between a hot-runner mould and a three-plate mould and has no runner system to re-grind. An insulated-runner mould is more difficult to start and operate than a three-plate mould, but it is a considerably easier than a hot-runner mould.

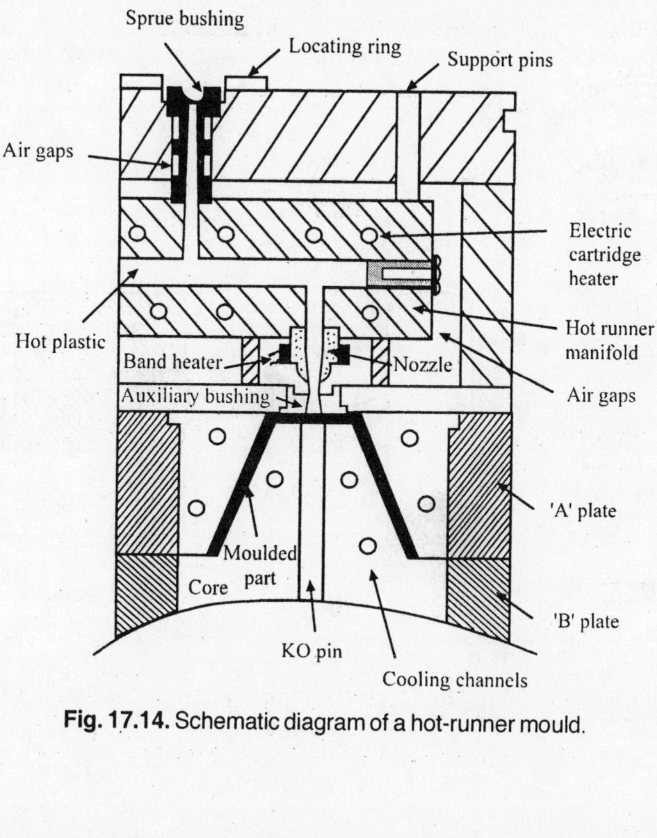

Fig. 17.14. Schematic diagram of a hot-runner mould.

Gating

The gate provides the connection between the runner and the mould cavity. It must permit enough material to flow into the mould to fill out the cavity. The type of the gate and its size and location in the mould strongly affect the moulding process and the quality of the moulded part. There are two types of gates: large and restricted. Restricted (pinpointed) gates are usually circular in cross-section and for most thermoplastics do not exceed 0.060 inch in diameter. The apparent viscosity of a thermoplastic is a function of shear rate—the viscosity decreases as the shear rate and, hence, the velocity increases. The use of the restricted gate is therefore advantageous, because the velocity of the plastic melt increases as it is forced through the small opening; in addition, some of the kinetic energy is transformed into heat, raising the local temperature of the plastic and thus further reducing its viscosity. The passage through a restricted area also results in higher mixing. The most common type of gate is the edge gate (Fig. 17.15a), where the part is gated either as a restricted or larger gate at some point on the edge. The edge gate is easy to construct and often is the only practical way of gating. It can be fanned out for large parts or when there is a special reason. Then it is called a fan gate (Fig. 17.15f). When it is required to orient the flow pattern in one direction, a flash gate (Fig. 17.15c) may be used. It involves extending the fan gate over the full length of the part but keeping it very thin.

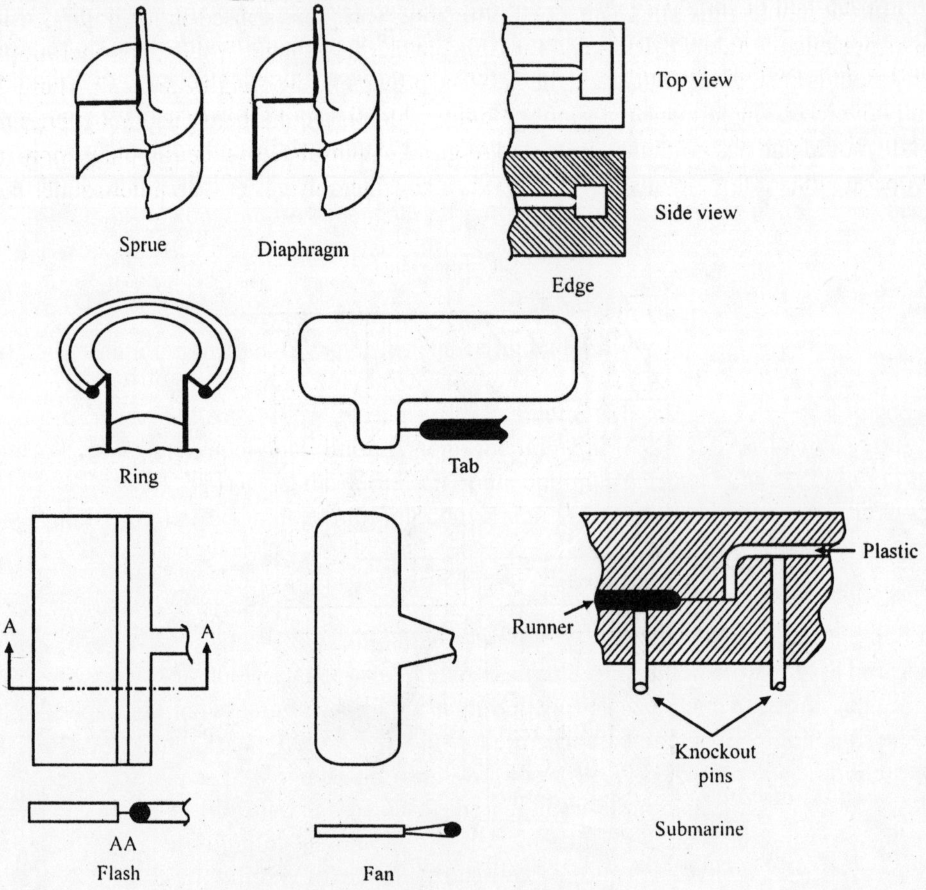

Fig. 17.15. Gating designs.

The most common gate for single-cavity moulds is the sprue gate (Fig. 17.15d). It feeds directly from the nozzle of the machine into the moulded part. The pressure loss is therefore minimum. But the sprue gate has the disadvantages of the lack of a cold slug, the high stress concentration around the gate area, and the need for gate removal. A diaphragm gate (Fig. 17.15e) has, in addition to the sprue, a circular area leading from the sprue to the piece. This type of gate is suitable for gating hollow tubes. The diaphragm eliminates stress concentration around the gate because the whole area is removed, but the cleaning of this gate is more difficult than for a sprue gate. Ring gates (Fig. 17.15f) accomplish the same purpose as gating internally in a hollow tube, but from the outside.

When the gate leads directly into the part, there may be surface imperfection due to jetting. This may be overcome by extending a tab from the part into which the gate is cut. This procedure is called tab gating (Fig. 17.15g). The tab has to be removed as a secondary operation. A submarine gate (Fig. 17.15h) is one that goes through the steel of the cavity. It is very often used in automatic moulds.

Venting

When the melted plastic fills the mould, it displaces the air. The displaced air must be removed quickly, or it may ignite the plastic and cause a characteristic burn, or it may restrict the flow of the melt into the mould cavity, resulting in incomplete filling. For venting the air from the cavity, slots can be milled, usually, opposite the gate. The slots usually range from 0.001 to 0.002 inch deep and from 3/8 to 1 inch wide. Additional venting is provided by the clearance between knockout pins and their holes. Note that the gate location is directly related to the consideration of proper venting.

Parting line

If one were inside a closed mould and looking outside, the mating junction of the mould cavities would appear as a line. It also appears as a line on the moulded piece and is called the parting line. A piece may have several parting lines. The selection of the parting line in mould design is influenced by the type of mould, number of cavities, shape of the piece, tapers, method of ejection, method of fabrication, venting, wall thickness, location and type of gating, inserts, post-moulding operations, and aesthetic considerations.

Cooling

The mould for thermoplastics receives the molten plastic in its cavity and cools it to solidity to the point of ejection. The mould is provided with cooling channels. The mould temperature is controlled by regulating the temperature of the cooling fluid and its rate of flow through the channels. Proper cooling or coolant circulation is essential for uniform repetitive mould cycling.

The functioning of the mould and the quality of the moulded part depend largely on the location of the cooling channel. Since the rate of heat transfer is reduced drastically by the interface of two metal pieces, no matter how well they fit, cooling channels should be located in cavities and cores themselves rather than only in the supporting plates. The cooling channels should be spaced evenly to prevent uneven temperatures on the mould surface. They should be as close to the plastic surface as possible, taking into account the strength of the mould material. The channels are connected to permit a uniform flow of the cooling or heating medium, and they are thermostatically controlled to maintain a given temperature. Another important factor in mould temperature control is the material the mould is made from. Beryllium copper has a high thermal conductivity, about twice that of steel and four times stainless steel. A beryllium copper cavity should thus cool about four times as fast as a stainless steel one. A mould made of beryllium copper would therefore run significantly faster then one of stainless steel.

Ejection

Once the moulded part has cooled sufficiently in the cavity, it has to be ejected. This is done mechanically by KO (knockout) pins, KO sleeves, stripper plates, stripper rings or compressed air, used either singly or in combination. The most frequent problem in new moulds is with ejection. Because there is no mathematical way of predicting the amount of ejection force needed, it is entirely a mater of experience.

Since ejection involves overcoming the forces of adhesion between the mould and the plastic, the area provided for the knockout is an important factor. If the area is too small, the knockout force will be concentrated, resulting in severe stresses on the part. As a result, the part may fail immediately or in later service. In material such as ABS and high-impact polystyrene, the severe stresses can also discolour the plastic.

Sticking in a mould makes ejection difficult. Sticking is often related to the elasticity of steel and is called packing. When injection pressure is applied to the molten plastic and forces it into the mould, the steel deforms; when the pressure is relieved, the steel retracts, acting as a clamp on the plastic. Packing is often eliminated by reducing the injection pressure and/or the injection forward time. Packing is a common problem in multicavity moulds and is caused by unequal filling. Thus, if a cavity seals off without filling, the material intended for that cavity is forced into other cavities, causing overfilling.

Standard mould bases

Standardisation of mould bases for injection moulding, is an important factor in the development of efficient mould making. Standard mould bases provide the mould maker with a mould base at lower cost and with much higher quality than if the base were manufactured by the mould maker. Replacement parts, such as locating ring and sprue bushings, loader pins and bushings, KO pins and push-back pins of high quality are also available to the moulder. Since these parts are common for many moulds, they can be stocked by the moulder in the plant and thus down time is minimised.

INJECTION MOULDING OF THERMOSETTING RESINS

Three methods were developed—jet moulding, straight injection moulding, and offset moulding. However, these techniques have only historical interest since they never really achieved significant commercial penetration.

Jet moulding involved the application of a special nozzle to a standard injection-moulding machine. Straight injection moulding (i.e., moulding in a straight line) was developed by altering standard plunger-type injection moulding machines to accommodate the thermosetting character of the resins. A jacketed cylinder, smoothly tapering to the orifices in the specially designed nozzles was incorporated, and the torpedoes or spreaders of standard injection machines were eliminated. Offset injection moulding may be described as a transfer-moulding operation carried out with an altered horizontal-injection-moulding machine in which the injection plunger displaces the heat-softened material into the heated transfer pot and from there into runners and cavities where rapid curing takes place. Offset moulds are generally used when parts which inserts are to be made. The advantages claimed for the offset process are fast cycles, large capacities, and low plunger pressure.

Screw-Injection Moulding of Thermosetting Resins

The foregoing three machines are basically plunger-type machines. But shortly after the development of screw-transfer machines, the concept of screw-injection moulding of thermosets, also known as direct screw transfer (or DST), was introduced. The potential of this technique for low-cost, high-volume production of moulded thermoset parts was quickly recognised, and today screw-injection machines are available in all clamp tonnages up to 1200 tonnes and shot sizes up to 10 lb. Coupled with this, there has been a new series of thermosetting moulding materials developed specifically for injection moulding. These materials have long life at moderate temperatures (approximately 200°F), which permits plastication in the screw barrel, and react (cure) very rapidly when the temperature is raised to 350 to 400°F (177 to 204°C), resulting in reduced cycle time.

A typical arrangement for a direct screw-transfer injection-moulding machine for thermosets is shown in Fig. 17.16. The machine has two sections mounted on a common base. One section constitutes the plasticising and injection unit, which includes the feed hopper, the heated barrel that encloses the screw, the hydraulic cylinder which pushes the screw forward to inject the plasticised material into the mould, and a motor to rotate the screw. The other section clamps and holds the mould halves together under pressure during the injection of the hot plastic melt into the mould.

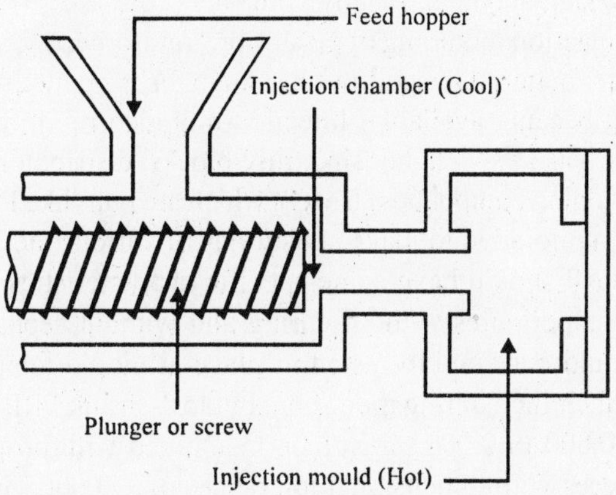

Fig. 17.16. Schematic of a direct screw-transfer moulding machine for thermosets.

The thermosetting material (in granular to pellet form) suitable for injection moulding is fed from the hopper into the barrel and is then moved forward by the rotation of the screw. During its passage, the material receives conductive heat from the wall of the heated barrel and frictional heat from the rotation of the screw. For thermosetting materials, the screw used is a zero-compression-ratio screw—i.e., the depths of flights of the screw at the feed-zone end and at the nozzle end are the same. By comparison, the screws used in thermoplastic moulding machines have compression ratios such that the depth of flight at the feed end is one and one-half to five times that at the nozzle end. This difference in screw configuration is a major difference between thermoplastic— and thermosetting-moulding machines.

As the material moves forward in the barrel due to rotation of the screw, it changes in consistency from a solid to a semi-fluid, and as it starts accumulating at the nozzle end, it exerts a backward pressure on the screw. The back pressure is thus used as a processing variable. The screw stops turning when the required amount of material—the charge—has accumulated at the nozzle end of the barrel, as sensed by a limit switch. (The charge is the exact volume of material required to fill the sprue, runners, and cavities of the mould). The screw is then moved forward like a plunger by hydraulic pressure (up to 20,000 psi) to force the hot plastic melt through the sprue of the mould and into the runner system, gates, and mould cavities. The nozzle temperature is controlled to maintain a proper balance between a hot mould (350–400°F), and a relatively cool barrel (150–200°F).

Moulded-in inserts are commonly used with thermosetting materials. However, since the screw-injection process is automatic, it is desirable to use post-assembled inserts rather than moulded-in inserts because moulded-in inserts require that the mould be held open each cycle to place the insets. A delay in the manual placement disrupts an automatic cyclic operations, affecting both the production rate and the product quality.

Tolerances of parts made by injection moulding of thermosetting materials are comparable to those produced by the compression and transfer methods described, earlier. Tolerances achieved are as low as ± 0.001 inch/inch, although ordinarily tolerances of ± 0.003 to 0.005 in./in. are economically practical.

Thermosetting materials used in screw-injection moulding are modified from conventional thermosetting compounds. These modifications are necessary to provide the working time-temperature relationship required for screw plasticating. The most commonly used injection-moulding thermosetting materials are the phenolics. Other thermosetting materials often moulded by the screw-injection process include melamine, urea, polyester, alkyd, and dially phthalate (DAP).

Since the mid-1970s the injection moulding of glass-fibre-reinforced thermosetting polyesters gained increasing importance as better materials (e.g., low shrinkage resins, pelletised forms of polyester/glass, etc.), equipment, and tooling became available. Injection-moulded reinforced thermoset plastics have thus made inroads in such markets as switch housing, fuse blocks, distributor caps, power-tool housings, office machines, etc. Bulk moulding compounds (BMC), which are puttylike FRP (fibrous glass-reinforced plastic) materials, are injection moulded to make substitutes of various metal die castings.

For injection moulding, FRP should have some specific characteristics. For example, it must flow easily at lower-than-mould temperatures without curing and without separating into resin, glass, and filler components, and it should cure rapidly when in place at mould temperature. A traditional FRP material shrinks about 0.003 inch/inch during moulding, but low-shrink FRP materials used for injection moulding shrink as little as 0.000 to 0.0005 inch/inch Combined with proper tooling, these materials thus permit production of pieces with dimensional tolerances of ± 0.0005 inch/inch.

Proper design of parts for injection moulding requires an understanding of the flow characteristics of material within the mould. In this respect, injection-moulded parts of thermosets are more like transfer-moulded parts than to compression moulded parts. Wall-section uniformity is an important consideration in part design. Cross-sections should be as uniform as possible, within the dictates of part requirements, since moulding cycles, and therefore costs, depend on the cure time of the thickest section. (For thermoplastics, however, it is the cooling time that is critical). A rule of thumb for estimating cycle times for a 1/4-inch wall section is 30 sec for injection-moulded thermosets (compared to 45 sec for thermoplastics). As a guideline for part design, a good working average for wall thickness is 1/8 to 3/16 inch, with a minimum of 1/16 inch.

STANDARDS FOR TOLERANCES ON MOULDED ARTICLES

Moulding materials shrink after they are taken out of the mould. The extent of shrinkage depends on the character of the material and the final mould temperature. A significantly large dimensional change takes place in cold-mould articles, because the cure or hardening is accomplished outside of the mould. Moreover, many of the conventional thermosetting and thermoplastic materials continue to shrink for many months after moulding. Such data on normal material shrinkage after moulding are commonly supplied by material suppliers.

The mould shrinkage described should not, however, be confused with tolerance. Tolerance is the variation in mould shrinkage rather than the shrinkage itself. Dimensional tolerances in a moulded article are the allowable variations, plus and minus, from a nominal or mean dimension. Charts were developed by the Society of Plastics Industry, to help obtain the magnitude of practical tolerances on the dimensions of articles moulded from a variety of thermosetting and thermoplastic materials, including phenolics, melamine and urea, epoxy, diallyl phthalate, alkyd, polyethylene, polypropylene, ABS, acetal, acrylic, polystyrene, polycarbonate, and flexible vinyl. The charts were based on data obtained from representative suppliers and moulders of plastics who responded to a questionnaire based on a hypothetical moulded articles.

EXTRUSION

The extrusion process is basically designed to continuously convert a soft material into a particular form. An oversimplified analogy may be a household meat grinder. However, unlike the extrudate from a meat grinder, plastic extrudates generally approach truly continuous formation. Like the usual meat grinder, the extruder (Fig. 17.17) is essentially a screw conveyor. It carries the cold plastic material (in granular or powdered form) forward by the action of the screw, squeezes it, and, with heat from external heaters and the friction of viscous flow, changes it to a molten stream. As it does this, it develops pressure on the material, which is highest right before the molten plastic enters the die. The screen pack, consisting of a number of fine or coarse mesh gauzes supported on a breaker plate and placed between the screw and the die, filter out dirt and unfused polymer lumps. The pressure on the molten plastic forces it through an adapter and into the die, which dictates the shape of the final extrudate. A die with a round opening as shown in Fig. 17.17, produces pipe; a square die opening produces a square profile, etc. Other continuous shapes, such as the film, sheet, rods, tubing, and filaments, can be produced with appropriate dies. Extruders are also used to apply insulation and jacketing to wire and cable and to coat substrates such as paper, cloth, and foil.

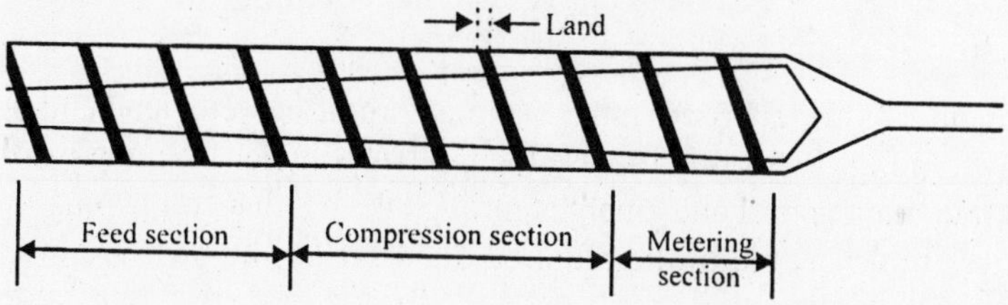

Fig. 17.17. Scheme for a typical single-screw showing extruding pipe.

When thermoplastic polymers are extruded, it is necessary to cool the extrudate below T_m or T_g to impart dimensional stability. This cooling can often be done simply by running the product through a tank of water, by spraying cold water, or, even, more simply, by air cooling. When rubber is extruded, dimensional stability results from cross-linking (vulcanisation). Interestingly, rubber extrusion for wire coating was the first application of the screw extruder in polymer processing.

Extruders have several other applications in polymer processing: in the blow moulding process they are used to make hollow objects such as bottles; in the blown-film process they are used for making wide films; they are also used for compounding plastics (i.e., adding various ingredients to a resin mix) and for converting plastics into the pellet shape commonly used in processing. In this last operation specialised equipment, such as the die plate-cutter assembly, is installed in place of the die, and an extrusion-type screw is used to provide plasticated melt for various injection-moulding process.

Multiple-Screw Extruders

Multiple-screw extruders (that is, extruders with more than a single screw) were developed largely as a compounding device for uniformly blending plasticisers, fillers, pigments, stabilisers, etc., into the polymer. Subsequently, the multiple-screw extruders also found use in the processing of plastics.

Multiple-screw extruders differ significantly from single-screw extruders in mode of operation. In a single-screw machine, friction between the resin and the rotating screw makes the resin rotate with the screw, and the friction between the rotating resin and the barrel pushes the material forward, and this also generates heat.

Increasing the screw speed and/or screw diameter to achieve a higher output rate in a single-screw extruder will therefore result in a higher buildup of frictional heat and higher temperatures. In contrast, in twin-screw extruders with intermeshing screws the relative motion of the flight of one screw inside the channel of the other pushes the material forward almost as if the machine were a positive-displacement gear pump which conveys the material with very low friction. In two-screw extruders, heat is therefore controlled independently from an outside source and is not influenced by screw speed. This fact becomes especially important when processing a heat-sensitive plastic like PVC. Multiple-screw extruders are therefore gaining wide acceptance for processing vinyls, although they are more expensive than single-screw machines. For the same reason, multiple-screw extruders have found a major use in the production of high-quality rigid PVC pipe of large diameter.

Several types of multiple-screw machines are available, including intermeshing co-rotating screws (in which the screws rotate in the same direction, and the flight of one screw moves inside the channel of the other), intermeshing counter-rotating screws (in which the screws rotate in opposite directions), and non-intermeshing counter-rotating screws. Multiple-screw extruders can involve either two screws (twin-screw design) or four screws.

Blown-Film Extrusion

The blown-film technique is widely used in the manufacture of polyethylene and other plastic films. A typical setup is shown in Fig. 17.18. In this case the molten polymer from the extruder head enters the die, where it flows round a mandrel and emerges through a ring-shaped opening in the form of a tube. The tube is expanded into a bubble of the required diameter by the pressure of internal air admitted through the centre of the mandrel.

The air contained in the bubble cannot escape because it is sealed by the die at one end and by the nip (or pinch) rolls at the other, so it acts like a permanent shaping mandrel once it has been injected. An even pressure of air is maintained to ensure uniform thickness of the film bubble.

The film bubble is cooled below the softening point of the polymer by blowing air on it from a cooling ring placed round the die. When the polymer, such as polyethylene, cools below the softening point, the crystalline material is cloudy compared with the clear amorphous melt. The transition line which coincides with this transformation is therefore called the frost line.

The ratio of bubble diameter to die diameter is called the blowup ratio. It may range as high as 4 to 5, but 2.5 is a more typical figure. Molecular orientation occurs in the film in the hoop direction during blowup, and orientation in the machine direction, that is, in the direction of the pinch rolls. The film bubble after solidification (at frost line) moves upward through guiding devices into a set of pinch rolls which flatten it. It can then be slit, gusseted, and surface-treated in line. (Vertical extrusion, shown in Fig. 17.18, is most common, although horizontal techniques have been successfully used.

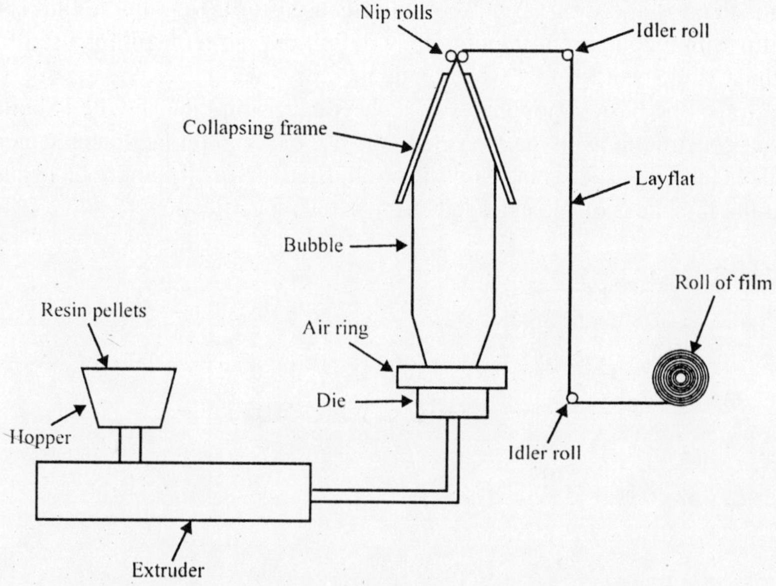

Fig. 17.18. Shematic of a typical blown-film extrusion setup.

Blown-film extrusion is an extremely complex subject, and a number of problems are associated with the production of good-quality film. Among the likely defects are variations in film thickness, surface imperfections (such as 'fish eyes', 'orange peel', haze), wrinkling, and low tensile strength. The factors affecting them are also numerous. 'Fish eyes' occur due to imperfect mixing in the extruder or due to contamination of the molten polymer. Both factors are controlled by the screen pack.

The blown-film technique has several advantages: the relative ease of changing film width and caliber by controlling the volume of air in the bubble and the speed of the screw; the elimination of the end effects (e.g., edge bead trim and non-uniform temperature that result from flat film extrusion); and the capability of biaxial orientation (i.e., orientation both in the hoop direction and in the machine direction), which results in nearly equal physical properties in both directions, thereby giving a film of maximum toughness.

After extrusion, blown-film is often slit and wound up as flat film, which is often much wider than anything produced by slot-die extrusion. Thus blow-films of diameters 7 ft or more have been produced, giving flat films of widths up to 24 ft. One example is reported of a 10 inch extruder with 5 ft diameter and a blowup ratio of 2.5, producing 1100 lb/hr of polyethylene film, which when collapsed and slit is 40 ft wide. Films in thicknesses of 0.004 to 0.008 inch are readily produced by the blown-film process. Polyethylene films of such large widths and small thicknesses find extensive uses in agriculture, and building.

Flat Film or Sheet Extrusion

In the flat-film process the polymer melt is extruded through a slot die (T-shaped or 'coat hanger' die), which may be as wide as 10 ft. The die has relatively thick wall sections on the final lands (as compared to the extrusion coating die) to minimise deflection of the lips from internal melt pressure. The die opening (for polyethylene) may be 0.015 to 0.030 inch—even for films that are less than 0.003 inch thick. The reason is that the speed of various driven rolls used for taking up the film is high enough to draw down the film with a concurrent thinning. (By definition, the term film is used for material less than 0.010 inch thick, and sheet for that which is thicker.

Fig. 17.19 illustrates the basic components of a sheet extrusion die. The die is built in two halves (that part) for easier construction and maintenance. The use of jack bolt facilitates separation of the die halves when the die is full of plastic. The die lip can be adjusted (across the entire lip length) to enable the processor to keep the thickness of the extruded sheet within specification.

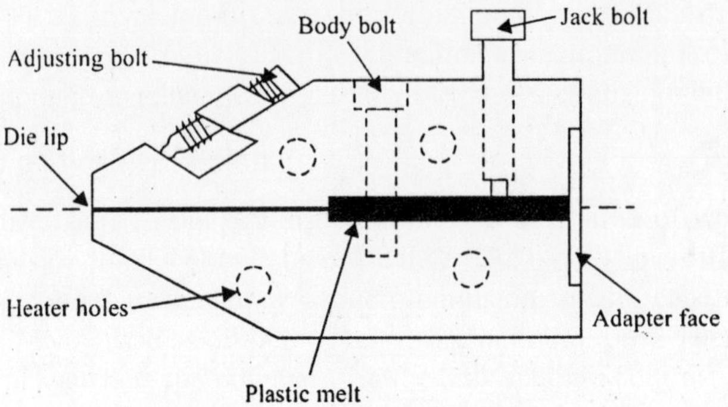

Fig. 17.19. Sheet extrusion die.

Following extrusion, the film may be chilled T_m or T_g by passing it through a water bath of over two or more chrome-plated chill rollers which have been cored for water cooling. A schematic diagram of a chill-roll (also called cast-film) operation is shown in Fig. 17.20. The polymer melt extruded as a web from the die is made dimensionally stable by contacting several chill-rolls before being pulled by the powered carrier rolls and wound up. The chrome-plated surface of the first roll is highly polished so that the product obtained is of extremely high gloss and clarity.

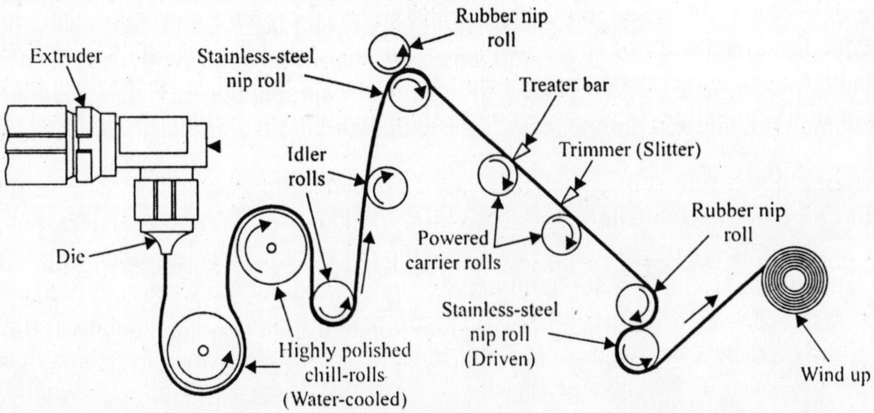

Fig. 17.20. Schematic diagram of chill-roll film extrusion.

In flat-film extrusion (particularly at high takeoff rates), there is a relatively high orientation of the film in the machine direction (i.e., the direction of the extrudate flow) and very low one in the traverse direction.

Biaxially oriented film can be produced by a flat-film extusion by using a tenter (Fig. 17.21). Polystyrene, for example, is first extruded through a slit die at about 190°C and cooled to about 120°C by passing between rolls. Inside a temperature-controlled box the moving sheet, re-warmed to 130°C, is grasped on either side by tenterhooks which exert a drawing tension (longitudinal stretching) as well as a widening tension (lateral stretching). Stretch ratios of 3:1 to 4:1 in both directions are commonly employed for biaxially oriented polystyrene film. Biaxial stretching leads to polymers of improved tensile strength. Commercially available oriented polystyrene film has a tensile strength of 10,000 to 12,000 psi (703 to 843 kg/cm^2), compared to 6,000 to 8,000 psi (422 to 562 kg/cm^2) for unstretched material.

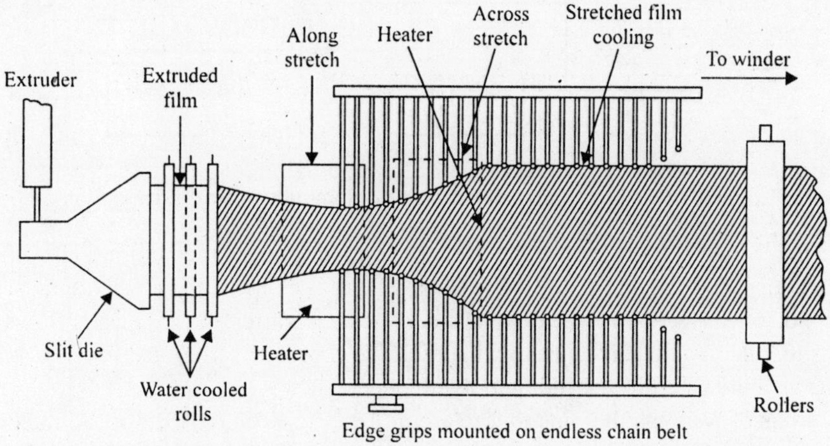

Fig. 17.21. Plax process for manufacture of biaxially stretched polystyrene film.

Biaxial orientation effects are important in the manufacture of films and sheet. Biaxially stretched polypropylene, poly(ethylene terephthalate) (e.g., Melinex) and poly(vinylidene chloride) (Saran) produced by flat-film extrusion and tentering are strong films of high clarity. In biaxial orientation, molecules are randomly oriented in two dimensions just as fibres would be in random mat; the orientation-induced crystallisation produces structures which do not interfere with the light waves. With polyethylene, biaxial orientation often can be achieved in blown-film extrusion.

Pipe

A typical pipe line consists of a single or a twin screw extruder, a die, equipment for inside and outside calibration, a cooling tank, a wall thickness measuring device, marking equipment, haul-off and automatic cutting and pallet equipment, or a windup unit for self-supporting pipe coils or lengths that are coiled on a drum (Fig. 17.22). Single screw extruders are generally used when processing PVC compound in granule form; twins handle powders of PVC. The adjustment and control of back pressure are very critical. PVC pipe is a big and very competitive market, so quality and profitability have been the most important requirements for years. Improving the equipment is almost of secondary importance because the equipment for good-quality products is already available.

The consumption of material, the main cost factor, can be gradually mininised by the use of current measuring the control systems. The goal is always tighter tolerance control to save material. Calibrating discs and pressure calibration methods of many different designs are used to meet various requirements. The operator's expertise in using these calibration systems is as important as controlling the complete line. A system of feedback and control of the extruder by a microprocessor is used to control wall thickness, combining ultrasonic gauging with gravimetric proportioning. Such new technologies are available but have to be debugged. Perhaps it can be said that any equipment from the 'past' is definitely non-competitive, based on all the new equipment that has been made available from upstream, through the extruder, and downstream. In the last few years, all the equipment has been significantly altered to increase profits.

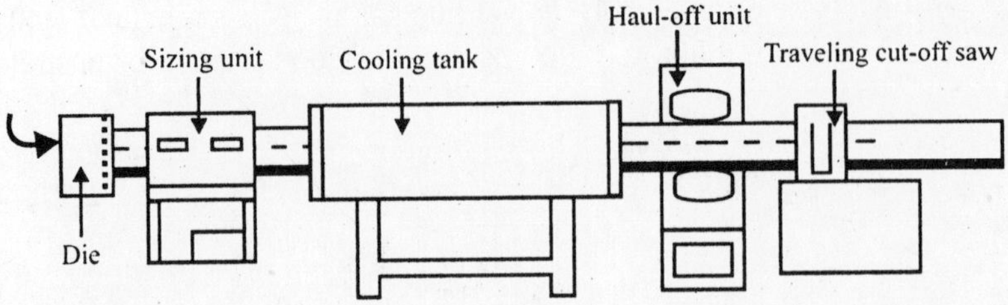

Fig. 17.22. Example of the important downstream equipment used in pipe and profile extrusion.

Plastic pipe is 1/2 inch or more in diameter. Smaller diameters are called tubing. The pipe die requires a central mandrel or core to support and shape the inside of the pipe wall. This core must be made as a spider with supporting arms to hold the core in the place concentrically. The plastic flows around these arms and welds together again before leaving the die. The core is water-cooled for PVC pipe in order to avoid degradation, and heated for extruding polyethylene.

Several methods of bringing the pipe to size are in use (Fig. 17.23). A European method is external sizing with a sizing sleeve. The plastic pipe is drawn through a water-cooled metal sleeve and held to the sleeve by air pressure inside the pipe. Alternatively, the pipe may be sized internally by a mandrel, in the extended mandrel method. The shrinkage of the cooling plastic pipe keeps a tight contact with the internal metal mandrel. The mandrel is water-cooled. This method is well suited to polyolefin plastics.

The sizing plate method is now less used than formerly. In this method the pipe takes its size by passing through a series of metal sizing plates. The newest and most used method of sizing is the vacuum trough. The pipe is fed through a long closed water-filled trough. Metal sizing rings in the trough give the pipe its desired diameter.

A vacuum is drawn over the water in the trough to reduce the pressure within the trough, and this reduced external pressure expands the pipe against the sizing rings. The inside of the pipe is at atmospheric pressure.

Wire and Cable Coverings

The covering or coating of wire and cable in continuous lengths with insulating plastics is an important application of extrusion, and large quantities of resin are used annually for this purpose. This application represented one of the first uses of extruders for rubber about 100 years ago. The wire and cable coating process resembles the process used for pipe extrusion (Fig. 17.22) with the difference that the conductor (which may be a single metal strand, a multiple strand, or even a bundle of previously individually insulated wires) to be covered is drawn through the mandrel on a continuous basis (Fig. 17.23). For thermoplastics such as polyethylene, nylon, and plasticised PVC, the coating is hardened by cooling below T_m or T_g by passing through a water trough. Rubber coatings, on the other hand, are to be cross-linked by heating subsequent to extrusion.

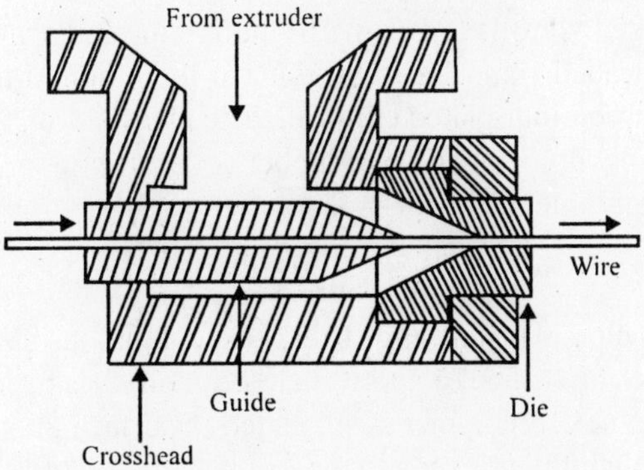

Fig. 17.23. Crosshead used for wire coating.

Extrusion Coating

Many substrates, including paper, paperboard, cellulose film, fibreboard, metal foils, or transparent films are coated with resins by direct extrusion. The resins most commonly used are the polyolefins, such as polyethylene, polypropylene, ionomer, and ethylene-vinyl acetate co-polymers. Nylon, PVC, and polyester are used to a lesser extent. Often combinations of these resins and substrates are used to provide a multilayer structure. [A related technique, called extrusion laminating, involves two or more substrates, such as paper and aluminium foil, combined by using a plastic film, (e.g., polyethylene) as the adhesive and as a moisture barrier].

Coatings are applied in thicknesses of about 0.2 to 15 mils, the common average being 0.5 to 2 mils, and the substrates range in thickness from 0.5 to more than 24 mils. The equipment used for extrusion coating is similar to that used for the extrusion of flat film. Fig. 17.24 shows a typical extrusion coating setup. The thin molten film from the extruder is pulled down into the nip between a chill roll and a pressure roll situated directly blow the die.

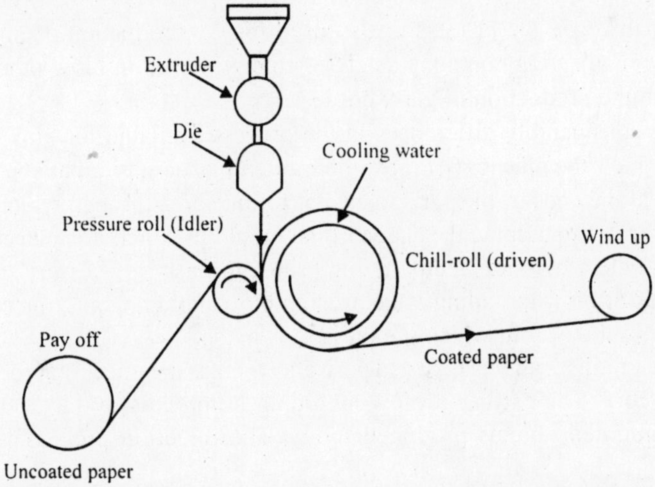

Fig. 17.24. Schematic diagram of paper coating for extrusion process.

The pressure between these two rolls forces the film on to the substrate while the substrate, moving at a speed faster than the extruded film, draws the film to the required thickness. The molten film is cooled by the water-cooled, chromium-plated chill roll. The pressure roll is also metallic but is covered with a rubber sleeve, usually neoprene or silicone rubber. After trimming, the coated material is wound up on conventional windup equipment.

Profile Extrusion

Profile extrusion is similar to pipe extusion (Fig. 17.22) except that the sizing mandrel is obviously not necessary. A die plate, in which an orifice of appropriate geometry has been cut, is placed on the face of the normal die assembly. The molten polymer is subjected to surface drag as it passes through the die, resulting in reduced flow through the thinner section of the orifice. This effect is countered by altering the shape of the orifice, but often this results in a wide difference in the orifice shape from the desired extrusion profile. Some examples are shown in Fig. 17.25.

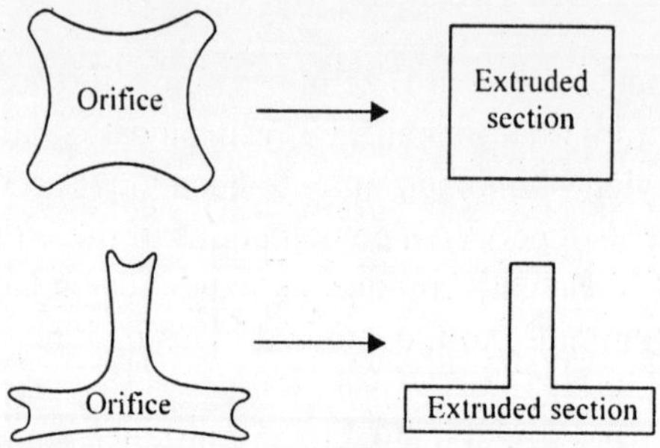

Fig. 17.25. Relationships between extruder die orifice and extruded section.

BLOW MOULDING

Basically, blow moulding is intended for use in manufacturing hollow plastic products, such as bottles and other containers. However, the process is also used for the production of toys, automobile parts, accessories, and many engineering components. The principles used in blow moulding are essentially similar to those used in the production of glass bottles.

Although there are considerably differences in the process available for blow moulding, the basic steps are the same: (i) melt the plastic; (ii) form the molten plastic into a parison (a tubelike shape of molten plastic); (iii) seal the ends of the parison except for one area through which the blowing air can enter; (iv) inflate the parison to assume the shape of the mould in which it is placed; (v) cool the blow-moulded part; (vi) eject the blow-moulded part; and (vii) trim flash if necessary.

Two basic processes of blow moulding are extrusion blow moulding and injection blow moulding. These processes differ in the way in which the parison is made.

The extrusion process utilises an unsupported parison, whereas the injection process utilises a parison supported on a metal core. The extrusion blow-moulding be process by far accounts for the largest percentage of blow-moulded objects produced today. The injection process is, however, gaining acceptance.

Although any thermoplastic can blow moulded, polyethylene products made by this technique are predominant. Polyethylene squeeze bottles form a large percentage of all blow-moulded products.

Extrusion Blow Moulding

Extrusion blow moulding consists basically of the extrusion of a predetermined length of parison (hollow tube of molten plastic) into a split die, which is then, closed, sealing both ends of the parison. Compressed air is introduced (through a blowing tube) into the parison, which blows up the fit to internal contours of the mould. As the polymer surface meets the cold metal wall of the mould, it is cooled rapidly below T_g or T_m. When the product is dimensionally stable, the mould is opened, the product is ejected, a new parison is introduced, and the cycle is repeated. The process affords high production rates.

In continuous extrusion blow moulding, a molten parison is produced continuously from a screw extruder. The moulds are mounted and moved. In one instance the mould sets are carried on the periphery of a rotating vertical wheel, in another on a rotating horizontal table. Such rotary machines are best suited for long runs and large volume applications. In the ram extrusion method the parison is formed in a cyclic manner by forcing a charge out from an accumulated molten mass, as in the pre-plasticiser injection-moulding machine. The transport arm cuts and holds the parison and lowers it into the waiting mould, where shaping under air pressure takes place.

A variation of the blow-extrusion process which is particularly suitable for heat-sensitive resins such as PVC is the cold preform moulding. The parison is produced by normal extrusion and cooled and stored until needed. The required length of tubing is then re-heated and blown to shape in a cold mould, as in conventional blow moulding. Since, unlike in the conventional process, the extruder is not coupled directly to the blow-moulding machine, there is less chance of a stoppage occurring, with consequent risk of holdup and degradation of the resin remaining in the extruder barrel. There is also less chance of the occurrence of 'dead' pockets and consequent degradation of resin in the straight-through die used in this process than in the usual crosshead used with a conventional machine.

Injection Blow Moulding

In this process the parison is injection moulded rather than extruded. In one system, for example, the parison is formed as a thick-walled tube around a blowing stick in a conventional injection-moulding machine. The parison is then transferred to a second, or blowing, mould in which the parison is inflated to the shape of the mould by passing compressed air down the blowing stick. The sequence is shown in Fig. 17.26. Injection blow moulding is relatively slow and is more restricted in choice of moulding materials as compared to extrusion blow moulding. The injection process, however, affords good control of neck and wall thicknesses of the moulded object. With this process it is also easier to produce unsymmetrical mouldings.

Blow Moulds

Generally, the blow mould is a cavity representing the outside of a blow-moulded part. The basic structure of a blow mould consists of a cast or machined block with a cavity, cooling system, venting system, pinchoffs, flash pockets, and mounting plate. The selection of material for the construction of a blow mould is based on the consideration of such factors as thermal conductivity, durability, cost of the materials, the resin being processed, and the desired quantity of the finished parts. Commonly used mould materials are beryllium, copper, aluminium, ampcoloy, A-2 steel, and 17-4 and 420 stainless steels.

Beryllium-copper (BeCu) alloys 165 and 25 are normally used for blow moulds. These materials display medium to good thermal conductivity with good durability. Stainless steels such as 17-4 and 420 are also frequently employed in blow moulds where durability and resistance to hydrochloric acid are required. Heat-treated A-2 steel is often used as an insert in pinchoffs where thermal conductivity is not a concern and high quality parts are required.

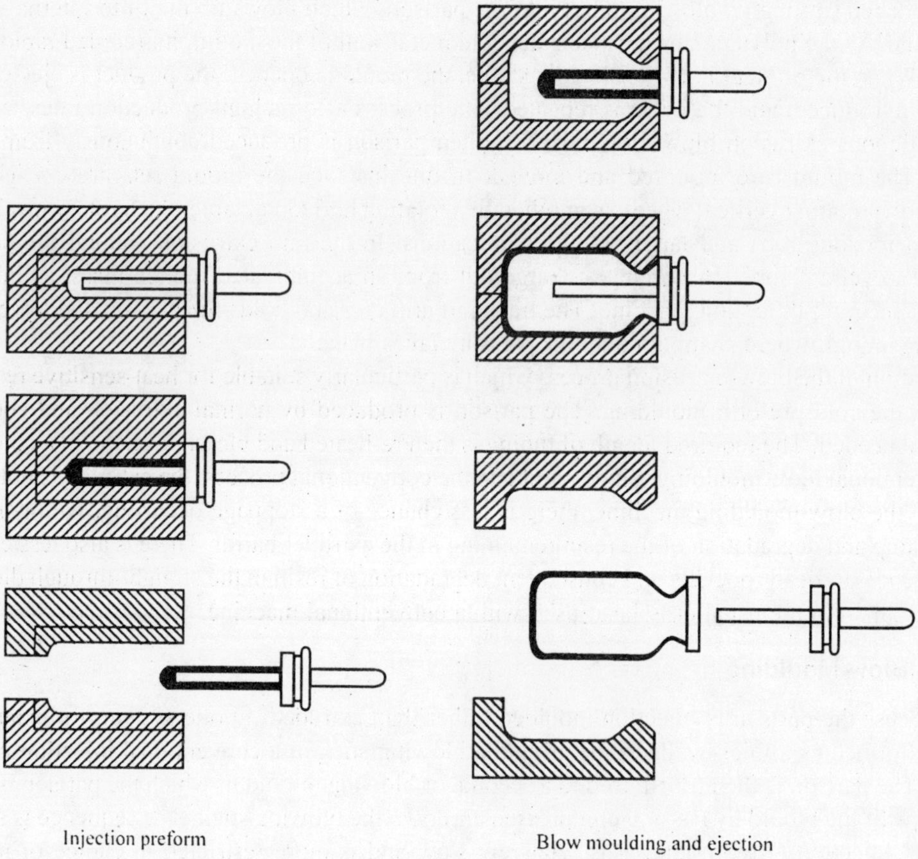

Fig. 17.26. Sketch of injection-blow-moulding process.

For blow moulding HDPE parts, aluminium is commonly employed for the base material, with BeCu or stainless steel inserts in the pinchoff area. For PVC parts BeCu, ampcoloy, or 17-4 stainless steel are used as the base material, with A-2 or stainless inserts in the pinchoff areas. For PET parts, the base mould is typically made of aluminium or BeCu, with A-2 or stainless steel pinchoffs.

The production speed of blow-moulded parts is generally limited by one of two factors: extruder capacity or cooling time in the mould. Cooling of mould is accomplished by a water circuit into the mould. Flood cooling and cast-in tubes are most common in cast moulds; drilled holes and milled slots are the normal in machined blow moulds.

In multiple-cavity moulds, series and parallel cooling circuits are used. Series cooling enters and cools one cavity, then moves to the next until all the cavities are cooled. The temperature of the water increases as it moves through the mould, and this results in non-uniform cooling. Parallel cooling, on the other hand, enters and exits all cavities simultaneously, thereby cooling all cavities at a uniform rate. Parallel cooling is thus the preferred method but it is not always possible due to limitations.

CALENDERING

The calendering process, used in the production of plastic film and sheet, is a continuous, synchronised method for converting raw materials, fusing them and passing them through nips of a series of co-operating rolls, into a sheet of specified thickness and width. This web may be polished or embossed, as required, and may be of either a rigid or a flexible character, with the 'hand' controlled by the formulation and the process.

It would appear, therefore, to be in direct competition with extrusion, but in fact it finds its main applications in the production of PVC sheet since with calendering there is much less risk of material degradation. Some film is also produced in this way from ABS, cellulosics, and toughened polystyrene.

Basically the process consists of a plastic formulation being passed between the nip of heated rolls. Usually four rolls with three nips are employed although a variety of combination is possible. The first nip controls the feed rate, the second nip acts as a metering unit, and the third nip controls the sheet thickness. The last roll may be matt, glossy, or embossed to provide the desired surface finish on the product. The other calender rolls must be highly polished and concentric to a very high degree of accuracy and hence, the equipment is expensive. Indeed since there is very little mixing during calendering the total plant required is very much more then the calender. A lot of preparatory work on the feedstock needs to be done before it reaches the calender rolls.

This includes mixing, blending, gelling and straining the formulation of resin, lubricants, plasticisers, stabilisers and pigments. Also since repairs to rolls can be very costly, it is essential to remove any foreign particles before the strip feed passes into the calender. Calendering is best known in the sheet formation of vinyl and ABS, and alloys of both.

Vinyl may be formulated to be soft/flexible, for low/high-temperature performance, oil/UV resistant, clear or opaque, Typically, the calender process gauge range is 0.002 to 0.045, and it has the advantage of higher throughput rates, compared to extrusion.

Unlike extrusion or injection moulding, a calender forms the web as a continuous extrusion between rolls and is not confined. A typical calender plant starts with the inventorying of raw materials in either silos for powders or tanks for liquids.

The only major ingredient not so handled is titanium dioxide, which has a tendency to cake if stored in bulk. Minor ingredients such as expensive stabilisers and pigments are manually or automatically handled. Then, based on the compounder's recipe, various proportions of the material are delivered to the mixer/blender.

BLENDING

Blending allows the distribution of all ingredients and, in the case of flexible formulations, permits the absorption of plasticisers into the resin to provide a free-flowing dry blend.

The blending of the recipe is critical, and two methods are employed to produce a uniform, coloured, stabilised blend of the desired flexibility, in powdered form.

1. High speed blending provides short cycles (5–10 minutes) and an intimate dispersion, suitable for both rigid and flexible compounding. Tandem vessels separate heating and cooling functions, with the latter sized adequately for blending two or more batches.
2. Ribbon blending has a lower capital cost but longer cycles (60 minutes) and usually accomplishes both heating and cooling in the same unit.

Selection of the unit is dependent upon cost and specific applications, such as product/colour changes.

The process temperature and time become an important part of the process. The residence time at elevated temperatures must be uniform and controlled throughout the calendering process.

The feed to the blenders may be either manual or automatic, and in some cases, even computerised to weigh in sequence automatically. Vessels are jacketed to accelerate heating or cooling, and a second unit may be necessary to permit product and colour changes.

Reground scrap is introduced in the ribbon blender or cooler, typically volumetrically.

After blending, the material is fed to a fluxing unit, where the powder is fused to a hot paste by mechanical/thermal energy in either a continuous or batch-type intensive mixer. Continuous mixers may be planetary design, single reciprocating screw, twin screw, or dual rotor mixers and will, if fed uniformly, deliver material at a uniform rate and stock temperature for direct feed to the calender (Fig. 17.27).

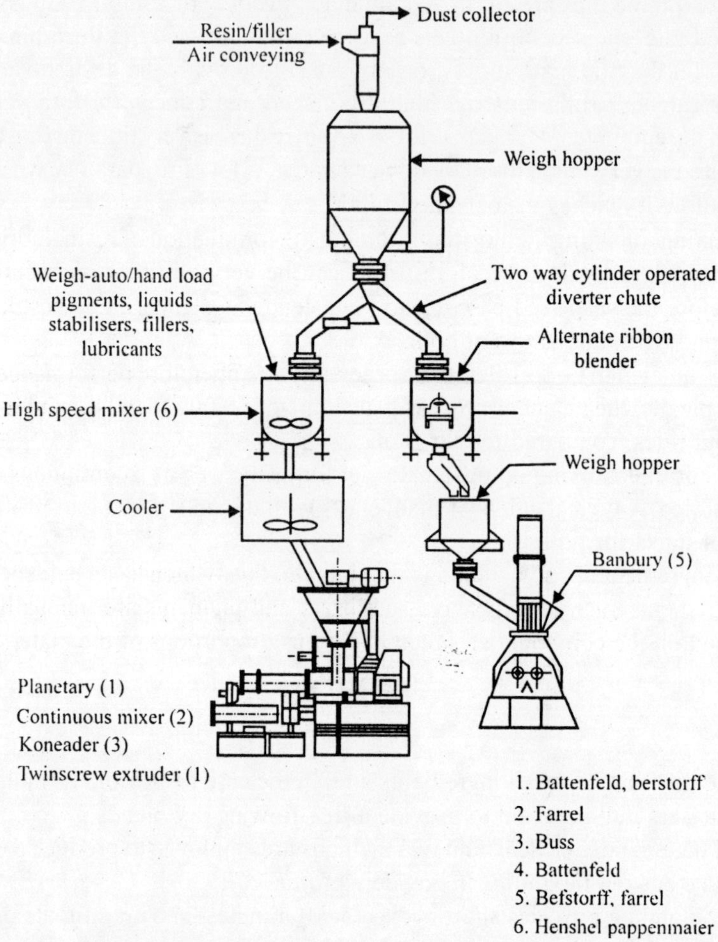

Fig. 17.27. Blending-fusion in a calender plant.

Batch type mixers are composed of a pair of rotors encased in a chamber, where the work is done between the rotor tips and the body or between the rotors (Fig. 17.28). Cycle times vary from 60 seconds for rigids to four times that for highly plasticised formulations. Batch sizes are established by the mixer capacity, based on the chamber volume.

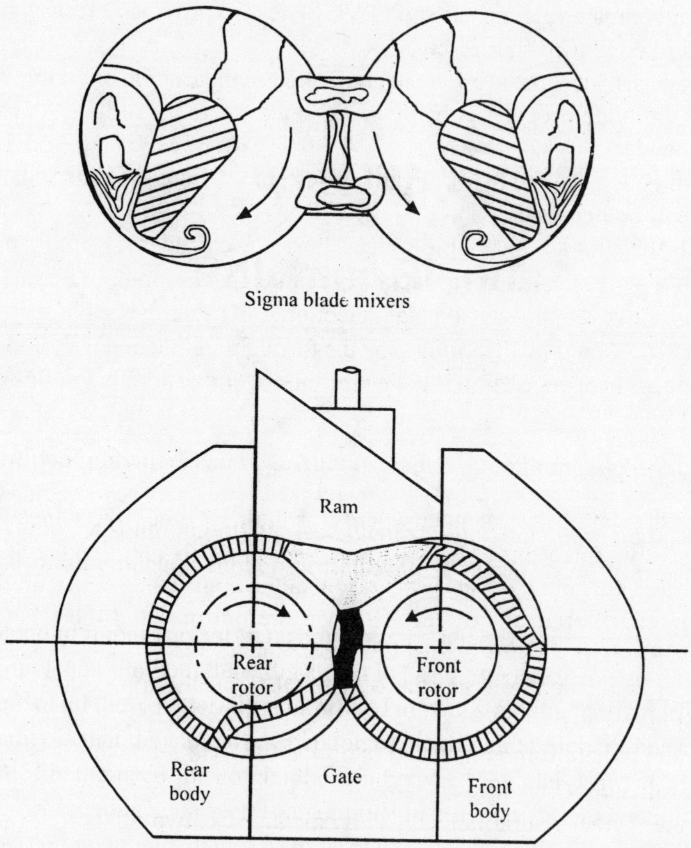

Fig. 17.28. Batch-type mixers.

Scrap, colour changes, and short runs are often better handled in the batch method; in continuous mixers, adequate size reduction of scrap is important to avoid problems caused by stratification and bridging.

SCRAP AND COLD TRIM

Scrap and cold trim handling from the product line poses one of the more difficult problems. Scrap and trim represent from 10 to 40 per cent of the product mix, depending on the trim width, rejects, and changes. A shorter flux rate and lower energy are required for re-melting the scrap versus virgin blend. The potential for decomposition contamination, colour drift, and gauge and flexibility variation exists whenever scrap is handled; yet for optimum uniformity, it is necessary to prepare the compound so that the blend of both new and old is fed uniformly to the ribbon blender, mixer-cooler, or fluxing unit. Careful control of the scrap percentage in the total mix is essential to a quality product and a reservoir or storage of scrap is necessary.

MILLS AND STRAINERS

The discharge from an intensive mixer may be fed to a mill, directly to a strainer, or directly to the calender for heat-sensitive materials such as rigid PVC. The mill provides a reservoir of material for the calender and can accept edge trim (hot or cold).

The mill has bored or drilled rolls, is oil- or steam-heated, with a single or double drive, and generally is a manual operation although progress is being made to automate the milling function and strip width control to feed the calender.

The size of the mill is determined by throughput rates, and if frequent order changes are required, a second or third unit will be necessary.

The mill then feeds the strainer, a variable speed, short *L/D* extruder. The molten material is cleaned by passing it through a wire mesh screen pack. To ensure that all metallic contamination has been removed, the plastic is conveyed through a metal detection unit as a rope or flat strip before entering the calender. The strainer size should be determined by the throughput rate and the temperature buildup that can be tolerated. Frictional heat generation is to be minimised, and generally the strainer temperature rise should be limited to 15°F.

CALENDER

The material is fed to the nip of the calender where, for the first time, it assumes the form of a sheet. The sheet is progressively pulled through two, three, or four subsequent banks in order to re-surface each of the two sides. Plastic calenders generally are made in four basic configurations:

1. The 'L' calender, wherein the offset roll is on the bottom, and the take-off is from the top roll (Fig. 17.29).
2. The inverted 'L' calender, wherein the offset roll is on the top, and the take-off is from the bottom roll (Fig. 17.30).
3. The 'F' calender, wherein the offset roll is on the top, and the take-off is from the middle roll, on the offset roll side (Fig. 17.31).
4. The 'Z' calender, either flat or inclined. This calender has two offset rolls, and the take-off may be made from either the top or the bottom offset roll, or from the back side of the stack roll (Fig. 17.32)

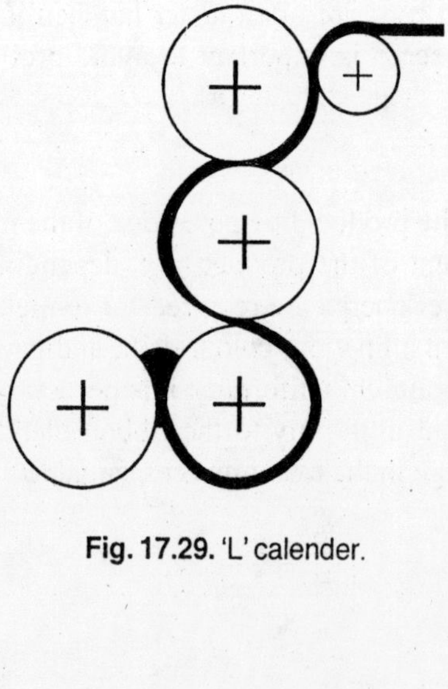

Fig. 17.29. 'L' calender.

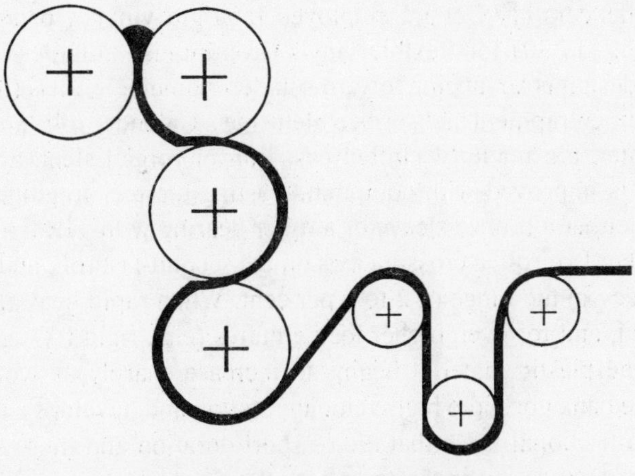

Fig. 17.30. Inverted 'L' calender.

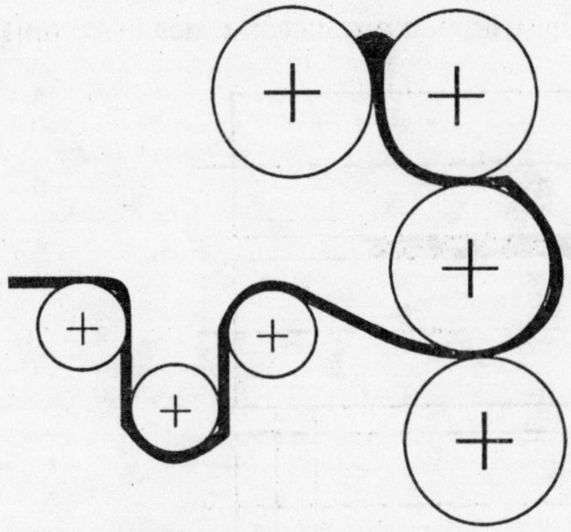

Fig. 17.31. 'F' calender.

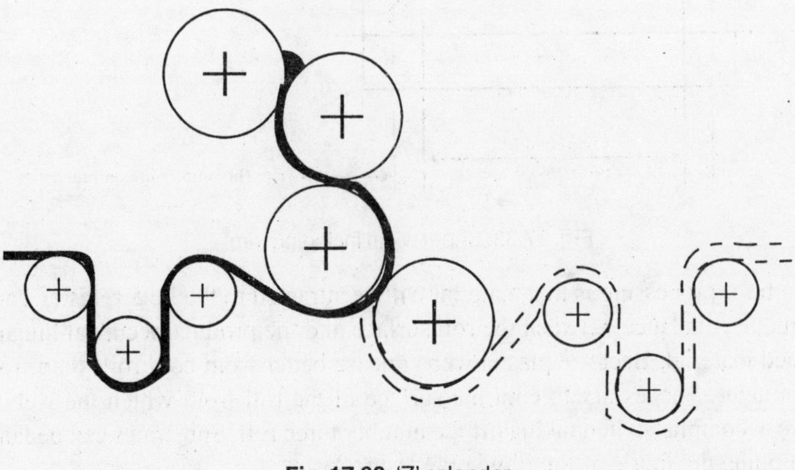

Fig. 17.32. 'Z' calender.

Typically, 'L' type calenders (17.29) are employed for rigid vinyls (for short feed transport) and inverted 'L' calenders (Fig. 17.30) for flexible vinyls (to minimise volatiles/condensation). Rolls are individually driven, and design is advancing toward smaller-diameter models (new calenders) to enable smaller banks, reducing air entrapment and surface blemishes. Calender rolls are drilled at the periphery for heating by oil or water, are made of chilled cast iron or forged steel, and are chrome-plated if necessary. Roll stiffness is improved with compound casting or steel forging to minimise deflection. Rolls are supported with either a bronze sleeve or a roller bearing with TIR targets 0.0003 inch runout. Roll bending on the number two roll, a crossing axis on the second-last roll, and roll bending on the last roll provide gauge accuracy in the range of 2 to 5 per cent. When rapid heat transfer is required, high-pressure hot water is used, and for even higher temperatures (e.g., rigid PVC), oil is employed.

The temperature of the plastic material begins to increase sharply as work is done between the several calender rolls. The bank goes into high-velocity rotation and develops a rolling action as it enters the nip. This causes high frictional heats that are of short duration and thus are tolerable. Should the temperature rise too high, degradation in the form of yellowing or burning begins to take place.

As the web moves through the calender nips, it widens out. The width of the sheet on the calender rolls is established by the final product trimmed width (Fig. 17.33). The sheet on the number two roll should be at least the width of the trimmed product in order to be uniform across the web.

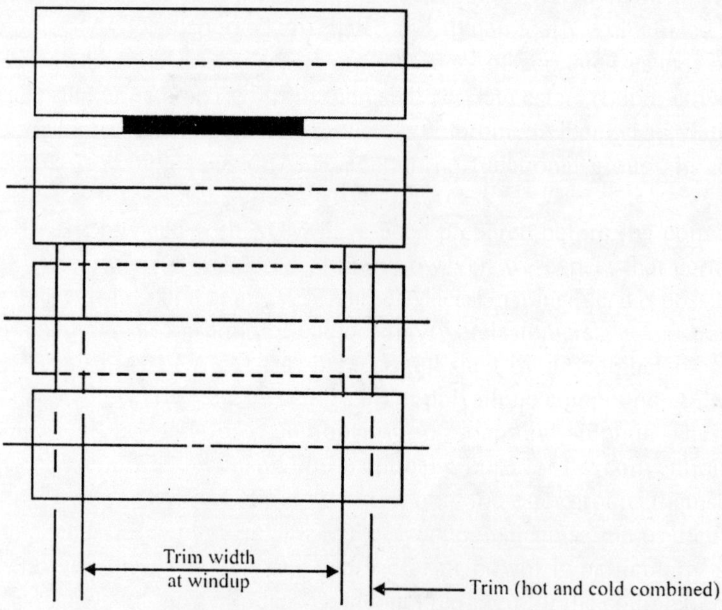

Fig. 17.33. Sheet width including trim.

A rule of thumb for processing is that material will be attracted to the hotter, faster, rougher roll. For rigid materials, the heat balance between the roll surface and the product occurs at much lower speeds than for plasticised material. Because plastic webs release better from cold rolls than from hot rolls, it frequently is desirable or necessary to cool the surface of the roll from which the web is 'taken off'. This is difficult to accomplish when taking off the number three roll. Substrates can be laminated in line and applied by nipping the fabric or the film at the last roll.

TAKE-OFF STRIPPER SECTION

Take-offs have been developed in Europe in recent years that improve on the older style, unheated, two- and three-roll units. These are temperature-controlled, multi-drive, smaller-diameter systems that provide annealing, improved dimensional stability for thin-gauge films and high film stretch. The increased stretch reduces calender surface speed, improves stock temperature control, and reduces air entrapment.

EMBOSSER

After leaving the stripping section, the web passes directly to cooling or is fed into an embossing unit, which can be arranged either vertically or horizontally. The embosser consists of an engraved steel roll pressing against a rubber roll, designed for a quick roll change (15 minutes) to minimise process interruption. The hardness and thickness of the rubber covering, together with the pressure applied by the steel roll, determine the depth of indentations or embossing in the plastic sheet. In order to keep the rubber cover cool, cooling water is circulated inside the steel arbour and outside the rubber covering, by means of cold roll contact or a water bath. The water on the rubber surface must be squeegeed off before contacting the plastic to prevent water marking of the film. The embossing roll is temperature-controlled, and this chilling effect sets the film surface.

From the embossing roll section, the web is annealed and is carried through a series of cooling drums and a beta ray gauge. The basic weight then can be correlated to the thickness, depending on the formulation used, and can, in turn, be used to control product thickness (final gap) and material distribution (roll bending/cross axis) automatically. SPC (statistical process control) and electronic maintenance data collection are options. Controlled and complete cooling is essential; if the film is wound up too hot, it will have a tendency to cool on standing, resulting in wrinkles, reduction in emboss pattern depth, 'snapback' and tight rolls, leading to printing/laminating difficulties.

Winder

The winder has a very demanding function—to wind thick/thin, wide/narrow widths; fast/should be at least 2 inches per side, and is collected or automatically ground.

SPINNING OF FIBRES

The term spinning, as used with natural fibres, refers to the twisting of short fibres into continuous lengths. In the modern synthetic fibre industry, however, the term is used for any process of producing continuous lengths by any means. (A few other terms used in the fibre industry should also be defined. A fibre may be defined as a unit of matter having a length at least 100 times its width or diameter. An individual strand of continuous length is called a filament. Twisting together filaments into a strand gives continuous filament yarn. If the filaments are assembled in a loose bundle, we have tow or roving. These can be chopped into small lengths (an inch to several inches long), referred to as staple. Spun yarn is made by twisting lengths of staple into a single continuous strand, and cord is formed by twisting together two or more yarns).

The primary fabrication process in the production of synthetic fibres is the spinning—i.e., the formation—of filaments. In every case the polymer is either melted or dissolved in a solvent and is put in filament form by forcing through a die, called spinneret, having a multiplicity of holes. Spinnerets for rayon spinning, for example, have as many as 10,000 holes in a 15-cm diameter platinum disc, and those for textile yarns may have 10 to 120 holes industrial yarns such as tyre core might be spun from spinnerets with up to 720 holes. Three major categories of spinning processes are melt, dry, and wet spinning. The features of the three processes are shown in Fig. 17.34 and the typical cross-sections of the fibres produced by them are shown in Fig. 17.35.

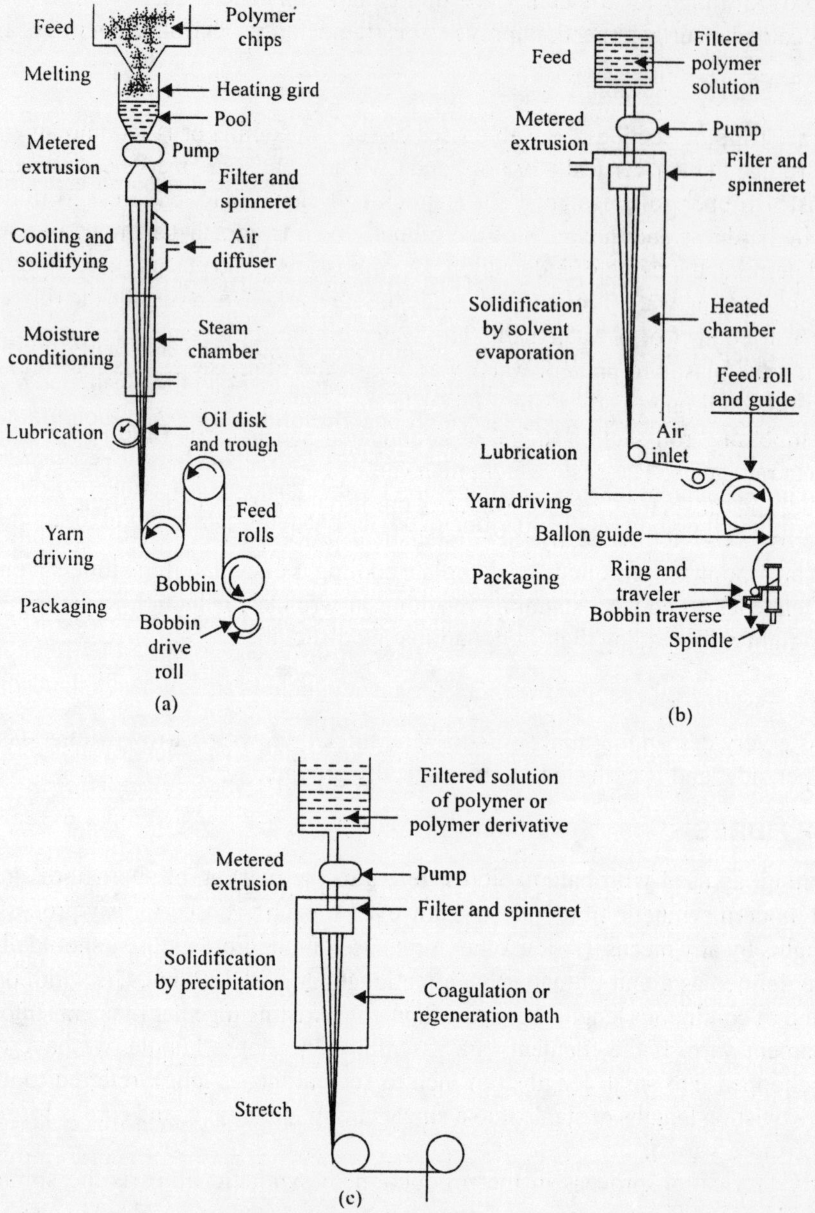

Fig. 17.34. Schematic of the three principal types of fibre spinning: (a) melt spinning; (b) dry spinning; and (c) wet spinning.

Melt Spinning

In melt spinning, which is the same as melt extrusion, the polymer is heated and the viscous melt is pumped through a spinneret. An inert atmosphere is provided in the melting chamber before the pump. Special pumps are used to operate in the temperature range necessary to produce a manageable melt (230 to 315°C). For nylon, for example, a gear pump is used to feed the melt to the spinneret (Fig. 17.34a). For a polymer with high melt viscosity such as polypropylene, a screw extruder is used to feed a heated spinneret. Dimensional stability of the fibre is obtained by cooling under tension.

Dry Spinning

In dry spinning, a polymer dissolved in a solvent is fed to a spinneret by using a pump. Dimensional stability of a fibre is achieved by evaporation of the solvent (Fig. 17.34b), which involves diffusion. The skin which forms first on the fibre by evaporation from the surface gradually collapses and wrinkles as more solvent diffuses out and the diameter decreases. The cross-section of a dry-spun fibre thus has an irregularly lobed appearance (Fig. 17.35). Recovery of the solvent used for dissolving the polymer is important to the economics of the process. Cellulose acetate dissolved in acetone and polyacrylonitrile dissolved in dimethylformamide are two typical examples.

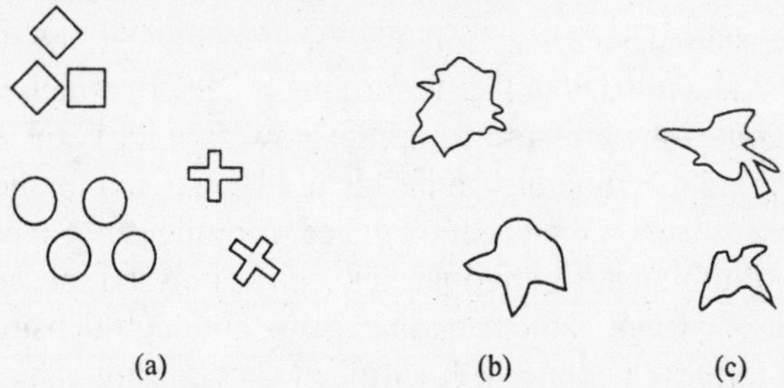

Fig. 17.35. Typical cross-section of fibres produced by different spinning processes: (a) melt-spun nylon from various shaped orifices; (b) dry-spun cellulose acetate from round orifice; and (c) wet-spun viscose rayon from round orifice.

Wet Spinning

Wet spinning also involves pumping a solution of the polymer to the spinneret. However, unlike dry spinning, dimensional stability is achieved by precipitating the polymer in a non-solvent (Fig. 17.34c). For example, polyacrylonitrile in dimethylformamide can be precipitated by passing a jet of the solution through a bath of water, which is miscible with the solvent but coagulates the polymer.

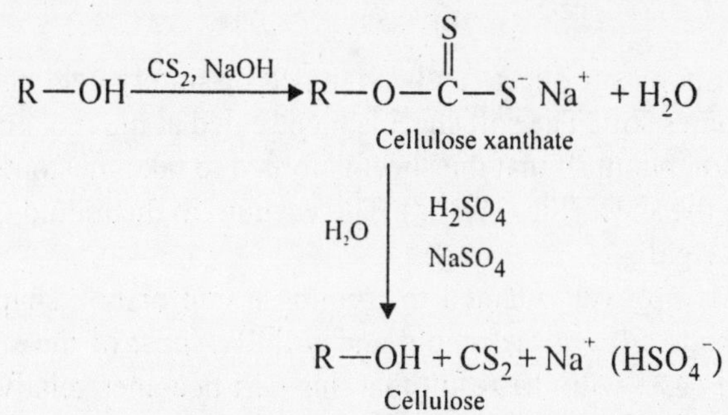

For wet-spinning cellulose triacetate a mixture of methylene chloride and alcohol can be used to dissolve the polymer, and a toluene bath can be used for precipitation of the polymer. In some cases the precipitation can also involve a chemical reaction. An important example is viscose rayon, which, is made by regenerating cellulose from a solution of cellulose xanthate in dilute alkali. If a slot die rather than a spinneret is used, the foregoing process would yield cellulose film (cellophane) instead of fibre.

Cold Drawing of Fibres

Almost all synthetic fibres are subjected to a drawing (stretching) operation to orient the crystalline structure in the direction of the fibre axis. Drawing orients crystallites in the direction of the stretch so that the modulus in that direction is increased and elongation at break is deceased. Usually the drawing is carried out at a temperature between T_g and T_m of the fibre. Thus, polyethylene ($T_g = -115°C$) can be drawn at room temperature, whereas nylon-6,6 ($T_g = 53°C$) should be heated or humidified to be drawn. T_g is depressed by the presence of moisture, which acts as a plasticiser. The drawing is accomplished by winding the yarn around a wheel or drum driven at a faster surface velocity than a preceding one.

THERMOFORMING

When heated, thermoplastic sheet becomes as soft as a sheet of rubber, and it can then be stretched to any given shape. This principle is utilised in thermoforming processes which may be divided into three main types: (i) vacuum forming; (ii) pressure forming (blow forming); and (iii) mechanical forming (e.g., matched metal forming), depending on the means used to stretch the heat softened sheet.

Since fully cured thermoset sheets cannot be re-softened, forming is not applicable to them. Common materials subjected to thermoforming are thermoplastics such as polystyrene, cellulose acetate, cellulose acetate butyrate, PVC, ABS, poly(methyl methacrylate), low- and high-density polyethylene, and polypropylene. The bulk of the forming is done with extruded sheets, although cast, calendered, or laminated sheets can also be formed.

In general, thermoforming techniques are best suited for producing mouldings of large area and very thin-walled mouldings, or where only short runs are required. Thermoformed articles include refrigerator and freezer door liners complete with formed-in compartments for eggs, butter, and bottles of various types, television masks, dishwasher housings, washing machine covers, various automobile parts (instrument panels, arm rests, ceilings, and door panels), large patterned diffusers in the lighting industry, displays in advertising, various parts in aircraft industry (windshields, interior panels, arm rests, serving trays, etc.), various housing (typewriters, dictaphones, and duplicating machines), toys, transparent packages, and much more.

Vacuum Forming

In vacuum forming, the thermoplastic sheet can be clamped or simply held against the rim of a mould and then heated until it becomes soft. The soft sheet is then sealed at the rim, and the air from the mould cavity is removed by a suction pump so that the sheet is forced to take the contours of the mould by the atmospheric pressure above the sheet (Fig. 17.36a). The vacuum in the mould cavity is maintained until the part cools and becomes rigid.

Straight cavity forming is not well adapted to forming a cup or box shape because as the sheet, drawn by vacuum, continues to fill out the mould and solidify, most of the stock is used up before it reaches the periphery of the base, with the result that this part becomes relatively thin and weak.

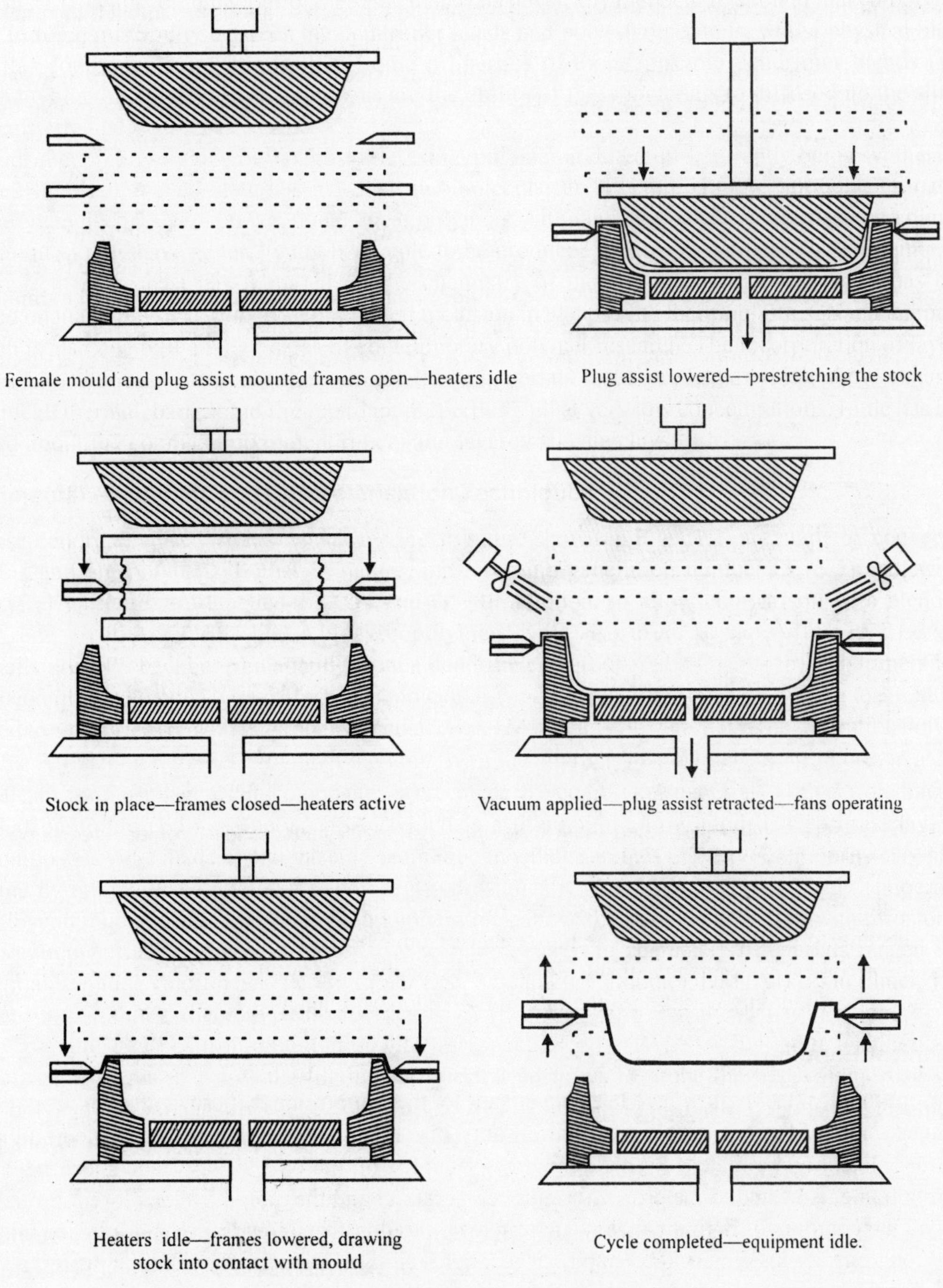

Fig. 17.36. Plug-assist forming using vacuum.

This difficulty is alleviated and uniformity of distribution in such shapes is promoted if the plug assist is used (Fig. 17.36b). The plug assist is any type of mechanical helper which carries extra stock toward an area where the part would otherwise be too thin. Plug-assist techniques are adaptable both to vacuum-forming and pressure forming techniques. The system shown in Fig. 17.36b. is thus known as plug assist vacuum forming.

Pressure Forming

Pressure forming is the reverse of vacuum forming. The plastic sheet is clamped, heated until it becomes soft, and sealed between a pressure head and the rim of a mould. By applying air pressure (Fig. 17.37), one forces the sheet to take the contours of the mould. Exhaust holes in the mould allow the trapped air to escape. After the part cools and becomes rigid, the pressure is released and the part is removed. As compared to vacuum forming, pressure forming affords a faster production cycle, greater part definition, and greater dimensional control.

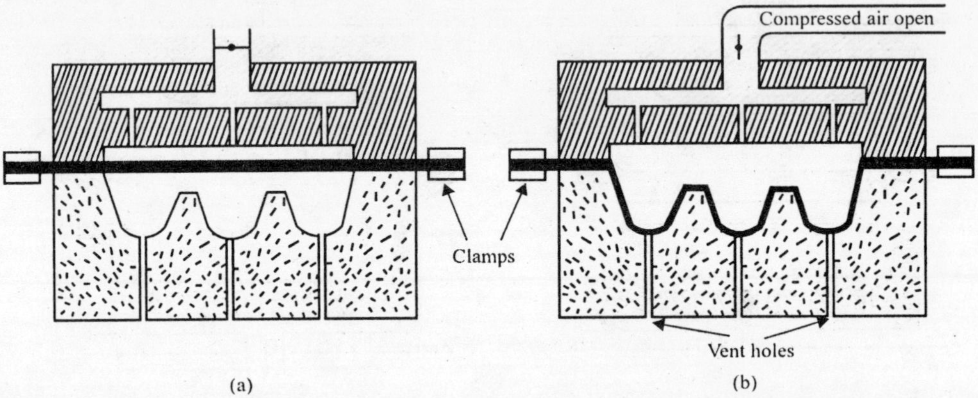

Fig. 17.37. Pressure forming: (a) heated sheet is clamped over mould cavity; and (b) compressed air pressure forces the sheet into the mould.

A variation of vacuum forming or pressure forming, called free forming or free blowing, is used with acrylic sheeting to produce parts that require superior optical quality (e.g., aircraft canopies). In this process the periphery is defined mechanically by clamping, but no mould is used, and the depth of draw or height is governed only by the vacuum or compressed air applied.

Mechanical Forming

Various mechanical techniques have been developed for thermoforming that use neither air pressure nor vacuum. Typical of these is matched mould forming (Fig. 17.38). A male mould is mounted on the top or bottom platen, and a matched female mould is mounted on the other. The plastic sheet, held by a clamping frame, is heated to the proper forming temperature, and the mould is then closed, forcing the plastic to the contours of both the male and the female moulds. The moulds are held in place until the plastic cools and attains dimensional stability, the latter facilitated by internal cooling of the mould. The matched mould technique affords excellent reproduction of mould detail and dimensional accuracy.

CASTING PROCESSES

There are two basic types of casting used in plastics industry: simple casting and plastisol casting.

Simple Casting

In simple casting, the liquid is simply poured into the mould without applying any force and allowed to solidify. Catalysts that cause the liquid to set are often added. The resin can be a natural liquid or a granular solid liquefied by heat. After the liquid resin is poured into the closed mould, the air bubbles are removed and the resin is allowed to cure either at room temperature or in an oven at low heat. When completely cured, the mould is split apart and the finished casting is removed. In the production of simple shapes such as rods, tubes, etc., usually two-piece metal mould with an entry hole for pouring in the liquid resin is used. For making flat-cast acrylic plexiglass or lucite sheets, two pieces of polished plate glass separated by a gasket with the edge sealed and one corner open are usually used as a mould.

Both thermosets and thermoplastics may be cast. Acrylics, polystyrene, polyesters, phenolics, and epoxies are commonly used for casting.

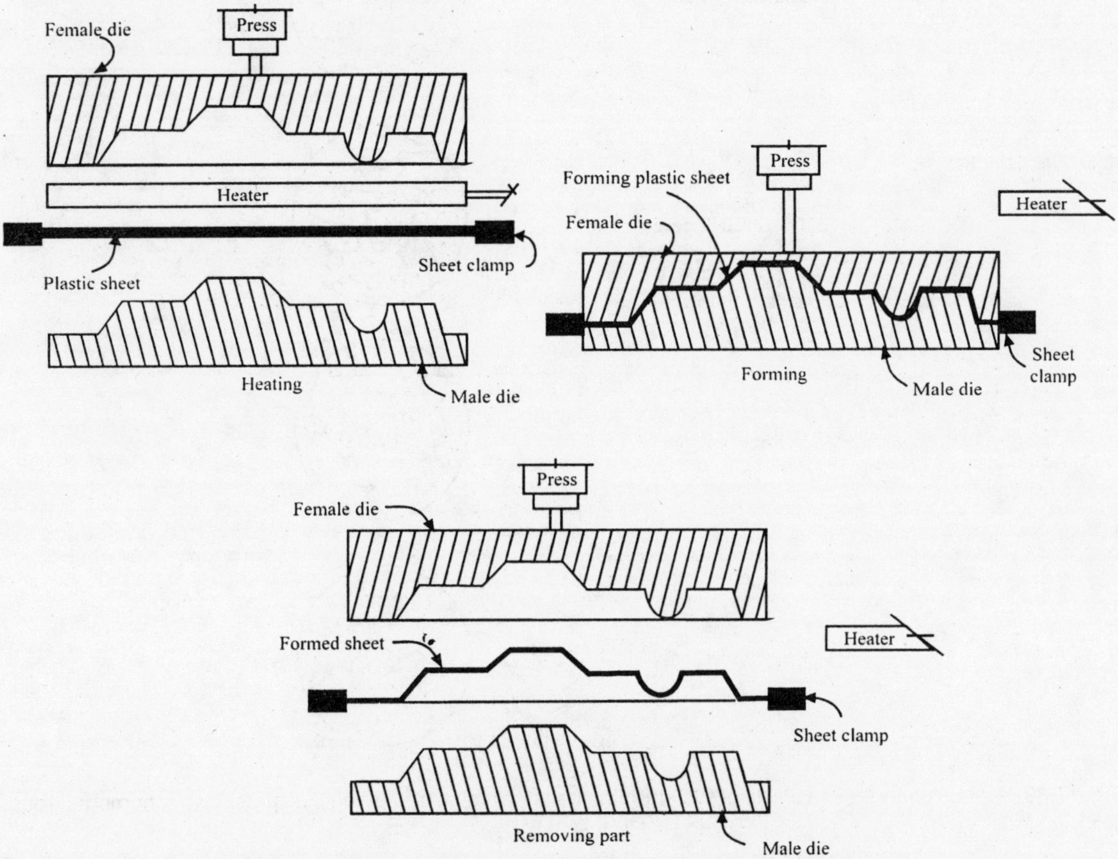

Fig. 17.38. Matched mould forming: (a) heating; and (b) forming.

Plastisol Casting

Plastisol casting, commonly used to manufacture hollow articles, is based on the fact that plastisol in fluid form is solidified as it comes in contact with a heated surface. A plastisol is a suspension of PVC in a liquid plasticiser to produce a fluid mixture that may range in viscosity from a pourable liquid to a heavy paste. This fluid may be sprayed onto a surface, poured into a mould, spread onto a substrate, etc. The plastisol is converted to a homogeneous solid ('vinyl') product through exposure to heat [e.g., 350°F (176°C)], depending on the resin type and plasticiser type and level. The heat causes the suspended resin to undergo fusion—that is, dissolution in the plasticiser (Fig. 17.39) so that on cooling, a flexible vinyl product is formed with little or no shrinkage. The product possesses all the excellent qualities of vinyl plastics.

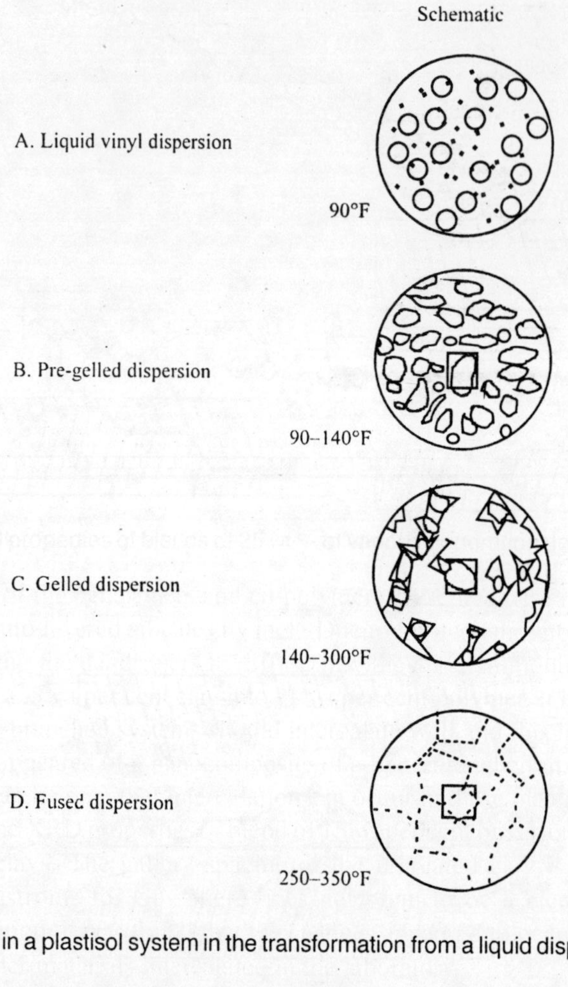

Fig. 17.39. Various changes in a plastisol system in the transformation from a liquid dispersion to a homogeneous solid.

Dispersion-grade PVC resins are used in plastisols. These resins are of fine particle size (0.1-2 micron in diameter), as compared to suspension type resins (commonly 75–200 micron in diameter) used in calender and extrusion processing. A plastisol is formed by simply mixing the dispersion-grade resin into the plasticiser with sufficient shearing action to ensure a reasonable dispersion of the resin particles. (PVC plasticisers are usually monomeric phthalate esters, the most important of them being the octyl esters based on 2-ethylhexyl alcohol and isooctyl alcohol, namely dioctyl phthalate and diisooctyl phthalate, respectively). The ease with which virtually all plastisol resins mix with plasticiser to form a smooth stable dispersion/paste is due to the fine particle size and the emulsifier coating on the resin particles. (The emulsifier coating aids the wetting of each particle by the plasticiser phase).

The liquid nature of the plastisol system is the key to its ready application. The plastisol may be spread onto a cloth, paper, or metal substrate, or otherwise cast or slashed into a mould. After coating or moulding, heat is applied, which causes the PVC resin particles to dissolve in the plasticiser and form a cohesive mass, which is, in effect, a solid solution of polymer in plasticiser. The various changes a plastisol system goes through in the transformation from a liquid dispersion to a homogeneous solid are schematically shown in Fig. 17.39. At 280°F (138°C) the molecules of plasticiser begin to enter between the polymer units, and fusion begins. If the plastisol were cooled after being brought to this temperature, it would give a cohesive mass with a minimum of physical strength. Full fusion occurs and full strength is accomplished when the plastisol is brought to approximately 325°F (163°C) before cooling. The optimum fusion temperature, however, depends on resin type and plasticiser type.

For coating applications it is common practice to add solvent (diluent) to a plastisol to bring down viscosity. This mixture is referred to as organosol. It may be applied by various coating methods to form a film on a substrate and then is heated to bring about fusion, as in the case of plastisol.

Unlike coating applications, there are some applications where it is desirable to have an infinite viscosity at low shear stress. For such applications, a plastisol can be gelled by adding a metallic soap (such as aluminium stearate) or finely filler as a gelling agent to produce a plastigel.

A plastigel can be moulded, placed on a pan, and heated to fusion without flow. The whole operation is like baking cookies.

A rigidgol is a plastisol of such formulation that it becomes a rigid, rather than a flexible, solid when fused. A very rigid product can be obtained when the plasticiser can be polymerised during or right after fusion. For example, a rigidsol can be made from 100 parts of PVC resin, 100 parts of triethylene glycol dimethacrylate (network forming plasticiser) and 1 part of di-tert-butyl peroxide (initiator). This mixture has a viscosity of only 3 poises compared with 25 poises for phthalate-based plastisol. However, after being heated for 10 min. at 350°F (176°C), the resin solvates and the plasticiser polymerises to a network structure, forming a hard, rigid glassy solid with flexural modulus of over 2.5×10^5 psi (1.76×10^4 kg/cm^2) at room temperature.

Three important variations of the plastisol casting, are dip casting, slush casting, and rotational casting.

Dip casting

A heated mould is dipped into liquid plastisol (Fig. 17.40a) and then drawn at a given rate. The solidified plastisol (with mould) is then cured in an oven at 350 to 400°F (176°C to 204°C). After it cools, the plastic is stripped from the mould. Items with intricate shapes such as transparent ladies' overshoes, flexible gloves, etc., can be made by this process.

The dipping process is also used for coating metal objects with vinyl plastics. For example, wire dish drainers, coat hangers, and other industrial and household metla items can be coated with a thick layer of flexible vinyl plastic by simply dipping in plastisol and applying fusion.

Slush casting

Slush casting is similar to slip casting (drain of ceramics. The liquid is poured into a preheated hollow metal mould, which is the shape of the outside of the object to be made (Fig. 17.40b). The plastisol in immediate contact with the walls of the hot mould solidifies. The thickness of the cast is governed by the time of stay in the mould. After the desired time of casting is finished, the excess liquid is poured out and the solidified plastisol with the mould is kept in an oven at 350 to 400°F (176 to 204°C). The mould is then opened to remove the plastic part, which now bears on its outer side the pattern of the inner side of the metal mould. Slush moulding is used for hollow, open articles. Squeezable dolls or parts of dolls and boot socks are moulded this way.

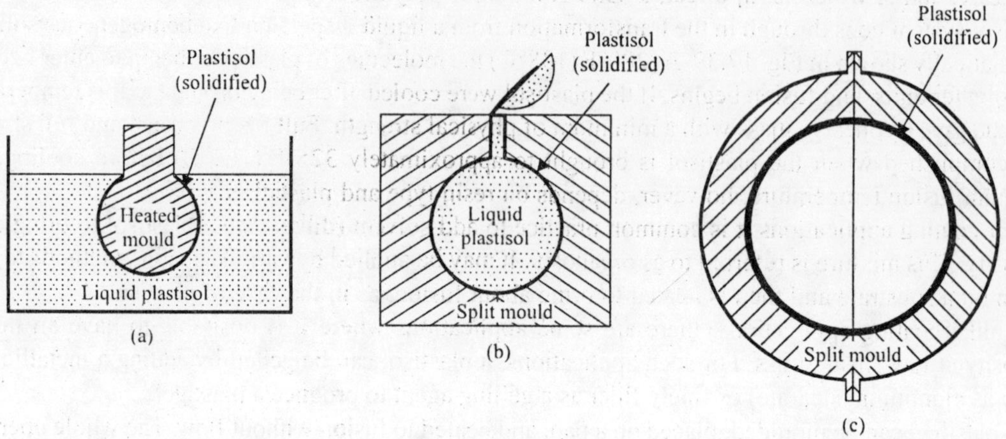

Fig. 17.40. Plastisol casting processes: (a) dip casting; (b) slush casting; and (c) rotational casting.

Rotational casting

In rotational casting a predetermined amount of liquid plastisol is placed in heated, closed, two-piece mould. The liquid is uniformly distributed against the walls of the mould in a thin uniform layer (Fig. 17.40c) by rotating the mould in two planes. The solidified plastisol in the mould is cured in an oven; the mould is then opened, and the part is removed. This method is used to make completely enclosed hollow objects. Doll parts, plastic fruits, squeeze bulbs, toilet floats, etc., can be made by rotational casting of plastisols.

REINFORCING PROCESSES

A reinforced plastic (RP) consists of polymeric resin strengthened by the properties of a reinforcing material. Reinforced plastics occupy a special place in the industry. They are at one and the same time both unique materials unto themselves and part and parcel of virtually every other segment of the plastics industry.

Reinforced plastics are composites in which a resin is combined with a reinforcing agent to improve one or more properties of the resin matrix. The resins may be either thermosetting or thermoplastic. Typical thermosetting resins used in reinforced plastics include unsaturated polyester, epoxy, phenolic, melamine, silicone, alkyd, and diallyl phthalate. In the field of reinforced thermoplastics, virtually every type of thermoplastic material can be, and has been, reinforced and commercially moulded. The more popular grades include nylon, polycarbonate, polypropylene, polyethylene, acetal, PVC, ABS, styrene-acrylonitrile, polysulphone, polyphenylene sulphide, and thermoplastic polyesters.

The reinforcement used in RP is a strong inert material bound into the plastic to improve its strength, stiffness, or impact resistance. The reinforcing agent can be fibrous, powdered, spherical, crystalline, or whisker, and made of organic, metallic, or ceramic material. Fibrous reinforcements are usually glass, although asbestos, sisal, cotton and high-performance fibres (discussed later) are also used. To be structurally effective, there must be a strong adhesive bond between the resin and the reinforcement. Most reinforcements are thus treated with sizes or finishes to provide maximum adhesion by the resins.

Although by definition, all reinforced plastics are composites (i.e., combinations of two materials—resin and reinforcement—that act synergistically to form a new third material RP with different properties than the original components) the term advanced composites, or high-strength composites, has taken on a special meaning. The term is applied to stiffer, higher modulus combinations involving exotic reinforcements such as graphite, boron, or other high-modulus fibres like aromatic polyamide fibres (Nomex and Kevlar) and extended-chain polyethylene fibres (Spectra ECPE), and resins like epoxy or some of the newer high heat-resistant plastics—polyamides, polyamideimide, polyquinoxalines, and polyphenylquinoxalines. Prime outlets for these materials are in the aerospace and aircraft industries.

Moulding Methods

As mentioned above, either thermosetting and thermoplastic resin can be used as the matrix component in reinforced plastics. Today, the reinforced thermoplastics (RTP) have become an accepted part of the reinforced plastics business, although smaller than the reinforced thermosets. Reinforced thermoplastics are generally made available to the processor in the form of injection moulding pellets (into which glass has been compounded) or concentrates (also a pellet but containing a much higher percentage of reinforcement, and designed to be mixed with non-reinforced pellets). It is also possible for the processor to do his own compounding of chopped glass and thermoplastic powder in injection moulding. In rotational moulding and casting techniques, the processor usually adds his own reinforcement. For thermoforming, glass-reinforced thermoplastic laminates are sold commercially. Structural foam moulding of glass-reinforced thermoplastics and reinforcing moulded urethane foams appeared as later developments in the field of reinforced thermoplastics.

Among thermosetting resins, unsaturated polyesters are by far the most widely used in reinforced plastics, largely because of their generally good properties, relatively easy handling, and relatively low cost. For special uses, however, other types are significant: epoxies for higher strength, phenolics for greater heat resistance, and silicones for their electrical properties and heat resistance. All these resins must be used in conjuction with a system of catalysts of curing agents in moulding thermoset composites. The type and amount strongly affect the properties, working life, and moulding characteristics of the resin. The polyester and epoxies are most often mixed with a catalyst just prior to moulding. The most widely used catalyst for polyesters is benzoyl peroxide.

Where heat is not available for curing, special catalyst-promoter systems can be used. With epoxies, an amine curing agent that reacts with the resin is most often used. However, there are many other types to choose from.

The polyester, epoxy, and thermosetting acrylic resins are usually thick liquid that become hard when cured. For this reason, they are most often combined with the reinforcement, by the moulder, by dipping or pouring. There are available, however, pre-impregnated reinforcement, (prepregs) for the moulder who wants to keep the operations as simple as possible.

Several methods are employed to make reinforced plastics. Although each method has the characteristics of either moulding or casting, the processes may be described as (i) hand lay-up; (ii) spray-up; (iii) matched moulding; (iv) vacuum-bag moulding; (v) pressure-bag moulding; (vi) continuous pultrusions; (vii) filament winding; and (viii) prepreg moulding.

Hand lay-up or contact moulding

A mould is first treated with a release agent (such as wax or silicone-mould release), and a coating of the liquid resin (usually polyester or epoxy) is brushed, rolled, or sprayed on the surface of the mould. Fibreglass cloth or mat is impregnated with resin and placed over the mould. Air bubbles are removed, and the mat or cloth is worked into intimate contact with the mould surface by squeegees, by rollers, or by hand (Fig. 17.41a). Additional layers of glass cloth or mat are added, if necessary, to build up the desired thickness. The resin hardens due to curing, as a result of the catalyst or hardener that was added to the resin just prior to its use. Curing occurs at room temperature, though it may be speeded up by heat. Ideally, any trimming should be carried out before the curing is complete, because the material will still be sufficiently soft for knives or shears. After curing, special cutting wheels may be needed for trimming.

Lowest-cost moulds such as simple plaster, concrete, or wood are used in this process, since pressures are low and little strength is required. However, dimensional accuracy of the moulded part is relatively low and, moreover, maximum strength is not developed in the process because the ratio of resin to filler is relatively high.

The hand lay-up process can be used for fabricating boat hulls, automobile bodies, swimming pools, chemical tanks, ducts, tubes, sheets, and housings, and for building, machinery, and autobody repairs.

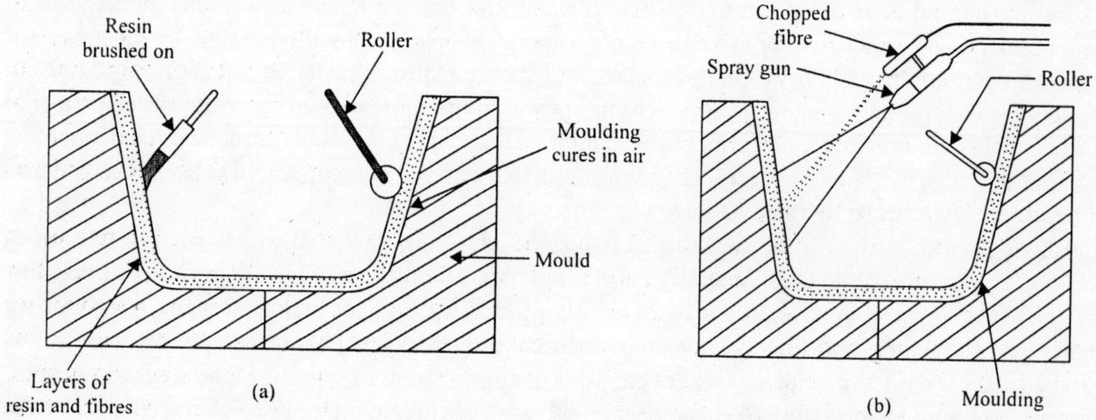

Fig. 17.41. (a) Basic hand lay-up method; and (b) spray-up technique.

Spray-up

A release agent is first applied on the mould surface, and measured amounts of resin, catalyst, promoter, and reinforcing material are sprayed with a multiheaded spray gun (Fig. 17.41b). The spray guns used for this work are different from those used for spraying glases, enamels, or paints. They usually consist of two or three nozzles, and each nozzle is used to spray a different material. One type, for example, sprays resin and promoter from one nozzle, resin and catalyst from another, and chopped glass fibres from a third.

The spray is directed on the mould to build up a uniform layer of desired thickness on the mould surface. The resin sets rapidly only when both catalyst and promoter are present. This method is particularly suitable for large bodies, tank linings, pools, roofs, etc.

Matched metal moulding

Matched metal moulding is used when the manufacture of articles of close tolerances and a high rate of production are required. Possible methods are preform moulding, sheet moulding, and dough moulding. In preform moulding the reinforcing material in mat or fibre form is preformed to the approximate shape and placed on one-half of the mould, which was coated previously with a release agent. The resin is then added to the preform, the second half of the mould (also coated previously with a release agent) is placed on the first half, and the two halves of the mould are then pressed together and heated (Fig. 17.42). The resin flows, impregnates the preform, and becomes hard. The cured part is removed by opening the mould. Because pressures of up to 200 psi (14 kg/cm^2) can be exerted upon the material to be moulded, a higher ratio of glass to resin may be used, resulting in a stronger product. The cure time in the mould depends on the temperature, varying typically from 10 minutes at 175°F (80°C) to only 1 minute at 300°F(150°C). The cure cycle can thus be very short, and a high production rate is possible.

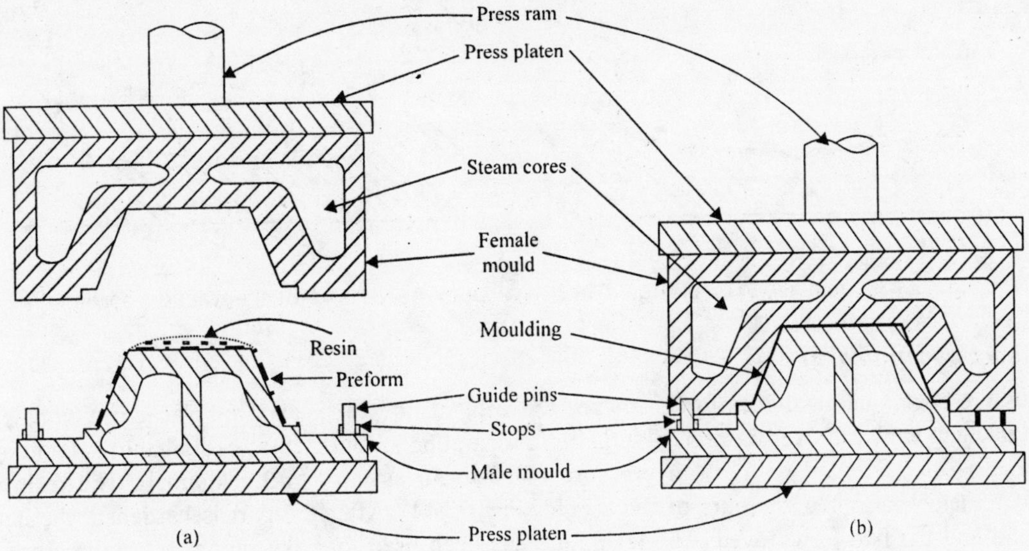

Fig. 17.42. Matched metal moulding: (a) before closing of die; and (b) after closing of die.

The moulding of sheet-moulding compounds (SMC) and dough-moulding compounds (DMC) is done 'dry'— i.e., it is not necessary to pour on resins. SMC, also called prepreg, is basically a polyester resin mixture (containing catalyst and pigment) reinforced with chopped strand mat or chopped roving and formed into a pliable sheet that can be handled easily, cut to shape, and placed between the halves of the heated mould.

The application of pressure then forces the sheet to take up the contours of the mould. DMC is a doughlike mixture of chopped strands with resin, catalyst, and pigment. The charge of dough, also called premix, may be placed in the lower half of the heated mould, although it is generally wise to preform it to the approximate shape of the cavity. When the mould is closed and pressure is applied, DMC flows readily to all sections of the cavity. Curing generally takes a couple of minutes for mould temperatures from 250 to 320°F (120 to 160°C). The method is used for the production of switch gear, trays, housings, structural and functional components.

Vacuum-bag moulding

In vacuum-bag moulding the reinforcement and the resin mixed with catalyst are placed in a mould, as in the hand lay-up method, and an airtight flexible bag (frequently rubber) is placed over it . As air is exhausted from the bag, atmospheric air forces the bag against the mould (Fig. 17.43). The resin and reinforcement mix now takes the contours of the mould. If the bag is placed in an autoclave or pressure chamber, higher pressure can be obtained on the surface. After the resin hardens, the vacuum is destroyed, the bag opened and removed, and the moulded part obtained. The technique has been used to make automobile body, aircraft component, and prototype moulds.

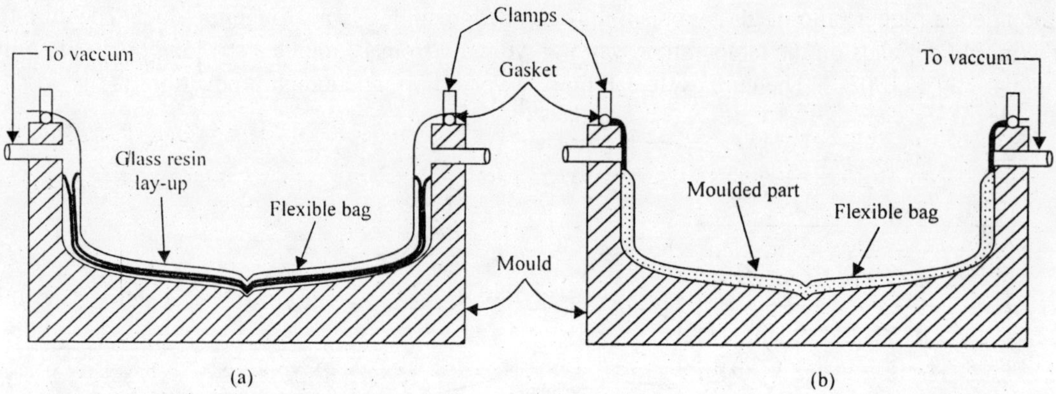

Fig. 17.43. Vacuum-bag moulding; (a) before vacuum applied; and (b) after vacuum applied.

Pressure-bag moulding

In pressure-bag moulding the reinforcement and the resin mixed with catalyst are placed in a mould, and a flexible bag is placed over the wet lay-up after a separating sheet (such as cellophane) is laid down. The bag is then inflated with an air pressure of 20 to 50 psi (1.4 to 3.5 kg/cm^2). The resin and reinforcement follow the contours of the mould (Fig. 17.44). After the part is hardened, the bag is deflated and the part is removed. The technique has been used to make radomes, small cases, and helmets.

Continuous pultrusion

Continuous strands, in the form of roving or other forms of reinforcement, are impregnated with liquid resin in a resin bath and pulled through a long, heated steel die which shapes the product and controls the resin content. The final cure is effected in an oven through which the stock is drawn by a suitable pulling device. The method produces shapes with high unidirectional strength (e.g., I-beams, rods, and shafts). Polyesters account for 90 per cent of pultrusion resin, epoxies for the balance.

Filament winding

In the filament-winding method, continuous strands of glass fibre are used in such a way as to achieve maximum utilisation of the fibre strength. In a typical process rovings or single strands are fed from a reel through a bath of resin and wound on a suitably designed rotating mandrel. Arranging for the resin-impregnated fibres to trasverse the mandrel at a controlled and predetermined (programmed) manner (Fig. 17.45) makes it possible to lay down the fibres in any desired fashion to give maximum strengths in the direction required. When the right number of layers have been applied, curing is done at room temperature or in an oven. For open-ended structures, such as cylinders or conical shapes, mandrel design is comparatively simple, either cored or solid steel or aluminium being ordinarily used for the purpose. For structures with integrally wound end closures, such as pressure vessels, careful consideration must be given to mandrel design and selection of mandrel material. A sand-poly(vinyl alcohol) combination, which disintegrates readily in hot water, is an excellent choice for diameters up to 5 ft.(1.5 m). Thus, a mandrel made of sand with water-soluble poly(vinyl alcohol) as a binder can be decomposed with water to recover the filament-wound part. Other mandrel materials include low-melting alloys, eutectic salts, soluble plasters, frangible or breakout plasters, and inflatables.

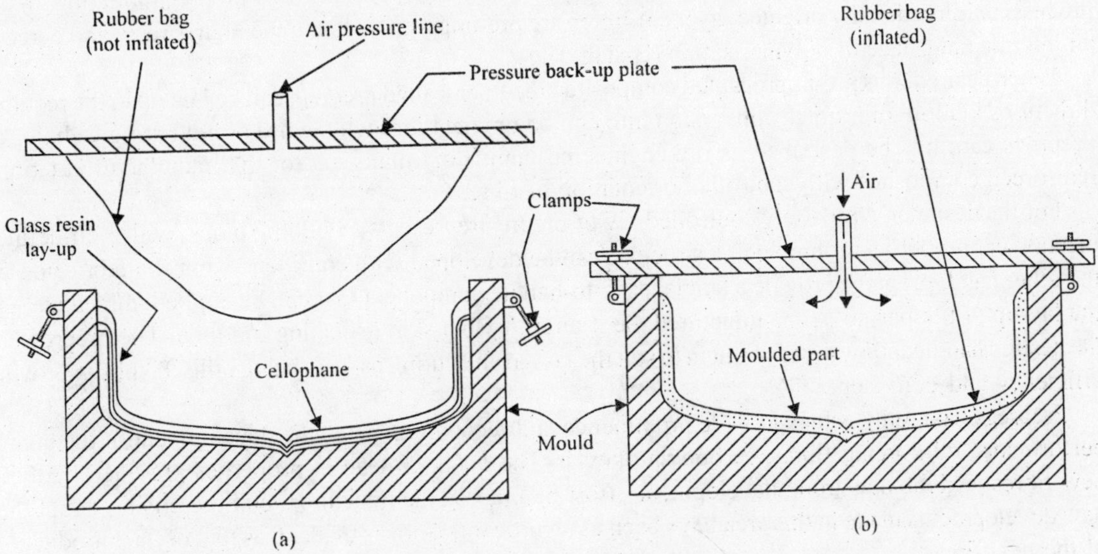

Fig. 17.44. Pressure-bag moulding: (a) during lay-up; and (b) during curing.

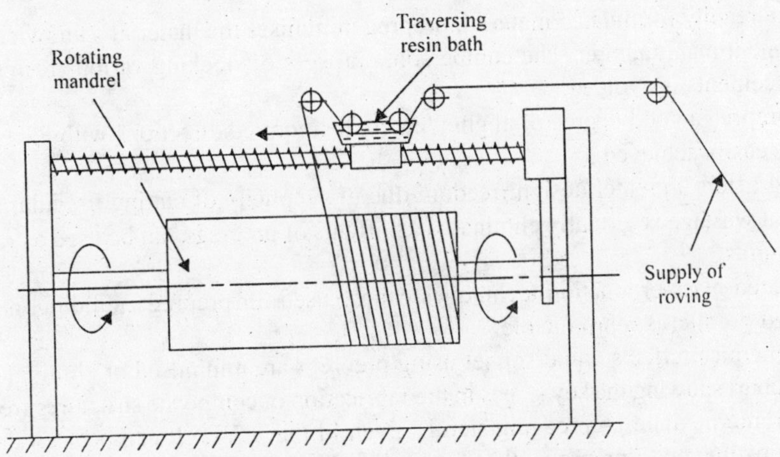

Fig. 17.45. Sketch of filament winding.

Because of high glass content, filament-wound parts have the highest strength-to-weight ratio of any reinforced thermoset. The process is thus highly suited to pressure vessels where reinforcement in the highly stressed hoop direction is important. Pipe installation, storage tanks, large rocker motor cases, interstage shrouds, high-pressure gas bottles, etc., are some of the products made of filament winding. The main limitation on the process is that it can only be used for fabricating objects which have some degree of symmetry about a central axis.

Prepreg moulding

Optimal strength and stiffness of continuous fibre-reinforced polymeric composites is obtained through controlled orientation of the continuous fibres. One means to achieve this is by prepreg moulding. In this process, unidirectionally oriented layers of fibres are pre-impregnated with the matrix resin and cured to an intermediate stage of polymerisation (B-stage).

When desired, this pre-impreganted composite precursor, called prepreg, can be laid up in the required directions in mould for quick conversion into end component through the use of hot curing techniques. Prepregs can thus be described as pre-engineered laminating materials for the manufacture of fibre-reinforced composites with controlled orientation of fibres.

For the designer, a precisely controlled ply of prepreg represents a building block, with well-defined mechanical properties from which a structure can be developed with confidence. For the fabricator, on the other hand, prepregs provide a single, easy-to-handle component that can be applied immediately to the lay-up of the part to be manufactured, be it aircraft wing skin or fishing rod tube. The prepreg has the desired handleability already built in to suit the lay-up and curing process being utilised, thus improving efficiency and consistency.

Prepregs have only achieved wide prominence and recognition since the development of the higher performance reinforcing fibres, carbon and kevlar. The quantum leap in properties provided by these new fibres generated a strong development effort by prepreg manufacturers and there is no doubt that new developments made in this area have been as significant, if not as evident, as that of the introduction of the new fibres.

The use of prepreg in the manufacture of a composite components offers several advantages over the conventional wet lay-up formulations:

1. Being a readily formulated material, prepreg minimises the material's knowledge required by a component manufacturer. The cumbersome process of stocking various resins, hardeners, and reinforcements is avoided.
2. With prepreg a good degree of alignments in the required directions with the correct amount of resin is easily achieved.
3. Prepreg offers a greater design freedom due to simplicity of cutting irregular shapes.
4. Material wastage is virtually eliminated as offcuts of prepregs can be used as random moulding compounds.
5. Automated mass-production techniques can be used for prepreg moulding, and the quality of moulded product is reproducible.
6. Toxic chemical effects on personnel using prepregs are minimised or eliminated.

A flowchart showing the key stages in the fabrication of composite structures from raw materials with an intermediate prepregging step is given in Fig. 17.46. It can be seen that there are two basic constituents to prepreg—the reinforcing fibre and the resin system. All of the advanced reinforcing fibres are available in continuous form, generally with a fixed filament diameter.

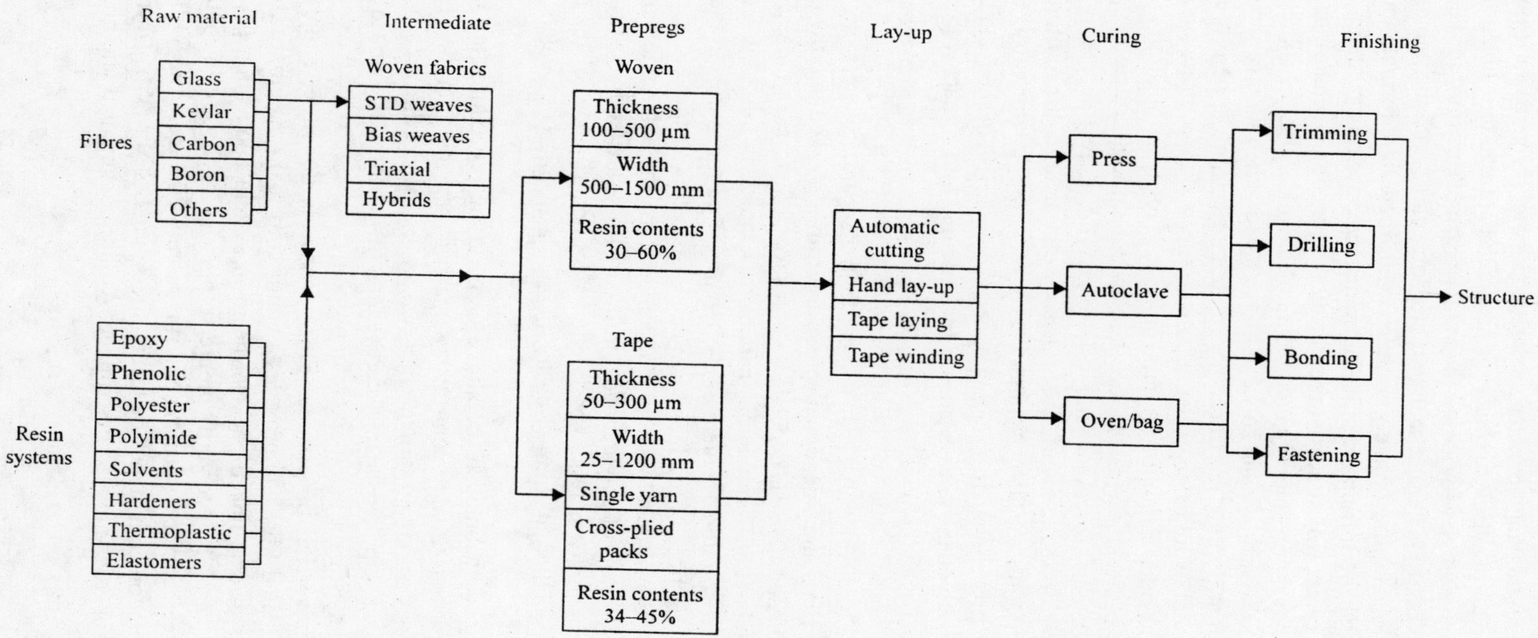

Fig. 17.46. Flow chart showing key stages in the fabrication of composite structures from raw materials by prepreg moulding.

Th number of filaments that the supplier arranges into a 'bundle' (or yarn) varies widely, and is an important determinant of the ability to weave fabric and make prepregs to a given thickness. In the majority of cases, the yarn is treated with a size to protect it from abrasion during the weaving or prepregging process. Often, the size is chosen such that it is compatible with the intended resin system.

Resin systems have developed into extremely complex multi-ingredient formulations in an effort to ensure the maximum property benefit from the fibre. Normally there are four methods of impregnation: (i) solution dip; (ii) solution spray; (iii) direct hot-melt coat; and (iv) film calendering.

The solution dip and solution spray impregnation techniques work with a matrix resin dissolved in a volatile carrier. The low viscosity of the resin solution allows good penetration of the reinforcing fibre bundles with resin. In solution dipping, the fibre, in yarn or fabric form, is passed through the resin solution and picks up an amount of solids dependent upon the speed of through-put and the solids level. With solution spraying, on the other hand, the required amount of resin formulation is metered directly onto the fibre. In both cases, the impregnated fibre is then put through a heat cycle to remove the solvent and 'advance' the chemical reaction in the resin to give the correct degree of tack.

Direct hot melt can be performed in a variety of ways. In one method, the reinforcing fibre web is dipped into a melt resin bath. A doctor blade, scraper bar, or metering roller controls the resin content. Alternately, the melt resin is first applied to a release paper, the thickness of the resin being determined by a doctor blade. The melt resin on the release paper is then brought into contact with a collimated fibre bundle and pressed into it in a heated impregnation zone.

In film calendering, which is a variation of the above method, the resin formulation is cast into a film from either hot-melt or solution and then stored. Thereafter in a separate process, the reinforcing fibre is sandwiched between two films and calendered so that the film is worked into the fibre.

The decision of which method to use is dependent upon several factors. Hot melt film processes are faster and cheaper, but certain resin formulations cannot be handled in this way, and hence solution methods have to be used. The solution dip method is often preferred for fabrics as the need to squeeze hot-melt and film into the interstices of the fabric can cause distortion of the weave pattern.

In a process based on a biconstituent tow impregnation concept, the polymeric matrix is introduced in fibrous form and a comingled tow of polymer and reinforcing fibres are fed into a heated impregnation zone (Fig. 17.47). In this method, it may be possible to effect better wetout of the reinforcing fibres with the matrix polymer, especially with high viscosity thermoplastic matrix polymers, through intimate comingling.

The machines necessary to accomplish the above prepregging procedures are many and varied. There are three distinct aspects to quality control: raw material screening, on-line control, and batch testing. All three are obviously important, but the first two are more critical.

Fibres and base resins are supplied against certificates of conformance and often property test certificates. On-line control during the manufacture of prepreg revolves around the correct ratio of fibre to resin. This is done by traversing Beta-gauge, which scans the dry fibre (either unidirectional or fabric) and then the impregnated fibre and provides a continuous real time plot of the ratio across the width of the material. This can be linked back to the resin system application point for continuous adjustment. Batch testing is carried out to verify prepreg properties, such as resin content, volatile level, and flow. The resin 'advancement' (chemical reaction) is monitored via a DSC (Differential Scanning Calorimeter) and the formulation consistency by testing the T_g via DSC or DMA (Dynamic Mechanical Analyser). The laminate properties are also determined. All are documented and quoted on a Release Certificate.

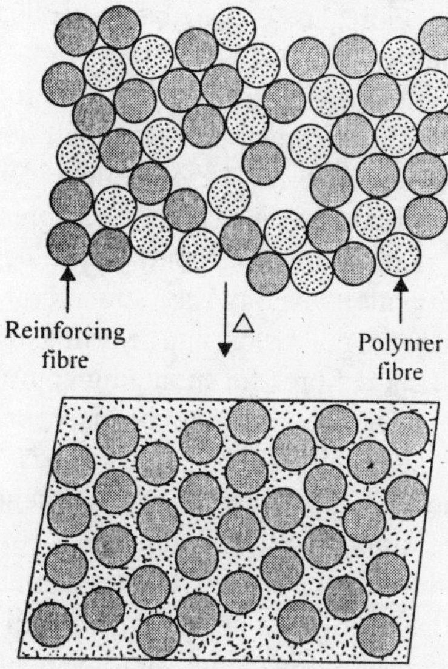

Fig. 17.47. Biconstituent tow impregnation using a comingled tow of polymer and reinforcing fibres.

The cured product is claimed to possess extraordinary fatigue life, no notch sensitivity, high ultimate strength, and superior corrosion resistance. The high ultimate strength and fatigue resistance of Scotchply laminate and relatively low modulus, make it about six times more efficient for spring use. The range of application of the reinforced plastic includes vibratory springs, sonar housings, landing gear, picker blades, snowmobile track reinforcement, helicopter blades, seaplane pontoons, missile casing, and archery bow laminate.

Fibrous Reinforcements

Although many types of reinforcements are used with plastics, glass fibres predominate. Fibrous glass reinforcements are available in many forms (described below). Asbestos is used in the form of loose fibre, paper, yarn, felt, and cloth. The two largest uses of asbestos in plastics are with poly(vinyl chloride) in vinyl asbestos tile and with polyesters and polypropylene.

Most natural and synthetic fibres do not have the strength required for a reinforced plastics part. However, when intermediate strengths are satisfactory, they can be used. In this category are nylon, rayon, cotton fabrics, and paper. Sisal fibres have also found use as a low cost reinforcing material in premix moulding compounds. High-modulus graphite and carbon fibres, aramid fibres and extended-chain polyethylene fibres are playing a more and more important role in reinforced plastics. Boron filaments, with outstanding tensile strengths, are usually used in the form of prepreg tapes and have been primarily evaluated for the aerospace and aircraft industry.

Glass fibres

A high-alkali 'A-glass' and a low-alkali 'E-glass' are used as reinforcements for polymer composites, the later being used most often. Since the modulus of E-glass is 10.5×10^6 psi and the tensile strength upwards of 250,000 psi, it is not surprising that the stiffness and strength of most plastics can be increased by compounding with glass. A more chemical resistant glass, sodium borosilicate (C-glass), and a higher-tensile-strength glass, S-glass, are also available. E-glass is a calcium alumino-silicate, and

S-glass is a magnesium aluminosilicate. Fibreglass is available as a collection of parallel filaments (roving), chopped strands, mat, and woven fabric.

Glass filaments are produced by melting a mixture of silica, limestone, and other reactants, depending on the type of glass and forcing the molten product through small holes (bushings). The hot filaments are gathered together and cooled by a water spray. These multiple glass filaments are gathered together into bundle, called a strand, which is wound up on a coil. Short fibres (staple) are produced by passing a stream of air across the filaments as they emerge from the bushings.

Rovings are rope-like bundles of continuous untwisted strands for use in such processes as preform press moulding, filament winding, spray-up, pultrusion, and centrifugal casting. They can also be converted into chopped strand mats or cut into short fibres for moulding compounds.

Chopped strands of glass 1/32 to 1/2 inch in length can be incorporated in thermoset or thermoplastic materials about as easily as the particulate fillers. Each strand may be made up to 204 individual filaments whose diameter is 2 to 7.5×10^{-4} inch. Chopped strands several inches long can be loosely bound as a mat that is porous and in which the strands are randomly oriented in two dimensions. This form is suitable for impregnation by a liquid polymer. After polymerisation or cross-linking (curing) under pressure, the composite will comprise a polymer-network matrix in which the individual strands are embedded. Chopped strand mats provide non-directional reinforcement (i.e., strengths in many directions, as contrasted to unidirectional forms which are continuous fibres, like roving, that provide strength in one direction). These mats are available in a variety of thicknesses, usually expressed in weight per square foot. In order to hold the fibres together, a resin binder is generally used, the type depending on the resin and moulding process. In some cases, the mats are stitched or needled, instead of using the resin binder. A woven glass fabric (cloth) might be used in place of the mat or in combination with it. In this case there will be a variation in strength with the angle between the axis of the fibres and the direction of stress. Twisted yarns are generally woven into fabrics of varying thickness and with tight or loose weaves, depending upon the application. Most are balanced weaves (i.e., equal amounts of yarn in each direction), although some are unidirectional (more fibres running in one direction). Although costlier, they offer a high degree of strength. Rovings can also be woven into a fabric that is less costly than the woven yarn fabrics, coarser, heavier, and easier to drape.

For optimum adhesion at the interface between the fibre surface (stationary phase) and the resin matrix (continuous phase), the glass fibres must be treated with coupling agents to improve the interfacial adhesion. The pioneer coupling agent (linking agent) was methacrylatochromatic chloride. This has been supplanted by organosilanes, organotitanates, and organozirconates. These coupling agents contain functional groups, one of which is attracted to the fibre surface and the other to resin (Fig. 17.48).

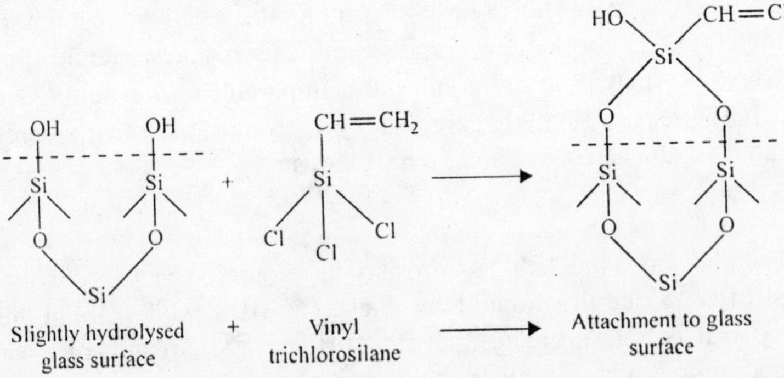

Fig. 17.48. Mechanism of functioning of a glass surface finish.

Graphite carbon fibres

Graphite carbon fibres are the predominant high-strength, high-modulus reinforcing agent used in the fabrication of high-performance polymer composites. In general, the term graphite fibre refers to fibres that have been treated above 1700°C (3092°F) and have tensile moduli of elasticity of 5×10^5 psi (3450 MPa) or greater. Carbon fibres are those products that have been processed below 1700°C (3092°F) and consequently exhibit elastic moduli up to 5×10^5 psi (3450 MPa). A further distinction is that the content of carbon fibres is 80 to 95 per cent carbon; of graphite, above, 99 per cent carbon. However, the industry has universally adopted the term 'graphite'. It will therefore be used to describe both product forms in this section.

These products exhibited relatively high tensile strengths of about 4×10^5 psi (2760 MPa) and were designed for rocket/missile ablative component applications. A significant event that led to the development of today's graphite industry was the utilisation of polyacrylonitrile (PAN) as a graphite precursor material. Subsequent work led to continued improvement of PAN-based graphite fibre properties by numerous researchers. These developments focused on stretching the PAN precursor to obtain a high degree or molecular orientation of the polymer molecules followed by stabilising it under tensile load, carbonisation, and graphitisation. PAN-based graphite fibres are now available with tensile moduli of up to 1.2×10^6 psi (8280 MPa) and tensile strengths above 8×10^5 psi (5516 MPa).

These fibres are made by melt spinning a low-cost isotropic molten (petroleum) pitch and then oxidising the filaments as they are spun. This step is followed by carbonisation at 1000°C (1832°F) in an inert atmosphere. Process modifications to improve the fibre properties evolved through the 1970s until pitch-based (including mesophase liquid crystal pitch) graphite fibres with tensile strengths up to 3.75×10^5 psi (2590 MPa) and tensile moduli to 1.2×10^6 psi (8300 MPa) were achievable.

Fibre manufacture

The pyrolysis of organic fibres used as graphite precursors is a multistage process. The three principal graphite precursors are PAN, pitch, and rayon, with PAN as the predominant product.

The commercial production of PAN precursor fibre is based on either dry or wet spinning technology. In both instances, the polymer is dissolved in either an organic or inorganic solvent at a concentration of 5–10 per cent by weight. The fibre is formed by extruding the polymer solution through spinneret holes into hot gas environment (dry spinning) or into a coagulating solvent (wet spinning). The wet spinning is more popular as it produces fibres with round cross-section, whereas the dry spinning results in fibrr with a dog bone corss-section. The wet-spun precursor making process includes three basic steps: polymerisation, spinning, and after treatments (Fig. 17.49). Acrylonitrile monomer and other co-monomers (methyl acrylate or vinyl acetate) are polymerised to form a polyacrylonitrile co-polymer. The reactor effluent solution, called 'dope', is purified, the unreacted monomers removed, and the solid contaminants filtered off. The spinning process next extrudes the purified dope through holes in spinnerets into a coagulating solution. The spun gel fibre then passes through a series of after-treatments such as stretching, oiling, and drying. The product is the PAN precursor.

In order to produce high-strength, high-modulus graphite fibres from the PAN order precursor, it is essential to produce preferred molecular orientation parallel to the fibre axis and then 'stabilise' the fibre against relaxation phenomena and chain scission reactions that may occur in subsequent carbonisation steps. A typical step-by-step PAN-based graphite manufacturing process beigns with the aforementioned precursor stabilisation which is followed by carbonisation, graphitisation, surface treatment, and sizing, as shown schematiclly in Fig. 17.50.

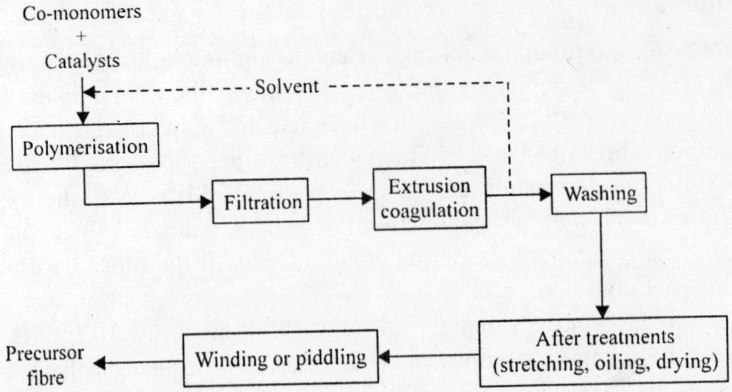

Fig. 17.49. Typical PAN precursor manufacturing steps based on solution polymerisation and wet spinning.

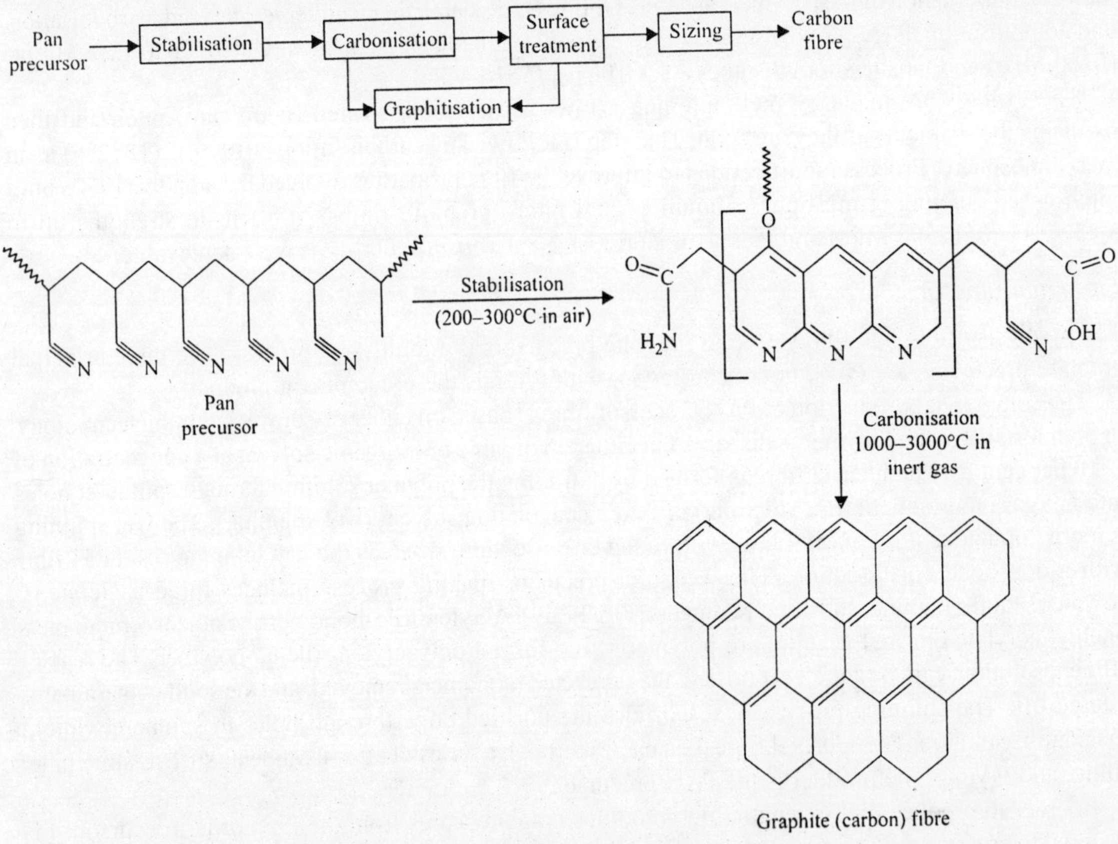

Fig. 17.50. Schematic of a typical step-by-step PAN-based graphite manufacturing process.

The stabilisation of the PAN precursor involves 'pre-oxidation' by heating the fibre in an air oven at 200–300°C (392–572°F) for approximately one hour while controlling the shrinkage/tension of the fibre so that the PAN polymer is converted into a thermally infusible aromatic ladder-like structure. The next step is the process of carbonisation, which pyrolyses the stabilised PAN-based fibres until they are transformed into graphite (carbon) fibres. The carbonisation treatment is done in an inert atmosphere (generally nitrogen) at temperatures greater than 1200°C (2192°F). This step removes hydrogen, oxygen, and nitrogen atoms from the ladderlike polymers whose aromatic rings then collapse into a graphitelike polycrystalline structure. It is during this stage that high-mechanical-property characteristics of graphite fibres are developed. The development of these properties is directly related to the formation and orientation of graphitelike fibres or ribbons within each individual fibre. Graphitisation performed at temperatures above 1800°C(3272°F) is an optional treatment. Its purpose is to improve the tensile modulus of elasticity of the fibre by improving the crystalline structure and orientation of graphitelike crystallites within each individual fibre. The higher heat-treatment temperatures used in graphitisation also results in a higher carbon content of that fibre. The final step in the process of producing graphite or carbon fibre is surface treatment and sizing prior to bobbin winding the continuous filaments. The surface treatment is essentially an oxidation of the fibre surface to promote wettability and adhesion with the matrix resin in the composite. Sizing improves handleability and wettability of the fibre with the matrix resin. Typical sizing agents are poly(vinyl alcohol), epoxy, and polyimide.

Pitch-based graphite fibres are produced by two processes. The precursor of one of these processes is a low-softening-point isotropic pitch and the process scheme includes the followings steps: (i) melt-spin isotropic pitch; (ii) thermoset at relatively low temperatures for long periods of time; (iii) carbonise in an inert atmosphere at 1000°C(1832°F); and (iv) stress graphitise at high temperatures 3000°C (5432°F). The high-performance fibres produced in this manner are relatively expensive because of the very long thermosetting times required and the need for high-temperature stress graphitisation.

The commercially more significant process for making pitch-based fibres is the mesophase process, which involves the following steps: (i) heat treat in an inert atmosphere at 400–450°C(752–842°F) for an extended period of time in order to transform pitch into a liquid-crystalline (mesophase) state; (ii) spin the mesophase pitch into fibres; (iii) thermoset the fibres at 300°C(572°F) for 2 1/2 hours; (iv) carbonise the fibres at 1000°C(1832°F) and (v) graphitise the fibres at 3000°C(5432°F). Since long thermosetting times and stress graphitisation treatment are not required, the high-temperature graphite fibres produced by this process are lower in cost.

The process by which rayon precursor is converted to graphite fibres includes four steps: (i) fibre spinning; (ii) stabilisation at 400°C(752°F) for long periods of time; (iii) carbonisation at 1300°C (2372°F); and (iv) stress graphitisation at high temperatures 3000°C (5432°F). The rayon based graphite fibres produced in this manner tend to be relatively expensive because of the very long stabilisation times required and the need for stress graphitisation at high temperatures.

Properties and applications

The excellent properties of graphite are directly attributable to the highly anisotropic nature of the graphite crystal. The standard-grade PAN-based graphite fibres, which make up the largest part of both the commercial and aerospace markets have tensile strengths ranging from 4.5×10^5 to 5.5×10^5 psi (3100 to 3800 MPa) and moduli of approximately 340,000 psi (2345 MPa). A family of intermediate-modulus/high-strain fibres with tensile strength up to 7×10^5 psi (4800 MPa) and modulus above 4×10^5 psi (2760 MPa) have been developed to meet high-performance aerospace requirements. The high-modulus fibres (both PAN- and pitch-based), used in high stiffness/low-strength applications, such as space hardware, have tension moduli ranging from 5×10^5 to 1.2×10^6 psi (3450 to 8280 MPa) and strain to failures generally greater than or equal to 1 per cent.

Graphite fibres are available to the user in a variety of forms: continuous filament for filament winding, braiding, or pultrusion; chopped fibre for injection or compression moulding; impregnated woven fabrics and unidirectional tapes for lamination.

The major markets for advanced graphite fibre composites are aerospace, marine, automotive, industrial equipment, and recreation. Military aerospace applications dominate the market and military consumptions are slated to increase rapidly as programs, which utilise a very high percentage of composites move from development to large-scale production. Graphite fibre usage in space applications is in a large measure linked to Space Station programs and production activities. Examples of non-aerospace military applications include portable, rapid deployment bridges for the army and propeller shafts for submarines. Fibre usage in the commercial aerospace sector is also growing. Commercial planes such as Boeing 767 and the Airbus A320 utilise two to three times the graphite fibre per plane that is used in older commercial models.

The biggest industrial market potential of graphite fibres is in the automotive sector. The graphite composite usage in this area should increase as lower-cost fibres become available. A major and growing use of chopped graphite fibres in the industrial market is as a reinforcement for thermoplastic-injection-moulding compounds. The advantages of such use include greater strength and stiffness, higher crepe and fatigue resistance, increased resistance to wear, higher electrical conductivity, and improved thermal stability and conductivity.

Aramid fibres

Aramid fibre is the generic name for aromatic polyamide fibres. As defined, an aramid fibre is a 'manufactured fibre in which the fibre forming substance is a long chain synthetic polyamide in which at least 85 per cent of the amide linkages are attached directly to two aromatic rings'.

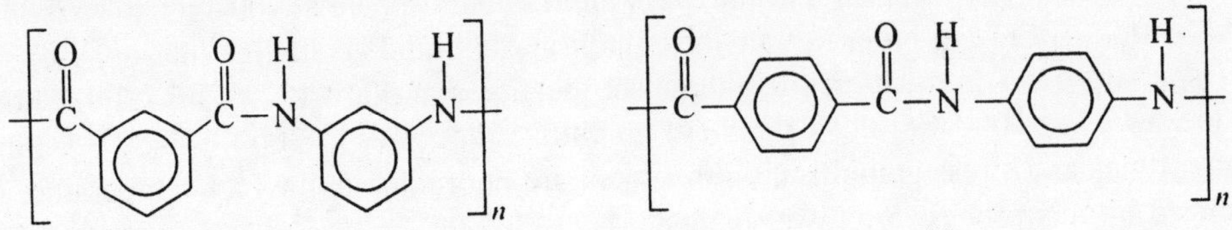

(I) (II)

Among the commercially available aramid fibres are Du Pont's Nomex (I) and Kelvar (II); in fact these trade names are commonly used in lieu of the generic name. Kevlar 49 is a high-modulus aramid fibre and is the most widely used reinforcing aramid fibre. Kevlar 29 has a lower modulus and Kevlar 149 has a higher modulus than Kevlar 49.

Aramid fibres can be used to advantage to obtain composites having lighter weight, greater stiffness, higher tensile strength, higher impact resistance, and lower notch sensitivity than composites incorporating E-glass or S-glass reinforcement. Weight savings over glass result from the lower specific gravity of aramid fibres, 90.4 lb/inch^3 (1.45 g/cm^3) versus E-glass, 159.0 lb/inch^3 (2.55 g/cm^3). Higher stiffnesses are reflected in a Young's modulus up to 19×10^6 psi (1.31×10^5 MPa) for Kevlar 49 and 27×10^6 psi (1.86×10^5 MPa) for Kevlar 149, compared to 10^7 psi (6.9×10^4 MPa) for E-glass and 12×10^6 psi (8.6×10^4 MPa) for S-glass. Aramid composites are more insulating than their glass counterparts, both electrically and thermally, more damped to mechanical and sonic vibrations, and are transparent to radar and sonar. Despite their outstanding mechanical properties, these high-modulus organic fibres have the processability normally associated with conventional textiles. This leads to wide versatility in the form of the reinforcement, e.g., yarns, rovings, woven and knit goods, felts and papers.

Applications

Fabrics woven from Kevlar 49 aramid fibre are often used as composite reinforcement, since fabrics offer biaxial strength and stiffness in a single ply. The mechanical properties of Kevlar 49 aramid are dependent on the fabric construction. The composite properties are functions of the fabric weave and the fibre volume fraction (typically 50 to 55 per cent with ply thickness 5 to 10 mils, depending on fabric construction). In 1987, Du Pont introduced high-modulus Kevlar 149. Compared to Kevlar 49 it has higher performance (47 per cent modulus increase) and lower dielectric properties (65 per cent decrease in moisture regain).

Over the past three decades, Kevlar has gained wide acceptance as a fibre reinforcement for composites in many end uses, such as tennis rackets, golf clubs, shafts, skis, ship masts, and fishing rods. The boating and aircraft industries make extensive use of advanced composites. Each Boeing 767 aircraft saves about 2,000 lb through the use of advanced composites. Cores of Nomex are used in the floor, ceilings, bulkheads, and side walls in the interior of the aircraft. Kevlar can be found in low-pressure cabin distribution ducts, cargo liners, and entry door panels. Composite facings of Kevlar are used over a honeycomb core of Nomex, with unidirectional carbon fibre reinforcement added selectively in critical areas.

For many applications, fabrics containing more than one fibre type offer significant advantages. Hybrids of carbon and Kevlar 49 aramid yield greater impact resistance over all-carbon construction and higher compressive strength over all-Kevlar construction. Hybrids of Kevlar 49 and glass offer enhanced properties and lower weight than constructions containing glass as the sole reinforcement and are less expensive than constructions using only Kevlar 49 reinforcement.

Extended-chain polyethylene fibres

Extended-chain polyethylene (ECPE) fibres are relatively recent entrants into the high-performance fibres field. ECPE fibres are arguably the highest modulus and highest strength fibres made. These are being utilised as a reinforcement in such applications as ballistic armour, impact shields, and radomes to take advantage of the fibre's unique properties.

Polyethylene is a flexible molecule that normally crystallises by folding back on itself. Thus fibres made by conventional melt spinning do not posses outstanding physical properties. ECPE fibres, on the other hand, are made by a process that results in most of the molecules being fully extended and oriented in the fibre direction, producing a dramatic increase in physical properties. Using a simple analogy, the structure of ECPE fibres can be described as that of a bundle of rods, with occasional entangled points that tie the structure together. Conventional PE, by comparison, is comprised of a number of short-length chain folds that do not contribute to material strength (Fig. 17.51). ECPE fibres are, moreover, made from ultrahigh molecular weight polyethylene (UHMWPE) with molecular weight generally 1 to 5 million that also contributes to superior mechanical properties. Conventional PE fibres, in comparison, have molecular weights in the range 50,000 to several hundred thousand. ECPE fibres exhibit a very high degree of crystalline orientation (95–99 per cent) and crystalline content (60–85 per cent).

High-modulus PE fibres can be produced by melt extrusion and solution spinning. The melt extrusion process leads to a fibre with high modulus but relatively low strength and high crepe whereas solution spinning in which very high-molecular-weight PE is utilised yields a fibre with both high modulus and high strength. The solution spinning process for a generalised ECPE fibre starts with the dissolution of polyethylene of approximately 1–5 million molecular weight in suitable solvent. This serves to disentangle the polymer chains, a key step in achieving an extended chain polymer structure. The solution must be fairly dilute to facilitate this process, but viscous enough to be spun using conventional melt spinning equipment. The cooling of the extrudate lends to the formation of a fibre that can be continuously dried to remove solvent or later extracted by an appropriate solvent. The fibres are generally postdrawn prior to final packaging.

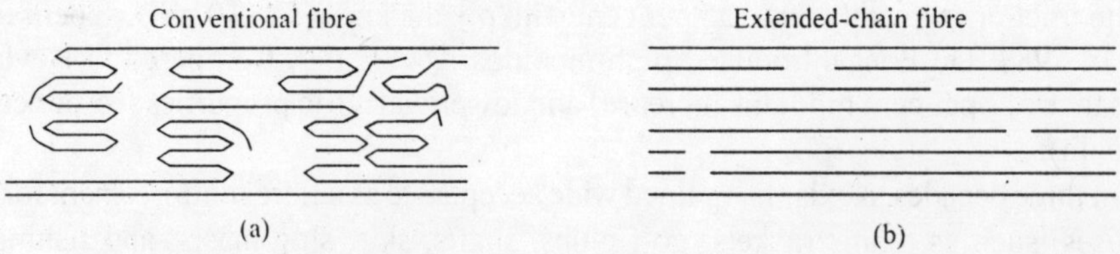

Fig. 17.51. Fibre morphology of polyethyelne. (a) Conventional PE fibre characterised by relatively low molecular weight, moderate orientation, and crystalline regions chain folded; and (b) extended-chain PE fibre characterised by very high molecular weight, very high degree of orientation, and minimum chain folding.

The solution spinning process is highly flexible and can provide an almost infinite number of process and product variations of ECPE fibres. Fibre strengths of (3.75–5.60) × 10^5 psi (2890–3860 MPa) and tensile moduli of (15–30) × 10^6 psi [(103–207) × 10^3 MPa] have been achieved. The properties are similar to other high-performance fibres; however because the density of PE is approximately two-thirds that of high modulus aramid and half that of high-modulus carbon fibre, ECPE fibres possess extraordinarily high specific strengths and specific moduli. Fig. 17.52 compares the specific strength versus specific modulus for currently available fibres.

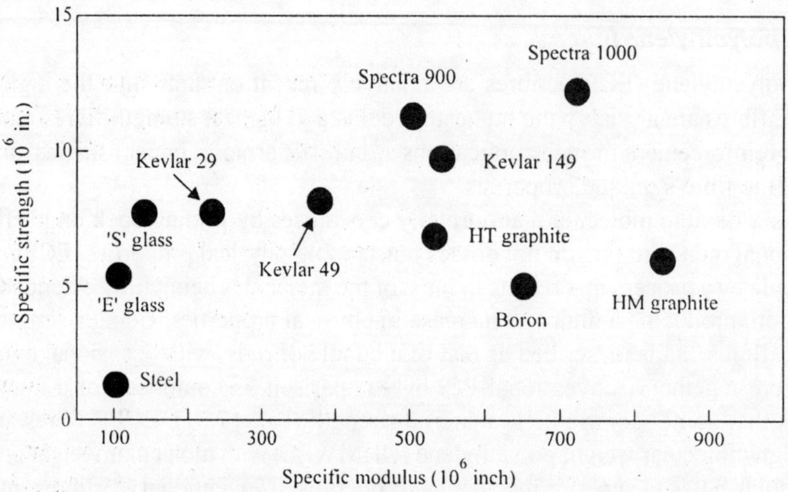

Fig. 17.52. Comparative properties of various reinforcing fibres.

Traditional binders and wetting agents are ineffective in improving resin adhesion to polyethylene. For ECPE fibres, this characteristic is actually advantageous in specific area. For instance, ballistic performance is inversely related to the degree of adhesion between the fibre and the resin matrix. However, for applications requiring higher levels of adhesion and wetout, it has been shown that by submitting ECPE fibre to specific surface treatment, such as corona discharge or plasma treatments, the adhesion of the fibre to various resins can be dramatically increased.

The chief application areas being explored and commercialised for ECPE fibres are divided between traditional fibre applications and hightech composite applications. The former include sailcloth, marine ropes, cables, sewing thread, nettings, and protective clothing. The latter includes impact shields, ballistics, radomes, medical implants, sports equipment, pressure vessels, and boat hulls.

ECPE fibres are well suited for high-performance yachting sails, offering, in addition, resistance to sea water and to typical cleaning solutions used in the sailing industry, such as bleach. The major sport equipment applications to date have been canoes, kayaks, and snow and water skis. Numerous other sport applications are under development.

The high-strength, lightweight, low-moisture absorption and excellent abrasion resistance or ECPE make it a natural candidate for marine rope. In marine rope applications, load, cycling, and abrasion resistance are critical. Thus a 12-strand ECPE braid, for example, is reported to withstand about eight times the number of cycles that cause failure in 12-strand aramid braid.

Specially toughened and dimensionally stabilised ECPE yarn has been used in a revolutionary new line of cut-resistant products. ECPE fibres are being used to produce cut-resistant gloves, arm guards, and chaps in such industries as meat packing, commercial fishing, and poultry processing and in sheet metal work, glass cutting, and power tool use.

ECPE's high strength and modulus and low specific gravity offer higher ballistic protection at lower density per area than is possible with currently used materials. The significant applications include flexible and rigid armour. Flexible armour is manufactured by joining multiple layers of fabric into the desired shape, the ballistic resistance being determined by the style of the fabric and the number of layers. Traditional rigid armour can be made by utilising woven ECPE fibre in either thermoset or thermoplastic materials. Ballistics are currently the dominant market segment. Products include helmets, helicopter seats, automotive and aircraft armour, armour radomes, and other industrial structures.

The radome (radome protective domes) market is also important for ECPE fibres. ECPE composite systems act as a shield that is virtually transparent to microwave signals, even in high-frequency regimes.

REACTION INJECTION MOULDING

In recent years, a new type of injection moulding called reaction injection moulding (RIM) has become important for fabricating thermosetting polymers. RIM differs from the conventional, injection moulding in that the finished product is made directly from monomers or low-molecular weight polymeric precursors (liquid reactants), which are rapidly mixed and injected into the mould even as the polymerisation reaction is taking place. Thus, synthesis of polymer prior to moulding is eliminated, and the energy requirements for handling of monomers are much less than those of viscous polymers. For RIM to be successful, the monomers or liquid reactants must be fast reacting, and the reaction rates must be carefully synchronised with the moulding process. Thus the polymers most commonly processed by RIM are polyurethanes and nylons though epoxies, and certain other polymers such as polycyclopentadiene have been processed by RIM. The process uses equipment that meters reactants to an accuracy of 1%, mixes them by high-pressure impingement, and dispenses the mixture into a closed mould. The mould, is in fact, a chemical reactor. The reaction in RIM takes place in a completely filled mould cavity. Reinforcing fillers are sometimes injected into the mould along with the reactants, a process called reinforced reaction injection moulding (RRIM). In the mould the functional groups of the liquid reactants react to form chemical linkages, producing solid polymers, which comprise polymeric chains or networks depending on the starting materials. The temperature of the mould plays a vital role in the polymerisation of reactants.

To produce a moulded part by the RIM process requires precise but realistic process control. Fig. 17.53 shows a simplified schematic of the RIM process. The important elements of the process are conditioning, metering, mixing, and moulding. All the liquid reactants require precise temperatures control. The flow property (viscosity) of the liquid reactants usually varies with temperature as does the density. For accurate metering the temperature must be controlled within very narrow limits. This is usually accomplished by recirculating reactants from conditioning or storage tanks designed to maintain raw material temperatures specified by the system supplier. These conditions are normally quite moderate (30–38°C) for polyurethanes. For some polymerisation systems, such as nylon which is processed at a high temperature, the machines are designed with heated lines and temperature control devices for pumps, mixers, and other components.

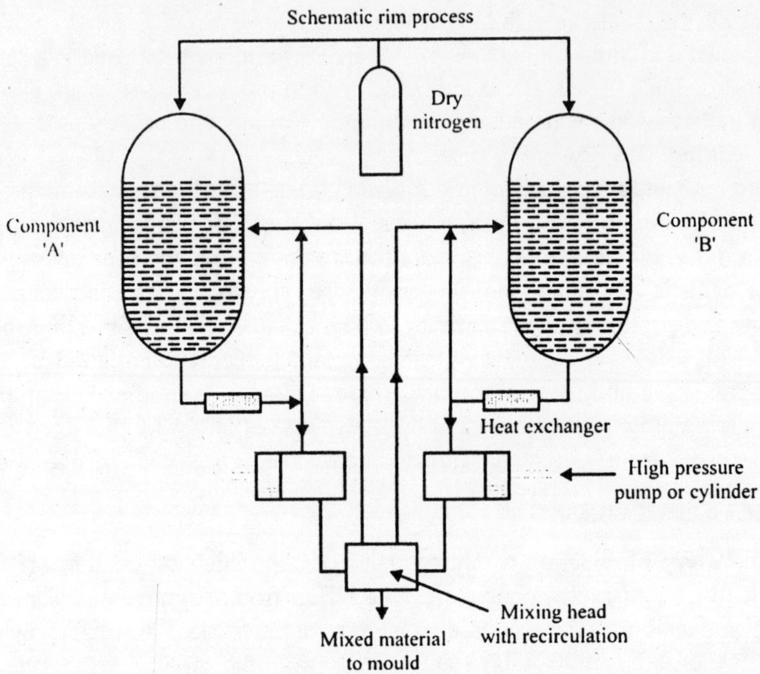

Fig. 17.53. Schematic of RIM process.

The moulds are also designed to control temperatures as the reaction characteristics of the RIM process are exothermic. A higher temperature increases the reaction rate, which results in decrease of cycle time. For polyurethane, however, speeding the reaction rate by operating the mould at elevated temperatures is to be avoided, as this changes the types of linkages produced. The automotive industries in the United States account for most of the commercial RIM production.

General Motors made the first production RIM parts moulded in the United States. The bumper cover for the Corvette was moulded experimentally and soon became a production part on a limited production car. The Company also moulded RIM nylon body panels that were used on selected production cars. Ford used RIM urethane bumper covers for its high-production Fairlane and Zephyr models, and moulded very large front end trim incorporating the bumper cover, headlight opening, and grille. Chrysler Omni and Horizon two-door models also used the multipurpose front end moulding. Another interesting application of RIM was the simulated wood trim used on the Chrysler station wagons and sport convertibles. A later development for RIM polyurethane, and to a lesser extent RIM nylons, is the application for housings of various instruments and appliances: computer housings, business machine housings, TV and radio cabinets, instrument cases, and similar electronic product enclosures. While elastomeric RIM is most commonly used in these applications, some housings are also moulded from RIM structural foam.

Though system suppliers do not always clearly differentiate between elastomeric and structural RIM, elastomeric RIM is moulded in thin cross-section (usually 0.125 inch) at high density while structural foam has an interior foam structure, a density about one-third that of elastomeric RIM, and is moulded in thicker cross-section (usually 0.375 inch).

Machinery

The three primary pieces of equipment necessary to make RIM parts are the RIM machine, the mould clamp, and the mould. The RIM machine performs four functions: (i) conditioning reactants; (ii) accurate metering; (iii) mixing by impingement; and (iv) dispersing mixed reactants directly into a closed mould.

Conditioning and temperature control are accomplished by recirculating reactants from storage tanks which are jacketed and/or contain tempering coils to maintain the process temperature required by the chemical system used. The tanks should be fitted with liquid-level-control devices to keep the raw material relatively constant. Recirculated material re-enters the conditioning tank through a dip tube, thereby controlling the amount of the dissolved gas in the circulating reactants. (Inert gas, such as dry nitrogen, is often introduced between the metering pump and the mixing head to provide a gas-bubble nucleus for the foaming reaction that finally fills and packs the mould).

Positive displacement pumps are commonly used in RIM machines for accurate metering of the reactants. The pumps can be adjusted to compensate for weight ratio differences characteristic of RIM chemical systems. There are two basic types of displacement machines. One uses metering pumps to pump hydraulic oil to the pistons, which actuate the reactant displacement cylinders. The other uses a single hydraulic pump to pressurise hydraulic oil and proportioning valves to determine flow rates to the hydraulic power pistons. A potentiometer linked directly to the piston system provides continuous monitoring of the flow.

Mixing heads perform three main functions of recycling reactants between shots, mixing by impingement, and cleaning the mixing chamber at the end of the shot. The most common designs use an angled mixing chamber that causes the mixed reactants to flow around a 90° bend in the head. Two clean-out pistons placed perpendicular to each other are required for cleaning out. A common mixing head uses bypass channels in the clean-out piston to recirculate raw materials.

RIM parts are moulded in metal moulds which shape the part by having a cavity cut to form a three-dimensional metal negative of the desired part shape. Most high-production RIM processes use machined steel moulds. These moulds have a reported working life of more than 1 million parts. Cast aluminium moulds have been used for textured parts most notably simulated-wood products such as cabinet doors. Aluminium moulds have a reported working life of more than 100,000 parts. Pressure is developed in the mould primarily by the foaming and packing done by R-11 vapourisation (discussed later). Clamping units for holding the mould closed are lighter and less expensive than pressures for injection moulding.

An expanding agent is vapourised to fill the mould (up to 70 per cent expansion for structured RIM) or pack the mould (possibly under 5 per cent expansion for elastomeric RIM). The solid part is formed as the concurrent chemical reaction forms the polymer chain or network that gives the part its ultimate physical properties.

RIM parts (especially polyurethanes) are usually removed from the mould before the chemical reactions that develop the physical properties are complete. The part is placed on a support jig that holds it in its final shape until it is fully cured. In some cases this is done by simply setting the supported part aside for 12 to 24 hour. More often the supported part is post-cured in an oven for several hours at temperatures of about 180°F (82°C). Nylons are completely reacted in the mould and postcuring is not necessary.

RIM polyurethanes made with aromatic isocyanates (such as pure or polymeric MDI) have a tendency to darken as a result of the effect of UV light on the chemical ring structure of the MDI component. Soft white limestone or fine carbon black is often used as filler to mask the effect of this colour change.

Polyurethanes manufactured with aliphatic isocyanates are light stable, and products are moulded in a wide range of bright colours. Especially interesting is the development of equipment to add colour concentrate, usually dispersed in a polyol, directly into the mixing head attached to a given mould. The basic urethane formula is adjusted to compensate for the additional reactive polyol. Using this technique with a multiclamp RIM line, it is possible to mould different colours using a single RIM machine. In some cases, aromatic RIM systems are moulded in colour, then painted the same colour. This technique eliminates the need to touch up every dent or scratch which would otherwise show up tan or white.

There are several systems for painting RIM parts. Usually, RIM parts are simply primed and painted. Before printing, however, the part is cleaned to remove mould-release agents. The most common mould releases are metal stearates (or soaps) that can be removed from the part by a water wash and paraffin waxes that are usually removed by solvent vapour degreasing. Silicone mould releases are to be avoided as they are very difficult to remove from the part, and paint will not stick to the silicone surface film.

Polyurethanes

The most common chemicals used in the RIM process for polyurethanes are isocyanates containing two or more isocyanate (—N=C=O) groups and polyols, which contain two or more hydroxyl (—OH) groups. These reactive end-groups, so named because they occur at the ends of the chemical structure, react chemically to form a urethane linkage:

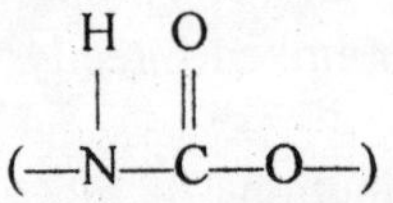

The chemical system, must be adjusted so that the number of isocyanates and hydroxyls balance and that all reactive end-groups are used in the formation of urethane linkages.

The number of polymer structures that can be formed using the urethane reaction is quite large. There are ways to produce polyurethanes having different physical properties. If linear polyols are reacted with diisocyanates, a flexible polyurethane will be formed. If a low boiling liquid, such as Refrigerant-11 (R-11), is incorporated into the system, the heat of reaction will produce a cellular structure. The resulting product will be a flexible polyurethane foam.

The physical properties of these materials can be varied by selecting polyols with shorter or longer polyol chains. The most common polyol or macroglycol chains are polyethers and polyesters. The composition of these thermoplastic chains also plays a role in the physical properties of the end product. These chain segments in the block co-polymer are often referred to as 'soft' blocks, or segments, while the polyurethane segments formed by the reaction of diisocyanate with glycol are referred to as 'hard' blocks or segments.

In addition to changing the chain composition and length, the physical properties can be varied by blending up to approximately 10 per cent of a long-chain triol (such as a triol adduct of ethylene oxide and propylene oxide with glycerol) into the basic resin system formulation. This produces branching in the 'soft' segment of the block co-polymer. Excessive triol modification may, however, diminish physical properties. Use of short-chain triols such as glycerol will produce cross-linking in the 'hard' segments in the polymer chain. The formation of hard blocks and cross-linking in the hard block tend to produce a stiffer, more rigid product. The hard blocks tend to be crystalline and reinforce the amorphous polymer, improving its strength.

Nylons

RIM nylons, like polyurethanes, form polymers very rapidly by the reaction of chemical end-groups. The linkages produced are as follows:

$$\text{Polyesteramide pre-polymer} + \underbrace{\overbrace{(CH_2)_5\,CONH}}_{\text{Caprolactam}} \rightarrow \text{Nylon block co-polymer}$$

Equipment used to manufacture RIM products must be extensively re-built to process RIM nylons. Because the viscosity of nylon RIM systems is low and the ingredients are quite reactive, leakage at the seals and the volumetric efficiency of the metering pumps (that is, the amount of material actually pumped divided by the volume displaced by the metering pumps) may cause problems, which require special attention. Hence, most manufacturers recommend having the machine designed specifically for nylon RIM systems.

Because of the excellent impact strength of nylon RIM, it has been used for bumper covers and automobile *fascia*. It also finds application in housings for business machines and electronics.

STRUCTURAL REACTION INJECTION MOULDING

Structural reaction injection moulding (SRIM) may be considered as a natural evolution of reaction injection moulding (RIM). It is a very attractive composite manufacturing process for producing large, complex structural parts economically. The basic concepts of the SRIM process are shown in Fig. 17.54. A preformed reinforcement is placed in a closed mould, and a reactive resin mixture that is mixed by impingement under high pressure in a specially designed mix head (like that in RIM) flows a low pressure through a runner system to fill the mould cavity, impregnating the reinforcement material in the process.

Once the mould cavity is filled, the resin quickly completes its reaction. A completed component can often be removed from the mould in as little as one minute. SRIM is similar to RIM in its intensive resin mixing procedures and its reliance on fast resin reaction rates. It is also similar to resin transfer moulding (RTM) (discussed later in this chapter) in employing preforms that are pre-placed in the cavity of a compression mould to obtain optimum composite mechanical properties. The term structural is added to the term reaction injection moulding to indicate the more highly reinforced nature of the composite components manufactured by SRIM.

The key to SRIM is the preform. It is a pre-shaped, three-dimensional precursor of the part to be moulded and does not contain the resin matrix. It can consist of fibrous reinforcements, core materials, metallic inserts, or plastic inserts. The reinforcements, cores, or inserts can be anything available that meets the economic, structural, and durability requirements of the parts. This tremendous manufacturing freedom allows a variety of alternative preform constructions.

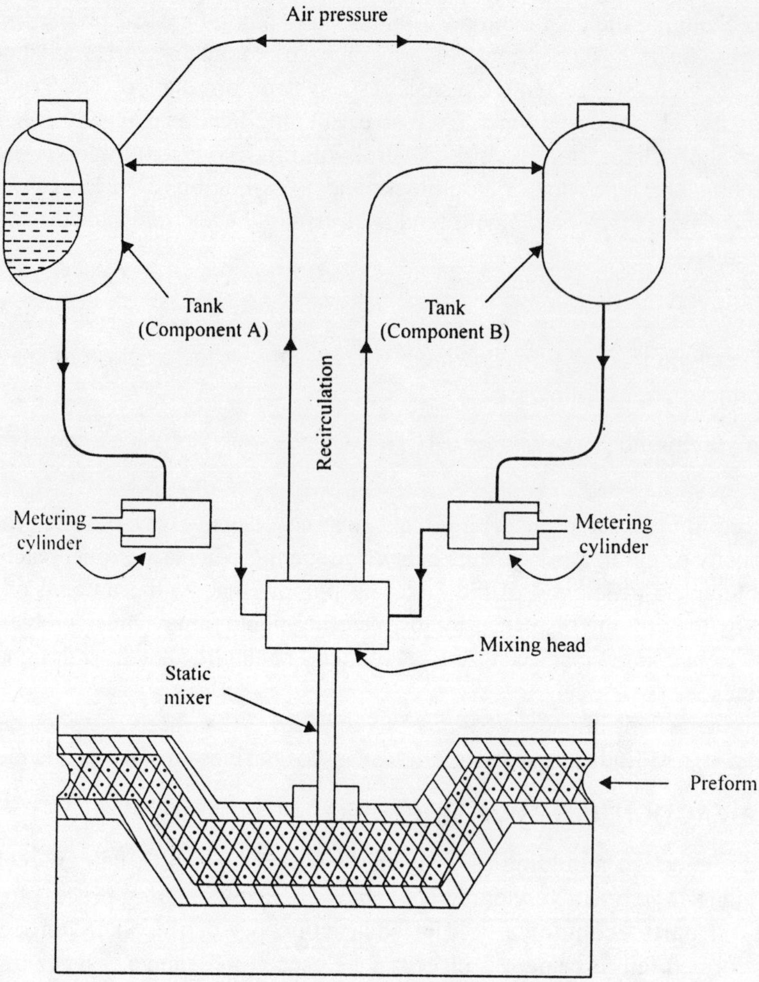

Fig. 17.54. Schematic of structural reaction injection moulding (SRIM).

Most commercial SRIM applications have been in general industry or in the automotive industry. The reinforcement material most commonly used has been fibreglass, due to the low cost. Fibreglass has been used in the form of woven cloth, continuous strand mat, or chopped glass.

Space-shaping cores can be used in the SRIM process to fabricate thick, three-dimensional parts with low densities. Specific grades of urethane-based foams, having densities of 6 to 8 lb/ft^3 and dimensional stability at SRIM moulding temperatures, are commonly used as moulded core materials. Fibreglass reinforcements and inserts can be placed around these cores, resulting in SRIM parts, moulded in one piece, that are very light-weight and structurally strong and stiff. Metallic inserts can be used in SRIM parts as local stiffeners, stressed attachment points, or weldable studs. The metallic material of choice is usually steel.

SRIM is a very labour-intensive process, and the consistency from preform to preform is usually poor. However, for very low manufacturing volumes this process can be cost-effective.

Most SRIM resins have several characteristics in common: their liquid reactants have room-temperature viscosity below 200 cps; their viscosity-cure curves are sigmoidal in shape, the typical mould-fill time being 10 to 90 sec; and their demould time is from 60 to 180 sec, varying with catalyst concentration. The low viscosity of SRIM resins and their relatively long fill times are crucial in allowing them to penetrate and flow through their reinforcing preforms.

The design of the gating and runner configuration (if any) is usually kept proprietary by the moulder. However, it appears that most SRIM parts are centre-gated, with vents located along the periphery of the part. This configuration allows the displaced air in the mould cavity to be expelled uniformly.

Applications

The ability of SRIM to fabricate large, lightweight composite parts, consisting of all types of precisely located inserts and judiciously selected reinforcements, is an advantage that other manufacturing processes find difficult to match. Moreover, large SRIM parts can often be moulded in 2 to 3 minutes, using clamping pressures as low as 100 psi. The capital requirements of SRIM are thus relatively low.

The first commercially produced SRIM part was the cover of the spare-tyre well in several automobiles produced by General Motors. Since then, SRIM automotive structural parts have included foamed door panels, instrument panel inserts, sunshades, and rear window decks. Non-automotive applications include satellite dishes and seat shells for the furniture market.

RESIN TRANSFER MOULDING

Resin transfer moulding (RTM) is similar to SRIM. In its common form, RTM is a closed-mould, low-pressure process in which dry, pre-shaped reinforcement material is placed in a closed mould and a polymer solution or resin is injected at a low pressure, filling the mould and thoroughly impregnating the reinforcement to form a composite part. The mould pressure in the RTM process is lower than that in both SRIM and RIM/RRIM and the moulding cycle time is much longer. The reinforcement and resin may take many forms, and the low pressure combined with the pre-oriented reinforcement package, affords a large range of component sizes, geometries, and performance options.

RTM is an excellent process choice for making prototype components. It allows representative prototypes to be moulded at low cost, unlike processes such as compression moulding and injection moulding, which require tools and equipment approaching actual production level. When prototyping with RTM, less reactive resins are generally used, allowing long fill times and easier control of the vents. Sizes can range from small components to very large, complex, three-dimensional structures. RTM provides two finished surfaces and controlled thickness, while other processes used for prototyping, such as hand lay-up and wet moulding, give only a single finished surface.

FOAMING PROCESSES

Plastics can be foamed in a variety of ways. The foamed plastics, also referred to as cellular or expanded plastics, have several inherent features which combine to make them economically important. Thus, a foamed plastic is a good heat insulator by virtue of the low conductivity of the gas (usually air) contained in the system, has a higher ratio of flexural modulus to density than when unfoamed, has greater load-bearing capacity per unit weight, and has considerably greater energy-storing or energy-dissipating capacity than the unfoamed material. Foamed plastics are therefore used in the making of insulation, as core materials for load-bearing structures, as packaging materials used in product protection during shipping, and as cushioning materials for furniture, bedding, and upholstery.

Among those plastics which are commercially produced in cellular form are polyurethane, PVC, polystyrene, polyethylene, polypropylene, epoxy, phenol-formaldehyde, urea-formaldehyde, ABS, cellulose acetate, styrene-acrylonitrile, silicone, and ionomers. However, note that it is possible today to produce virtually every thermoplastic and thermoset material in cellular form. In general, the basic properties of the respective polymers are present in the cellular products except, of course, those changed by conversion to cellular form.

Foamed plastics can be classified according to the nature of cells in them into closed-cell type and open-cell type. In a closed-cell foam each individual cell, more or less spherical in shape, is completely closed in by a wall of plastic, whereas in an open-cell foam individual cells are inter-connecting, as in a sponge. Closed-cell foams are usually produced in processes where some pressure is maintained during the cell formation stage. Free expansion during cell formation typically produces open-cell foams.

Most foaming processes, however, produce both kinds. A closed-cell foam makes a better buoy or life jacket because the cells do not fill with liquid. In cushioning applications, however, it is desirable to have compression to cause air to flow from cell to cell and thereby dissipate energy, so the open-cell type is more suitable. Foamed plastics can be produced in a wide range of densities—from 0.1 lb/ft^3 (0.0016 g/cm^3) to 60 lb/ft^3 (0.96 g/cm^3)—and can be made flexible, semi-rigid, or rigid.

A rigid foam is defined as one in which the polymer matrix exists in the crystalline state or, if amorphous, is below its T_g. Following from this, a flexible cellular polymer is a system in which the matrix polymer is above its T_g. According to this classification, most polyolefins, polystyrene, phenolic, polycarbonate, polyphenylene oxide, and some polyurethane foams are rigid, whereas rubber foams, elastomeric polyurethanes, certain polyolefins, and plasticised PVC are flexible. Intermediate between these two extremes is a class of polymer foams known as semi-rigid. Their stress-strain behaviour is, however, closer to that of flexible systems than that exhibited by rigid cellular polymers. The group of rigid cellular polymers can be further sub-divided according to whether they are used: (i) for non-load-bearing applications, such as thermal insulation; or as (ii) load-bearing structural materials, which require high stiffness, strength and impact resistance.

The description of cellular foams as low, medium or high density is very common in practice. This is, however, not exact as the different density ranges which correspond to each of these items are not strictly defined. The following figures can, however, serve as a rough general guide:

	lb/ft³	*kg/m³*
Low density	0.1–3	2–50
Medium density	3–21	50–350
High density	21–60	350–960

(Note that the density of a polymer foam refers to its bulk density, defined by the ratio of total weight/total volume of the polymer and gaseous component. Obviously the gas phase contributes considerably to the volume of the end product, while the solid component contributes almost to the entire weight).

Obtained forms of foamed plastics are blocks, sheets, slabs, boards, moulded products, and extruded shapes. These plastics can also be sprayed onto substrates to form coatings, foamed in place between walls (i.e., poured into the empty space in liquid form and allowed to foam), or used as a core in mechanical structures. It has also become possible to process foamed plastics by conventional processing machines like extruders and injection-moulding machines.

Polymer foams may be homogeneous with a uniform cellular morphology throughout or they may be structurally anisotropic. They may have an integral solid polymer skin or they may be multicomponent in which the polymer skin is of different composition to the polymeric cellular core. Schematic representations of the different physical forms of cellular polymers are given in Fig. 17.55. Some special types of foams, namely, structural foams, reinforced foams, and syntactic foams are represented by Figs. 17.55c to f. These are described in a later section.

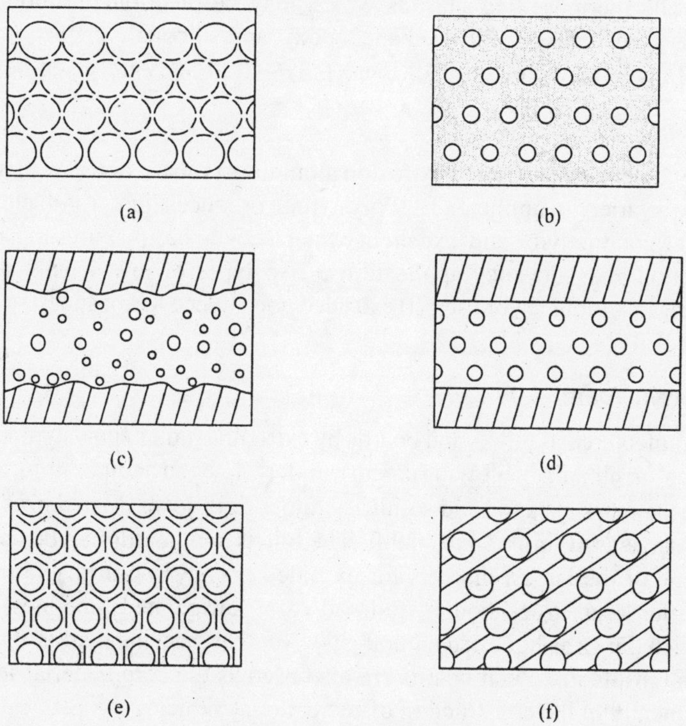

Fig. 17.55. Schematic representations of section through different types of cellular polymer. (a) Low-density open-cell foam; (b) high-density closed-cell foam; (c) single-component structural foam with cellular core and integral solid skin; (d) multicomponent structural foam; (e) fibre-reinforced closed-cell foam; and (f) syntactic foam.

Foaming of plastics can be done in a variety of ways. Most of them typically involve creating gases to make foam during the foaming cycle. Once the polymer has been expanded or 'blown', the cellular structure must be stabilised rapidly; otherwise it would collapse. Two stabilisation methods are used. First, if the polymer is a thermoplastic, expansion is carried out above the softening or melting point, and the form is then immediately cooled to below this temperature. This is called physical stabilisation. The second method—chemical stabilisation—requires the polymer to be cross-linked immediately following the expansion step. Common foaming processes follow:

1. Air is whipped into a dispersion or solution of the plastic, which is then hardened by heat or catalytic action or both.
2. A low-boiling liquid is incorporated in the plastic mix and volatilised by heat.
3. Carbon dioxide gas is produced within the plastic mass by chemical reaction.
4. A gas, such as nitrogen, is dissolved in the plastic melt under pressure and allowed to expand by reducing the pressure as the melt is extruded.
5. A gas, such as nitrogen, is generated within the plastic mass by thermal decomposition of a chemical blowing agent.

6. Microscopically small hollow beads of resin or even glass (e.g., microballoons) are embedded in a resin matrix.

Foams can be made with both thermoplastic and thermosetting plastics. The well known commercial thermoplastic foams are polystyrene, poly(vinyl chloride), polyethylene, polypropylene, ABS co-polymer, cellulose acetate. The thermosetting plastics which may be mentioned, among others, are phenol-formaldehyde, urea-formaldehyde, polyurethane, epoxy, and silicone.

The methods of manufacture of some of these polymeric foams are given below.

Polystyrene Foams

Polystyrene, widely used in injection and extrusion moulding, is also extensively used in the manufacture of plastic foams for a variety of applications. Polystyrene produces light, rigid, closed-cell plastic foams having low thermal conductivity and excellent water resistance, meeting the requirements of low-temperature insulation and buoyancy applications. Two types of low-density polystyrene foams are available to the fabricator, moulder, or user: (i) extruded polystyrene foam; and (ii) expandable polystyrene for moulded foam.

Extruded polystyrene foam

This material is manufactured as billets and boards by extruding molten polystyrene containing a blowing agent (nitrogen gas or chemical blowing agent) under elevated temperature and pressure into the atmosphere where the mass expands and solidifies into a rigid foam. Many sizes of extruded foam are available, some as large as 10 in. × 24 in. × 9 ft. The billets and boards can be used directly or cut into different forms. One of the largest markets for extruded polystyrene in the form of boards is in low-temperature insulation (e.g., truck bodies, railroad cars, refrigerated pipelines, and low-temperature storage tanks for such things as liquefied natural gas). Another growing market for extruded polystyrene boards is residential insulation. Such boards are also used as the core material for structural sandwich panels, used prominently in the construction of recreational vehicles.

Expandable polystyrene

Expandable polystyrene is produced in the form of free-flowing pellets or beads containing a blowing agent. Thus, pellets chopped from an ordinary melt extruder or beads produced by suspension polymerisation are impregnated with a hydrocarbon such as pentane. Below 70°F (21°C) the vapour pressure of the pentane dissolved in the polymer is low enough to permit storage of the impregnated material (expandable polystyrene) in a closed container at ordinary temperature and pressure. Even so, manufacturers do not recommend storing for more than a few months.

The expandable polystyrene beads may be used in a tubular blow-extrusion process (Fig. 17.56) to produce polystyrene foam sheet, which can subsequently be formed into containers, such as egg cartons and cold-drink cups, by thermoforming techniques.

Expandable polystyrene beads are often moulded in two separate steps: (i) pre-expansion or pre-foaming of the expandable beads by heat; and (ii) further expansion and fusion of the pre-expanded beads by heat in the enclosed space of a shaping mould.

Steam heat is used for pre-expansion in an agitated drum with a residence time of a few minutes. As the beads expand, a rotating agitator prevents them from fusing together, and the pre-expanded beads, being lighter, are forced to the top of the drum and out of the discharge chute. They are then collected in storage bins for ageing prior to moulding. The usual lower limit of bulk density for bead pre-expansion is 1.0 lb/ft^3 (0.016 g/cm^3), compared to the virgin bead bulk density of about 35 lb/ft^3 (0.56 g/cm^3).

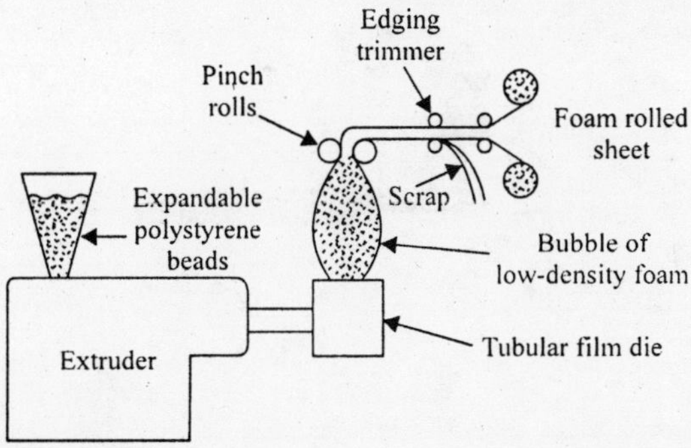

Fig. 17.56. Tubular blow extrusion of production of low-density polystyrene foam sheet.

Moulding of pre-expanded (pre-foamed) beads requires exposing them to heat in a confined space. In a typical operation pre-foamed beads are loaded into the mould cavity, the mould is closed, and steam is injected into the mould jacket.

The pre-foamed beads expand further and fuse together as the temperature exceeds T_g. The mould is cooled by water spray before removing the moulded article. Packages shaped to fit their contents (small sailboats, toys, drinking cups, etc.) are made in this way. Special machines have been designed to produce thin-walled polystyrene foam cups.

Very small beads at a pre-foamed density of approximately 4 to 5 lb/ft^3 (0.06 to 0.08 g/cm^3) are used, which allow easy flow into the moulding cavity and produce a cup having the proper stiffness for handling.

Structural foams

Structural foam is the term usually used for foam produced in an injection moulding press and made of almost any thermoplastic resin. Structural foam is always produced with an integral skin on the outer surfaces and a cellular core in the interior, and is used almost exclusively for production of moulded parts.

The process is thus ideally suited for fabrication of parts such as business machine housings (commonly of ABS), and similar parts or components in which lightweight and stiffness are required.

The structural foam injection moulding process (Fig. 17.57), by which a product with a cellular core and a solid skin can be moulded in a single operation, gets its name from the application of its product rather than the mechanism of the process itself. In a manner directly opposite to the vented extruder, a blowing agent, often nitrogen, is injected with gas, is then forced into the accumulator where it is maintained at a pressure and temperature high enough to prevent foaming (Fig. 17.57a). When a sufficient charge has accumulated it is transferred into the mould (Fig. 17.57b). The melt foams and fills the mould at a relatively low pressure (1.3–2.6 MPa) compared to the much higher pressure in the accumulator. The lower operating pressures of the moulds make the moulds less expensive than those used for conventional injection moulding. However, the cycle times are longer because the foam being a good insulator, takes longer time to cool.

Structural foams can also be made using a chemical blowing agent (discussed later) rather than an inert gas. In that case, a change in pressure or temperature on entering the mould triggers gas formation. Today structural foam injection moulding is a very fast-growing polymer processing technique that can be used to modify the properties of thermoplastics to suit specific applications.

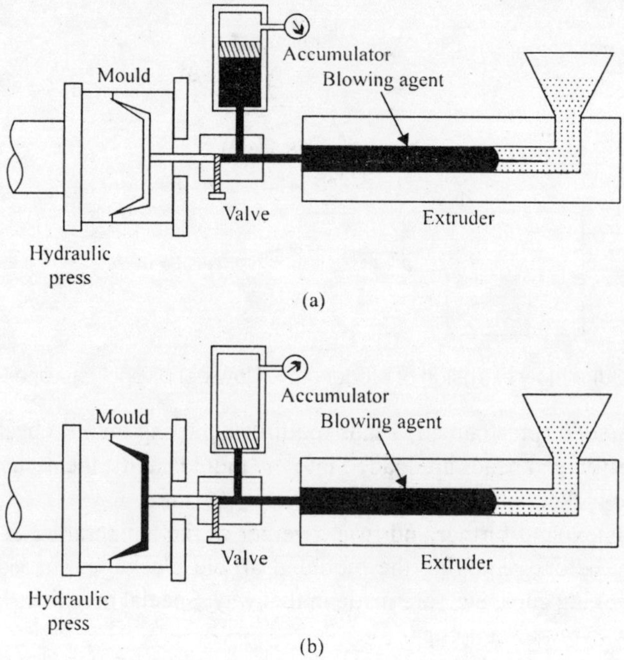

Fig. 17.57. Structural foam process. (a) Filling the accumulator. The blowing agent (usually nitrogen) is injected into the melt in the extruder before it is passed into the accumulator; (b) filling the mould. The accumulator ram injects the melt into the mould where the reduced pressure allows the gas to foam the resin.

Polyolefin Foams

Polyolefin foams can be produced with closely controlled density and cell structure. Generally the mechanical properties of polyolefins lies between those of a rigid and a flexible foam. Polyolefin foams have a very good chemical and abrasion resistance as well as good thermal insulation properties. Cross-linking improves foam stability and polymer properties. A variety of foams can be produced from various types of polyethylenes and cross-linked systems having a very wide range of physical properties, and foams can be tailor-made to a specific application. Polypropylene has a higher thermostability than polyethylene. The production volume of polyolefin foams is not as high as that of polystyrene, polyurethane, or PVC foams. This is due to the higher cost of production and some technical difficulties in the production of polyolefin foams. The structural foam injection moulding process, described previously (Fig. 17.57) for polystyrene, is also used for polyethylene and polypropylene structural foams.

Cellular polyethylene is produced in the form of planks, rods, or billets by extrusion. In this technique, a blowing agent is mixed with the polyethylene pellets, and during the mixing and shearing in the barrel, the pellets melt and the blowing agent decomposes. The liberated gas remains dissolved in the molten resin under high pressure. On extrusion, the gas expands and forms the cellular material. Polyethylene foams are found in packaging, marine bumpers, sporting goods padding, and other products.

Polyurethane Foams

Polyurethane foams, also known as urethane foams or U-foams, are prepared by reacting hydroxyl-terminated compounds called polyols with an isocyanate. Isocyanates in use today include toluene diisocyanate, known as TDI, crude methylenebis(4-phenyl-isocyanate), known as MDI, and different types of blends, such ass TDI/crude MDI. Polyols, the other major ingredient of the urethane foam, are active hydrogen-containing compounds, usually polyester diols and polyether diols.

It is possible to prepare many different types of foams by simply changing the molecular weight of the polyol, since it is the molecular backbone formed by the reaction between isocyanate and polyol that supplies the reactive sites for cross-linking which in turn largely determines whether a given foam will be flexible, semi-rigid, or rigid. In general, high-molecular-weight polyols with low functionality produce a structure with a low amount of cross-linking and, hence, a flexible foam. On the other hand, low-molecular-weight polyols of high functionality produce a structure with a high degree of cross-linking and, consequently, a rigid foam. Of course, the formulation can be varied to produce any degree of flexiblility or rigidity within these two extremes.

The reactions by which urethane foam are produced can be carried out in a single stage (one-shot process) or in a sequence of several stages (pre-polymer process and quasi-pre-polymer process). These variations led to 27 basic types of products or processes, all of which have been used commercially. In the one-shot process all of the ingredients—isocyanate, polyol, blowing agent, catalyst, additives, etc.—are mixed simultaneously, and the mixture is allowed to foam. In the pre-polymer method a portion of the polyol is reacted with an excess of isocyanate to yield a pre-polymer having isocyanate end groups. The pre-polymer is then mixed with additional polyol, catalyst, and other additives to cause foaming. The quasi-pre-polymer process is intermediate between the pre-polymer and one-shot processes.

Flexible polyurethane foams

The major interest in flexible polyurethane foams is for cushions and other upholstery materials. Principal market outlets include furniture cushioning, carpet underlay, bedding, automotive seating, crash pads for automobiles, and packaging. The density range of flexible foams is usually 1 to 6 lb/ft^3 (0.016 to 0.096 g/cm^3). The foam is made in continuous loaves several feet in width and height and then sliced into slabs of desired thickness.

One-shot process

The bulk of the flexible polyurethane foam is now being manufactured by the one-shot process using polyether type-polyols because they generally produce foams of better cushioning characteristics. The main components of a one-shot formulation are polyol, isocyanate, catalyst, surfactant, and blowing agent. Today the bulk of the polyether polyols used for flexible foams are propylene oxide polymers. The polymers prepared by polymerising the oxide in the presence of propylene glycol as an initiator and a caustic catalyst are diols having the general structure

$$\underset{\displaystyle OH}{\overset{}{CH_3-\underset{|}{CH}}}-CH_2-O\!\left(CH_2-\underset{\underset{\displaystyle CH_3}{|}}{CH}-O\right)_n CH_2-\underset{\underset{\displaystyle OH}{|}}{CH}-CH_3$$

The polyethers made by polymerising propylene oxide using trimethyl propane, 1,2,6-hexanetriol, or glycerol as initiator are triols of the following general type:

$$HO\text{-}(\text{-}C_3H_6O-)_nCH_2-CH(OH)-CH_2(C_3H_6O)_nOH$$

The higher hydroxyl content of these polyethers leads to foams of better loadbearing characteristics. Molecular weight in the range 3000 to 3500 is found to give the best combination of properties.

The second largest component in the foam formulation is the isocyanate. The most suitable and most commonly used isocyanate is 80:20 TDI—i.e., 80:20 mixture of tolylene-2,4-diisocyanate and tolylene-2,6-diisocyanate.

One-shot processes require sufficiently powerful catalysts to catalyse both the gas evolution and chain extension reaction. Use of varying combinations of an organometallic tin catalyst (such as dibutyltin dilaurate and stannous octoate) with a tertiary amine (such as alkyl morpholines and triethylamine), makes it possible to obtain highly active systems in which foaming and cross-linking reactions could be properly balanced. The surface active agent is an essential ingredient in formulations. It facilitates the dispersion of water in the hydrophobic resin by decreasing the surface tension of the system. In addition, it also aids nucleation, stabilises the foam, and regulates the cell size and uniformity. A wide range of surfactants, both ionic and non-ionic, have been used at various times. Commonly used among them are the water-soluble polyether siloxanes.

Water is an important additive in urethane foam formulation. The water reacts with isocyanate to produce carbon dioxide and urea bridges. An additional amount of isocyanate corresponding to the water present must therefore be incorporated in the foaming mix. The more water that is present, the more gas that is evolved and the greater number of active urea points for cross-linking. This results in foams of lower density but higher degree of cross-linking. So when soft foams are required, a volatile liquid such a trichloromonofluoromethane (bp 23.8°C) may be incorporated as a blowing agent. This liquid will volatilise during the exothermic urethane reaction and will increase the total gas in the foaming system, thereby decreasing the density, but it will not increase the degree of cross-linking. However, where it is desired to increase the cross-link density independently of the isocyanate-water reaction, polyvalent alcohols, such as glycerol and pentaerythritol, and various amines may be added as additional cross-linking agents. A typical formulation of one-shot urethane foam system is shown in Table 17.3.

Table 17.3. Urethane one-shot foam formulation.

Ingredient	*Parts by weight*
Poly(propylene oxide), mol. wt. 2000 and 2 OH/molecule	35.5
Poly(propylene oxide) initiated with trifunctional alcohol, mol. wt. 3000 and 3 OH/molecule	35.5
Toluene diisocyanate (80:20 TDI)	26.0
Dibutyltin dilaurate	0.3
Triethylamine	0.05
Water	1.85
Surfactant (silicone)	0.60
Trichloromonofluoromethane (CCl_3F)	12.0
Final density of foam = 1.4 lb/ft^3 or 0.022 g/cm^3 (2.0 lb/ft^3 or 0.032 g/cm^3 if CCl_3F is omitted)	

Most foam is produced in block form from Henecke-type machines or some modification of them. In this process, several streams of the ingredients are fed to a mixing head which oscillates in a horizontal plane. In a typical process, four streams may be fed to the mixing head: e.g., polyol and fluorocarbon (if any); isocyanate; water, amine, and silicone; and tin catalyst. The reaction is carried out with slightly warmed components. Foaming is generally complete within a minute of the mixture emerging from the mixing head. The emergent reacting mixture runs into a trough, which is moving backward at right angles to the direction of traverse of the reciprocating mixing head. In this way the whole trough is covered with the foaming mass.

Other developments of one-shot flexible foam systems include direct moulding, where the mixture is fed into a mould cavity (with or without inserts such as springs, frames, etc.) and cured by heat. In a typical application, moulds would be filled and closed, then heated rapidly to 300 to 400°F (149 to 204°C) to develop maximum properties. A good deal of flexible urethane foam is now being made by the cold-cure technique. This involves more reactive polyols and isocyanates in special foaming formulations which would cure in a reasonable time to their maximum physical properties without the need for additional heat over and above that supplied by the exothermic reaction of the foaming process. Cold-cure foaming is used in the production of what is known as high-resilient foams having high sag factor (i.e., ratio of the load needed to compress foam by 65 per cent to the load needed to compress foam by 25 per cent), which is most important to cushioning characteristics. True-cold-cure foams will produce a sag factor of 3 to 3.2, compared to 2 to 2.6 for hot-cured foams.

Pre-polymer process

In the pre-polymer process the polyol is reacted with an excess of isocyanate to give an isocyanate-terminated pre-polymer which is reasonably stable and has less handling hazards than free isocyanate. If water, catalysts, and other ingredients are added to the product, a foam will result. For better load-bearing and cushioning properties, a low-molecular-weight triol, such as glycerol and trimethylolpropane, is added to the polyol before it reacts with the isocyanate. The triol provides a site for chain branching.

Although the two-step pre-polymer process is less important than the one-shot process, it has the advantage of low exotherms, greater flexibility in design of compounds, and reduced handling hazards.

Quasi-pre-polymer process

In the quasi-pre-polymer process a pre-polymer of low molecular weight and hence low viscosity is formed by reacting a polyol with a large excess of isocyanate. This pre-polymer, which has a large number of free isocyanate groups, is then reacted at the time of foaming with additional hydroxy compound, water, and catalyst to produce the foam.

The additional hydroxy compound, which may be a polyol or a simple molecule such as glycerol or ethylene glycol, also functions as a viscosity depressant. The system thus has the advantage of having low-viscosity components, compared to the pre-polymer process, but there are problems with high exotherms and a high free-isocyanate content. Quasi-pre-polymer systems based on polyester polyols and polyether polyols are becoming important in shoe soling, the former being most wear resistant and the latter the easiest to process.

Rigid and semi-rigid foams

The flexible foams discussed in previous sections have polymer structures with low degrees of cross-linking. Semi-rigid and rigid forms of urethane are products having higher degrees of cross-linking. Thus, if polyols of higher functionality—i.e., more hydroxyl groups per molecule—are used, less flexible products may be obtained, and in the case of polyol with a sufficiently high functionality, rigid foams will result.

The normal density range for rigid and semi-rigid foams is about 1 to 3 lb/ft^3 (0.016 to 0.048 g/cm^3). Some packaging applications, however, use densities down to 0.5 lb/ft^3 (0.008 g/cm^3); for furniture applications densities can go as high as 20 to 60 lb/ft^3 (0.32 to 0.96 g/cm^3), thus approaching solids. At densities of from 2 lb/ft^3 (0.032 g/cm^3) to 12 lb/ft^3 (0.19 /cm^3) or more, these foams combine the best of structural and insulating properties.

Semi-rigid (or semi-flexible) foams are characterised by low resilience and high energy-absorbing characteristics. They have thus found prime outlet in the automotive industry for applications like safety padding, arm rests, horn buttons, etc. These foams are cold cured and involve special polymeric isocyanates. They are usually applied behind vinyl or ABS skins. In cold curing, the liquid ingredients are simply poured into a mould in which vinyl or ABS skins and metal inserts for attachments have been laid. The liquid foams fills the cavity, bonding to the skin and inserts. Formulations and processing techniques are also available to produce self-skinning semi-rigid foam in which the foam comes out of the mould with a continuous skin of the same material.

Rigid urethane foams have outstanding thermal insulation properties and have proved to be far superior to any other polymeric foam in this respect. Besides, these rigid foams have excellent compression strength, outstanding buoyancy (floatation) characteristics, good dimensional stability, low water absorption, and the ability to accept nails or screws like wood. Because of these characteristics, rigid foams have found ready acceptance for such applications as insulation, refrigeration, packaging, construction, marine uses, and transportation.

For such diverse applications several processes are now available to produce rigid urethane foam. These include foam-in-place (or pour-in-place), spraying, moulding, slab, and laminates (i.e., foam cores with integral skins produced as a single unit). One-shot techniques can be used without difficulty, although in most systems the reaction is slower than with the flexible foam, and conditions of manufacture are less critical. Pre-polymer and quasi-pre-polymer systems were also developed in the United States for rigid and semi-rigid foams, largely to reduce the hazards involved in handling TDI where there are severe ventilation problems.

In the foam-in-place process a liquid urethane chemical mixture containing a fluorocarbon blowing agent is simply poured into a cavity or metered in by machine. The liquid flows to the bottom of the cavity and foams up, filling all cracks and corners and forming a strong seamless core with good adhesion to the inside of the walls that form the cavity. The cavity, of course, can be any space, from the space between two walls of a refrigerator to that between the top and bottom hull of a boat. However, if the cavity is the interior of a closed mould, the process is known as moulding.

Rigid urethane foam can be applied by spraying with a two-component spray gun and a urethane system in which all reactants are incorporated either in the polyol or in the isocyanate. The spraying process can be used for applying rigid foam to the inside of building panels, for insulating cold-storage rooms, for insulating railroad cars, etc.

Rigid urethane foam is made in the form of slab stock by the one-shot technique. As in the Henecke process the reactants are metered separately into a mixing head where they are mixed and deposited onto a conveyor. The mixing head oscillates in a horizontal plane to insure an even deposition. Since the foaming urethane can structurally bond itself to most substrates, it is possible (by metering the liquid urethane mixture directly onto the surface skin) to produce board stock with integral skins already attached to the surface of the foam. Sandwich-construction building panels are made by this technique.

Foamed Rubber

Although foamed rubber and foamed urethanes have many similar properties, the processes by which they are made differ radically. In a simple process a solution of soap is added to natural rubber latex so that a froth will result on beating. Antioxidants, cross-linking agents, and a foam stabiliser are added as aqueous dispersions. Foaming is done by combined agitation and aeration with automatic mixing and foaming machines. The stabiliser is usually sodium silicofluoride (Na_2SiF_6). The salt hydrolyses, yielding a silica gel which increases the viscosity of the aqueous phase and prevents the foam from collapsing. A typical cross-linking agent is a combination of sulphur and the zinc salt of mercaptobenzothiazole (accelerator). Cross-linking (curing or vulcanisation) with this agent takes place in 30 minutes at 100°C. When making a large article such as a mattress, a metal mould may be filled with the foamed latex and heated by steam at atmospheric pressure. After removing the foamed rubber article from the mould, it may be dewatered by compressing it between rolls or by centrifuging and by drying with hot air in a tunnel dryer. In foamed rubber formulation a part of the natural rubber latex can be replaced by a synthetic rubber latex. One such combination is shown in Table 17.4.

Table 17.4. Foamed rubber formulation.

Ingredient	*Parts by weight*
Styrene-butadiene latex (65% solids)	123
Natural rubber latex (60% solids)	33
Potassium oleate	0.75
Sulphur	2.25
Accelerators:	
Zinc diethyldithiocarbamate	0.75
Zinc salt of mercaptobenzothiazole	1.0
Trimene base (reaction product of ethyl chloride, formaldehyde, and ammonia)	0.8
Antioxidant (phenolic)	0.75
Zinc oxide	3.0
Na_2SiF_6	2.5

Poly(Vinyl Chloride) Foams

A number of methods have been devised for producing cellular products from poly(vinyl chloride) (PVC), either by a mechanical blowing process or by one of several chemical blowing techniques. PVC foams are produced in rigid or flexible forms. The greatest interest in rigid PVC foam is in applications where low-flammability requirements prevail. It has an almost completely closed cell structure and therefore low water absorption. The rigid PVC foam is used as the cellular layer of some sandwich and multilayer panels.

Plastisols are the most widely used route to flexible expanded PVC products. Dolls, gaskets, and resilient covers for tool handles, for example, are produced from expandable plastisol compounds by moulding, while varied types of upholstery, garment fabrics, and foam layer in coated-fabric flooring are made from coatings with such compounds. Fig. 17.58 shows a schematic representation of the German Trovipor process for producing flexible, mainly open cell, and low to medium density (60–270 kg/m^3, 3.75–16.87 lb/ft^3) PVC foam. It is normally produced in the form of continuous sheet.

Many chemical blowing (foaming) agents have been developed for cellular elastomers and plastics, which generally speaking, are organic nitrogen compounds that are stable at normal storage and mixing temperatures but undergo decomposition with gas evolution at reasonably well-defined temperatures. Three important characteristics of a chemical blowing agent are the decomposition temperature, the volume of gas generated by unit weight, and the nature of the decomposition products. Since nitrogen is an inert, odourless, non-toxic gas, nitrogen-producing organic substances are preferred as blowing agents. Several examples of blowing agents especially recommended for vinyl plastisols are shown in Table 17.5; in each case the gas generated is nitrogen.

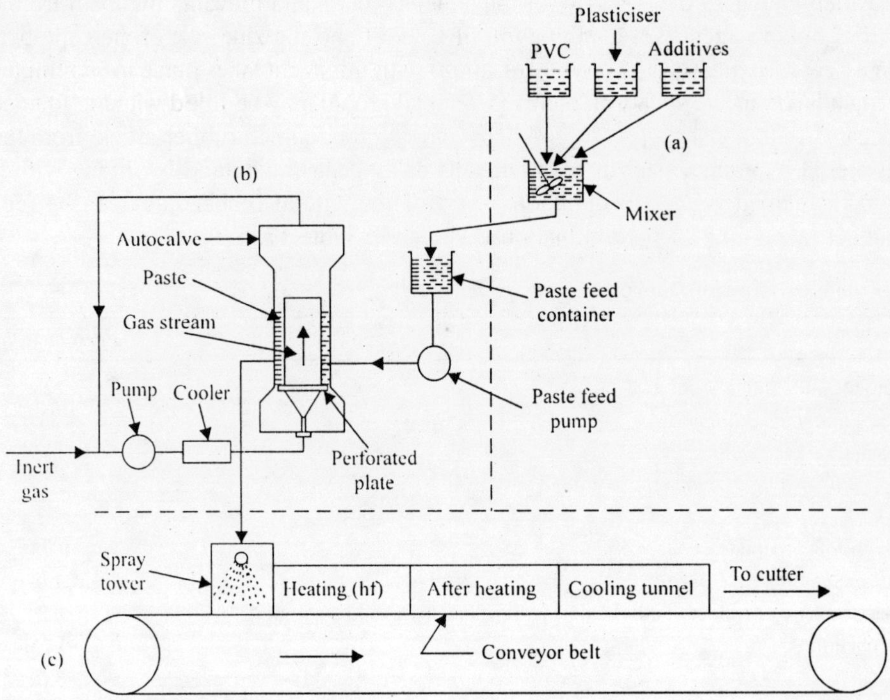

Fig. 17.58. Schematic representation of the Trovipor process. (a) PVC paste preparation; (b) Gasification of PVC paste; and (c) Spraying and fusion.

Table 17.5. Commercial blowing (foaming) agents.

Chemical type	*Decomposition temperature in air (°C)*	*Decomposition position range in plastics (°C)*	*Gas yield (mL/g)*
Azo compounds			
Azobisformamide (azodicarbonamide)	195–200	160–200	220
Azobisisobutyronitrile	115	90–115	130
Diazoaminobenzene	103	95–100	115
N-Nitroso compounds			
N,N'-Dimethyl-*N,N'*-dinitrosoterephthalimide	105	90–105	126
N,N'-Dinitrosopentamethylenetetramine	195	130–190	265
Sulphonyl hydrazides			
Benzenesulphonylhydrazide	> 95	95–100	130
Toluene-(4)-sulphonyl hydrazide	103	100–106	120
Benzene-1,3-disulphonyl hydrazide	146	115–130	85
4,4'-oxybis (benzenesulphonylhydrazide)	150	120–140	125

To produce uniform cells, the blowing agent must be uniformly dispersed or dissolved in the plastisol and uniformly nucleated. It should decompose rapidly and smoothly over a narrow temperature range corresponding to the attainment of a high viscosity or gelation of the plastisol system. The gelation involves solvation of the resin in plasticiser at 300 to 400°F (149 to 204°C), the temperature depending on the ingredients employed in the plastisol. The foam quality is largely determined by the matching of the decomposition of the blowing agent to the gelation of the polymer system. If gelation occurs before gas evolution, large holes or fissures may form. On the other hand, if gas evolution occurs too soon before gelation, the cells may collapse, giving a coarse, weak, and spongy product. Among the blowing agents listed in Table 17.5 azobisformamide (ABFA) is the most widely used for vinyls because it fulfills the requirements efficiently. ABFA decomposition can also be adjusted through proper choice of metal organic activators so that the gas evolution occurs over a narrow range within the wide range given in Table 17.5.

Closed-cell foams result when the decomposition and gelation are carried out in a closed mould almost filled with plastisol. After the heating cycle, the material is cooled in the mould under pressure until it is dimensionally stable. The mould is then opened, and the free article is again subjected to heat (below the previous moulding temperature) for final expansion. Protective padding, life jackets, buoys, and floats are some items made by this process.

The blowing agents given in Table 17.5 can be used to make foamed rubber. A stable network in this product results from the cross-linking reaction (vulcanisation), which thus corresponds to the step of fusion in the case of plastisols. Some thermoplastics also can be foamed by thermal decomposition of blowing agents even though they do not undergo an increase in dimensional stability at an elevated temperature. In this case the viscosity of the melt is high enough to slow down the collapse of gas bubbles so that when the polymer is cooled below its T_m a reasonably uniform cell structure can be built in. Cellular polyethylene is made in this way.

Special Foams

Some special types of foams are: (i) structural foams; (ii) syntactic foams and multifoams; and (iii) reinforced foams. Structural foams (Figs. 17.54c and d), which possess full-density skins and cellular cores, are similar to structural sandwich constructions or to human bones, which have solid surfaces but cellular cores. Structural foams may be manufactured by high pressure processes or by low-pressure processes (Fig. 17.57). The first one may provide denser, smoother skins with greater fidelity to fine detail in the mould than may be true of low-pressure processes. Fine wood detail, for example, is used for simulated wood furniture and simulated wood beams. Surfaces made by low-pressure processes may, however, show swirl or other textures, not necessarily detracting from their usefulness. Almost any thermoplastic or thermosetting polymer can be formulated into a structural foam.

In the case of syntactic foams (or spheroplastics), instead of employing a blowing agent to form bubbles in the polymer mass, hollow spherical particles, called microspheres, microcapsules, or microballoons, are embedded in a matrix of unblown polymer. (In multifoams, microspheres are combined with a foamed polymer to provide both kinds of cells). Since the polymer matrix is not foamed, but is filled mechanically with the hollow spheres, syntactic materials may also be thought of as reinforced or filled plastics, with the gas-containing particles being the reinforcing component. Synthetic wood, for instance, is provided by a mixture of polyester and small hollow glass spheres (microspheres).

The cellular structure of the syntactic foam depends on the size, quantity, and distributive uniformity of the microspheres. Since the microspheres have continuous shells, the final material will, as a rule, have completely enclosed cells, and thus can be called absolute foamed plastics or 'absolute' closed-cell foams. This, together with the absence of microstructural anisotropy (because the microspheres have practically all the same size and are uniformly distributed in the matrix), gives a syntactic material its valuable properties. They have better strength-to-weight ratios than conventional foamed plastics, absorb less water, and can withstand considerable hydrostatic pressures. Using hollow sphere means that the final material is lighter than one containing a compact filler, such as glass powder, talc, kaolin, quartz meal, or asbestos.

Fig. 17.59a and Fig 17.59b are graphical presentations of syntactic foam structure in which the two components, namely, microspheres and resin fill completely the whole volume (no dispersed voids) and the density of the product is thus calculated from the relative proportion of the two. Measured density values often differ from the calculated ones due to the existence of some isolated or interconnected irregularly shaped voids, as shown in Fig. 17.59c. The voids are usually an incidental part of the composite, as it is not easy to avoid their formation. Nevertheless, voids are often introduced intentionally to reduce the density below the minimum possible in a close-packed two-phase structure (Fig. 17.59a).

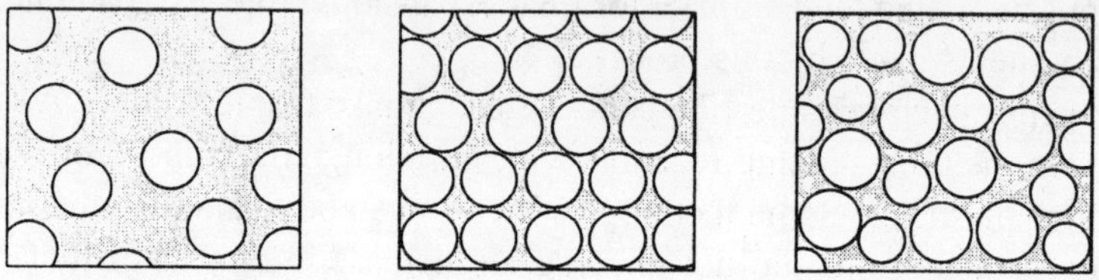

(a) (b) (c)

Fig. 17.59. Graphical representations of syntactic foam structures. (a) Two-phase composite with random dispersion of spheres; (b) two-phase composite with hexagonal close-packed structure of uniform sized spheres (74 per cent by vol.); and (c) three-phase composite containing packed microspheres, dispersed voids and binding resin.

Syntactic foams exhibit their best mechanical behaviour in the compressive mode. The spheres themselves are an extremely strong structure and hence can withstand such stresses very well. Syntactic materials consisting of hollow glass microspheres in epoxy resin are used for sandwich structures and as potting compounds for high-density electronic modules and other units likely to encounter hydrostatic pressures. Hollow glass microspheres and powdered aluminium in resin are used as core materials for sandwich construction and radomes. Hollow glass microspheres in aluminium matix are used for aerospace and extreme hydrostatic pressure (oceanographic) applications in view of low weight and high compressive strength.

Polymer foams may be reinforced, usually with short glass fibres, and also other fibres such as asbestos or metal, and other reinforcements such as carbon black. The reinforcing agent is generally introduced into the basic components and is blown along with them, to form part of and to reinforce the walls of the cells (Fig. 17.54e). When this is done, it is not unusual to obtain increases in mechanical properties of 400–500 per cent with fibreglass content up to 50 per cent by weight, especially in thermosettings.

The principal advantages of reinforcement, in addition to increased strength and stiffness, are improved dimensional stability, resistance to extremes of temperature and resistance to crepe.

Two processes for the manufacture of glass-reinforced foam laminates used as building materials, namely, free-rise process and restrained rise process are presented in Figs. 17.60 and 17.61. The glass fibre reinforcement is a thin (0.25–1.25 mm) mat supplied in roll form.

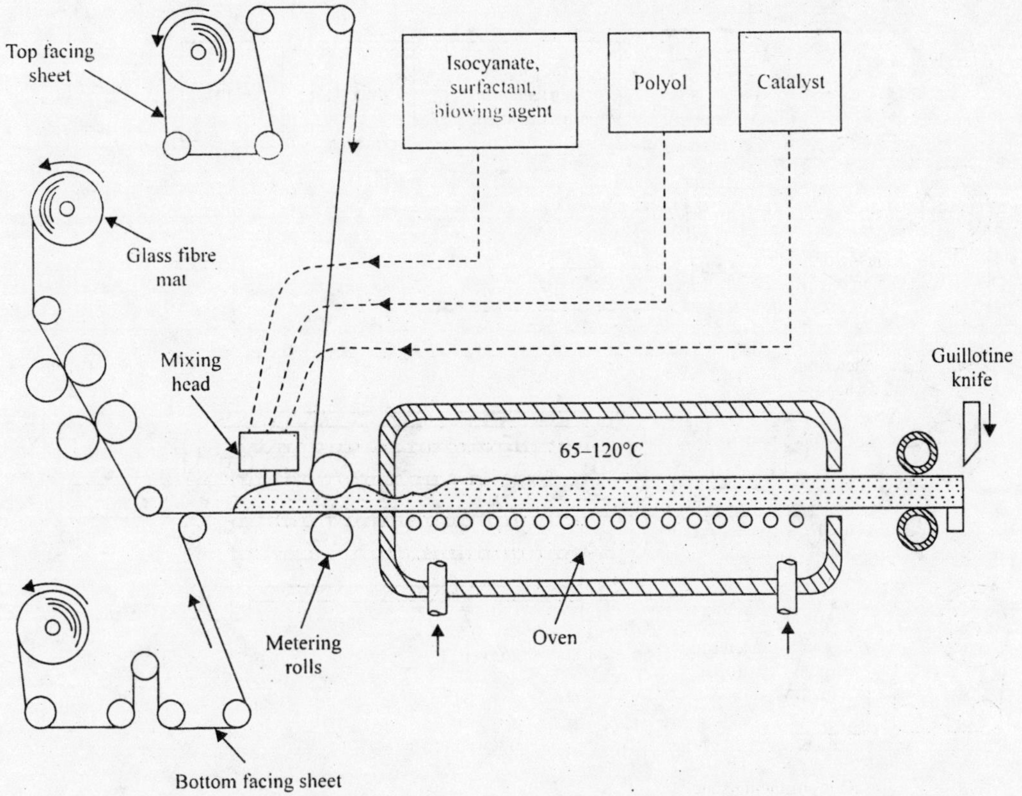

Fig. 17.60. Schematic of a free-rise process for manufacture of glass-reinforced foam laminates.

It consists of layers of relatively long (1.5–4 m) glass fibres, the fibres in one layer being at an acute angle to the fibres in each next adjacent layer.

A small amount of silane-modified polyester, or other binder is present, at a level of 2–10 per cent by weight.

This type of glass mat is relatively porous to the passage of liquids and is also capable of expanding within a mixture of rising foam chemicals to provide a uniform, three-dimensional reinforcing network within the final foamed laminate. The glass fibre reinforcement is functionally effective when used at levels of 4 to 24 g per board foot of the laminate. Various types of facing sheets may be used, such as aluminium foil for building insulation products, asphalt-saturated felts for roof insulation or any other material, e.g., paper and plastic films.

COMPUTER INTEGRATED INJECTION MOULDING

The ultimate result of Computer Integrated Injection Moulding (CIIM) in software packages is to translate the results of computer simulation of the moulding of a specific part into machine setting for specific microprocessor-controlled machines (Fig. 17.62). CIIM automates the entry of a large number of set points in microprocessor-controlled machines and maximises their efficiency, on the basis of extensive development.

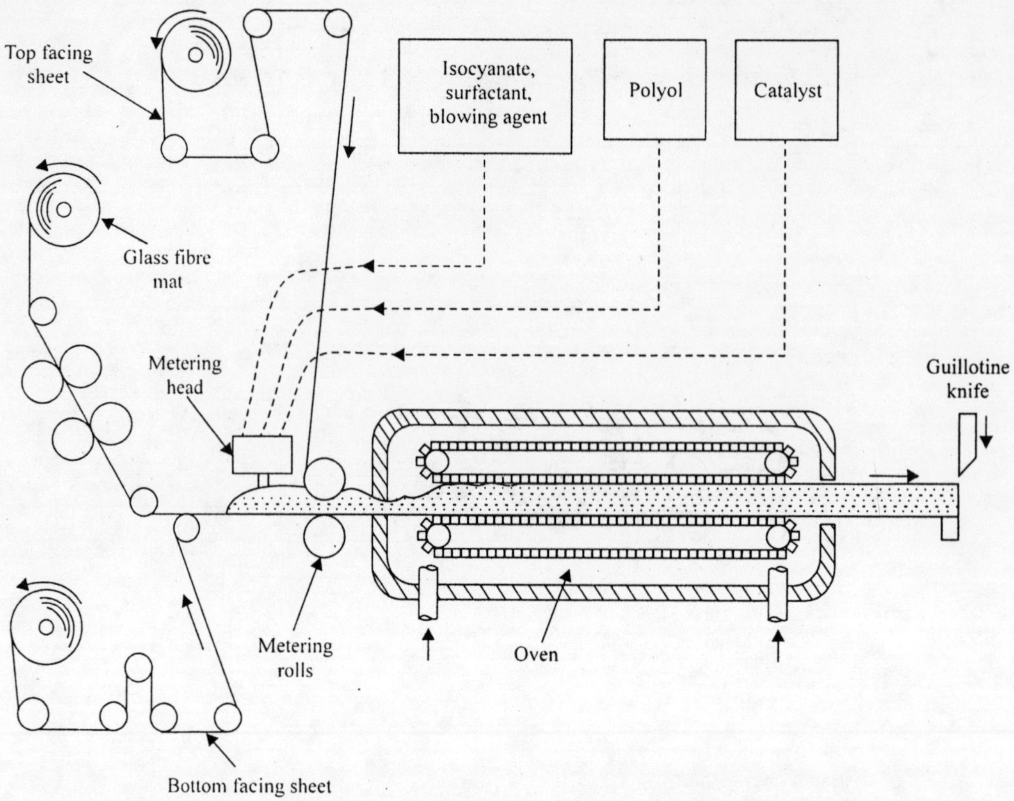

Fig. 17.61. Schematic of restrained rise process for manufacture of glass-reinforced foam laminates.

Microprocessor Control Systems

Microprocessor control systems (MCS) make it possible to completely automate an IM plant. They control machines, automatically, enabling them to achieve high quality and zero defects. These systems readily adapt to enhancing the ability of processing machines. There are many mouldings that would be difficult, if not impossible, to produce at the desired quality level without this feature.

Once processing variables are optimised through computer simulation, these values are entered in computer programmes in the form of a large number of machine settings. Establishing the initial settings during start-up is inherently complex and time-consuming. The many benefits of these systems are well recognised and accepted, but it is evident that self-regulation of IM can be effective only when the design of the part and the mould is optimised, and when the correct processing conditions for the operation have been predetermined. Otherwise, a self-regulating machine is confused and can provide conflicting instructions. The results could be disastrous, damaging the machine or the mould. Therefore, the efficient utilisation of microprocessor-control systems depends on the success of utilising correct and optimum programmes.

Computerised Process Simulation of IM

The simulation approach replaces the traditional trial-and-error method. Programmes, such as the Barnnardt TM Concept, are packaged for the complete moulding process, including material selection, moulding and cost optimisation, flow analysis, computerised shrinkage evaluation, and mould thermal analysis. The programmes include mould filling, packing, and so forth, which accurately model the performance of microprocessor programmed injection. Major 3D CAD systems for part and mould design, as well as for structural and flow analysis, are integrated with these systems.

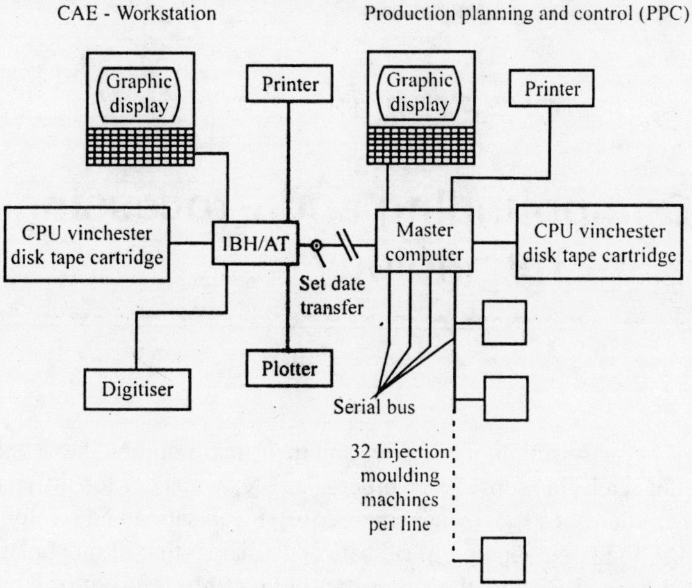

Software packages (plastics and computer)
Inconcept expert system for moulding
MS - Material selection
MCO - Moulding and cost optimisation
FA - Flow analysis
CSE - Computerised shrinkage evaluation
PP - Post processor

Software packages (netstal)
PPP - Production planning package
SDM - Set data management
PDM - Production data management

Fig. 17.62. Overview of Computer Integrated Injection Moulding (CIIM).

CHAPTER 18

Rubber Compounding and Processing Technology

INTRODUCTION

The term compounding describes the selection of additives and their incorporation into a polymer so as to give a homogeneous mixture ready for subsequent processing steps. Except for differences in the nature of the compounding ingredients, especially as required for vulcanisation and reinforcement and described under those headings, the compounding of rubbers is similar to that of most plastics.

In the case of natural rubber, SBR, and the other synthetic rubbers produced by emulsion polymerisation, the polymer is first available in the form of a latex. The normal procedure in compounding is to coagulate and dry the latex, and then to masticate and compound the rubber on mills or in other equipment. With SBR, however, two important compounding steps, *oil extending* and *masterbatching*, are commonly carried out before coagulation of the latex.

When designing a mixing formulation for specific end use, a compounder must take account not only of those vulcanisate properties essential to satisfy service requirements but cost of the raw materials involved and the production processes by which these will be transformed into the final product.

Compounding Ingredients

No rubber becomes technically useful if its molecules are not cross-linked, at least partially, by a process known as curing or vulcanisation. For natural rubber and many synthetic rubbers, particularly the diene rubbers, the curing agent most commonly used is sulphur. But sulphur curing takes place at technically viable rates only at a relatively high temperature (>140°C) and, moreover, if sulphur alone is used, optimum curing requires use of a fairly high doses of sulphur, typically 8–10 parts per hundred parts of rubber (phr), and heating for nearly 8 hr at 140°C. Sulphur doses has been substantially lowered, however, with the advent of organic accelerators. Thus, incorporation of only 0.2–2.0 phr of accelerator allows reduction of sulphur dose from 8–10 phr to 0.5–3 phr and effective curing is achieved in a time scale of a few minutes to nearly an hour depending on temperature (100–140°C), and type of the selected accelerator. The low sulphur doses required in the accelerated sulphur vulcanisation has not only eliminated bloom (migration of unreacted sulphur to the surface of the vulcanisate), which was a common feature of the earlier non-accelerated technology, but also has led to the production of vulcanisates of greatly improved physical properties and good resistance to heat and ageing.

The selection of the accelerator depends largely on the nature of the rubber taken, the design of the product, and the processing conditions. It is important to adopt a vulcanising system that not only gives a rapid and effective cross-linking at the desired vulcanising temperatures but also resists premature vulcanisation (scorching) at somewhat lower temperatures encountered in such operations as mixing, extrusion, calendering, and otherwise shaping the rubber before final cross-linking. This may require the use of delayed-action type accelerators as exemplified by sulphenamides. Other principal types of accelerators with different properties are guanidines, thiazoles, dithiocarbamates, thiurams, and xanthates. Accelerators are more appropriately classified according to the speed of curing induced in their presence in NR systems. In the order of increasing speed of curing, they are classified as slow, medium, semiultra, and ultra accelerators.

The problem of scorching or premature vulcanisation is very acute with ultra or fast accelerators. Rubber stocks are usually bad conductors of heat and therefore flow of heat to the interior of a vulcanising stock from outside is very slow. As a result, in thick items the outer layers may reach a state of overcuring before the core or interior layers begin to cure. For such thick items, a slow accelerator is most suitable.

For butyl and EPDM rubbers, which have very limited unsaturations, slow accelerators are, however, unsuitable and, fast accelerators should be used at high temperatures for good curing at convenient rates. Since butyl rubber is characterised by reversion, a phenomenon of decrease of tensile strength and modulus with time of cure after reaching a maximum, duration of heating at curing temperatures must be carefully controlled, and prolonged heating must be avoided.

For rubbers with highér degree of unsaturations, an ideal accelerator is one that is stable during mixing, processing, and storage of the mix, but that reacts and decomposes sharply at the high vulcanisation temperature to effect fast curring. These requirements or demands are closely fulfilled by the delayed-action accelerators. Some typical examples of organic accelerators are given in Table 18.1.

Table 18.1. Organic vulcanisation accelerator.

Name of accelerator	*Symbol*
Mercapto-accelerators	
2-Mercaptobenzothiazole	MBT
Zinc-2-mercaptobenzothiazole	ZMBT
Dibenzothiazyl disulphide	MBTS
Sulphenamide-accelerators	
N-cyclohexyl-2-benzothiazylsulphenamide	CBS
N-tert-butyl-2-benzothiazylsulphenamide	TBBS
2-Benzothiazyl-N-sulphenemorpholide	MBS
N,N-dicyclohexyl-2-benzothiazylsulphenamide	DCBS
Thiuram accelerators	
Tetramethylthiuram disulphide	TMTD
Tetramethylthiuram monosulphide	TMTM
Tetraethylthiuram disulphide	TETD
Dimethyldiphenylthiuram disulphide	MPTD

(Contd ...)

Name of accelerator	*Symbol*
Dipentamethylenthiuram tetrasulphide	DPTT
Dithiocarbamate accelerators	
Zinc dimethyldithiocarbamate	ZDMC
Zinc diethyldithiocarbamate	ZDEC
Zinc dibutyldithiocarbamate	ZDBC
Zinc pentamethylenedithiocarbamate	Z5MC
Zinc ethylphenyldithiocarbamate	ZEPC
Zinc dibenzyldithiocarbamate	ZBEC
Piperidine pentamethylenedithiocarbamate	PPC
Sodium dimethyldithiocarbamate	NaDMC
Sodium dibutyldithiocarbamate	NaDBC
Selenium dimethyldithiocarbamate	SeDMC
Tellurium dimethyldithiocarbamate	TeDMC
Lead dimethyldithiocarbamate	PbDMC
Cadmium dimethyldithiocarbamate	CdDMC
Cadmium pentamethylenedithiocarbamate	Cd5MC
Copper dimethyldithiocarbamate	CuDMC
Copper dibutyldithiocarbamate	CuDBC
Bismuth dimethyldithiocarbamate	BiDMC
Dithiocarbamylsulphenamide	
N-oxydiethylenedithiocarbamyl-N'-oxydiethylene sulphenamide	OTOS
Xanthate accelerators	
Zinc isopropylxanthate	ZIX
Zinc butylxanthate	ZBX
Sodium isopropylxanthate	NaIX
Guanidine accelerators	
Diphenylguanidine	DPG
Di-o-tolylguanidine	DOTG
o-Tolylbiguanidine	OTBG
Amine accelerators	
Butyraldehydeaniline	BAA
Tricrotonylidenetetramine	TCT
Hexamethylenetetramine	HEXA
Polyethylenepolyamines	PEP
Cyclohexylethylamine	CEA
Dibutylamine	DBA
Thiourea accelerators	
N,N'-ethylenethiourea (=2-mercaptoimidazoline)	ETU

(Contd ...)

Name of accelerator	*Symbol*
N,N-diphenylthiourea (= thiocarbanilide)	DPTU
N,N'-diethylthiourea	DETU
Dithiophosphate accelerators	
Zinc dibutyldithiophosphate	ZDBP
Copper diisopropyldithiophosphate	CuIDP
Sulphur donors	
2-Benzothiazole-N-morpholyldisulphide	MBSS
Dimorpholine disulphide	DTDM
Vulcanisation retarders	
Cyclohexylthiophthalimide	CTP
Phthalic anhydride	PTA
Benzoic acid	BES
Salicylic acid	SCS
N-nitrosodiphenylamine	NDPA

The spectacular effects of modern organic accelerators in sulphur vulcanisation of rubber are observed only in the presence of some other additives known as accelerator activators. They are usually two-component systems comprising a metal oxide and a fatty acid. The primary requirement for satisfactory activation of accelerator is good dispersability or solubility of the activators in rubber. Oxides of bivalent metals such as zinc, calcium, magnesium, lead, and cadmium act as activators in combination with stearic acid. A combination of zinc oxide and stearic acid is almost universally used. (Where a high degree of transparency is required, the activator may be a fatty acid salt such as zinc stearate). Besides speeding up the rate of curing, activators also bring about improvements in physical properties of vulcanisates. This is highlighted by the data given in Table 18.2.

Table 18.2. Effect of activator on vulcanisation.

Time of cure, min	*Tensile strength, psi (MPa)* *ZnO, phr*	
	0.0	*5.0*
15	100 (0.7)	2300 (16)
30	400 (2.8)	2900 (20)
60	1050 (7.2)	2900 (20)
90	1300 (9.0)	2900 (20)

Base compound: NR (pale crepe) 100; sulphur 3; mercaptobenzothiazole (MBT) 0.5.
Temperature of vulcanisation 142°C.

A sulphur-curing system thus has basically four components: a sulphur vulcanising agent, an accelerator (sometimes combinations of accelerators), a metal oxide, and a fatty acid. In addition, in order to improve the resistance to scorching, a pre-vulcanisation inhibitor such as *N*-cyclohexylthiophthalimide may be incorporated without adverse effects on either the rate of cure or physical properties of the vulcanisate.

The level of accelerator used varies from polymer to polymer. Some typical curing system for the diene rubbers (NR, SBR, and NBR) and for two olefin rubbers (IIR and EPDM) are given in Table 18.3.

Table 18.3. Components of sulphur vulcanisation systems.

Additive (phr)	*Rubber*				
	NR	*SBR*	*NBR*	*IIR*	*EPDM*
Sulphur	2.5	2.0	1.5	2.0	1.5
Zinc oxide	5.0	5.0	5.0	3.0	5.0
Stearic acid	2.0	2.0	1.0	2.0	1.0
TBBS	0.6	1.0	–	–	–
MBTS	–	–	1.0	0.5	–
MBT	–	–	–	–	1.5
TMTD	–	–	0.1	1.0	0.5

In addition to the components of the vulcanisation system, several other additives are commonly used with diene rubber. Rubber in general, and diene rubbers in particular, are blended with many more additives than is common for most thermoplastics with the possible exceptions of PVC.

The major additional classes of additives are:

1. Antidegradants (antioxidants and antiozonants).
2. Processing aids (peptisers, plasticisers, softeners and extenders, tackifiers, etc.).
3. Fillers.
4. Pigments.
5. Others (retarders, blowing agents).

Processing Aids

Peptisers are added to rubber at the beginning of mastication and are used to increase the efficiency of mastication. They act chemically and effectively at temperatures greater than 65°C and hasten the rate of breakdown of rubber chains during mastication. Common peptisers are zinc thiobenzoate, zinc-2-benzamidothiophenate, thio-β-naphthol, etc. Processing aids other than the peptiser and compounding ingredients (additives) are added after the rubber attains the desired plasticity on mastication.

Common process aids, besides the peptiser, are pine tar, mineral oil, wax, factice, coumarone-indene resins, petroleum resins, rosin derivatives, and polyterpenes. Their main effect is to make rubber soft and tacky to facilitate uniform mixing, particularly when high loading of carbon black or other fillers is to be used.

Factice (vulcanised oil) is a soft material made by treating drying or semi-drying vegetable oils with sulphur monochloride (cold or white factice) or by heating the oils with sulphur at 140–160°C (hot or brown factice). The use of factice (5–30 phr) allows efficient mixing and dispersion of powdery ingredients and gives a better rubber mix for the purpose of extrusion.

Ester plasticisers (phthalates and phosphates) that are used to plasticise PVC are also used as process aids, particularly with NBR and CR. Polymerisable plasticiser such as ethylene glycol dimethacrylate are particularly useful for peroxide curing rubbers. They act as plasticisers or tackifiers during mixing and undergo polymerisation by peroxide initiation during cure.

The diene hydrocarbon rubbers are often blended with hydrocarbon oils. The oils decrease polymer viscosity and reduce hardness and low temperatures brittle point of the cured product. They are thus closely analogous to the plasticisers used with thermoplastics but are generally know as softeners. Three main types of softeners are distinguished: aliphatic, aromatic, and naphthenic. The naphthenics are preferred for general all-round properties.

Natural rubbers exhibit the phenomenon known as tack. Thus when two clean surfaces of masticated rubber are brought into contact the two surfaces strongly adhere to each other, which is a consequence of interpenetration of molecular ends followed by crystallisation. Amorphous rubbers such as SBR do not display such tack and it is necessary to add tackifiers such as rosin derivatives and polyterpenes.

Fillers

Major fillers used in the rubber industry are classified as: (i) non-black fillers such as china clay, whiting, magnesium carbonate, hydrated alumina, anhydrous, and hydrated silicas and silicates including those in the form of ground mineral such as slate powder, talc, or french chalk, and (ii) carbon blacks.

Rather peculiar to the rubber industry is the use of the fine particle size reinforcing fillers, particularly carbon black. Fillers may be used for 50 phr to as high as 100–120 phr or even higher proportions. Their used improves such properties as modulus, tear strength, abrasion resistance, and hardness. They are essential with amorphous rubbers such as SBR and polybutadiene that has little strength without them. They are less essential with strain-crystallising rubbers such as NR for many applications but are important in the manufacture of tyres and related products.

Carbon blacks are essentially elemental carbon and are produced by thermal decomposition or partial combustion of liquid or gaseous hydrocarbons to carbon and hydrogen. The principal types, according to their method of production, are channel black, furnace black, and thermal black.

Thermal black is made from natural gas by the thermatomic process in which methane is cracked over hot bricks at a temperature of 1600°F (871°C) to form amorphous carbon and hydrogen. Thermal black consists of relatively coarse particles and is used principally as a pigment. A few grades (FT and MT referring to fine thermal and medium thermal) are also used in the rubber industry.

Most of the carbon black used in the rubber industry is made by the furnace process (furnace black), that is, by burning natural gas or vapourised aromatic hydrocarbon oil in a closed furnace with about 50 per cent of the air required for complete combustion. Furnace black produced from natural gas has an intermediate particle size, while that produced from oil can be made in wide range of controlled particle sizes and is particularly suitable for reinforcing rubbers. Quite a variety of grades of furnace blacks are available, e.g., fine furnace black (FF), high modulus (HMF), high elongation (HEF), reinforcing (RF), semi-reinforcing (SRF), high abrasion (HAF), super abrasion (SAF), intermediate super abrasion (ISAF), fast extruding (FEF), general purpose (GPF), easy processing (EPF), conducting (CF), and super conducting furnace black (SCF).

Channel black is characterised by lower pH, higher volatile content, and high surface area. It has the smallest particle size of any industrial material. A few grades of channel black (HPC, MPC, or EPC corresponding to hard, medium, or easy processing channel) are used in the rubber industry.

For carbon-black fillers, structure, particles size, particle porosity, and overall physico-chemical nature of particle surface are important factor in deciding cure rate and degree of reinforcement attainable. The pH of the carbon black has a profound influence. Acidic blacks (channel blacks) tend to retard the curing process while alkaline blacks (furnace blacks) produce a rate-enhancing effect in relation to curing, and may even give rise to scorching.

Another important factor is the particle size of the carbon black filler. The smaller the particle size, the higher the reinforcement, but the poorer the processability because of the longer time needed for dispersion and the greater heat produced during mixing. Blacks of the smallest particle size are thus unsuitable for use in rubber compounding.

For carbon black fillers the terms structure is used to represent the clustering together and entanglement of fine carbon particles into long chains and three-dimensional aggregates. High-structure black produce high-modulus vulcanisates as high shear forces applied during mixing break the agglomerates down to many active free radical sites, which bind the rubber molecules, thereby leading to greater reinforcement. In non-structure blacks the aggregates are almost non-existent.

Most of the non-black fillers used in rubber compounds are of non-reinforcing types. They are added for various objectives, the most important being cost reduction. Precipitated silica (hydrated), containing about 10–12 per cent water with average particle size ranging from 10–40 nm, produce effective reinforcements and are widely used in translucent and coloured products. Finely ground magnesium carbonate and aluminium silicate also induce good reinforcing effects. Precipitated calcium carbonate and activated clacium carbonate (obtained by treating calcium carbonate with a stearate) are used as semi-reinforcing fillers.

Short fibres of cotton, rayon, or nylon may be added to rubber to enhance modulus and tear and abrasion resistance of the vulcanisates. Some resins such as 'high styrene resins' and novolac-type phenolic resins mixed with hexamethylene tetramine may be used as reinforcing fillers or additives. Whereas SBR has a styrene content of about 23.5 per cent and is rubbery, styrene-butadiene co-polymers containing about 50 per cent styrene are leatherlike whilst with 70 per cent styrene the materials are more rigid thermoplastics but with low softening points. Both of these co-polymers are known in the rubber industry as high styrene resins and are usually blended with a hydrocarbon rubber such as NR and SBR. Such blends have found use in shoe soles, car wash brushes and other mouldings, but in recent years have suffered increasing competition from conventional thermoplastics and thermoplastic rubbers.

Mastication and Mixing

A deficiency of natural rubber, compared with the synthetics, is its very high molecular weight, which makes mixing of compounding ingredients and subsequent processing by extrusion and other shaping operations difficult. For natural rubber it is thus absolutely necessary, while for synthetic rubbers it is helpful, to subject the stock to a process of breakdown of the molecular chains prior to compounding. This is effected by subjecting the rubber to high mechanical work (shearing action), a process commonly known as mastication. Mastication and mixing are conveniently done using two-roll mills and internal mixers. The oxygen in air plays a critical role during mastication.

Rubber that has been masticated is more soft and flows more readily than the unmasticated material. Mastication also allows preparation of solutions of high solids content because of the much lower solution viscosity of the degraded rubber. Rubber is also rendered tacky by mastication, which means that the uncured rubber sticks to itself readily so that articles of suitable thickness can be built up from layers of masticated rubber or rubberised fabric without the use of a solvent.

Open mill

The mainstays of the rubber industry has been the two-roll (open) mill and the Banbury (internal) mixer. Roll mills were first used for rubber mixing. The plastics and adhesives industries later adopted these tools. The two-roll mill (Fig. 18.1) consists of two opposite-rotating rollers placed close to one another with the roll axes parallel and horizontal, so that relatively small gap or nip (adjustable) between the cylindrical surface exists. The speeds of the two rolls are usually different, the front roll having a slower speed. For natural rubber mixing, a friction ratio of 1:1.2 for the front to back roll may be used. For some synthetic rubbers or highly filled NR mixes, friction ratios close to 1.0 produce good results.

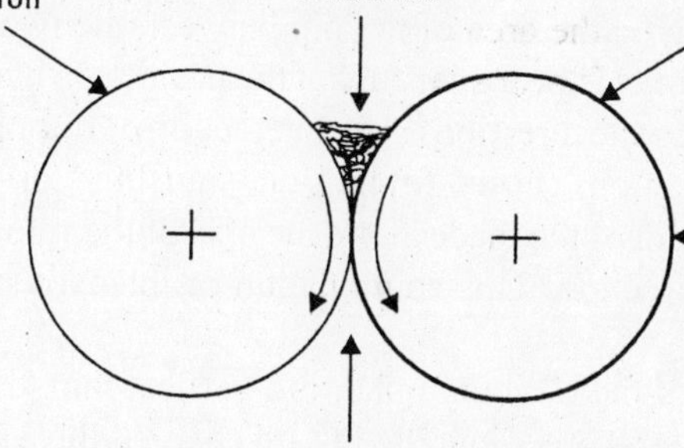

Fig. 18.1. Section showing the features of a two-roll open mill.

The nip is adjusted so that when pieces of rubber are placed between the rolls, they are deformed by shearing action and squeezed through the nip to which they are returned by the operator. On repeated passage through the nip, the separate pieces of rubber soon join into a single mass to form a band around the front roll and a moving 'bank' above the nip. As the rolls keep on rotating, the operator uses a knife to cut through the band on the front roll, removes the mass in parts from time to time, and places it in a new position to ensure uniform treatment and mixing through the nip.

With most rubbers other than natural rubber, addition of different compounding ingredients may be normally started soon after a uniform band is formed on the roll and a bank is obtained. For natural rubber, however, milling is usually continued to masticate the rubber to the desired plasticity, and mixing of compounding ingredients is started only after the adequate mastication. Mixing is effected by adding the different ingredients onto the bank. They are gradually dispersed into the rubber, which is cut at intervals, rolled over, and recycled through the nip of the moving rolls to produce uniform mixing. Since the rate of mastication is a function of temperature, time and temperature of mastication have to be controlled or kept uniform from batch to batch in order to get the desired uniform products from different batches.

Roll mills vary greatly in size from very small laboratory machines with rollers of about 1 inch. in diameter and driven by fractional horsepower motors to very large mills with rollers of nearly 3 ft in diameter and 7 or 8 ft in length and driven by motors over 100 hp.

Internal batch mixers

Internal batch mixtures are widely used in the rubber industry. They are also used for processing plastics such as vinyl, polyolefins, ABS, and polystyrene, along with thermosets such as melamines and ureas because they can hold materials at a constant temperature.

The principle of internal batch mixing was first introduced in 1916 with the development of the Banbury mixer (Fig. 18.2). A Banbury-type internal mixer essentially consists of a cylindrical chamber or shell within which materials to be mixed are deformed by rotating blades or rotors with protrusions. The mixer is provided with a feed door and hopper at the top and a discharge door at the bottom. In most cases a shell actually consists of two adjacent cylindrical shells, partly intersecting each other. In each shell there is a rotor describing an arc concentric with the shell section. The ridge between the two cylindrical chamber sections helps force mixing, and the acute convergence of the rotors and the chamber walls imparts shearing. As the rubber or mix is worked and sheared between the two rotors and between each rotor and the body of the casing, mastication takes place over the wide area, unlike in a open mill where it is restricted only in the area of the nip between the two rolls.

The rotor blade of the Banbury mixer is pear shaped, but the projection is spiral along the axis and the two spirals interlock and rotate in opposite directions. The interaction of rotor blades between themselves, in addition to producing shearing action, causes folding or 'shuffling' or the mass, which is further accentuated by the helical arrangement of the blade along the axis of the rotor, thereby imparting motion to the mass in the third, or axial, direction. This combination of intensive working produces a highly homogeneous mix.

An important and novel feature of the Banbury mixer is a vertical ram to press the mass into contact with the two rotors. Rubbers, fillers, and other ingredients are charged through the feed hopper and then held in the mixing chamber under the pressure of the hydraulic (or manual) ram. As a result, incorporation of solids is more rapid. Both the cored rotors and the walls of the mixing chamber can be cooled or heated by circulating fluid. Because of the large power consumption of such a machine (up to 500 hp) the cylinder walls are usually water-cooled by sprays.

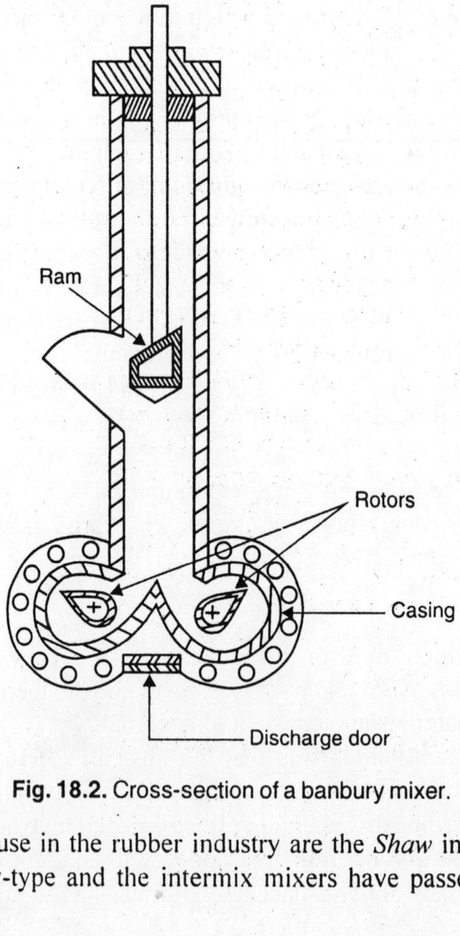

Fig. 18.2. Cross-section of a banbury mixer.

Other major machines in use in the rubber industry are the *Shaw* intermix and the *Baker-Perkins* shear-mix. Both the Banbury-type and the intermix mixers have passed through modifications and refinements at different points in time. Mechanical feeding and direct oil injection in measured doses into the mixing chamber through a separate oil injection port are notable features of modern internal mixers. Higher rotor speeds and, in the Banbury type, the higher ram pressures, are used for speedy output. In Banbury-type mixers, the rotors run at different speeds, while in intermix mixers the rotor speeds are equal but the kneading action between the thicker portion of one rotor and the thinner portion of the other produces a frictional effect.

The operation of internal mixers is power intensive, and a given job is performed at a much higher speed and in a much shorter time than on a two-roll open mill. However, the major vulcanising agent (such as sulphur) is often added later on a two-roll mill so as to eliminate possibilities of scorching. And even if this is not practiced, the mix, after being discharged from the internal mixer, is usually passed through a two-roll mill in order to convert it from irregular lumps to a sheet form for convenience in subsequent processing.

Reclaimed Rubber

The use of reclaimed rubber in a fresh rubber mix not only amounts to waste utilisation but offers some processing and economic advantages to make it highly valued in rubber compounding. Though waste vulcanised rubber is normally not processable, application of heat and chemical agents to ground vulcanised waste rubber leads to substantial depolymerisation whereby conversion of the rubber to a soft, plastic processable state is effected. Rubber so regenerated for re-use is commonly known as *reclaimed rubber* or simply as *reclaim*. Reclaimed rubber can be easily re-vulcanised.

Worn-out tyres and scraps and trimmings of other vulcanised products constitute the raw material for reclaimed rubber. Therefore a good reclaiming process must not only turn the rubber soft and plastic but also must remove reinforcing cords and fabrics that may be present. There are a number of commercial processes for rubber regeneration: (i) alkali digestion process; (ii) neutral or zinc chloride digestion process; (iii) heater or pan process, and (iv) reclaimator process.

Tyres are most commonly reclaimed by digestion processes. For processes (i) and (ii), debeaded tyres and scraps, cut into pieces, are ground with two-roll mills or other devices developed for the purpose. Two-roll mills generally used for grinding tyres turn at a ratio of about 1:3, thus providing the shearing action necessary to rip the tyre apart. The rubber chunks are screened, and the large material is recycled until the desired size is reached. The ground rubber is then mixed with a peptiser, softener, and heavy naphtha, and charged into spherical autoclaves with requisite quantities of water containing caustic soda for process or zinc chloride for process. The textile is destroyed and mostly lost in the digestion process. Steam pressure and also the amount of air or oxygen in the autoclave greatly influence the period necessary for rubber reclaiming. On completion of the process, the pressure is released, the contents of the autoclave discharged into water, centrifuged, pressed to squeeze out water, and dried. The material is finally processed through a two-roll mill during which mineral fillers and oils may be added to give a product of required specific gravity and oil extension.

Butyl and natural rubber tubes and other fibre-free scrap rubbers are reclaimed by means of the heater or pan process. Brass tube fittings and other metal are removed from the scrap. The scrap is mechanically ground, mixed with reclaiming agents, loaded into pans or devulcanising boats, and autoclaved at steam pressures of 10–14 atm (1.03–1.40 MPa) for 3–8 hr. The reclaim is finally processed much the same way as in the digester process.

The reclaimator process is more attractive than the above processes. The reclaimator is essentially a high-pressure extruder that devulcanises fibre-free rubber continuously. Ground scrap is mechanically treated in hammer mills to remove the textile material, mixed with reclaiming oils and other materials, and then fed into the reclaimator. High pressure and shear between the rubber mixture and the extruder barrel walls effectively reclaim the rubber mixture. Devulcanisation occurs at 175–205°C in a few minutes and turns the rubber into reclaim that issues from the machine continuously. The whole regeneration, which is a dry process, may be completed in about 30 min.

The reclaiming oils and chemicals are complex wood and petroleum derivatives that swell the rubber and provide access for breaking the rubber bonds with heat, pressure, chemicals, and mechanical shearing. Approximately 2–4 parts of oils are used per 100 parts of scrap rubber. Some examples of reclaiming oils include monocyclic and mixed terpenes, i.e., pine-tar products, saturated polymerised petroleum hydrocarbons, aryl disulphides in petroleum oil, cycloparafinic hydrocarbons, and alkyl aryl polyether alcohols.

Reclaimed rubber contains all the fillers present in the original scrap or waste rubber. It shows very good ageing characteristics and is characterised by less heat development during mixing and processing as compared to fresh rubber. The use of reclaim in a fresh rubber mix is advantageous not on consideration of physical mechanical properties but essentially for smooth processing and reduced cost.

Some Major Rubber Products

The most important applications of rubbers is in the transport sector, with tyres and related products consuming nearly 70 per cent of the rubber produced. Next in importance is the application in belting for making flat conveyor and (power) transmission belts and V-belts (for power transmission), and in the hose industry for making different hoses. Rubber also has a large outlet in cellular and microcellular products. Other important and special applications of rubber are in the areas of adhesives, coated fabrics, rain-wear, footwear, pipes and tubing, wire insulation, cables and sheaths, tank lining for chemical plants and oil storage, gaskets and diaphragms, rubber mats, rubber rollers, sports goods, toys and balloons, and a wide variety of moulded mechanical and miscellaneous products.

Tyres

Tyre technology is a very specialised area, and a tyre designer is faced with the difficult task of trying to satisfy all the needs of the vehicle manufacturer, the prime factors of consideration, however, being safety and tread life.

Fig. 18.3 shows the constructions of a standard bias (diagonal) ply tyre and a radial ply tyre. The major components of a tyre are: bead, carcass, sidewalls, and tread. In terms of material composition, a tyre on an average contains nearly 50 per cent of its weight in actual rubber; for oil extended rubbers (typically containing 25 parts of aromatic or cycloparaffinic oils to 75 parts of rubber), it is less. The remainder included carbon black, textile cord, and other compounding ingredients plus the beads.

The bead is constructed from a number of turns or coils of high tensile steel wire coated with copper and brass to ensure good adhesion of the rubber coating applied on it. The beads function as rigid, practically inextensible units that retain the inflated tyre on the rim.

The carcass forms the backbone of the tyre. The main part of the carcass is the tyre cord. The cords consist of textile threads twisted together. Rayon, nylon, and polyester cords are widely used. Steel cords are also used. While the practice of laying rubber-impregnated cords in position is still followed, the use of woven fabric is more widespread. To promote a good bond with rubber, the fibre or cord is treated with an adhesive composition such as water-soluble resorcinol-formaldehyde resin and aqueous emulsion of a co-polymer of butadiene, styrene, and vinyl pyride (70:15:15). The resin-coated cord or fabric is dried and coated with a rubber compound by calendering, whereby each cord is isolated from its neighbour. For conventional tyres, the rubber-coated fabric is then cut to a pre-determined width and bias angle. The bias-cut plies are joined end-to-end into a continuous length and batched into roll form, interleaved with a textile lining to prevent self-adhesion.

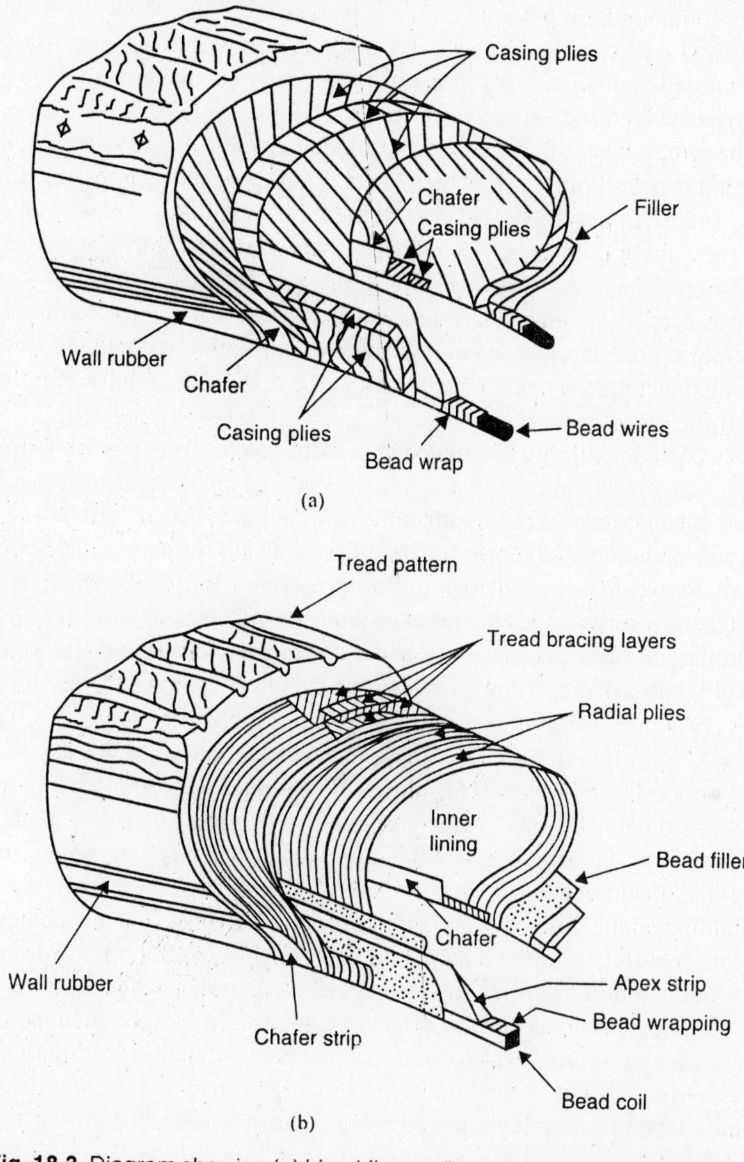

Fig. 18.3. Diagram showing (a) bias (diagonal) ply tyre and (b) radial ply tyre.

The sidewall is a layer of extruded rubber compound that protects the carcass framework from weathering and from damage from chafing. Together with the tread and overlapping it in the buttress region, the sidewall forms the outermost layer of the tyre. It is the most highly strained tyre component and is susceptible to two types of failure—flex cracking and ozone cracking.

The tread is also formed by extrusion, but different rubber compounds are used for sidewall and tread. When making the sidewall and the tread in two separate extrusion lines, it is useful to take the extruded sidewall to the second extruder, which would deliver the tread to the sidewall. To simplify the tyre-building operation, the sidewall and the tread may also be produced as a single unit by simultaneous extrusion of the two compounds in a single band—the tread over the sidewall—with two extruders arranged head-to-head with a common double die. The band is rapidly cooled to avoid scorching and then cut to the appropriate length. The tread receives its characteristic pattern from the mould when the subsequently built tyre is vulcanised.

While building the tyre, a layer of specially compounded cushion rubber may be used to keep heat development on flexing to a minimum and achieve better adhesion between the tread and the carcass. One or more layers of fabric, known as breakers or bracing layers, may be placed below the cushion. The bracing layers raise the modulus of the tread zone and level out local blows to the tread as it contacts the road.

The tyre building is carried out on a flat drum, which is rotated at a controlled speed. The plies of rubber-coated cord fabric are placed in position, one over the other, and rolled down as the drum rotates, the inner lining being placed next to the drum and the ends of the drum being flanged to suit the bead configuration of the tyre.

The plies of rubber-coated textile are assembled in three basic constructions—bias (diagonal), radial, and bias belted.

Bias tyres have an even number of plies with cords at an angle of 30–38° from the tread centre line. Passenger-car bias tyres commonly have two or four plies, with six for heavy duty service. Truck tyres are often built with six to twelve plies, although the larger earthmover types may contain thirty or more.

In the radial-ply tyre, one or two plies are set at an angle of 90° from the centre line and a breaker or belt or rubber-coated wire or textile is added under the tread. This construction gives a different tread-road interaction, resulting in a decreased rate of wear. The sidewall is thin and very flexible. The riding and steering qualities are noticeably different from those of a bias-ply tyre and require different suspension systems.

The bias-belted tyre, on the other hand, has much of the tread wear and traction advantage of the radial tyre, but the shift from bias to bias-belted tyres requires less radical change in vehicle suspension systems and in tyre building machines. These features make the bias-belted tyre attractive to both automobile and tyre manufacturers.

When the tyre building on the drum is complete, the drum is collapsed and the uncured ('green') tyre is removed. The cylindrical shapes of the uncured tyre obtained from the building drum is transformed into a toroidal shape in the mould resulting in a circumferential stretch of the order of 60 per cent. For shaping and curing, the uncured tyre is pressed against the inner face of the heated mould by an internal bag or bladder made of a pre-cured heat-resistant rubber and inflated by a high-pressure steam or circulating hot water. Different types of moulding presses are in use.

In the Bag-O-Matic type of press, the uncured tyre is placed over a special type of bag or bladder, and the operation of shaping, curing, and ejection of the cured tyre are accomplished by an automatic sequence of machine operation.

Pneumatic tyres require an air container or inner tube of rubber inside the tyre. The manufacture of inner tubes is done essentially in three steps: extrusion, cutting into length, and insertion of valve and vulcanisation. The ends of the cut length of the tube are joined after insertion of the valve prior to vulcanisation.

Tubeless tyres have, instead of an inner tube, an inner liner, which is a layer of rubber cured inside the casing to contain the air, and a chafer around the bead contoured to form an airtight seal with the rim.

Belting and hoses

Uses of flat conveyor and (power) transmission belts and V-belts (for power transmission) are to be found in almost all major industries. V-belts for different types cover applications ranging from fan belts for automobiles, belts for low-power drives for domestic, laboratory, and light industrial applications, and high-power belts for large industrial applications.

Textile cords or fabric and even steel cords constitute an important part of all rubber belting and hoses. The various types of cords used in the tyres industry are also in use in the belting industry.

Essential steps in making conveyor belts are: (i) drying the fabric; (ii) frictioning of the hot fabric with a rubber compound and topping to give additional rubber between plies and the outer ply and cover, using a three- or four-roll calender; (iii) belt building; and (iv) vulcanisation.

The process of belt building essentially consists of cutting, laying, and folding the frictioned and coated fabric to give the desired number of plies. The construction may be straight-laid or graded-ply types (Fig. 18.4) with the joints in neighbouring plies being staggered to eliminate weakness and failure. Finally, the cover coat is applied by calendering.

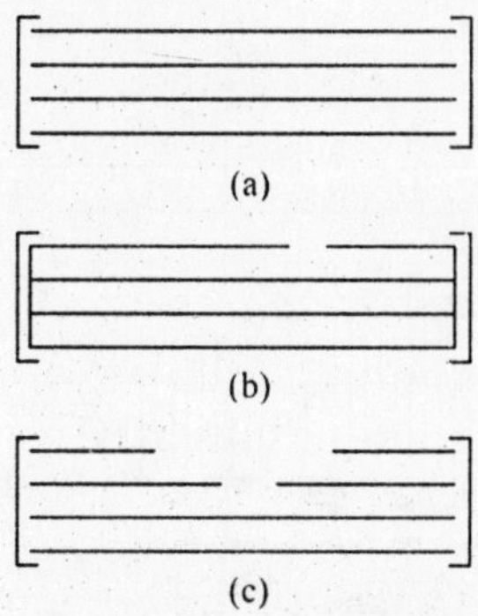

Fig. 18.4. Construction of conveyor belts: (a) straight laid, (b) folded jacket, and (c) graded ply.

Vulcanisation of conveyor belts may be carried out in sections using press cure or continuously by means of a Rotocure equipment. In press cure, vulcanisation is done by heating in long presses, the belt being moved between successive cures by a length less than the length of the press platens. Since in this process the end or overlap zones receive an additional cure, it is desirable to minimise damage or weakness due to overcure by using flat cure mixes and allowing water cooling at each end section of the platen. In each step of vulcanisation, the section of the belt to be vulcanised is gripped and stretched hydraulically to minimise or eliminate elongation during use. The difficulties of press cure may, however, be avoided by adopting continuous vulcanisation with a Rotocure equipment in which the actual curing operation is carried out between an internally steam heated cylinder and heated steel band. Rotocure is also useful for the vulcanisation of transmission belts and rubber sheeting.

The most common feature of V-belts is their having a cross-section of the shape of a regular trapezium with the unparallel sides at an angle of 40°C (Fig. 18.5). The V-belts usually consist of five sections: (i) the top section known as the tension section; (ii) the bottom section, called the compression section; (iii) the cord section located at the neutral zone; (iv) the cushion section on either side of the cord section; and (v) one or two layers of rubberised fabric, called the jacket section covering the whole assembly.

Three different rubber compounds are required for use in the above construction of a V-belt: (i) a base compound, which is the major constituent on a weight basis; (ii) a soft and resilient cushion compound required for surrounding and protecting the reinforcing cord assembly; and (iii) a friction compound used for rubberising the fabric casing of the belt.

Relatively short length V-belts are built layer by layer on rotatable collapsible drum formers. The separate belts are then cut out with knives and transferred to a skiving machine that imparts the desired V-shape. The fabric jacket is then applied, and the belts are vulcanised in open steam using multi-cavity ring moulds for smaller belts. Long belts are built similarly on V-groove sheave pulleys using weftless cord fabric in place of individually wound cord. They are vulcanised endlessly by moulding in a hydraulic press under controlled tension.

A rubber hose has three concentric layers along the length. While the innermost part consisting of a rubber lining or tube is required to resist the action of the material that would pass through the top layer is meant to play the role of a protective layer to resist weathering, oils, chemicals, abrasion, etc. Between the inner lining and the outer cover is given a layer of reinforcement of textile yarn or steel wire applied by spiralling, knitting, braiding, or circular loom weaving. A cut woven fabric wrapped straight or on the bias may also be used to reinforce the inner lining or tube.

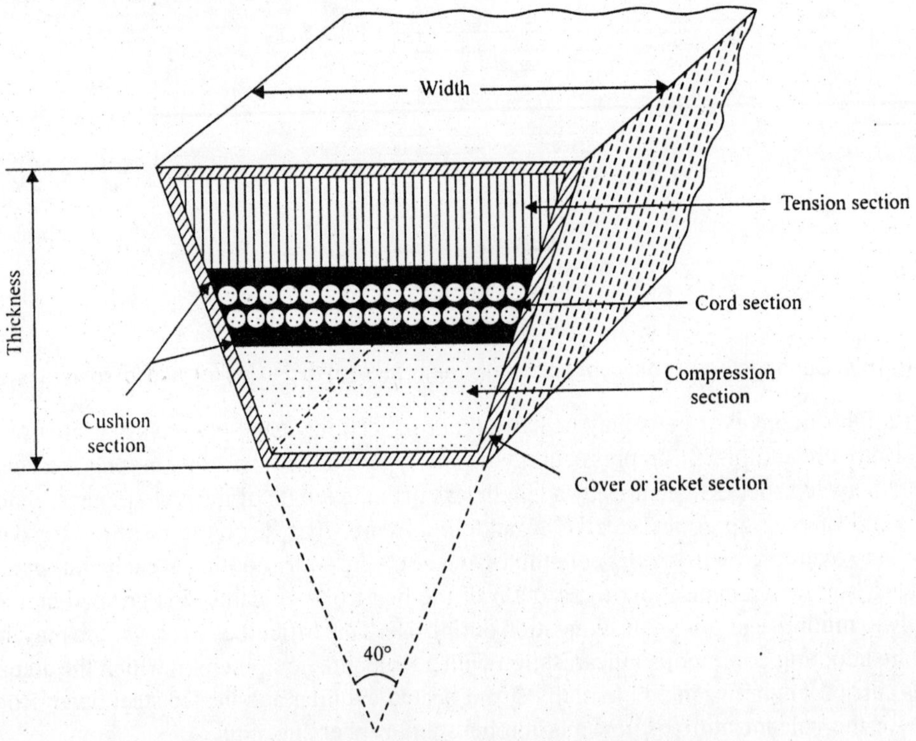

Fig. 18.5. Cross-section showing V-belt construction.

Essentially, the process of hose building consists of extruding the lining or tube, braiding or spiralling the textile around the cooled tube, and applying an outer cover of rubber to the reinforced hose using a cross-head extruder. Several methods are employed for vulcanisation. In one process, the built hose is passed through a lead press or lead extruder to give a layer of lead cover to the hose. The hose is then wound on a drum, filled with water or air, and the ends are sealed. The whole assembly is then heated to achieve vulcanisation. The water or air inside expands, and the vulcanising hose is pressed against the lead acting as the mould. On completion of cure, the sealed ends are cut open, the lead cover is removed by slitting lengthwise in a stripping machine, and the cured hose is coiled up.

Cellular rubber products

Cellular rubber may be described as an assembly of a multitude of cells distributed in a rubber matrix more or less uniformly. The cells may be interconnected (open cells) as in a sponge or separate (closed cells). Foam rubber made from a liquid starting materials such as latex, described earlier, is of open-cell type. Cellular products made from solid rubber are commonly called sponge (open cell structure) and expanded rubber (closed cell structure).

The technology of making cellular products from solid rubber is solely dependent on the incorporation of a blowing agent, usually a gas such as nitrogen or a chemical blowing agent, into the rubber compound. The most widely used chemical blowing agent for this application is dinitroso pentamethylene tetramine (DNPT). The curing is carried out either freely using hot air or steam or in a mould that is only partially filled with the moulding compound. Synthetic rubbers, particularly SBR, are preferred as they allow precise control over level of viscosity required for obtaining consistent product quality. The sponge and expanded rubber product include carpet backing, sheets, profiles, and mouldings.

The development of microcellular rubber has brought a revolution in footwear technology. Microcellular rubber with an extremely fine non-communicating cell structure and very comfortable wearing properties, is the lightest form of soling that can be produced. Density of soling as low as 0.3 g/cm^3 may be obtained with a high dose (8–10 phr) of DNPT at a curing temperature of 140–150°C. For common solings, the density normally varies between 0.5 and 0.8 g/cm^3.

High hardness and improved abrasion resistance along with low density, desired in microcellular soling, can be achieved by using SBR and high styrene resins with NR in right proportions. Higher proportions of high styrene resins give products of higher hardness and abrasion resistance and lower density. Silicious fillers such as precipitated silica and aluminium or calcium silicate also give higher hardness, abrasion resistance, and split tear strength. Microcellular crumbs can be used in considerable quantity along with china clay and whiting to reduce the product cost. Higher proportions of stearic acid (5–10 phr) are normally used in microcellular compounds in order to bring down the decomposition temperature of DNPT type blowing agents. Post-cure oven stabilisation of the microcellular sheets typically at 100°C for 4hr reduces the delayed shrinkage after cure to a minimum.

CHAPTER 19

Miscellaneous Processing Techniques

INTRODUCTION

Polymers in their pure form, as obtained from the manufacturing plants after isolation and purification, are called 'virgin' polymers or 'virgin' resins. Barring a few (such as polystyrene, polyethylene or polypropylene), virgin polymers, as such, may not be amenable for procesing straightaway. Virgin PVC, for example, is a material of horny texture and cannot be moulded without making it soft by the addition of a plasticiser. Natural rubber similarly requires a vulcanising agent to render it mouldable. Most polymers are protected from thermal, oxidative and photo-degradation by incorporating appropriate stabilisers. Many articles are obtained in attractive colours by adding suitable dyes and pigments to the polymeric material prior to its moulding. Lubricants and process aids are added to most of the polymers to reduce friction and improve flow within processing machineries. Fillers are added to impart special properties and reduce the finished product cost. The process involving incorporation of ingredients such as plasticisers, vulcanising agents, curatives, stabilisers, fillers, colouring agents, flame retardants and lubricants into the virgin resin is known as 'compounding' of the resin.

Before compounding procedures for the various specific vulcanisate properties are described, some general remarks will be made about the requirements for mixing and processing characteristics. This is necessary as a compromise has often to be made to permit a formulation, which has been developed for particular vulcanisate properties, to be mixed and processed in a given manner.

COATING PROCESSES

A coating is a thin layer of material used to protect or decorate a substrate. Most often, a coating is intended to remain bonded to the surface permanently although there are strippable coatings which are used only to afford temporary protection. An example of the latter type is the strippable hot-melt coating with ethyl cellulose as the binder, which is used to protect melt pieces such as drill bits or other tools and gears from corrosion and mechanical abrasion during shipping and handling.

Two of the principal methods of coating substrates with a polymer, namely extrusion coating and calendering have already been dealt with in this chapter. Other methods of coating continuous webs include the use of dip, knife, brush, and spray. Dip coating, as applied to PVC, has already been described on plastisols. In knife coating the coating is applied either by passing the web over a roll running partly immersed in the coating material or by running the coating material onto the face of the web while the thickness of the coating is controlled by a sharply profiled bar (or knife). This techniques, also referred to as spreading, is used extensively for coating fabrics with PVC. The PVC is prepared in the form of a paste, and more than one layer is usually applied, each layer being gelated by means of heat before the next layer is added.

Lacquers are a class of coatings in which film formation results from mere evaporation of the solvent(s). The term 'lacquer' usually connotes a solution of a polymer. Mixtures of solvents and diluents (extenders which may not be good solvents when used alone) are usually needed to achieve a proper balance of volatility and compatibility and a smooth coherent film or drying. Some familiar examples of lacquers are the spray cans of touch-up paint sold to the auto owner. These are mostly pigmented acrylic resins in solvents together with a very volatile solvent [usually dichlorodifluoromethane (CCl_2F_2)] which acts as a propellant. A typical lacquer formulation for coating steel surfaces contains polymer, pigment, plasticiser (non-volatile solvent) and volatile solvents.

Latex paints or emulsion paints are another class of coatings which form films by loss of a liquid and deposition of a polymer layer. The paints are composed of two dispersions: (i) a resin dispersion, which is either a latex formed by emulsion polymerisation or resin in emulsion form, and (ii) a dispersion of colorants, fillers, extenders, etc., obtained by milling the dry ingredients into water. The two dispersion are blended to produce an emulsion paint. Surfactants and protective colloids are added to stabilise the product. Emulsion paints are characterised by the fact that the binder (polymer) is in a water-dispersed form, whereas in a solvent paint it is in solution form. In emulsion systems the external water phase controls the viscosity, and the molecular weight of the polymer in the internal phase does not affect it, so polymers of high molecular weight are readily utilised in these systems. This is an advantage of emulsion paints. The minimum temperature at which the latex particles will coalesce to form a continuous layer depends mainly on the T_g. The T_g of a latex paint polymer is therefore adjusted by co-polymerisation or plasticisation to a suitable range. The three principal polymer latexes used in emulsion paints are styrene-butadiene co-polymer, poly(vinyl acetate), and acrylic resin.

Although the term 'paint' has been used for latex-based systems as well as many others, traditionally it refers to one of the oldest coating systems known—that of a pigment combined with a drying oil, usually a solvent. Drying oils (e.g., linseed, tung), by virtue of their multiple unsaturation, behave like polyfunctional monomers which can polymerise ('dry') to produce film by a combination of oxidation and free-radical propagation. Oil-soluble metallic soaps are used to catalyse the oxidation process.

Combinations of resin with drying oils yield oleoresinous varnishes, whereas addition of a pigment to a varnish yields an enamel. The combination of hard, wear-resistant resin with softer, resilient, drying oil films can be designed to give products with a wide range of durability, gloss, and hardness. Another route to obtaining a balanced combination of these properties is the alkyd resin, formed from alcohols and acids (and hence the 'alkyd'). Alkyds are actually a type of polyester resin and are produced by direct fusion of glycerol, phthalic anhydride, and drying oil at 410 to 450°F (210 to 232°C). The process involves an esterification reaction of the alcohol and the anhydride and transesterification of the drying oil. A common mode of operation today is thus to start with the free fatty acids from the drying oil rather than with the triglycerides.

Fluidised Bed Coating

Fluidised bed coating is essentially an adaptation of dip coating and designed to be used with plastics in the form of a powder of fine particle size. It is applied for coating metallic objects with plastics. The uniqueness of the process lies in the fact that both thermoplastics and thermosetting resins can be used for the coating. Uniform coating of thicknesses from 0.005 to 0.080 inch. (0.13 to 2.00 mm) can be built on many substrates such as aluminium, carbon steel, brass, and expanded metal. The coating is usually applied for electrical insulation and to enhance the corrosion resistance and chemical resistance of metallic parts.

Low-melting polymers are most appropriate for fluidised beds. Higher-melting polymers must have a sufficiently low melt viscosity that the particles can flow and fuse together to form a continuous coating. Thermoplastic polymers in common use include nylon, poly(vinyl chloride), acrylics, polyethylene, and polypropylene. Other possible materials are thermoplastic urethane, silicones, EVA, polystyrene, or any other low-melting, low-viscosity polymer. Thermosetting polymers are limited largely to epoxy and epoxy/polyester hybrids since other thermosets, such as phenolic and urea-formaldehyde resins, give off volatile by-products that can create voids in the coating.

The powder resin particles range in size from about 20 to 200 microns. Particles larger than 200 microns are difficult to suspend. Particles smaller than 20 microns may create excessive dusting and release of particles from the top of the bed.

The actual coating process is uncomplicated; however, achievement of a uniform coating requires considerable skill. The metal object to be coated is heated to well above the fusion point of the resin and dipped into a bed of powdered resin fluidised by the passage of air through a porous plate. The bed is not heated; only the surface of the object to be coated is hot. As the powder contacts the hot substrate, the particles adhere, melt, and flow together to form a continuously conforming coating. The object is removed from the bed when the desired coating thickness is obtained. On cooling, the coated resin resumes its original characteristics. In the case of a thermosetting resin, additional time at elevated temperature may be required to complete the cure.

The ability of the fluidised bed to continuously conform to and coat parts having unusual shapes and sizes permits a high degree of flexibility in application. This is invaluable to the processor who wishes to coat one-of-a-kind products. In general, the coatings are smooth and glossy, with excellent adhesion to the substrate, providing the hermetic seal necessary for proper maintenance. Fluidised bed coating is certainly the most efficient method of applying a thick coating in a single step. Thicknesses of 0.1 inch. (2.5 mm) or greater are easily attained. Probably the single biggest advantage of powder coating is the nearly 100 per cent utilisation of the coating resin, without the hazard or expense of solvents.

Almost all the coatings applied by this process have a definite function, chiefly electrical insulation, but it can be used for applications that simply require a thick coating with powder. Examples of electrical applications include small motor stators and rotors, electronic components (capacitors or resistors), transformer casings, covers, laminations, and busbar. Other items coated using fluidised beds include valve bodies used in chemical industries, refinery equipment, and appliance and pump parts.

Spray Coating

Spray coating is especially useful for articles that are too large for dip coating or fluidised bed coating. The process consists of blowing out fine polymer powders through a specially designed burner nozzle, which is usually flame heated by means of acetylene or some similar gas, or it can be heated electrically. Compressed air or oxygen is used as the propelling force for blowing the polymer powder.

Electrostatic Spraying

The electrostatic spraying of polymer powders utilises the principle that oppositely charged particles attract, a principle that has been used for many years in spraying solvent-based paints. In electrostatic spraying, polymer powder is first fluidised in a bed to separate and suspend the particles. It is then transferred through a hose by air to a specially designed spray gun. As the powder passes through the gun, direct contact with gun and ionised air applies an electrostatic charge to the particles of powder. The contact area may be a sleeve that extends the length of the gun or merely small pins that extend into the passageway of the powder. For safety the gun is designed for high voltage but low amperage.

The part to be coated is electrically grounded, attracting the charged particles. This produces a more even coating and reduces overspray. Parts to be coated may be pre-heated, thereby forming the coating immediately, or they may be coated cold as the electrostatic charge will hold the particles in place until heat is applied. Once heat is applied, the particles melt and flow together forming a continuous protective coating. Coatings 50–75 microns thick can be applied electrostatically to cold objects; coatings up to 250 microns thick to hot objects. The polymers used in spraying of powders are the same as those used in fluidised beds. The key characteristic for any polymer, thermoplastic or thermoset, applied as a powder is low melt viscosity, which enables polymer particles to flow together and form a continuous coating.

The chief characteristic of electrostatically applied powder polymer is the ability to produce a thin coating. It is the preferred method of producing a coating of 0.001–0.002 inch. (0.025–0.050 mm) thickness. It is a continuous process and suited to automated assembly-line production. With a well designed recovery system and fully enclosed spray booth, the process permits full utilisation of the powder. The particle size of sprayed powders is smaller than that of powders used in fluidised bed coating. The average particles size is 30–60 microns.

Much of what is termed decorative powder coating is applied by electrostatic spray. Appliances, laboratory instruments, transformer housings, engine parts, and chain-link fences are among the many types of products so coated for decorative purposes. Coatings that provide electrical insulation are also applied by electrostatic spray. Objects coated include electrical motor armatures and stators, electrical switchgear boxes, and magnet wire. Corrosion-resistant coatings are designed to prevent the corrosion of the underlaying substrates. These include pipe, fencing, concrete reinforcing bars, valves, conduit, and pumps. Polymers used for corrosion protection are usually thermosetting, particularly epoxies because of their superior adhesion characteristics.

POWDER MOULDING OF THERMOPLASTICS

Static (Sinter) Moulding

The process is often used with polyethylene and is limited to making open-ended containers. The mould which represents the exterior shape of the product is filled with powder. The filled mould is heated in an oven, causing the powder to melt and thus creating a wall of plastics on the inner surface of the mould. After a specific time to build the required wall thickness, the excess powder is dumped from the mould and the mould is returned to the oven to smooth the inner wall. The mould is then cooled, and the product is removed. The product is strain free, unlike pressure-moulded products. In polyethylene this is especially significant if the product is used to contain oxidising acids.

Rotational Moulding

Rotational moulding (popularly known as rotomoulding) is best suited for large, hollow products, requiring complicated curves, uniform wall thickness, a good finish, and stress-free strength. It has been used for a variety of products such as car and truck body components (including an entire car body), industrial containers, telephone booths, portable outhouses, garbage cans, ice buckets, appliance housing, light globes, toys, and boat hulls. The process is applicable to most thermoplastics and has also been adapted for possible use with thermosets.

Essentially four steps are involved in rotational moulding: loading, melting and shaping, cooling, and unloading. In the loading stage a pre-determined weight of powdered plastic is charged into a hollow mould. The mould halves are closed, and the loaded mould is moved into a hot oven where it is caused to simultaneously rotate in two perpendicular planes. A 4:1 ratio of rotation speeds on minor and major axes is generally used for symmetrically shaped objects, but wide variability of ratios is necessary for objects having complicated configurations. On most units the heating is done by air or by a liquid of high specific heat. The temperature in the oven may range from 500 to 900°F (260 to 482°C), depending on the material and the product. As the moulds continue to rotate in the oven, the polymer melts and forms a homogenous layer of molten plastic, distributed evenly on the mould cavity walls through gravitational force (centrifugal force is not a factor). The moulds are then moved, while still rotating, from the oven into the cooling chamber, where the moulds and contents are cooled by forced cold air, water fog, or water spray. Finally, the moulds are opened, and the parts removed. (Rotational moulding of plastisol, is similar to that described here. However, in this case the plastic is charged in the form of liquid dispersion, which gels in the cavity of the rotating mould during the heating cycle in the oven).

The most common rotational moulding machine in use today is the carousel-type machine. These are three-arm machines consisting of three cantilevered arms or mould spindles extending from a rotating hub in the centre of the unit. In operation, individual arms are simultaneously involved in different phases (loading, heating, and cooling) in three stations so that no arms are idle at any time (Fig. 19.1).

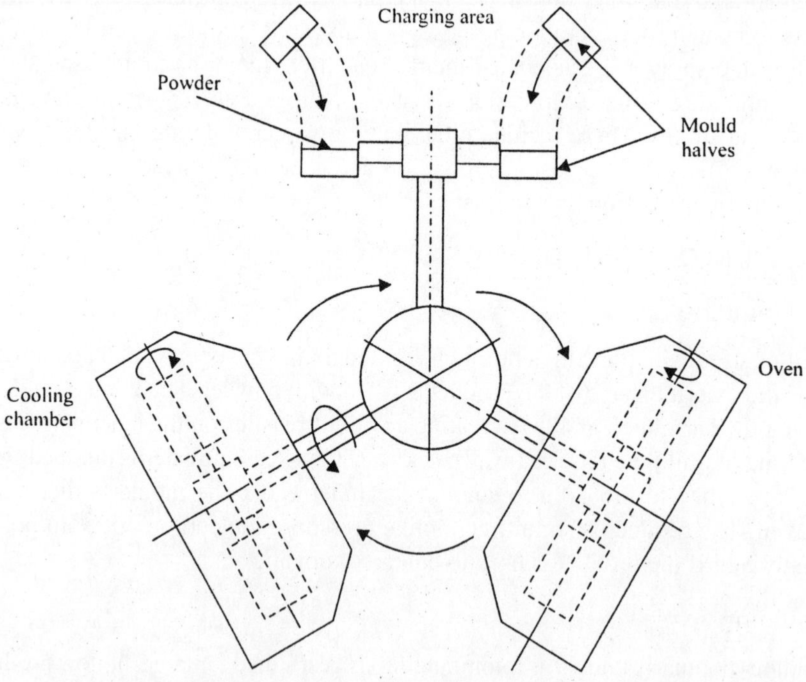

Fig. 19.1. Basics of continuous-type three-arm rotational moulding machine.

Modern rotational-moulding machines enable large parts to be moulded (e.g., 500-lb car bodies and 500-gal industrial containers). For producing small parts an arm of a carousel-type machine may hold as many as 96 cavities.

Centrifugal Casting

Centrifugal casting is generally used for making large tube forms. It consists of rotating a heated tube mould which is charged uniformly with powdered thermoplastic along its length. When a tubular molten layer of the desired thickness builds up on the mould surface, the heat source is removed and the mould is cooled. The mould, however, continues to rotate during cooling, thus maintaining uniform wall thickness of the tube. Upon completion of the cooling cycle, the plastic tube, which has shrunk away from the mould surface, is removed, and the process is repeated. Usual tube sizes moulded by the process range from 6 to 30 inch. in diameter and up to 96 inch. in length.

ADHESIVE BONDING OF PLASTICS

Adhesives are widely used for joining and assembling of plastics by virtue of low cost and adaptability to high-speed production. They can be sub-divided into solvent or dope cements, which are suitable for most thermoplastics (not thermosets), and monomeric or polymerisable cements which can be used for most thermoplastics and thermosets.

Solvent cements and dope cements function by attacking the surfaces of the adherends so that they soften and, on evaporation of the solvent, will join together. The dope cements, or bodied cements, differ from the straight solvents in that they contain, in solution, quantity of the same plastic which is being bonded. On drying, these cements leave a film of plastic that contributes to the bond between the surfaces to be joined.

Monomeric or polymerisable cements consist of a reactive monomer, identical with or compatible with the plastic to be bonded, together with a suitable catalyst system and accelerator. The mixture polymerises either at room temperature or at a temperature below the softening point of the thermoplastic being joined. In order to hasten the bonding and to reduce shrinkage, some amount of the solid plastic may also be initially dissolved in the monomer. Adhesives of this type may be of an entirely different chemical type than the plastic being joined. A typical example is the liquid mixture of epoxy resins and hardener which by virtue of the chemical reactivity and hydrogen bonding available from the epoxy adhesive provide excellent adhesion to many materials.

In joining with solvents or adhesives, it is very important that the surfaces of the joint be clean and well matched since poor contact of mating surfaces can cause many troubles. The problem of getting proper contact is aggravated by shrinkage, warpage, flash, marks from ejector pins, and non-flat surfaces.

Solvent Cementing

Solvent cementing or solvent welding basically involves softening the bonding area with a solvent or a solvent containing small quantities of the parent plastic, referred to as dope or bodied cement, generally containing less than 15 per cent resin. The solvent or cement must be of such composition that is will dry completely without blushing. Light pressures must be applied to the cemented joint until it has hardened to the extent that there is no movement when released. Structural bonds of up to 100 per cent of the strength of the parent material are possible with this type of bonding.

Table 19.1 gives list of typical solvents selected as useful for cementing of plastics. A key to selection of solvent in this table is how fast they evaporate: a fast-evaporating solvent may not last long enough for some assemblies, while evaporation that is too slow can hold up production. It may be noted that solvent bonding of dissimilar materials is possible when the materials can be bonded with the same solvents (Table 19.1).

Table 19.1. Typical solvents for solvent cementing of plastics.

Plastics	*Solvent*
ABS	Methylene chloride, tetrahydrofuran, methyl ethyl ketone, methyl isobutyl ketone
Acetate	Methylene chloride, chloroform, acetone, ethyl acetate, methyl ethyl ketone
Acrylic	Methylene chloride, ethylene dichloride
Cellulosics	Acetone, methyl ethyl ketone
Nylon	Aqueous phenol, solutions of resorcinol in alcohol, solutions of calcium chloride in alcohol
PPO	Methylene chloride, chloroform, ethylene dichloride, trichloroethylene
PVC	Cyclohexane, tetrahydrofuran, dichlorobenzene
Polycarbonate	Methylene chloride, ethylene dichloride
Polystyrene	Methylene chloride, ethylene dichloride, trichloroethylene, methyl ethyl ketone, xylene.
Polysulphone	Methylene chloride

The solvent is usually applied by the soak method in which pieces are immersed in the solvent, softened, removed, quickly brought together, and held under light pressure for some time. Areas adjoining the joint area should be masked to prevent them from being etched. For some applications where the surfaces to be cemented fit very closely, it is possible to introduce the cementing solvent by brush, eyedropper, or hypodermic needle into the edges of the joint. The solvent is allowed to spread to the rest of the joint by capillary action.

Adhesive Bonding

Adhesive bonding is a process in which the adhesive acts as an agent to hold two substrates together (as opposed to solvent cementing where the parent materials actually become an integral part of the bond) and the adhesion is chemical.

Table 19.2 gives a list of various adhesives and typical applications in bonding of plastics. This table is not complete, but it does give a general idea of what types of adhesives are used and where. It should be remembered, however, that thousands and thousands of variations of standard adhesives are available off the shelf. The computer may shape up as an excellent selection aid for adhesives. Selection is made according to the combination of properties desired, tack time, strength, method of application, and economics (performance/cost ratio).

The form that the adhesive takes (liquids, mastics, hot melts, etc.) can have a bearing on how and where they are used. Thus liquid adhesives lend themselves very well to hand-type techniques such as tube, spatula, brush, roller, and squeeze bottle, and also high-speed processing techniques such as spray and automatic roller coating. They may be solvent or water based; the anaerobics (cyanoacrylate-based bonding agents that cure in the absence of air) also fall into this class, as well as a number of epoxies. The anaerobics, which can give some very high bond strength and are usable with all materials except polyethylene and fluorocarbons, are dispensed by the drop. A thin application of the anaerobics is said to give better bond strength than a thick application.

Table 19.2. Typical adhesives for bonding plastics.

	ABS	*Acetal*	*Acrylic*	*Cellulosics*	*Fluoro carbons*	*Nylon*	*PVC*	*PC*	*PE*	*PP*	*PS*	*MF*	*PF*	*Polyesters*	*PU*
Metal	10	3.10	2,3	2,3	9,10	2.10	2.3,7.10.13.14	10	2,12	1	12	3.14	2	4	3,4
Paper	–	3,10	–	–	9.10	3	–	13	–	1	4.12.13	–	–	–	4.13
Wood	10	10	2,3	3	10	2.3	3.10.13	10,13	2	1	12.13	2.3	2	2	13
Rubber	10	3	1–4	1–4	10	2	3,4,7	4.13	2	1	5	2.3.14	2.3	1.4	4,13
Ceramics	–	10	2,3	3	10	3,10	3,4	10,13	2	1	–	2	2	2	3
ABS	10.14	2,4,6,10	2,4,6,10	–	–	8,10	10	4,10	10	–	–	–	–	–	–
Acetal	2,4,6,10	3,10	10	–	10	10	10	–	–	–	–	–	10	–	–
Acrylic	2,4,6,10	2,4,6,10	15	–	10	2,3,7,9	3,4,6,10	–	–	–	–	–	4,8,10	10	4,10
Cellulosic	–	–	–	3,4,13	10	2,3,7,9	10	–	–	–	–	–	10	–	–
Fluoro carbons	–	10	–	–	9,10	–	10	–	10	–	10	–	–	–	10
Nylon	8,10	8,10	2,3,7,9	2,3,7,9	–	2,9,10,13	–	–	–	–	–	8,10	8,10,14	8,10	–
PVC	10	10	3,4,6,10	–	10	–	3,4,6,13	–	–	–	–	–	4,7	10	–
PC	10	–	–	–	–	–	3,4,6,13	15	–	–	–	–	–	–	–
PE	10	–	–	–	10	–	–	–	10,12	–	–	–	–	–	–
PP	–	–	–	–	–	–	–	–	–	10,12	–	–	–	–	–
PS	–	–	–	–	10	–	–	–	–	–	4,5,10,12,13	–	4,10,14	–	–
MF	–	–	–	–	–	–	–	–	–	–	–	2,3,10,12,13	–	–	–
PF	10,14	14	4,8,10	4,8,10	10	–	4,7	–	–	–	4,10	3,8,10,11	2–4,10,12,14	–	–
Polyester	–	–	10	10	–	–	10	–	–	–	–	–	–	3,10,12,13	–
PU	–	–	4,10	–	10	–	–	–	–	–	–	–	–	–	3,4,10,13

Elastomeric: 1, Natural rubber. 2, Neoprene. 3, Nitrile. 4, Urethane. 5, Styrene-butadiene. Thermoplastic: 6, Poly(vinyl acetate). 7, Polyamide. Thermosetting: 8, Phenol-formaldehyde. 9, Resorcinol, Phenol-resorcinol/formaldehyde. 10, Epoxy. 11, Urea-formaldehyde. Resin: 12, Phenolic-poly(vinyl butyral. 13, Polyester. Other: 14, Cyanoacrylate. 15, Solvent.

The more viscous, mastic-type cements include some of the epoxies, urethanes and silicones. Epoxies adhere well to both thermosets and thermoplastics. But epoxies are not recommended for most polyolefin bonding. Urethane adhesives have made inroads into flexible packaging, the shoe industry, vinyl bonding. Polyester-based polyurethanes are often preferred over polyether systems because of their higher cohesive and adhesive properties. Silicones are especially recommended where both bonding and sealing are desired.

Hot melts—100 per cent solids adhesives that are heated to produce a workable material—are based on polyethylene, saturated polyester or polyamide in chunk, granule, pellet, rope, or slug form. Saturated polyesters are the primary hot melts for plastics; the polyethylenes are largely used in packaging; polyamides are used most widely in the shoe industry. Application speeds of hot melts are high and it pays to consider hot melts if production requirements are correspondingly high.

Pressure sensitives are contact-bond adhesives. Usually rubber based, they provide a low-strength, permanently tacky bond. They have number of consumer applications (e.g., cellophane tape), but they are also used in industrial applications where a permanent bond is not desirable or where a strong bond may not be necessary. The adhesive itself is applied rapidly by spray. Assembly is merely a matter of pressing the parts together.

Film adhesives require an outside means such as heat, water, or solvent to reactivate them to a tacky state. Among the film types are some hot melts, epoxies, phenolics, elastomers, and polyamides. Film adhesives can be die cut into complicated shapes to ensure precision bonding of unusual shapes. Applications for this type of adhesive include bonding plastic bezels onto automobiles, attaching trim to both interiors and exteriors, and attaching nameplates on luggage.

While most plastics bond without trouble once the proper adhesive has been selected, a few notably polyolefins, fluorocarbons and acetals, require special treatment prior to bonding. Untreated polyolefins adhere to very few substrates (a reason that polyethylene is such a popular material for packaging adhesives). Their treating methods include corona discharge, flame treating (especially for large, irregular-shaped articles), and surface oxidation by dipping the articles in a solution of potassium dichromate and sulphuric acid. A rubber-based adhesive that remains permanently tacky is generally recommended. Fluorocarbons can be prepared for bonding by cleaning with a solvent such as acetone and then treating with special etching solution. For acetals, one pre-treatment method involves immersing articles in a special solution composed mainly of perchloroethylene, drying at 120°C (250°F), rinsing, and then air drying.

Joining of Specific Plastics

The same basic handling techniques apply to almost all thermoplastic materials. In the following section, however, a few thermoplastics will be treated separately, with mention of the specific cements most suitable for each. The section on acrylics should be read in connection with the cementing of any other thermoplastics.

Cast Acrylic Sheeting

Articles of considerable size and complexity can be fabricated from methyl methacrylate plastics by joining sections together by solvent welding. The technique described here applies to cast sheeting. With articles made from methyl methacrylate moulding powders or extruded rod, tubing or other shapes, joining is generally not as satisfactory as a with cast sheeting.

With care and practice, the transparency of acrylic resin can be retained in joints with the formation of a complete union of the two surfaces brought into contact. Usually one of the two surfaces to be joined is soaked in the cementing solvent until a soft, swollen layer (cushion) has been formed upon it. This soft surface is then pressed against the surface to be attached and held in contact with it so that the excess solvent contained in the soaked area softens it also. The role of the cushion formed by soaking the surface in the solvent is solely to enable the two surfaces to be brought into complete conformity, so as to exclude air bubbles and to compensate the solvent, the softening action should be confined to the immediate region of the joint, by masking the rest of the surface with a tape.

For some purposes, it may be desirable to dissolve clean savings of methyl methacrylate resin in the solvent in order to raise its viscosity so that it can be handled like glue.

Only very slight pressure should be applied at first after bringing the soaked surface into contact with the surface to which it is to be joined. Then after about 15 to 30 sec, the cementing jig must be tightened to apply pressure.

Most of the conventional types of joint construction can be employed, such as overlap, butt, rabbet, mitter, and scarf, depending on the service requirement of the article. The area of the joint must be large enough to develop sufficient strength.

Methylene chloride, a solvent which produces joints of medium strength, requires a soaking time of 3 to 10 min and sets very quickly. Ethylene dichloride is slightly slower and is apt to produce cloudy joints. The most universally applicable type of solvent cement is the polymerisable type, comprising a mixture of solvent and catalysed monomer. These are mobile liquids, volatile, rapid in action, and capable of yielding strong sound bonds. An example of these is a 40–60 mixture of methyl methacrylate monomer and methylene chloride. Before using this cement, 1.2 grams of benzoyl peroxide per pint of solvent, should be added. Heat treatment or annealing of joints made with solvent cements is highly desirable because it greatly increases the strength of the joint.

Cellulosics

The cements used with cellulosic plastics are of two types: (i) solvent type, consisting only a solvent or a mixture of solvents; (ii) dope type, consisting of a solution of the cellulosic plastic in a solvent or mixture of solvents.

Acetone and mixtures of acetone and methyl 'cellosolve' are commonly used as solvent cements for cellulose acetate. Acetone is a strong solvent for the plastic, but evaporates rapidly. The addition of methyl "cellosolve" retards the evaporation, prevents blushing, and permits more time for handling the parts after application of the cement.

A cement of the dope type leaves upon drying a film of plastic that forms the bond between the surfaces to be joined. These cements are generally used when an imperfect fit of the parts requires filling. A typical composition of the dope-type cement for cellulose acetate is given below:

Ingredients	*Parts by weight*
Cellulose acetate	130
Acetone	400
Methyl 'cellosolve'	150
Methyl 'cellosolve' acetate	50

Other cellulosics, cellulose acetate butyrate and propionate are cemented in accordance with the technique described for cellulose acetate. In the case of dope cements, the plastic to be dissolved in solvents is cellulose propionate. Similarly for ethyl cellulose plastics, the strongest bonds are made by solvents or by solvents bodied with ethyl cellulose plastic.

Nylon

The recommended cements for nylon-to-nylon bonding are generally solvents, such as aqueous phenol, solutions of resorcinol in alcohol, and solutions of calcium chloride in alcohol, sometimes 'bodied' by the inclusion of nylon in small percentages.

Aqueous phenol containing 10–15 per cent water is the most generally used cement for bonding nylon to itself. The bond achieved by use of this cement is water resistant, flexible, and has strength approaching that of the nylon. If the parts fit well, they can be assembled immediately after applying the cement with a brush or cotton swab to the mating surfaces. However, if the fit is poor or loose at the mating surface, a waiting period of 2 to 3 min to soften the surface will help in obtaining a satisfactory joint. The parts are fastened together by means of spring clamps, 'C' clamps or toggle clamps in such a way that a uniform pressure is exerted on the bonded areas. If the joint is exposed to moist air as it cures, it may turn white.

Curing the assembly overnight in a dry atmosphere at the ambient temperature or for 1 hr in an oven at 40°C, will prevent the joint from whitening. The assembly is next immersed in boiling water in order to remove the excess phenol and to ensure a strong bond.

Calcium-chloride-ethanol solution bodied with nylon is recommended for nylon-to-nylon joint where there is possibility of contact with foods or where phenol or resorcinol would be otherwise objectionable.

For bonding nylon to metal and other materials, various commercial adhesives, especially those based on phenol-formaldehyde and epoxy resins, are sometimes used. Epoxy adhesives (in two-parts system), for example, have been used to produce satisfactory joints between nylon and metal, wood glass and leather.

Polycarbonate

Solvent cementing of parts of polycarbonate may be effected by the use of a variety of solvents or light solutions of polycarbonate in solvents. Methylene chloride, a 1 to 5 per cent solution of polycarbonate in methylene chloride, and a mixture of methylene chloride and ethylene dichloride (with a maximum of 40 per cent ethylene dichloride) are commonly recommended. Solvent should be applied to only one of the bonding surfaces while the other half remains dry and ready in the clamping fixture. As soon as the two parts have been put together, pressure should be applied immediately. Pressure between 200 and 600 psi is suggested for best results. Holding time in the pressure fixture is approximately 1 to 5 min, depending on the size of the bonding area.

For bonding moulded parts of polycarbonate to other plastics, glass, wood, aluminium, brass, steel, and other materials, a wide variety of adhesives can be used. Generally, the best results are obtained with solventless materials, such as epoxies and urethanes.

Polyethylene

The good solvent resistance of polyethylene and other olefins precludes the use of solvent-type cements. Several commercial rubber-type adhesives produce moderate adhesion with polyethylene that has been surface treated.

One technique for surface treatment is to dip polyethylene is a chromic acid bath (made up of concentrated sulphuric acid 150 parts by weight , water 12 parts, and potassium dichromate 7.5 parts) for about 30 sec at 70°C. The parts are rinsed with water after this treatment. Still another effective surface treatment for producing cementable surfaces on polyethylene is electrical discharge. The open oxidising flame method is also used extensively for this purpose.

Polystyrene

Complex assemblies of polystyrene, usually moulded in section, may be joined by means of solvents and adhesives. Polystyrene is soluble in a wide variety of solvents. According to their relative volatilities they may be divided into three groups—fast drying, medium drying, and slow drying. Methylene chloride, ethylene dichloride, and trichloroethylene are some of the fast-drying solvents that produce strong joints. However, they are unsatisfactory for transparent articles of polystyrene because they cause rapid crazing. 'Medium-drying' solvents such as toluene, perchloroethylene, and ethyl benzene that have higher boiling temperature are less apt to cause crazing. High-boiling or 'slow-drying' solvents such as amylbenzol and 2-ethylnaphthalene often require excessive time for development of sufficient bond strength, but they will not cause crazing to appear so quickly. Up to 65 per cent of a fat- or medium-drying solvent may be added to a slow-drying solvent to speed up the development of initial tack without greatly reducing the time before crazing appears.

A bodied, or more viscous, solvent may be required by certain joint designs and for producing airtight or watertight seals. These are made by dissolving a usually 5 to 15 per cent of polystyrene by weight in solvent. Numerous commercial adhesives, mostly of the solvent-dispersed, lacquer of rubber-based type, are also available for bonding polystyrene plastics. Solvent-based contact cements provide the strongest bond between polystyrene and wood. These adhesives all have a neoprene (polychloroprene) base and a ketonic-aromatic solvent system. Use of excessive amounts of these adhesives can, however, cause crazing since they contain polystyrene-active solvents—ketones and aromatics.

Poly (Vinyl Chloride) and Co-polymers

Cements for co-polymer resins are usually of two types: solvent cements and dope-type cements. On account of the relative insolubility of poly(vinyl chloride) and the markedly increased effect of solvents with the increasing content of vinyl acetate in the co-polymer resins, there exists among the vinyl chloride-acetate co-polymer system a greater diversity of composition and of ability to be cemented by solvents than in most of the other plastics that are cemented. Because of the greater solubility of the vinyl acetate component of the co-polymer, the cements which depend upon solvent action for strength will be more effective with co-polymers containing the greater percentage of vinyl acetate.

The co-polymer resin are most rapidly dissolved by the ketone solvents, such as acetone, methyl ethyl ketone, methyl isobutyl ketone, and cyclohexanone. Propylene oxide also is a very useful solvent in hastening solution of copolymer resins, especially those of high molecular weight, and of straight poly (vinyl chloride). This solvent penetrates the resins very rapidly and, in amounts up to about 20 to 25 per cent, improves the 'bite' into the resin. The chlorinated hydrocarbons also are excellent solvents for the vinyl chloride-vinyl acetate co-polymer resins, and are suitable for use in cements.

Thermosetting Plastics

Adhesive bonding is, for various reasons, the logical method of fastening or joining cross-linked and reinforced thermoset plastics to themselves, or to other materials. Since most thermoset plastics are quite resistant to solvents, and heat, heat-curing solvent-dispersed adhesives may be used. Such adhesives consist of reactive or thermosetting resins (e.g., phenolics, epoxies, urea-formaldehydes, alkyds, and combinations of these), together with compatible film-formers such as elastomers or vinyl-aldehyde condensation resins. Isocyanates are frequently added as modifiers to improve specific adhesion to surfaces that are difficult to bond. These adhesives may be applied not only in solvent-dispersed form, but also in the form of film, either unsupported, or supported on fabric, glass-mat, and so forth.

A great many of outstanding adhesive formulations are based on epoxy resins. A broad spectrum of adhesive formulations with a wide range of available properties have resulted from the use of polymeric hardeners such as polyamides and polyamines, phenolics, isocyanates, alkyds, and combinations of amines with polysulphide elastomers, and the "alloying" of the epoxy with compatible polymeric film-formers, such as poly(vinyl acetate) and certain elastomers.

In cemented assemblies of thermoset plastics and metals, where structural strength is generally desired, the adhesive must be more rigid than those used for bonding plastic to plastic, i.e., one with modulus, strength, and coefficient of thermal expansion between those of the plastic and the metal. In many cases, such adhesives are stronger than the plastic itself.

WELDING

Often it is necessary to join two or more components of plastics to produce a particular setup or to repair a broken part. For some thermoplastics solvent welding is applicable. The process uses solvents which dissolve the plastic to provide molecular interlocking and then evaporate. Normally it requires close-fitting joints. The more common method of joining plastics, however, is to use heat, with or without pressure. Various heat welding processes are available. Those processes in common commercial use are described here.

Hot-Gas Welding

Hot-gas welding, which bears a superficial resemblance to welding of metals with an oxyacetylene flame, is particularly useful for joining thermoplastics sheets in the fabrication of chemical plant items, such as tanks and ducting. The sheets to be joined are cleaned, beveled, and placed side by side so that the two bevealed edges form a V-shaped channel. The tip of a filler rod (of the same plastic) is placed in the channel, and both it and the adjacent area of the sheets are heated with a hot-gas stream (200–400°C) directed from an electrically heated hot-gas nozzle (Fig. 19.2(a), which melts the plastics. The plastics then fuse and unite the two sheets.

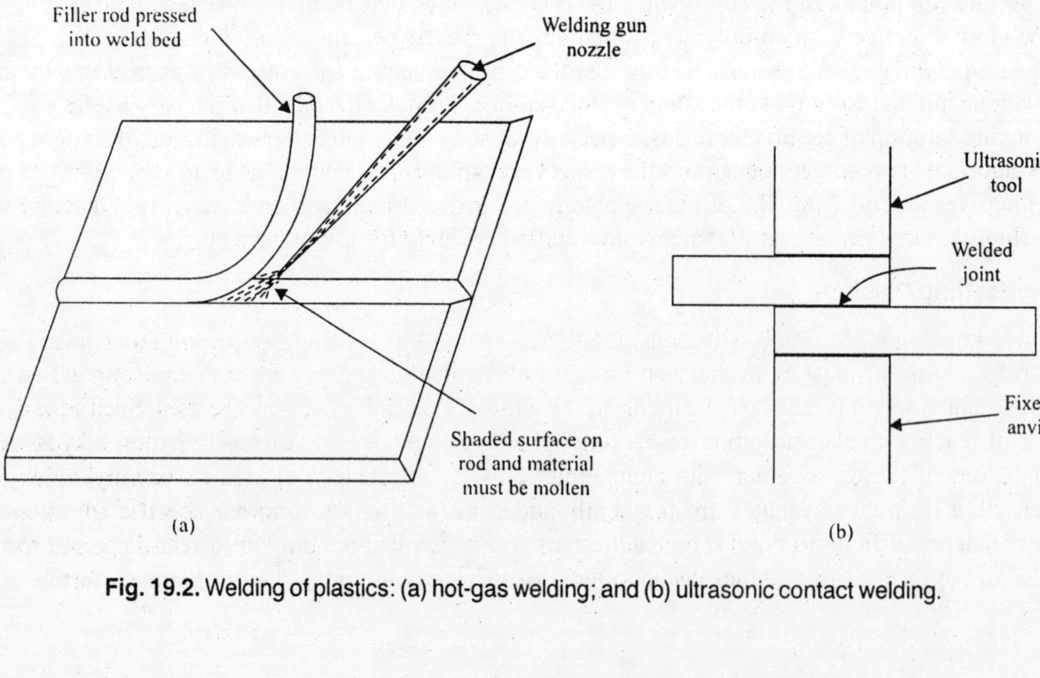

Fig. 19.2. Welding of plastics: (a) hot-gas welding; and (b) ultrasonic contact welding.

The hot gas may be air in PVC welding, but for polyethylene an inert gas such as nitrogen must be used to prevent oxidation of the plastics during welding.

Fusion Welding

Fusion or hot-tool welding is accomplished with an electrically heated hot plate or a heated tool (usually of metal), which is used to bring the two plastic surfaces to be joined to the required temperature. The polyfusion process for joining plastic pipes by means of injection-moulded couplings is an example of this type of welding. The tool for this process is so shaped that one side of it fits over the pipe while the other side fits into the coupling. The tool is heated and used to soften the outside wall of the pipe and the inside wall of the coupling. The pipe and coupling are firmly pressed together and held until the joint cools to achieve the maximum strength of the weld. The tool is chrome plated to prevent the plastic sticking to its surfaces.

Friction Welding

In friction or spin-welding of thermoplastics one of the two pieces to be jointed is fixed in the chuck of a modified lathe and rotated at high speed while the other piece is held against it until frictional heat causes the polymer to flow. The chuck is stopped, and the two pieces are allowed to cool under pressure. The process is limited to objects having a circular configuration. Typical examples are dual-coloured knobs, moulded hemispheres, and injection-moulded bottle halves.

High-Frequency Welding

Dielectric or high-frequency welding can be used for joining those thermoplastics which have high dielectric-loss characteristics, including cellulose acetate, ABS, and PVC. Obviously, polyethylene, polypropylene, and polystyrene cannot be welded by this method. The device used for high-frequency welding is essentially a radio transmitter operated at frequencies between 27 and 40 MHz. The energy obtained from the transmitter is directed to electrodes of the welding apparatus. The high-frequency field causes the molecules in the plastic to vibrate and rub against each other very fast, which creates frictional heat sufficient to melt the interfaces and produce a weld.

Ultrasonic Welding

In ultrasonic welding the molecules of the plastic to be welded are sufficiently disturbed by the application of ultrahigh-frequency mechanical energy to create frictional heat, thereby causing the plastics to melt and join quickly and firmly. The machinery for ultrasonic welding consists of an electronic device which generates electrical energy at 20/50 kHz/sec and a transducer (either magnetostrictive or piezoelectric) to convert the electrical energy to mechanical energy. In the contact-welding method (Fig. 19.2b) the ultrasonic force from the transducer is transmitted to the objects (to be welded) through a tool or 'horn', generally made of titanium. The amplitude of the motion of the horn is from 0.0005 to 0.005 in. (0.013 to 0.13 mm) depending on the design. The method is generally used for welding thin or less rigid thermoplastics, such as films, or sheets of polyethylene, plasticised PVC, and others having low stiffness.

CHAPTER 20

Decoration of Plastics

INTRODUCTION

Commercial techniques for decorating plastics are almost as varied as plastics themselves. Depending on end-use applications or market demands, virtually any desired effect or combination of effects, shading of tone, and degree of brightness can be imparted to flexible or rigid plastics products.

The primary decorating technique is raw-materials colouring achieved at the compounding stage. Although most thermoplastics are produced in natural white or colourless transparent form, colour is usually added by directly blending colorants into the base resin prior to the processing stage. These colorants (or colour concentrates) are available in a wide range of stock shades with precise tinctorial values. Colours can also be matched to exact customer specifications and these specifications kept in computer memory to ensure batch-to-batch or order-to-order consistency. Colour blending can also be utilitarian, as in colour-coded wire and cable sheathing.

Besides basic raw-materials colouring, mentioned above, designers have a large palette of decorating media at their disposal. Plastics can be decorated in various ways, which include painting processes, direct printing, transfer decoration, in-mould decoration, embossing, vacuum metallising, sputtering, and electroplating. Most of these processes require bonding other media, such as inks, enamels, and other materials to the plastics to be decorated. Some plastics, notably polyolefins and acetals, are however, highly resistant to bonding and need separate treatment to activate the surface. Commonly used treatment processes are flame treatment, electronic treatments such as corona discharge and plasma discharge, and chemical treatment.

In *flame treatment*, plastic objects such as bottles and film are passed through an oxidising gas flame. Momentary contact with the film causes oxidation of the surface, which makes it receptive to material used in decorating the product.

In the *corona discharge* process the plastic film to be treated is allowed to pass over an insulated metal drum beneath conductors charged with a high voltage. When the electron discharge ('corona') between the charged conductors and the drum strikes the intervening film surface, oxidation occurs and makes the surface receptive to coatings. Moulded products are also treated in a similar manner, often by fully automatic machinery.

In the *plasma process*, air at low pressure is passed through an electric discharge, where it is partially dissociated into the plasma state and then expanded into a closed vacuum chamber containing the plastic object to be treated. The plasma reacting with the surfaces of the plastic alters their physicochemical characteristics in a manner that affords excellent adhesion to surface coatings. The process can be used for batch processing of plastics products, including films which may be unreeled in the vacuum chamber for treatment.

Acetal resin products are surface treated by a chemical process consisting of subjecting the product to a short acid dip that results in an etched surface receptive to paint.

PAINTING

Virtually all plastics, both thermoplastic and thermosetting, can be painted, with or without priming or other preliminary preparation procedures. The process, however, requires special consideration of the resin-solvent system to achieve adhesion, adequate covering, and chemical resistance. Painting operations have the advantage of being as simple or as sophisticated as the application may dictate.

Plastics parts or materials can be coated manually by brushing, dipping, hand-spray painting, flow coating or roller coating; they can be automatically spray-painted with rotating or reciprocating spray guns, and electrostatically painted using a conductive precoating procedure. It is also possible to paint areas selectively, through a variety of masking options. Three-dimensional effects can be achieved by placing multicolour painting intaglio designs on the under surface of clear plastic products.

Painting operations have the advantage of offering almost unlimited colour options as well as great variety of surface finishes and final surface properties to meet such needs as gloss, UV resistance, abrasion resistance, and chemical resistance. Limitations of painting on plastics are usually dictated by limitations of the base materials themselves. For example, to achieve extremely high abrasion resistance requires coatings that are cured at elevated temperatures that may exceed the temperature resistance of the base plastics.

Total surface coverages as well as masked operations in painting can be adapted to almost unlimited geometries. Parts as small as knobs, pushbottons, delicate nameplates, and medallions are readily decorated by the painting process either by manual options (for low-volume production) or automatic systems (for higher volume production). Painting is routinely done on huge parts such as TV cabinets and exterior automotive parts such as hoods and deck lids. Production painting of larger parts is often a completely automated operation involving robotized spraying arms that have been programmed to apply paint in a carefully orchestrated pattern to achieve maximum speed and efficiency in part coating.

PRINTING

The primary printing presses used in plastics are gravure printing, flexography, silk-screen printing, and pad printing.

Gravure Printing

Gravure printing is a process that requires the use of an engraved metal cylinder or roller. *Rotogravure* is thus an appropriate title for this printing process. The engraving or etching process on the surfaces of the metal cylinder results in recessed areas that pick up ink or liquid coatings from a reservoir, the excess ink or coatings being wiped from the foreground areas of the cylinder by a doctor blade or a squeegee action. The ink-laden cylinder is then rolled over the medium to be printed. The etching process in producing the gravure cylinders is such that very fine tones and shading results can be achieved. With proper formulation of printing ink, the gravure process can be applied to a great variety of plastic substrates. Virtually all thermoplastic film or sheet applications are printable by this process.

A good example of the capabilities of the gravure process is the printing of woodgrain patterns on carrier foil for use in hot-stamping applications (described later). Woodgrain patterns may require the application of several coating to achieve the proper effect. Several engraved cylinders can be used in sequence for continuous printing.

Flexography

Flexography uses a flexible printing plate, typically a metal-silicone rubber-bonded combination with the rubber surface processed to leave the printing surface raised over the background area. The raised and recessed areas on the surface can be fabricated through photographic etching and/or engraving. After transferring the ink from a reservoir through a roller-doctor blade system onto the curved flexible plate, the ink is transferred off the raised portions to the material to be printed.

Fabricating the silicone rubber-metal printing plate is less expensive than engraving or etching metal roller dies. Flexography is adaptable to shorter volume production and the material to be printed can be fed one at a time.

The process is suitable for a variety of applications, ranging form simple label film to decoration on moulded parts such as plaques, medallions or wall tile. However, the flexible printing plates used in flexograph do not permit the very fine detail that can be achieved on metal surfaces such as used in gravure printing. There are also limitations to the size and shape of articles that can be printed.

Screen Process Printing

The process derived its name from the use of silk cloth or silk screen in the transfer of printing ink to articles to be printed. Integral to the process is the use of a suitable open-weave cloth or screen (silk is still commonly but not exclusively used) stretched over a frame work. Screens made of nylon or other synthetic materials are often employed, as also stainless steel or other metallic screens. The stretched screen is selectively coated through the use of a stencil corresponding to an art copy of the image to be printed; this coated (closed) area resists the passage of printing ink, which can only penetrate through the uncoated (open) areas of the screen. There are various ways to prepare the screen for printing, other than stencilling.

The screen printing process involves the use of a rubber squeegee or metal flood bar to carry the ink over the open mesh areas of the screen. The squeegee action also forces ink through the open mesh areas by pressure onto the surface to be printed. This very nature of the process suggests some geometry limitations. It is not adaptable to complex shapes, but is effective on flat surfaces or the peripheral areas of cylinders.

Silk screening is a versatile process, with a lot of flexibility in inks that can be accommodated, therefore allowing it to be used for a wide variety of plastic substrates. Silk screening can also generate good design detail. Tooling for screen printing usually involves only the feed or holding devices required to position the substrate under the screen.

The process does not require a great deal of auxiliary equipment. The process cost is also offset somewhat by the relatively low cost of producing screens, and its adaptability to manual screening for low-volume needs and to full automation where volumes demand is anticipated. High-volume applications may, however, dictate that screen be made of woven stainless steel threads rather than the less expensive (but less durable) silk, nylon, or polyester. The detail of printing can be adjusted through proper selection of the screen media. The finer the mesh weaver the greater detail that will result, which is consideration when delicate graphics or patterns are required.

Polyolefins such as polyethylene and polypropylene must be surface treated before being printed. The most effective way is in an integrated machine where surface treatment takes place right before printing. A time lapse will means that the treatment will lose some of its effect. Three methods are used: flame treatment, corona discharge, and chemical treatment. Flame treatment is considered the most practical and most widely used.

Screen printing is performed manually for small lots at the rate of 200 to 600 pieces per hour. Large-volume products, such as detergent bottles, are printed by automatic screening at rate approaching 2000 to 3000 impressions per hour, depending on the nature of the product. Printing of multicolour impressions by this process requires the use of a screen for each colour and adequate drying time for each colour before the next colour is applied. Split screens enable simultaneous printing of several colours. However, split screens can only be used if colours are not mixed in the design.

Pad Printing

Pad printing uses printing principles and techniques from letterpress and flexography. The uniqueness of the process has to do with the use of a smooth silicone pad that picks up ink impression from an engraved or etched plate and transfers it to the product to be decorated. The engraved plate, known as a *cliche*, is produced in a manner similar to that of printing plates for offset of gravure roller printing. The silicone pick-up pad can be designed to meet almost any shape and configuration of the product part. This ability has prompted tremendous growth in pad printing. An additional capability of the process is that it can print several colours and impressions within one cycle of operation. Coatings can be layered when wet to accomplish multicolour design with very accurate registration and impression quality.

Since the silicone pad used in pad printing process conforms to the surface of the individual item being printed, the process is extremely valuable for transferring imprints to grossly irregular surfaces incorporating concave and convex features. An example is the imprinting of graphics onto the contoured surface of key caps such as are used in business machine keyboards. The silicone pad, moreover, has the ability to adapt to part changes form cycle to cycle. This is a very important attribute when production runs produce changing tolerances that result in slightly inconsistent imprint area surfaces.

Pad printing is also efficient for imprinting very delicate designs and/or graphics. But for graphics or designs to be imprinted on irregular surfaces, it is important to understand the relationship between picking an imprint off a flat surface and transferring it onto a contoured surface, that is, the imprint on the flat surface will not look or dimension the same as the final result as viewed or measured on the contoured product. Thus, depending on the degree of change of the surface geometry, it may be necessary to extrapolate the dimensions from the contoured surface to the flat surface.

The process is adaptable to short runs with manual loading and unloading as well as to high-volume applications involving multiple colours and automatic part feeding. The silicone pads utilised in the process do wear out. Dependent upon the contours and the impression produced, the pad may only last for 20,000–40,000 cycles.

HOT STAMPING

Hot stamping is one of the original methods of decorating plastics materials. Though familiarly known as hot stamping, the terminology 'coated foil transferring' might be more appropriate since in this process the printed coating on a carrier film is transferred onto a plastic surface. Secure adhesion is accomplished with the use of heat, pressure, and time under controlled conditions. The key to the process is the use of a carrier film (usually a polycarbonate, polyester, or cellophane) upon which various coatings provide the desired decorative effect. The coated foil is placed over the plastic to be decorated, and a heated die forces the foil onto the plastic. The proper control of heat, pressure, and time transfers the coating off the carrier foil onto the plastic.

The hot-stamping process is a versatile tool for plastics decoration. A wide variety of coatings can be deposited on the carrier film which allows the process to be used on almost any thermoplastic material and many thermosets. Coatings on the carrier film can be formulated specifically for different plastics materials to enhance adhesion and to react to proper heat exposure as dictated by the plastic substrate. Coatings can be virtually of any colour; metallic effects can be imparted by depositing microthin coatings of gold or silver or chrome; multiple coatings can be applied to the carrier film to achieve such effects as woodgraining, marbleizing, or multicoloured designs. Three-dimensional decorative effects can also be achieved by embossing the surface of the carrier medium coating.

The hot-stamping process also has some limitations. To achieve a good finish quality, it is necessary to ensure even contact of the die through the carrier medium onto the plastic surface in order to control the heat, time, and pressure variables on tall surfaces of the coating. The geometry of the surfaces to be decorated must, therefore, be considered when evaluating the use of hot stamping. Even so, with proper tooling it is possible to decorate surfaces other than flat areas.

Simplicity of operation is a key advantage of the hot-stamping process, which finds many end-use applications ranging from selective imprinting, such as on the windows of microwave-oven doors or covering large areas, such as woodgrained surfaces of television cabinets. A wide variety of hot-stamping equipment is available. This ranges from hand-held heated roller dies utilised in prototype work or sample to huge roller presses such as for laying finishes on continuous 4-ft wide rolls of ABS thermoforming stock. Equipment is also available to do peripheral stamping such as decorating perimeters of rollers or dials, such as odometer wheels.

IN-MOULD DECORATING

As the name implies, in-mould decorating is a process in which a predecorated overlay (film), or *decal*, is placed in the mould, where the decorated element is fused to the moulded part during the heating/cooling cycles of the moulding operation. Since the decorated coating is bonded between the plastic part and the film (which will be the exterior surface), thus forming an integral part of the product, it produces one of the most durable and permanent decorations. High quality melamine dinnerware is decorated by this method, and so are a host of other household and hardware plastics goods.

In-mould decoration can be done with either injection moulding of thermoplastics or compression moulding of thermosets. Thermosetting plastics are decorated with a two-stage process. For melamine products, for example, the mould is loaded with the moulding powder in the usually manner and closed. It is opened after a practical cure, and the decorative 'foil' or overlay is placed in position. The mould is then closed again, and the curing cycle is completed. The overlay consists of a cellulose sheet having printed decoration and covered with a thin layer of partially cured clear melamine resin. During the moulding cycle the overlay is fused to the product and becomes a part of the moulding. The process relatively inexpensive, especially when a multicolour decoration is required.

For in-mould decoration of thermoplastic products, single-stage process is used. The foil or overlay is thus placed in the mould cavity prior to the injection of the polymer. It is held in place in the mould by its inherent static charge. Shifting is prevented during moulding by inducing an additional charge by passing the wand of an electronic static charging unit over the foil after it is properly positioned. The overlay, in all cases, is printed or decorated film (0.003–0.005 in. thick) of the same polymer. Thus, polystyrene film is used for a polystyrene product, and polypropylene film for a polypropylene product. A similar procedure may also be used for decorating blow-moulded products.

MOULDED-IN LETTERING

Moulded letters, figures, and decorated designs are often used on moulded parts to identify and to decorate the product. Letters may be raised or depressed, or they may face up from a depressed panel. Raised lettering is by far the cheaper because it requires engraved letters on the mould surface which can be cut or stamped into the cavity of the mould. Depressed lettering, on the other hand, requires raised mould letters to create the depressed letters on the product. Such moulds are more expensive to make. If mould cavities are hobbed, the hob is engraved with the depressed letters so that the hob forms raised letters in the mould cavity. The moulded part thus becomes a replica of the hob and contains depressed letters. The depressed letters are filled with paint, and the surface is wiped, which produces an attractive appearance.

Raised or depressed lettering usually does not exceed 0.03 in. (0.76 mm). For lettering which must be applied to the parting line or side walls, it is desirable to provide sufficient draft for the letters to 'draw' out the mould.

Moulded-in lettering offers unusual decorative possibilities for transparent thermoplastics materials. Thus lettering may be cut in on the underside of the product, and contrasting colours may be used for the background and lettering to produce an attractive three-dimensional effect.

EMBOSSING

Embossing is used for producing a tactile texture or patterns on plastics sheet or film. As the process involves the use of heat and pressure to texture a semifinished substrate, embossing is largely limited to thermoplastics materials. However, it can be adapted to thermoset composites, such as melamine-impregnated sheet stock.

Embossing is most commonly done with a two-roller system, in which one roller carries the embossing pattern and the other provides the essential pressure backup and feeding actions. Texture or pattern can be applied to the embossing or surface roller through a variety of processes, including conventional engraving, chemical engraving, etching, and lesser cutting. The backup or forming roller can be made of various materials, depending on the degree of texturing or patterning required. While rollers are typically made of smooth steel, formed steel, or rubber-coated steel, matched-steel rolls are also used for deep or very detailed embossing. The latter involve the use of two engraved rollers of male-female configuration, with the backup roller having projections to match the depressions in the embossing roller. The matched roller process provides maximum pattern definition at high roller speeds.

Embossing can also be performed without rollers. In this process, textured aluminium foil is placed over the substrate, such as sheet stock or laminates, and subjected to heat and pressure in a compression press or laminating press. The combined action of heat and pressure transfers the pattern from the aluminium foil to the laminate; the foil is subsequently removed and discharged. Instead of using foil, stainless steel plates with engraved textures can be used in the press cycle time and again, offering multiple impression economies.

Each of these embossing process has advantages and disadvantages. Roller systems are advantageous for handling continuous film or sheet materials. On the other hand, the foil or engraved steel laminating pressure approach is desirable for individual sheet stock preparation.

Embossing is most frequently used as a method of decorating nonslip packaging materials, vinyl wall coverings, furniture laminates, building panel laminates, textured foil for hot stamping, and other applications where the innate quality of three-dimensional printing is of value.

ELECTROPLATING

Electroplating is a chemical process for depositing heavy metals on plastics to achieve decorative effects and/or upgraded functionality. Since plastics are nonconductors of electricity, electroplating requires that the surface be properly conditioned and sensitized to receive metallic coatings. The principle of electroplating is to electrically conduct metal atoms such as copper, nickel and chrome off anodes placed within the plating baths through the plating solutions and onto the plastic production part. The target, i.e., the production part, acts as a cathode via connection to conductive plating racks, the part being attached to the plating rack with metal holding devices, spring-loaded contacts, or prongs. The point of contact between the plating rack and the plastic part forms the continuity of the current flow from anode through the solution onto the plastic part.

The process of electroplating begins with the plastic part attached to the plating rack being introduced into the preplate section. This involves a partcleaning operation, a chemical action to provide minute undercuts for good adhesion, and catalysation to prepare the surface for a deposit of electroless nickel or copper. The catalysation is accomplished by absorbing on the surface a sensitiser, such as acidified stannous chloride, which is oxidised in subsequent nucleation step (typically, treatment with an acidified palladium chloride solution) to deposit a catalytic film on the plastic surface. When the catalytic surface is next introduced into an electroless plating bath, the reaction between the metal salt and the reducing agent, both present in the bath, takes place only on this surface and deposits a layer of metal. The above preplate procedure is designed to create a surface on the plastic parts that will develop a bond between the plastic and the first nickel or copper deposit. These initial deposits are extremely thin, in the micron (10^{-6} mm) range. This first deposit is designed to increase conductivity uniformly over the plastic surface without building up stress or concentrated conductive areas.

When preplating is completed (and the plastic articles have a conductive coating), it is possible to proceed to the electroplating operation, which is very similar to conventional electroplating on metal. The preplated plastic part proceed through conventional heavy metal plating bath, interspersed with rinse and cleaning baths. A 0.5 to 1 mil (0.01 to 0.25 mm) deposit of ductile copper comes first, followed by a somewhat thinner nickel coating, followed in turn by a final chrome plating, applied in fractions of a mil (0.025 mm). The parts are then again put through a rinsing and cleaning operation to complete the electroplating process. Precise control of each plating thickness is necessary to achieve the essential performance properties, such as adhesion, perspiration resistance, gloss and thermocycling. The copper-nickel-chrome process is designed to meet most commercial plating demands.

Electroplating of plastic products provides the high-quality appearance and wear resistance of metal combined with the light weight and corrosion resistance of plastics. Physical properties such as tensile strength, flexural strength, and heat-deflection temperature are also enhanced in the plated plastics.

Plating is done on many plastics, including phenolic, urea, ABS, acetal, and polycarbonate. Many automotive, appliance, and hardware used of plated plastics include knobs, instrument cluster panels, bezel, speaker grilles, and nameplates. In marine searchlights zinc has been replaced by chrome-plated ABS plastics to gain lighter weight, greater corrosion resistance, and lower cost. An advantage of plastics plating is that, unlike metal die castings, which require buffing in most cases after plating, plastics do not ordinarily require this extra expensive operation. The use of plated plastics also affords the possibility of obtaining attractive texture contrasts.

Fundamental Concepts of Electroplating

The fundamental concepts and production techniques for the electroplating of plastics involves the following steps:

1. Moulding of the plastic.
2. Conditioning treatments for adhesion.
3. Activation of the surfaces with a catalytic material.
4. Electroless plating for conductivity.
5. Electroplating to specifications.

This relatively simple process has replaced almost completely the older procedures based on batch mechanical treatments and has allowed the use of automatic plating equipment in large-scale production operations.

Using the sequence mentioned above it is possible to electroplate a large number of commercial plastics. The only step specific to each plastic is the conditioning treatment by which an adequate metal-to-plastic bond is obtained. This treatment in some cases is a simple chemical etching process, whereas for plastics with high chemical resistance (like the fluoropolymers) it can involve complicated multistep techniques.

Polymers

The principal properties for making a plastic adequate for electroplating are the quality of the surfaces and the relative ease by which it can be treated for adhesion. The first polymer that offered adequate properties from both standpoints was acrylonitrile–butadiene–styrene co-polymer and this material is still by far the most widely used (about 90 per cent) by the industry for electroplating. This dominant position is now being challenged by other polymers. Special types of poly (oxyphenylene) and polypropylene have been developed which can be plated in the same production lines used for acrylonitrile–butadiene–styrene co-polymer parts. This has allowed an increased flexibility in the plating operations.

The surface structure of platable plastics has been investigated and the relation between surface porosity and the performance of the plated parts is now better understood. To obtain parts that will cover uniformly and have an adequate adhesion, a deep, uniform porosity is needed. In the case of acrylonitrile–butadiene–styrene plastic, this porosity is produced by removing the butadiene phase from the surface by chemical etching. For other platable plastics, additives (such as titanium dioxide, talcum, and silicon dioxide) are used to provide surface porosity.

An area of recent development is the plating of glass or paper filled epoxy, phenolic, or polyester laminates for the electronics industry. Following essentially the same techniques as used in the plating of acrylonitrile–butadiene–styrene co-polymer, conductive patterns can be deposited for the production of printed-circuit boards by additive processing. In additive printed-circuit production, the polymer laminates are conditioned, activated, and then plated with electroless copper to the desired thickness. Using plating resists or selective activation, it is possible to restrict the electroless copper deposition to the areas of the electric circuit (pattern plating). By additive techniques, very fine line patterns can be produced at competitive costs. In some of these processes, the metal adhesion is improved by coating with additives, topographic modification, or by the use of resin-rich boards.

Moulding and conditioning

Moulding will determine, to a large degree, the internal stresses and surface characteristics of the parts that affect the outcome of the plating operations. The most important development in moulding for electroplating has been the success of the industry in producing large moulding free of stresses and with surfaces of a remarkable quality. This is particularly important for the automotive industry where large chrome-plated parts are required to replace metal with lighter materials. As already mentioned, the present trend in the industry is to use plastics that can be processed through the same cycle as used for acrylonitrile–butadiene–styrene copolymer. Mixtures of chromic and sulphuric acids are universally used for the etching of plastics. Research on improvements is being done on chrome-free chemical treatments, to reduce the waste-treatment problems associated with the use of hexavalent chromium. Several patents have been issued to this effect.

The use of gas and vapour-phase treatment have also been studied, and low-temperature gas plasmas have been reported as being effective. Nevertheless, at the present time, the industry has not proceeded actively to eliminate chromic acid etches. In most operations, chrome-plating baths in the electroplating line have already made the treatment of chromic wastes mandatory, so that facilities are available for disposing of the etching materials.

Several new techniques have been reported to improve the metal-to-plastic adhesion and performance of the finished parts. These treatments are used either before or after the chromic–sulphuric etch.

Activating

There are several techniques for depositing a catalystic layer on the surface of the plastic and rendering it active for electroless plating. Of these processes, the combined palladium–tin nucleation is mostly used at the present time. This type of activation has the advantage of stability and also causes much less plating on the racks used to hold the parts. The catalysts are obtained by reacting stannous and palladium salts under carefully controlled conditions. The catalystic behaviour of the material will be influenced to a large extent by the reaction conditions.

Several patents have been issued covering the manufacture of the tin–palladium catalyst and their nature has been the subject of several studies. At present, the active species is considered to be either a palladium–tin complex or a palladium colloid.

Plating

Electroless plating baths based on the autocatalytic chemical reduction of copper and nickel are used for metallising the plastics. During the past years electroless nickel solutions have become widely used for making the plastic surfaces conductive and have displaced electroless copper, to a large degree. Electroless nickel solutions are very stable and exhibit less tendency to plate out on the plating racks. Still, at present there is a renewed interest in using electroless copper, since this metal appears to have an advantage over nickel for outdoor exposure conditions; acrylonitrile–butadiene–styrene co-polymer parts plated with electroless nickel exhibit problems of lateral corrosion and loss of adhesion.

In almost all cases, a layer of copper is used as the first electrodeposit. Bright acid copper electroplating baths are used almost universally. This copper deposit has two functions: due to the brightness and leveling properties, it will cover scratches and surface defects and make mechanical polishing unnecessary. Also, it is ductile and conductive and improves the thermal characteristics of the plated parts. Copper thicknesses of 10–20 μm are used, depending on specifications and the final use of the parts.

Nickel and chrome deposits are commonly used over the copper layer. For applications where the parts are exposed to corrosion, duplex nickel and microcracked chrome deposits have become standard. Lately, a strong trend has developed towards plating selected areas only, instead of the total surface of the parts. The development of very effective 'stop-off coating' has made this process possible, and design engineers are finding decorative and cost advantages for appliance and automotive parts.

Applications

The major markets for plated plastic parts are automotive components, plumbing fixtures, builders hardware and appliance parts. For these applications, acrylonitrile–butadiene–styrene co-polymer is by far the most commonly used plastic.

Poly(oxyphenylene) plastics are used to some extent for special outside exposure in automotive parts. Polysulphone resins are used to a much lesser degree for high-performance applications, like photographic components.

The electronic industry has become a major market for plated plastics by adopting the additive process for the manufacturing of printed-circuit boards. At present, mainly glass-reinforced epoxy and paper-filled phenolic laminates are used.

The general increase in raw material and commodity prices has affected the costs of electroplated plastics as well as those of the competitive materials.

An additional cost of electroplating is waste disposal. New capital investments and higher operating costs are needed to meet the current Environmental Protection Agency regulations for plating effluents. More stringent regulations, would affect even more adversely the economics of plating on plastics.

Specifications and test methods

There are still no official standards for the industry, but the American Society for Electroplated Plastics has issued guidelines. Standard Specifications for Decorative Electroplated Coatings of Copper, Nickel, and Chrome on Plastics are being established at present by the Section B08.03.06 of the ASTM.

Many major corporations in the automotive and appliance fields have issued specifications for internal use and for suppliers. The major consideration of these specifications is the end-use of the plated parts. With increased experience, more rigid standards are being applied regarding the type and thickness of the electrodeposited coatings. One of the major causes for exfoliation and lack of adhesion is in-sufficient thickness of the metal deposits. In the automotive industry, a total metal thickness of 1.5 mil (38μ) for outside applications and 0.8 mil (20μ) for inside applications are the usual requirements.

Adhesion, thermal cycling and corrosion are still the main tests for plated plastics. The interpretation and testing procedure have been changed. Adhesion measured as peel strength by the Jacquet test has lost some of its practical significance, since peel strength and field performance cannot readily be correlated. On the other hand, peel testing can be a valuable research tool for optimising surface-conditioning processes.

Thermal cycling is the most widely accepted tool for testing the mechanical performance of plated plastics. There is no agreement yet as to how the test should be performed and the trend is to vary the conditions of the test to simulate the relative severity of the final use conditions. Corrosion is being tested mostly by the copper accelerated salt spray test (CASS).

Experience in the automotive field indicates that this is the accelerated corrosion method that reproduces most closely the failures found in the field.

Future outlook

The electroplating of plastics has now become an established an mature industry and therefore, its short-range change will be governed mostly by economic factors. Plastics other than acrylonitrile–butadiene–styrene co-polymer will probably gain an increased share of the market and new polymers are expected to become available in plating-grade compounds. The outlook for circuit boards made by additive processing is excellent.

VACUUM METALLISING

Vacuum metallising is a process whereby a bright thin film of metal is deposited on the surface of a moulded product or film under high vacuum. The metal may be gold, silver, or most generally, aluminium. The process produces a somewhat delicate surface compared to electroplating. The metallising process can be used on virtually all properly (surface) prepared thermoplastic and thermosetting materials.

Small clips of the metal to be deposited are attached to a filament. When the filament is heated electrically, the clip melts and through capillary action, coat the filament. An increased supply of electrical energy then causes vaporisation of this metal coating, and plating of the plastic product takes place. To minimise surface defects and enhanced the adhesion of the metal coating, manufactures initially give the plastics parts lacquer base coat and dry in an oven. The lacquered parts are secured to a rack fitted with filaments, to which are fastened clips of metal to be vaporised. The vaporisation and deposition are accomplished at high vacuum (about 1/2 micron). The axles supporting the part holding the fixtures are moved so as to rotate the parts during the plating cycle to promote uniform deposition. The thickness of the coating produced is about 5×10^{-6} inch. (127 nm). After the deposition is completed, the parts are removed and dipped or sprayed with a top-coat lacquer to protect the metal from abrasion. Colour tones, such as gold, copper, and brass may be added to this coating if desired.

Vacuum metallising of polymer films, such as cellulose acetate, butyrate, and Mylar, is performed, in essentially the same way. Film rolls are unreeled and rewound during the deposition process to metallised the desired surface. A protective abrasion-resistant coating is then applied to the metallised surface in an automatic coating machine.

Central to the vacuum metallising process is the use of a high vacuum chamber, which can range from very small 2 by 2 ft (0.6×0.6 m) laboratory units to high vacuum production units measuring 6 by 6 ft (1.8×1.8 m) or more. The function of the chamber is to hold and rotate the plastic parts to be decorated, to fixture the heat sources necessary to vapourise the coating metal, and to provide vacuum pumps for reducing internal pressures to about 1/2 micron of mercury, required for achieving a clean transfer to the plastic surfaces.

Vacuum metallising is a versatile process used in a great variety of applications. Examples range from highly decorative cosmetic closures to automotive grilles and instrument clusters. Vacuum metallised plastic parts can replace metal parts with large saving in manufacturing cost and weight. The process can also serve functional needs, such as lamp reflectors or diffusion grids for overhead fluorescent lighting. The aluminium colour looks like chrome. Since it will not corrode like chrome plating on metal, it has replaced such parts as automobile grills and boat parts. Vacuum metallising on interior surfaces of computer or communication equipment provides a degree of radio frequency interference shielding.

SECTION VI

Special Topics

CHAPTER 21

Polymer Blends

INTRODUCTION

The commercial utility of polymers can be greatly extended through the use of blends and alloys. Some important examples are:

1. Impact polystyrene—a solution polymerised blends of polybutadiene and polystyrene.
2. ABS—A blend of an emulsion rubber-grafted with styrene/acrylonitrile—in a matrix of styrene/acrylonitrile.
3. Noryl®—A blend of impact polystyrene and polyphenylene oxide.
4. Impact polypropylene—Polyethylene dispersed in polypropylene.

A common theme of all these materials is improved impact strength, particularly at low temperatures. A second commonality is that they are all dispersion of a particulate rubbery phase in a polymer matrix. A third characteristics is the need for good adhesion between the phases.

Common means for creating such blends are emulsion polymerisation, precipitation polymerisation, solution blending and compounding. In the last three of these techniques, a nonlinear diffusion phenomenon known as spinodal decomposition plays an important role in determining the blends morphology. The polymers to be blended are dissolved in a good solvent to form a single phase solution. Then, the solvent is suddenly removed by flash devolatilisation.

With millisecond, the system goes from homogeneous, single phase to deep within the two-phase region. Phase separation is spontaneous and quick, although the separated domains may experience subsequent rippening.

Computer-aided design is being applied to polymer blends. Excellent progress has been made for the prediction of morphologies achieved by solution blending (compositional quenching) and good progress has been made for the case of conventional compounding (twin screw extrusion).

RUBBER—PLASTIC BLENDS

Dynamic vulcanisation has been used to prepare thermoplastic rubber compositions. *Fischer* used the process to prepare compositions containing in partially vulcanised rubber. It has since been found that improved, very strong elastomeric compositions of EPDM and polypropylene could be prepared by dynamic vulcanisation provided that the rubber was completely vulcanised.

The dynamic vulcanisation process for thermoplastic elastomers can be described as follows:

After sufficient melt-mixing of plastic and rubber, vulcanising agent are added. Vulcanisation of the rubber phase occurs as mixing continues. After removal from the mixer, the cooled blend can be chopped, extruded, pelletised, injection-moulded, etc. Such a composition is described as a dispersion of very small particles of vulcanised rubber in a thermoplastic resin matrix.

Dynamic vulcanisation gives the following improvements, in comparison with blends which have not been dynamically vulcanised: reduced set, improved ultimate properties, improved fatigue resistance, improved resistance to attack by hot oils, greater stability of melt-phase morphology, greater melt strength, and so on.

EPDM—Polyolefin Compositions

The dynamic vulcanisation of blends of EPDM rubber with polyolefins (PP or PE) has been described. The rubber-plastic proportions and the extents of vulcanisation were varied. In a few instances the rubber was first press cured and then ground to various particle sizes. The ground rubber particles were then mixed with molten polypropylene. It was found that the ultimate properties varied inversely with rubber particle size, as the smallest particle sizes of vulcanised rubber were obtained by dynamic vulcanisation.

Only a small amount of cross-link formation is required for a large improvement in tension set; however, tensile strength improves rather continuously as the cross-link density of the rubber phase is increased. Compositions can be vulcanised by accelerated sulphur, methylolphenolic materials (e.g., catalysed by $SnCl_2$), or other curatives.

As the concentration of the polyolefin resin increases the compositions become less like rubber and more like plastic. Modulus, hardness, tension set, and strength increases.

NBR—Nylon Compositions

Excellent elastomeric NBR-nylon compositions have also been prepared by dynamic vulcanisation during the melt-mixing of intimidate blends of NBR with various nylons. In this case, the effect of curatives was complicated by the fact that some nitrile rubbers tend to self-cure at temperatures of mixing. Sulphur, phenolic, maleimide, or peroxide curatives can be used. The thermoplastic elastomeric compositions prepared by the dynamic vulcanisation of nylon-NBR blends are highly resistant to hot oil. As in the case of the EPDM-polyolefin blends, increases in the amount of rubber in the composition reduce stiffness but increase resistance to permanent set.

Other Elastomer Compositions Prepared by Dynamic Vulcanisation

In addition to EPDM-polyolefin and NBR-nylon combinations, a large number of other rubber-plastic combinations have been used to prepare thermoplastic vulcanisates by dynamic vulcanisation.

The best compositions are prepared when the surface energies of the rubber and plastic material are matched, when the entanglement molecular length of the rubber molecule is small, and when the plastic material is crystalline. It is also necessary that neither the plastic nor the rubber decomposes in the presence of the other at temperatures required for melt-mixing. Also, in each case, a curing system appropriate for the rubber under the conditions of melt-mixing is required.

Technological Compatibilisation of NBR–Polyolefin Blends by Rubber–Plastic Graft Formation

Were it not for their gross mutual incompatibility, a combination of a polyolefin resin with NBR would produce a good oil-resistant thermoplastic elastomeric composition by dynamic vulcanisation. The presence of a small amount of a block co-polymer can, however, technologically compatibilise blends, provided that the block co-polymer contains blocks which are compatible with each of the blend components. The compatibilising block co-polymer can even form *in situ*; e.g., polypropylene can be modified by the action of maleic anhydride in the presence of a decomposition organic peroxide. If part of the polypropylene in a NBR-polypropylene blend is maleic-modified, and if part of the NBR is amine-terminated, then compatibilising amounts of a NBR-polypropylene block co-polymer can form *in situ* during melt-mixing. Such a composition, after dynamic vulcanisation (of the NBR), can exhibit strength-related properties almost as good as those of the more nearly thermodynamically compatible EPDM-polypropylene compositions.

Technological Applications

The lower cost of thermoplastic processing is the motivation for the development of thermoplastic elastomers; however, failure in the achievement of truly rubberlike properties has impeded the acceptance of thermoplastic elastomer technology. Nevertheless, relatively recently commercialised compositions based on polypropylene and completely vulcanised EPDM have many of the excellent properties of the polyurethane and co-polyester-type thermoplastic elastomers and even improved set and fatigue resistance. Applications of these materials can be listed as follows: caster wheels, convoluted bellows, diaphragms, gaskets, seals, tubing, mounts, bumpers, glazing seals, shields, suction cups, torque couplings, vibration isolators, plugs, connectors, rollers, oil-well injection lines, handles, grips, hose covers, vacuum tubbing, bushings, grommets, protective sleeves, shock isolators, ducts, various hoses (e.g, hydraulic, agricultural spray, paint spray, plant air-water, mine hose), wire and cable insulation and strain relief, jacketing.

Rubber-Rubber Blends

If all of the polymeric components of the blend are curable elastomers, the dynamic vulcanisation process can be used to produce new elastomeric, thermoset compositions. As stated earlier, dynamic vulcanisation is the vulcanising or cross-linking of one polymer during its molten-state mixing with another polymer or with other polymers. The polymers are first thoroughly mixed and then, during further mixing, one of the polymers is obliged to become cross-linked, whereas the remaining polymeric material remains uncross-linked. After completion of the dynamic vulcanisation process, such a mixture is still millable, shapable, and mouldable if the composition comprises enough uncross-linked polymer. The dynamically vulcanised blend can then be processed much in the same way that conventional millable, vulcanisable elastomers are processed. Additional curative for the uncured component or components of the dynamically vulcanised blend is added at a processing temperature below the temperature for facile vulcanisation. Other additives such as fillers, extenders, plasticisers, and antidegradants can also be added at this time. Then the final mix is vulcanised again, but this time with the material in its final form, e.g., in a mould.

The dynamic vulcanisation process can be accomplished in several ways:

1. At a temperature suitable for vulcanisation but in the absence of vulcanising ingredients or curatives for any of the polymers, the polymers are thoroughly mixed together to form an intimate blend. Then with continued mixing, curative (or curatives) for only one of the polymers is added. Vulcanisation of only one of the elastomeric polymers then occurs. The mixing and mastication are continued until completion of the cross-linking reaction.
2. The polymers can be thoroughly mixed together in the presence of a curing system for only one of the elastomers at a temperature below that which permits facile cross-linking within the processing

time frame. Then the temperature is increased, with continued mixing, until a high enough temperature is reached to permit cross-linking to occur within a reasonable length of time. Mixing is continued until the dynamic vulcanisation process is completed.

3. The polymers can be thoroughly mixed together in the presence of a curing system for only one elastomer at a temperature below that which permits cross-linking within the processing time frame. Then the blend is removed from the mixing equipment and stored for later use. The stored material (possibly in the form of milled strips or talc-coated pellets) is introduced into hot mixing-processing equipment, and again, with continued mixing and mastication, dynamic vulcanisation of only one of the blended elastomer polymers occurs while mixing is continued until the completion of the dynamic vulcanisation process.

It should be noted that in these three cases of dynamic vulcanisation, the polymers, polymer proportions, and curing ingredients must be selected such that only part of the blended polymeric components can be cross-linked during the dynamic vulcanisation process. After the dynamic vulcanisation process, the composition must contain enough uncross-linked polymer (raw rubber) to permit the masticative incorporation of additional curing ingredients and other additives and to permit formability before the final vulcanisation of the remaining raw rubber (in heated mould).

The dynamic vulcanisation of three types of rubber-rubber blends is considered here: dynamically vulcanised acrylated rubber (ACM) dispersed in nitrile rubber (NBR), dynamically vulcanised rubbery acrylate ester polymer (ACR) dispersed in NBR, and dynamically vulcanised ACR dispersed in fluoroelastomer (FKM). Both ACM and ACR are polymers of acrylated esters; however, ACM, the conventional acrylate rubber, is prepared from the usual acrylate esters (butyl, ethyl, ethoxyethyl, etc.) and at least one curing-site-containing monomer such as chloromethylbenzyl acrylate, glycidyl acrylate, or acrylic acid, whereas the molecules of ACR contain no functional groups in addition to the ester groups themselves.

Compositions of NBR blended with ACM have been prepared before by conventional techniques (without dynamic vulcanisation of the ACM phase); however, we expect that certain advantages will result from the dynamic vulcanisation of the ACM phase. At least in principle, it is possible to independently control the cross-linking or vulcanisation of each of the two elastomers.

The dynamic vulcanisation of ACM can be accomplished by a number of chemistries such as aminolysis of C–Cl bond of a curing site by using a di- or polyamino compound. ACM polymers are also vulcanised by using small amounts of a combination of soap and sulphur. These methods are not applicable to ACR polymers, which contain no functional groups for the conventional vulcanisation of ACM. The functional groups of the ACR polymers are only the ester groups themselves.

We have now applied a different chemistry to the dynamic vulcanisation of the ACR. Tin-catalysed ester interchange between the ACR polymer and a diol or polyol is particularly suited for dynamic vulcanisation, as the removal of the ACR-derived alcohol required for the completion of the interchange reaction can readily be accomplished in mixing equipment operated at elevated temperatures under reduced pressure or under the flow of an inert gas such as nitrogen. The chemistry is a variant of that used for the alcoholysis of vinyl acetate polymers. The ester-interchange technology can be applied to compositions of either NBR and ACR or FKM and ACR. It is especially applicable to the FKM compositions because the excellent thermal stability of FKM readily permits the use of the higher temperatures needed to best accomplished the ester-interchange dynamic vulcanisation.

It is excepted that there might be end-use-related advantages associated with the use of dynamic vulcanisation to cross-link a depressed phase. Some advantages could be related to phase morphology.

The dynamic vulcanisation process would force the ACM or ACR to be a particulated phase dispersed in the NBR or FKM. The morphology would be stabilised because the particles, having been cross-linked, would be unable to flow or coalesce. The resulting millable elastomeric alloy could then be reproducibly processed into uniform finished parts. The reactively large and highly shaped domains frequently associated with conventionally mixed blends can act as stress-concentrating flaws. It is hoped that this will be greatly reduced in the alloys prepared by dynamic vulcanisation.

It was indicated from experimental work that the dynamic vulcanisation of an ACM or of an ACR during its mixing with NBR or FKM, produces a new type of millable speciality elastomeric alloys. In the case of the ACR, dynamic vulcanisation by the application of the tin-catalysed ester-interchange reaction enables the cross-linking of the ACR polymer even though its molecules contain no special sites for cross-linking.

In the case of the NBR blends, dynamic vulcanisation of either dispersed ACM or ACR gives good mechanical properties and resistance to hot oil, but without the usual poor ozone resistance associated with NBR. In the case of the KFM-ACR dynamically vulcanised blends, compositions are produced which perform nearly as well as do compositions containing only the FKM elastomer. This could have important economic consequences, as in the case of the composition presented here, the FKM has been diluted with an equal volume of ACR which should cost a small fraction of the cost of the FKM elastomer.

Blends of Elastomers and Thermoplastics

General background about elastomeric alloys

Thermoplastics that contain plasticiser (such as plasticised PVC) can be regarded as a type of precursor or cheap version of TPE, although these are really elastified or impact-modified thermoplastics. Similar, but technically more effective results are obtained when thermoplastics are blended with non-cross-linked rubber. They can have the same effect as macro-molecular plasticisers but, by definition, they are not yet elastomers. This is the case, for instance, with pairs of polymers and blend ratios in which the remaining crystal structures of the thermoplastics guarantee the strength and the still non-cross-linked rubber segments provide the elastic properties. These must therefore be regarded as elastified or impact-modified thermoplastics. Such blends do not contain any elastomer, by definition, and hence cannot be strictly classified as TPE.

The elastifying effect becomes even more pronounced when rubber/thermoplastic blends of this have the rubber phase dynamically cross-linked *in situ*, making the mix into a true thermoplastic/elastomer blend, or when rubber molecules are grafted on to thermoplastic molecules, or when styrene triblock copolymers are blended with thermoplastics. Blends of these types are also called elastomeric alloys or true types of TPE made of elastomers and thermoplastics.

Elastomeric alloys, such as those made up of thermoplastics and cross-linked elastomers, whose properties depend on the level of cross-linking of the elastomer segments, display different behaviour. There are many possibilities when it comes to manufacturing alloys of this type. A number of the more frequently cited polymer pairs will be listed by way of example: NR/PP; EPDM/PP, EPDM/PS, EPDM/SAN, EPDM/PA 6, EPDM/PA 66; EVA/PS, EVA/SAN, EVA/PA 11; EVA/PVDC; NBR/PP, NBR/PS, NBR/SAN, NBR/PA 6 to 9; NBR/PVC. Recently also blends of SBS (SIS) and PP come on the market. These obviously lead to very different classes of Thermoplastic elastomers (TPE). This also explains the broad range of properties found in TPE based on elastomeric alloys, particularly when consideration is given to the type of bonding, i.e., grafting, covalent cross-linking and ionomer cross-linking. The class of elastomeric alloys thus covers a particularly wide hardness range.

In the soft range (Shore A hardnesses of 55 to 85), elastomeric alloys display good elastic properties and are capable of replacing covalent cross-linked elastomers within certain temperature limits in the same way as styrene triblock co-polymers. At higher hardness level they fill the gap between elastomers and thermoplastics, in competition with other classes of TPE, such as thermoplastic polyurethanes, co-polyesters and polyetheramides. At the top end of the hardness scale they constitute impact-modified thermoplastics.

The service temperatures of these TPE classes are obviously conditioned by the T_g values of the elastomers used and the melting point of the thermoplastics. At present there is no indication that TPE of this type can be used at temperatures higher than 100 to 120°C. The high melt temperature calls for processing on plastics processing machinery.

EPDM/PP-blends (TPE-O, resp. TPO)

Blends of elastomers and thermoplastics, particularly EPDM and PP, are used *inter alia* for bumpers on automobiles and for cable sheathing. Since it is possible to blend EPDM and PP, for example, in any ratio, there is theoretically a continuous spectrum from elastified PP to EPDM reinforced with thermoplastics. Where the PP component predominates, which is by far the chief filed of application, PP constitutes the continuous phase with uniformly and finely dispersed rubber segments. With a very high EPDM component, by contrast, the structure reverses and the material is a PP—reinforced EPDM. The properties of blends containing a large quantity of PP—the impact-resistant PP grades which are available under a large number of brand names—depend on the quantity of EPDM and on the uniformity and size of the micro-heterogeneous rubber phase. They have only a relatively low level of elasticity (above all, they have virtually no entropy elasticity), display a low elongation at break and high values for compression set. They have only a low-level elastomeric character in the service temperature range, making it difficult to classify them as TPEs in the true sense of the term.

In contrast, through dynamic cross-linking *in situ* of EPDM with PP or PP/PE-blends during mixing, elastomeric alloys (TPOs) can be made with rubber-elastic behaviour over a wide range of temperatures (–40°C to +125°C). The rubber-elastic behaviour of such elastomeric alloys is quite different from that of non-cross-linked EPDM/PP blends. The morphology of *in situ* cross-linked polymer blends differs from those of non-cross-linked ones. The distribution of rubber particles is much finer and, therefore, the heterogenous phase distribution is much better for the TPOs. The properties of the fully cross-linked elastomer particles do not change even at processing temperatures up to 180–225°C. In contrast, the melt temperature of the applied polyolefins influences the processing and application temperatures. The maximum application temperature can be expected to be in the range between 100°C and 125°C. The high melt temperatures of TPOs call for processing (e.g., calendering, extrusion, injection moulding etc.) thermoplastics processing machines.

In general, the properties of each individual TPO type depend, apart from EPDM and the polyolefine types, also on the blend ratios, the degree of cross-linking and kind of compounding of EPDM phases, and, of course, of the micro-heterogenous phase distribution of EPDM in polyolefines. The well-known TPO sortiments from various producers e.g., vary in each case in wide hardness ranges from approximately Shore A of 55 up to Shore D of 75 with good elastic behaviour of the softer ranges. These products can replace covalently cross-linked elastomers in a lot of applications and they fill the gap between elastomers and thermoplastics. They compete with TPUs, thermoplastic co-polyester and with TPE-As. In the highest hardness range they compete also with impact resistant thermoplastics.

The UV and ozone resistance of TPOs is excellent and they are resistant in most inorganic chemicals as well as in a lot of polar organic solvents like break liquids, but not in aliphatic and in aromatic solvent, in which they have great volume swell. TPOs also have good abrasion resistance and the electrical insulation properties are excellent. The compression set resistance between low minus degrees up to approximately 70°C is low and it will be enlarged above 100°C rapidly, coming at 125°C near to 100 per cent. The stress crack propagation as well as the migration of plasticisers and oils is low.

Under the additional application of primers finished TPO parts can be varnished with normal laquer systems. Because of the low density of TPOs, the short processing times under avoidance of vulcanisation processes, low rate of waste etc. TPO parts can be produced cheaper than those made out of EPDM.

NR/PP-blends (TPE-NR)

In the same way as described above compounded NR can also be blended with PP or PP/PE blends. As *in situ* cross-linking by peroxides during mixing cycles results in thermoplastic NR (TPE-NR). Blends of *Guayule* rubber with PP are also possible. Because the olefinic component TPE-NR is stiffer than normal NR elastomers and the stiffness increases of course with increasing polyolefin share. The hardness, the modulus of elasticity and the tensile strength increase too. The higher stiffness limits the low temperature applicability approximately –30°C and the application under heat is limited to approximately +70°C. At higher temperatures the compression set increases rapidly and the strength properties will be reduced drastically. TPE-NR has a high tensile strength at room temperature of approximately 20 MPa in flow direction and 17 MPa in the opposite direction.

The weather and ozone resistance of TPE-NR is much better than that of NR-elastomers and the dynamic properties are also reported to be quite good. The electrical insulation properties are excellent. The finished TPE-NR parts can be varnished with normal lacker systems. The applicability of TPE-NR (–30°C up to +70°C) is limited compared with that of NR elastomers. But because of the high stiffness TPE-NR can also be applied in such areas in which higher hardness is required, for instance in car body parts, shoes or boats parts etc. beside cables and a lot of other applications.

For commercial calculations the same arguments as for TPOs are valid. TPE-NR can be processed on rubber processing machinery, which is of advantages for the rubber industry. Therefore, TPE-NR will be an attractive material for technical as well as for commercial reasons. The products are too new yet for having a measurable market share.

NBR/PP-blends (TPE-NBR)

In the same way as described for TPOs, compounded NBR can also be blended with PP or PP/PE blends. And *in situ* cross-linking during the mixing cycle results in thermoplastic NBR (TPE-NBR). Micro-heterogenious phase distribution can be obtained by cross-linked NBR in PP, which can be processed like other TPEs. TPE-NBR is rubber-elastic in the application temperature range (–20°C up to 100°C), and the hardness range ranks on Shore A of 80 to Shore D of 50.

The good swelling resistance of TPE-NBR in automotive oils and fuels is combined with an excellent ozone resistance, contrary to covalently cross-linked NBR. Also the acid and alkali resistance is good. The compression set resistance is acceptable even at temperatures of 100°C. Therefore, there are some possibilities to exchange covalent cross-linked NBR against the cheaper processable TPE-NBR for the same applications. Cost savings up to 50 per cent, based on a finished part, have been mentioned for TPE-NBR versus covalently cross-linked NBR elastomers.

NBR/PVC thermoplastic blends

By blending NBR grades (mainly X-NBR) with PVC, another kind of thermoplastic NBR can be prepared. With increasing NBR share the blend gets stiffer. The hardness of the described products ranges from

Shore A of 67 up to 75 with tensile strength of 10 up to 15 and elongation of break of 300 per cent up to 470 per cent. The properties vary with the kind of NBR grade, their molecular weight, acrylonitrile content, carboxylating rate, curing rate, blend ratio etc. These products seem to be experimental at the time being.

Polyvinyl Chloride—Nitrile Rubber Blend

Blends of PVC and nitrile rubbers are outstanding additives in rubber compounding which substantially improve ozone resistance, flame retardant characteristics and over all physical properties. Properly formulated PVC/NBR compounds are comparable to products based on neoprene, hypalon rubbers.

Most of the PVC/NBR compounds are translucent, light straw colour which is a pre-fluxed homogenised polyblend manufactured under highly controlled shear conditions using special stabilisers, antioxidants, and other additives. PVC/NBR blends are well protected against degradation due to oxygen, ozone or heat and against gelling on storage.

Typical properties of PVC/NBR compounds:

Tests at 23°C	*50:50 Blends*	*30:70 Blends*
Specific gravity	1.13–1.15	1.01–1.1
Hardness (Shore A)	65–75	50–60
Raw Mooney viscosity (ML 1 + 4 at 100°C)	95–115	95–115

Advantages of PVC/NBR polyblend

PVC/NBR blend gives products similar to traditionally formulated NBR compounds, but with significant improvements in the following properties. (i) ozone resistance; (ii) plane retardancy; (iii) abrasion resistance; (iv) processability; and (v) gloss and finish characteristics, etc.

The overall improvement of ozone resistance is so good as compared to traditionally NBR compounds that former compounds are equivalent to neoprene compounds. The superior oil and solvent resistance of PVC/NBR compounds allow them even to substitute polysulphide compounds in selected applications. The overall characteristics of suitably formulated PVC/NBR blend also make it possible to substitute Hypalon type speciality rubber in limited applications.

Processing technique

PVC/NBR compounds can be conveniently milled, extruded, compression moulded, using traditional processing equipments for NBR.

Areas of application

Suitably formulated PVC/NBR compounds can find uses in petroleum hose and hose covers, LPG tubing and hoses, moulded goods, printing roll covers, textile coats and aprons, floor tiles, cable sheating, footwear, flame resistant conveyor belting, channel and weather resistant strips, etc.

THERMOPLASTIC BLENDS AND NANOCOMPOSITES OF DENDRITIC MACROMOLECULES

A wide range of properties of polybenzyl ether dendrimers, novel linear-dendrimer co-polymers or in physical blends with differing polymers have been examined for miscibility, free volume/density and rheological properties. In particular, the effects of covalent connectivity found in a co-polymer have been contrasted with combinations of the same materials physically blended. It is found that co-polymers appear to force miscibility between the dendrimer heads and poly(styrene) tails, whilst physical blends of similar components are immiscible. Some properties of more miscible dendrimer blends (with poly(phenyl methacrylate) are examined, as are the ability of these materials to diffuse into the slits of organically-treated nanoclays.

Dendrimers represent one of the most interesting polymer architectures currently being synthesised. They are highly branched materials in which the molecular architecture is more 'globular' in nature, than the entangled model of conventional linear polymers. Although they can be synthesised by a number of techniques, they have generally not been able to be produced in large quantities. Therefore much of the research in the past has focussed on the chemistry of their synthesis. Far less work has been reported on many physical properties or on their inclusion in blends with thermoplastics. Nanocomposite research is also one of the major areas of contemporary polymer research. The incorporation of layered silicates into thermoplastics and thermosets allows materials to be produced which show enhanced mechanical, thermal, barrier and fire retardant properties – all at very low concentrations. Little has been reported about the combination of dendrimers and layered silicate clays.

Experimental – Materials and Characterisation Techniques

The base dendrimers used in this work have the structure shown in Fig. 21.1, are made by convergent synthesis and are polybenzyl ether in nature, have varying generation number (3 to 5) and, with a hydroxyl at the core, are labelled G3, G4 and G5. In addition, to allow comparison with blends of polystyrene, hybrid dendritic-linear block copolymers were also used, as shown in Fig. 21.2 and essentially are a PS backbone emanating from a dendritic end group. All blends of the dendrimers were combined either with polystyrene of various molecular weights (ranging from MW 10,716 to 33,280, all obtained from Aldrich) or with poly(phenyl methacrylate) MW=165,230, DP=2.05 from Scientific Polymer Products. All blends were made by melt blending with the appropriate polymer for 5 minutes in a 2 – 4 gm scale in a mini-mixer. Nanocomposites were made by combining materials with an organically-modified bentonite (B34 – Rheox, USA) which is a dimethylditallow modified natural clay. The nanocomposites were made by the 'biscuit' technique in which samples (usually) of majority clay phase are made by mixing clay and polymer powder, hydrostatically compressing them at room temperature into a 'biscuit' and annealing in the oven at temperatures above the polymer T_g, to allow indiffusion for periods ranging from 5 mins to 4 hours. The techniques used to characterise materials ranged from differential scanning calorimeter (Perkin Elmer DSC7, scan 10°C/minute), DMTA (Perkin Elmer, 1 Hz, scan rate 5°C/minute), rheology (Rheometrics RDA II, 2 mm gap), density (Micromeritics Accupyc Gas Pycnometer) to positron annihilation lifetime spectroscopy (PALS). This latter technique is a radioactive technique that uses the lifetime of injected positrons to probe the free volume size scale within polymers on an angstrom level. Only a very few results from all these systems and techniques can be shown in this section.

Results

The polymers of generation 3, 4 and 5 shown in Fig. 21.1 were found to have glass transitions (from DSC) of 36, 42 and 44 °C, respectively, due to the longer dendrimer arms, clearly the effect plateauing at generation 4 – 5. The total free volume fraction from PALS measured at 22 °C was found to be very similar for all 3 materials, with a slight decrease in free volume size with increasing generation, possibly due to backfolding or burying of heads within the dendrimer. The bulk density from pycnometry was appropriately found to increase slightly with generation, in agreement with the free volume dependence. All materials showed essentially a Newtonian-behaviour in the melt (although there appeared to be some structure in the lowest generation 3 material, which could be removed on preheating).

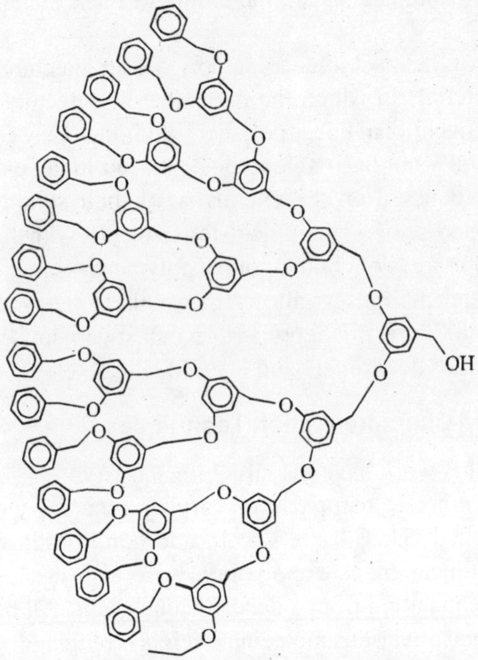

Fig. 21.1. 4th Generation polybenzyl ether dendron.

In order to determine the effects of connectivity or otherwise on dendrimer properties, all dendrimers were blended with PS in a range of concentrations, including that of the hybrid copolymers shown in Fig. 21.2. Thus a comparison in properties can be made in terms of physical mixtures (blends) and covalent connectivity (co-polymers). An example is shown for T_g and density of a generation 4 polymer, blended as a function of PS content (Fig. 21.3). [Note: the PS used in the blend has a similar molecular weight to that of the tail in the comparative copolymer]. It can be seen that despite the large number of aromatic groups on the periphery of the dendrimer, there is very low miscibility between the dendrimer and the PS (2 T_gs observed). In comparison, only 1 T_g was found (using DSC and DMTA) for the co-polymer, indicating that the dendritic head group is well dispersed within material being covalently coupled, and the T_g is lower than that of the PS phase alone. It can also be seen that the density of the blends lie just below the rule-of-mixtures line, indicating that a poor interface leads to excess free volume. Indeed the size of the free volume sites (data not shown here) from PALS indicates that there is a positive deviation in size from additivity, due to this extra free volume at the poor interface. This has been found by us in the past in purely thermoplastic blends to be a good indicator of miscibility (negative deviation in size of a blend implying miscibility, with an overshoot in free volume size indicative of immiscibility). The same system can be seen in terms of rheology in which the rheological properties (complex viscosity) are determined for a range of materials. All rheological measurements are undertaken at 150°C, except the G4 dendrimer which, because of its very low viscosity, is measured at 90°C. It can be seen in this case that the dendrimer physically blended with the PS (BG4PS1) is not as effective as the co-polymer of similar PS tail molecular weight (G4PS1) at reducing viscosity. This is interesting since, in the solid state, it appeared as though the T_g of the PS tail of the co-polymer was influenced in an intimate-mixing sense with the covalently-bonded head groups. It appears as though this intimacy of mixing is more efficient at reducing viscosity than the micron-sized domains found in the immiscible PS – G4 blends (for example). One application often stated for dendrimers is the possibility that they could be used as viscosity reducers. Interestingly the size of the free volume pores in the co-polymer G4PS1, is lower than the physical G4 + PS1 blend of the same composition (confirming closer packing in the solid state), whilst the bulk densities are the same (within experimental error).

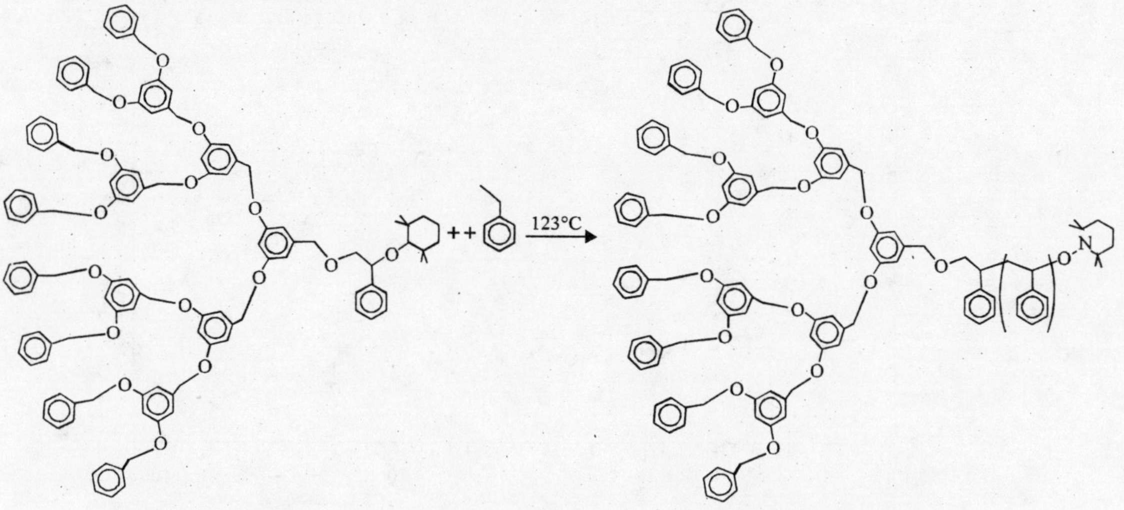

Fig. 21.2. Hybrid dendritic-linear polymer polymerisation of styrene initiated by 3rd.

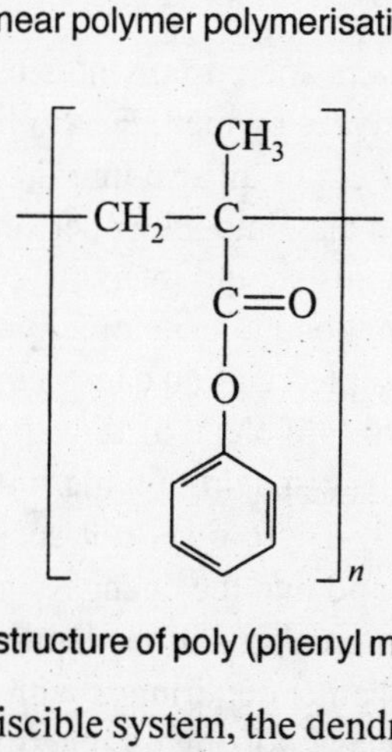

Fig. 21.3. Chemical structure of poly (phenyl methacrylate) PPMA.

In order to try and examine a more miscible system, the dendrimers were combined with poly(phenyl methacrylate) (PPMA) (Fig. 21.3) since it was proposed that its aromatic pendant groups would show favorable interactions with the terminal groups of the dendrimers. This was found to be the case. Interestingly, the generation of dendrimer was important as to whether blends were fully miscible, seen in DSC thermograms (as for G3 + PPMA) or partially-miscible (G4 + PPMA). Interestingly, in both blend systems, there was a slight positive deviation in density (not shown here) indicative of close packing, and this was slightly stronger for the more miscible G3 + PPMA blend. As expected, the size of the free volume showed the most negative undershoot in terms of blend composition for the G3 + PPMA blend, indicative of miscibility. It may be that the smaller G3 molecules, with more rigid arms, are better able to pack into the coiled PPMA thermoplastic, or perhaps have a greater number of interactions. The interaction is particularly evident in the melt, where addition of much lower viscosity dendrimers of all generations *increased* the viscosity of the neat PPMA (Fig. 21.4).

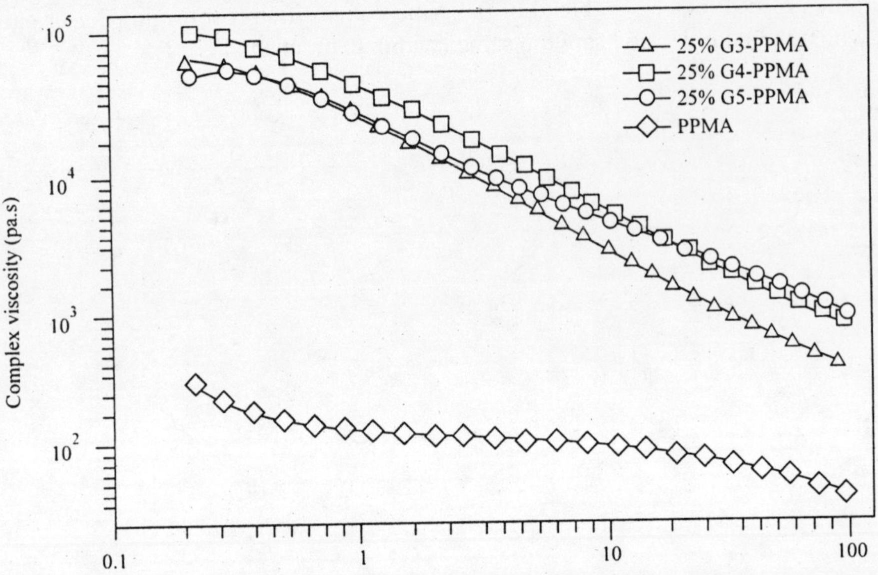

Fig. 21.4. Rheological properties of blends of 25 wt % of various generation dendrimers on PPMA.

The final modification of the dendrimers and co-polymers presented here is a brief investigation of their ability to intercalate into layered silicates by melt. Due to limited amount of sample, majority clay samples were used (allowing the dendrimers to diffuse into the alkyl ammonium-treated clay slits), as opposed to melt blending ca. 5 wt per cent clay into 95 wt per cent polymer. It had been mentioned that, thermodynamically, highly branched systems should intercalate well and this has not yet been reported (Note: a recent report has appeared of a nanocomposite of a hyperbranched material and nanoclays, by solution intercalation). It can be seen that intercalation can occur in these materials (all held for 4 hrs at 120°C). Fig. 21.5. Shows the XRD properties of blend of 15 wt per cent of various generation dendrimers with organo-treated nanoclays. The gallery spacing of the untreated clay is 31 angstroms, and this increases to about 45 angstroms for G3. There does not seem to be a clear trend with increasing generation number. In addition, it was found that the changes appeared to occur in the first 10 minutes, very rapid compared to other materials intercalated in the laboratory.

Multicomponent blends of polybenzyl ether dendrimers with other polymers have been investigated, as well as with organically-treated nanoclays – using a range of techniques which probe thermal, free volume and rheological properties. In particular, a comparison has been made between blends with poly(styrene) and dendrimer-poly(styrene) co-polymers. The linear-dendritic co-polymers appear to force a higher degree of miscibility than physical blends and this leads to a lower viscosity. It was found that dendrimers were miscible (for lower generations) or partially-miscible (for higher generations) in blends with poly(phenyl methacrylate), due to the range of secondary interactions, leading to more viscous materials. Dendrimers appeared able to intercalate a particular organo-ion treated nanoclay, although no dependence on dendrimer generation was observable at this stage.

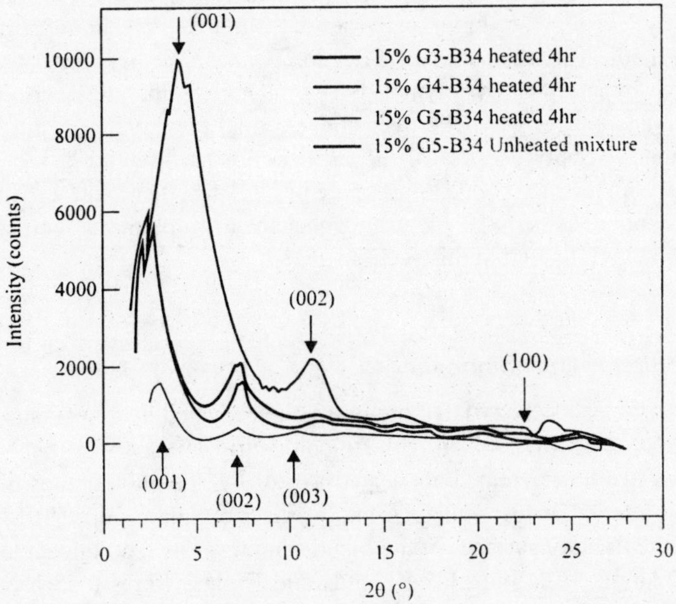

Fig. 21.5. XRD properties of blends of 15 wt % of various generation dendrimers with organo-treated nanoclays.

BLENDS OF DENDRITIC HYPERBRANCHED POLYMER AND HIGH TEMPERATURE CURING EPOXY FOR COMPOSITE APPLICATIONS

In this section, a series of blends of an epoxy functionalised hyperbranched polymer (HBP) and high temperature curing epoxy were investigated. The phase separation behaviour of the blends was investigated by a hot-stage microscopy and dynamic mechanical analysis. The HBP is miscible with epoxy at 120°C and undergoes phase separation during curing leading to the formation of a two-phase micro-structure. Finally the blends were investigated as matrix for glass fibre reinforced composite. It was found that the epoxy/HBP blends based composites shows significantly higher impact strength compared to unmodified epoxy based composites. Over the last two decades, the use of epoxy based fibre reinforced plastic (FRP) composites, in structural applications, has increased tremendously and this dramatic growth is expected to continue in the future. The composites possess many useful properties like high specific stiffness and strength, dimensional stability, adequate electrical properties and excellent corrosion resistance, which make them attractive to the automobile and aerospace industries. However, the thermoset based FRP composites are known to be highly susceptible to internal damage caused by a low velocity impact, which may lead to severe safety and reliability problems. One of the successful methods to enhance the toughness of FRP composite is matrix toughening. Epoxy matrixes are most effectively toughened by blending with reactive liquid rubbers like carboxyl-terminated co-polymer of butadiene and acrylonitrile (CTBN). The improvement in fracture toughness is generally achieved without a significant reduction of thermal and mechanical properties of the cross-linked epoxy resin.

However, the main deficiency of CTBN is the high level of unsaturation in their structure, which provides sites for degradation reaction in oxidative and high temperature environment. Secondly, there remains a possibility that traces of free acrylonitrile, which is carcinogenic, might exist and limits the use of these materials. It has been found that acrylate based liquid rubbers, can be successfully used as an alternative to CTBN. The modification of epoxy with linear liquid rubber results in significant increase in prepolymer viscosity. The increase in viscosity of the resin hinders resin impregnation into the fibres resulting in poor mechanical performance of FRP composites. We have investigated here an epoxy functionalised dendritic hyperbranched polymer (HBP) and high temperature curing blends as matrix for FRP composites. The advantages of HBPs over conventional toughening agents are that HBPs offer much lower prepolymer viscosity because of their spherical structure and lack of chain entanglement and stronger adhesion with the matrix due to presence of high density of surface functional groups. The phase separation behaviour of epoxy/HBP blend and mechanical properties of their composites reinforced with glass fibres will be investigated in this section.

Experimental

Characterization of blends and composite

The phase separation process of epoxy/HBP blend was investigated by a hot stage optical microscope. The epoxy/HBP blend (100/15) was mixed with stoichiometric amount of hardener and a drop of resin mixture taken between two transparent films heated to 120°C. The photographs of the resin film were taken at different time interval. A low voltage scanning electron microscope (SEM) (JEOL, JSM 840) was used to examine the fracture surfaces of the blend samples. The Izod impact test for the composite sample (thickness ~2.5mm) made out of the blends and E glass fibres (15 layers) was carried out according to ASTM D-256 using an impact tester (Tinius Olsen, Model 892 T).

Results and Discussions

The epoxy/HBP blend with stichiometric amount of DETDA was subjected to hot-stage optical microscopic analysis. It was found that HBP is initially miscible with epoxy at 120°C and undergo phase separation with the advancement of curing reaction. DMA studies shows two peaks; one at high temperature (200°C) for epoxy and other at subambient temperature c.a. –35°C for HBP (Fig. 21.6).

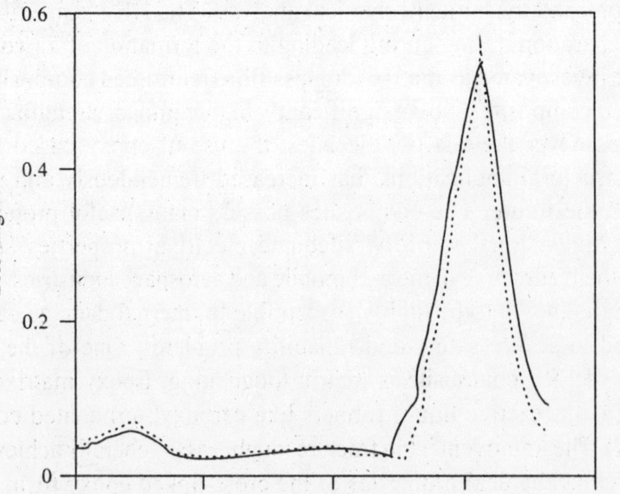

Fig. 21.6. Loss factor vs. plots of epoxy/HBP blends:(—) 100/0 (-----) 100/15.

This indicated phase separation between epoxy and HBP. The phase separation is also confirmed by SEM studies. The SEM microphotograph contains HB blend for 15 per cent. This is in the form of spherical HBP particles (3–5 μm) which are dispersed in continuous epoxy matric. Finally the glass fibre reinforced composites based on the blends were evaluated for the Izod impact strength. The composites based 15 per cent HBP containing blend shows impact strength 2000 J/m compared to 900 J/m for unmodified epoxy based composite. The increase in impact strength is believed to be due to increase in toughness of the matrix as a result of dissipation of mechanical energy by rubber cavitations and shear yielding.

ETHYLENE PROPYLENE DIENE TERPOLYMER (EPDM)- CHLOROPRENE RUBBER(CR) BLENDS AS ROCKET MOTOR INSULATION MATERIAL

EPDM rubber based rocket motor insulation has better thermal properties and density advantage. Polarity improvement of the rubber matrix by incorporation of Chloroprene rubber was experimented in order to improve the propellant bondability. EPDM -CR blends with lower CR proportions were evaluated for insulation characteristics. Physical properties of a chosen blend ratio were optimised by filler loading. A 6 per cent CR blended EPDM formulation containing 20-24phr silica resulted as a potential rocket motor insulation material.

Insulation, liner and inhibition systems are three important *inert* subsytems in a solid rocket motor that contribute to safety, structural integrity and combustion design. Though these do not contribute to thrust generation, reduction in the mass of the insulation would lead to increased payload capacity and lower launch costs.

Insulator is a thick elastomeric material inter-posed between the rocket case and propellant to provide thermal insulation of the case. While thermal, erosion and mechanical properties are the governing factors in insulation lay up design, insulator to propellant interface bonding characteristics become significant in material selection.

The currently used insulation systems have specific gravity values in the range 1.1 to 1.4. Reduction of density by 15 per cent or so would yield tremendous pay off , especially in gigantic solid booster motors where the insulation weight contribution is about 1.3 to 1.5 per cent of the propellant tonnage. Of the various elastomers used as insulator, nitrile rubber (NBR) is the most common base polymer. This is being replaced by EPDM in advanced solid boosters. EPDM as a polymer, has certain advantages over NBR like better thermal characteristics, better low temperature flexibility and a wider range of compounding choices. The same non polar nature that bestows these advantages poses potent drawbacks like inferior interface properties. This section also discusses how polarity improvement of the insulator matrix was attempted by blending EPDM with CR not so common a blend combination.

Experimental

Materials

Herlene H502 and Skyprene B30 were blended in a two roll mill. Precipitated silica (Pre-o-sil) powder of less than 1 micron average particle size was the reinforcing filler and a sulphur-methyl TUAD- NA 22 system was used for vulcanisation reaction. Varying Skyprene content in the blend, in proportions given by 3 per cent, 6 per cent, 8.7 per cent and 11.5 per cent, property variations were studied with respect to general requirements of a rocket motor insulator material. Filler content was then varied in one promising formulation among these. The objective was to develop a low density insulation system.

Methods

Cure characteristics of the formulations were determined using a moving Die Rheometer, RheoTECH-MD at 120°C and 145°C. Mechanical and interface properties were determined using a universal testing machine(Instron 4469) as per ASTM D 416.

Interface characteristics namely the tensile bond strength(TBS) and the wheel peel strength(PS), both indicating the rubber to propellant bond strength were measured at a cross head speed of 50 mm/min. Samples for TBS and PS were cut from a carton that simulated an integrated motor interior, prepared by lining the wall of a cylindrical plastic vessel with a properly abraded and dried rubber sheet and then by casting propellant slurry into it to bond with a polyurethane bonding agent that has been spray coated earlier onto the rubber sheet and preserved moisture free overnight. This carton after oven curing for few days can be cut into wheel peel discs and TBS blocks. Peel strength determination is done using wheel pull method, a 90° peel test.

Bond strength in the axial mode is determined through the tensile bond strength, where in, one cubic inch propellant blocks with insulator interface at one end and bonded axially to two metal blocks of similar dimensions are tested axially in the Instron. Density measurements were performed using the classical buoyancy technique. Thermal conductivity and specific heat capacity values were determined using a thermal conductivity detector (Holomatrix Inc. USA) at 80°C and a Differential Scanning Calorimeter (Mettler). Erosion rate was determined using a test set up (Plasma arc jet facility) in which high heat flux levels can be simulated. Cylindrical specimen of dia 10mm and length 100mm were used for erosion tests done at a heat flux of 300 W/sq. cm in air.

Results and Discussion

Characterisation of EPDM-CR blends with varying blend composition

Mechanical properties

In this series of experiments, keeping the formula specific gravity in the range 0.98-1.02, CR content in the blend was varied from 3.1 to 11.5 per cent. It is observed that the tensile strength increases and the elongation decreases with increase in CR per cent where as the modulus indices viz., M-100and M-300 do not show much variation. The increase in tensile strength with percentage of CR validates the fact that EPDM and CR tend to exist as distinct phases and increased continuity of the CR phase is a possible reason for this observation. The state of cure in the mixes has been steady as indicated by the both the modulus indices. The variation in ultimate elongation values along the series can be attributed to the behaviour pattern of incompatible blends. Perhaps this phenomena can be due to morphological changes occurring in the blended rubber matrix. The role of curative migration leading to morphological changes has been studied. The effect of blend ratio on mechanical properties are given in Table 21.1.

Table 21.1. Effect of blend ratio on mechanical properties

Properties	*CR content %*			
	3.1	6.0	8.7	11.5
Tensile strength (ksc)	36	43	56	112
Elongation %	700	610	550	980
M-100 (ksc)	12.3	12.8	12.7	12.8
M-300 (ksc)	20.5	20.7	20.3	19.6
Hardness(Shore A)	56	56	56	55

Thermal properties

Since the blends studied had the purpose of usefulness as solid rocket insulator, it was imperative that addition of CR should not increase thermal conductivity or decrease specific heat capacity drastically. However no major deviation of thermal properties were observed. With the specific heat values of the order of 0.52, and the thermal conductivities in the range of 3 to 4 $\times 10^{-4}$, the results are within their desirable limits of a good insulation system for solid propellant motors. Erosion rates of these formulations did not follow any specific trend with the more polar CR content.

Table 21.2 gives the thermal properties ie; specific heat, thermal conductivity and arc jet erosion rate for three typically chosen compositions (3 per cent, 6 per cent, 8.7 per cent).

Table 21.2. Thermal properties at different EPDM to CR blend ratios.

Properties	*CR content %*			
	3.1	6.0	8.7	11.5
Specific heat (cal/g/°c)	0.56	0.52	0.53	–
Thermal conductivity (cal/sec/cm/°c)	4.6×10^{-4}	4.4×10^{-4}	4.3×10^{-4}	–
Erosion rate (mm/sec)	0.16	0.18	0.19	–

Interface properties

Peel strength values more than 0.8 kg/cm and tensile bond strength more than 6 kg/cm^2 were obtained for all formulations except 3.1 per cent CR. These values are adequate for establishing a good bond between the propellant and the insulator. Peel values were lower than 0.4 kg/cm with EPDM rubber alone. Thus incorporation of CR into the non polar matrix has served the purpose.

The formulation with 6 per cent CR content that was found to be meeting the requirements was then chosen for filler variation studies in order to improve the physical properties.

Effect of variation in Pre-O-sil loading

As expected mechanical properties could be improved with filler loading and Table 21.3 gives quantitative data presented as a mean of six readings in each case.

Thus it is inferred that pre-o-sil loading at 20–24 phr would result in a formulation with specific gravity around 1 and acceptable physical properties.

Table 21.3. Effect of Pre-o-sil on mechanical properties.

Properties	*Filler loading (phr)*			
	0	10	30	60
Tensile strength (ksc)	33.8	43	134	151
Elongation %	420	610	1060	1120
M-100 (ksc)	9.1	12.8	15.4	18.0
M-300 (ksc)	14.1	20.7	25.4	31.6
Specific gravity	0.913	0.953	1.04	1.19
Hardness (shore A)	55	56	65	70

Within the limitations of the experiments the following conclusions were drawn. The insulation based on EPDM-CR blend with silica content as filler is a good potential candidate as an insulator for using in solid propellant motors. This class of insulation has good mechanical properties, low density, desirable thermal properties and interface properties.

A typical formulation with 6 per cent Chloroprene rubber loading and reinforced with 20 to 30 phr precipitated silica meets the specific requirements except calling for minor improvement in erosion resistance for which factors such as char retention rate also has to be considered and not merely the flame inhibition aspect. Scope of erosion resistance improvement has to be looked into by incorporation of short fibers like kevlar or melamine.

CHAPTER 22

Polymerisation Reaction Engineering: Some Current Problems

INTRODUCTION

Two problems have been described which give a glimpse of the exciting possibilities which exist in the area of polymerisation engineering. In the first of these, multiobjective optimal solutions have been obtained for an industrial, semibatch nylon 6 reactor using an adapted version of (nondominated sorting) genetic algorithm. This was made possible by the availability of a good model. The optimal solutions indicate significant improvements in the operation of the reactor, and these have been (partially) implemented in industry with some degree of success. Work on an experimental study leading to the implementation of an on-line optimising control scheme on methyl methacrylate (MMA) polymerisation, is described next. Some preliminary results have been presented on the bulk polymerisation of MMA which suggest that viscosity of the reaction mass can be used as a software sensor in control studies. The techniques and concepts developed are quite general in both the cases, and are applicable to other polymerising systems as well.

The field of polymer reaction engineering has matured considerably since *Flory* first postulated the equal-reactivity hypothesis (now associated with his name) in 1936. Several experimental studies followed this work and the rate constants of the main reactions were evaluated for important polymerisation using data obtained in small glass ampoules under isothermal conditions. In the late sixties, the emphasis shifted and the rate constants of important side reactions were evaluated using newly developed analytical techniques like HPLC, gas chromatography, spectroscopy (UV, NMR, IR), etc. In addition, several chemical engineers entered the field and started to study complex polymerisation, e.g., polypropylene, polyphenylene oxide, polymides, etc., as well as reported work on various aspects of heat and mass transfer. Studies on the optimisation of *ideal* reactors started to appear in the open literature to obtain interesting qualitative justifications of existing industrial reactors.

For example, it was found that one should use high concentrations of water initially in the polymerisation of ε-caprolactam, followed by removal of water to very low levels, in order to minimise the reaction/residence times for nylon 6 polymerisation while producing polymer of number average chain lengths, μ_n, of around 150. Only in the late eighties, studies were reported in the open literature on the simulation and optimisation of industrial polymerisation reactor with plant data appearing in slightly disguised forms. These include work of *Hamielec* and *Kiparissides* on LDPE reactors, as well as work on nylon 6 systems. This paved the way for several industrially relevant and useful studies. Some of the recent work on the simulation, optimisation and control of polymerisation reactors have been described here, to give a glimpse of the fascinating possibilities in these areas.

SIMULATION AND OPTIMISATION OF AN INDUSTRIAL NYLON 6 REACTOR

A semibatch nylon 6 reactor having a capacity of about 7000–10000 kg per batch, has been modeled and optimised. It is shown schematically in Fig. 22.1. The reactor is equipped with a low-speed anchor or ribbon agitator for mixing the reaction mass. A liquid mixture of caprolactam, C_1, and water, W, is heated by condensing vapours at a constant temperature, T_j, in the jacket. Polymerisation takes place in the liquid reaction mass as its temperature, T, goes above 220°C, and some water and caprolactam vapourise.

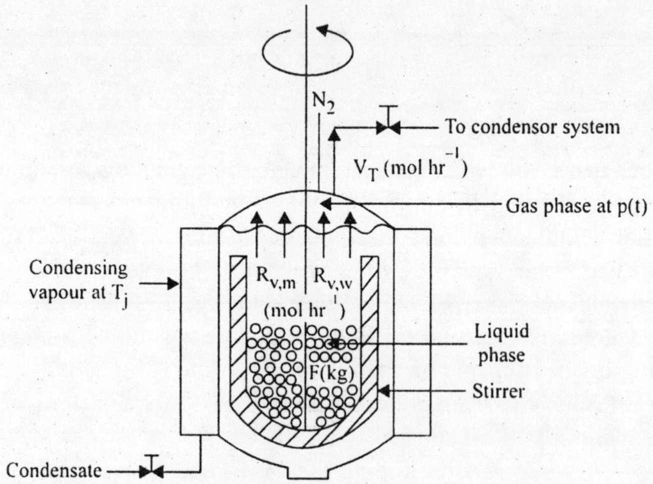

Fig. 22.1. Schematic representation of the industrial semibatch nylon 6 reactor.

The pressure, p, above the liquid reaction mixture is manipulated by a control valve which releases the vapour mixture of nitrogen (inert), monomer and water, to a condenser at a prescribed rate, V_T mol/hr. This helps maintain a desired pressure history, $p(t)$, in the reactor. The pressure histories, $p(t)$, used presently are shown in Fig. 22.2 (curves marked 'current') for three different industrial runs, corresponding to three different initial values of the water concentration, $[W]_0$ (2.52, 3.45 and 4.43 wt per cent). The following dimensionless variables are being used in this study:

$$\Pi \equiv (p - p_0)/(p_{\max,\text{ref}} - p_0); \quad \theta \equiv \frac{(T - T_{0,\text{ref}})}{T_{j,\text{ref}} - T_{0,\text{ref}}}; \tau \equiv t/t_{f,\text{ref}} \qquad \text{... (22.1)}$$

where p_0, the initial pressure, is 1 atm, $p_{\max,\text{ref}}$ is the maximum pressure used currently (indicated by subscript 'ref') while $t_{f,\text{ref}}$ is the reaction time used currently.

A detailed mathematical model is developed first, accounting for the heat transfer from the jacket fluid to the reaction mass, vapourisation of water and ε-caprolactam from the reaction mass, build up of pressure and maintenance of a prescribed pressure history above the liquid reaction mass in the reactor, change in the viscosity of the reaction mixture during polymerisation and its effect on the heat mass transfer rates, and the various reactions leading to the formation of the polymer and side products. The model involves five unknown parameters, four of which describe the activity coefficients of the monomer and water, and one which describes the inside film heat transfer coefficient. The values of these parameters have been obtained by curve-fitting one set ($[W]_0$ = 3.45 per cent) of industrial data (entire temperature history, pressure history in the initial period, and the final value of μ_n) using the Box complex technique. Fig. 22.3 (curves b) shows model results for the dimensionless temperature and pressure as a function of τ for $[W]_0$ = 3.45 per cent.

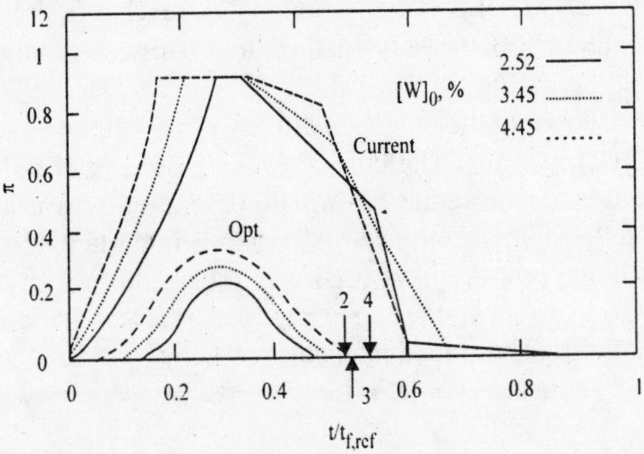

Fig. 22.2. Dimensionless pressure histories used currently to produce three grades of nylon 6, using different values of $[W]_0$. Optimal histories corresponding to points, O_i, in the Pareto optimal solutions also shown. Arrows indicate the values of $t_f/t_{f,ref}$ for the three $[W]_0$. Values 2,3 and 4 on the arrows indicate $[W]_0$ = 2.52, 3.45 and 4.43 per cent respectively.

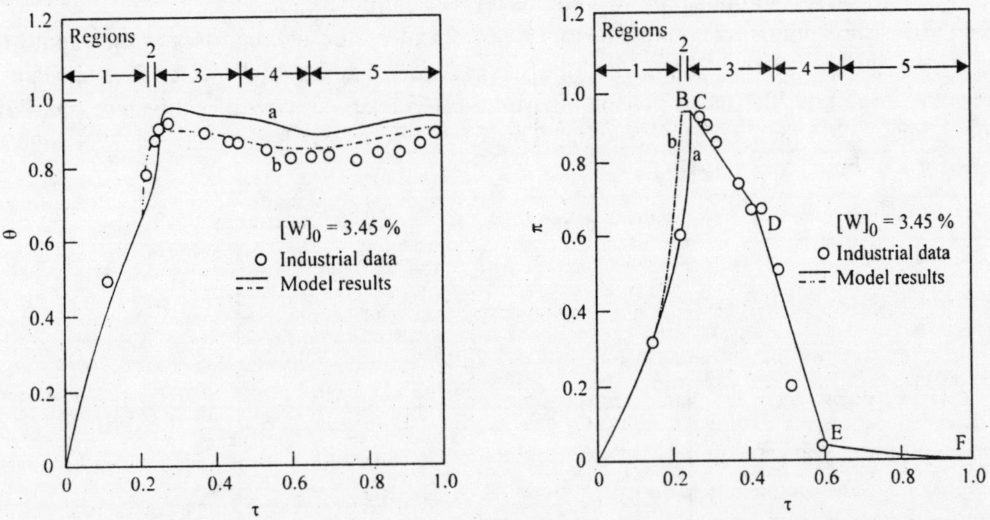

Fig. 22.3. Variation of the dimensionless temperature, θ, and pressure, π, with dimensionless time, τ, for $[W]_0$ = 3.45 %. Circles show industrial data. Curves (b) show model results using the best-fit values of the parameters.

Fairly good agreement between industrial data and simulation results is observed. The computed values of μ_n at $\tau = 1$ ($t = t_{f,ref}$) is 151.76, which agrees well with the industrial values of 152.5 for the product. These parameter values are then used to predict (without further tuning) the results for the other two values of $[W]_0$ for which industrial data is available. The agreement is found to be good. This confirms that the model accounts for most of the important physicochemical phenomena which play an important role in the reactor.

With a good reactor model available, the stage is set for optimising its performance. Two objective functions are selected for minimisation:

1. The dimensionless reaction time, $t_f/t_{f,ref}$, where the total reaction time, t_f, is determined by the time required for the degree of polymerisation, μ_n, to reach a desired value, $\mu_{n,ref}$ (selected as the value obtained currently in the industrial reactor). This ensures the quality of the product in terms of its physical properites.
2. The non-dimensional concentration of the undesirable cyclic dimer, C_2, in the polymer produced, i.e., we minimise $[C_2]_f/[C_2]_{f,ref}$. The cyclics cause problems in polymer processing and hot water extraction is used to remove them. Since this operation is expensive and energy-intensive, minimisation of the final cyclic dimer concentration is important.

While optimisation, the final monomer conversion, $x_{m,f}$, is constrained to lie at desired (current) value, $x_{m,ref}$. This ensures no increase in downstream recycling costs. The optimisation problem being solved in the present study is, thus, described mathematically by

$$\underset{u(t)}{\text{Min}} \quad I \equiv [t_f/t_{f,ref},\ [C_2]_f/[C_2]_{f,ref}] \qquad \text{... (22.2a)}$$

subject to

$$x_{m,f} = x_{m,ref} \qquad \text{... (22.2b)}$$

$$\mu_{n,f} = \mu_{n,ref} \text{ (decides } t_f) \qquad \text{... (22.2c)}$$

Mass and energy balance equations ... (22.2d)

Any additional physical limits on the control variables ... (22.2e)

where $u(t)$ is the optimising variable history. The control vector, u, comprises of the vapour-release rate, $V_T(t)$, which is a *function of time*, and the jacket fluid temperature, T_j, which is a constant *value*.

The solution of Eq. 22.2 using genetic algorithm (GA) leads to a set of equivalent (non-inferior) optimum points, rather than a unique solution. This is referred to as a Pareto set. Fig. 22.4 shows these solutions for the three values of $[W]_0$.

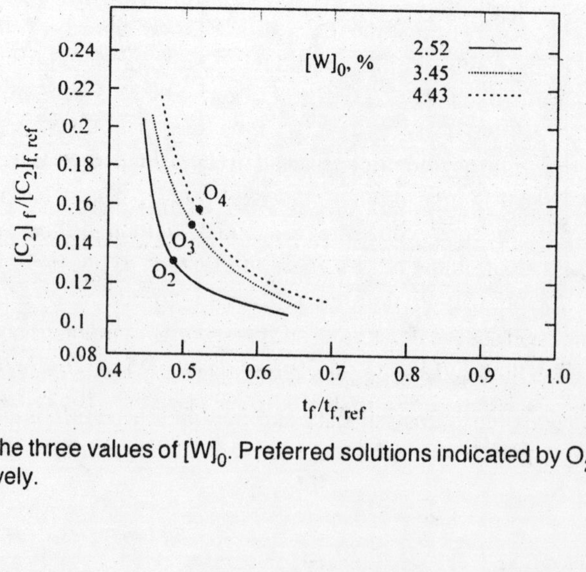

Fig. 22.4. Pareto sets for the three values of $[W]_0$. Preferred solutions indicated by O_2, O_3, and O_4 for $[W]_0$ 2.52, 3.45 and 4.43%, respectively.

A characteristic of any two points on a Pareto set is that when one goes from one to the other, this leads to an upgradation of one (or more) objective function, but simultaneously leads to a worsening of the other (or of at least one other) objective function. It is obvious that the Pareto set generated helps channelise the thinking of the decision-maker since it narrows down his choices substantially.

Each point on the Pareto sets in Fig. 22.4 corresponds to a different V_T history as well as T_j. Fig. 22.5 shows the optimal values of T_j corresponding to the Paretos for all three values of $[W]_0$, while Fig. 22.2 shows the pressure histories (corresponding to the optimal V_T histories) for three points on the paretos of Fig. 22.4, indicated by points $O_2 - O_4$ (called preferred solutions since they are nearest to the point of intersection of the asymptotes). It is found that the optimal solutions are far superior to the current operation. The optimal solutions have been implemented on the industrial reactor and have lead to substantial improvements in the operation.

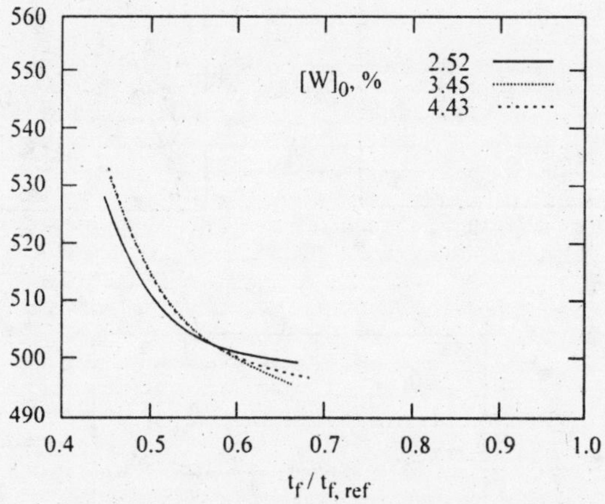

Fig. 22.5. Jacket fluid temperatures corresponding to different points on the three Paretos shown in Fig. 22.4.

ON-LINE OPTIMISING CONTROL OF FREE-RADICAL BULK POLYMERISATIONS

We now review some of recent work on the opimising control of addition polymerisations which exhibit (micro) diffusional limitations [referred to as the cage, gel (or Trommsdorff) and glass effects]. Even though we have taken a sample system, homopolymerisation of methyl methacrylate to study this, the techniques and concepts are applicable to several other systems, e.g., polystyrene (PS), copolymers, etc. The first step is to obtain a good model describing the kinetics of free radical polymerisations. Table 22.1 lists the three most important reactions taking place.

Table 22.1. Kinetic scheme of polymerisation of MMA.

Initiation	$I \xrightarrow{(f)k_d} 2R$
	$R + M \xrightarrow{k_i = k_p} P_1$
Propagation	$P_n + M \xrightarrow{k_p} P_{n+1}$
Termination by disproportionation	$P_n + P_m \xrightarrow{k_{td}} D_n + D_m$

The gel or Trommsdorff effect represents the effect of decreasing diffusivity of macroradicals, P_n (due to increasing viscosity of the reaction mass), on the termination rate constant, k_{td}. It is manifested as a sudden increase in the rate of polymerisation and the weight-average molecular weight, M_w, after some time (this necessitates the use of model-based predictive control for reactors). The glass and cage effects represent the decrease of k_p and the initiator efficiency, f, respectively, and are manifested experimentally by the observation that the reaction stops short of complete monometer conversion (Fig. 22.6).

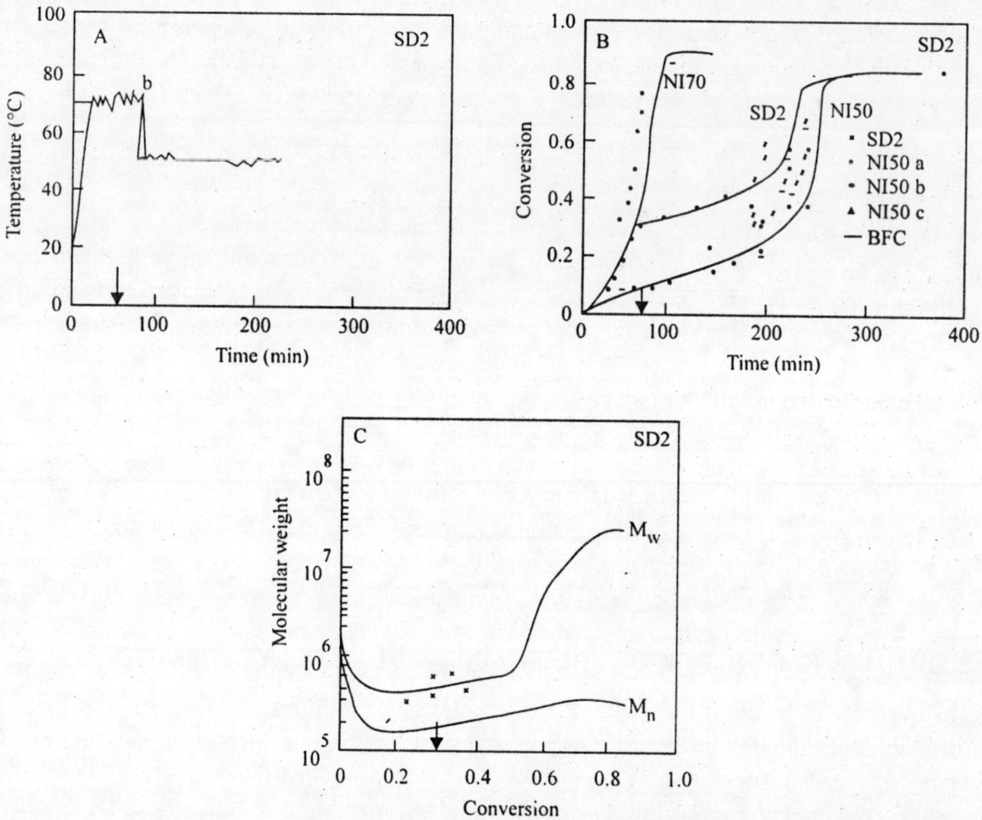

Fig. 22.6. Experimental results on one step decrease (SD2) run for MMA-AIBN. Theoretical predictions using the measured temperature history in the model are shown by solid curves (BFC). $[I]_0 = 25.8$ mol/m^3. Arrows indicate the time at which the step change is effected. Near-isothermal (NI50 and NI70) results for x_m are shown for comparison.

The model so developed is then tested against experimental data. Bulk polymerisation of MMA is carried out in a PC-interfaced, one litre, stainless steel, Parr® reactor under near-isothermal and non-isothermal conditions mimicking industrial operations. Samples of the reaction mass are withdrawn periodically, quenched and analysed for monomer conversion (gravimetry) and for average molecular weight, M_v (dilute solution viscometry). One set of results is shown in Fig. 22.6. Reasonable accord between model predictions (using parameter values obtained from isothermal results on small ampoules–(see BFC curves) and experimental data are observed. Similar agreement has been obtained for several other non-isothermal cases as well as for cases involving step-increases in the initiator concentration. These studies validate our model.

An important problem in the control of free radical bulk polymerisation is the estimation of the state of the system at any time. Densitometers and gel permeation chromatography have been used in some experimental studies involving solution polymerisations for estimating $x_m(t)$ and $\mu_n(t)$ [or $\mu_w(t)$. However, these experimental techniques cannot be used conveniently for bulk polymerisations, and inferential state estimation techniques (also called software sensors) have to be used.

One possible strategy is to use experimental values of the viscosity, $\eta(t)$, of the reaction mass for inferential state estimation. The feasibility of such a scheme for model-based inferential state estimation using $T(t)$ and $\eta(t)$, has been demonstrated recently using theoretically generated 'experimental' data (with addition of noise to model-generated values).

A few 'experimental' points in the estimation horizon, $t_1 - N_1\,(\Delta t) \le t \le t_1$, at any time, t_1, are used to estimate the parameters in the model.

These parameters can then be used to predict the polymerisation at future time, $t_1 \le t \le t_1 + N2(\Delta t)$. It is found that if the estimation horizon lies much before the gel-effect, the short-term ($N_2 = 5$) predictions are quite good.

However, *long-term* predictions ($N_2 >> 5$) using the present technique could be either good or poor, depending on the points used in the estimation horizon. Long range predictions can be significantly improved if a large estimation horizon (higher N_1) is used. These results indicate the 'experimental' data on $T(t)$ and $\eta(t)$ can, indeed, be used with a model to estimate the state of the system [$x_m(t)$, $\mu_n(t)$, $\mu_w(t)$]. In fact, our studies indicate that the reverse prediction of x_m and μ_n using T and η is unique, provided a good model is used.

A reactor-viscometer assembly has been set up recently in a laboratory which can be used to measure, on-line, the viscosity of the MMA-PMMA reaction mixture polymerised under any general temperature history. Samples can also be taken out at different times and analysed for monomer conversion and average molecular weight. Some preliminary experimental data on the viscosity of the reaction mass, monomer conversion and average molecular weight, have been collected under isothermal conditions. These are shown in Fig. 22.7.

Good reproducibility is observed. Also, the monomer conversions obtained on this assembly match well with those taken by *Balke* and *Hamielec* in small glass ampoules. These results can be used to develop correlations between viscosity of the reaction mass and conversion, average molecular weight and temperature. Also, the reactor-viscometer assembly can easily be used for experimental on-line optimising control studies.

A possible scheme for on-line optimising control can now be described. The polymerisation process could be started using an off-line optimised temperature trajectory. However, the trajectory can usually be followed only approximately because of experimental limitations (or disturbances entering the process, or material being added to or removed from the reactor, or equipment failure occurring temporarily). All these call for modifications of the remainder of the temperature trajectory.

Experimental data on $T(t)$ and $\eta(t)$ up to time, t_1, could be used to calculate the model parameters and, thus, to estimate the state of the system.

A fresh optimisation (on-line) could then be carried out, starting with the state at t_1, as estimate above. This trajectory could then be implemented until the end of polymerisation or till a new trajectory is required to be found.

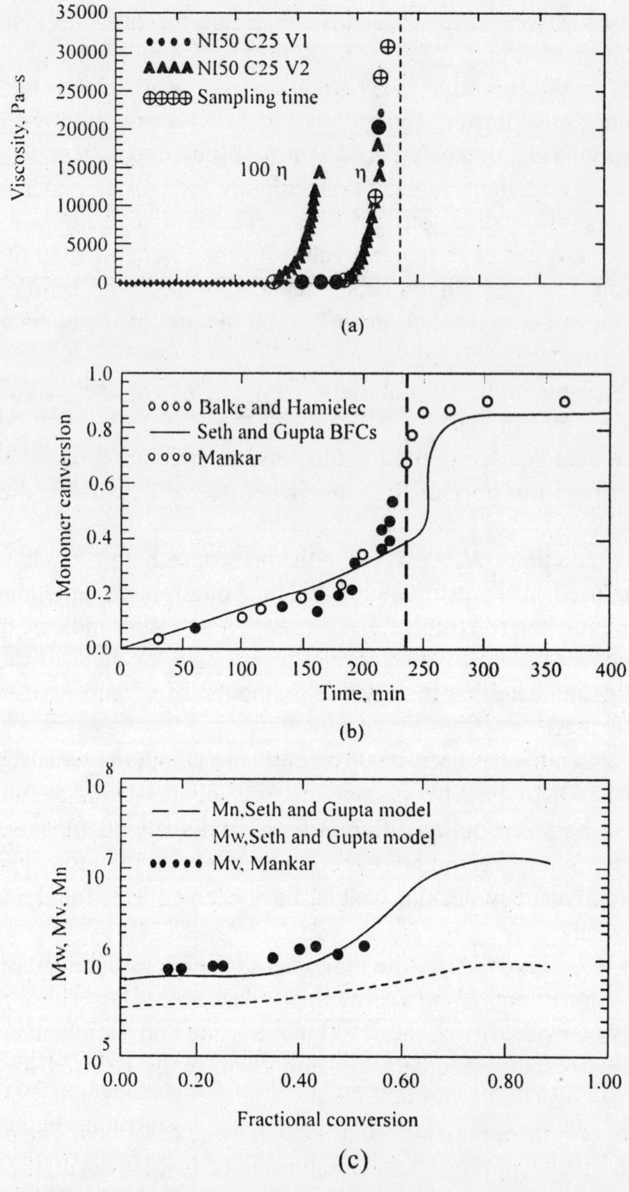

Fig. 22.7. (a) Experimental viscosity data for duplicate runs N150 C25 V1 AND N150 C25 V2 at 50°C using $[I]_0 = 25.48$ mol/m^3. (b), (c): Data points of Balke and Hamielec (o) as well as samples taken from the viscometer (⊕) also shown. Solid curves represent model results using best-fit correlations (BFCs).

NOMENCLATURE

Ci	Caprolactam (i = 1) and cyclic dimer (i = 2)
D_n	Dead polymer molecule having n repeating units
f	Initiator efficiency

I	Initiator
$\boldsymbol{I}$	Vector of objective functions, I_m; $m = 1, 2$
k_d, k_i, k_p, k_{td}	Rate constants for the reactions in Table 22.1 at any time (s^{-1} or $m^3\ mol^{-1}\ s^{-1}$)
M	Monomer
M_n	Number average molecular weight [= $(MW_m)\ (\lambda + \mu_1)/(\lambda_0 + \mu_0)$] (kg mol^{-1})
M_v	Average molecular weight as measured by intrinsic viscosity (kg mol^{-1})
M_w	Weight average molecular weight [= $(MW_m)\ (\lambda_2 + \mu_2)/(\lambda_0 + \mu_1)$] (kg mol^{-1})
N_1	Number of points in the curve fit or estimation horizon
N_2	Number of points in the prediction horizon
O_i	Preferred solutions for *i*th case ($i = 2, 3, 4$ indicating $[W]_0 = 2.52, 3.45$ and 4.43 %)
p	Pressure (kPa or atm)
P_n	Growing polymer radical having n repeat units
R	Primary radical
t	Time (s)
t_1	Time at which state estimation is made (s)
T	Temperature (K)
$\boldsymbol{u}$	Vector of control variables
V_T	Vapour realease rate (mol/hr)
x_m	Monomer conversion (molar)

Greek Letters

θ_f, θ_t, θ_p	Adjustable parameters in the model for cage, gel and glass effects, respectively ($m^3\ mol^{-1}$, s, s)
λ_k	*k*th ($k = 0, 1, 2, ...$) moment of live (P_n) polymer radicals $\left[\equiv \sum_{n=1}^{\infty} n^k P_n\right]$
μ_k	*k*th ($k = 0, 1, 2, ...$) moment of dead (D_n) polymer chains $\left[\equiv \sum_{n=1}^{\infty} n^k D_n\right]$
μ_n	Number average chain length at time t [= $(\lambda_1 + \mu_1)/(\lambda_0 + \mu_0)$]
η	Viscosity of the reaction mass (Pa s)

Subscripts/Superscripts

d	Desired value
f	Final (value for the product)
j	Jacket
o	Feed/initial conditions
ref	Reference (value used in industrial reactor currently)
t_f	Value at $t = t_f$

CAGE EFFECT

Many polymerisation reactions are initiated by radical fragments formed from thermal or photochemical dissociation of special molecules. When a molecule does acquire enough energy to dissociate a bond, the product radical will necessarily be separating from each other at the conclusion of the primary dissociation process. If the dissociation takes place in a dilute gas, the mean free path is so much greater than molecular dimensions that the specific pair of radicals will never encounter each other again after they separate.

If the dissociation takes place in liquid phase, the surrounding molecules will interfere with escape of the fragments and make it probable that the product radical will recombine with each other rather than react with other species. *Rabinowitch* first called attention to this situation and referred to the fragments as being tapped in a 'cage'. He called the general process *primary recombination. Zimmerman* and *Noyes* distinguished between primary recombination of radicals that never become separated by as much as a molecular diameter and *secondary recombination* of radicals that attain separations of this magnitude but subsequently diffuse together. These two processes are not easily distinguishable by experiment, and *Hamill* has proposed the term *geminate recombination* for the total recombination of original pairs of fragments by both processes.

Because cage effects interfere with separation of fragments from dissociation they reduce the efficiency with which radicals can be produced for processes such as initiation of polymerisation. Because the radicals are produced in pairs rather than singly, their subsequent reactions with scavengers and with each other exhibit some unconventional kinetic features. Both of these effects are discussed below.

Efficiency of Dissociation

A very simplified model of the dissociation process has been developed by *Noyes*. Let two spherical particles each of mass m and radius a separate with each having initial speed u_0. If the solvent can be regarded as a hydrodynamic continuum of viscosity η, each sphere will traverse a distance $mu_0/6\pi\eta a$ before it is brought to rest by frictional forces.

If two particles having a finite initial separation undergo random diffusion in a three-dimensional continuum, the probability is less than unity that they will ever encounter each other. For initial separations much larger than the dimensions of the particles this probability is inversely proportional to the initial separation, and numerical calculations have indicated that such a relationship is approximately valid for shorter initial separations. If the particles are formed from a compound with bond energy ε by photochemical dissociation with light of frequency v, and if ϕ is the probability that they will escape the cage and be available for other reactions, the relationship shown in equation 22.3 exists.

$$\frac{1}{1-\phi} = 1 + \frac{\sqrt{m(hv-\varepsilon)}}{6\pi\eta a^2} \quad \text{... (22.3)}$$

If the particles are formed by a thermal dissociation but are initially separated by a distance s (such as the radicals from dissociation of an azo compound R–N=N–R), the even simpler form is given in equation 22.4.

$$\phi = 1 - 2a/s \quad \text{... (22.4)}$$

Of course, the derivations of these equations assume that the particles are virtually sure to recombine if they encounter each other. The equations given here neglect a multiplicative term used in a previous treatment because it is usually too near unity to be of practical significance.

Equation 22.3 expresses in quantitative form the obvious conclusion that per quantum absorbed photochemical dissociation will be more efficient with light of short wavelength and in solvents of low viscosity.

Since real solvents consist of discrete molecules, the assumption of a hydrodynamic continuum must be a great oversimplification. The predictions of the simple theory can be tested quantitatively because all of the quantities in equation 22.3 are measurable or estimable for some systems. Table 22.2 illustrates the agreement of theory and experiment for photochemical dissociation of iodine at 25°C in a number of hydrocarbon solvents of different viscosities.

Table 22.2. Quantum yields for Iodine dissociation in hydrocarbons.

Solvent viscosity, η, poise	*Wavelength, λ, Å*	*Quantum yield, ϕ* Calculated	Observed
0.0029	4074	0.54	0.83
	4358	0.52	0.66
	5461	0.46	0.46
	5790	0.44	0.36
	6430	0.40	0.14
	7350	0.31	0.11
0.017	4358	0.145	0.18
0.54	4358	0.0052	0.086
0.8	4358	0.0035	0.048
1.8	4358	0.0016	0.038
3.8	4358	0.0007	0.036

Table 22.2 shows that equation 22.3 clearly predicts quantum yields of the right order of magnitude except in very viscous solvents and at long wavelengths where the particles have little excess kinetic energy; these are the conditions where failure of the theory is most to be expected. Increasing viscosity beyond a few tenths of a poise does not decrease quantum yield much further because the dissociation process interacts only with segments of the large molecules present. In solvent of low viscosity, equation 22.3 predicts too great a quantum yield at long wavelengths and perhaps gives a somewhat low prediction at very short wavelengths. Some specific solvent effects are indicated also because quantum yields in chlorinated solvents are lower than those in hydrocarbon solvents of comparable viscosity.

Although equation 22.3 must be used with some caution, it has considerable value for estimating the quantum efficiency for photochemical initiation of radical polymerisations.

Equation 22.4 is gross simplification that predicts no solvent effect on the efficiency of initiation by thermal dissociation of an azonitrile. The studies of *Hammond* and *Boozer* clearly demonstrate that these efficiencies are distinctly influenced by changes in solvent.

Kinetics of Scavenging

For purposes of this topic, any species in a solution is called a *scavenger* if it can react with radicals formed in a dissociation. The scavenger may be a polymerisable monomer whose reaction is desired, or it may be an impurity whose presence is undesirable.

If a scavenger is present at concentrations less than 0.01 mole/l, it can scarcely react with many radicals except those that have already escaped geminate recombination. However, at higher concentrations a very reactive scavenger can compete with the recombination process.

A mathematical treatment of this situation has been developed by *Noyes*. Let two radicals be produced with a small initial separation as the result of thermal or photochemical dissociation, and let $h(t)dt$ be the probability that they will react with each other between time t and $t + dt$ afterward. For random diffusion in a solvent continuum, the theory of random flights predicts the relationship shown in equation 22.5, where a is a parameter having dimensions time$^{1/2}$ and β' is the total probability that the pair will ever react with each other.

$$h(t) = \frac{ae^{-\pi a^2/\beta'^2 t}}{t^{3/2}} \qquad \text{... (22.5)}$$

Table 22.3. Examples of cage effects.

Substance decomposing	*Method of decomposition*	*Solvent*	*Temperature, °C*	*% Geminate reaction*
$(CH_3COO)_2$	Thermal	iso-C_8H_{18}	65	24
$(C_6H_5(CH_2)_4COO)_2$	Thermal	CCl_4	55	55
$(CH_3N{=})_2$	Photochemical	iso-C_8H_{18}	6	80
			25	77
			50	73
			75	66
			95	62
$(CF_3N{=})_2$	Photochemical	iso-C_8H_{18}	65	27
$[(CH_3)_2C(CN)N{=}]_2$	Thermal	CCl_4	62	56
		C_6H_6	62	39
		C_6H_5Cl	62	40
		$C_6H_5NO_2$	62	25
		CH_3NO_2	62	25
		Br_2	62	0
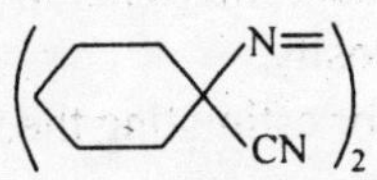	Thermal	C_6H_5Cl	80	39
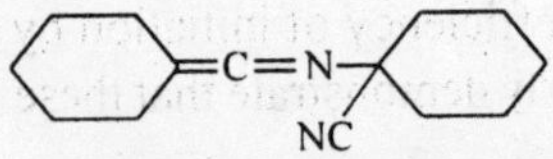	Thermal	C_6H_5Cl	80	73

If the solution contains a scavenger at low concentration, it may react with all radicals that escape geminate recombination but yet have negligible chance of reacting with radicals that would not have escaped the cage in which they were formed. Under such conditions, the amount of reaction with the scavenger may be virtually independent of scavenger concentration over a considerable range.

If the scavenger is more concentrated, it may react with radicals that would otherwise have undergone geminate recombination. The competition between scavenging and geminate recombination is complicated both because of the form of equation 22.5 and because the scavenging reaction cannot be described by a conventional rate constant that is independent of time. However, the treatment predicts that, to a first approximation, the scavenging of radicals that would otherwise have undergone geminate recombination is proportional to the square root of scavenger concentration. A similar conclusion was reached independently by *Hamill* and *Williams*.

At still higher concentrations of scavenger, the additional scavenging should increase more rapidly than the square root of the concentration of scavenger, but the effect should approach saturation when most of the geminate radicals are being scavenged.

The degree of validity of this interpretation is still somewhat observed. Many Scientists have observed situations in which efficiency of radical productions is less than unity yet is independent of scavenger concentration over a wide range; these systems obviously involve negligible competition with geminate recombination.

At the other extreme, *Trapp* and *Hammond* have observed that liquid bromine is such an efficient scavenger that it can trap all of the radicals produced by thermal dissociation of 2,2′-azobisisobutyronitrile; geminate recombination is completely suppressed in this unusual solvent.

Observations on several other systems can best be interpreted if scavengers are indeed competing with geminate recombination. The competition often does seem to vary as the square root of scavenger concentration, but *Waits* and *Hammond* have established in their elegant experiments that the first power of that concentrations gives a better fit to their data over a wide range of conditions. They interpret their results in terms of a model based on the probability that a radical will have a scavenger molecule for a nearest neighbour at the moment it is formed. Although the best theoretical description of the effect is still somewhat in doubt, there is general agreement that scavengers at concentrations above 0.01M can compete with geminate recombination, and this competition seems to vary as the square root of scavenger concentration when only a few per cent of the geminate radicals are being scavenged.

Cage Effects in Polymerisation Initiators

Polymerisation of many monomers can be initiated by species that dissociate to produce pairs of radicals. These radicals may then undergo geminate reactions that reduce their efficiency at initiating the desired polymerisation.

Azoisobutyronitrile (AIBN), $(CH_3)_2C(CN)–N=N–C(CN)(CH_3)_2$, is a commonly used initiator that exhibits a significant cage effect even though the two radicals are separated by a molecule of nitrogen at the time they are formed. Geminate recombination leads chiefly to tetramethylsuccinonitrile, $(CH_3)_2C(CN)–C(CN)(CN)(CH_3)_2$, but a significant amount of ketenimine, $(CH_3)_2C=C=N–C(CN)(CH_3)_2$, is also formed. This compound dissociates into radicals at a rate comparable to that of the original azonitrile. Because these radicals are formed without a nitrogen molecule to separate them, a higher percentage undergo geminate recombination than is the case with the azonitrile.

Diacyl peroxides, $(RCOO)_2$, are also used as initiators for radical polymerisations. The oxygen-oxygen bond breaks first, but the resulting $RCO_2•$ radical readily loses carbon dioxide. The products of geminate reaction depend upon the nature of the R group. *Herk*, *Feld*, and *Szwarc* found no ethane when diacetyl peroxide ($R = CH_3$) was decomposed in the presence of gas-phase scavengers, but considerable ethane was formed by geminate reaction in solution. They also observed geminate formation of methyl acetate, CH_3COOCH_3, indicating a finite life for the acetate radical. Since they were unable to scavenge this radical with any reagent not present in the cage, they estimated that the reaction $CH_3CO_2• \longrightarrow CH_3• + CO_2$ has a rate constant of about 10^{10} sec^{-1} at 65°C and that the activation energy is about 5 kcal/mole. On the other hand, *DeTar* and *Lamb* found that moist iodine could scavenge a considerable number of $RCO_2•$ radicals that escaped geminate reaction during the decomposition of δ-phenylvaleryl peroxide (R = $C_6H_5(CH_2)_4$). *Nozaki* and *Bartlett* found abundant evidence for nongeminate reactions of benzoate radicals during the decomposition of benzoyl peroxide (R = C_6H_5). Products of these peroxide decompositions obviously depend upon the nature of the R group.

Table 22.3 summarises the fractions of geminate reaction by radicals produced in decompositions under various conditions. Several competing geminate reactions are often involved, and the original references should be consulted for details. Only those radicals that do not undergo geminate reactions escape the cage and become available for other processes. The data illustrate the efficiencies of radial production that are apt to be encountered and suggest the effects to be expected from changes in solvent, temperature, radical type, and nature of molecule (such as N_2 or CO_2) separating caged radicals at the time of formation. Attempts to predict these effects quantitatively must obviously be undertaken with extreme caution.

If radicals differ in their effectiveness at initiating a reaction, these geminate processes are obviously of importance. However, most chemists employing such initiators are primarily interested in the efficiency with which radicals are produced rather than in the nature of those radicals. If radicals react by geminate processes, the expensive initiator is destroyed without causing any of the desired chain reactions such as polymerisation or air oxidation.

If v_i is the rate of initiation of radical chains, and if chains are terminated by reactions of radical pairs, the total amount of reaction is proportional to $\sqrt{v_i}$ and the chain length is proportional to $1 \sqrt{v_i}$. Hence cage effects that reduce the efficiency of radical production will reduce the total amount of reaction resulting from initiator decomposition. However, for a polymerisation without chain transfer, these same cage effects may serve to increase slightly the average molecular weight of the polymer that is produced.

The above discussion indicates that useful radicals are formed slightly more efficiently if the temperature is raised and distinctly more efficiently if the viscosity of the medium is reduced. If the reaction is initiated photochemically, efficiency per quantum absorbed will also be improved by decreasing the wavelength of the light. Cage effects are only important when initiating reactions produce two or more radicals simultaneously. Radicals may also be produced by reactions such as $Fe^{2+} + H_2O_2 \longrightarrow FeOH^{2+} + \bullet H$. However, since this redox reaction is virtually irreversible and only a single radical is produced, no cage effect will operate to reduce efficiency. Of course polymerisation can also take place by cationic and anionic mechanisms. The chains are usually initiated singly rather than in pairs, and no cage effects are to be anticipated.

Cage Effects in Radiation Chemistry

The preceding discussion was confined to dissociation processes that produce two radicals in a cage. When high-energy ionizing radiation is absorbed in a medium, large local, concentrations of radicals may be produced. With strongly ionising particles like α rays and other positive ions, these radicals can be assumed to be formed in small cylindrical regions that expand as diffusion and recombination take place. When β rays and the electrons ejected by γ rays traverse a solution, they produce secondary electrons that are each assumed to produce small numbers of radicals initially confined to spherical 'spurs'. The description of these systems is much more difficult than when radicals are produced in pairs.

CHAPTER 23

Development of a Software Sensor for Control of Bulk Polymerisation of Methyl Methacrylate

INTRODUCTION

Most industrial reactors for the manufacture of commodity polymers operate under non-isothermal, semibatch reactor conditions. The operating conditions are usually decided semiempirically and possibly have some optimality criteria as their basis. Often, these criteria are intuition based. There is a scope for improvement of the operating conditions in most cases, and the vast knowledge base now available in polymer reaction engineering can be exploited to considerable advantage. Some modifications with relatively small investments can lead to a substantial increase in the profitability and viability of these units. The basic kinetic scheme characterising several important free radical polymerisations [e.g., polymethylmethacrylate (PMMA), polystyrene (PS), polyvinylchloride (PVA), etc.] is shown in Table 23.1. The three most important rate constants, k_d, k_p and k_t ($\cong k_{td}$ for PMMA, since combination is unimportant) are associated with diffusional limitations,. The gel or Trommsdorff effect represents the effect of decreasing diffusivity of macroradicals, P_n (due to increasing viscosity of the reaction mass), on the termination rate constant, k_{td}, while the glass and cage effects represents similar effects on k_p and k_d (through the initiator efficiency, f), respectively. The gel effect is exhibited as a sharp increase in the rate of polymerisation after some time, and the glass effect is exhibited as a stoppage of the reaction before complete monomer conversion. Several scientists have attempted to model the gel, glass and cage effects theoretically. *Chiu* presented a molecular theory which led to a spate of research activity. Several interesting studies with an engineering bias followed the publication of this model, focusing on such aspects as parametric sensitivity, optimisation optimal parameter estimation, etc. Recently a new model has been developed which is valid for both batch as well as semibatch reactors operating under more general non-isothermal conditions. The theoretical predictions for non-isothermal operation as well as with intermediate addition of initiator have recently been confirmed experimentally on bulk polymerisation of MMA, and so this model can now be used for on-line optimising control of semibatch, non-isothermal reactors.

Very few experimental studies have been reported in the last few years on the control of free-radical polymerisation reactors. A scheme has been implemented which tracked a specified optimal temperature history in a solution polymerisation (PMMA) batch reactor using non-linear (globally linearising) control. *Ellis* used an on-line size-exclusion chromatograph to measure the molecular weight distribution (MWD) in an experimental reactor in which solution polymerisation is carried out. Temperature monomer conversion and simultaneous temperature change and monomer addition were used as manipulated variables to achieve a specified weight average molecular weight, M_w, at the end. *Crowley* used a densitometer for on-line measurement of density to infer monomer conversion in a batch solution MMA polymerisation reactor and showed that the reactor can be better controlled with an on-line state estimator.

Table 23.1. Kinetic scheme for polymerisation of methyl methacrylate.

Initiation	I	$\xrightarrow{k_d}$	2R
	R + M	$\xrightarrow{k_i}$	P_1
Propagation	$P_n + M$	$\xrightarrow{k_p}$	P_{n+1}
Termination by disproportionation	$P_n + P_m$	$\xrightarrow{k_{td}}$	$D_n + D_m$

All these studies on the control of MMA polymerisation reactors have been carried out under experimental conditions where in diffusional limitations are not manifested. As such, these studies are somewhat limited in scope, at least in so far as (some) industrial systems are concerned. Control of polymerisation reactors in the presence of severe diffusional limitations is expected to pose a formidable challenge, since these reactors are associated with extremely fast increase in the monomer conversion, x_m, with time, t. Such rapid changes in $x_m(t)$ leave almost no choice to the control engineer but to use theoretical models having excellent predictive capabilities. Model based control is, thus, essential for the proper control of such reactors, since often, one has to take control action (like changing the temperature) much before diffusional effects manifest themselves.

The control of free-radical bulk polymerisation reactors poses yet another problem, namely, lack of suitable measuring instruments which can be used online. Use of gel permeation chromatography/HPLC to obtain the molecular weight distribution, as well as the use of available densitometers for estimating monomer conversion is limited to solution polymerisations, and cannot be used easily for bulk polymerisations. The slow progress in the automatic control of polymerisation reactors has been attributed to the lack of robust, on-line measuring devices. Indeed, *Embiricu* have recently investigated on advanced control of polymerisation reactors, and have mentioned that very few studies have been reported on property estimation techniques for the control of bulk polymerisations. In a recent review, *Kiparissides* and *Morris* have similarly emphasised the growing need of reliable predictive models and new in-line control algorithms to meet tighter constraints on the product quality of polymers. They point out that robust analytical instrumentation and sensors to provide on-line measurements of polymer properties are generally not available, and offline analysis can result in long and variable response times. Hence, there is an urgent need for both online and inline sensors for polymer quality monitoring and control. The associated use of easy to measure secondary variables to provide reliable inferential measurements (called software sensors) is an important strategic and adjunct requirement. One possibility first suggested is the use of experimental values of the viscosity, $\eta(t)$, of the reaction mass (or related parameter like torque on stirring motor, etc.) along with the measured values of the temperature, $T(t)$, for such purposes. The feasibility of such a model-based inferential state estimation using $T(t)$ and $\eta(t)$, has been studied recently using theoretically generated 'experimental' data, and it has been demonstrated that this scheme can be used experimentally for online optimising control of bulk polymerisation reactors. For such a model-based inferential state estimation technique, one needs, in addition to a good model for the reactor, an equation relating the viscosity of the reaction mass and the temperature, T, to the state of the system (monomer conversion, x_m, and weight average molecular weight, M_w). *Mortiz* has suggested the use of the *Lyons-Tobolsky* equation. This relation has two curve-fit parameters, k_H and b, which must be obtained using experimental data on viscosity of MMA-PMMA systems under different conditions. In this section we report data on the viscosity of the reaction mass as a function of time during bulk polymerisation of MMA under isothermal conditions in a Haake® viscometer modified to work as a viscometer-cum-reactor. The data generated can be used to obtain values of k_H and b using an optimal parameter estimation technique. The experimental set up developed in this study has an added advantage that it can easily be used to study, experimentally, on-line optimising control of bulk MMA polymerisation in the future by developing appropriate software for periodic state estimation and optimisation.

EXPERIMENTAL SYSTEM

Modification of Viscometer

The experimental setup (Fig. 23.1) consists of a Haake® viscometer, (M5 osc, Haake Mess-Technik GmbH, Germany) having a stainless steel cup-and-bob (Haake SV400), which can be heated electrically in a controlled manner through a temperature controlled (Haake TP500) so as to track a specified set-point temperature history (not requiring fast cooling). The existing viscometer is modified to work simultaneously as a reactor by providing a stainless steel cover with appropriate input-lines for the liquid reaction mixture as well as for argon, at the top of the cup and bob assembly.

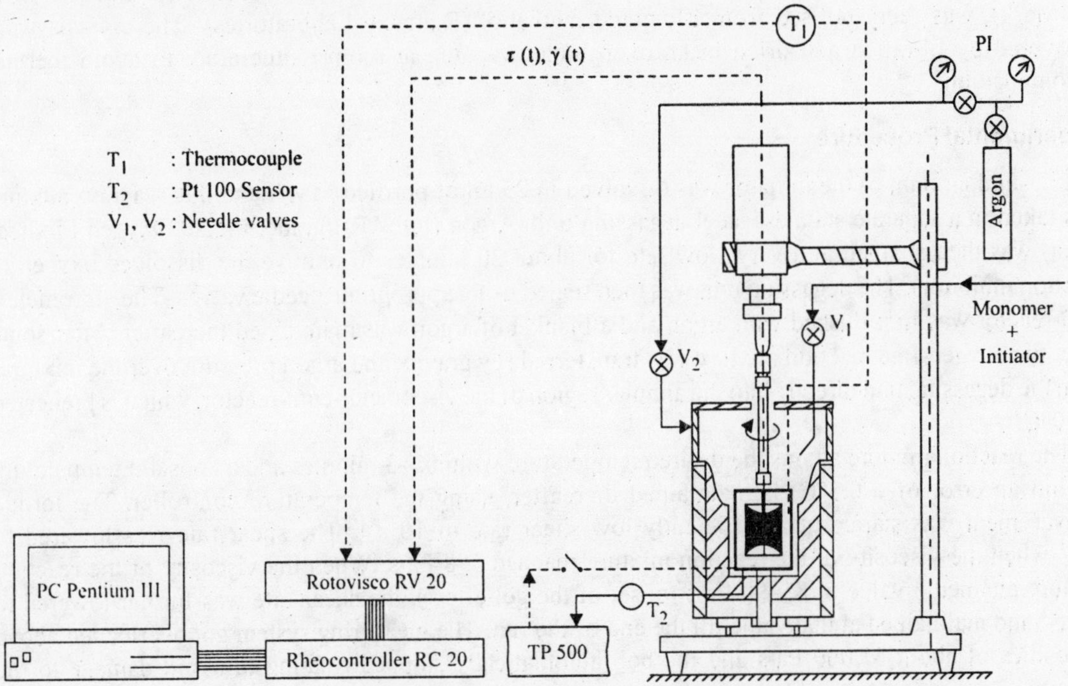

Fig. 23.1. Schematic diagram of the viscometer-cum-reactor assembly.

This enables to flush out air as well as maintain a blanket of inert gas (argon) over the reaction mass inside the annular region, while not interfering with the rotating bob. A small amount of argon leaks out continuously from the thin gap between a nylon 6 gasket and the rotating bob. This gap is required so that the rotating bob does not experience any torque at contact. The viscometer-cum-reactor sends the signal to a measuring system (Haake RV20) which in interfaced with a PC Pentium III over a RS-232 serial line through the rheocontroller (Haake RC20). The viscosity of the reaction mass is measured at desired time intervals. The computer records the values of the torque, τ, and the shear rate, γ, which in turn, are used to calculate the viscosity of the reaction mass as a function of time. The polymerisation is, thus, carried out in the annular region between the cup-and-bob, while simultaneously measuring the viscosity *on-line*.

Calibration of Viscometer-Cum-Reactor

The viscometer-cum-reactor should be calibrated prior to its use. Standard viscosity test fluids (E6000 and E40000) supplied by Gebruder Haake GmbH, Germany, are used for this purpose.

Purification of Monomer and Initiator

LR grade MMA (E. Merk) was purified prior to use. The monomer was washed three times with equal volumes of 5 per cent NaOH (E. Merck) solution in double distilled water to remove the hydroquinone inhibitor. Traces of NaOH were removed by washing thrice with double distilled water. The monomer was then dried by passing it through beds of silica gel (mesh 3–8) and molecular sieves (type 30–541, Linde division, Union Carbide, USA). The monomer thus obtained was distilled under vacuum (~50 mm Hg) at about 30°C. The final yield of monomer was about 60 per cent. The initiator, LR grade AIBN (E. Merk), was recrystallised from LR grade methanol (Ranbaxy Laboratories). The crystals were recovered by filtration and dried in an oven under vacuum at room temperature to avoid thermal decomposition.

Experimental Procedure

The desired amount of the initiator was dissolved in 25 ml of purified MMA and this reaction mixture was taken in a separate stainless steel degassing unit. Argon (IOLAR-I grade, Indian Oxygen Limited, India) was then sparged at a very slow rate for about 20 minutes to remove the dissolved oxygen (a reaction inhibitor). The degassing unit was then sealed using appropriate needle valves. The viscometer-cum-reactor was first flushed with argon and a blanket of argon was maintained thereafter. After some time, the oxygen-free reaction mixture was transferred (by gravity and argon pressure over the mixture) from the degassing unit directly into the annular region of the viscometer-cum-reactor, which is preheated to 50°C.

The reaction mixture attains the desired temperature within 2–3 minutes and a constant temperature (within an error of ± 0.5°C) is maintained thereafter, using the temperature controller. The torque measurement was started at a sufficiently low shear rate of 10 s^{-1}. The shear rate was lowered to 2 s^{-1} when the viscosity of the reaction mixture reached 5–8 Pa-s. When the viscosity of the reaction mixture attained a value of 85–90 Pa-s (onset of the gel-effect), the shear rate was further lowered to 0.1 s^{-1} and maintained at that value till the end of the run. The measuring system got overloaded above viscosities of about 31,000 Pa-s and the bob automatically stopped rotating to avoid damage to the system.

The rotating bob worked as a stirrer during the polymerisation. The shear rate, torque and viscosity, were recorded by the PC Pentium III as a function of time. The temperature of the reaction mass was measured with the help of a thermocouple dipped in the reaction mass. The temperature history was recorded by the PC Pentium III through the rheocontroller.

Determination of Monomer Conversion and Viscosity Average Molecular Weight

Samples of the reaction mixture were taken out from the viscometer-cum-reaction at desired times. The viscosity measurement was stopped at the sampling instant and the sample was taken out and transferred to bottles containing LR grade acetone with small amount of inhibitor, hydroquinone to stop polymerisation instantly. Thus, one experimental run was required to be carried out to get one sample, as it was not possible to take out the samples from the viscometer-cum-reactor while measuring the viscosity. The monomer conversion was determined gravimetrically. The viscosity average molecular weight, M_v, of the polymer sample was determined by dilute solution viscometry with an Ubbelohde viscometer [501 01 (Schott Geräte GmbH, Germany] using well established techniques.

RESULTS AND DISCUSSION

Bulk polymerisation of MMA was carried out in the viscometer-cum-reactor at two different initiator lodgings ($[I])_0$ = 15.48 and 25.8 mol/m^3) under isothermal conditions (50°C). As the reaction proceeds, the reaction mixture becomes more and more viscous. The viscosity of the reaction mass was found to increase dramatically in the gel-effect region, when the monomer conversion as well as M_v, increased sharply. We were able to measure and record several data points in the gel-effect region. This happens to be the main region of interest for on-line optimising control of bulk polymerisation reactors planned for the near future. Fig. 23.2 shows the temperature history for one experimental run (temperature isothermal 50°C, initial concentration of initiator = $[I]_0$ = 25.48 mol/m^3 {run No. TI50 C25a} α, β denotes the duplicate runs) during the polymerisation.

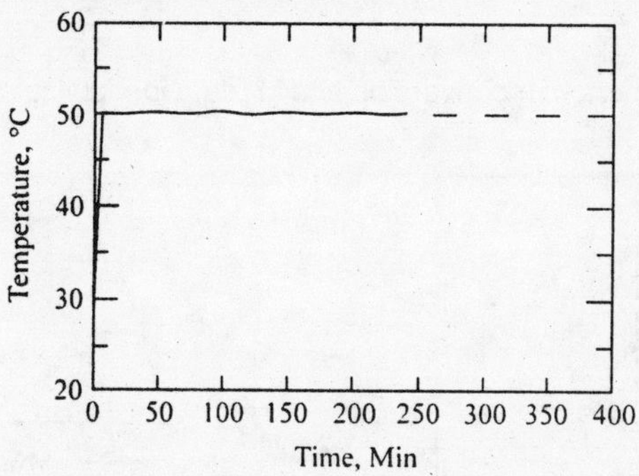

Fig. 23.2. Temperature history for run TI50 C25a.

It can be seen that isothermal conditions were, indeed, maintained. The temperature histories for the other runs were quite similar and are not included here for the sake of brevity. Duplicate runs were carried out, at times, to ensure reproducibility. Fig. 23.3a shows viscosity as a function of time for the two duplicate runs, TI50 C15a and TI50 C15b. The theoretically predicted monomer conversion obtained from the model using globally-optimised best-fit correlations (BFCs) are compared with our experimental data on monomer conversion Fig. 23.3b. Our experimental data on M_v, alongwith the model predicted values of M_n and M_w, are shown in Fig. 23.3c. Fig. 23.4a-c shows similar data for another set of duplicate runs, TI50 C25a and TI50 C25b. Our monomer conversion data are found to compare well with experimental results of *Balke* and *Hamielec* for both cases. For both initiator loadings, a definite correspondence is observed between the dramatic rise in the viscosity as well as that for monomer conversion. The suitability of the computer-interfaced experimental setup made in laboratory is thus demonstrated. Values of k_H and b at 50°C can now be easily obtained by curve-fitting the data.

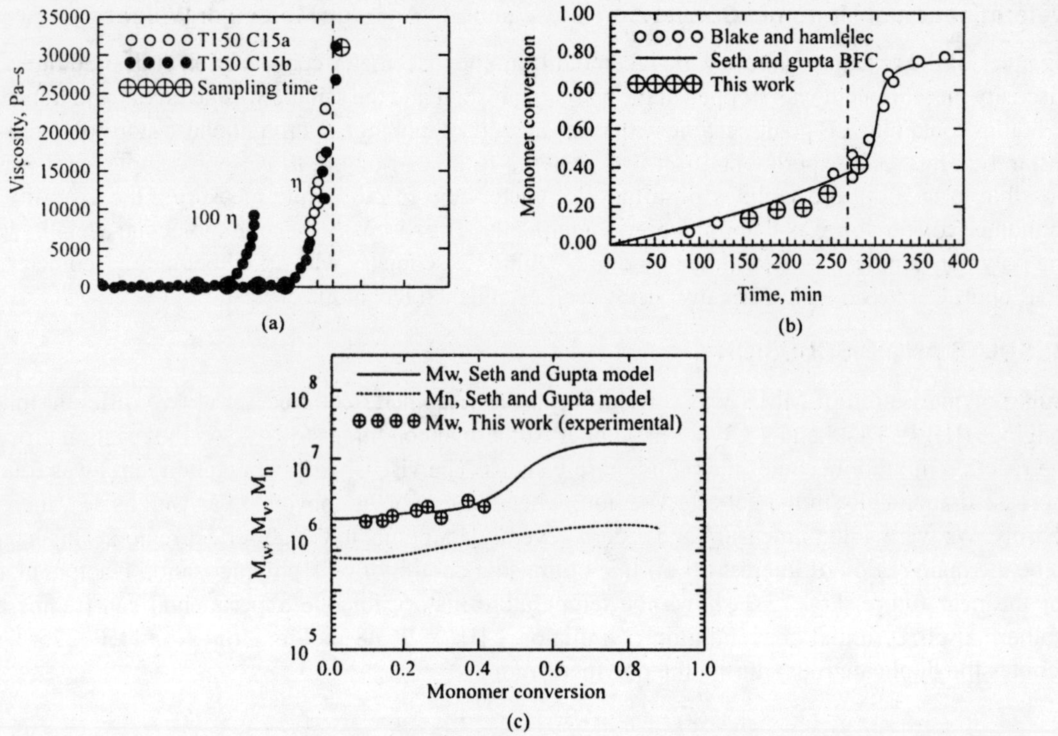

Fig. 23.3. Experimental viscosity, monomer conversion and M_v data for duplicate runs TI50 C15 and TI50 C15b.

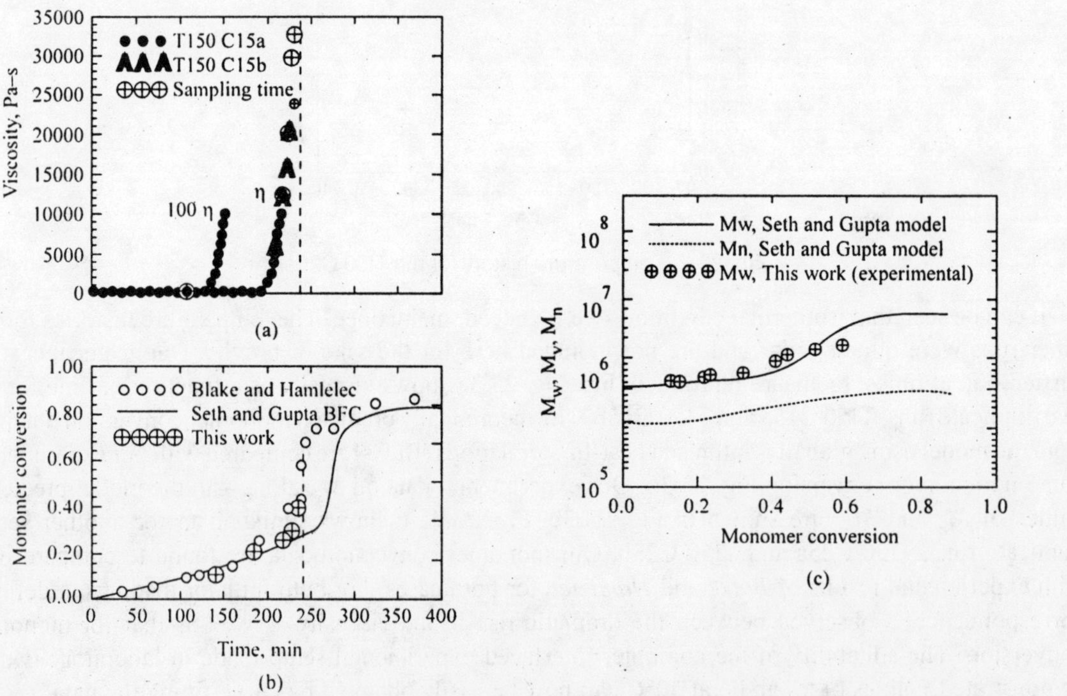

Fig. 23.4. Experimental viscosity, monomer conversion data and M_v data for duplicate runs TI50 C25a and TI50 C25b.

This work is in progress. A Haake viscometer has been modified to act simultaneously as a polymerisation reactor. Development of a software sensor have been demonstrated. This is probably the first attempt of its kind to measure the viscosity of a polymerising mixture (during bulk polymerisation) and record the viscosities as high as 31,000 Pa-s, in the open literature. Further work on on-line optimising control of this system using the experimental setup developed in this work is in progress. The knowledge gained on this system can easily be used for other free-radical homo- and co-polymerisation which exhibit gel, glass and cage effect (e.g., PS).

CHAPTER 24

Water Soluble Polymers

INTRODUCTION

Synthetic polymers are known to exhibit broad and varied activities with biological systems. Many of these include antibacterial, antineoplastic, blood extending, anticlotting, interferon-inducing, antiviral, enzymatic, herbicidal, and antifungal activities. Since biological polymer applications are so extensive, this chapter will be concerned primarily with some of the synthetic polymers that are soluble in aqueous media.

Water-soluble polymers generally fall into one of three major classes: (i) neutral species which include polyols, polyethers, and polymers with amine oxide groups; (ii) polyanions such as carboxylates, sulphates, and phosphates; and (iii) polycations including polymeric quaternary ammonium and sulphonium salts.

Water-soluble polymers can distribute themselves in a living system by blood or lymphatic circulation, by cellular transport through the involvement of mobile phagocytic cells, or by absorption on cell surfaces. The role of polymers in the biology of organisms relates to their soluble or insoluble forms in tissue and tissue fluid, some of which resemble blood protein or nucleotides. Blood is a heterogenous non-Newtonian type fluid composed of cellular elements such as erythrocytes, leucocytes, and platelets which are suspended in the plasma.

The compatibility of a polymer or any substance for that matter, in blood, requires that it should not cause thrombosis or destroy cellular elements, alter plasma proteins, deplete electrolytes, or cause acute or delayed toxic or immune-allergic tissue damage. Therefore, clinically one usually searches for biologically active polymers that are similar to blood proteins.

The water solubility of polymer is significant not only for transport but also for systemic administrations, since injection of suspension into blood vessels can cause 'colloido-clasmic shock' or 'macro-molecular syndrome' with major hypersensitive clinical toxicity. Similarly, the interaction of the polymer with blood proteins and cell surfaces is essential, not only for distribution, but also for specific reactivity since they readily interact with other macro-molecules to form supermolecular structures. This association phenomenon occurs with neutral macro-molecules because of weak van der Walls interactions, and more importantly, hydrogen bonding. Polyanions and polycations form much stronger electrostatic interactions through ionic bonds.

Complex formations with polyelectrolytes are sensitive to charge density, ionic strength, concentration, and stereochemistry of both the polymer and the substrate. Since the complexes formed by these polymeric interactions differ in stability, it should be possible for polymers to have an inherent selectivity and specificity. Detailed evidence, however, of these facts and how they relate to biological systems is presently lacking.

A further essential criterion that must be met by biological useful polymers is the absence of lethal toxicity in both the polymer itself and its metabolites. Consequently, there must be an appreciable difference between toxic and therapeutic doses. Also, the polymer should be completely eliminated from the organism since retention may cause histocytic reactions, which can produce tumorlike growths that could be carcinogenic.

For example, water-soluble polymers have been successfully employed clinically as plasma extenders. Their main role is to maintain osmotic pressure to increase or restore the effective circulating blood volume. Molecular weight is important here, since blood volume restoration must be retained for several hours. Therefore, a polymer of sufficient molecular weight (greater than 30,000) must be employed so as to delay clearance, but still be below the kidney threshold which is approximately 50,000. Consequently, the molecular-weight distribution is more critical than the average molecular weight since very large molecules can also cause erythrocyte aggregation, and changes in platelet or leukocyte distribution or morphology. Prolonged storage *in vivo* can be avoided by selecting polymer with appropriate molecular dimension so as to facilitate their clearance of the liver, kidneys, and other organs. Another approach to this problem is to fabricate biodegradable polymers that can retain their biologically active structure long enough to perform their task, and then are converted to harmless by-products that are readily eliminated. This latter concept has not been widely attempted, except in the development of degradable coatings for sustained drug release and prolonged drug activity.

CATIONIC POLYMERS

Cationic polymers are positively charged polyelectrolytes that have several biomedical applications which include antiviral, antibacterial, and antineoplastic activities. These activities are primarily related to their ability to bind to surfaces with negative electrostatic charges. Cell surfaces have a large excess of negative charges due to the acid anion moieties on the cell membrane. Each type of cell exhibits its own characteristic electronegative charges, e.g., the spleen cells, lymph cells, and tumor cells have a greater negative charge than thymocytes and macrophages. Difficulty is encountered, however, in exploiting these slight differences in electrostatic phenomena of cell surfaces therapeutically because of the lack of specificity of presently known polycations.

Antibacterial Activity

Several quaternary ammonium polymers are known as potential bacteriostatic and bactericidal polyelectrolytes. Ionene polymers, a specific type of quaternary ammonium polymer have also been bactericidal. Ionene polymers are particularly active against a variety of gram-positive and gram-negative bacteria. Their antimicrobial activity, which is usually only a few ppm, varies with the distributions of the positive charge; e.g., the 6,10-ionene bromide polymer has the highest inhibitory effect on *E. coli*. Low-molecular-weight (less than 1000) polymeric ammonium salts do not inhibit bacterial growth. Vinyl alcohol–vinylaminosalicylate co-polymers have been reported to exhibit in vitro antitubercular activity. Agglutination of bacteria has been observed for polylysine, perhaps the most actively biologically studied polycation.

Although these cationic polymers are very active against a number of microbial organisms, they have proven to be for the most part too toxic for practical clinical use, at least by systemic administration.

Adriamycin, a cationic antibiotic that intercalates into deoxyribonucleic acid, has been complexed with deoxyribonucleic acid with enhancement of clinical tumor activity. It was designed to increase endocytic uptake with subsequent lysozomal breakdown of deoxyribonucleic acid in tumor cells which would result in the release of the active adriamycin. This concept of complexing a biologically active cation to a biodegradable polyanion to effect changes in pharmacokinetics has great clinical potential for both detoxification and therapeutic specificity.

Growth Inhibition

Several polycations have been shown to inhibit preferentially the growth of malignant cells. Many cationic polyelectrolytes have been found to inhibit the growth of allogeneic and syngeneic tumors in mice at nontoxic levels to the host. Antitumor activity is apparently related to specific characteristics of the polycations, such as molecular weight and zeta potential, and to the mode of administration. Quaternary polyethylenimine, polypropylenimine, and polyvinylimidazoline are cytotoxic and show specificity of binding to tumor cells. It has been reported that ionene polymers that have been covalently bonded to polymeric microspheres exhibit specific cytotoxicity to leukemic murine cells. It has also been found that normal human diploid WI-38 cells adhere at a lower rate than Simian virus transformed WI-38 cells. This event has been related to the fact that tumor cells exhibit different surface-cell properties, such as anodic mobility, which binds them preferentially to polyionene-treated surfaces. Further, lymphocytes pretreated with polycations have been observed to increase cytotoxicity to solid tumor cells. Several synthetic polycations, particularly polyvinylimidazoline, have shown strong adjuvant activity with diphtheria and tetanus toxoids. In addition, immunoadjuvant properties for this and related polycations have been observed with experimental tumors in mice. Polylysine, for example, when added to synthetic cationic polymers such as diethylaminoethyldextran, enhances the interferon-inducing activity of the polyamines.

ANIONIC POLYMERS

Polyanions are polyelectrolytes with negative charges that generally reside on pendant groups attached to the backbone of the polymers chain. Both natural (eg., heparin and heparinoids) and synthetic [eg, poly(acrylic acid) and 'pyran'] polyanions produce a variety of biological activities. Some of these effects are: induction of interferon, antiviral activity including inhibition of Maloney sarcoma, Rauscher leukemia, Friend leukemia, polyoma, vesicular stomatitis virus, Mengo, encephalomyocarditis, and foot-and-mouth disease. Further, they enhance host resistance to bacteria and fungi, enhance the immune response, inhibit adjuvant arthritis and, depending on polymer size, either depress or stimulate the functional phagocytic activity of the reticuloendothelial system.

Polyanions that enter into biological functions through distribution throughout the host are similar to certain proteins, glycoproteins, or polynucleotides that modulate a variety of biological responses related to host-defence reactions, including resistance to viral, bacterial, protozoal, and fungal infection. In relation to these, immunologic and hormonal responses, inflammation, would repair, blood clotting, and tissue damage are subject to the action of these macromolecules.

The toxicology of polyanions is of prime importance to their medical applicability. They can cause anemia, leukocytosis, hepatosplenomegaly, liver enzyme induction, and can sensitize biological systems to the lethal action of bacterial endotoxins. Further, they interfere with Type I and Type II drug metabolism mediated by liver microsomal systems by prolonging high levels of barbiturates and aminopyrenes through blocking of detoxification. Specific molecular weight fractions polyanions that diminish these side effects and antiviral effects are presently under study.

Anticlotting activity

Thrombosis is the complex plugging of blood vessels in capillaries, veins, and arteries which impedes tissue perfusion and/or results in the release of tissue-damaging by-products. This problem is encountered following surgery, clinical sepsis, abortion, infection, hemolysis resulting from malaria, and with mismatched blood transfusions. It is also experienced in whole-blood storage, with the use of heart-lung machines, and artificial kidneys.

Heparin, heparinoids, and mucopolysaccharides are among the most effective water-soluble polymeric anticoagulants. Heparin is a naturally-occurring, sulphated polysaccharide found in trace amounts in certain mammalian tissue. This polyanion has been in clinical use for the prevention of blood thrombosis, or clotting. Heparinoids are sulphated polysaccharides related to heparin which are prepared by partial degradation of natural polysaccharides followed by sulphonation or phosphorylation. Their physical properties are similar to heparin with high sulphate or phosphate anion content and water solubility. Mucopolysaccharides are also naturally-occurring carbohydrates, like heparin, but with fewer sulphate groups and weaker anticoagulant activity. Heparin and mucopolysaccharides are important structural materials of connective tissue in contrast to neutral polysaccharides (starches and glycogens) which are storage oriented. Heparin and heparinoids, although primarily developed clinically for their anticoagulant and lipolytic activity, have several other biological activities.

Several synthetic sulphonate polymers and co-polymers have been prepared for anticoagulant-activity studies. The sulphonate group is unique in that it apparently does not bind calcium or magnesium ions, contrary to carboxylates and phosphates. For example, a typical sulphonate polymer, such as poly(styrenesulphonic acid), is not precipitable by calcium or even barium ions. The ability of these polyanions to maintain a high negative charge in the presence of calcium and/or magnesium ions is the apparent basis of the effectiveness of heparin and other polysulphonates as anticoagulants. Ethylenesulphonic acid polymers, including co-polymers, in particular have marked advantages over other types of synthetic materials in this regards and are being clinically utilised. However, poly(ethylenesulphonate) and poly(styrenesulphonate) have also been shown to displace histones bound to nucleic acid, resulting in stimulation of protein synthesis and ribonucleic acid polymerase activity.

Anionic heteropolymolybdates have physiological properties similar to heparin. Although they are inorganic polyelectrolytes, they have a relatively high molecular weight (3,000). They have been shown to have *in vitro* anticoagulant and lipolytic properties, to activate pyrophosphatase, and to inhibit virus, replication through effects on inhibition of reverse transcriptase.

Growth inhibition

The action of polyanions as mitotic inhibitors and their functional role in growth processes have been reviewed as has the role of polynucleotides in immunology and virus resistance. The inhibitory effect of heparin and other anticoagulants on tumor growth and spreading by tumor metastasis as well as the penetration of chemotherapeutic agents into the tumor environment has also recently been reviewed. A possible mechanism for this activity of polyanions may be related to coupling of the polyanion to tumor antigen. However, the action of polyanions on a wide range of enzymes, such as alteration of the isoelectric point of proteins, displacement of nucleohistone, and antiviral action, all indicate alternative concepts of antitumor action.

The anticoagulant effect of heparin and its relation to calcium binding and enzyme inhibition was evaluated as a potential impediment of tumor growth. Tumor growth and normal regeneration is associated with changes in polyanionic stromal or surface polysaccharides. Many clinical side effects noted with heparinoids represent inhibitions of rapidly growing epithelium, such as hair loss, and alterations in fingernail growth. Other side effects are alopecia, ulcerations of the gastrointestinal tract, and osteoporosis leading to pathologic fractures. Heparin, heparinoids, and synthetic polysulphonates have all been found to inhibit bone and cartilage growth, with resultant fracture of weight-bearing surfaces, or pseudoarthrosis.

Pyran, a carboxylic hydrolyzate of divinyl ether–maleic anhydride co-polymer, exhibits a broad spectrum of antitumor activity against transplanted tumors and appears to have a good therapeutic index. Pyran induces the production of interferon in both mice and humans. It has minimal *in vitro* toxicity to tumor cells in contrast to alkylating or other chemotherapeutic agents. Pyran administration, prior to inoculation of mice with Friend leukemia virus, Lewis lung tumor, allogenic sarcoma 180, or Ehrlich ascites tumor, significantly inhibits subsequent tumor growth. Similar inhibition of Friend leukemia virus splenomegaly is obtained with methylvinyl ether–maleic anhydride and styrene–maleic anhydride co-polymers. Pyran also inhibits tumor induced by Rauscher leukemia virus, polyoma virus, and Gross leukemia virus in normal or thymectomised, immunosuppressed mice.

Pyran administration after tumor implantation also inhibits the growth of first-generation transplant of mammary carcinoma, a chemically induced tumor, and delays or reduces the onset of dimethyl-benzanthrancene-caused tumors in the hamster cheek pouch. This antitumor action of pyran is not correlated with phagocytic stimulation or interferon-inducing capacity, but *in vitro* work indicates that the antitumor action may correlate with an increased macrophage-killing capacity.

The route of pyran administration is important because intravenous injection can stimulate Friend leukemia virus activity, in contrast to intraperitoneal administration, which is inhibitory. Paradoxically, certain murine sarcoma virus-induced tumors are stimulated by pyran as well as by other interferon inducers. This stimulation may be related to the ability of pyran to induced hepatosplenomegaly or splenomegaly in normal mice.

The mechanism of action of pyran is being examined in a number of systems. Only prophylactic treatment with pyran was effective against systemic infection of mice with encephalomyocarditis virus, or against intravenous or intravaginal infection of mice with herpes simplex virus type. The investigations in various systems have indicated that activated macrophages may be the common effector of the action of pyran. For example, the enhancement of reticuloendothelial function was observed after pyran treatment; this effect was correlated with enhanced resistance of mice to infection with micro-organisms such as staphylococcus and pneumococcus. Further, peritoneal exudate cells have been obtained from pyran-treated animals that exhibited cytotoxicity for tumor cells and inhibited virus growth *in vitro*.

The antitumor mechanism of action appears to involve an early cytostasis phase that is nonspecific and a delayed cytotoxic phase that is specific for tumor cells. Several investigators have combined immunomodulators with tumor cell vaccine in an effort to potentiate vaccine effectiveness. Marked protection occurred following immunoprophylaxis with a vaccine consisting of the synthetic immunomodulator pyran and irradiated L1210 tumor cells. Moreover pyran, in combination with chemotherapy, enhanced the production of long-terms survivors in a murine leukemia. Recently, it was found that pyran treatment, as an adjuvant to chemotherapy, significantly prolonged the life of animals bearing the Lewis lung carcinoma and LSTRA lymphoma.

The toxicology of pyran is as complex as its therapeutic advantages. Hepatic and renal damage represent the major toxicity observed at high dosages in the dog and monkey. In mice, ^{14}C-labeled pyran with an average molecular weight of 17,000 is lost from the blood within 24 hours; at 5 weeks, there is 30 per cent residual activity shared between liver and spleen. At 3 months, less than 10 per cent of its original activity is present. A sample with an average molecular weight of 17,000 to 22,500 has been in clinical trial in human subjects, but its use is limited due to acute vascular effects. Thrombocytopenia, which is reminiscent of the action of poly(ethylenesulphonate), is also observed and hemolysis has been reported in one case. Cytoplasmic inclusion following pyran administration have been found in all the nucleated elements of the blood, as well as in the phagocytic reticuloendothelial cells of liver, spleen, and bone marrow. The anticoagulation effects are due to interference with fibrin formation, but bleeding has not been observed as a problem. Four different types of anionic polymers were prepared and their activity compared directly to the activity of pyran. These were acrylic acid–maleic anhydride co-polymer, maleic anhydride homopolymer, furan–maleic anhydride co-polymer, and aspartic acid–glutamic acid co-polymer.

Acrylic acid–maleic anhydride co-polymer with a viscosity slightly higher than pyran exhibited hepatosplenomegaly but not to the same extent as pyran. This co-polymer sensitizes to endotoxin, exhibits antitumor activity equal to that of pyran, possesses antiviral activity against encephalomyocarditis virus, and shows slight activity against the microsomal enzymes as indicated by the prolongation of hexobarbital-induced sleeping time. Maleic anhydride homopolymer slowed the growth of the primary Lewis lung tumor, but it did not cause a significant increase in life span. It also provided a moderate degree of protection from encephalomyocarditis virus. Maleic anhydride homopolymer caused slight hepatosplenomegaly, but did not alter the hexobarbital sleeping time. Recently, several different molecular-weight fractions of maleic anhydride homopolymer ranging from 1,000–100,000 were prepared and tested. These material administered interperitoneally produced 100 per cent inhibition of abnormal peripheral blood leukocytes. Bone marrow cells, spleen cells, and serum provided no protection. These and other studies show this polymer to be very similar in biological activity to pyran but it appears to be less toxic.

Better therapeutic indexes appear to be obtained for polyanions having discrete, narrow molecular weights. The clinical pyran, for example, has an average molecular weight of 17,000–22,500, but it is a mixture of a variety of both high-and low-molecular-weight fractions. Pyran, maleic anhydride homopolymer, and acrylic acid-maleic anhydride co-polymer fractions of lower molecular weights appear to maintain antitumor activity with better therapeutic index. Toxicity as manifested by weight loss, anemia, leukocytosis, hepatosplenomegaly, prolongation of type I (aminopyran) barbiturate metabolism, and endotoxin sensitisation can be reduced or eliminated. Most importantly, the biphasic phagocytic effect (phagocytic inhibition followed by stimulation) produced by the clinical pyran appeared to be a summation effect of the distribution of both high-and low-molecular-weight fractions included in the parent sample. Similar results have been observed for fractionated samples of acrylic acid–maleic anhydride co-polymer and for hydrolysed poly(maleic anhydride).

Poly(acrylic acid) has been found to be effective against polyoma virus-induced tumor in mice and rats. Compared to other related polyanions such as dextran sulphate, poly(strenesulphonate), poly(vinyl sulphate), and poly(phloroglucinol phosphate), poly(acrylic acid) was found to be significantly more antivirally active and less cytotoxic. Further, poly(acrylic acid) was found to be a good interferon inducer as well. Glucan and poly(methyacrylate) mobilize macrophages and produced tumor necrosis on local injection into tumors which is similar to what has been found for live and fixed vaccines.

Chlorite-oxidised oxypolysaccharides, a type of polyacetal carboxylic acids known as COAM, were developed because they have a polyanionic chain resembling poly(acrylic acid), an essentially biodegradable backbone, and a slow degradation rate in the tissue so that the compounds persist long enough to provide protection for a long period of time. The polymers are prepared by a two-step oxidation of polysaccharides. First, the anhydropyranose units are cleaved with periodate and then the resultant aldehyde groups are oxidised with sodium chlorite to carboxylic groups. This structure involves a polymer chain with both polycarboxylic and some polyhydroxyl functionality depending on the degree of oxidation. Chlorite-oxidised oxy-polysaccharides inhibit the cytopathic effect of vesicular stomatitis virus in the mouse embryo. Antiviral activity was only observed when at least 64 per cent of the glucopy -ranose units were oxidised and a very high-molecular weight (2×10^6) polymer was used. Intraperitoneal injection was found to protect mice against Mengo, vaccinia, Semliki Forest, and influenza viruses. It also inhibits spontaneous mammary carcinoma. This polymer has effectively inhibited tumor growth in mice with Moloney strain of murine sarcoma virus. Furthermore, chlorite-oxidised oxy-polysaccharides administered with the interferon inducer statalon, demonstrates a synergistic effect in inhibition of Friend leukemia virus.

In general, polyanions have been found to be fairly stable but not very susceptible to bio-degradation, which may account for their prolonged activity. Pyran and poly(acrylic acid), particularly, are poorly excreted and appear not to be enzymatically attacked. Although chlorite-oxidised oxy-polysaccharides have been claimed to be more bio-degradable than poly(acrylic acid), no direct evidence has been reported.

The future of synthetic polyanions in clinical or veterinary medicine will reside in close attention to bio-degradability and/or control of molecular weight in relation to the polyanions or polycations used systemically.

Antiinflammatory Agents

The reduction of internal and external inflammation poses a difficult medical problem. Water-soluble sodium salts of *N*-antipyrinylmethylacrylamide–*N*-(o-carboxylphenyl)methacrylamide co-polymers have reduced inflammation to 68 per cent after intraperitoneal injection. 19-Hydroxysteroids esterified with a co-polymer of vinylpyrrolidone and a carboxylic monomer such as maleic anhydride, acrylic acid, acryloyl chloride, and methyl acrylate produced water-soluble materials with molecular weights of 10,000–40,000. When administered systemically, these drugs suppressed inflammation and prolonged action was observed when the drugs was applied directly to the site. A serious side effect is an allergic shock which is probably due to the high molecular weight of the polymer.

NEUTRAL POLYMERS

Neutral polymers are generally not very water soluble, but the few that are have some interesting biological applications.

Plasma Extenders

Netural water-soluble polymers are employed clinically to maintain osmotic pressure to increase or restore an effective circulating blood volume. The molecular-weight distribution is important in this application as a high molecular weight (above 30,000) is necessary to delay clearance in order to maintain blood-volume restoration, but the molecular weight must also be low enough to clear the kidneys (below 50,000).

Poly(vinylpyrrolidone) is clinically the most widely used plasma extender at the present time. Preparations of this polymer have been found to be nonantigenic and with no apparent toxic side effects. However, appreciable storage in the liver, spleen, and reticuloendothelial system has been reported with no apparent irreversible toxicity. In addition, poly(vinylpyrrolidone) is an active detoxifier. The osmotic diuresis produced by the right molecular weight accelerates the excretion of barbiturates, uric acid, and other toxicants. Further, this polymer's osmotic diuretic effect can prevent renal tubular damage caused by deposition of urate, haemoglobin, and other proteins an in renal tubule. Gum arabic (acacia), molecular weight 1,200–2,400, also has good osmotic and retention characteristics, but severe antigenic reactions have been reported. Methylcellulose in 0.25 per cent solution is effective in the treatment of experimental shock, but is retained in the liver, kidney, and spleen. Other polymeric plasma extenders include dextrans, mannitol, pectin, and polyglucose. These vary in their effectiveness, but all suffer from prolonged periods of storage in the body, or sthenoplastic effects, that alter the characteristics of platelets or red blood cells and cause changes in blood coagulation.

Other biological applications

Poly(vinylpyridene-*N*-oxides) have been employed as antisilicosis agents. The polymer inhibits siliocolic fibrosis and significantly increases the self purification of the lungs of quartz dust. Protection appears to be due to electrostatic binding of the polyelectrolyte to the electronegative surface of silica particles in the lung and clearance may be attributed to macrophage stimulation to engulf the silica particles. Water-soluble polyoxyalkylenes exhibit good laxative properties, particularly in the 1,000–11,000 molecular-weight range. These polymers have also been employed in the treatment of hypercholestermic conditions, with substantial reductions reported in human blood-serum cholesterol levels.

Trends

The most immediate clinical place for polyanions is an immunoadjuvants to enhance the biological effectiveness of relatively inactive vaccines. This concept has been shown to be useful in the treatment of tumors, rabies, and hoof-and-mouth disease in rodents. Clinically, the concomitant adjuvant use of vaccines with synthetic polyanions is feasible as the quantity of polymer necessary is below the toxic threshold.

Both synthetic polyanions and polycations are providing us with useful information as to controlling factors in host lymphocyte and macrophage response. An understanding of the mechanism of polyanion action on plutonium liver affinity could provide us with more effective means of coping with this environmental hazard. The future for these agents is not only their direct effects on cell systems, i.e., in viral neutralisation, enzyme, and cell-component binding, but is also in the area of host stimulation to protect against viral, bacterial, and fungal pathogens, as well as tumor growth. The future of synthetic polymers in clinical or veterinary medicine will reside in close attention to bio-degradability and/or control of molecular weight, particularly for the case of the polyanions or polycations used systemically.

ELECTRO-KINETIC BEHAVIOUR OF SOME WATER SOLUBLE POLYMERS AND THEIR BLENDS

Polyelectrolytes are polymers, which undergo electrolytic dissociation when dissolved in the proper polar solvents. In solution, the polyelectrolyte macro-molecule is the polyion surrounded by equivalent amounts of small oppositely charged ions called counter ions, which are smaller in size than polyions, by several orders.

Polyelectrolytes are classified into polyacids, polybases and polyampholytes depending on their ionized forms being anion, cation and both anion and cation, respectively. They attract much attention to the investigators due to their specific properties such as solubility in water, ability to interact charge particles and surfaces and they are widely used as flocculants, coagulants, sensors and so forth. Various properties of polyelectrolytes were studied in the past by several investigators. One of these properties is the conductivity that has been observed here to compare the electro-kinetic properties of individual polymers as well as their blends with high and low molecular weight (MW).

Experimental

Bi-distilled water used for the preparation of this solutions was investigated. Polyacrylamide (PAA) was synthesised from the aqueous solution of acrylamide (AA) monomer with γ-ray using Co-60 source and PVP. The prepared PAA has molecular weight (MW) is about 10^7, whereas MW of PVP was of 40,000.

During preparation of the solutions, it has been found that in case of PAA the concentration greater than 0.2 per cent was not homogeneous. Hence, it was initially prepared at a concentration of 0.2 per cent for all samples and thereafter the solution was diluted successively to 0.1, 0.075, 0.05, 0.025 and 0.0125 per cent by adding required amount of bi-distilled water. The blended solutions PP's with different proportions of (PVP: PAA) per cent were (95:5) per cent, (80:10) per cent and (85:15) per cent, which are hereinafter abbreviated as PP1, PP2 and PP3, respectively. The conductivity of polyelectrolytes with different concentrations was measured using a WTW LF-330 conductivity meter (Germany) fitted with a LetracoR-325 electrode that was initially calibrated by standard potassium chloride solution. The electrode was immersed in various polyelectrolyte solutions and the conductivity was directly recorded from the digital display. The conductivity of the bi-distilled water used in the experiment was found 3 μS/cm. The equivalent conductance was estimated by the relation: $\chi_c=(\chi_r-\chi_o)/C$, where, χ_c, χ_r, and χ_o are the equivalent conductance of solution, conductance of solution and conductance of the distilled water, respectively. C denotes the concentration of the solution.

Thus to measure the conductivity PAA, PVP and their blends in aqueous solution were the main aim of this investigation. Increasing the concentrations reveals the increase of conductivity but decrease in equivalent conductivity up to a certain range. Conductivity has been found to be more in PAA than PVP, which is explained by introducing a statistical consideration and the molecular conformation. By introducing high MW polyelectrolytes in to low MW ones, the conductivity could be enhanced.

NOVEL METHODS FOR THE SYNTHESIS OF SOLUBLE POLYMERS FROM DIVINYL MONOMERS

Free radical polymerisation of divinyl monomers leads to cross-linking. Selective polymerisation of one of the vinyl group would overcome this limitation. Such polymers will find diverse applications in latex paints, controlled release systems, microlithography etc. Synthesis of soluble divinyl polymers using anionic polymerisation technique was reported by *Hiller* and *Funke*. But anionic polymerisation technique limits the type of monomer that can be used and the scale of synthesis. Cyclodextrins (CD) are well known cyclic species that solubilise hydrophobic molecules in aqueous media. In this section the method of protection of one of the vinyl group of divinyl monomer by forming supramolecular complex with β-CD has been discussed. Homopolymerisation of these complexes and co-polymerisation using various monomers is also reported.

Experimental

Supramolecular complex preparation

Equimolar amount of the divinyl monomer was added with stirring to a solution of β-CD in water at room temperature. The inclusion complex precipitated and was filtered, washed with water and MeOH and dried in a vacuum oven at room temperature. The yield of the complexes was 60–75 per cent.

Polymerisation

1. Homopolymerisation [poly (ethylene glycol dimethyl acrylate)]: 1gm(0.75mmole) ethylene glycol dimethyl acrylate/β-CD complex was dissolved in 5ml N, N dimethyl formamide.10 mg

(0.07mmole) azoisobutyronitrile (AIBN) was added and flushed with nitrogen for 15 min. The polymerisation was carried out for 20 hrs at 65°C. The polymer was precipitated in water; filtered and dried under vacuum at room temperature. The yield of the polymer was 68 per cent.

2. Co-polymerisation: [poly (methyl methacrylate-co-ethylene glycol dimethyl acrylate)]: 2 gm (0.02 moles) methyl methacrylate, 5.3 gm (3.9 mmoles) ethylene glycol dimethyl acrylate/β-CD complex and 20mg (0.14mmoles) AIBN were dissolved in N, N dimethyl formamide. Nitrogen was purged for 15 min. The polymerisation was carried out for 16 hrs at 65°C. Polymer was precipitated in water, filtered and dried under vacuum at room temperature. The yield of the polymer was 74 per cent. Homopolymers of different divinyl monomer complexes and co-polymers using various monomers was also carried out using above procedure. Soluble homopolymers and co-polymers of divinyl monomers are shown in Table 24.1.

Table 24.1. Soluble homopolymers and co-polymers of divinyl monomers.

Complexes of β-CD with	*Polymer*	*Molecular Wt.(Mw)*	*Polydispersity index*
Ethylene glycol dimethyl acrylate (EGDMA)	Poly EGDMA	1,00,754	4.8
	P(MMA-co-EGDMA)	42,000	2.1
	P(AA-co-EGDMA)	–	–
	P(HEMA-co-EGDMA)	–	–
Trimethalol propane trimethacrylate (TMP)(1:1)	P (MMA-co-TMP)	40,214	9.2
Trimethalol propane trimethacrylate (TMP)(1:2)	P (MMA-co-TMP)	72,960	12.8
Vinyl methacrylate (VMA)	P (VA-co-VMA)	4464	1.7
Divinyl benzene(DVB)	P (Sty-co-DVB)	11,770	2.82

Note: (a) Methyl methacrylate (MMA), Acrylic acid (AA), and Vinyl acetate (VA) Styrene (Sty).
(b) Molecular wts were measured by GPC using THF as a mobile phase.

Results and Discussion

Inclusion complexes of β-CD with divinyl monomers were synthesised by mixing the monomer with β-CD solution in water followed by stirring at room temp. for 24 hrs. All mixtures became gradually turbid, an indication of the formation of crystalline inclusion complexes between monomer and β-CD. The complex was isolated by filtration and washed with water to remove uncomplexed β-CD and then with MeOH to remove free monomer. The complexes were characterised using IR, NMR, and TGA.

FTIR is a very useful tool to prove the presence of both guest and host components in a supramolecular complex. The spectrum contains distinct peaks of the complex e.g. in EGDMA/β-CD complex the peak at 1720 cm^{-1} shows the stretching mode of carbonyl and 1654 cm^{-1} double bond of divinyl monomer. The broad hydroxyl band of β-CD at 3263 cm^{-1} is shifted to a higher frequency at 3298 cm^{-1} and it becomes sharper.

The structure of the complexes was studied by ^1H NMR spectroscopy. It showed presence of vinyl unsaturation. The thermal properties of the complexes were investigated by TGA. For e.g. complex of EGDMA/β-CD decomposes at 354.5°C while the decomposition temp of β-CD is 344.4°C.The complex shows higher decomposition temperature. Thus the formation of supramolecular complex can improve thermal stability of CD. Because the guest monomer is in the cavity of β-CD molecule, it should decompose after the decomposition of host molecule. A free radical azo initiator, AIBN was used to initiate polymerisation of complexed monomers in solvents such as N, N dimethyl formamide, dimethyl sulfoxide etc. During the polymerisation β-CD slips from the divinyl monomer and gives the polymers containing free unsaturated groups, which are soluble in organic solvents. These soluble polymers containing unsaturation can be further polymerised using azo, redox, or peroxide type initiator. By NMR spectroscopy, it was proved that all the polymers are free from β-CD.

The polymers formed are linear in nature as confirmed by Intrinsic Viscosity measurements. It shows that as the molecular wt increases, intrinsic viscosity also increases linearly (Table 24.2).

Table 24.2. Intrinsic Viscosity's of Poly (MMA-co-EGDMA).

MolecularWt.(Mn)	*Polydispersity Index*	*[η] dl/gm*
3.026	3.4	0.0419
3.924	7.4	0.0643
7.968	4.9	0.0970
12.175	4.0	0.1096
36.753	5.4	0.2129

Note: (a) All polymerisations were carried out varying initiator concentrations to get polymers of various molecular wts.
(b) Intrinsic Viscosities were measured for the polymer in THF at 25°C.

Thus formation of supramolecular complex helps to achieve selective polymerisation of divinyl monomers. Such polymers can be further cast into films, made into microspheres or any desired shape and further converted into insoluble forms in the second stage.

Polymers which are solvent soluble, thermally fusible and which would be latter converted into product having enhanced mechanical/thermal/solvent resistance properties have applications in the field of latex paints, controlled release systems, microlithography, microelectromechanical systems.

CHAPTER 25

Polymers That Resist High Temperature

INTRODUCTION

Polymer science is the science of 'macromolecules', i.e., the molecules of large dimensions, having molecular weights in thousands. Molecules of this type are built up of very large numbers of simple repeating units called monomers. These monomer units are linked together in long, chain-like structures. Today, we have many types of polymers which have been designed and tailored by chemists' skill.

A new class of polymers, namely, high temperature polymers, has appeared in the last decade. A considerable amount of research work is being carried out for the development of these polymers. They, unlike other polymers, are extraordinarily resistant to heat. We can use them at high temperatures. The main stimulus has come from the development and characterisation of new materials for space exploration.

Construction of a spacecraft or a satellite, for instance, needs materials that can withstand high surface temperatures, are light in weight and have high dimensional stability. Modern military and civil aircraft use a large percentage of such materials. Conventional materials like metals, glass and ceramics are being fast replaced.

APPLICATIONS

High temperature polymers cover a wide range of applications. Fibre reinforced plastics, based on high temperature polymers, are made by incorporating tough fibres in a thermoplastic or a thermosetting polymer. Thermosetting polymers become very tough on application of heat. Reinforced plastics are very light in weight and even stronger than metals.

The density of reinforced plastics lies between 1.0 to 1.5, whereas that of metals between 3.0 to 7.5. Also, they are resistant to heat. Reinforced plastics find application in the building of ships, planes, cars and bridges, etc. The fibres used in the making of reinforced plastics are called structural fibres.

These structural fibres are tough and strong; many of them have been based on high temperature polymers. Aero-space adhesives are also made from these high temperature polymers. In space systems, such adhesives offer the advantage of a light weight structure and can join dissimilar materials.

These adhesives have moisture resistance, fatigue resistance and high temperature resistance. Many other applications of these polymers are in electrical insulations, protective coatings and for various kinds of films and plastics, etc.

WHAT IS A HIGH TEMPERATURE POLYMER?

The present day commercial polymers, which include films, fibres, laminates and plastics, etc., have one major limitation. They have low melting points and therefore cannot operate at elevated temperatures. Majority of these polymers decompose or readily oxidise at temperatures below 250°C. The current demand is for polymers which retain useful properties for a period of about 10,000 hours at 300°C or above or 1,000 hours at much higher temperatures. Such polymers should possess high thermo-oxidative stability, i.e., they should withstand degradation of properties over long periods at high temperature in air.

GENERAL PRINCIPLES

What type of polymer structure should be chosen for heat-stable materials? The question has stimulated scientists to understand basic principles which govern the heat-stability of polymers and their resistance to oxidative degradation. One basic factor which determines the stability is the strength of the bonds which join different groups in the polymer backbone. If the connecting links are stronger, the resulting polymer chain would be more stable. For instance, a polymer chain with Si—O bonds would be more stable than one with C—C bonds, because Si—O has a higher bond energy than C—C bond. Also, a chain containing phenylene units in backbone would be more stable than a chain with alkylene backbone (Fig. 25.1). The phenylene rings are stiffer and more resistant to deformation.

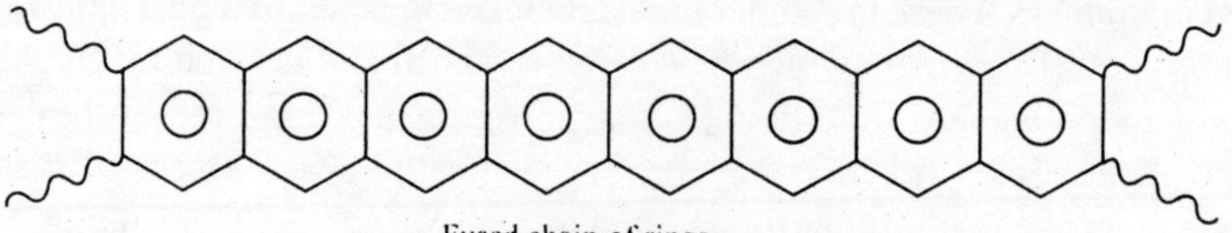

Fused chain of rings
(A double stranded polymer, an ideal structure for a high temperature polymer)

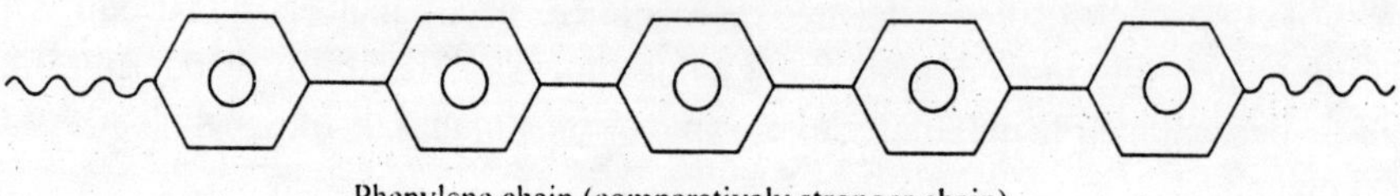

Phenylene chain (comparatively stronger chain)

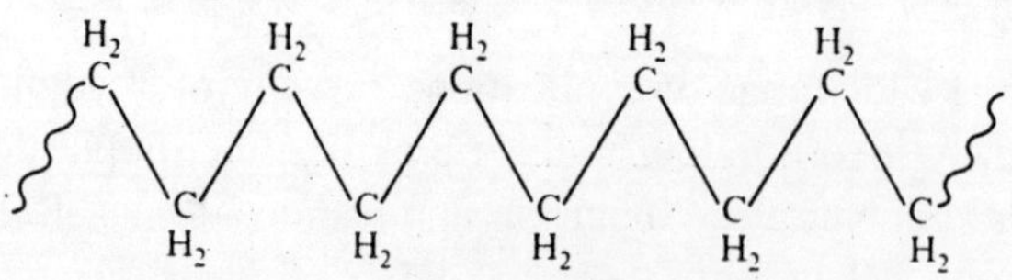

Weak alkylene chain

Fig. 25.1. Polymer-chain showing joining of different groups with two strands.

In fact, it has been observed that placement of phenylene or heterocyclic rings in the backbone improves thermal stability (heat-stability). These ring structures do not decompose noticeably below 600°C–650°C. Similarly, a polymer chain, in which different groups are joined with two strands (Fig. 25.1) rather than one, obviously, would be stronger. Chains having double stand have been prepared by fusing different rings in a linear fashion. Based on above arguments, several approaches have been attempted to obtain stable backbone of the polymer.

A wealth of polymers has been prepared, the chemical composition has been tailored to obtain a particular balance of physical properties, tractibility and thermal stability. For this purpose, various kinds of connecting links have been used for the phenylene moities in the polymer backbone (Fig. 25.2). The most promising seems to be:

$$X = \text{—S—},\ SO_2\text{—},\ \text{—CO—},$$
$$\text{—COO—},\ \text{CONH—},\ \text{—N=N—},$$
$$\text{—CH=CH—},\ \text{—}CF_2\text{—},\ \text{—}CH_2\text{—}.$$

The main advantage achieved by the incorporation of these groups is decrease in rigidity of the polymer chain. These connecting links increase the flexibility and processability of the product, without decreasing too much the thermal stability of the polymers. In this way the temperature of the polymers is raised to 400°C or above.

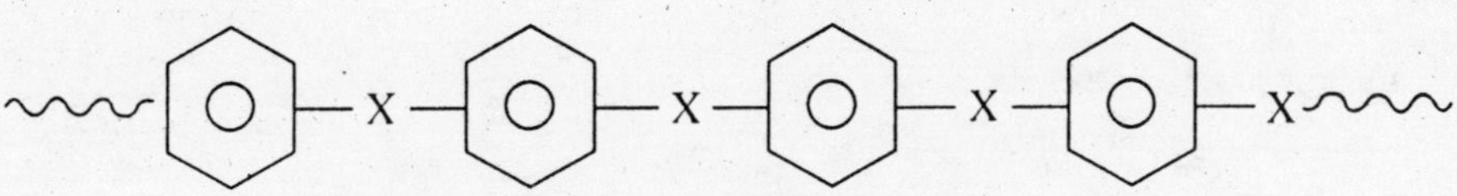

A model structure, X stands for different linking units for the phenylene ring in the polymer back-bone

Fig. 25.2. Phenylene moities in the polymer backbone.

POLYMERS WHICH RESIST HIGH TEMPERATURE

Some of the high temperature resistant polymer are discussed below:

Aromatic Polyamide

The first commercial product of the thermally stable type was a polyamide, viz., Du Pont's wholly aromatic nylon, known as Nomex (Fig. 25.3). It forms the basis of an important thermally stable fibre. This material can be produced in a simple manner, i.e., by means of solution polycondensation of two monomers, *m*-phenylene diamine and isophthaloyl chloride. Solution polycondensation is a method of polymerisation in which two monomers are reacted in the solution form.

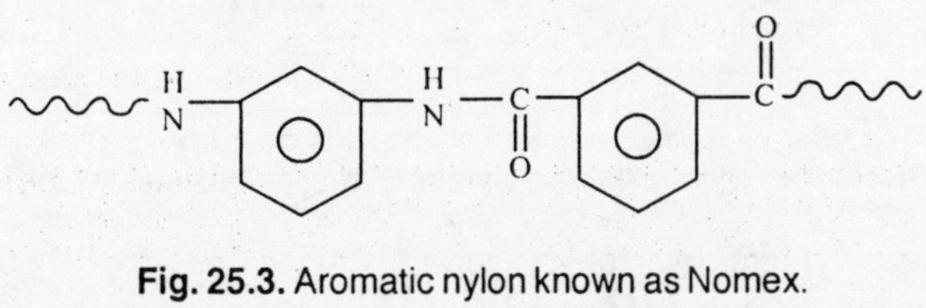

Fig. 25.3. Aromatic nylon known as Nomex.

The monomers condense to form a high polymer. Slight changes in the structure of the monomers show a marked influence on physical properties of the polymer obtained. Phthaloyl chloride gave low melting polymers, while isophthaloyl chloride and terephthaloyl chloride gave high melting polymers (Table 25.1).

This increase in softening point from 200°C to in excess of 360°C has been attributed to the influence of steric factors. *Para* and *meta*-structures have higher interchain attraction and can crystallise better than *ortho* structures. Moreover, *ortho*-structures have more flexibility than *para* and *meta*-structures. A flexible chain forms a low melting polymer than a rigid chain.

Table 25.1. Structure and softening range of aromatic polyamide.

Structure of the polymer	*Softening range C*
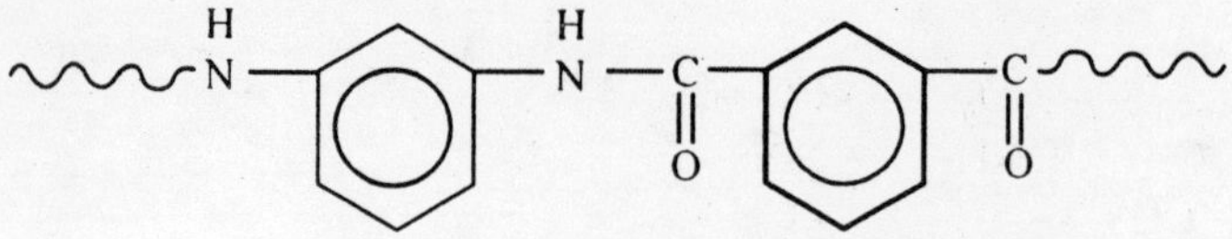 (*meta* linkage)	>360
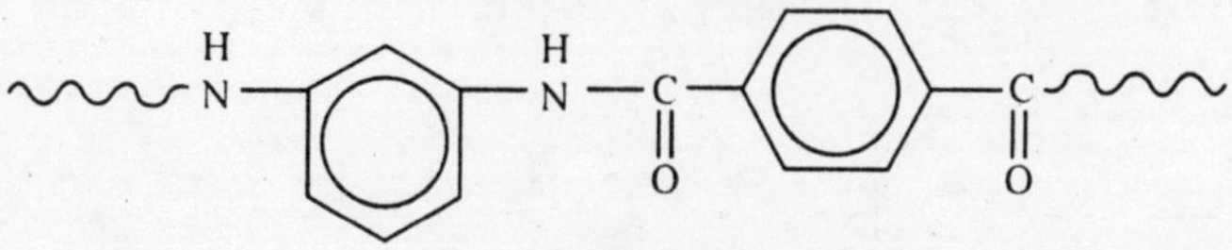 (*para* linkage)	>360
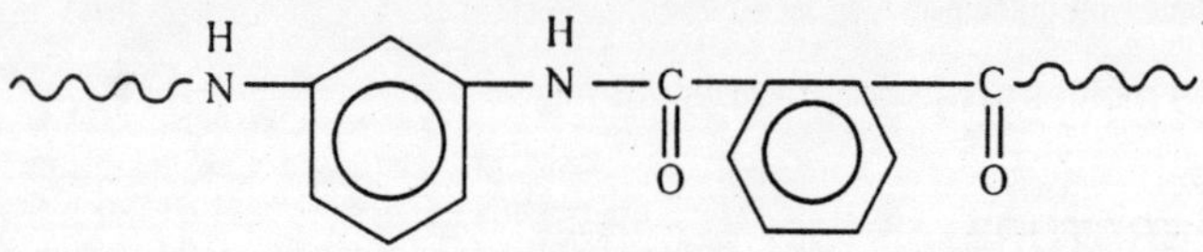 (*ortho* linkage)	~ 200

Polyphenylene Oxides

One more promising class of polymers which has found commercialisation is polyphenylene oxide type (Fig. 25.4). These polymers are synthesised by oxidative coupling of 2,6-substituted phenols.

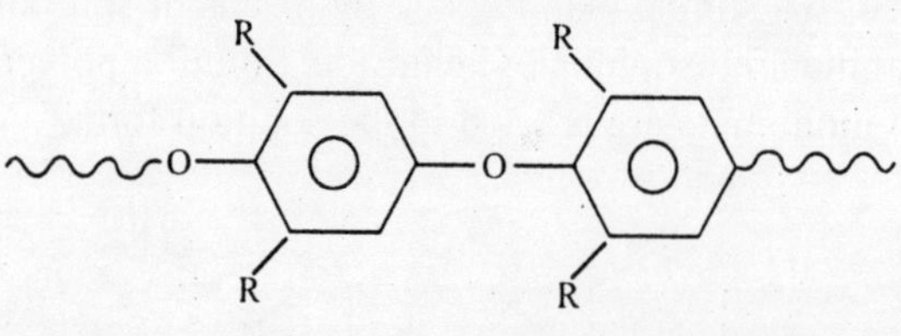

Fig. 25.4. Polyphenylene oxide polymer, R represents an alkyl group.

The advantage of these polymers, over other thermally stable polymers is that they can be melt-fabricated and have melting temperatures of the order of 350°C. They can be processed on conventional moulding and extrusion equipment, particularly notable physical property is its low thermal expansion rate. This property makes it fit for use where high dimensional tolerance is the prime requirement.

Polyimides

A new generation of polymers with inherently rigid chains has been developed. These materials are prepared by incorporating highly stable and rigid heterocyclic ring systems into the polymer chain. One such polymer system developed of high commercial importance is polyimide (Fig 25.5). Polyimides are available for use as plastics, films, laminating resins, insulating coatings, and high temperature structural adhesives. Based on these are the tough and rigid fibres which have been used in fibre reinforced plastics in some experimental tests. Thermal stability of polyimide is attributed to the presence of inert imide ring and high interchain attraction, i.e., high cohesion between the polymer chains. These polymers are generally not meltable and decompose only above 500°C. On similar lines many heterocyclic ring structures have been used for synthesis of high temperature polymers. Some important ones are shown in Fig. 25.6.

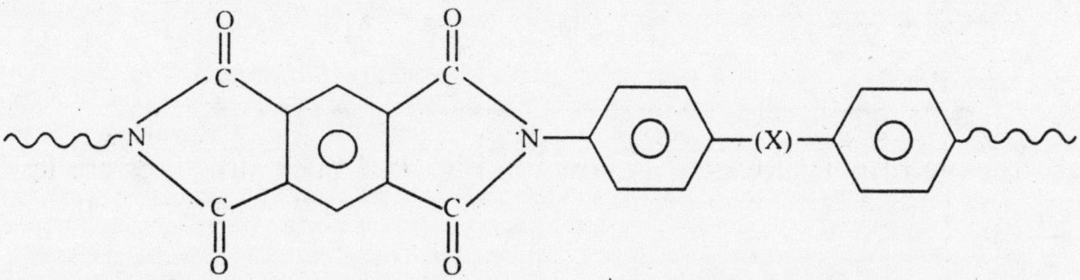

Fig. 25.5. Polyimide polymers 'X' can be any flexible linking like –O–, –CO–, etc.

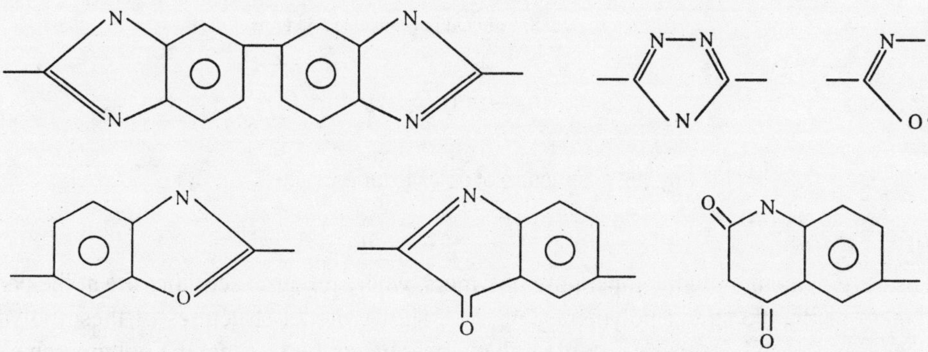

Fig. 25.6. Some heterocyclic ring structure which form the basis of many polymers.

Organo-silicon Polymers

Many bonds like Si-N, Si-O, B-N, B-O have higher bond energy than C-C bond (Table 25.2). These bonds with higher strength will impart further stability to the polymer chain. This fact prompted the synthesis of many organic-inorganic polymers. Out of a large variety of such polymers synthesised, organo-silicon polymers have found a great potential for diferent high temperature commercial products.

Of many other applications which silicone polymers offer, silicone rubbers is the most important one. Ordinary rubber has a tendency to become brittle at very low temperatures. However, silicone rubbers possess good heat stability and maintain physical properties over a wide temperature range of –90°C to 250°C.

Because of this superiority, silicone rubbers have been used in the tyres of fighter aircrafts. Embrittlement of ordinary rubber at high altitudes (about –80°C in the stratosphere) causes many accidents at the time of landing. The unique thermal properties of silicone rubbers have provided invaluable in the manufacture of hot airducting used in de-icing of aircraft wings. Aircraft wings get covered with ice at high altitudes. Another important application of silicone rubbers is the encapsulation of fragile components, especially in the electronics field. Other applications of silicone polymers are in high temperature silicone grease, and silicone release agents, etc.

Table 25.2. Bond energies of different bonds which can be used for the building of different polymers.

Bond type	*Bond energy Kcal/mole*
C–C	80.0
Si–O	89.0
Si–N	110.3
B–N	104.3
B–O	119.3

The general structure of polysiloxanes is shown in Fig. 25.7 Such structures are fairly resistant to oxidation and chemical attack.

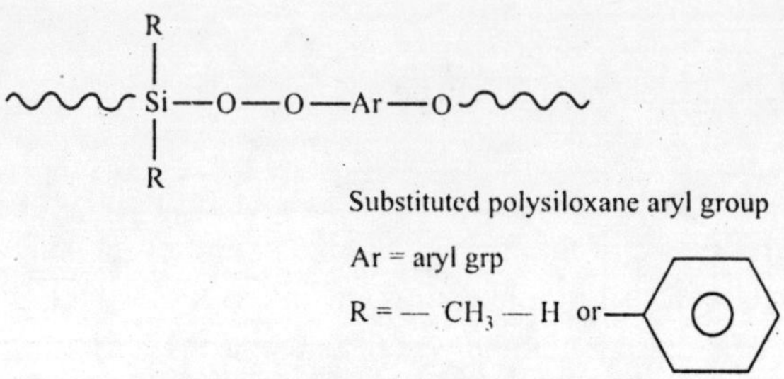

Fig. 25.7. Structure of polysiloxanes.

Ladder Polymers

Another field of development of high temperature polymers, where research activities are at the peak, is of ladder polymers or double strand polymers. Fig. 25.8 shows model structures of these polymers. The interest in ladder polymers as thermo-stable polymers is due to the fact that the polymer chain can not be broken by single bond breaking reaction.

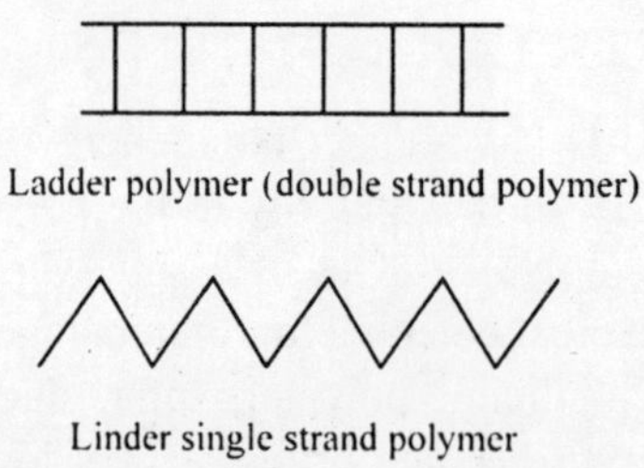

Fig. 25.8. Model structure of ladder polymers.

When a break occurs at any point along a double strand or ladder polymer, the second strand keeps the entire polymer system intact, and maintains overall chain integrity. This is not so in single strand polymers.

Chain scissioning in different cases is shown in Fig. 25.9. In ladder polymers, the close proximity of the fragmented ends makes recombination probable. To sever a polymer chain, cleavage of two or more adjacent bonds should occur. Thus a double strand or ladder polymer will be more thermally stable due to this effect.

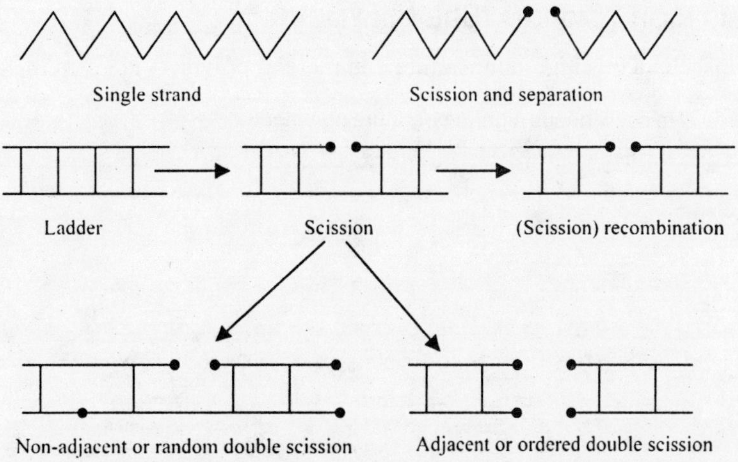

Fig. 25.9. Chain scissoning in different cases of ladder polymers.

The remarkable thermal stability of 'black orlon' (Fig. 25.10), which is obtained by regulated pyrolysis of polyacrylonitrile, has attracted much interest. Fibres of black orlon are highly tough and have found an important application in the manufacture of fibre reinforced plastics.

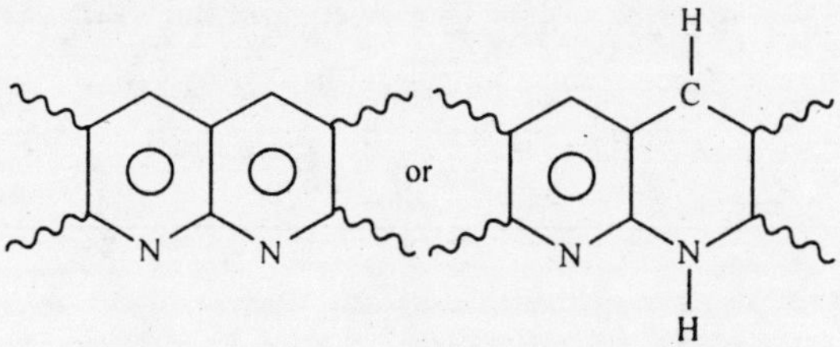

Fig. 25.10. Thermal stability of 'black orlon'.

The stability of black orlon led to attempts to synthesise such structures with higher molecular weights. The synthesis of such a double strand aromatic heterocyclic structure is shown in Fig. 25.11. Such polymers are highly rigid and infusible. They are soluble only in sulphuric acid and are stable in nitrogen upto 600°C.

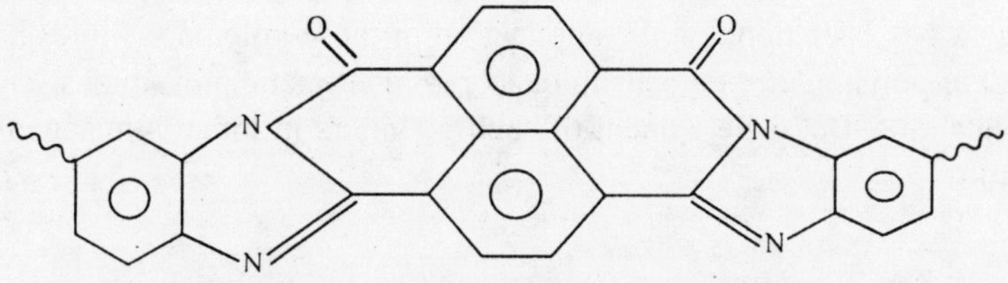

Fig. 25.11. Synthesis of a step ladder polymer.

One more polymer of importance with incomplete ladder structure (also called step ladder structure) has been prepared and fibres produced from it. BBB, as it is called, has fulfilled much of the theoretical expectations from it. Newer researches on ladder structures seem to reveal the possibility that products of this type can be fabricated into films and fibres which will show good thermal stability and oxidative stability at temperatures 100°C above that of non-ladder structures.

MISCELLANEOUS HIGH TEMPERATURE POLYMERS

The structure of miscellaneous high-temperature commercial polymers is given in Table 25.3.

Table 25.3. Structures of miscellaneous high-temperature polymers.

Polymer	Structure
Polybenzimidazole	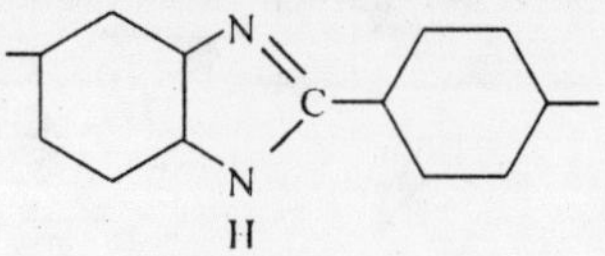
Polyarylate	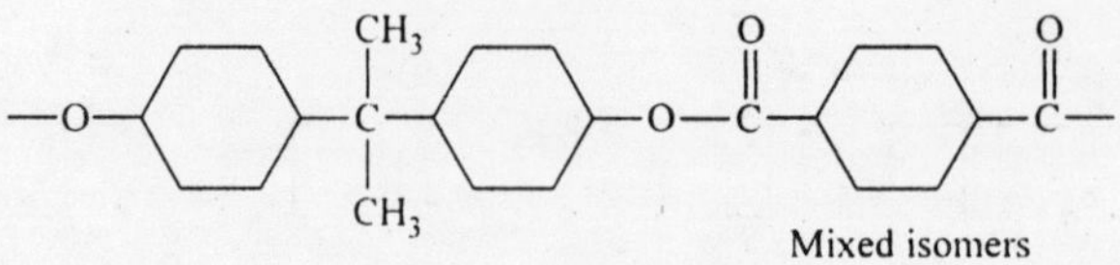
Polyaryletherketone	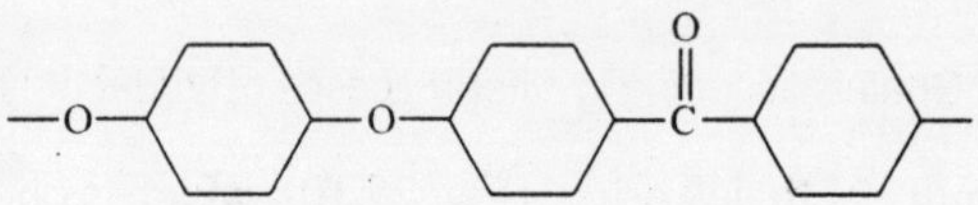
Polyarylethersulphone	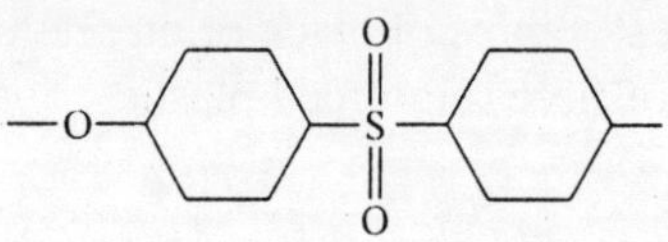
Polynorbornene	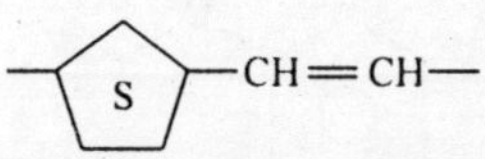

Progress in India

In India, the commercial exploitation of high temperature polymers is still a far way off, although the importance of these polymers has been realised by defence and other research organisations. A definite headway has been made forward for the development of such polymers in some research laboratories such as the National Chemical Laboratory, Pune and National Aeronautical Laboratory, Bangalore. At the Indian Institute of Technology, New Delhi, work has been in progress in various fields of high temperature polymers. Apart from fundamental work, which is being extensively carried out on these polymers, a good amount of work has been done on the application of these polymers. Fibres have been prepared from polymers such as polyimides, polybenzimidazoles and aromatic polyamides. These fibres find use in reinforcing applications. But development of such polymers in India is indispensable for its space research programmes.

References

Alfrey, T., *Testing of Polymers,* Interscience, New York, USA.

Baer, E., *Polymer Degradation,* Reinhold, New York. USA.

Beck, R. D., *Polymer Rheology,* Van Nostrand Reinhold, USA.

Bernhardt, E. C., *Co-polymerisation,* Pergamon, UK.

Bruins, P. F., *Polyblends and Composites,* Wiley Interscience, New York.

Brydson, J.A., *Principles of Polymer Chemistry,* Elsevier Applied Science, London.

Ceresa, R. J., *Block and Graft Co-polymerisation,* Wiley Interscience, New York.

Craig, A.S., *Polymer Applications and Technology,* Philosophical Library, New York.

Crochet, M. J., *Numerical Simulation of Non-Newtonian Flow,* Elsevier, USA.

Culbertson, B. and **Pittman, C.,** *New Monomers and Polymers,* Noyes, UK.

Dealy, J. M., *Kinetics and Mechanisms of Polymerisation Reactions,* Van Nostrand Reinhold, USA.

Donnet, J.B., *Molecular Weight Distributions in Polymers,* Marcel Dekker, New York.

Dorgham, M. A., and **Rosato, D. V.,** *Mechanism of Emulsion Polymerisation,* Interscience Enterprises, Geneva, Switzerland.

Du Bois, J. H. and **John, F. W.,** *Polymer Chemistry,* Van Nostrand Reinhold, USA.

Du Bois, J. H. and **Pribble, W. I.,** *Principles of Polymerisation,* Van Nostrand Reinhold, USA.

Dym, J. B., *Polymers Applications and Technology,* Van Nostrand Reinhold, USA.

Dym, J. B., *Thermodynamics of Polymerisation,* Industrial Press, UK.

Ehrenstein, G. W. and **Erhard, G.,** *Block and Graft Polymers,* Hanser, USA.

Ferry, J. D, *Viscoelastic Properties of Polymers,* Wiley Interscience, New York.

Fisher, E. C., *Contemporary Polymer Chemistry,* The Plastics and Rubber Inst., UK.

Fisher, E. G. and **Whitfield, J. L.,** *Emulsion Polymerisation,* Wiley Interscience, New York.

Frados, J., *Plastics Engineering Handbook,* Van Nostrand Reinhold, USA.

Gacher, R., *Plastics Additives,* Henser, USA.

Griff, A. A., *Plastics Extrusion Technology,* Van Nostrand Reinhold, USA.

Harper, C. A., *Handbook of Plastics and Elastomers,* McGraw-Hill, New York.

Hertzberg, R. W. and **Manson, J. A.,** *Suspension Polymerisation,* Academic Press, UK.

Kraus, G., *Fundamental Principles of Condensation Polymerisation,* Interscience, New York.

Kusy, P. F., *Trends in Polymer Characterisation*, Noyes, UK.

Levy, S. and **DuBois, J. H.,** *Electronic Properties of Polymers*, Van Nostrand Reinhold, USA.

MacDermott, C., *Selecting Thermoplastics for Engineering Applications*, Marcel Dekker, Reinhold, USA.

Mark-Bikales-Overberger-Menges, *Encyclopedia of Polymer Science and Engineering*, Vols. 1-15, Wiley Interscience, New York.

Martelli, F. G., *'Mass Spectrometry'*, Van Nostrand Reinhold, USA.

Matthews, G. A. R., *Polymer Mixing Technology*, Elsevier, USA.

McKelvey, J. M., *Polymer Processing*, Wiley Interscience, New York.

Middleman, S. S., *Fundamentals of Polymer Processing*, McGraw-Hill, New York.

Parker, S. P., *Free Radicals in Solution*, McGraw-Hill, USA.

Rauwendaal, C., *Polymer Extrusion*, Van Nostrand Reinhold, USA

Richardson, P. A., *Physical Chemistry of Polymers*, Noyes, UK.

Rosato, D. V., *Rheology–Theory and Applications*, Van Nostrand Reinhold, USA.

Rosato, D.V. and **Rosato, D.V.,** *Organic Polymers*, Van Nostrand Reinhold, USA.

Rosato, D.V. and **Rosato, D.V.,** *Polyamides*, Van Nostrand Reinhold, USA.

Rubin, I. I., *Crystallisation of Polymers*, Wiley Interscience, New York.

Schenkel, G., *Synthetic Polymers* , Elsevier, USA.

Schnabel, W., *Polymer Degradation; Principles and Applications*, Pergamon, UK.

Schwartz, S.S., and **Goodman, S.,** *Polymer-blends*, Van Nostrand Reinhold, USA.

Sors, L., *Physical Chemistry of High Polymers*, Van Nostrand Reinhold, USA.

Throne, J. L., *Plastics Processing Engineering*, Noyes, UK.

Titow, W. V., *Polymer Chemistry,* Elsevier, USA.

Tobin, W. J., *Principles of Polymer Morphology*, Noyes, UK.

Treloar, L.R., *Kinetics of Chain Polymerisation,* Clarendon Press, London.

Turi, E. A., *Thermal Characterisation of Polymer Materials*, Reinhold, USA.

Wienert, L.A., *Polymer Structure Properties and Applications,* Reinhold, New York. USA.

Winolow, R.M., *Environmental Effects on Polymeric Materials,* Interscience, New York, USA.

Williams, J. G., *Stress Analysis of Polymers*, Wiley, Interscience, New York.

Wiley, R.E., *Testing of Polymers,* Interscience, New York, USA.

Index